Theoretical Methods in Medium-Energy and Heavy-Ion Physics

NATO ADVANCED STUDY INSTITUTES SERIES

A series of edited volumes comprising multifaceted studies of contemporary scientific issues by some of the best scientific minds in the world, assembled in cooperation with NATO Scientific Affairs Division.

Series B: Physics

RECENT VOLUMES IN THIS SERIES

Volume 29 – Electron–Phonon Interactions and Phase Transitions
edited by Tormod Riste

Volume 30 – Many Degrees of Freedom in Field Theory
edited by L. Streit

Volume 31 – Many Degrees of Freedom in Particle Theory
edited by H. Satz

Volume 32 – Electron and Ion Spectroscopy of Solids
edited by L. Fiermans, J. Vennik, and W. Dekeyser

Volume 33 – Microscopic Structure and Dynamics of Liquids
edited by J. Dupuy and A. J. Dianoux

Volume 34 – Path Integrals and Their Applications in Quantum, Statistical, and Solid State Physics
edited by G. Papadopoulos and J. T. Devreese

Volume 35 – Correlation Functions and Quasiparticle Interactions in Condensed Matter
edited by J. Woods Halley

Volume 36 – Strongly Coupled Plasmas
edited by Gabor Kalman

Volume 37 – Coherence in Spectroscopy and Modern Physics
edited by F. T. Arecchi, R. Bonifacio, and M. O. Scully

Volume 38 – Theoretical Methods in Medium-Energy and Heavy-Ion Physics
edited by K. W. McVoy and W. A. Friedman

This series is published by an international board of publishers in conjunction with NATO Scientific Affairs Division

A Life Sciences **B Physics**	Plenum Publishing Corporation London and New York
C Mathematical and Physical Sciences	D. Reidel Publishing Company Dordrecht and Boston
D Behavioral and Social Sciences	Sijthoff International Publishing Company Leiden
E Applied Sciences	Noordhoff International Publishing Leiden

Theoretical Methods in Medium-Energy and Heavy-Ion Physics

Edited by
K.W. McVoy
and
W.A. Friedman
University of Wisconsin – Madison

PLENUM PRESS • NEW YORK AND LONDON
Published in cooperation with NATO Scientific Affairs Division

Library of Congress Cataloging in Publication Data

Nato Advanced Studies Institute on Theoretical Nuclear Physics, University of Wisconsin–Madison, 1978.
Theoretical methods in medium-energy and heavy-ion physics.

(Nato advanced study institutes series: Series B, Physics; v. 38).
Includes bibliographical references and index.
1. Heavy ion collisions – Congresses. 2. Pions – Scattering – Congresses. 3. Heavy ions – Congresses. I. McVoy, K. W. II. Friedman, William Albert, 1938- III. Title. IV. Series.
QC794.6.C6N37 1978 539.7 78-11583
ISBN 0-306-40062-6

Proceedings of the NATO Advanced Studies Institute on Theoretical Nuclear Physics held at the University of Wisconsin, Madison, Wisconsin, June 12–23, 1978

A Division of Plenum Publishing Corporation
227 West 17th Street, New York, N.Y. 10011

Printed in the United States of America

PREFACE

A NATO Advanced Studies Institute was held June 12-23, 1978, at the University of Wisconsin in Madison, Wisconsin. It was a topical Institute in theoretical nuclear physics and had the somewhat novel feature of focussing not on a single topic but on two closely allied ones: pion-nucleus and heavy-ion physics. These two fields, both dedicated to the investigation of short-wavelength properties of nuclei, have many techniques and concepts in common, and essentially become one in the topic of relativistic heavy-ion physics. The purpose of including both in a single Institute was to encourage the practitioners in each of these fields to learn from those in the other; to judge from the liveliness of the questioning which ensued, the purpose was well-served indeed.

Because the Institute was viewed as one which served both educational and research ends, the lecturers took particular pains to develop their subjects in a careful, coherent sequence. The result is a compendium of advanced techniques and current results in these two rapidly-expanding fields of nuclear theory which should serve interested physicists as an ideal introduction to the fields.

In addition to the support provided by the Scientific Affairs Division of NATO, substantial financial assistance was provided by the U.S. National Science Foundation and the Graduate School of the University of Wisconsin.

We wish to express our gratitude to all these agencies for their generosity; to the speakers at the Institute for so conscientiously having their manuscripts available for the participants at the beginning of the meeting; to Mrs. Betty Perry for invaluable secretarial assistance; and to our long-suffering wives for their good cheer and moral support.

W.A. Friedman
K.W. McVoy

CONTENTS

PART I: HEAVY ION REACTIONS

PART I

HEAVY ION REACTIONS

THE CURRENT EXPERIMENTAL SITUATION IN HEAVY-ION REACTIONS

David K. Scott

Lawrence Berkeley Laboratory
University of California
Berkeley, California 94720

INTRODUCTION

Let us begin on a grandiose note by comparing heavy-ion collisions, which occur on the shortest scales of time and space in the Universe (10^{-23} sec and 10^{-13} cm), with the collisions of galaxies (then both exponents are positive!). Figure 1.1 shows the spectacular NGC 5194 spiral nebula in Canes Venatici,[1] with the satellite nebula NGC 5195. The analysis of this type of cosmological event uses a simple potential model with gravitational forces folded over the mass density distributions.[2,3] The collision of two equal mass

Figure 1.1

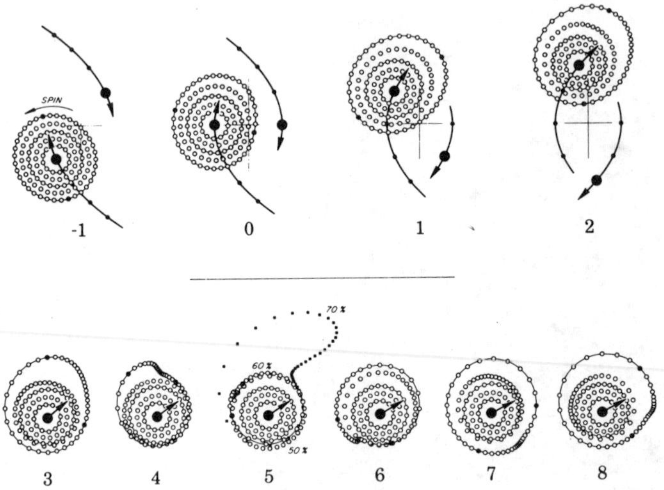

Figure 1.2

galaxies, where one has some initial symmetric distribution counter to the parabolic orbit of the incident galaxy, is shown in Fig. 1.2. As time passes, we see the build-up of a tidal wave which eventually spews out mass in the "target fragmentation region," leaving behind some hot, residual system which seeks a stable mode. Now compare the collision of ^{20}Ne on ^{238}U at incident relativistic energies of 250 MeV/nucleon and 2.1 GeV/nucleon in Fig. 1.3; these pictures were generated by solving the hydrodynamic equations,[4] and show nuclear rather than galactic matter streaming out, as the wounded

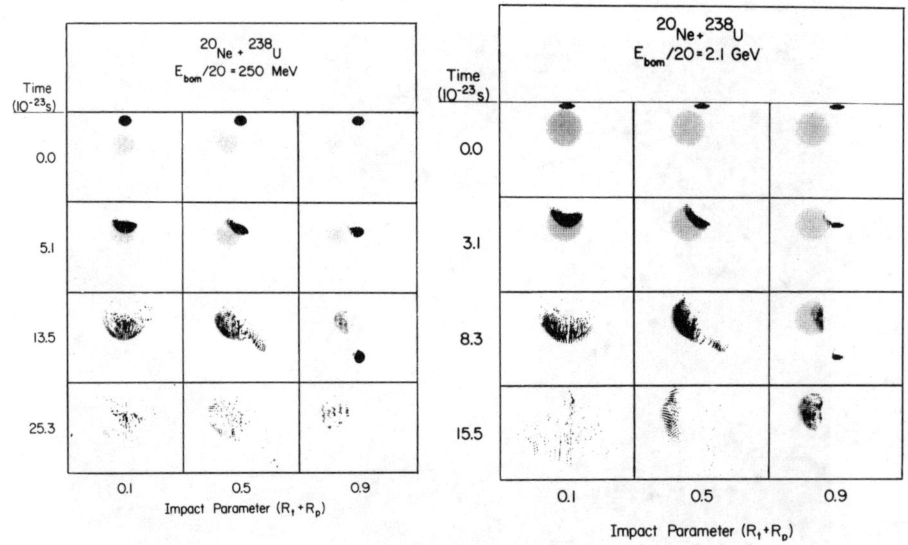

Figure 1.3

nuclei try to recover. (The hydrodynamic equations have also been solved for star-star collisions.[5,6])

The relevance of heavy-ion collisions to cosmological events may be even more profound. In Fig. 1.4 is shown the temperature reached in the nuclear fireball (the region of matter dispersed between the target and the projectile in Fig. 1.3) as a function of the incident energy of two colliding ions,[7] for two assumptions about the hadronic mass spectrum. The curve labeled "experimental" corresponds to a mass spectrum containing essentially the known particles, while that labeled "Hagedorn" corresponds to the bootstrap hypothesis of an exponential growth of hadrons. In this model the temperature limits at ≈140 MeV (and such a limit may have been observed[8]), a temperature approaching the limit reached at the earliest recognizable moments of our Universe, in the Cosmic Big Bang.[9] After this beginning to our lectures, let us hope that we do not end with a whimper!

These examples demonstrate that there is considerable interest throughout the whole of physics in the collisions of structured objects, especially insofar as the phenomena may be explained in the context of a microscopic theory. In the most general sense, this motivation justifies the enormous effort and expense poured into providing heavy-ion beams as massive as uranium up to energies of 2 GeV/nucleon for the study of nuclear interactions. (Useful sources on developments in the field are contained in Refs. 10-30.) A more specific motivation becomes evident when we take a panoramic view of the stability diagram[31] for nuclear species in Fig. 1.5. There are 300 stable nuclear species. During the last half century only some 1300 additional radioisotopes have been identified and studied. It is estimated that in the interaction of U + U, 6000 new

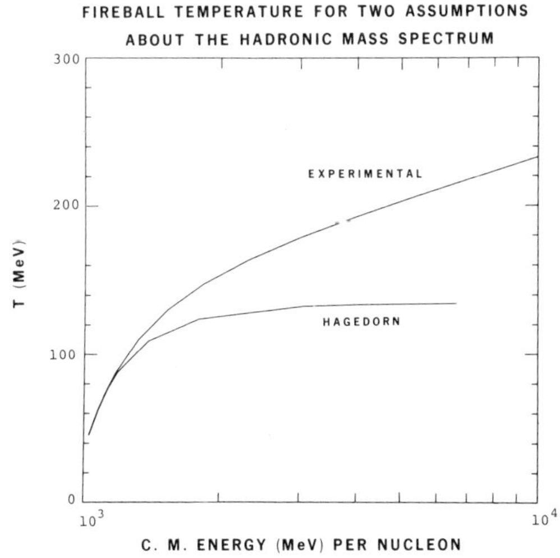

Figure 1.4

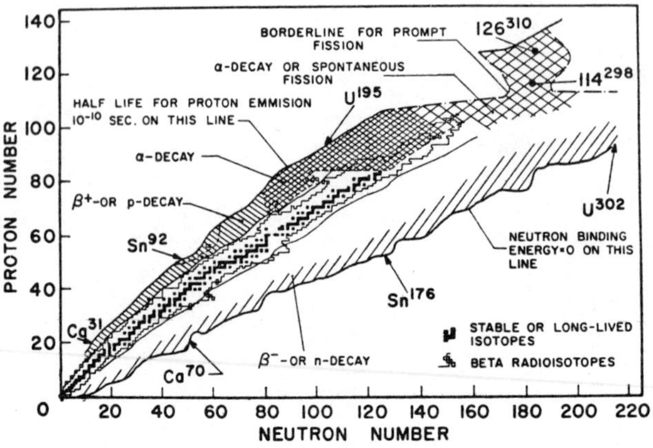

Figure 1.5

species could be formed. The historic role of heavy-ion physics, through the study of these nuclei, will be to relax the limitations that have been imposed on the study of nuclear physics over its 60 year history — limitations of nuclear charge and mass number, limitation of spherical shape, limitations of "normal" temperatures and pressures and reaction mechanisms. The influence of very heavy-ion accelerators is already beginning to be felt in theoretical chemistry, in atomic physics, and quantum electrodynamics as well as in nuclear physics itself. Over the last few years, a wave of enthusiasm has caused nuclear physicists to focus on research with heavy ions, and the view both near and far is one of increasing excitement which has pervaded the conference halls and the research laboratories, dominated the research proposals and preoccupied the funding agencies. It shows no signs of abatement.

In these lectures I shall attempt to give a survey of the present experimental situation in Heavy-Ion Physics. I shall draw heavily from a similar course of lectures delivered last year,[30] updated by the many new trends which have emerged since that time — or which were unknown to me then! In order to chart a navigable course through the vast territory of heavy-ion literature, I shall make a division into three continents, named (a) Microscopia, (b) Macroscopia, and (c) Asymptotia, which will deal in turn (a) with the simple excitation of discrete states in elastic scattering, transfer and compound nuclear reactions; (b) with more drastic perturbations of the nucleus high in the continuum through fusion, fission and deeply-inelastic scattering; and (c) with the (possibly) limiting asymptotic phenomena of relativistic heavy-ion collisions. However, it will be one of the goals of these lectures — and my selection of material is so guided — to show that there are definite signs of a *Continental drift*, with a merging of the microscopic, macroscopic and asymptotic approaches. When they finally become a Trinity, no doubt we shall find Utopia, but I am afraid we shall not reach it in these lectures. However, the very fact that we are gathered here to discuss both heavy-ion *and* pion physics is also an indication of the reunification of the many branches into which nuclear physics has become divided. Perhaps we could do well to reflect on Benjamin Franklin's injunction to his colleagues, "Gentlemen, let us all hang together, or we may all hang separately." In other words, make out of necessity a golden opportunity to strike down artificial barriers in physics, providing a better perspective on many aspects of nuclear dynamics.[32]

1. MICROSCOPIA

We shall begin by defining some of the parameters of heavy-ion reactions, and then use this knowledge to describe the characteristic features of elastic scattering. The status of optical potentials is then treated, followed by their incorporation into the DWBA formalism for simple transfer reactions. A survey of more complicated multinucleon transfer leads us to heavy-ion compound nuclear reactions, from which most of our knowledge of new types of states excited in heavy-ion collisions is presently being gleaned. Throughout this, and the subsequent lectures, the emphasis will be on heavy-ion collisions at energies well above the barrier, since this region is the wave of the future.

1.1 Characteristics of Heavy-Ion Collisions

In the collision of nuclei with charge and mass numbers Z_1, A_1 and Z_2, A_2, some useful quantities are defined in Fig. 1.6

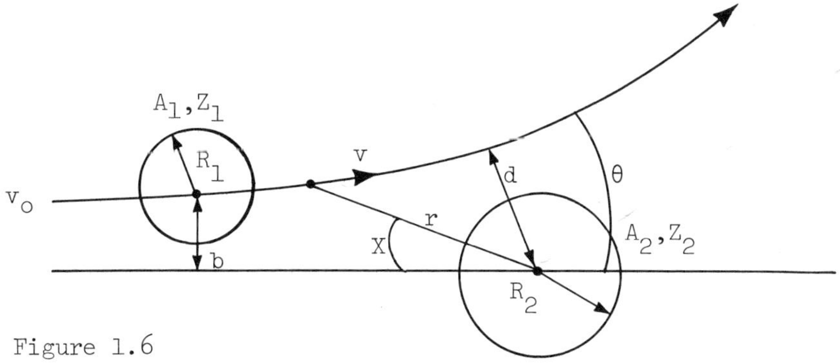

Figure 1.6

Reduced mass $\mu = \dfrac{mA_1A_2}{A_1+A_2}$, m = nucleon mass. (1.1)

Relative velocity = v,

$$\frac{v}{c} = \sqrt{\frac{E_{lab}}{469\,A_1}}\;, \qquad E \text{ in MeV}\;. \tag{1.2}$$

$$\text{Wave number}\quad k = \frac{1}{\bar{\lambda}} = \frac{\mu v}{\hbar} = \frac{4{\cdot}8\;A_1A_2}{A_1 + A_2}\left(\frac{v}{c}\right)\;. \tag{1.3}$$

Kinetic energy of relative motion $E = \frac{1}{2}\mu v^2$. (1.4)

Half distance of closest approach in head-on collision

$$a = \frac{Z_1Z_2e^2}{\mu v^2} = \frac{Z_1Z_2}{2E_{cm}}\left(\frac{e^2}{\hbar c}\right)\hbar c\;. \tag{1.5}$$

$$\text{Sommerfeld parameter}\quad \eta = ka = \frac{Z_1Z_2e^2}{\hbar v} \tag{1.6}$$

Classical impact parameter = b.

Associated angular momentum = kb = ℓ (partial wave).

Scattering angle = θ.

Strong interaction radius $R = R_1 + R_2 = r_o(A_1^{1/3} + A_2^{1/3})$.

For a Rutherford orbit,

$$\begin{aligned} d &= a(1 + \operatorname{cosec}\theta/2) \\ &= a + \sqrt{a^2 + b^2} \\ &= \eta/k\;(1 + \sqrt{1 + (\ell/\eta)^2} \end{aligned} \tag{1.7}$$

Critical scattering angle θ_g or θ_c when $d = R$.

$$\sin\frac{\theta_c}{2} = \frac{a}{R - a} \tag{1.8}$$

$$b_c = R\sqrt{1 - 2a/R} \tag{1.9}$$

$$\ell_c = kb_c = kR(1 - 2\eta/kR) \tag{1.10}$$

Heavy-ion reactions are characterized by large values of $kR = R/\bar{\lambda} \gg 1$. Such considerations lead us to the concept of a semi-classical trajectory, associated with different impact parameters. Indeed the very features that complicate *numerical* calculations for heavy-ion interactions, high orbital angular momenta $\ell = kR$ and large Sommerfeld parameter η, are just those that may be turned to advantage in semi-classical *analytical* computations. Referring to Fig. 1.6, we can write for a given point on the orbit, by conservation of angular momentum and energy:

$$\mu r^2 \dot{\chi} = \ell = \mu v_o b \tag{1.11}$$

$$\tfrac{1}{2}\mu \dot{r}^2 + \tfrac{1}{2}\mu r^2 \dot{\chi}^2 + V(r) = E = \tfrac{1}{2}\mu v_o^2 \tag{1.12}$$

Then

$$\frac{d\chi}{dr} = \frac{d\chi/dt}{dr/dt} = \frac{\dot{\chi}}{\dot{r}} = \frac{\ell}{r^2} \frac{1}{\sqrt{E - V(r) - \ell^2/2\mu r^2}} \,. \tag{1.13}$$

and we can calculate the scattering angle

$$\theta(\ell) = \pi - 2\int_d^\infty \frac{\ell}{r^2} \frac{dr}{\sqrt{2\mu E - 2\mu V(r) - \ell^2/r^2}} \tag{1.14}$$

since $\theta = \pi - 2\chi$. Here $V(r)$ is the total potential, comprising Coulomb + nuclear. Equation (1.14) enables us to construct a scattering diagram and a deflection function diagram, which typically looks like Fig. 1.7.

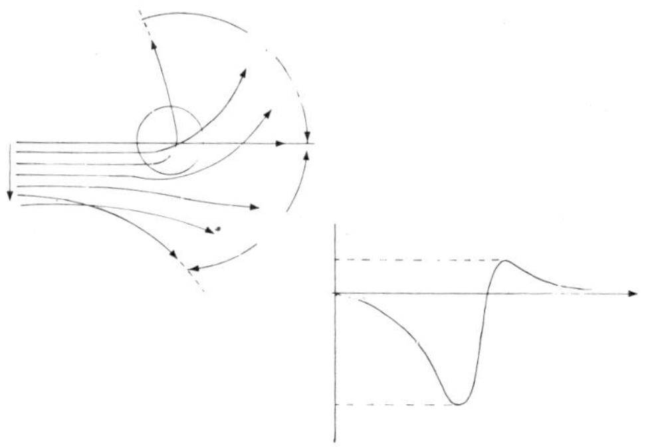

Figure 1.7

For large impact parameter b the trajectory follows a Coulomb orbit, and as b decreases θ initially decreases. At smaller impact parameters the attractive nuclear potential pulls the trajectory forward so there is a maximum scattering angle $\theta_r^{(C)}$, called the Coulomb rainbow angle, beyond which scattering is forbidden classically. The attraction pulls the trajectories round to a maximum *negative* angle, after which still smaller impact parameters again scatter to smaller angles. This negative maximum is called the nuclear rainbow angle, $\theta_r^{(N)}$. The trend is concisely represented in the deflection function diagram at the bottom. One of the contrasts between light- and heavy-ion scattering is the prominence of nuclear rainbows in the former and Coulomb rainbows in the latter.[33] These considerations lead us to predict an elastic scattering distribution (Fig. 1.8).

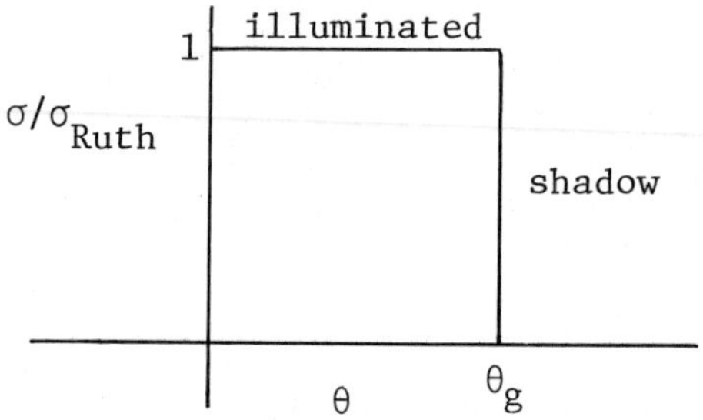

Figure 1.8

The scattering follows the Rutherford pattern up to the grazing trajectory. Beyond that is the shadow region, where classically no particles penetrate. Note, however, that a similar picture can be generated by ***strong absorption*** inside the grazing trajectory. Then the shadow is generated by an imaginary rather than the real potential.[34]

We compare these zeroth order predictions with the two standard forms occurring experimentally in Fig. 1.9, which shows the scattering of ^{16}O of 10 MeV/nucleon on ^{208}Pb and ^{12}C.

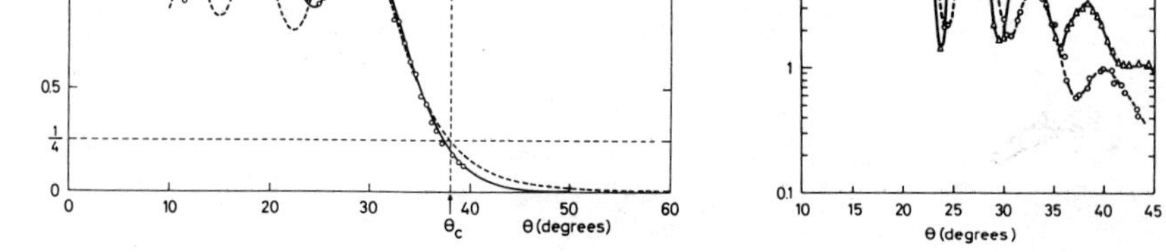

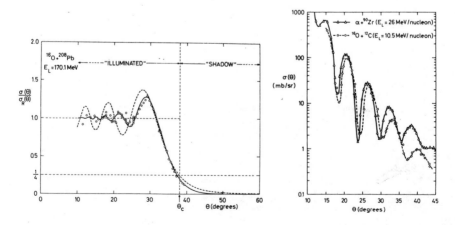

Figure 1.9

These are examples of Fresnel and Fraunhoffer diffraction. In the case of $^{16}O+^{208}Pb$, the scattering is Coulomb dominated and the *average trend* is indeed as in Fig. 1.8. An interpretation of the *diffraction patterns* is possible in the semiclassical picture by introducing *complex trajectories*,[35,36] and is discussed by R. Schaeffer in this lecture series.

1.2 More Formal Treatment of Elastic Scattering

The scattering amplitude can be written

$$f(\theta) = \frac{1}{ik} \sum_{\ell} (2\ell+1) P_{\ell}(\cos\theta) (e^{2i\delta_{\ell}} - 1) \tag{1.15}$$

Using semi-classical ideas:[37,38]

a) Replace ℓ by continuous variable L, $\ell + \frac{1}{2} \rightarrow L$.
b) Assume continuous variation of phase shift $\delta(L)$ with L.
c) Replace $P_{\ell}(\cos\theta)$ by an asymptotic form for large L.
d) Replace Σ by $\int$.

Then

$$f(\theta) \approx \frac{1}{ik} \int L \, dL \, J_0(L \sin\theta) \left(e^{2i\delta(L)} - 1\right) \tag{1.16}$$

is valid if $\theta \lesssim \pi/6$.

If we set

$$\begin{aligned} e^{2i\delta(L)} &= 1, \qquad L > L_c \\ &= 0, \qquad L < L_c \; , \end{aligned} \tag{1.17}$$

(i.e., no scattering if $L > L_c$, complete absorption if $L < L_c$), the integral can be evaluated to give the diffractive cross section

$$\sigma_D(\theta) \approx (kR^2)^2 \left[\frac{J_1(kR\theta)}{kR\theta}\right]^2 \tag{1.18}$$

where L = kR. This diffraction cross section has a characteristic oscillatory behavior with spacing

$$\Delta\theta_D \approx \pi/kR \; . \tag{1.19}$$

In order to discover the predicted trend of differential cross sections we tabulate some values of parameters in Table 1.1. We see that the $^{16}O + ^{12}C$ reaction at 168 MeV has a small Sommerfeld parameter η and has similar values of η, kR, $\lambda\!\!\!^{-}$, a, R, θ_c to the reaction $\alpha + ^{94}Zr$ at 104 MeV. There is therefore nothing mysterious about the almost exactly similar differential cross sections shown in Fig. 1.9(b), of the predicted Fraunhoffer diffraction spacing, 4.6°.

TABLE 1.1. Interaction radius computed as $r_o(A_1^{1/3} + A_2^{1/3})$ with r_o = 1.6 fm.

Ions	a	R (fm)	λ	E (MeV)	v/c	η	θ_c	kR	$\Delta\theta_o$
$\alpha + {}^{94}Zn$	0.577	9.81	0.231	104	0.235	2.49	7.17	42	4.24
${}^{16}O + {}^{12}C$	0.479	7.69	0.203	168	0.150	2.34	7.62	38	4.74
${}^{16}O + {}^{208}Pb$	1.628	13.51	0.069	312	0.204	23.5	17.75	197	0.92
${}^{238}U + {}^{238}U$	5.114	19.83	0.033	2380	0.053	1166	40.67	600	0.30

A reaction such as ${}^{16}O + {}^{208}Pb$ is characterized by a large η parameter and is Coulomb-dominated, leading to Fresnel scattering (the Fraunhoffer scattering would be difficult to observe experimentally since $\Delta\theta_D \approx 0.3°$).

In this case we make the large angle approximation in $f(\theta)$:

$$f(\theta) \approx \frac{1}{ik}\int_0^\infty L\,dL\left(\frac{2}{\pi L\sin\theta}\right)^{\frac{1}{2}} \cos\left(L\theta - \frac{\pi}{4}\right)\left(e^{2i\delta(L)} - 1\right) \quad (1.20)$$

At a scattering angle θ, the main contribution to the integral comes from values of L near L_θ given by

$$2\left(\frac{d\delta(L)}{dL}\right)_\theta = \pm\theta \quad (1.21)$$

[Note: This is an equation for L_θ: for Coulomb phase shifts gives $L_\theta = \eta\cot(\theta/2)$.]

Expand $\delta(L)$ about L_θ:

$$\delta(L) = \delta(L_\theta) + \left(\frac{d\delta}{dL}\right)(L - L_\theta) + \tfrac{1}{2}\left(\frac{d^2\delta}{dL^2}\right)(L - L_\theta)^2 + \ldots. \quad (1.22)$$

$$\therefore 2\delta(L) = 2\delta(L_\theta) + \theta(L - L_\theta) + \tfrac{1}{2}\left(\frac{d\theta}{dL}\right)(L - L_\theta)^2 + \ldots. \quad (1.23)$$

Taking out slowly varying functions, and replacing the lower limit of integration by L_c (i.e. sharp cut-off model):

$$f(\theta) \approx \frac{1}{k}\sqrt{\frac{L_\theta}{2\pi\sin\theta}}\; e^{i\alpha(\theta)} \int_{L_c}^{\infty} dL \exp\left[\frac{i}{2}\left(\frac{d\theta}{dL}\right)_\theta (L - L_\theta)^2\right] \tag{1.24}$$

This is just the Fresnel integral (compare Fig. 1.9(a)).

Introducing a new variable x by

$$\pi x^2 = \left(\frac{d\theta}{dL}\right)_\theta (L - L_\theta)^2 \tag{1.25}$$

$$f(\theta) = \frac{1}{k}\sqrt{\frac{L_\theta (dL/d\theta)_\theta}{2\sin\theta}}\; e^{i\alpha(\theta)} \int_{x_c}^{\infty} dx \exp \frac{i\pi}{2} x^2 \tag{1.26}$$

The integral can be evaluated, replacing $x_c \to -\infty$, i.e. $L_c \ll L_\theta$, as $\sqrt{2}\, e^{i\pi/4}$. Then

$$f(\theta) = \frac{1}{k}\sqrt{\frac{L_\theta (dL/d\theta)}{\sin\theta}}\; e^{i\bar{\alpha}(\theta)} \quad \text{where} \quad \bar{\alpha} = \alpha + \frac{\pi}{4}\left(\frac{d\theta}{dL}\right)_\theta \tag{1.27}$$

and

$$\sigma(\theta) = |f(\theta)|^2 = \frac{1}{\sin\theta}\left(\frac{b\,db}{d\theta}\right) \tag{1.28}$$

where $L_\theta = kb_\theta$, which is just the classical scattering formula.

Now we note that if x_c is set equal to zero, i.e., $L = L_c$, we have the simple result that at the *critical angle* θ_c,

$$\frac{\sigma(\theta)}{\sigma_R(\theta)} = \frac{1}{4} \tag{1.29}$$

which is the origin of the famous "quarter-point" recipe.[39] We shall see that this point (and others closely related) dominate most heavy-ion elastic scattering experiments. To make further progress we either have to introduce more elaborate parameterizations of the phase shifts[38] (which *can* be done, e.g. smooth cut-off instead of sharp cut-off) or resort to the common practice of dressing everything up by an *optical potential*.

1.3 Optical Model Analysis of Elastic Scattering

Most analyses have used a Saxon-Woods nuclear optical potential. (The Coulomb and centrifugal potentials must also be included.)

$$U(r) = -V(e^{x}+1)^{-1} - iW(e^{x'}+1)^{-1} \tag{1.30}$$

where

$$x = (r-R)/a \qquad R = r_o(A_1^{1/3} + A_2^{1/3})$$

$$x' = (r-R')/a' \qquad R' = r_o'(A_1^{1/3} + A_2^{1/3})$$

Most often the four-parameter form, $R = R'$ and $a = a'$, is used.

The most coherent picture would be that of quoting a global set of parameters, but we are not quite there yet. There are tremendous ambiguities associated with the potentials for the scattering of strongly absorbed particles, which are sensitive only to the extreme tail of the potential.

As an example, consider data for the reduction $^{16}O + {}^{208}Pb$ at 192 MeV shown in Fig. 1.10(a) (similar to that shown in Fig. 1.9(a)). The analysis with Saxon-Woods potentials in Fig. 1.10(b) illustrates three potentials which fit the 192 MeV data equally well.[40] *Only the value of the potential at 12.5 fm is well determined.* Note that the actual value of the nuclear potential at this point (≈ 1 MeV) is very small compared to the Coulomb (≈ 75 MeV). The cross-over point is called the *sensitive radius* (R_S) and has the same significance as the Fresnel ¼-point discussed previously.[41] In fact, from Fig. 1.10(a), $\theta_{1/4} = 31.4°$. Then,

$$L_{1/4} = \eta \cot(\theta_{1/4}/2) = 105 \;, \qquad \eta = 29.9 \tag{1.31}$$

and

$$R_{1/4} = \eta/k\left(1 + \sqrt{1+(1/\eta)^2}\right) = 12.2 \text{ fm} \tag{1.32}$$

which is close to the 12.5 fm of the cross-over. The point also coincides with the radius associated with the ℓ-value at which the optical model transmission coefficient drops to ½, ($R_{1/2}$), and $L_{1/2} = 106$ in the above example. This distance is typically 2 or 3 fm *larger* than the sum of the radii of the two ions, at which their densities fall to one-half of the central value.[41] Even when absorption is almost complete, only the 10% regions overlap. From classical perturbation theory it can be shown[42] that elastic scattering mainly determines the real part of the optical potential at a point slightly inside the distance of closest approach for a

J. B. BALL *et al.*

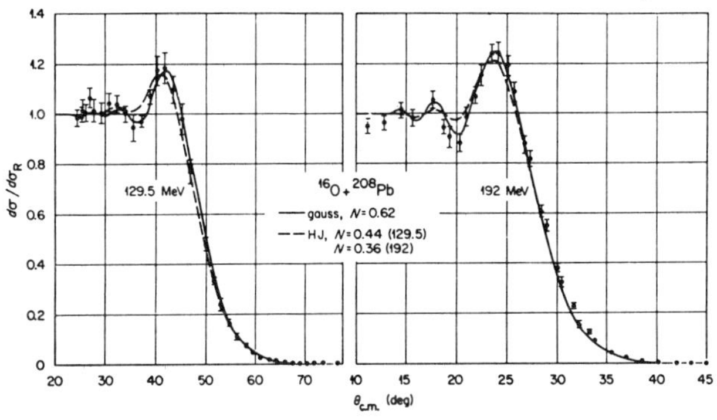

Figure 1.10(a)

ELASTIC SCATTERING

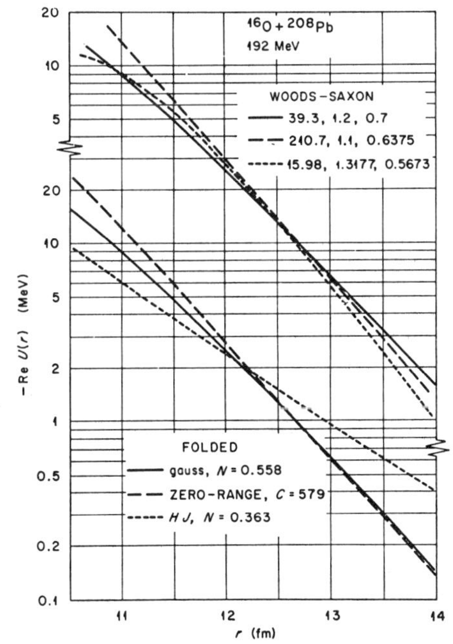

Figure 1.10(b)

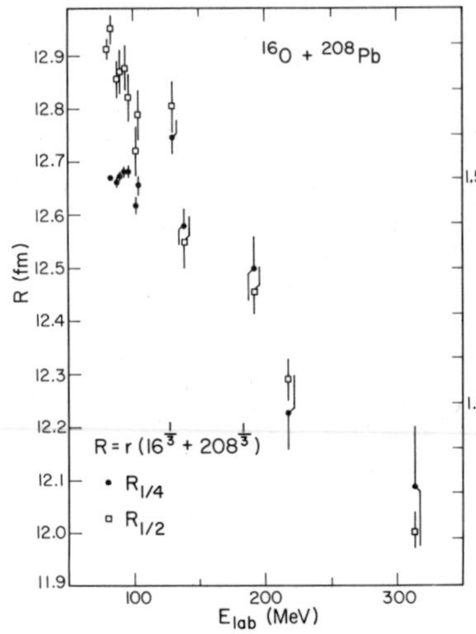

Figure 1.11

trajectory leading to a rainbow angle, and this distance should become constant at high energies. A detailed analysis of the data for the $^{16}O + ^{208}Pb$ system[34] shows that from 90 MeV to 190 MeV, the scattering is indeed refractive, with $R_{1/2}$ roughly constant. Recently the elastic scattering has been extended to 315 MeV (see Fig. 1.11) suggesting rather that the distance continues to decrease, and that higher energies may be able to prove the potential deeper inside the nucleus.[43]

Higher bombarding energies have been used in an attempt to resolve the ambiguities in the $^{16}O + ^{28}Si$ system.[44,45] The data at 215 MeV are shown in Fig. 1.12. The idea is to take data beyond the rainbow angle, where an exponentially decreasing cross section will be observed if the real potential is sufficiently weak. Too

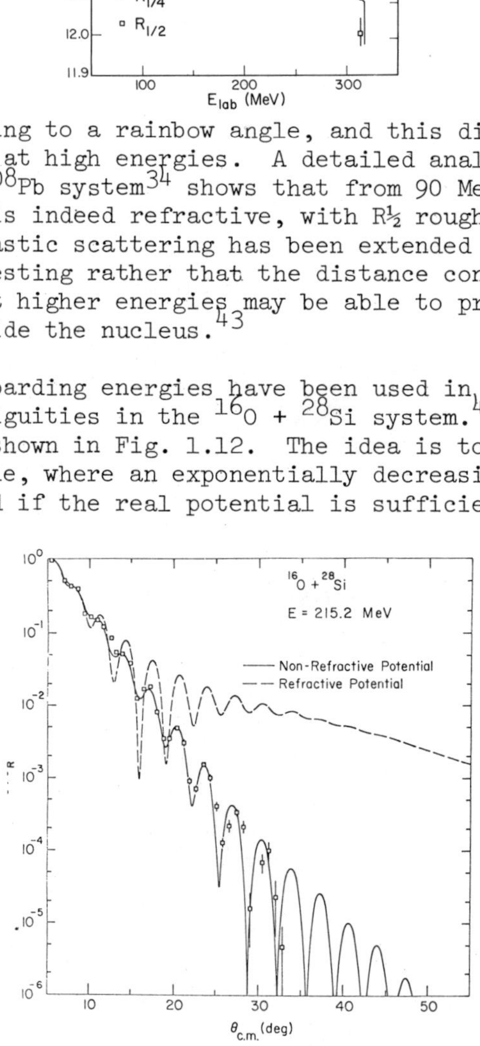

Figure 1.12

much absorption will always give rise to a *diffractive* pattern. The data are clearly diffractive, and call for potentials with V/W < 0.5 (in contrast to those for light ions for which V/W ≈ 5.0), assuming an energy independence; this is expected to be small for heavy-ions.[46] The solid curve is for V = 10, W = 23, r_o = 1.35, r_o' = 1.23, a = 0.618 and a' = 0.552, whereas the dashed curve is for a deep potential of 100 MeV. The potentials extracted for $^{12}C + ^{28}Si$ are quantitatively very similar.[47]

Given the abrupt change in character of potentials for light ions (e.g., alpha particles) and heavy ions as light as ^{12}C, obviously one must look in between, say at ^{6}Li. In fact the results[47] in Fig. 1.13 have a pronounced nuclear rainbow similar to α-scattering, completely at variance with shallow 10 MeV diffractive potentials, but unable nonetheless to pin down the real potential to better than between 150 and 200 MeV (with W ≈ 40 MeV in both cases). Now the search is on with ^{9}Be, and no doubt Mother Nature will be clever enough to hide the sudden transition between light and heavy ions in the nucleus ^{8}Be! The suddenness of the transition is a challenge to fundamental theoretical derivations of heavy-ion potentials and we end our discussion of elastic scattering with a catalogue of some of these approaches.

1.4 More General Approach to Heavy-Ion Potentials

As we have seen, the study of heavy-ion potentials is hampered in general by the insensitivity of elastic scattering to all but the value of the potential at the strong interaction radius. It is natural therefore that both experiment and theory should turn to methods which determine the potential at closer distances. Another distance where the nucleus-nucleus interaction is established can be estimated from the liquid drop model. This is the distance

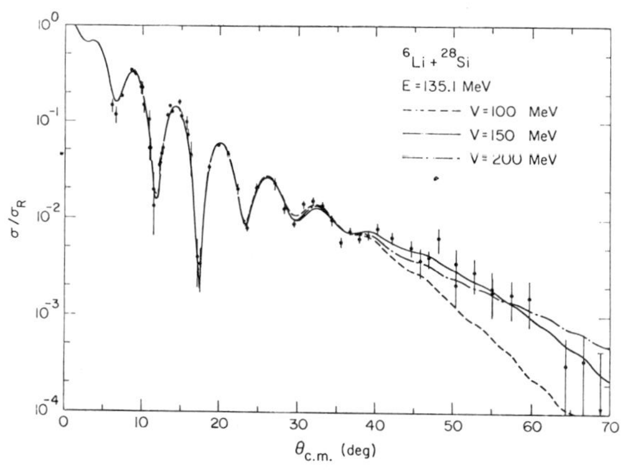

Figure 1.13

corresponding to the sum of the half-density radii R_1 and R_2 where the attractive force is:[48]

$$F = 4\pi\gamma \frac{R_1 R_2}{R_1 + R_2} , \quad R_1 + R_2 = R_0 \tag{1.33}$$

where $\gamma \approx 0.95$ MeV·fm^{-2} is the surface tension coefficient. The previously determined sensitive radius and the value of the potential at this point, together with the value of the force:

$$\left(\frac{dV}{dr}\right)_{r=R_0} = \frac{V}{4a} = 4\pi\gamma \frac{R_1 R_2}{R_0} \tag{1.34}$$

determine the two parameters V and a. The sum of the half density radii R_1+R_2 can be evaluated using expressions of the form:[49]

$$R_1 = 1.12\ A^{1/3} - 0.86\ A^{-1/3} \tag{1.35}$$

(The deviation from strict proportionality to $A^{1/3}$ comes from purely geometrical considerations of a spherical distribution with a diffuse surface.) Using these equations, the nuclear potential can be calculated for any target projectile combination, and lead typically to potentials 60 MeV deep, of diffuseness 0.85 fm.

These simple considerations have been generalized by the *Proximity Force Theorem* which states:[50]

> *"The force between rigid gently curved surfaces is proportional to the potential per unit area between flat surfaces."*

For frozen, spherical density distributions, the force between two nuclei as a function of distance s between their surfaces is

$$F(s) = 2\pi \frac{R_1 R_2}{R_1 + R_2} e(s) \tag{1.36}$$

where e(s) is the potential energy per unit area, as a function of the distance between flat surfaces. The touching of two flat surfaces results in a potential energy gain per unit area equal to twice the surface energy coefficient,

$$\therefore\ e(0) = -2\gamma$$

leading to the same maximum force as above. (The force becomes repulsive as the two density distributions overlap.)

For the potential we obtain,

$$U(s) = 2\pi \frac{R_1 R_2}{R_1 + R_2} \int_s^\infty e(s')ds' \tag{1.37}$$

where

$$s = r - (R_1 + R_2) \ .$$

The interaction is given in terms of a universal function e(s); once known or calculated for one pair of nuclei, we immediately have information about other pairs. Although based on a liquid drop model, the formula is actually very general. Suppose that the interaction energy is represented by a folding formula with a δ-function interaction:

$$U = A\ \rho_1(r_1)\rho_2\ (r - r_1)d\underline{r_1} \tag{1.38}$$

If the densities ρ_1, ρ_2 have Saxon-Woods shapes

$$\rho = \frac{\rho_0}{[1 + \exp\left(\frac{r - R}{a}\right)]} \tag{1.39}$$

then the integral can be evaluated:[51]

$$U(s) = 2\pi A\ \rho_0^2 \frac{R_1 R_2}{R_1 + R_2} \int_s^\infty \frac{s'ds'}{\exp \frac{s'}{a-1}} \tag{1.40}$$

where $s = r - (R_1 + R_2)$, and has the proximity form with a particular expression for e(s). This result begins to link for us the *microscopic* and *macroscopic* approaches to potentials.

To compare with experiment, we write U(s) in the form

$$U = 4\pi\gamma \frac{R_1 R_2}{R_1 + R_2} b\phi(\zeta) \tag{1.41}$$

where $\zeta = s/b$, $b = 1$ fm, and $\gamma \approx 0.95\ \text{MeV}\cdot\text{fm}^{-2}$. The universal function ϕ has been evaluated using the nuclear Thomas-Fermi method. We find:

$$\begin{aligned} \phi(\zeta < 1.25) &= -\tfrac{1}{2}(\zeta - 2.54)^2 - 0.85(\zeta - 2.54)^3 \\ \phi(\zeta > 1.25) &= -3.437\ \exp(-\zeta/0.75) \end{aligned} \tag{1.42}$$

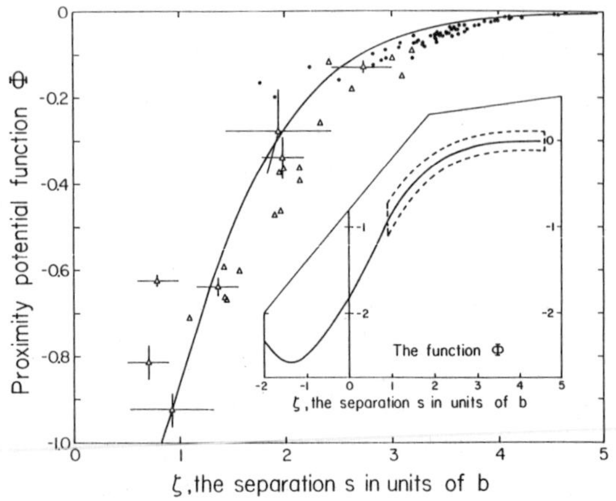

Figure 1.14

and is plotted in Fig. 1.14.[52]

The theoretical proximity function $\phi(\xi)$ in the extreme tail region has been compared with nuclear potentials deduced from an analysis of elastic scattering data, leading to values of ϕ from 0 to -0.16, and are reproduced in the figure by circles. We see (as expected) that elastic scattering tests the potential over ϕ at large values of ζ, i.e., radial distances near the strong absorption radius.

As we shall see in later sections, *inelastic* processes probe the potential to much smaller radii.[34] Values derived in this way are shown as triangles. The theoretical proximity potential is in good agreement with the data over the entire range of distances. A similar global comparison is discussed in Ref. 53, where the potential is tested at distances where friction effects are important, but this subject leads us into Macroscopia.

Many other approaches are taken to the theoretical derivation of heavy-ion potentials; for example, the folding model,[54-56] and the energy density formalism.[57,58] Perhaps it is appropriate to conclude with a comparison[42] in Fig. 1.15 of some of these potentials, evaluated at the sensitive radius with the Saxon-Woods potential for a wide range of interacting systems. Equally good agreement is produced by the *empirical potential* of proximity type:

$$V(r) = 50 \frac{R_1 R_2}{R_1 + R_2} \exp\left(\frac{r - R_1 - R_2}{\alpha}\right)$$

with $R_1 = 1.233\, A_i^{1/3} - 0.978 A_i^{-1/3}$ and a = 0.63 fm.

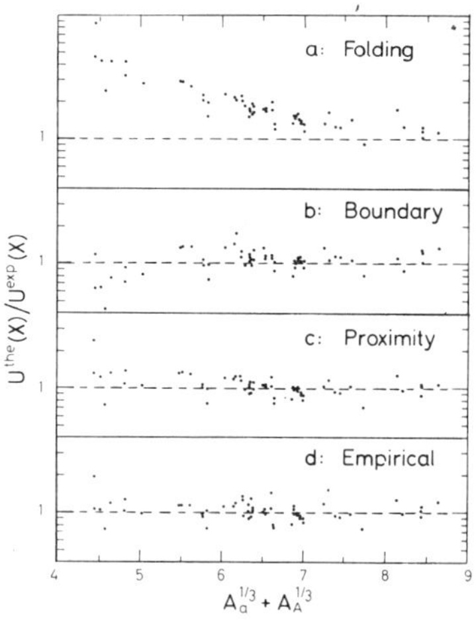

Figure 1.15

1.5 Transfer Reactions

The resurgence of interest in microscopic heavy-ion reactions around 1970 was largely (and rightly) triggered by the great hope that multinucleon transfers (which are possible only via heavy ion reactions) would reveal a rich spectrum of new types of states in nuclei, e.g., nuclear quartets.[12,59] The ideal scenario is to take the optical potentials from the elastic scattering studies of the previous sections, compute distorted waves in the initial and final channels, plug them into the DWBA transfer amplitude to get the cross sections for transfer. Since 1970, however, many studies of one, two, three and four nuclear transfers[60-62] (some of which are also possible with light ions!) indicate that the mechanisms are complicated by high order coupled channels and multistep effects. The whole subject has become bogged down in a welter of computational details. Let me try to show that the situation is not quite as black as it is often painted, and that heavy-ion reactions can still make an attack on nuclear structure problems.[64]

Look at a nucleus such as ^{20}Ne in which the spherical-basis shell model generates rotational like spectra described as (2s,1d).[4] A clear "rotational band" is predicted in agreement with experiment (Fig. 1.16), not only for level positions but also for E2 transition strengths (those in brackets are collective model, the others are shell model). It seems that the shell model is winning, because of the fall-off of E2 strength for the higher spin states. The shell model also predicts that the band should terminate at J=8, whereas the collective model, as classically conceived, goes on forever, to states of 10, 12 .... If the band *did* run on, it would be a triumph for the collective model, but it would not be the end of the shell model. We would argue that as the excitation increases, so does the tendency to loosen the ^{16}O core so that the configurations such as $1p^{-2}(2s,1d)^6$ creep in, bringing higher angular

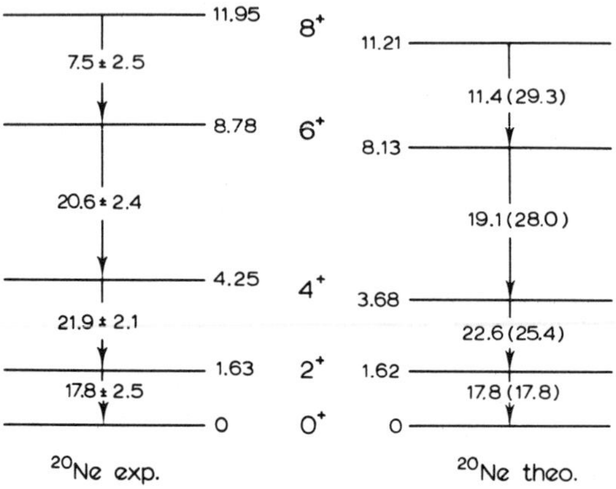

Figure 1.16

momentum. (Such merging of single particle and collective aspects will be taken up in our discussion of much higher angular momenta in nuclei, in the lecture on Macroscopia). If the band stops at J=8, the argument for the truth of the shell model as against the classical rotational model becomes very strong.

The states of the band should be strongly populated by attaching an α-particle to the ^{16}O core, and the same is true for the configurationally equivalent case in ^{16}O, by α-transfer on ^{12}C into the band beginning at 6.05 MeV. Now take a look at the spectrum[65] for the $^{12}C(^{11}B,^{7}Li)^{16}O$ reaction at 114 MeV in Fig. 1.17. We imagine the α-particle popped onto the ^{12}C surface, bringing in an angular momentum of several units due to its linear motion in the ^{11}B. The striking feature of the spectrum is the *extreme selectivity*. Only a few states appear up to 21 MeV excitation which can be identified with members of the rotational band

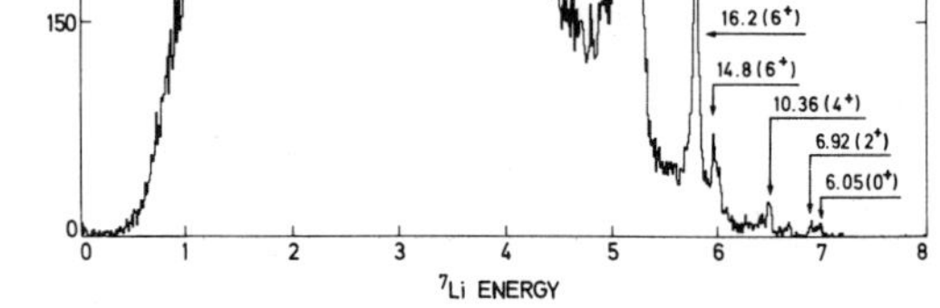

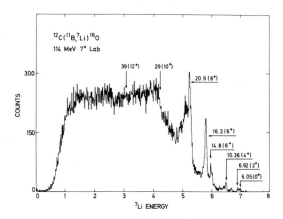

Figure 1.17

up to 8^+ (and also a negative parity band up to 7^-). Remember that the level density in ^{16}O around 20 MeV is many tens of levels/MeV. There is little sign of 10^+ and 12^+ levels which the $E_J = \hbar^2/2J\, I(I+1)$ rotational scheme would place around 29 and 39 MeV. So this simple spectrum, *almost by inspection*, already strengthens our feeling that the shell model is probably an excellent first order description of nuclear structure and that the collective models are probably to be regarded as much more convenient representations of some aspects of the shell model, but secondary to it, rather than models that contain truths *beyond* those to be distilled from shell model wavefunctions.[64] However, we do need a quantitative theory of the reaction dynamics to predict the strengths of the states in Fig. 1.17. Let us begin with a simple, semiclassical one.

This model[66,67] assumes that the particles move on classical trajectories, as illustrated in Fig. 1.18. (The transfer is dealt with quantum-mechanically.) There are three kinematical conditions to be satisfied if the transfer probability of the cluster m (a nuclear or group of nuclears) is to be large. (We shall return to this theory in Lecture 3 on Deeply-Inelastic Scattering.) The cluster starts in an initial state $(\ell_1\lambda_1)$ and ends in $(\ell_2\lambda_2)$.

$$\Delta k = k_0 - \frac{\lambda_1}{R_1} - \frac{\lambda_2}{R_2} \approx 0 \tag{1.43}$$

$$k_0 = \frac{mv}{\hbar}$$

where v is the speed of the particle at the transfer point.

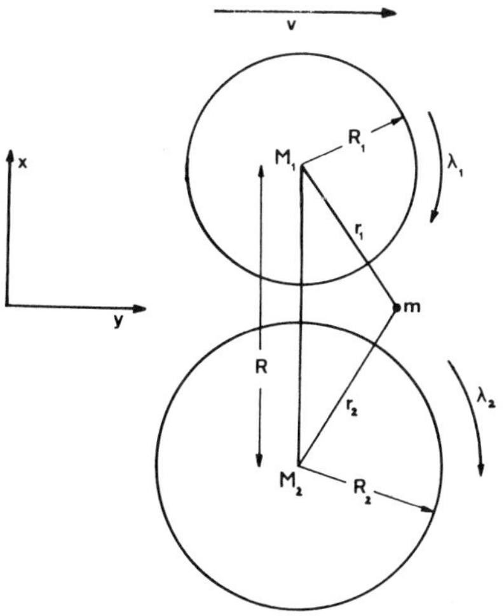

Figure 1.18

$$\Delta L = \lambda_2 - \lambda_1 + \tfrac{1}{2}k_0(R_1 - R_2) + Q\text{eff}\,\frac{R}{\hbar v} \approx 0 \qquad (1.44)$$

$$Q\text{eff} = Q - (z_1^f z_2^f - z_1^i z_2^i)\, e^2/R \qquad (1.45)$$

$$\ell_1 + \lambda_1,\ \ell_2 + \lambda_2 \text{ even.}$$

These conditions imply, respectively: conservation of the y-component of angular momentum of the transferred nucleon; conservation of angular momentum; and confinement of the transfer to the reaction plane, i.e., the angles θ in the spherical harmonics of the single particle wave functions are $\approx \pi/2$. An approximate expression for the transition probability is:

$$P(\lambda_2\lambda_1) \approx S_1 S_2 P_0(R) \left| Y_{\ell_1}^{\lambda_1}\left(\frac{\pi}{2},0\right) Y_{\ell_2}^{\lambda_2}\left(\frac{\pi}{2},0\right)\right|^2 \times \exp\left[-\left(\frac{R\Delta k}{\sigma_1}\right)^2 - \left(\frac{\Delta L}{\sigma_2}\right)^2\right] \qquad (1.46)$$

where $P_0(R)$ is determined by the radial wave functions at the nuclear surface, and σ_1, σ_2 measure the spreads in Δk, ΔL from zero as allowed by the uncertainty principle. The total transition probability is then calculated by summing over the final magnetic substates and averaging over the initial substates, weighted by angular momentum coupling coefficients and the spectroscopic factors (S_1, S_2) for finding the cluster in the initial and final states. However, the localization and semi-classical aspects of the transfer usually mean that the reaction is "well matched" for a restricted range of λ_1, λ_2 and ℓ_1, ℓ_2. The spectroscopic amplitudes in the rotational band are very simple to calculate in the SU(3) model. They are just proportional to the intensities of the SU(3) [80] representation in each state, which are equal for all members of the band (at about 0.36). A comparison of the experimental and theoretical cross sections[65] for the positive parity band are given in Table 1.2. (Theory and experiment are normalized for the 6^+ state.) There is still some uncertainty about the location of the 8^+ state[68] but it is more likely to be associated with the broad structure at 22 MeV excitation rather than at 20.9 MeV, which appears rather to be the 7^- member of the negative parity band. (Since the two states have roughly equal cross section, this ambiguity does not affect our discussion of Table 1.2.) By continuing this type of study to higher incident energies,[69] so that possible 10^+ and 12^+ states are *definitely* not disfavored by the reaction dynamics, it may still be possible to make interesting statements about nuclear structure, with only a *skeletal* reaction theory.

By comparing one, two, three and four nucleon transfers on different targets, all leading to the same *final* nucleus, it is possible to bootstrap one's way up through a hierarchy of simple

TABLE 1.2. Experimental and Theoretical Cross Sections for the Reaction $^{12}C(^{11}B, ^{7}Li)^{16}O$.

E_x, J^{π}	6.05,0^+	6.92,2^+	10.35,4^+	16.23,6^+	20.9,8^+?
$\frac{d\sigma}{d\Omega}$ (EXPT) mb/sr	≈0.0	0.006	0.019	0.250	0.224
$\frac{d\sigma}{d\Omega}$ (TH)	0.000	0.003	0.038	0.250	0.178

stretched, cluster configurations in light nuclei.[67,70] Indeed these experiments have already led to the formulation of literal cluster models by convoluting an α-particle with the core as a function of their separation, adding up all the nuclear-nuclear interactions to generate them from an effective α-core potential.[71]

Another impressive demonstration that few nucleon transfer reactions can proceed by simple α-transfer comes from a comparison of it with the presumed inverse process, α-decay. Nuclei in the lead region are ideally suited to this test. For example, it is possible to derive a "reduced α-width" rate for ^{212}Po (0.727 MeV, 2^+) and ^{212}Po(gs) states from their decay to ^{208}Pb, from the formula,

$$\delta^2 = h/\tau P$$

where τ is the mean life and P the penetrability. Then, $\delta^2(2^+)/\delta^2(0^+) = 0.61$, in excellent agreement with the spectroscopic factor ratio $S(2^+)/S(0^+) = 0.64$, deduced from a direct reaction analysis of $^{208}Pb(^{16}O,^{12}C)^{212}Po$, leading to the conclusion that the basic quantities measured in alpha transfer and decay are homologous.[72,73] (There is, however, an intriguing problem that absolute values of the decay widths are underestimated by the shell model by a factor of 1000--which may indicate substantial clustering of alphas in the surface region,[74,75] and therefore surface phenomena not presently described by the shell model.) However, one is encouraged to look for other alpha particle strengths,[76] e.g., alpha vibrations,[77] analogous to pairing vibrations, so far with a mystifying lack of success.[76]

This type of stimulus is surely what we should expect and demand of heavy-ion transfer reactions. After all we do not need heavy-ions to study one and two nucleon transfers! Many interesting possibilities remain, so far almost completely untapped. Three and four *neutron* transfers are available *only* by heavy-ion reactions but even today there has only been a handful of studies.[78-81] Such reactions enable us to *locate* not only new configurations in nuclei, but also *new nuclei themselves*. Frequently, just the knowledge that a nucleus exists, stable against decay by strong interactions, together with the ground state mass-excess, can lead to new nuclear structure information. A striking case is the Na isotopes, which extend from ^{19}Na to ^{33}Na, the widest range of (N-Z)/A known to man (apart from He isotopes). This information led[82] to the prediction of a sudden shape change from spherical to deformed in the Na isotopes. Perhaps we should be devoting at least as much time to exploring these possibilities of testing our nuclear structure theories on exotic nuclei, as we spend on studying all the complexities of the reaction mechanism. Nevertheless, we must now spend some time looking at these complexities!

The formal quantal evaluation of heavy-ion direct reactions uses the DWBA. Symbolically the reaction can be written[83]

$$(a + c) + b \rightarrow (b + c) + a$$

where a, b, are the heavy-ion cores and c is the transferred particle. Then

$$T_{fi}^{DWBA} = \langle \chi_f \, \phi_{b+c} \, \phi_a \, | V_{ac} | \chi_i \, \phi_{a+c} \, \phi_b \rangle \tag{1.47}$$

where χ_f, χ_i are distorted waves, the scattering eigenfunctions, and ϕ are the eigenfunctions of nuclear Hamiltonians (see Fig. 1.19). The interaction V_{ac} (or V_{bc}) causes the transition (as usual one assumes that the core-core interaction V_{ab} cancels the potential in the initial channel).

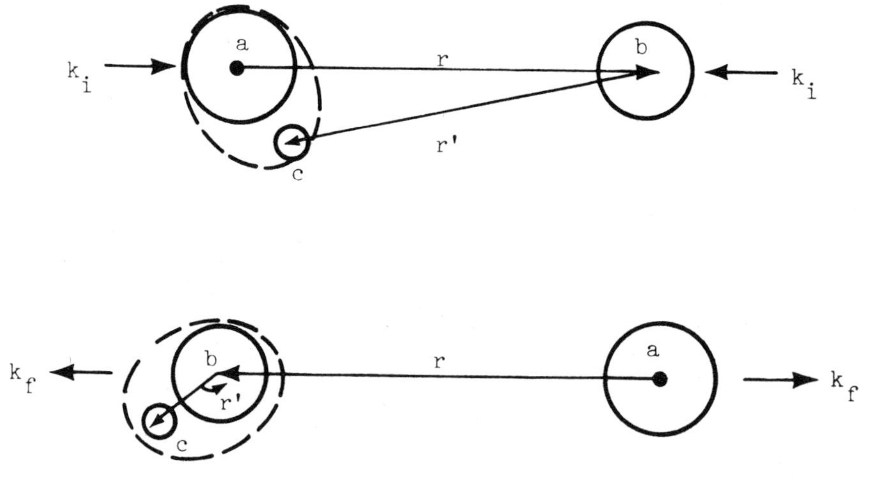

Figure 1.19

Using the coordinates of Fig. 1.19,

$$T_{fi} = \int d^3r \int d^3r' \, \chi_f^{(-)*}\left(k_f; r - \frac{r'}{A_f}\right) u_f^*(r') \, V_{ac}(r + r')$$
$$U_i(r + r') \, \chi_i^{(+)}\left(k_i; \frac{A_i - 1}{A_i} r - \frac{r'}{A_i}\right) \qquad (1.48)$$

where u_i, u_f are bound-state wave functions for c in the initial and final states, and $A_i = m_a + m_c/m_c$, $A_f = m_b + m_c/m_c$. This integral can be evaluated exactly and the correct procedure for calculating transfer reactions is: determine the distorted waves from an analysis of elastic scattering where the potential is fixed by some prescription such as that of Section 1.3, and then use them in the transfer integral.[84] This prescription has had many successes, but we wish here to concentrate on *failures*. Therefore, it is instructive to disentangle the various contributions to the six-dimensional integral.

A great simplification occurs if "recoil effects" are dropped, i.e., r'/A_f and r'/A_i are removed from the distorted waves. Then:

$$T_{fi} = \int d^3r \, \chi_f^{(-)*}(\underline{k}_f; \underline{r}) \quad \chi_i^{(+)}\left(\underline{k}_i; \frac{A_i - 1}{A_i} \underline{r}\right) G_{if}(r)$$

$$G_{if}(r) = \int d^3r' \, u_f^*(\underline{r}') \, V_{ac}(\underline{r} + \underline{r}') \, u_i(\underline{r} + \underline{r}') \qquad (1.49)$$

and we have two 3-D integrals. If, in addition, we make the "zero range" approximation:

$$G_{if}(r) = u_f^*(-r) \, u_i(0)$$

and

$$T_{fi} \propto \int d^3r \, \chi_f^*(k_f, r) \, \chi_i(k_i, r) \, u_f^*(r) \; . \qquad (1.50)$$

As an example, take an initial state where (a+c) and b are in $\ell = 0$ while in final state c is bound to b with orbital angular momentum L. The angular momentum transfer is L. Thus $u_f^* \propto \psi_L^*(r) \, Y_L^M(\hat{r})$.

Simplifying still further to a ring locus model (strong absorption) with *plane waves* $e^{i\underline{k}\cdot\underline{r}}$, and if the z-axis is chosen perpendicular to the annulus, $\Theta = \pi/2$ in the spherical harmonics, then

$$T_{fi}^{LM} \propto P_L^M(\pi/2) \int_0^{\pi} d\phi \; \exp[i(\underline{k}_i - \underline{k}_f)\cdot\underline{r}] \; \exp(im\phi)$$

$$\propto P_L^M(\pi/2) \int_0^{2\pi} d\phi \; \exp(iqR\cos\phi + im\phi)$$

$$= 2\pi \; P_L^M(\pi/2) \; J_M(qR) \qquad (1.51)$$

When the cross section is summed over all M-substates, the Legendre function requires L+M even, and therefore even L transfer will have oscillatory angular distributions characterized by:

$$\sum_M [J_M(2kR \sin\Theta/2)]^2 \qquad (1.52)$$

with even M; likewise odd L-transfer will have only odd M and we arrive at the well-known phase rules.

It is found that the main contribution at low energies is associated with $|M| = L$. Classically this corresponds to the transferred particle making the transition between orbits which are nearly perpendicular to the reaction plane; furthermore, as Fig. 1.20 shows,[60] if the initial value of m is $+\ell_i$, the final value will be $-\ell_f$ and the transfer is likely to occur with a large change in the component of L along the z-axis.

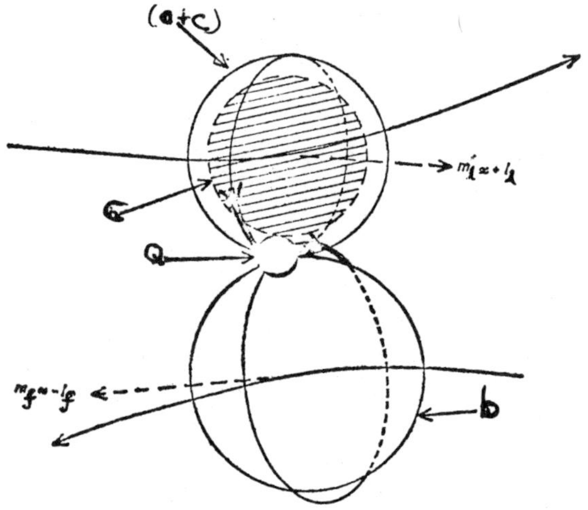

Figure 1.20

The period of the angular oscillations (as usual) is $\approx \pi/kR$ at small angles. Take for example the stripping and pick-up reactions $^{40}Ca(^{13}C,^{12}C)^{41}C$ and $^{40}Ca(^{13}C,^{14}N)^{39}K$ which have been studied at 68 MeV. The data for both reactions[85] shown in Fig. 1.21 have oscillatory angular distributions of period $\pi/kR \approx 5°$ ($k \sim 4.97\ f^{-1}$ and $R \sim 8$ fm). For the stripping reaction, the DWBA (dashed line) works perfectly, but for pick-up (which should be mainly L = 1 transfer) the oscillations are exactly out of phase-- in fact, they fit with M = 0, rather than M = 1 in contradiction to our derived rules, and in contradiction to any reasonable attempts at rectification by the usual parameter juggling of optical model and bound-state parameters. An impressive array of escape routes have been brought to bear on this problem, which certainly do credit to the imagination of the theoreticians! Amongst the possible explanations are helicity spin flip,[86] molecular orbital approach[87] in which the interaction of the transferred particle with both cores is treated explicitly during the entire process. (See also polarization phenomena with the two-center shell model wave functions in heavy-ion transfer;[88] whether such processes are important in heavy-ion reactions depends on the ratio

$$\frac{\text{transit time}}{\text{nuclear period}} \approx \left(\frac{E_{\text{Fermi}}}{E/A}\right)^{1/2} \approx 3$$

for this reaction). An even more formidable explanation may be in the coupled channels approach to heavy-ion reactions.

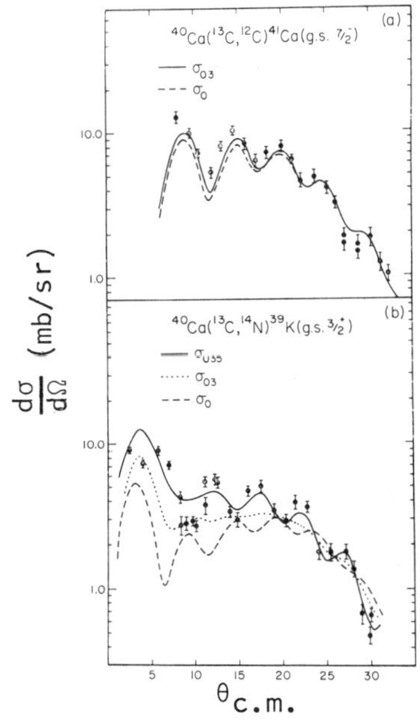

Figure 1.21

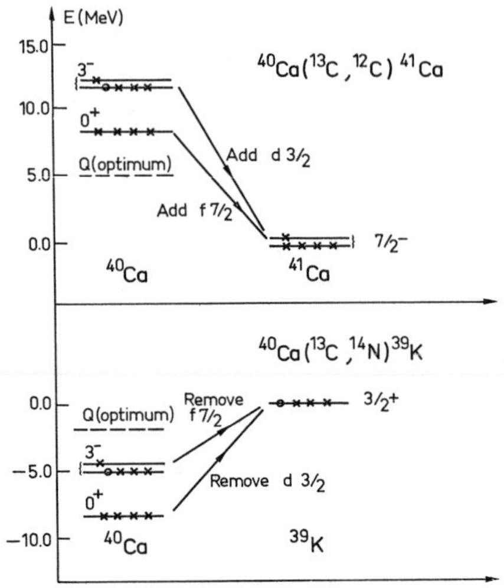

Figure 1.22

Coupled Channel Effects

It has been suggested that in addition to transferring the particle between the *ground states*, other routes may be important through, for example, pre-excitation of the ^{40}Ca prior to transfer.[89] (Such processes are two-step and go beyond the first-order perturbative treatment of the DWBA.) Some possibilities are illustrated in Fig. 1.22. For the stripping reaction the ^{40}Ca gs can be reached by adding an $f_{7/2}$ particle to ^{40}Ca (a transition from $(\ell_i - \frac{1}{2})$ in ^{13}C to $(\ell_f + \frac{1}{2})$ in ^{41}Ca) *or* by adding a $d_{3/2}$ particle to the pre-excited ^{40}Ca, 3^- state $((\ell_i + \frac{1}{2})$ to $(\ell_f + \frac{1}{2}))$. Remember, by our earlier arguments the latter is disfavored; it is further inhibited by the optimum Q-value ($Q_{opt} \approx -\frac{1}{2}mv^2 + \Delta V_C$) which is not very negative for neutron transfer, where $\Delta V_c = 0$. (This expression for Q_{opt} can be derived easily from equs. 1.43, 1.44 by assuming $\lambda_1 \simeq 0$ on the average, evaluating λ_2 from equ. 1.43 and substituting in equ. 1.44.) Therefore the inclusion of these routes does not have much effect on the stripping reaction (see Fig. 1.21).

Both arguments are reversed for pick-up, and we see that inclusion of 3^- and 5^- excitations improve the agreement of the phase of the oscillations.[90] This situation is not very satisfactory, because there are many other routes that could be included, and in fact inclusion of them all would far exceed present computational techniques. Furthermore, the strength required for the inelastic routes appear to exceed those observed experimentally.[91] However, they are still *too few* to produce the average couplings that we know how to handle via an absorptive potential.

The effects of coupled channels not only introduce additional transition routes to the final state; through the inelastic transitions they also modify the optical model wave functions of relative motion. The influence is quite subtle, as illustrated by inelastic scattering[92]

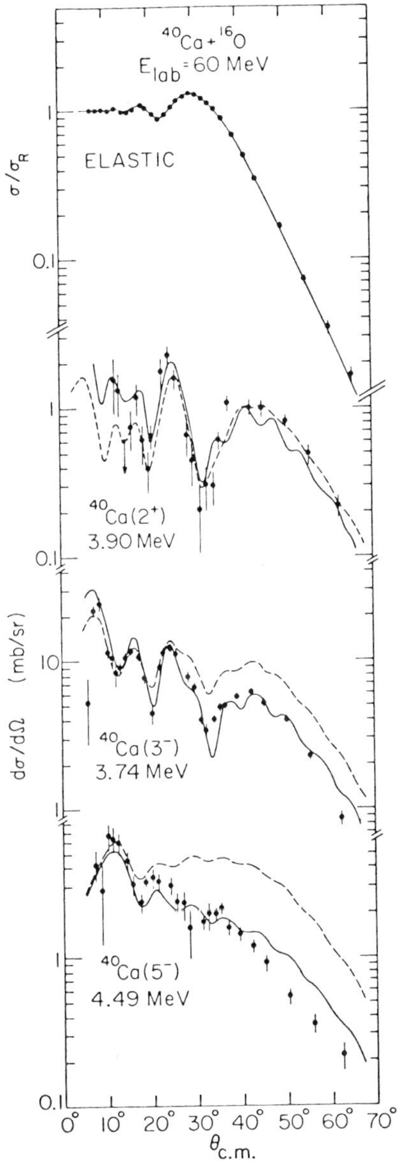

Figure 1.23

of ^{16}O or ^{40}Ca in Fig. 1.23. The DWBA (dashed line) cannot fit all transitions simultaneously. Since the 3^- state is strongly coupled to the ground state, the imaginary part of the optical potential was modified to reproduce the elastic scattering in a coupled channels calculation involving only the ground and the 3^- state. The CCBA calculations then work for all the states (solid line), in particular the 5^- which is not directly coupled. This behavior is in contradiction to the assumption generally made in DWBA that in a direct reaction calculation for a given transition all other non-directly coupled channels can be treated through an average absorptive potential. These observations may account for the failure of DWBA in many heavy-ion transfer calculations, and hopefully will dispense with the ad hoc changes made in optical model parameters.

The previous section may have conveyed the impression that the present status of heavy-ion transfer reactions is a little bit like opening Pandora's Box. Nonetheless, it may be just in these complexities that some of the unique, interesting heavy-ion physics lies for nuclear spectroscopy. Let us look at a striking example. Consider two-neutron transfer, stripping, and pick-up reactions, as illustrated in Fig. 1.24. In pick-up to the 2^+ state, route 2 is direct, and in stripping, 3 is direct. Routes 1 and 4 are branches of indirect routes which can also contribute to transfer via inelastic scattering in the initial and final states. For *vibrational* nuclei the sign of the amplitudes 2 and 3 is opposite and leads to opposite interference patterns with the indirect routes--destructive in stripping and constructive in pick-up.[93,94] A further refinement is introduced by the contribution of Coulomb and nuclear terms to the indirect routes, which enter with opposite signs, and interfere differently with the direct routes.

In the pick-up reaction $^{76}Ge(^{16}O,^{18}O)^{74}Ge$, a very weak interference dip is observed[95] for the 2^+ of $^{74}Ge^*$ but not of $^{18}O^*$. It turns out that the direct transition to the 2^+ of ^{74}Ge is negligible, corresponding to the removal of two neutrons from the gs BCS superfluid vacuum of ^{76}Ge, leaving ^{74}Ge in the 2^+ particle-hole vibration. The main population is from the two-step process, first by the removal of a neutron pair to the gs of ^{74}Ge, followed by the creation of a quasi-particle pair of the 2^+. The dip is then caused by

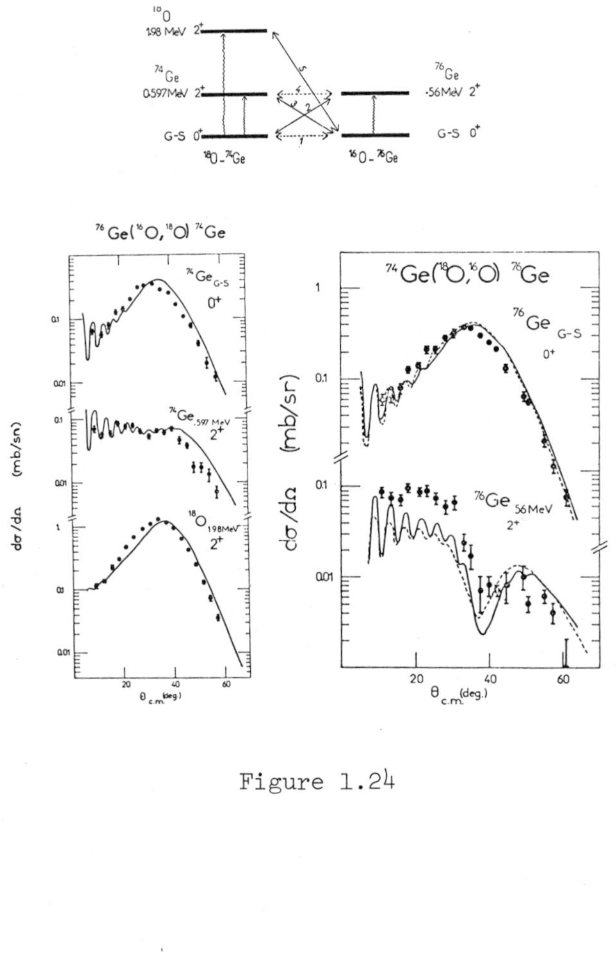

Figure 1.24

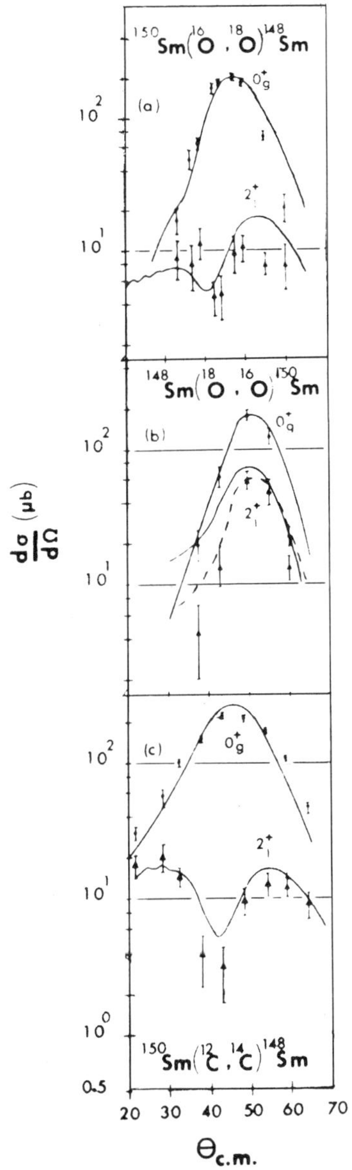

Figure 1.25

Coulomb-nuclear interference in the inelastic scattering section. For the stripping reaction, on the other hand, the direct transition to the 2^+ of ^{76}Ge is strong, and interferes destructively with the nuclear amplitude of the indirect routes, giving rise to a pronounced modification of the characteristic bell-shaped differential cross sections. The ground state transitions are of course identical in the two reactions, since they correspond roughly to time-reversed processes. The theoretical calculations shown require as input optical model parameters for the initial and final channels, deformation parameters for the inelastic excitation, detailed spectroscopic amplitudes for all the states involved in the coupling. The success of the theory is an encouraging indicator that this field--almost unique to heavy-ion transfer--could become important for unravelling sensitive details of the structure of collective states.[96-98]

The data for a similar pair of transfer reactions on Samarium isotopes at 100 MeV are shown in Fig. 1.25. Here the interference is the *opposite* sign from the Sm isotopes.[99] The theoretical curves are the first attempt to incorporate the *dynamic deformation method* with the CCBA formalism. This method is to be contrasted with an alternative attempt[100] to explain these data with the *boson expansion method*. In this latter theory the nuclear deformation effects arise as a result of complex mixing of a large number of spherical bosons whereas in the DDM method the nuclear deformations are introduced in the single particle basis, and further the deformations are treated as dynamic variables (in β and γ). The striking shape differences between the 2_1^+ distributions are however still not satisfactorily explained.

As an illustration of the scope for imagination in the study of heavy-ion reactions, it is fascinating to note that the interference phenomena due to multistep processes can be described in a Regge pole parameterization.[101] There occur two poles found at positions of the barrier-top resonances of the entrance and exit channels, in a well matched reaction. If the poles for the transfer are very different from these, it is a clear sign that intermediate channels are important, indicating a multistep process. Another example comes from the old question of whether surface transparent imaginary potentials are necessary to fit the interference oscillations in two particle transfer reactions.[102] These diffractive oscillations are usually attributed to interferences between a peripheral Coulomb-dominated orbit on one side of the target nucleus and a slightly penetrating orbit on the far side. Too strong an absorption reduces the penetrating flux and extinguishes the interference pattern. However, it is also possible that the Coulomb dominated orbit can be weakened by multistep effects, and the final resolution is a very delicate balance.

There are severe technical problems both in the measurement and the computation of two nucleon transfer reactions of the type described above. To resolve the low lying collective states and identify the two neutron transfer products from elastic scattering is difficult. To calculate the absolute magnitude of two neutron transfer, complicated by problems such as simultaneous v. successive transfer,[103] is also no mean feat. We have only to look at the *quality* of both the data and the theory to wonder if our tools[104] would not be of much poorer quality without the challenge of heavy ions.

However, problems are also showing up in the much simpler *one* nucleon transfer reactions. Recently it has become possible to study heavy-ion transfer reactions over a wide energy range from sub-Coulomb up to 20 MeV/A. An example is the $^{208}Pb(^{16}O,^{15}N)^{209}Bi$ reaction. Because of the variety of low-lying single particle states outside the doubly-magic ^{208}Pb, this reaction has almost become a standard for testing reaction theories.[105]

Techniques for evaluating the finite-range, recoil DWBA are available and have been applied to the ^{16}O + ^{208}Pb data as a function of energy.[105] Such a study is an ideal test of the reaction model, compared to data at a single or closely spaced energies, where deficiencies may be masked by the extreme sensitivity to extraneous details, e.g., the wave functions used to describe the initial and final bound states.

The calculations used optical parameters, V = 51, r_V = 1.11, W = 51, r_W = 1.11, a_V = 0.79, and a_W = 0.74. The bound states were generated in Saxon-Woods wells with the depth adjusted to reproduce the binding energy: for ^{208}Pb + p, r_V = 1.28, a_V = 0.76, $V_{spin-orbit}$ = 6 MeV, r_{so} = 1.09, and a_{so} = 0.60; for ^{15}N + p, r_V = 1.20, a_V = 0.65, v_{so} = 7 MeV, r_{so} = 1.20 and a_{so} = 0.65. The resultant spectroscopic factors, normalized to unity for the ground state are shown in Table 1.3 and compared with other reactions and with theory. The satisfactory agreement is typical of the other beam energies *when each set of data is treated in isolation.*

When we compare experiment and theory as a function of energy (using the theoretical spectroscopic factors with their absolute values, when $S(h_{9/2})$ = 0.95) a failure of the theory by almost a factor of 10 is encountered from the sub-Coulomb energy of 69 MeV

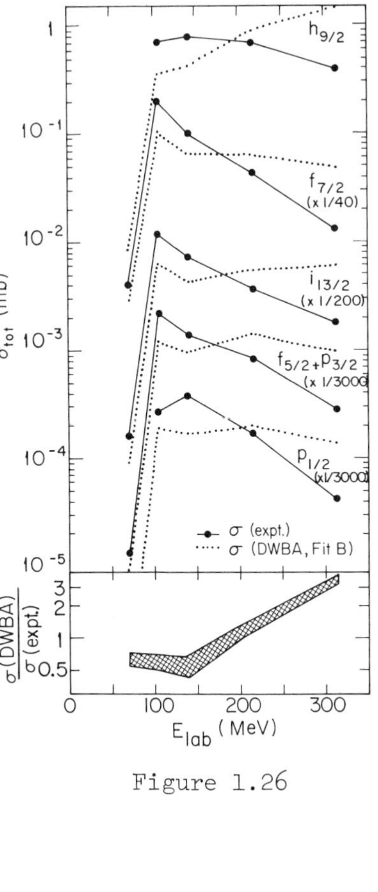

Figure 1.26

TABLE 1.3 Spectroscopic factors for $^{208}Pb(^{16}O,^{15}N)^{209}Bi$ data at 312.6 MeV.

State	E	$S(^{16}O,^{15}N)$	$S(^{12}C,^{11}B)$	$S(^{3}He,d)$	S(Theory)
$1h_{9/2}$	0.00	1.00	1.00	1.00	1.00
$2f_{7/2}$	0.90	0.85	0.96	0.67	0.89
$1i_{13/2}$	1.61	0.77	0.89	0.48	0.74
$2f_{5/2}$	2.84	0.77	0.64	0.75	0.69
$3p_{3/2}$	3.12	0.74	0.82	0.57	0.78
$3p_{1/2}$	3.64	0.69	--	0.38	0.57

up to 312.6 MeV (see Fig. 1.26). Of course such disagreements could be patched up, energy by energy, by ad hoc variations of bound state parameters and optical potentials, sacrificing if necessary the qualitative relationship of the bound state potentials to the nucleon-nucleon optical potential, as well as the quality of the optical model fits to the elastic scattering. Such strategems miss the spirit of the model and even worse have no predictive power. Rather we should say that the method has failed and look for possible causes, as yet unknown.

1.6 Compound Nuclear Reactions

It may have come as a surprise that our discussion of transfer reactions had nothing to say about multinucleon transfers of more than four nucleons. It was discovered that such reactions usually proceed by the formation of a compound nucleus,[106] with subsequent evaporation of a complex fragment. These reactions also have some striking characteristics. For example, the differential cross sections are symmetric about 90° with a form $1/\sin\theta$, characteristic of emission from a high spin compound nucleus[106]:

$$\left(\frac{d\sigma}{d\Omega} \rightarrow \frac{d\sigma}{d\theta} \cdot \frac{d\theta}{d\Omega} \Rightarrow 1/\sin\theta, \text{ since } \frac{d\sigma}{d\theta} \text{ is constant}\right).$$

Sometimes the spectra show a highly selective excitation of high spin states (reminiscent of a direct reaction) and often they are entirely featureless. Compare for example the reactions $^{14}N(^{14}N,\alpha)^{24}Mg$ and $^{10}B(^{12}C,d)^{20}Ne$ in Fig. 1.27.[107,108]

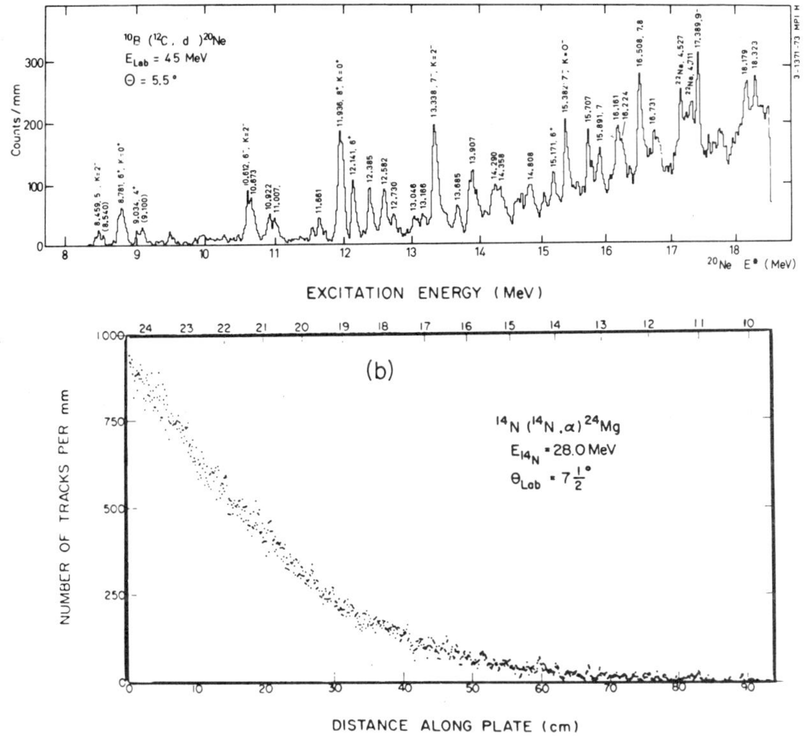

Figure 1.27

It turns out that both the formation and decay of the compound nucleus are dominated by a few partial waves close to the grazing value, and therefore it is plausible that only those levels located inside or near the curve defined by $L_{inc}^{grazing}$ and $L_{out}^{grazing}$ (which is a function of the Q-value and excitation energy of the reaction, i.e. $E_f = E_{CM} + Q - E_X$ and $L_{out}^{grazing} \approx R_f\sqrt{2M_fE_f}$) will be strongly excited. The shape of the spectrum is determined by the overlap between this curve and the yrast line of the final nucleus, the lowest excitation possible in the nucleus for a given J. Above this locus the level density increases exponentially. So one expects for example, from Fig. 1.28, that the $(^{12}C,d)$ reaction would be selective[109] and the $(^{14}N,\alpha)$ reaction not,[110] which is just the experimental observation.

Figure 1.28

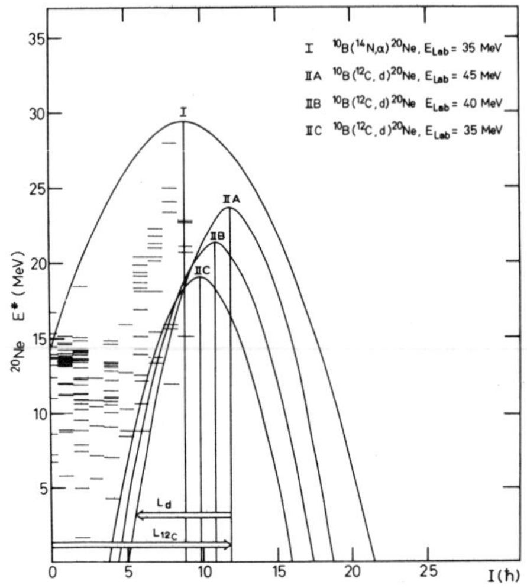

For a detailed quantitative treatment, Hauser-Feshbach calculations are necessary,[111] with many attendant technical and philosophical difficulties. In the formation of the compound nucleus, the summation over angular momentum may have to be truncated, because the compound nucleus is unable to support large amounts before fission. The spin cut-off and level density parameters have to be determined. It turns out that the calculations of the *ratio* of two cross sections is relatively stable against all these multifarious uncertainties. The fits of the ratio of the statistical theory cross sections for states at E* = 11.92 and 12.14 in

Figure 1.29

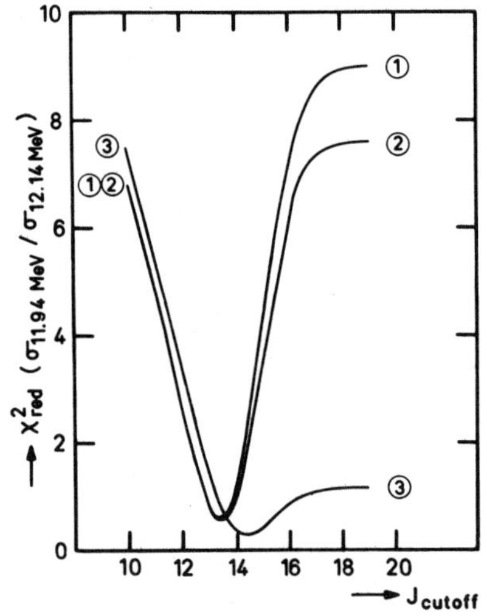

^{20}Ne to the ratio of the experimental cross sections for different choices of the level density parameter "a" (curves 1 and 2 average "a" over shell effects (a ~ A/6); curve 3 takes into account the final nucleus shell effects) are shown[112] in Fig. 1.29, as a function of the angular momentum cut-off, J_{crit}. Clearly this quantity can be deduced with high accuracy(14 ± 1) for this $^{10}B(^{12}C,d)^{20}Ne$ reaction at 45 MeV. (We shall discuss the origins of J_{crit} in the next lecture.) But clearly, having determined it for states of *known* spin, the procedure can be turned around, and new spin assignments made from the observed relative cross sections. (For a more detailed discussion see my lecture notes in Ref. 30.)

Now, we go *beyond conventional spectroscopy* and we discuss the evidence for nuclear molecular states, which are formed by the two colliding ions rotating in a dumb-bell configuration.[113,114] These have manifested themselves as resonances in the excitation functions of heavy-ion elastic scattering[115] and of reactions. For $^{12}C + ^{12}C$ and $^{16}O + ^{16}O$ elastic scattering the resonances are shown[116] in Fig. 1.30. There are wild oscillations which continue unabated to high energies (the equivalent excitation energy in ^{24}Mg for the $^{12}C + ^{12}C$ system is E_{CM} + 13.93 MeV). At the lower energies the resonances have been interpreted as shape resonances and fitted[117] with a potential of the form shown in Table 1.4.

The fits obtained have the correct characters (see Fig. 1.31) and at certain energies are almost pure $[P_L(\cos\theta)]^2$. The values of

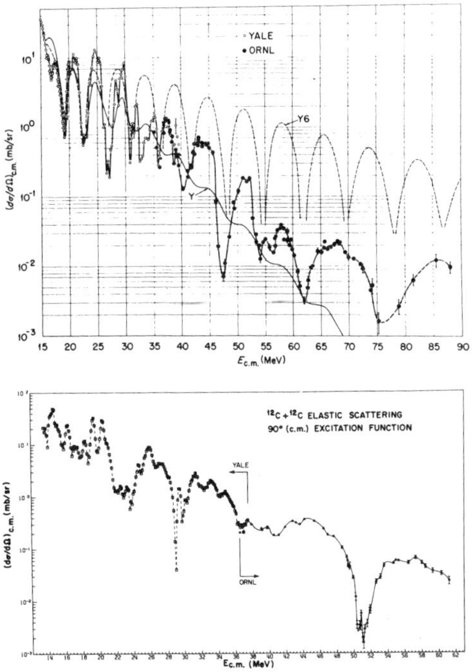

Figure 1.30

TABLE 1.4

System	V	R	a	W	R_I	a_I
$C^{12} + C^{12}$	14	6.18	0.35	$0.4 \pm 0.1E$	6.41	0.35
$O^{16} + O^{16}$	17	6.8	0.49	$0.8 \pm 0.2E$	6.40	0.15

L are shown at the right. At these energies the phase shifts are close to $\pi/2$. The quality of fit for the $^{16}O + {}^{16}O$ system up to high energies with the above potential appears in Fig. 1.30 as the curve Y6. Weak absorption is essential for a description of the resonance width ($\approx 2W$); only if the surface regions remain transparent can the interacting nuclei retain their identity for a sufficiently long period to make a molecular description meaningful. (For a microscopic justification of this transparency, see refs. 118, 119.)

Closer examination of the excitation functions reveals that in addition to the potential shape resonances there is a superimposed fine structure of ≈ 100 KeV width. Such structure has been discovered in the excitation functions of many reaction channels. A good example is the $^{12}C(^{12}C,p)^{23}Na$ reaction illustrated in Fig. 1.32

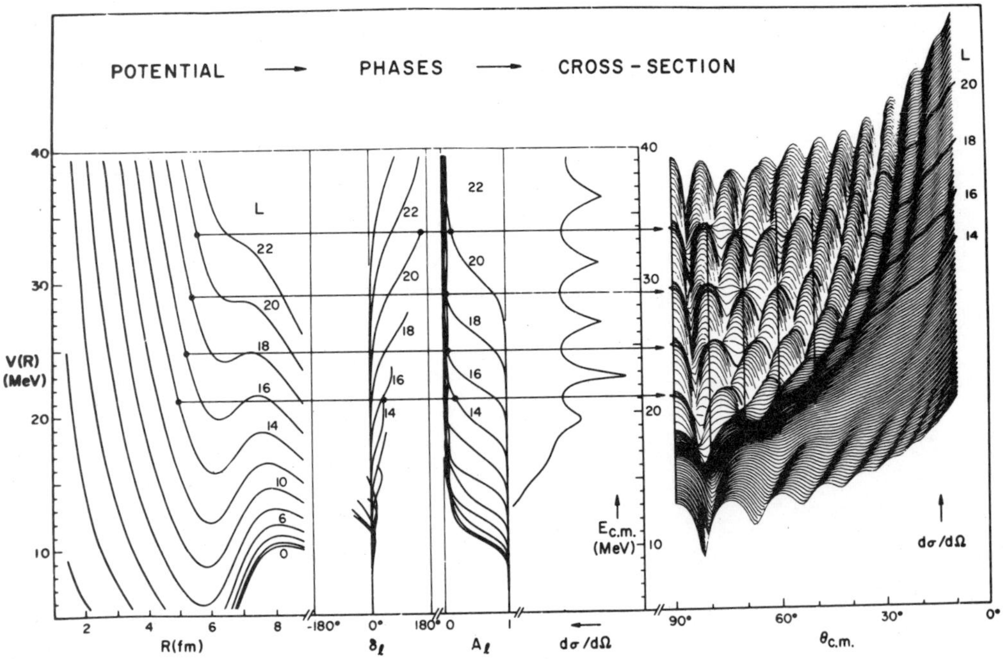

Figure 1.31

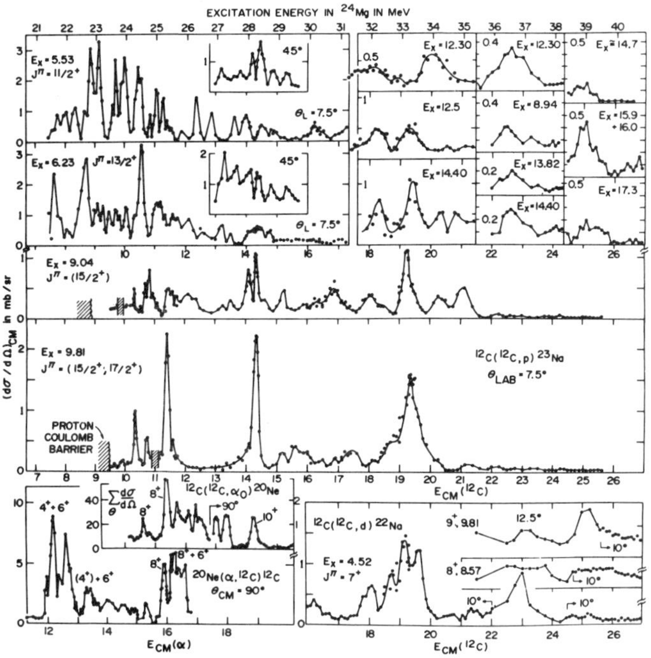

Figure 1.32

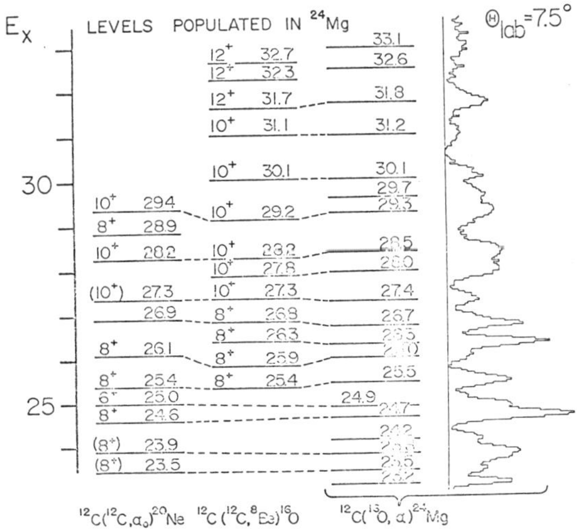

Figure 1.33

for several different residual states in ^{23}Na, and compared with other outgoing α, d channels.[120] The equivalent excitation energies of the compound ^{24}Mg system is shown at the top. There exist pronounced narrow resonances at 11.4, 14.3 and 19.3 MeV which are strongly correlated in different channels. By comparing branching ratios, spins of 8^+, 10^+ and 12^+ were assigned.

Another example is the $^{12}C(^{16}O,\alpha)^{24}Mg$ reaction[121] for which the energy spectrum, averaged over incident energies from 62-100 MeV, is shown in Fig. 1.33, and compared with other "α-particle" channels. Possible correspondences in the spectra are indicated by the dashed lines. Because of the differing non-resonant background which can interfere with the resonant amplitude, the energy of the resonance is not necessarily the same in all channels; however the shift cannot be much larger than the width (note that in contradistinction to our discussion of this type of reaction earlier, there is evidence for direct aspects in the observed selectivity--e.g., there is a preponderance of positive parity levels, whereas positive and negative natural parity states in the J = 6 to 12 ħ region are expected on the compound picture; these multinucleon transfers may therefore also be useful for populating states of particular structure in a direct process). We notice that the levels appear to be grouping themselves into clusters of a given J^π.

A summary of all reported resonances[114] appears in Fig. 1.34; the groups fall on a line constituting a Regge trajectory[122], or quasi-molecular rotational band, where

$$E_J \; \alpha \; \frac{\hbar^2}{2\mathcal{J}} \; J(J+1) \tag{1.53}$$

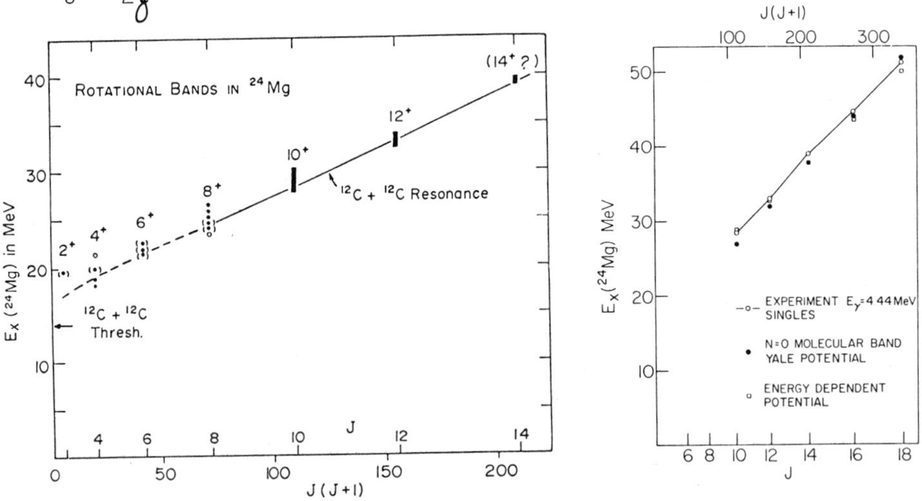

Figure 1.34

The resonances correspond to pockets in the potential for the different partial waves (see Fig. 1.31). The slope of the line in Fig. 1.34 corresponds to the $\hbar^2/2\mathcal{J} \approx 100$ KeV, just the value we calculate for two carbon nuclei in dumbell rotation at the grazing distance (see Fig. 1.35). (For comparison, the $\hbar^2/2\mathcal{J}$ of the ground state band is ≈ 200 KeV, i.e. a lower moment of inertia $\approx \frac{2}{5} MR^2$. Extrapolation of the band to the 0^+ member on the vertical axis shows that the band begins almost at the threshold for $^{12}C + {}^{12}C$ in 24 Mg, as predicted in a cluster molecular model.[123] Pushing the picture still further, we obtain the value 2.6×10^{21} sec^{-1} for the frequency of rotation corresponding, e.g. to the 8^+ resonance at $\approx$25 MeV, and considering the envelope of all the 8^+ resonances ($\approx$3 MeV) as the width of the molecular resonance, we obtain a life-time of 4×10^{-22} sec. Thus the two ^{12}C nuclei would perform $\approx 1/10$ of a full rotation before either coalescing a splitting into the $^{12}C + {}^{12}C$ exit channel.[124]

The fact that the resonances of a given spin group and secondly that their centroids fall close to the value of the Yale potential (Table 1.4) suggests that, because of the gross structure, windows exist for the specific angular momenta. These windows permit the carbon nuclei to be in close contact, to interact and thereby to fragment into a number of narrow doorway state resonances. This interaction must be weak, because a strong one would have moved the resonances out of the window. Also the summed widths of a resonance of given J is an appreciable fraction of the gross structure width. Several models of this fragmentation exist,[114] one of which involves the excitation of the ^{12}C nucleus to its 2^+, 4.43 MeV level, or the double excitation of both nuclei.[125,126] A resonance occurs at an

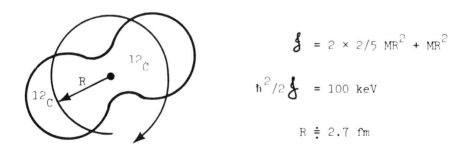

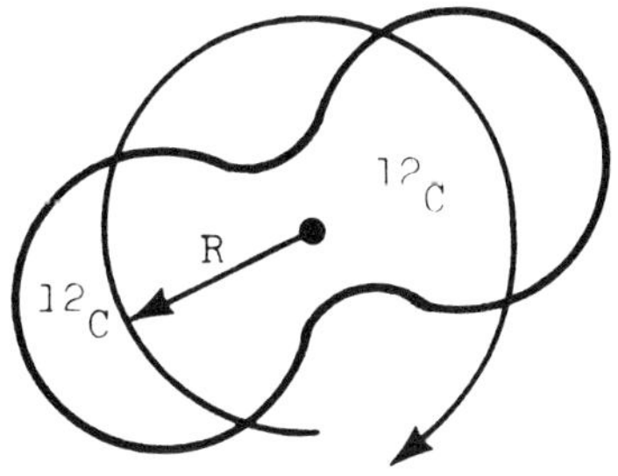

$$\mathcal{J} = 2 \times 2/5\ MR^2 + MR^2$$

$$\hbar^2/2\mathcal{J} = 100 \text{ keV}$$

$$R \doteq 2.7 \text{ fm}$$

Figure 1.35

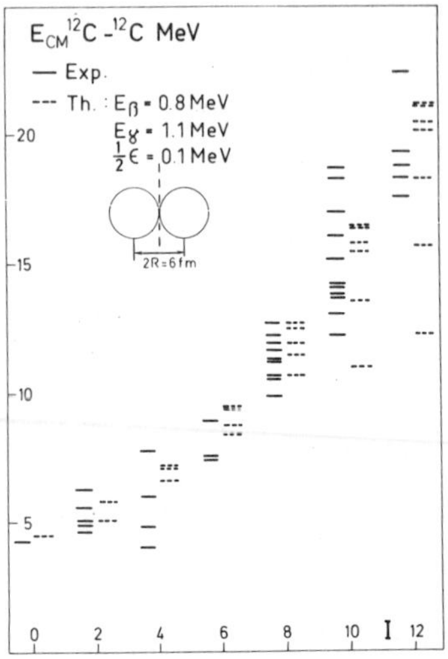

Figure 1.36

energy such that after the excitation of the nuclei, they are in a quasi-bound state of the appropriate angular momentum. Thus the doorway state consists of excited ^{12}C nuclei trapped in a potential well pocket. Another approach[124] lets the shock of the initial collision lead to surface vibrations in the system, similar to β,γ vibrations. These split up the wide rotational resonance. Applying the first order rotation-vibration model[127] leads to a rather satisfactory agreement with the data (Fig. 1.36).

Support for the first picture of the resonances comes from a recent experiment [128,129] on the integrated cross sections for the reactions $^{12}C(^{12}C,{}^{12}C^*)\,^{12}C^*$ where either of the final ^{12}C can be excited into the 2^+ level at 4.43 MeV. Figure 1.37 shows that both

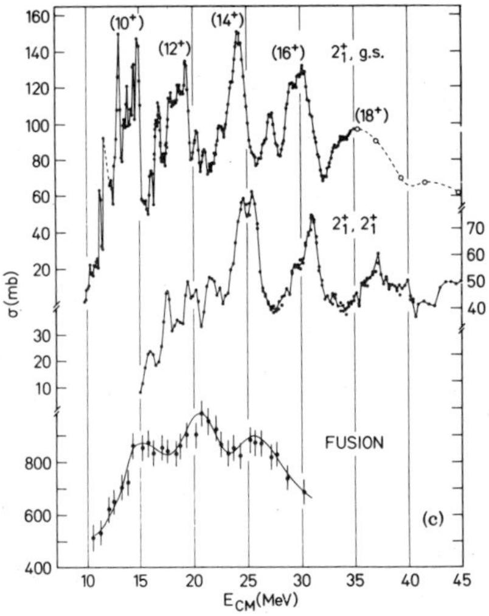

Figure 1.37

the double and single excitation functions are dominated by broad resonances *and* underlying fine structure. The upper three resonances fall nicely on the continuation of the molecular band, with the same moment of inertia, and with suggested spins 14^+, 16^+, 18^+ (see Fig. 1.34). The resonances also appear to line up with data on the fusion cross section. A partial width decomposition for the $J^\pi = 10^+, 12^+, 14^+$ gross structure resonance is made by assuming that the experimental total width is given by:

$$\Gamma = \Gamma_C + \Gamma_{2^+} + \Gamma_{2^+,2^+} + \Gamma_{cn}$$

and that, 1.54

$$\sigma_i = 2(2J + 1)\pi\lambda\!\!\!^{-2} \frac{\Gamma_c\Gamma_i}{(\Gamma/2)^2}$$

(with $i = 2^+$, $2^+\cdot 2^+$ and cn) relates the resonant total cross sections δ_i and the various partial widths. The compound nucleus cross section σ_{cn} and width Γ_{cn} are identified with the resonant component of the fusion cross section. One of the resultant solutions of the quadratic equations is given in Table 1.5, and compared with the predicted total width of the quasi-molecular model.

TABLE 1.5

J^π	Ex ^{24}Mg	Γ_{TOT}	Γ_C	Γ_{2^+}	$\Gamma_{2^+2^+}$	Γ_{cn}	Molecular Band Ex.	Molecular Band Γ_{tot}
10^+	28.5	1.8	1.35	0.11	$\leqslant$0.01	0.13	28.6	1.1
12^+	33.0	3.0	2.41	0.22	$\leqslant$0.04	0.33	32.8	2.0
14^+	39.0	2.5	1.94	0.27	0.13	0.16	37.8	3.4

The extracted widths are somewhat less than those of the quasi-molecular rotational band, indicating the intermediate structural nature of the states. It is also true that this type of intermediate structure, believed once to be almost unique to the $^{12}C + {}^{12}C$ system, is also emerging[130-132] in the $^{12}C + {}^{16}O$ and, more excitingly, in much heavier systems, as we now discuss.

Recall the system $^{16}O + {}^{28}Si$ which we discussed earlier (Section 1.3.1) as an example of elastic scatterings over a wide energy range to determine the optical potential. Recently[133] angular distributions have been extended into the backward hemisphere (Fig. 1.38), and reveals an oscillatory pattern, which is quite distinct from the forward angle Fresnel and Fraunhoffer diffraction patterns. In fact a continuation to backward angles of the angular

Figure 1.38

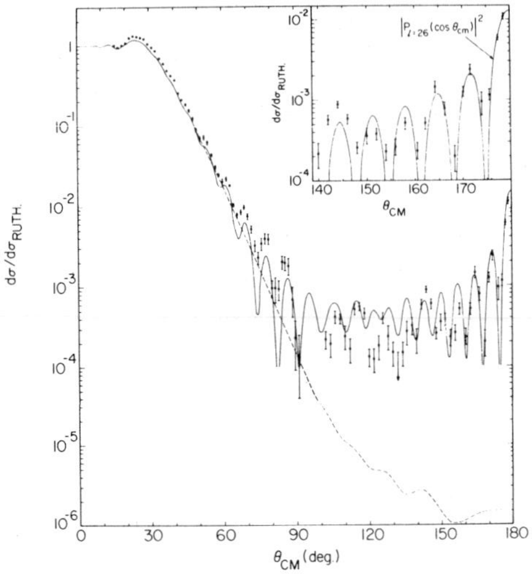

distributions predicted by the "unique" potentials established in section 1.3 leads to the dashed curve. The oscillations are characteristic of $|P_\ell = 26(\cos\theta)|^2$, with $\ell = 26$ close to the grazing partial wave, and *may* find a natural explanation in terms of a surface Regge pole resonance.[101,134]

It is therefore perhaps no surprise to find that excitation functions for $^{16}O + ^{28}Si$ and $^{12}C + ^{28}Si$ also give rise to resonance structure[135,136] very similar to the lighter systems we have been discussing, as the examples in Fig. 1.39 show.[135] At each of the resonances, the differential cross sections have a fairly pure $|P_\ell(\cos\theta)|^2$ form, and for the peaks in $^{16}O + ^{28}Si$ at E_{CM} = 21, 26, 32 and 35 MeV, the ℓ values are 9, 17, 22 and 24 ℏ. The irregular

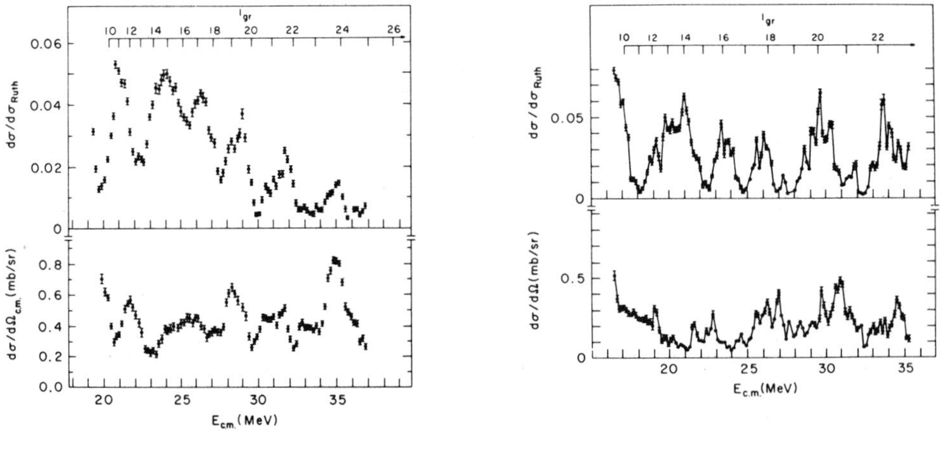

Figure 1.39

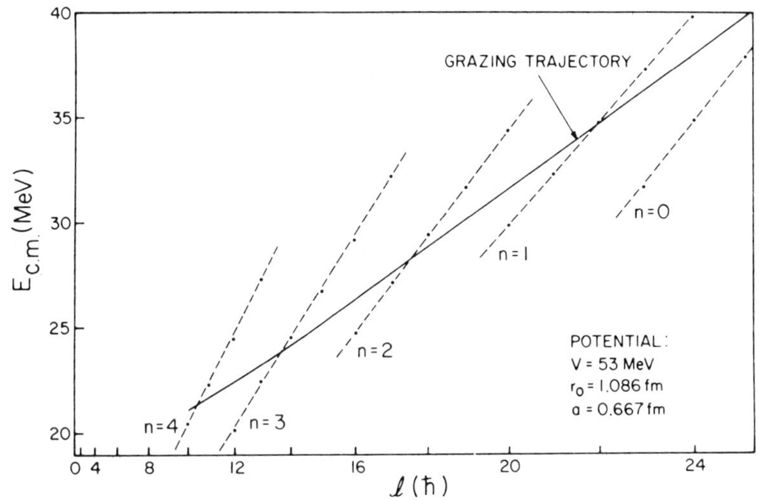

Figure 1.40

spin sequence is very difficult to reconcile with the Regge molecular band,[137] which follows the grazing trajectory. However a calculation of shape resonances using a folding model potential leads to several rotational bands, all with moments of inertia smaller than the grazing trajectory. The observed irregular sequence could be due to the intersection of the grazing trajectory (see Fig. 1.40) with rotational bands of different principal quantum numbers.[135] It would

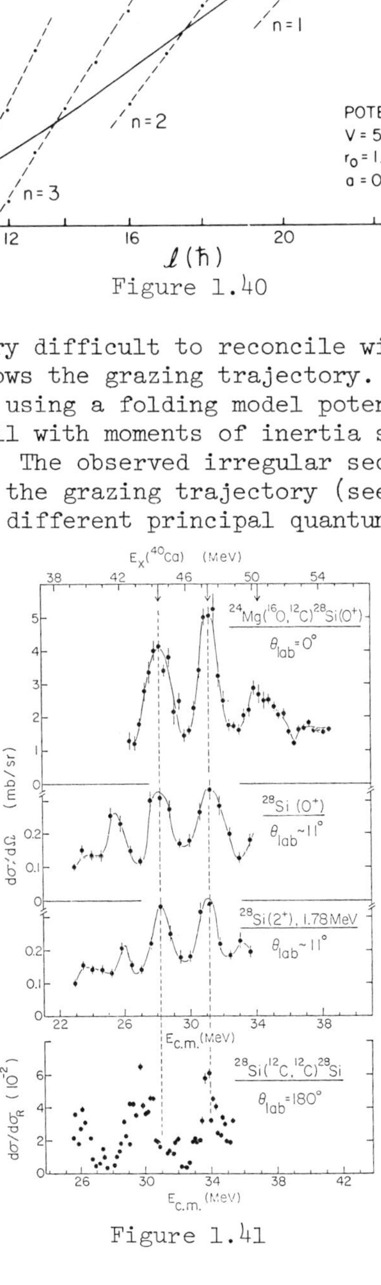

Figure 1.41

also be interesting to know whether interferences between the reflected waves from the inner and outer potential barrier as have been recently discussed[137(a)] might produce the structures. A very recent explanation has been given in terms of a parity-dependent potential.[137(b)]

Since we have primarily discussed elastic scattering and transfer reactions in this lecture, it is appropriate to end with a synthesis of the two, which gives a new direction towards the understanding of these resonances. If these phenomena indeed occur in the grazing partial waves, then similar effects might show up at forward angles in *transfer* reactions, where the contributing ℓ-waves are also strongly surface peaked. The excitation function[138] for the ^{24}Mg (^{16}O, ^{12}C) ^{28}Si reaction appears in Fig. 1.41. (Here the *exit* channel is one in which resonances in elastic scattering are observed.) There is indeed strong resonant behavior, which coincides with, elastic and inelastic channels. Are these also shape resonances, generated by surface transparent potentials, or are they evidence for more subtle effects in the structure of ^{40}Ca at high angular momenta? Perhaps the α-transfer plays a special role, and therefore many other channels have to be tested. It seems clear however that even complicated systems at very high excitation are revealing a most unexpected simplicity.

There is hope that this simplicity can be treated in a microscopic model which describes the fragments by displaced oscillator shell model wave functions.[450] For ^{16}O + ^{16}O and + ^{40}Ca the minima of the energy expectation values for various angular momenta are in good agreement with the experimental resonance energies, confirming the concept of an underlying quasimolecular structure. A first test of this interpretation is provided by the fact that the intrinsic state of such a nuclear molecule has mixed parity. Whereas shell model states show a gap of $\approx \hbar\omega$ between positive and negative parity states, a nuclear molecule should have positive and negative parity states in a common band. Hence, if the concept of a nuclear molecule is applicable one should find little or no splitting between bands of positive and negative parity. For the system α + ^{40}Ca the experimental splitting is less than 0.6 MeV. The microscopic description also yields resonances in the ^{16}O + ^{40}Ca system and therefore they appear to be a widespread feature of heavy-ion systems both experimentally and theoretically. The microscopic treatment shows that a description in the framework of a simple optical potential must be non-local and energy dependent. This fact may explain the recent spurt of activity which "explains" the resonances in the ^{16}O + ^{28}Si system by a variety of unusual potentials, e.g., a parity dependent potential,[137(b)] or an energy dependent, surface transparent potential.[451]

Only a short time ago, the resonances in the $^{12}C + ^{12}C$ system were believed to be unique, giving us only a glimpse of shape resonances and also the next stage in the hierarchy of increasing complexity of doorway states. The carbon nuclei avoided both the Scylla of being too easily polarizable and the Charybdis of not being polarizable at all.[114] Now we are through these straits, and the whole ocean lies ahead to explore for years to come. This exploration can be made with the low-energy, Tandem Accelerators scattered around the world. Compared with the mighty ocean-going Titanic of the Berkeley Bevalac, these "outboard motor boats" are inexpensive to run, and it is exciting that they continue to reveal fundamental aspects of the nucleus. Hopefully the Berkeley Bevalac will lead to its own fundamental discoveries, but that subject must wait until the last lecture. In the next lecture, we move on to much higher perturbations of the nucleus, beyond the region of discrete excitations, which has dominated our discussion of Microscopia.

2. MACROSCOPIA (FUSION AND FISSION)

The last lecture ended on a hopeful note. By means of heavy-ion reactions, the possibility is at hand of observing nuclei under unusual conditions of rotation and shape. Already discrete states of spin 18 $\hbar$ have been observed in nuclei at excitation energies of over 50 MeV. The theoretical description of this state of motion presents a challenge comparable to understanding the rotation of homogeneous masses as idealized representations of planets and stars back in Newton's days. It is a challenge that has been met in a remarkable series of experimental and theoretical developments. In this lecture we convey some idea of violent changes of shape undergone by the nucleus as more angular momentum is added to the fused system. Eventually the nucleus cannot sustain the centrifugal forces and it flies apart in fission. This behavior has an important bearing on the problem of synthesizing superheavy elements, once regarded as the prime motivation for the construction of heavy-ion accelerators.

2.1 Nuclei at High Angular Momentum

Before embarking on a discussion of nuclei subjected to these extreme stresses, we should note that the determination of nuclear matter and charge distributions of nuclei near their ground states has long been an important stimulus to the development of nuclear structure theories. Information on the moments of the nuclear charge distribution comes from experiments with electromagnetic probes, whereas the nuclear matter distributions come from hadronic scattering experiments. The availability of high energy, heavy-ion beams has expanded the horizons for inelastic excitation by hadrons, because they display interesting interference effects between

Coulomb and nuclear excitation. In the DWBA, the excitation of a collective level is described in the interaction form factor $F_L^C(r) + F_L^N(r)$, where

$$F_L^C(r) = \frac{eZ_1 \; 4\Pi_1 \sqrt{B(EL)}}{2L+1} \; \frac{1}{r^{L+1}} \tag{2.1}$$

$$F_L^N(r) = \beta_L^N (V_R R_R \frac{df}{dr} + i \; W_I R_I \frac{dg}{dr})$$

Here L is the multipolarity of the transition and F^C and F^N are Coulomb and nuclear excitation forces. The latter is proportional to the derivative of the optical potential. β_L^N is the potential deformation. Since V_R is usually attractive, while the Coulomb potential is repulsive, there result minima in the scattering angular distribution of excitation functions.

From vast and beautiful literature on this subject,[139,140] we select an example from the collision of very heavy nuclei,[141] Kr + Th and Ar + U (Fig. 2.1). The excitation functions for back-scattered particles in coincidence with the de-excitation γ-ray cascade are shown. The solid line is the prediction of pure Coulomb excitation (using a semi-classical approach[142]), which agrees with the low spin data. But there is a rich variety of interference phenomena due to Coulomb-Nuclear interference; the sign, strength and energy for onset are state dependent. The solid and dashed-dot lines use proximity nuclear potentials[143] of

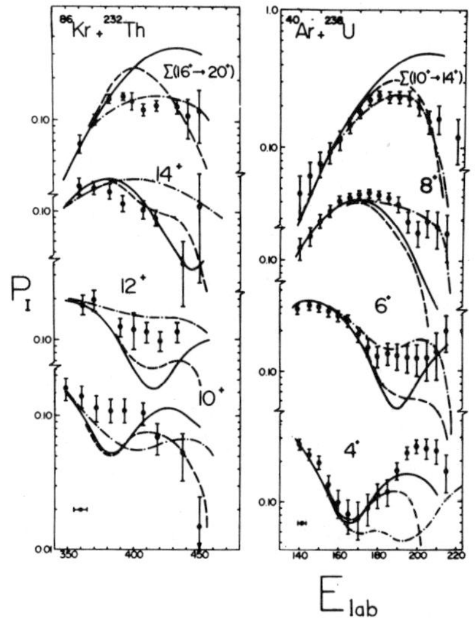

Figure 2.1

the type we discussed in Lecture 1. Since these potentials fit some states but not others we infer that inelastic excitation carries information about the nuclear potential *beyond* that contained in elastic scattering. It may therefore be possible to probe the nuclear surface directly, and we may learn even more about the delicate shapes of nuclei such as ^{234}U, at present known to carry both quadrupole (β_2) and hexadecupole (β_4) deformations (Fig. 2.2).[139] There are also some remarkable experiments on Coulomb excitation of low lying states in Pt with Xe projectiles, that suggest rigid triaxial shapes,[144] contradicting theories of γ-soft nuclear potential surfaces.[145]

Another recent development in the study of deformations describes the Coulomb excitation of collective states by a long range imaginary potential.[146] The remarkable merit of this approach is that a nontrivial theory with no free parameters can be evaluated without a computer, and gives specific cross-section predictions.[147] Indeed the beauty of *both* the above methods is the reliance on semiclassical, analytical methods, originally touted as the great virtue of heavy-ion collisions, but which fell into disrepute for a few years, to return now with renewed vigor.

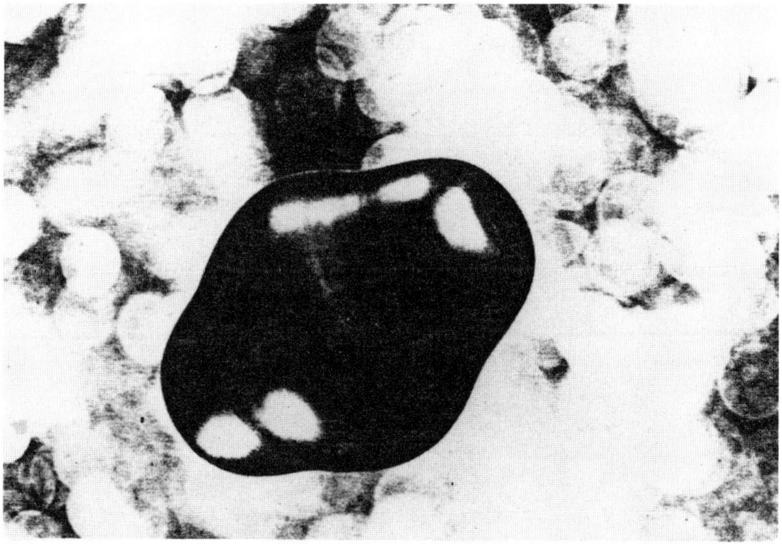

Figure 2.2

The discovery in 1971 of a pronounced irregularity around spin $16\hbar$ (called backbending) in the otherwise very regular behavior of the rotational sequence of even-even rare earth nuclei, has opened up a vigorous research field in the study of high angular momentum in nuclei.[149,150] An illustration of the backbending phenomenon appears in Fig. 2.3.

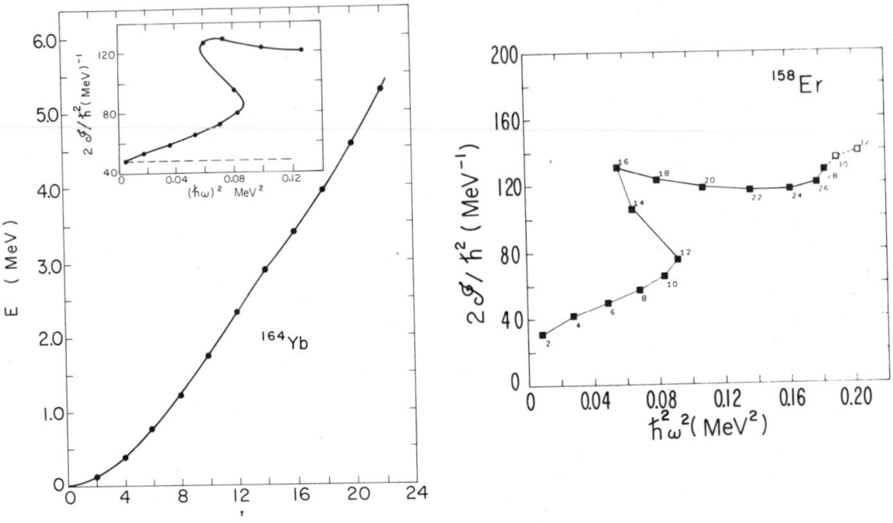

Figure 2.3

A slight discontinuity is evident in the plot of:

$$E_J \propto \frac{\hbar^2}{2\mathcal{J}} J(J + 1) \tag{2.2}$$

at J = 14. On the Variable Moment of Inertia model[151] we write:

$$E_J = \frac{\hbar^2}{2\mathcal{J}} J(J + 1) + \tfrac{1}{2} C \left[\frac{\mathcal{J}}{\hbar^2} - \frac{\mathcal{J}_0}{\hbar^2} \right]^2 \tag{2.3}$$

and

$$\frac{\mathcal{J}}{\hbar^2} = \frac{\mathcal{J}_0}{\hbar^2} + \frac{3}{4C} (\hbar\omega)^2 \tag{2.4}$$

Therefore a plot of moment of inertia versus the rotational frequency squared should yield a straight line. The Inset in Fig. 2.3 shows a marked departure from this trend, with a sudden increase in the moment of inertia.

Three effects have been given serious consideration as the causes for backbending. These are:[150]

- a collapse of the pairing correlations;[152]
- a shape change, i.e. change of deformation;[153]
- an alignment of the angular momenta of two high j nucleons with that of the rotating core.[154]

The fact that the moments of inertia of a most deformed nuclei are about one-half of the rigid body value is attributed to pairing correlations, which partly prevent the nucleons from following the rotation. It now appears more likely that backbending is due to the breaking of one pair rather than total pairing collapse (the gradual reduction of pairing appears rather to account for the variable moment of inertia up to the backbend). The physical process involved in breaking the pair is the Coriolis force which forces the angular momentum vector j of the particle to decouple from the deformation (symmetry) axis and align with the rotation axis. In the $i_{13/2}$ orbit, for example, this effect gives a total of $12\hbar$ which can replace an equal amount of core rotational angular momentum.

On this model, at still higher angular momenta, additional pairs of high-j nucleons will tend to be aligned, and just such a discontinuity appears to be observed[155] in the $^{122}Sn(^{40}Ar,4n)^{158}Er$ reaction at 166 MeV, in which large amounts of angular momenta are deposited (Fig. 2.3). Here the second discontinuity at $J = 28\hbar$ appears to make a further step towards the formation of an oblate nucleus in which all the angular momenta is carried by aligned particles.[156] At the first backbend, two different rotational bands cross. Below the crossing, the levels belong to the ground state band, and above they belong to a superband with a larger moment of inertia. Another explanation of the *second* discontinuity operates from the assumption that if the superband is really based on an aligned two-particle (high j)^2 configuration, then the superband should cross the ground state band not once but *twice*.[157] In this case, (see Fig. 2.4) beyond the second crossing, the lowest band is again the ground state band. A test of this model would be to follow the groud state band beyond the first crossing to see how the energies of these levels compare with the prediction.

The existence of two bands has been demonstrated *directly* in some cases by following the ground state band *beyond the backbending* region. Such is the case in ^{164}Er for which the γ-deexcitation spectra following Coulomb excitation with a ^{136}Xe beam, and the

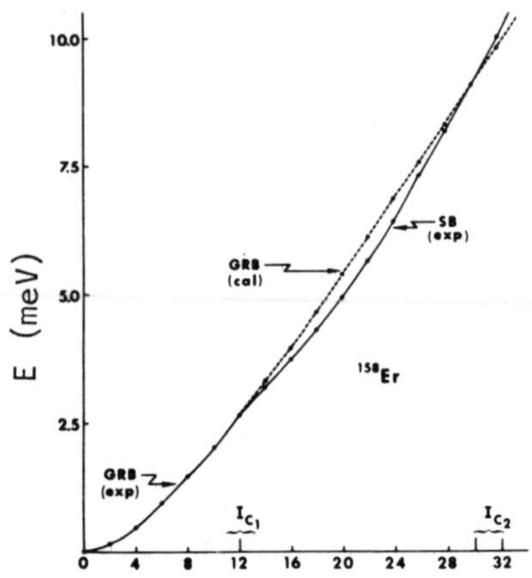

Figure 2.4

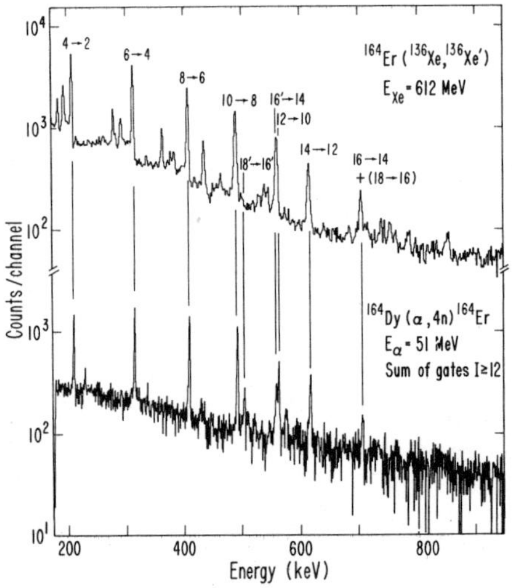

Figure 2.5

$^{164}Dy(\alpha,4n)$ reaction, are compared[158] in Fig. 2.5. The spectrum for $(\alpha,4n)$ demonstrates how backbending manifests itself experimentally, when a gate is set on a certain (high-J) transition and the coincidence E2 cascade to the lower levels is observed. It is clear that the transitions labelled 16'-14 and 18'-16' are "out of sequence" compared to the regular spacing of the 4-2, 6-4, 8-6 etc. transitions. Note, however, that in the upper part of the spectrum from Coulomb excitation there are, in addition, regularly spaced transitions 16-14, 18-16 which are the continuation of the ground state band *beyond* the J = 16 backbending region (compare Fig. 2.3). Only recently have sufficiently heavy beams become available to Coulomb excite very high spin states.

The rotation-alignment model actually predicts a *series of similar* rotation-aligned superbands. The lowest one, discussed above has only even spin members, and evolves (in ^{164}Er) from a $K = 0^+$ band (at spin 0) to a structure at $I \geqslant 16$ which is mainly two $i_{13/2}$ quasineutrons coupled to J = 12, aligned with the core rotation. The next two superbands are predicted to start out as a single $K = 4^+$ band, evolving into the lowest odd spin (yrast odd) and the *second* lowest even spin (yrare even) rotation-aligned bands. They still have a dominant $(i_{13/2})$ configuration at high spin, and in the extreme limit, the rotation-alignment model predicts that yrast even-spin (I), the yrast odd-spin (I-1) and the yrare even-spin (I-2) states all have the same rotational energy. The structure of the superbands can be probed by studying their interactions with the ground and γ-vibrational bands. The higher lying γ-band is an excellent probe because it intersects both the even *and* the odd-spin states of the superbands. All these bands have been sorted out by a variety of (H·I,xn) coincidence experiments[159] (Fig. 2.6); an excellent and truly remarkable agreement between experiment (a) and theory (b) is observed.

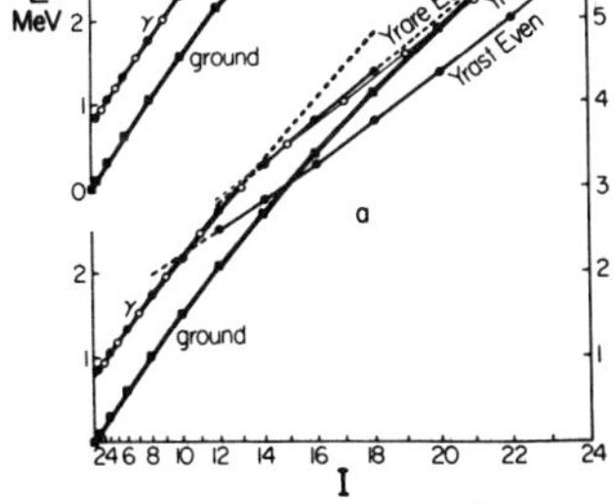

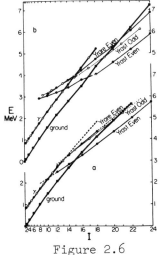

Figure 2.6

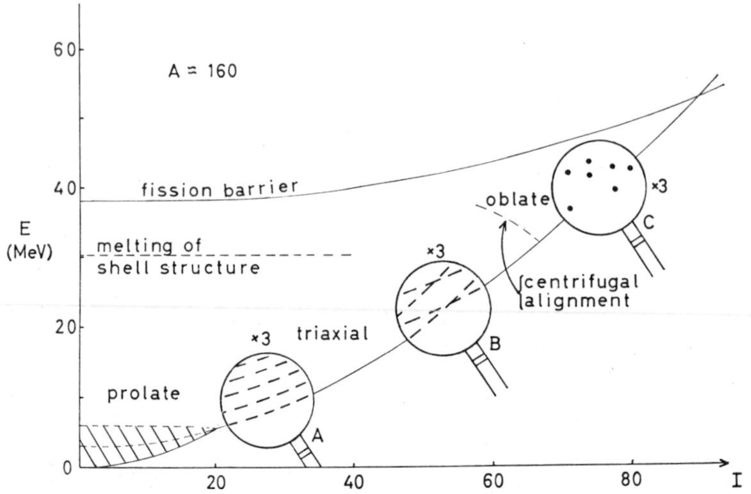

Figure 2.7

Guided by this introduction to high spin phenomena, let us now speculate[160,161] on the possible behavior of nuclei as even large amounts of energy and angular momentum are deposited (Fig. 2.7). The lower, approximately parabolic, line is the yrast line so there are no levels in the nucleus below this. The upper line gives the fission barrier, which sets an upper limit to the study of levels of the nucleus. The intersection of the two gives the effective maximum angular momentum for the nucleus. Nuclei in the rare earth region have prolate shapes near the ground state as a result of shell structure, and they have strong pairing correlations. The hatched region indicates where pairing correlations exist, which terminate as we have seen, around I = 20, where the two bands cross.

Some insight into the region above I = 20 comes from the liquid drop model. A rigidly rotating charged drop prefers an oblate shape until shortly before fission. The large moment of inertia of oblate shapes minimizes the total energy. Although the nucleus cannot rotate about a symmetry axis, it has been shown[162] that for a Fermi gas the states obtained by aligning the angular momenta of individual particles along the symmetry axis is the same as would be obtained by rigid rotation about that axis. These deformation-aligned states in oblate nuclei therefore generally are lower than the rotation-aligned states in prolate nuclei. At high angular momentum the nucleus becomes oblate and the angular momentun is carried by aligned individual nucleons (region C in the figure). This region may be identified by the occurrence of isomeric states,[160] due to the absence of smooth rotational band structure. At the very highest spins the nucleus may become triaxial before fission. The increase in deformation and moment of inertia is predicted to be so rapid that the rotational frequency will decrease, leading to a "super-backbend." Between the prolate and oblate regions, nuclei are also expected to become triaxial. Wobbling motion is then

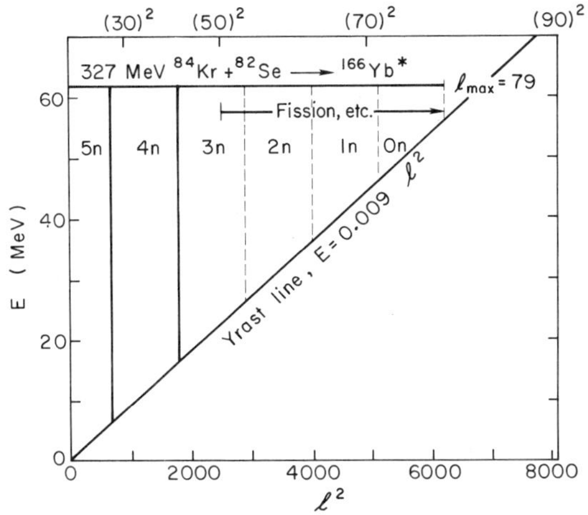

Figure 2.8

possible in addition to rotation about the axis with largest moment of inertia, and could give rise to a series of closely spaced parallel bands.[163] (Note that two aligned high-j orbits represent a triaxial bulge in prolate nuclei.)

How do we get an experimental handle on these new modes of motion of the nucleus? The problem is to learn about high spin states above I = 20, as discussed above, especially those along the yrast line, where the nucleus is thermally cool and does not have a high density of states. The remarkable feature of the (HI,xn) reaction is that it can locate us along different regions of the yrast line.[164-165] This works as follows: in Fig. 2.8, the compound nucleus ^{166}Yb is formed with an angular momentum distribution from $J = 0$ to $J = \ell_{max}$ at excitation energy $E_{CM} + Q \approx 60$ MeV by the partial cross sections:

$$\sigma_\ell = \pi \chi^2 (2\ell + 1)\, T_\ell \qquad (2.5)$$

The successive evaporation of x neutrons from these states is assumed to remove practically no angular momentum and an average of 2 MeV kinetic energy plus the binding energy of ≈ 8 MeV. Neutron evaporation continues until the available energy above the yrast line is less than 10 MeV. Since

$$E_y = \frac{\hbar^2}{2\mathcal{J}} \ell(\ell + 1) \qquad (2.6)$$

a given value of x occurs in the sharply defined "bin" ℓ_i to ℓ_f where:

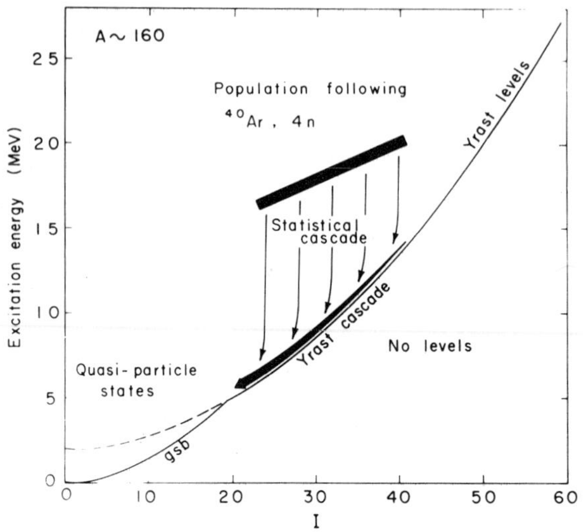

Figure 2.9

$$E_y(\ell_i) + 10 = E_{CM} + Q - 10x$$
$$E_y(\ell_f) = E_{CM} + Q - 10x \qquad (2.7)$$

The partial cross section for the evaporation of x neutrons is then:

$$\sigma_x = \pi \lambda\!\!\!^{-2} \sum_{\ell_i}^{\ell_f} (2\ell + 1)T_\ell \approx \pi \lambda\!\!\!^{-2}[\ell_f(\ell_f + 1) - \ell_i(\ell_i + 1)] \qquad (2.8)$$

As long as $0 < \ell_i < \ell_f < \ell_{max}$, it follows that

$$\sigma_x = \pi \lambda\!\!\!^{-2} \frac{2\mathcal{J}}{\hbar^2} \cdot 10, \textit{ independent of } x \; . \qquad (2.9)$$

(The largest and smallest bins can be truncated due to the limits $\ell_i = 0$, $\ell_f = \ell_{max}$.) Furthermore, the mean angular momentum $\bar{\ell}$ of the states on which the neuron evaporation chains terminate is predicted for each bin:

$$\bar{\ell} = \frac{2}{3} \frac{\ell_f^2 + \ell_f \ell_i + \ell_i^2}{\ell_f + \ell_i} \qquad (2.10)$$

Channels corresponding to different numbers of evaporated neutrons have different angular momentum ranges and the highest angular

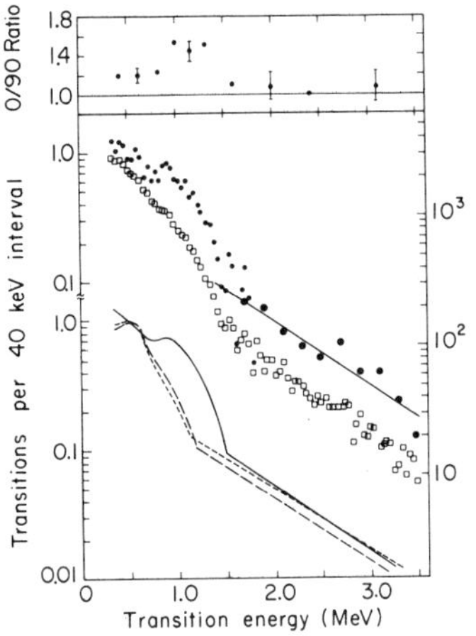

Figure 2.10

momenta are in the channels with the fewest evaporated neutrons. These results have been demonstrated experimentally.[166]

A specific application is shown[165] in Fig. 2.9 for the initial production of A ~ 160, with an Argon beam of 170 MeV. The initial excitation is 70 MeV and the 4n channel drops down to roughly 10 MeV above the yrast line, without removing much angular momentum. There is still a high density of levels, and there follows a high-energy statistical cascade of dipole transitions, which still do not carry off much angular momentum. Approaching the yrast line the level density becomes small; and the most likely mechanism is then stretched E2 transitions along the yrast collective bands. Eventually these run into the discrete levels of the ground state band (like Fig. 2.5). By setting gates on the lines corresponding to the 4n channel one can look at the corresponding spectrum in several NaI counters placed around the target.

The observed continuum spectrum for the $^{126}Te(^{40}Ar,4n)^{162}Yb$ reaction is shown in Fig. 2.10, by the hollow squares.[167] The dots show the corrected spectrum after efficiency unfolding. The exponential tail is associated with the statistical dipole emission, and the lower energy bump with the E2 cascade (confirmed by the anisotrophy shown at the top of the figure, obtained by comparing the spectra at 0° and 90°). The integral of the bump gives the number of gamma rays.

Then we determine the average angular momentum $\bar{\ell}$ carried in the cascades as

$$\ell = 2(\bar{N}_{\gamma} + \delta)$$

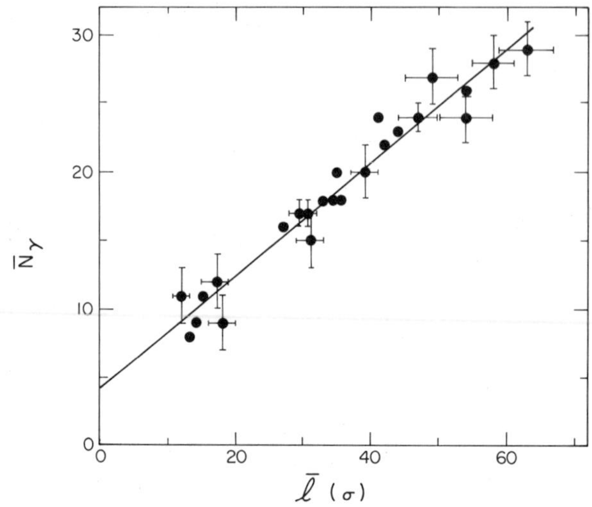

Figure 2.11

where δ is the number of statistical γ-rays removing no angular momentum. (Note however that some very recent measurements indicate that a dipole component is present in the yrast cascade, the precise origin of which is not understood.[168]) Our earlier theorems about the bins and the associated $\bar{\ell}$ of the different xn reactions, now enable the construction[165] of Fig. 2.11. The slope is not *exactly* one half, but close at 0.43. If we also associate the bump edge with transitions from the states of highest spin in the bin, we can determine the moments of inertia at these high spins from the relation:

$$E_\gamma = \frac{\hbar^2}{2\mathcal{J}}(4I - 2) \tag{2.11}$$

describing the transition energies in a rotor.

The results are shown[165,167] in Fig. 2.12 for ^{162}Yb, plotted in the backbending fashion of Fig. 2.3. Since ^{162}Yb has not been tracked completely through a backbend, ^{160}Yb is also shown (open circles). At the highest rotational frequency, the moment of inertia approaches the rigid sphere value with A = 162, $\mathcal{J} = 2/5\ MR^2$, $2\mathcal{J}/\hbar^2 \approx 140\ \text{MeV}^{-1}$. The last point on the plot is associated with the (^{40}Ar,4n) reaction, which as we saw earlier, originates from angular momentum $\approx 35\hbar$. Since the deformed moment of inertia would be a little larger (by 10%) and since the measured values fall below this line, some residual pairing correlations may still persist even at this high angular momentum.

A great deal of experimental ingenuity is presently invested in methods for unravelling the information about nuclear shapes at still higher rotational angular momentum. A promising technique[169,170]

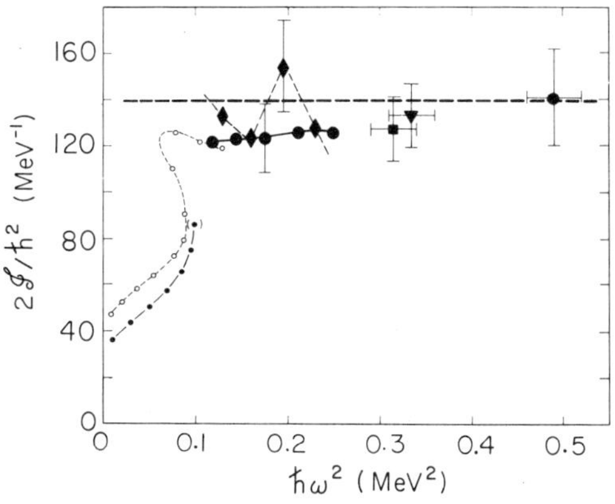

Figure 2.12

is to look at the multiplicity (the number of γ-rays) associated with each transition in the continuum. If there is some relationship, like:

$$E_\gamma = \frac{\hbar^2}{2\mathcal{J}} I(I+1)$$

at work this will be reflected as structure in the spectrum and imply a *prolate* nuclear shape. The absence of structure, on the other hand, is an indication of non collective motion and hence spherical or *oblate* shapes. An array of NaI counters is placed around the beam axis and the spectrum in another detector is unfolded in coincidence with one, two, three....counters of the array. Examples of coincidence and singles spectra for three reactions are shown in Fig. 2.13. (The singles spectrum shows the yrast and statistical cascades just as in Fig. 2.10.) In coincidence the yrast cascade yields a bump in the E_γ v multiplicity curve, the upper edge of which moves to higher energies as more angular momentum is brought in at higher incident energies (remember $E_\gamma \propto I(I+1)$). The spectrum is well reproduced in (b) with a cascade of I/2 rotational transitions from spin I to 0 whose energies are,

$$E_\gamma = \frac{\hbar^2}{2\mathcal{J}} (4I-2).$$

The data determine the moment of inertia to be 95% of the rigid sphere value. By contrast, the $^{100}Mo + ^{48}Ca$ example leads to nuclei in the N = 82 closed shell region, and the absence of structure in the multiplicity spectrum remains up to high spins. The rotational competition starts only at 50ħ, implying that this system is still oblate up to this spin, and then becomes prolate. These trends are actually in agreement with detailed calculations of potential energy

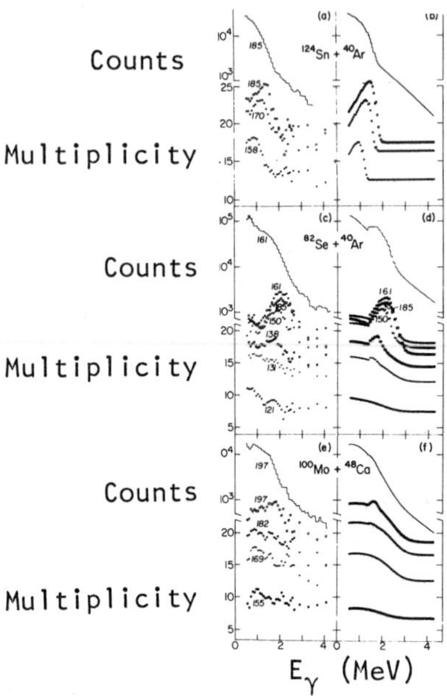

Figure 2.13

surfaces over the full (β,γ) plane, which use cranked modified-oscillator potentials with a Strutinsky-type normalisation to the liquid drop![171] Clearly we are on the way to finding out about the dynamics of nuclear rotation at very high spins indeed.

For a nucleus with oblate shape and with the angular momentum oriented in the direction of the symmetry axes, we encounter a form of rotational motion which is radically different from the usual prolate rotation. In the oblate case, the average density and potential remain static. (See Fig. 2.14.)

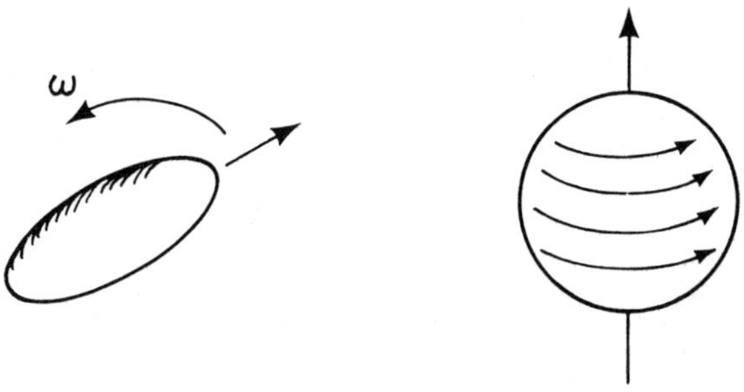

Figure 2.14

Each single particle orbit contributes a quantized angular momentum in the direction of the rotation axis. The transitions from one state to the next along the yrast line involve successive rearrangements in the filling of single particle orbits, and the energies along the yrast line exhibit irregularity, although on the average the yrast states have a rotational dependence of energy on spin with a mean effective moment of inertia equal to that of a rigid body rotating about the oblate axis of symmetry. The deviations from the mean, enhanced by shell effects, may cause large irregularities in the yrast sequence, and the nucleus may become trapped in isomeric states[160] with lifetimes orders of magnitude longer than rotational transitions. A systematic search for such yrast traps has been undertaken with beams of Ar, Ti and Cu in a hundred different target-projectile combinations.[172] The γ-emission from the recoiling compound nuclei were studied by detectors selecting high multiplicity (see above). The survey identified an island of high-spin isomeric states centered around neutron number 84 with lifetimes in the region of a few to several hundred nanoseconds. The interpretation will be quite speculative until the spin and decay schemes are pinned down, but it is fascinating to note that several theoretical calculations[171,173,174] point to this region of isotopes as especially favorable for the occurrence of yrast traps based on the oblate coupling scheme.

So far we have concentrated on γ-emission for transmitting information about nuclei at high angular momentum. However, once formed the compound nucleus has to decay by particle emission, from which important properties of the compound system become accessible, such as the temperature, distribution of angular momenta, moments of inertia, and degree of equilibration. Analysis of the data requires a comparison with the predictions of a statistical evaporation code. Remarkable progress has been made in refining the

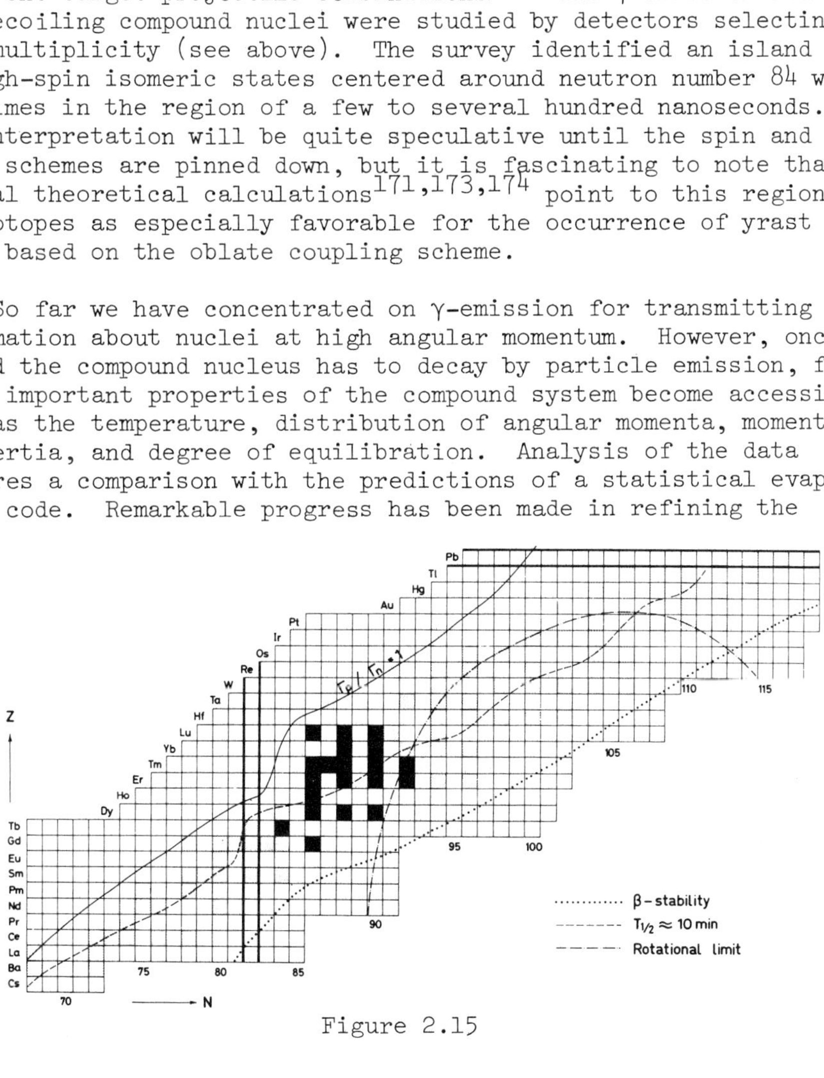

Figure 2.15

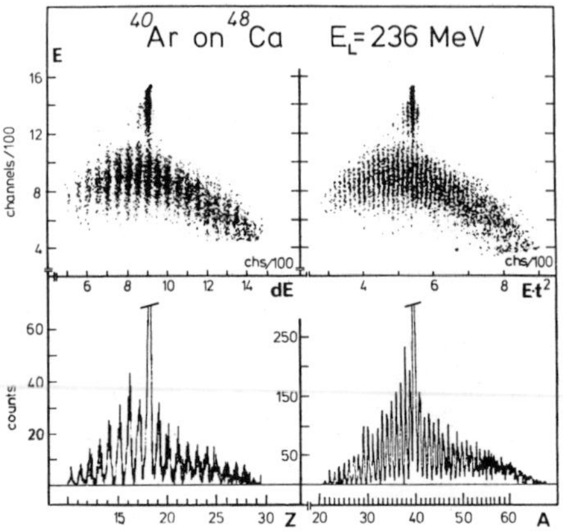

Figure 2.16

calculational[175,176] and experimental techniques. Experimental data and evaporation residues (the remnant of the compound nucleus after particle decay) can be obtained for individual A,Z by an apparatus which measures ΔE, E (to determine Z) and time-of-flight (to determine $A \propto Et^2$). A "state of the art" example is shown in Fig. 2.16 for ^{40}Ar + ^{48}Ca at E = 236 MeV. In this particular experiment[177] evaporation residues were not being measured, but the figure demonstrates that it is possible to resolve individual Z up to 30 (in fact, up to 65 has been achieved) and individual A up to 60.

The measured evaporation residues in the reaction ^{19}F + ^{27}Al at 76 MeV are compared with statistical calculations[176] in the bottom part of Fig. 2.17. The upper sections decompose the calculation into contributions from different angular momenta in the compound nucleus. It is clear that increased α-particle emission is associated with higher angular momentum and therefore these residues probe the region of the energy-angular momentum space closest to the yrast line of the compound nucleus. A reconstruction of the "decay scheme" of the compound nucleus is shown in Fig. 2.18. (It is clear from this figure that our earlier discussion of the particle emission down to the yrast line producing the γ-cascades was oversimplified for light nuclei--see Fig. 2.9.

An important input to the statistical calculations is the level density in the nucleus at (in this example) excitations up to 70 MeV, and angular momenta up to $40\hbar$. Nuclei in this region are likely to behave like liquid drops, and the influence of individual shell structure of a nucleus on the level density and pairing energy vanishes. Based on theoretical predictions[178] one assumes in those calculations[179,180] that above a given excitation energy, U (liquid drop model) ≈ 15 MeV, the shell effects disappear. An

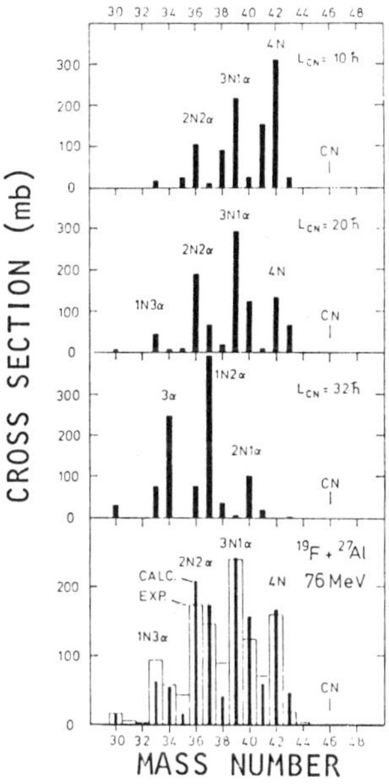

Figure 2.17

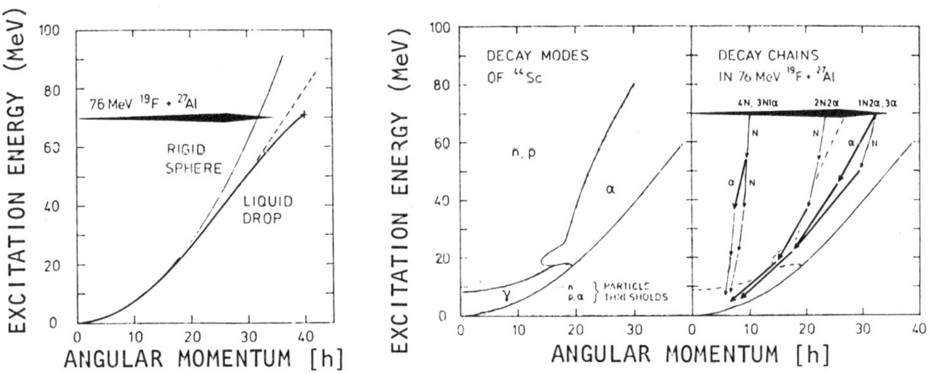

Figure 2.18

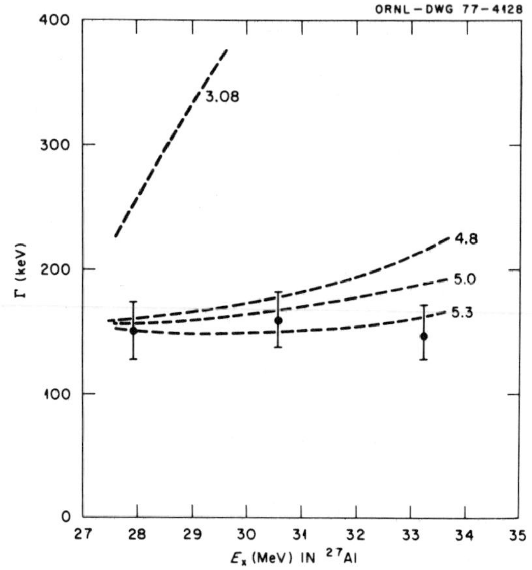

Figure 2.19

appropriate allowance for the deformability under rotation is made by using:

$$\mathcal{J} = \mathcal{J}_o \, (1 + \delta L^2) \qquad (2.12)$$

In this way we obtain an yrast line deviating from that of a sphere, as shown in the third section of Fig. 2.18. Because of the connection between the shape of the yrast line and the shape of the nucleus itself, information on the latter may be forthcoming from measuring the ratio between nucleon and α-particle emission (see the left-hand sections of the figure). The quantitative analysis yielded $2 \times 10^{-4} < \delta < 5 \times 10^{-4}$, which is to be compared with the prediction of 2.5×10^{-4} for the detailed shape calculation[181,182] (to be discussed in the next section).

Yet another method for extracting moments of inertia at high excitation and angular momentum is the measurement of coherence widths Γ. These can be evaluated in terms of the number of open channels (Hauser-Feshbach denominated) at a given compound nucleus J and of the level density in the compound system of (Ex,J) at excitation Ex. The slope of Γ versus Ex is primarily a function of the effective moment of inertia, via the spin cut-off parameter $\sigma^2 = \mathcal{J} T/\hbar^2$, with T the nuclear temperature.[183] For the $^{12}C(^{15}N,\alpha)^{27}A\ell$ system[184] a comparison of Γ v Ex in $^{27}A\ell$ is shown in Fig. 2.19 for different statistical model predictions labelled by $\mathcal{J}/\hbar^2$. The moment of inertia $\mathcal{J}/\hbar^2$ of 5.3 MeV^{-1} greatly exceeds[185] that extracted by fitting the low lying member of the ground state rotational band (≈ 3 MeV^{-1}). It will certainly be exciting to learn more about the predicted exotic shapes that nuclei, under the influence of heavy-ion collisions, will assume from experiments such as those described in the section. Since our whole discussion presupposed the formation of the compound nucleus, we must now check this assumption.

2.2 To Fuse or Not to Fuse

That is certainly a question at the forefront of much modern research with heavy ions. It is well known that if a deformable fluid mass is set spinning it will flatten and eventually fly apart.[182] To discuss the equilibrium shapes of a rotating nucleus we set up an effective potential energy and look for configurations that are stationary:

$$P.E = E_{Coul} + E_{nuc} + E_{rot} \tag{2.13}$$

where

$$E_{rot} = \frac{\hbar^2 \ell(\ell+1)}{2\mathcal{J}(\alpha_2\alpha_3\alpha_4)} \tag{2.14}$$

It is convenient to introduce two dimensionless numbers, specifying the relative sizes of the three energy components.[182,186,187] Choose the surface energy of a spherical drop as a unit:

$$E_S^{(0)} = 4\pi R^2\gamma = C_2 A^{2/3} \tag{2.15}$$

with $C_2 \approx 17.9$ MeV. Then specify the amount of charge on the nucleus by

$$x = \frac{\frac{1}{2}E^o_C}{E^o_S} \approx \frac{1}{50}\frac{Z^2}{A} \tag{2.16}$$

For the angular momentum, specify

$$y = \frac{E^{o}_{rot}}{E^{o}_{S}} \approx \frac{\frac{1}{2}\hbar^{2}\ell^{2}}{\frac{2}{5}MR^{2}} \cdot \frac{1}{C_{2}A^{2/3}} \approx \frac{2\ell^{2}}{A^{7/3}} \qquad 2.17$$

In terms of these parameters, Fig. 2.20 illustrates some shapes, in each case for the ground-state (stable) shape and the saddle point (unstable shape) — labeled H and PP respectively. As the rotation speed increases, the ground state flattens and the saddle point thickens its neck. In the bottom figure the ground-state pseudospheroid loses stability and becomes triaxial, resembling a

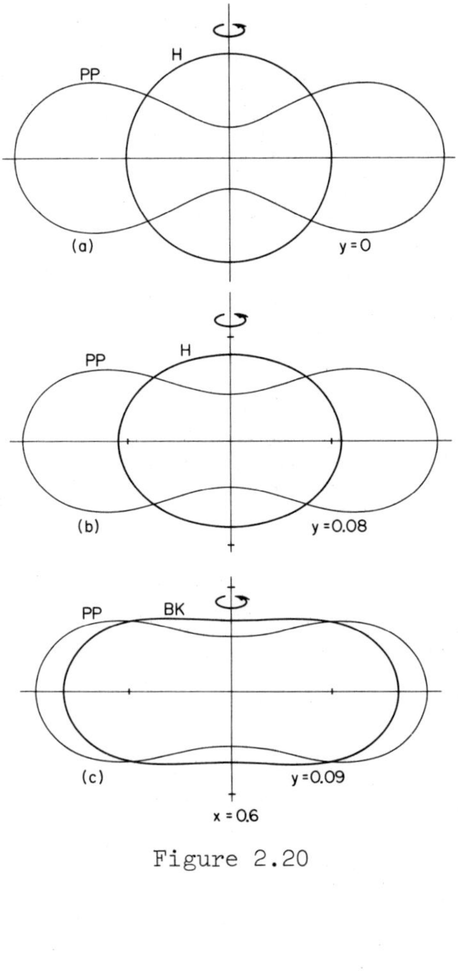

Figure 2.20

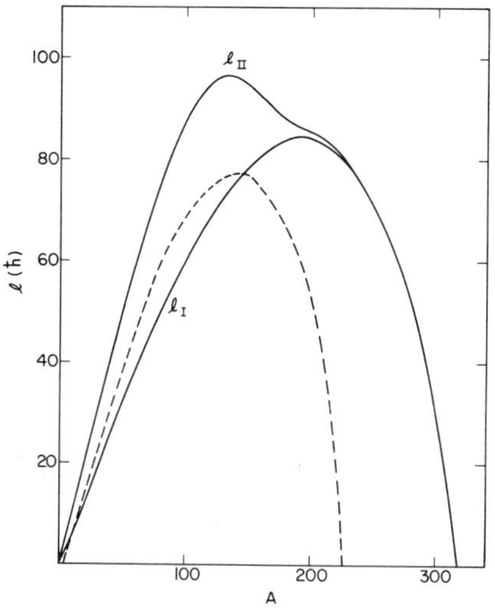

Figure 2.21

flattened cylinder with rounded edges, beginning to merge with the saddle shape. At slightly higher angular momenta the stable and unstable families merge and the fission barrier vanishes.

This behavior can be translated into an angular momentum plot versus mass (Fig. 2.21). For vanishing of the fission barrier the resultant curve is ℓ_{II}. No nucleus can support more than 100ħ, and neither light nor heavy nuclei can support very many units. The dashed curve shows the angular momentum required to lower the fission barrier to 8 MeV; this curve is indicative of the maximum the nucleus could support and still survive the risk of fission in the de-excitation process.

By conservation of energy and angular momentum, it follows that the closest distance of approach of projectile and target is given by r_{min}, where for impact parameter b,

$$\left(\frac{b}{r_{min}}\right)^2 = \left(1 - \frac{V}{E}\right) \qquad 2.18$$

which, for given r_{min}, is a hyperbola for b^2 versus E. If r_{min} is chosen as the strong interaction radius (R_1+R_2), this curve divides the plane (b v E) into two regions: distant collisions where the nuclei pass each other without appreciable interaction, and close collisions where the corresponding πb^2 gives the reaction cross section. Because of diffuseness, this region is given some width in Fig. 2.22. The curves are constructed for R_1+R_2+d. The plane can be further subdivided by curves corresponding to the locus of fixed angular momentum ℓ:

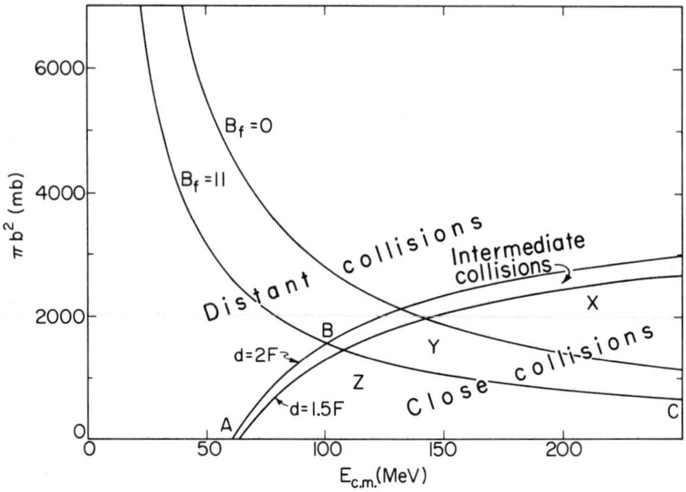

Figure 2.22

$$\ell = b\sqrt{2\mu E} \qquad b^2 = \frac{\ell^2}{2\mu}\ \frac{1}{E} \qquad \mu = \text{reduced mass}$$

The value of y (or ℓ) at which the fission barrier vanishes can be inserted to construct the additional curves on Fig. 2.22 (both for zero fission barrier and where it has become equal to the binding energy of a nucleon, which marks where the de-excitation mode changes to nucleon emission and the compound nucleus would be detectable).

To the left of B_f=0, a compound nucleus could form, and to the left of B_f=11 it would definitely survive. We shall see later however that the prediction of the *formation* of a compound nucleus

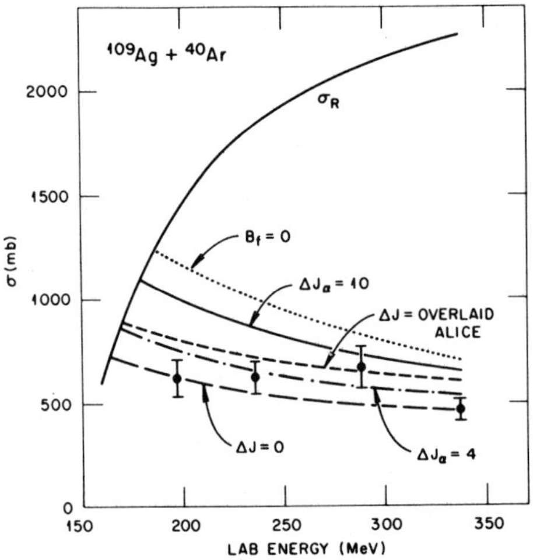

Figure 2.23

is a dynamical question, beyond the scope of these considerations. Only if this critical curve lies *totally* above ABC, can the curve ABC represent the cross section for formation and survival of the compound nucleus. The figures are constructed for $^{20}Ne+^{107}Ag$. Now we compare with actual data[188] for a much heavier system, $^{40}Ar+^{109}Ag$.

The fusion products are experimentally identified by detecting evaporation residues after evaporation of nucleons and alpha particles[189] and are shown in Fig. 2.23; the trend follows that of Fig. 2.22. The line $B_f = 0$ is marked and also more precise calculations using the computer code ALICE, which deals more properly with particle evaporation, and in particular with the angular momentum they carry off (represented by $\Delta J = 10$ etc). Detailed discussions of the fusion of heavy systems are given in the reviews of Refs. 190 and 191.

In many cases we find that the fusion cross section is much less than the reaction cross section, although the fission barrier has still not disappeared. It appears that the ions have to reach a critical *distance* of overlap of nuclear matter before fusion sets in.[192,193] To take into account the effects of a critical distance we write[194] for the fusion and the total reaction cross sections:

$$\sigma_f = \pi\lambda\!\!\!^-{}^2 \sum_0^\infty (2\ell+1)\, P_\ell \tag{2.20}$$

$$\sigma_R = \pi\lambda\!\!\!^-{}^2 \sum_0^\infty (2\ell+1) \tag{2.21}$$

where P_ℓ are the probabilities that fusion takes place after the barrier is passed. For P_ℓ we assume:

$$P_\ell = \begin{cases} 1 & \ell \leqslant \ell_{cr} \\ 0 & \ell > \ell_{cr} \end{cases} \tag{2.22}$$

Then the summation in Eq. 2.20 leads to

$$\sigma_f(E) = \pi\lambda\!\!\!^-{}^2(\ell_{cr}+1)^2 \approx \pi\lambda\!\!\!^-{}^2\ell_{cr}{}^2 \tag{2.23}$$

The turning point for the partial wave $\ell = \ell_{cr}$ is deduced from the expression:

$$E = V(R_{cr}) + \frac{\hbar^2 \ell_{cr}(\ell_{cr}+1)}{2\mu R_{cr}^2} \qquad 2.24$$

Substituting for ℓ_{cr} in Eq. 2.23, gives

$$\sigma_f = \pi R_{cr}^2 \left(1 - \frac{V(R_{cr})}{E}\right) \qquad 2.25$$

This expression is just equivalent to the usual formula for the reaction cross section (see for example Eq. 2.18) with R_{cr} replaced by the interaction barrier radius R_B:

$$\sigma_R = \pi R_B^2 \left(1 - \frac{V(R_B)}{E}\right) \qquad 2.26$$

It turns out that $R_{cr} \approx 1.00\ (A_1^{1/3} + A_2^{1/3})$ for a wide range of ions. This interpenetration distance corresponds to the overlap of the half density radii of the nuclear matter distributions.[195] The radius is marked[196] on Fig. 2.24 for $^{16}O + ^{48}Ca$. Up to a certain critical energy, for all partial waves that surmount the outer barrier, the two ions succeed in interpenetrating to the critical distance (assuming there is not too much radial friction near the barrier top -(dashed line) and fuse. Above this critical energy, however, the increasing centrifugal barrier does not allow

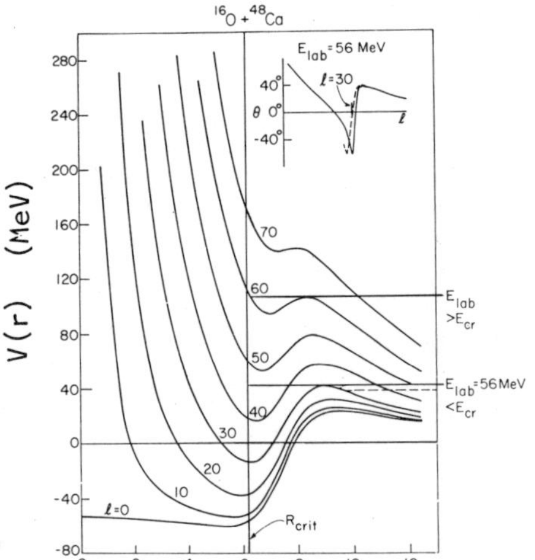

Figure 2.24

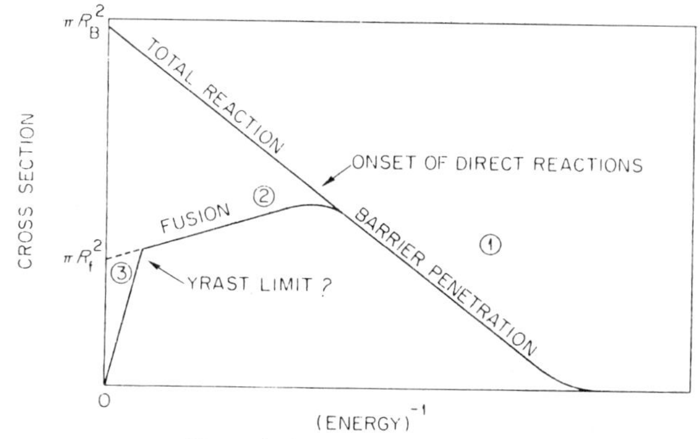

Characteristic Energy Regions for Fusion.

Figure 2.25

the ions to penetrate for all partial waves, and the fusion cross section becomes smaller than σ_R. (This scheme is valid when the dynamical path for fusion lies inside the saddlepoint, a situation which is not usually fulfilled for heavy systems — see the discussion in Ref. 30).

From these equations we generate the schematic representation[197] of fusion and total reaction cross sections as a function of 1/E in Fig. 2.25. In region 2, the critical energy is passed and the fusion cross section changes slope — it may increase, stay constant or decrease, depending on the value of $V(R_{cr})$ at this point. In region 3, the limit of maximum angular momentum in the compound system is surpassed. Just these features appear to be observed[198] in $^{14}N+^{12}C$ system shown in Fig. 2.26. If the data are represented

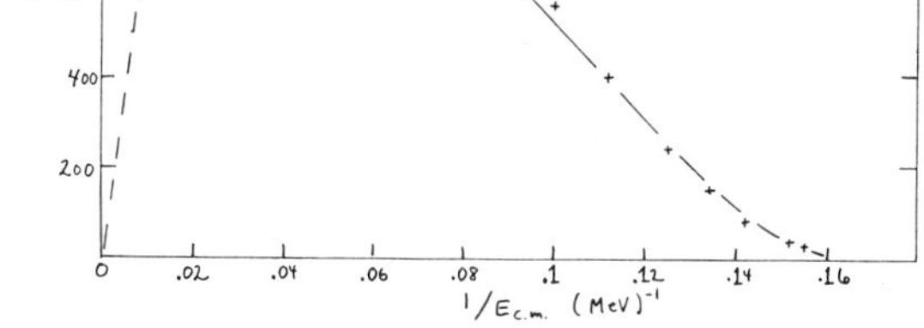

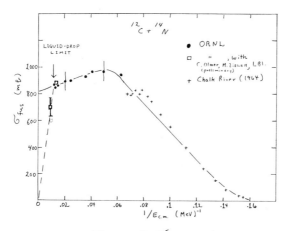

Figure 2.26

in terms of the critical angular momentum, as in Eq. 2.23, then the value $\ell_{cr}(\ell_{cr}+1) = 734\hbar^2$ does indeed correspond to the limit of $26.6\hbar$ expected from Fig. 2.21 for $A \approx 26$. The predicted shape is that of a very deformed, triaxial nucleus with $R_{max} \approx 2R$ and $R_{min} \approx 0.4R$, with R the radius of the spherical ground state. In view of these extreme shapes, it is perhaps more realistic to consider[199] a critical deformation, or moment of inertia, which determines whether fusion occurs or not; in a more formal derivation[194] R_{cr} is introduced via the equation $\mathcal{J}_{cr} = \mu R_{cr}^2$. The study of much heavier systems, beyond the liquid drop fission limit, should soon be possible with the higher energies becoming available.[200]

Since the slope and intercept beyond the critical energy determine V_{cr} and $R_{cr} \approx 1.0(A_1^{1/3} + A_2^{1/3})$, these measurements can be used to determine the potential at much smaller distances than is possible from elastic scattering[52] (we call $R_{1/4}$ and R_s in Lecture 1), and indeed were used to construct some of the points in Fig. 1.14. A thorough analysis of potentials, synthesizing information from the total reaction cross section, the fusion cross section, elastic and transfer reactions is given in Ref. 34; such an approach may help to remove the ambiguities we discussed for $^{16}O+^{28}Si$ in Lecture 1. However if there is significant radial friction (and the next lecture will show that there is) then our earlier equations should contain $(1 - \frac{V+E_F}{E})$ rather than $(1-\frac{V}{E})$, where E_F is the energy loss due to friction on that portion of the trajectory leading up to the barrier. Roughly we can see that neglect of friction produces an underestimate of the potential.[52] At the critical distance where frictional dissipation is very strong the whole method of analysis presented here becomes questionable. Nevertheless a variant of this analysis,[53] using a proximity potential has been used to extract potential depths down to values of s (in Fig. 1.14) which are negative, i.e., very strong overlap of the nuclear matter. A questionable assumption in many of these treatments is the sudden approximation, i.e., a potential which conserves the structure of each nucleus.[201] At the opposite extreme is the adiabatic approach, which allows a continuous change of potential.[202] Ultimately a full dynamical calculation is required, in which the fusion cross section depends not only on the static shapes but also on coupling to internal degrees of freedom.[203-206] In the classical limit this approach leads to an equation of motion with frictional forces.[207] Then it becomes possible to describe in complete fusion events, or deeply-inelastic scattering; in the next lecture we shall see that these processes consume the missing cross section[208] of region 2 in Fig. 2.25.

2.3 More Microscopic (and Speculative) Aspects

In our introduction to these lectures we mentioned that the

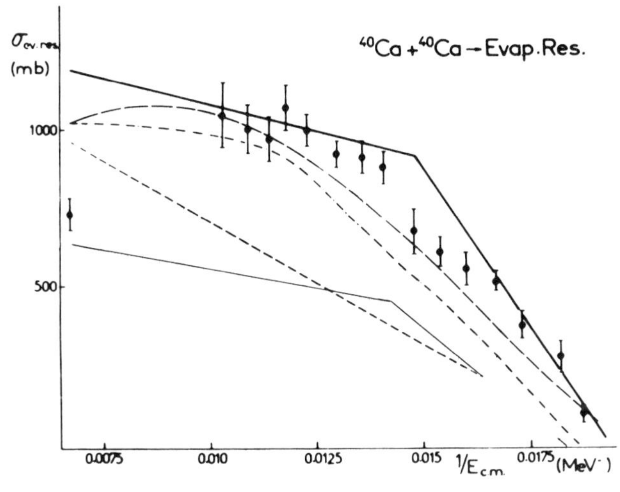

Figure 2.27

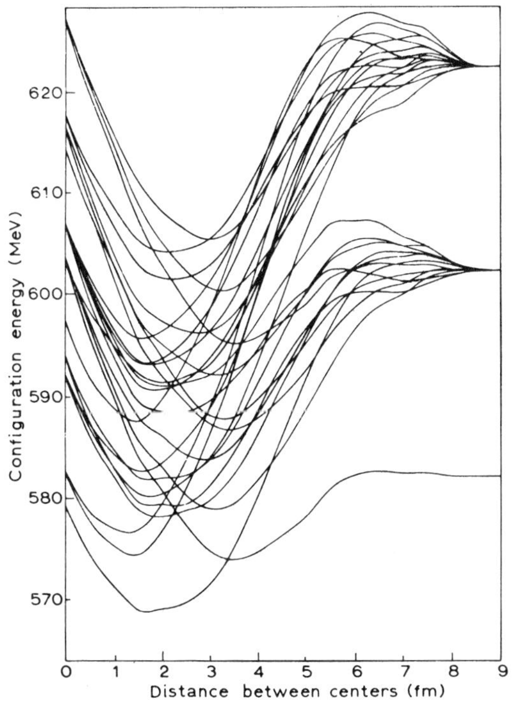

Figure 2.28

microscopic, and the macroscopic were not really distinct subjects, but so far in our discussion of fusion processes we have ignored any effects of individual nucleons, the fundamental constituents of nuclei. In Fig. 2.27 is a plot of the $^{40}Ca+^{40}Ca$ fusion cross section,[209] plotted in our familiar framework. In the notation of Eqs. 2.25, 2.26, the solid line uses the parameters,

$$V(R_B) = 51.5 \text{ MeV}, \quad R_B = 1.49\,(A_1^{1/3} + A_2^{1/3}) = 10.2\text{fm}$$

$$V(R_{cr}) = 24 \text{ MeV}, \quad R_{cr} = 0.97\,(A_1^{1/3} + A_2^{1/3}) = 6.65\text{fm}$$

The critical potential is positive, which classifies the system as "heavy" (compare Fig. 2.26 where it is negative). Since this system comprises two closed-shell nuclei, the tightness associated with shell effects could manifest itself by a decrease of the radius parameter, compared with neighboring systems; such a comparison could give some information on the role of individual nucleons in the fusion process. The dashed curve in fact corresponds to a calculation with a smaller critical distance determined from Hartree-Fock densities for ^{40}Ca. One physical interpretation of the critical radius comes from the two-center shell model.[210] This is illustrated in Fig. 2.28 for $^{16}O+^{16}O$; at distances less than 3.4 fm the lowest configuration becomes the ground state of ^{32}S and at large distances it is the $^{16}O+^{16}O$ ground state. At the level crossing, strong energy losses should occur. It would appear from Fig. 2.27 that there is no evidence for this closed-shell effect. However, another doubly magic system $^{48}Ca+^{208}Pb$ does indicate[211] a lowered fusion cross section.

In lighter nuclei, there is some evidence for a shell effect. For example, in Fig. 2.29 the valence nucleons of a new oscillator shell appears to cause a discontinuous jump in cross section of 200 mb at the sd-shell. Unfortunately the systematic trend is broken[212] by one point for ^{15}N,[213] and also by some recent results with ^{20}Ne.[213] The microscopic aspects, and the exact manner in which the valence nucleons affect the fusion cross section via the complexity of Fig. 2.28 is still poorly understood. An even more challenging observation is the presence of oscillations in the fusion cross sections of light, closed shell systems,[212,214] such as $^{12}C + ^{12}C$ and $^{12}C + ^{16}O$, whereas the $^{12}C + ^{18}O$ system behaves according to the systematics we have described in section 2.2. Some examples are shown in Fig. 2.30. The story becomes even more subtle with the observation that these oscillations are correlated with the resonances appearing in the excitation functions $^{12}C + ^{12}C \rightarrow ^{12}C(2^+) + ^{12}C(2^+)$, which, as we discussed in the first lecture, have been attributed to molecular shape resonances.[128-130] (See Fig. 1.37) (For a detailed discussion, see Ref. 211.) Even the association with closed shell systems, and/or even partial

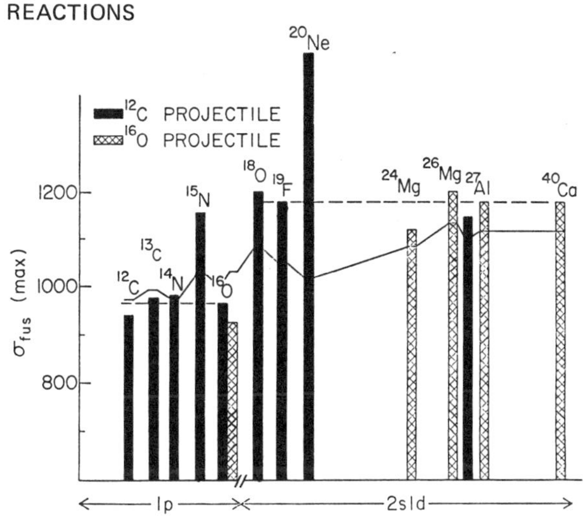

Figure 2.29

waves, appears dubious with the recent observation[215] that the oscillations are *also* present in $^{18}O + ^{16}O, ^{12}C$.

To return to the $^{40}Ca + ^{40}Ca$ system, which did not exhibit shell effects, it is conjectured that fast, collective excitations[216] could provide the first step in overcoming the shell gap before adiabatic effects (such as level crossings) become important. In this model, the two nuclei move on trajectories constrained by the proximity potential, and at the turning point of the radial motion dissipate energy into vibrational and intrinsic motion. Fusion happens predominantly close to the orbiting angular momenta, and a satisfactory description of the data in Fig. 2.27 is obtained.

Figure 2.30

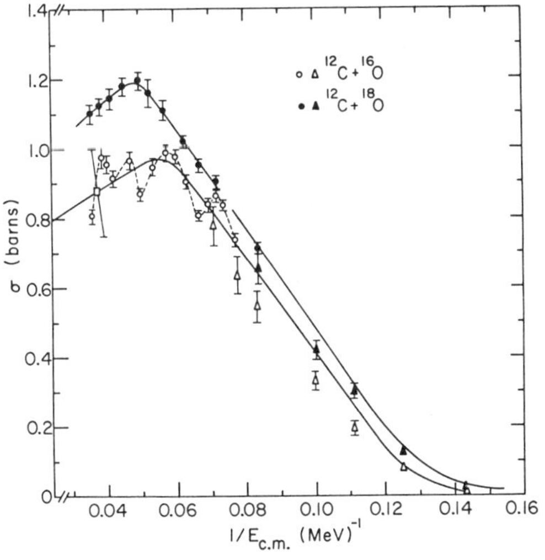

The smaller impact parameters tend to make the ions bounce off one another, a feature which is also present in the time dependent Hartree-Fock Model.[217-219] So far we have been largely concerned with the "macrophysics" of nuclear matter. Of course this is not a new subject, since fission has been with us for a long time. But there have not been many studies of the *dynamics* of fission. It has mostly been an attempt to understand the energetics and other properties of the fission barrier. Is it possible to get some understanding of all these processes in some microscopic framework? A convenient starting point is the mean field or Hartree-Fock approximation, which has enjoyed great success in the static case.[220] This works because the density is high, the effective forces are strong, and the Pauli principle inhibits collisions. In a time-dependent generalization the rate of change of the mean field must be small enough so that it does not produce large excitations of the independent particles in a short time. The kinetic energy per nucleon should not be too large compared to the Fermi energy (≈30 MeV). The last lecture will carry us beyond this regime.

The TDHF equations for the single particle wave functions ψ_n are given by

$$i \frac{\partial}{\partial t} \psi_n(\underline{r},t) = H(t)\psi_n(\underline{r},t)$$

$$H(t) = -\frac{h^2}{2m} \nabla^2 + V(t) \qquad (2.27)$$

and V(t) is an integral over the two-body interaction calculated self-consistently with the single particle wave functions. At each instant of time one has to calculate a mean field produced by the influence of all other particles. As the solutions are stepped in time, the self-consistent field is simply the Hartree-Fock potential at the previous step. The initial systems are represented by a product of single particle wave functions calculated in a moving potential; after the collision, one needs a mixture of both sets of wave functions.

A computer display of the density distributions of these calculations for $^{40}Ca + ^{40}Ca$ at 8 MeV/nucleon in a head-on collision is shown in Fig. 2.31(a), as a function of time.[218] (Because of the symmetry the complete picture should be visualized with an identical pattern below the bottom axis and to the left of the vertical axis.) The contour stripes mark density intervals of 0.04 nucleons/fm^3. We see that taking these calculations at face value (which is premature regarding the state of the art) the nuclei do not fuse, but separate after 0.65×10^{-21} sec oscillating in a predominantly octupole mode. In earlier stages of the diagram all the aspects of fission dynamics, including the neck formation

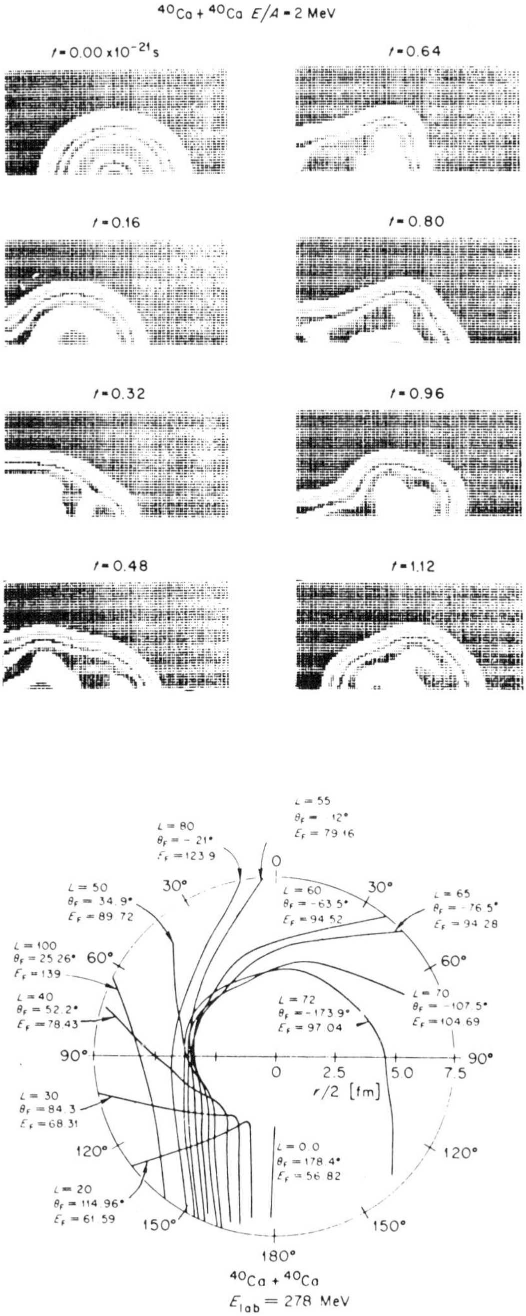

40Ca + 40Ca E/A = 2 MeV
t = 0.00 x10^-21 s
t = 0.64
t = 0.16
t = 0.80
t = 0.32
t = 0.96
t = 0.48
t = 1.12
L = 55
θF = -12°
EF = 79.16
L = 80
θF = - 21°
EF = 123.9
0
30°
30°
L = 50
θF = 34.9°
EF = 89.72
L = 60
θF = -63.5°
EF = 94.52
L = 65
θF = -76.5°
EF = 94.28
L = 100
θF = 25.26°
EF = 139
60°
60°
L = 40
θF = 52.2°
EF = 78.43
L = 72
θF = -173.9°
EF = 97.04
L = 70
θF = -107.5°
EF = 104.69
90°
90°
0
2.5
5.0
7.5
r/2 [fm]
L = 30
θF = 84.3
EF = 68.31
120°
120°
L = 0.0
θF = 178.4°
EF = 56.82
L = 20
θF = 114.96°
EF = 61.59
150°
150°
180°
40Ca + 40Ca
Elab = 278 MeV

Figure 2.31

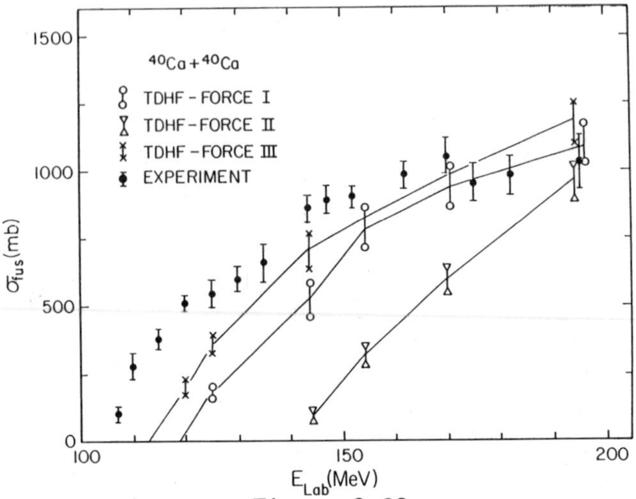

Figure 2.32

and scission, are in evidence. In Fig. 2.31(b), a "trajectory diagram" is constructed showing the final energy and scattering angle for different partial waves. The small waves "bounce" backwards up to ℓ = 30. Some waves fuse and others go into partial orbiting with deflection to negative angles. (This diagram is considerably more sophisticated than our sketch in Fig. 1.6, but it contains the same information.) As shown in Fig. 2.32, the calculation using TDHF Force III gives a reasonable description[221] of the Ca + Ca fusion data.

The possibility that low partial waves do not fuse (i.e., that there is a lower cut-off in partial wave as well as an upper) is an

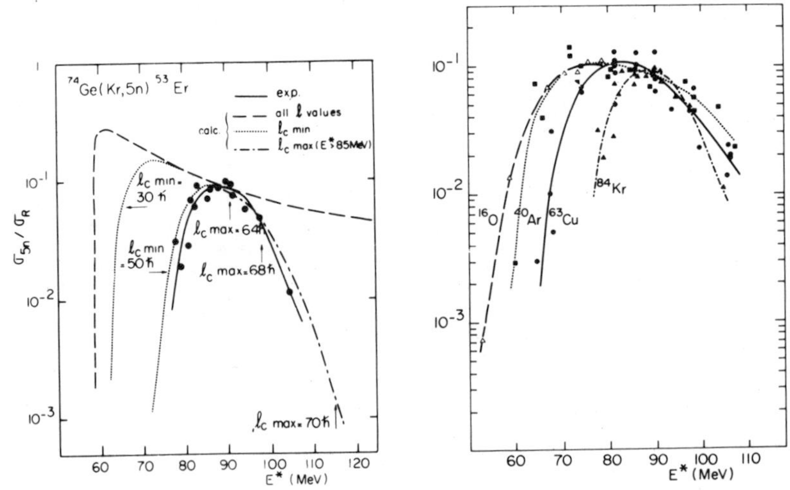

Figure 2.33

idea that has been around in the literature for some time.[191,222-224] A detailed study was made of evaporation residues from the formation of the compound system ^{158}Er by comparing the results of different formation experiments $^{16}O + {}^{142}Nd$, $^{40}Ar + {}^{118}Sn$, $^{84}Kr + {}^{74}Ge$, and $^{63}Cu + {}^{96}Zr$ (the last giving a slightly different compound nucleus). The excitation functions for a particular evaporation channel (5n) are shown in Fig. 2.33. We recall from the "bin diagram" of Fig. 2.8 that this channel should be associated with the *same excitation energy region* of the compound nucleus, regardless of how it was formed, but the evidence in Fig. 2.33 clearly indicates a shift in the onset of this decay channel for the heavier projectiles. (The thresholds are indeed found to be identical for different light projectiles, C, O, and Ne.) Figure 2.8 also reminds us that the lower energy part of the curve must be associated with the low angular momentum population, since all the available excitation energy has to be removed by five neutrons. For the Kr case in Fig. 2.33 then, the large shift in the threshold implies a *lower limit in the J population of the compound nucleus*. The quantitative calculation in Fig. 2.33(b) indicates that $J_{lower} \approx 50\hbar$ would do the trick; as usual there is a J_{upper} of $68\hbar$ in this example (expected from Fig. 2.21).

The implied lower ℓ cut-off is at variance with experiments which measure the fusion cross section via average quantities such as the evaporation residues, or the number of γ-rays in the deexcitation of the residues.[225] One suggestion for the resolution of this paradox is that the apparent thresholds for specific xn channels in the Ar and Kr bombardments may be confused by the presence of non-equliibrium effects in the decay of the evaporation residues. There is certainly evidence for sprays of forward peaked light particles accompanying fusion[226-228] at low angular momenta. Some cross section from low ℓ values is then shifted from xn to (yp, xn) channels. The question of a lower cut-off is still not finally resolved.

2.4 Superheavy Elements

The subtleties of heavy-ion fusion have been discussed, and it will come as no surprise that fusion into superheavy elements has presented enormous obstacles. Schiffer has described the problem in a picturesque metaphor.[229] Suppose you lived in the age of Columbus and you were convinced that his idea of finding an easy trade route to the East Indies was wrong. There were plenty of clever geographers who calculated that the earth was much larger, or that it was flat....and he would never make it to India. Should he have been discouraged? He would certainly have missed something interesting that turned up on the way! Without the elusive goal of Superheavy Elements, perhaps we would have missed some of the discoveries described in this lecture--and the next.

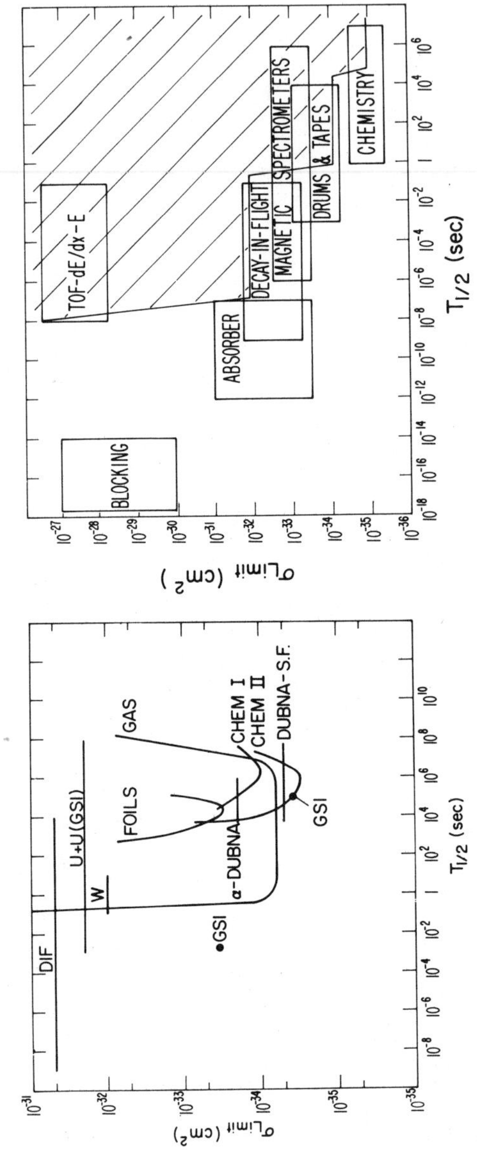
TOF-dE/dx-E
BLOCKING
ABSORBER
DECAY-IN-FLIGHT
MAGNETIC SPECTROMETERS
DRUMS & TAPES
CHEMISTRY
σLimit (cm²)
T1/2 (sec)
DIF
U+U(GSI)
W
FOILS
GAS
CHEM I
CHEM II
DUBNA-S.F.
α-DUBNA
GSI

Figure 2.34

Despite intensive searches in major laboratories in the U.S., U.S.S.R., Germany and France, no evidence for superheavy elements in nuclear reactions have been found (for a recent review see Refs. 230, 231). (Brief successes[232] in Monazite inclusions were short-lived.[233]) Upper limits for the cross sections are shown on the left side of Fig. 2.34. Most of the limits are obtained by failing to detect any spontaneous fission activity; one event would correspond to the quoted cross sections, and it is questionable whether the methods would make us believe one event. Some other experimental techniques, and their attainable limits, are illustrated on the right. It seems clear that one must turn to methods capable of exploring shorter lifetimes and (preferably) yielding higher cross sections particularly since the most favorable[234] combination ^{48}Ca + ^{248}Cf has come up with negative results.[235,236]

There may be a way, by bombarding Uranium with Uranium. Fig. 2.35 is a plot of the total kinetic energy loss suffered in the collision as a function of the Z of the observed fragment.[237] The two horizontal lines indicate the total available kinetic energy and the Coulomb repulsion energy for spherical fragments; the

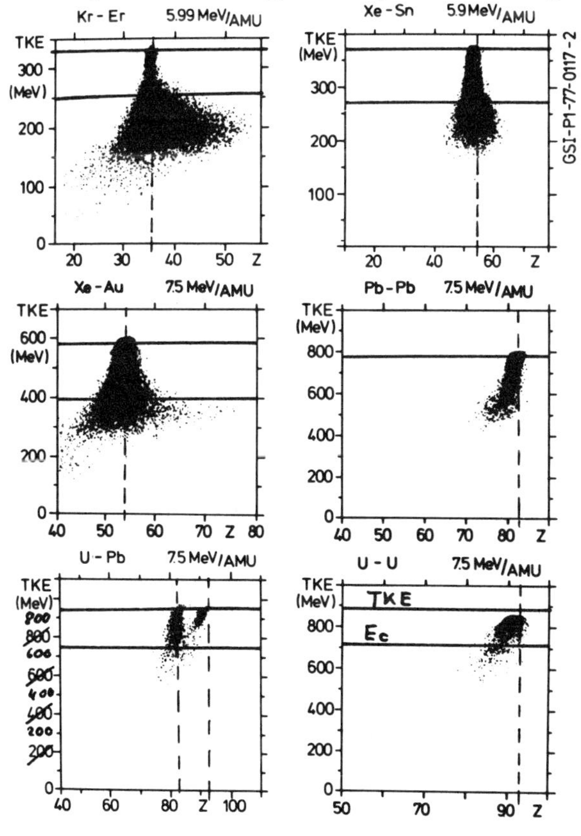

Figure 2.35

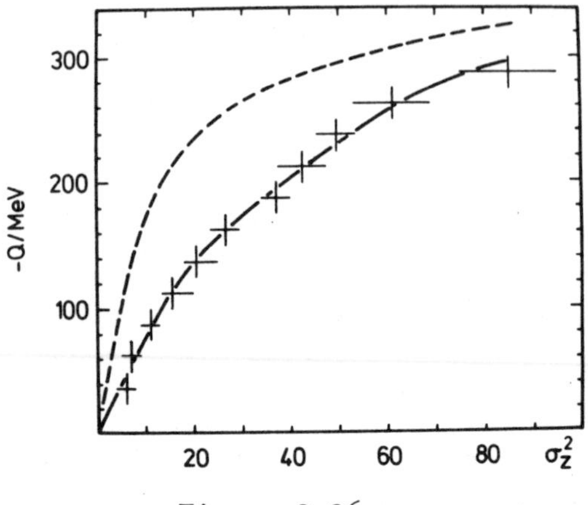

Figure 2.36

vertical line marks the projectile z. The principle feature of the first three reactions is that the width of the Z distribution increases with energy loss. This trend can be explained in a diffusion model as an increased interaction time[238] (to be discussed in the next lecture). The functional relationship for systems such as Xe + Au is shown in Fig. 2.36 by the dashed line. The encouraging feature is that this curve does not apply to U + U (solid line) which has a smaller energy loss for a given σ_Z^2, and therefore less likelihood for the product to fission. To set a scale, the difference between the two curves is 16 orders of magnitude in the production cross section for element 114, but the calculated cross section is still two orders of magnitude smaller than the present limit (10^{-32} cm^2). We devote the whole of the next lecture to a discussion of this type of reaction, known as Deeply-Inelastic Scattering.

3. MACROSCOPIA (DEEPLY-INELASTIC SCATTERING)

For a long time, nuclear reactions—induced by either light or heavy-ions—were classified into two extremes. In quasi-elastic direct reactions, a few nucleons were exchanged between target and projectile, the loss of kinetic energy was small, and the effect of the interactions was a minor perturbation of the system. In fusion reactions, on the other hand, the projectile was completely swallowed up by the target, and all the initial kinetic energy was transformed into intrinsic excitation energy, or carried away by secondary light particles. We know now that such a treatment is too Procrustean; a whole hierarchy of intermediate processes exists. In light-ion reactions, there is a well-developed theory of doorway states to describe reactions intermediate between direct and compound.[239,240] In heavy-ion reactions, a similar intermediate reaction process is called deeply-inelastic scattering.[241,242] We can hope that the formal theoretical approaches developed to describe this phenomenon will lead one day to a generalized theory of nuclear reactions, light or heavy, at high or low incident energy. When the history of nuclear physics in the seventies is written, surely the study of deeply-inelastic scattering will emerge as one of the major new directions in nuclear physics. It is a subject that has captured the imaginations of nuclear chemists, physicists and theorists all over the world, who have abandoned their traditional research areas in droves. They have created a monument which by sheer size alone inspires awe. There are many names for the new reaction process—quasi-fission, damped collisions, partially-relaxed phenomena, deeply-inelastic scattering and, because they are Characteristic of Heavy-Ion Processes,[219] CHIPS. Last year I suggested a new compromise[30] which captured the spirit of these processes, of FISSION CHIPS, but it does not appear to have been accepted. Recently I noticed a misprint[169] in a Physical Review Letter, which introduced a new word to the English language, FUSSION. Since deeply-inelastic scattering is betwixt and between fusion and fission, I now suggest a new name, FUSSION CHIPS (an often heard dialectic variant of this English delicacy!)

3.1 The Phenomenon

To see why "fussion chips" is an appropriate terminology, let us look at some of the paths taken by heavy-ion collisions[191] as they have emerged in the first two lectures (see Fig. 3.1). The time scale is in units of $\tau = 10^{-22}$ sec. It shows how the composite system may proceed towards compound nuclear formation, preceded and succeeded by particle emission, and possibly ending in symmetric fission. *But there is also a new path,* where the composite system *never* fuses completely; rather, it separates on a relatively short time scale into two fragments, reminiscent of the initial ions which went into partial orbit. (Is there a connection with the quasimolecular states?) This is the new process. A schematic division of the reaction cross section as a function of ℓ is also given in Fig. 3.1. The sloping line represents the unitarity limit $\sigma_\ell = \pi\lambda\!\!\!^{-2}(2\ell+1)$. For high partial waves (associated with the strong interaction radius) direct reactions such as transfer and inelastic scattering occur. Then at closer collisions come the CHIP processes, until ℓ_{crit} allows the onset of fusion. For the upper ℓ-values, the system decays primarily by fission; for the lower values the compound nucleus survives to emit particles and leads to evaporation residues. Finally there may exist a ℓ_{lower} boundary where the nuclei do not fuse, but result (possibly) in other fussion chips processes.

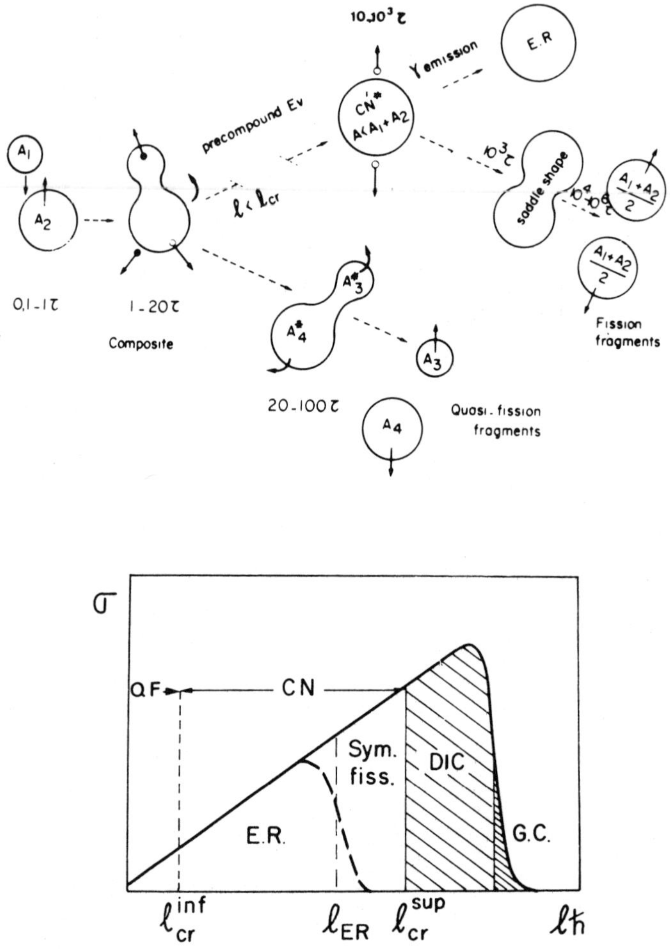
10-10³ τ
γ emission
E.R
precompound Ev
CN*
A<A₁+A₂
A₁
A₂
ℓ<ℓcr
10³ τ
saddle shape
(A₁+A₂)/2
(A₁+A₂)/2
0,1-1τ
1-20τ
Composite
A₃*
A₄*
A₃
Fission fragments
20-100τ
A₄
Quasi-fission fragments
σ
QF
CN
Sym. fiss.
DIC
E.R.
G.C.
ℓcr inf
ℓER
ℓcr sup
ℓħ

Figure 3.1

A powerful means of getting a global picture of the reaction processes comes from analyzing the mass distribution of reaction products after irradiation, using radiochemical techniques. An example[211] for $^{48}Ca + ^{208}Pb$ is shown in Fig. 3.2. The mass distribution is decomposed into several components. The "rabbit ears" close to the projectile and target masses are few—nucleon transfers and quasi-elastic processes. The broad curve A, centered at half the combined mass of target and projectile, is symmetric fusion-fission, whereas curve B is the transfer induced sequential fusion of the target. The regions C and D represent the intermediate deeply-inelastic processes.

The energy spectra for a similar experiment $^{40}Ar + ^{232}Th$ at 388 MeV, taken with solid-state counter telescopes, is shown in Fig. 3.3(a) for a selection of different elements as a function of angle. These data were some of the first to address the new phenomenon of deeply-inelastic scattering,[243] but they illustrate very clearly the characteristic features. The spectra in general have two components: a high-energy component—referred to as "quasi-elastic"—with an energy not too far removed from $(E_{inc} + Q)$, and a low-energy component, peaking close to the Coulomb barrier for the appropriate exit channel. (For example, Cl at 45° peaks at ≈ 140 MeV, and V_c ≈ 155 MeV.) This observation is the origin of the term "deeply-inelastic" or "strongly damped," because all the initial energy has been dissipated, and the fragments emerge with the Coulomb energy (like fission fragments). However, we notice that as we move towards forward angles, there is *a tendency for the two components to move together and merge into one.*

The corresponding angular distributions, shown in Fig. 3.3(b) for products with masses in the vicinity of the projectile have a grazing angle peak characteristic of the peripheral collisions we

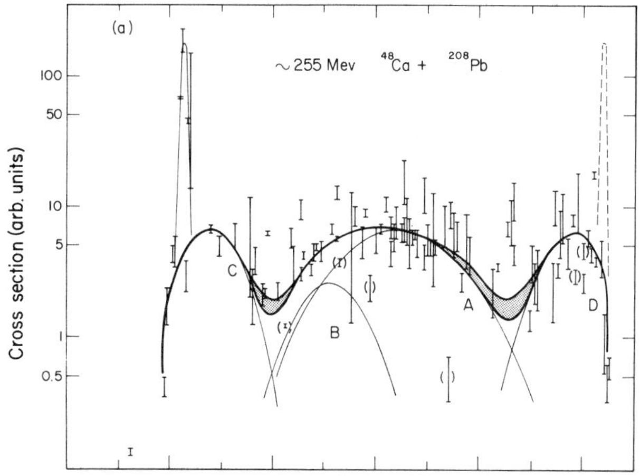

Figure 3.2

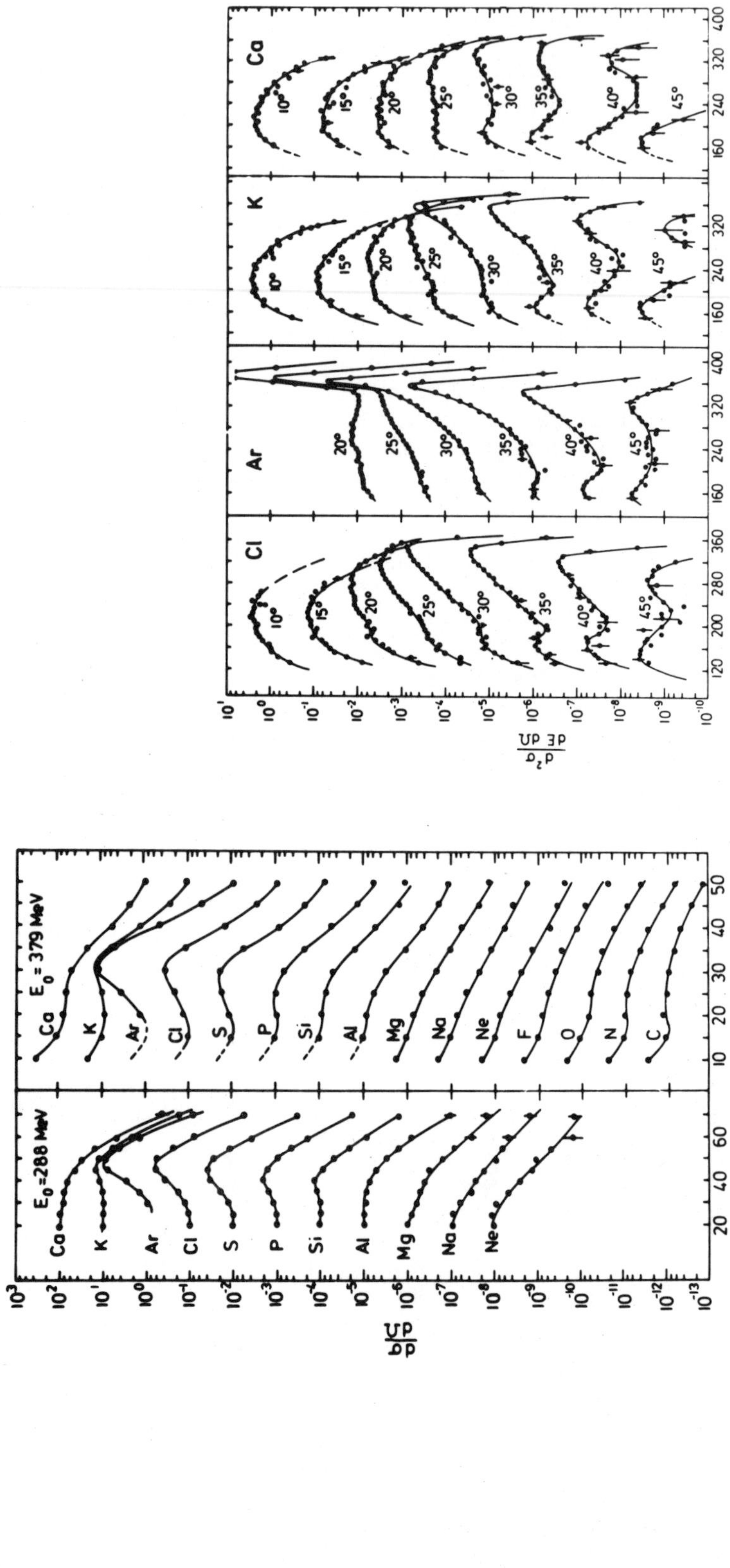

Figure 3.3

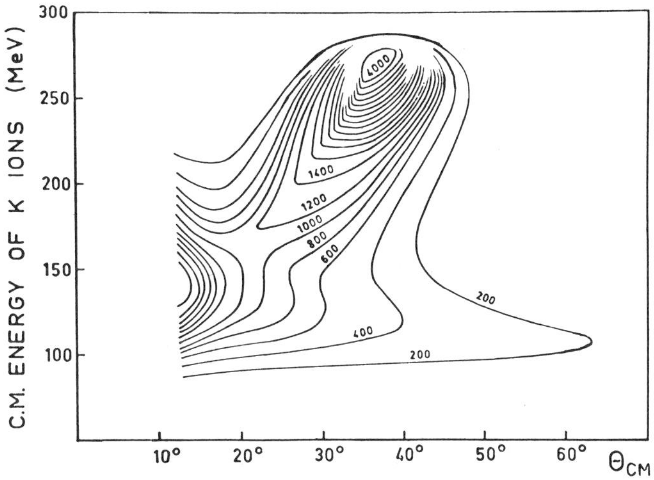

Figure 3.4(a)

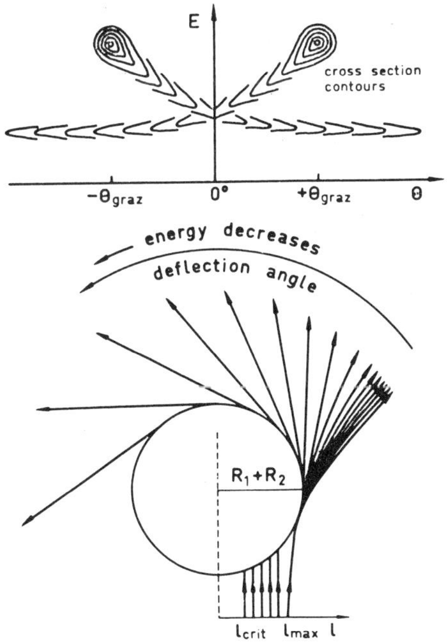

Figure 3.4(b)

discussed in Lecture 1. Products far removed from the beam have much flatter distributions and, in many experiments, approach the $1/\sin\theta$ form characteristic of heavy-ion compound nuclear reactions (Lecture 2). The information from Figs. 3.3(a) and (b) can be combined into the contour plot of Fig. 3.4(a), which shows a pronounced mountain at high energy close to the grazing angle, with a ridge running down to lower energies and smaller angles.[244] It merges with another ridge which *increases* in energy from large angles to more forward angles.

The presently accepted interpretation is illustrated schematically in Fig. 3.4(b). (Alternative interpretations based on a double rainbow cannot be excluded in all cases.[245]) The higher partial waves undergo grazing collisions, in the region of the maximum of the deflection function (compare Fig. 1.6) and lead to the large quasi-elastic peak. Between the grazing and the critical partial waves, the orbits are pulled around by the attractive potential and the two ions form a *temporary dimolecular system*. They do *not* fuse to form a compound nucleus, which happens for waves smaller than ℓ_{crit} (compare the discussion in the last Lecture). The low-energy ridge in the contour plot is then naturally associated with *negative angle scattering*. The longer the association and the greater the angle of rotation, the more the angular distributions approach the form $1/\sin\theta$. Figure 3.3 then suggests nucleon transfers will occur during the time the two nuclei are in contact and the magnitude of the exchange is correlated with the kinetic energy dissipation and with increasing angle rotation. A comparison of the contour plot in Fig. 3.4 with those in Fig. 2.35 shows that the detailed behavior depends strongly on the Z_1Z_2 product of the colliding system. For U + U for example the angular distributions become broader with significant fractions *backwards* of the grazing angle, indicative of a grazing angle, Coulomb dominated deflection.[246] A critical parameter is the reduced Sommerfeld parameter,

$$\eta' = \frac{Z_1 Z_2 \ell^2}{\hbar v'} .$$

Here v' is the velocity of the two ions at the interaction barrier). The quantity is roughly the ratio of the Coulomb force $Z_1Z_2e^2/R^2$ and the frictional force (responsible for dissipating the initial kinetic energy) which is proportional to the velocity and the product of the nuclear densities $\propto 1/R^2$. Systems with $\eta' \lesssim 150$-200 give rise to orbiting whereas those with $\eta' \gtrsim 250$-300 do not.[247] Another important parameter defining the characteristic behavior is the ratio E/B of the center of mass energy to the Coulomb barrier.[248] (Some of the many extensive reviews on the subject of deeply-inelastic scattering are given in Refs. 249-254.)

Before proceeding further with the logical analytical predictions of the rotating, dinuclear model, we must describe some experiments relating to direct experimental evidence for its validity. An important aspect is that these reactions are basically binary processes, and this has been established by coincidence measurements of the projectile and target-like fragments (see e.g. Ref. 255).

Another consequence of Fig. 3.4 is that the direction of rotation of the quasi-elastic (positive angle) and deeply-inelastic (negative angle) fragments should be opposite. Further, in a classical picture of a peripheral collision, we expect the angular momentum to be oriented perpendicular to the reaction plane. For the quasi-elastic transfer, the semi-classical model discussed in Lecture 1 gives some predictions[66,67] of the polarization. Evaluating λ_2 from the Eq. 1.43 and substituting Eq. 1.44 gives:

$$Q_{eff} \approx \frac{\lambda_1 \hbar v}{R_1} - \frac{\hbar v}{R}\,\frac{k_o R}{2} \approx \frac{\lambda_1}{R_1}\hbar v - \tfrac{1}{2} m v^2 \qquad 3.1$$

Since the incident nucleus is left in a hole state of the transferred particles, the sign of its polarization should just be opposite to λ_1. Vanishing polarization is predicted at the "optimum Q-value", best satisfying the semi-classical matching conditions:

$$Q_{opt} = -\tfrac{1}{2}\, m v^2 + (Z_1^f Z_2^f - Z_1 Z_2^i) e^2/R \qquad 3.2$$

If $Q > Q_{opt}$, the polarization is negative and if $Q < Q_{opt}$, it is positive. (For a more detailed investigation using DWBA theory see ref. 256).

Just these features have been studied[257,258] in the reaction ^{100}Mo (^{14}N, ^{12}B)^{102}Ru at 90 MeV by measuring the β-decay asymmetry of the ^{12}B($J^\pi = 1^+$, $E_{\beta max} = 13.37$ MeV, $t_{\frac{1}{2}} = 20.3$ ms). The angular distribution of β rays with respect to the polarization $\underline{P}$ is given by $W(\theta) = 1 - P\cos\theta$. The apparatus is sketched in Fig. 3.5(a). The beam irradiation was cyclic and on alternate cycles the spin direction of the ^{12}B was reversed with an RF field to eliminate instrumental asymmetries. The results are shown in Fig. 3.5(b). We see that the energy spectrum of the ^{12}B (determined for N_β) peaks in the continuum at a Q-value of ≈ -23 MeV, compared to $Q_{opt} - \tfrac{1}{2}m\,v^2 + \Delta V_c \approx -21$ MeV. Calculations for both the energy spectrum and the polarization are shown. The value of P indeed reaches zero at the peak of the distributions, but for more negative Q-values the values of P do not become sufficiently positive. Therefore, for these larger energy losses, we need a process tending to give additional polarization of the opposite sign, e.g. deeply-inelastic scattering.

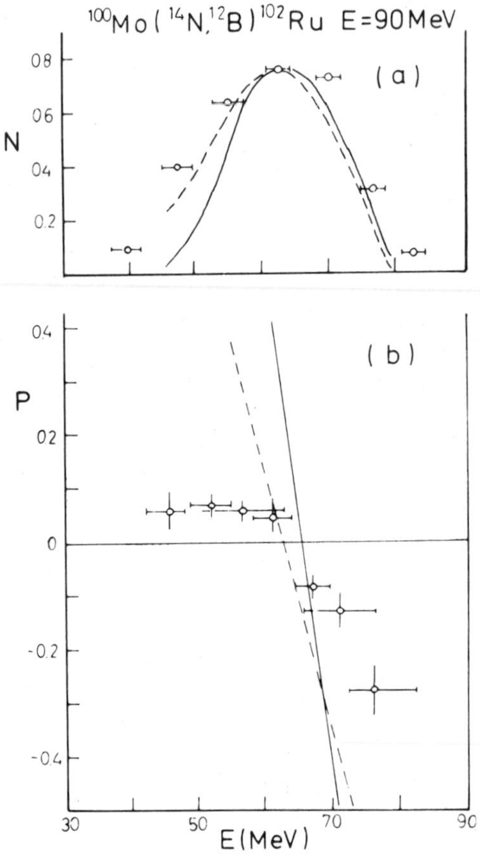

Figure 3.5

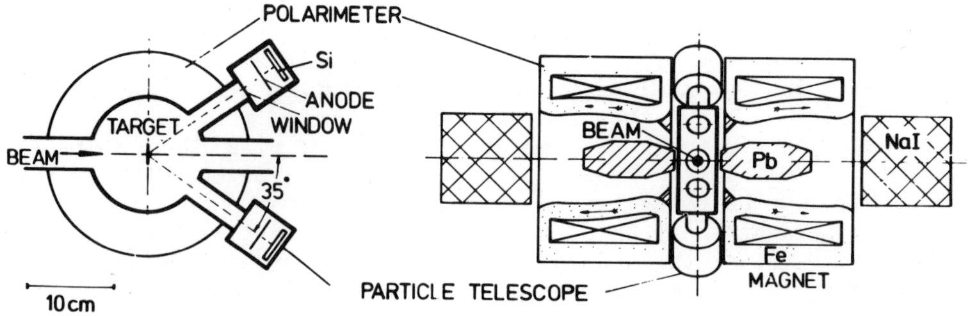

Figure 3.6

For the ^{40}Ar + Ag system at 300 MeV quasi- and deeply inelastic processes are clearly separated. The polarization of the fragments (*both* projectile and target fragment spin in the same direction) has been determined[259] from the circular polarization of the subsequent de-excitation γ-radiation. The direction of polarization can be measured by scattering the emitted γ-rays from the polarized electrons in magnetized iron. (Remember the classic experiment on measuring parity violation in weak interactions.) The experimental apparatus, sketched in Fig. 3.6, used two polarmeters in a symmetric configuration normal to the scattering plane defined by the two heavy-ion counters at 35°, which detected the fragments. The observed polarizations are consistent with the dinuclear positive and negative angle scattering model, although the absolute values are less than expected. I have described both these experiments in some detail because they are characteristic of the sophistication of techniques in current use.

One of the striking characteristics of deeply-inelastic scattering is the rapid dissipation of the initial kinetic energy into internal excitation on a time scale comparable to direct reactions. The complete equilibration hypothesis implies that the composite system reaches a common temperature and divides into fragments with excitation energies proportional to their thermal capacities (masses). In a cunning analysis of the ^{86}Kr + ^{124}Sn reaction at 440 and 720 MeV incident energies, this division has been convincingly demonstrated.[260] The bottom part of Fig. 3.7 shows the observed average mass of each projectile-like element emitted (upper triangles for 440 Mev, lower for 720 MeV). Since we are interested in the difference in mass numbers at each Z, the ordinate has been expanded by making zero equal a mass division having the same Z/A as the composite system (achieved by subtracting 201/86 Z_f). If the observed products *really* represented the primary fragments, these

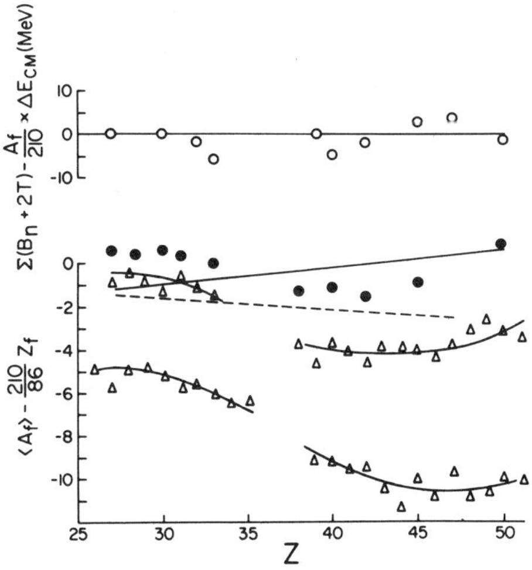

Figure 3.7

points would all line up at zero for complete relaxation; indeed the solid points are an attempt at reconstruction of the primary distribution for the lower energy, and the straight lines are other theoretical reconstructions. The fact that the data are displaced from zero is simply a reflection of the subsequent neutron emission of the excited prefragments, and these displacements therefore contain information on this excitation energy. The displacement of the two curves is just related to the difference in partition of the extra 165 MeV (difference of the center of mass energies) of internal excitation. If this energy is divided in proportion to the masses, the fragment A_f should receive the fraction $(A_f/210)\Delta E$. As the upper part of Fig. 3.7 shows, this quantity is indeed equal to $\Sigma(Bn + 2T)$, where the summation is over the observed neutron difference seen in the bottom part of the figure, Bn is the neutron binding energy and 2T is the average kinetic energy of the emitted neutron. The temperature T was estimated from complete equilibration of the energy. Clearly the clustering of the points around zero, confirms the division of energy in the ratio of the masses.

The examples we have discussed in this section should indicate the scope for inventiveness and imagination in the study of heavy-ion reactions. Now we proceed to discuss the underlying mechanisms responsible for the large and fast damping of the energy, and the relaxation of other degrees of freedom.

3.2 Energy Dissipation in Deeply Inelastic Scattering

Two hypotheses are currently in vogue to explain the description of large amounts of the initial kinetic energy into intrinsic excitation of the dinuclear complex. Since they approach the problem from the antithetical viewpoints of one-body dissipation and collective dissipation, it is likely that the ultimate truth will involve a synthesis, just as our understanding of nuclear structure involves single particle and collective aspects. We should recall that the transition between direct and compound nuclear processes is not unique to heavy-ion reactions. A typical energy spectrum for a light-ion reaction (e.g., p, p') appears[261,262] in Fig. 3.8. The energy abscissa can also be regarded as a time variable; the low energy compound region corresponds to a long interaction time and the high energy direct region to a short time. There is a continuous evolution from direct and multistep direct reactions to compound nuclear reactions. The relationship to heavy-ion reactions becomes plausible if we compare the evolution of the differential cross sections in Fig. 3.3 and those of Fig. 3.9, which pertains to direct and multistep reactions leading to *discrete final states*.[196,263] In this type of reaction the microscopic techniques we discussed in Lecture 1 are well developed and there is hope that similar approaches can be applied to heavy-ion reactions.[264-266]

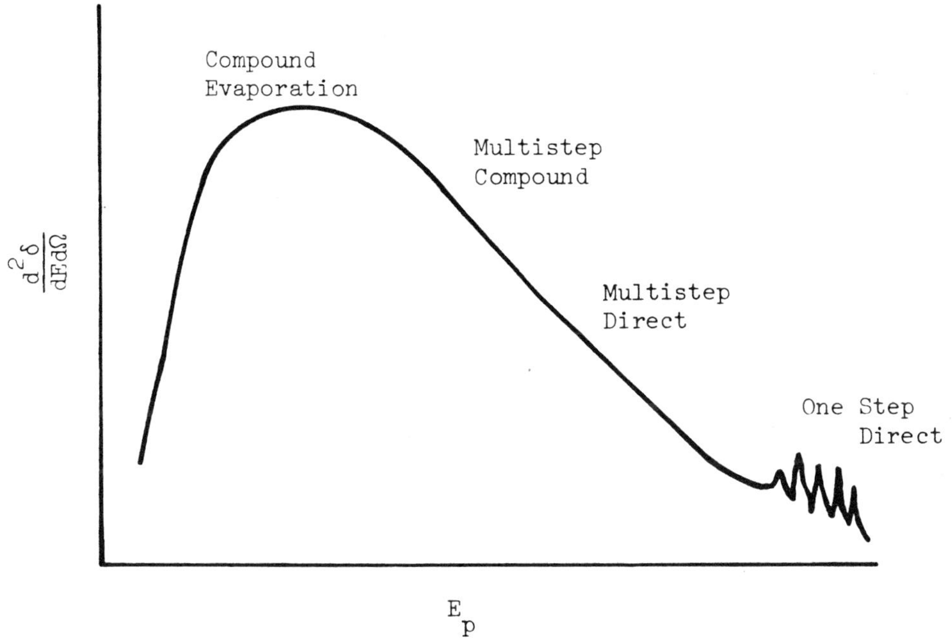

Figure 3.8

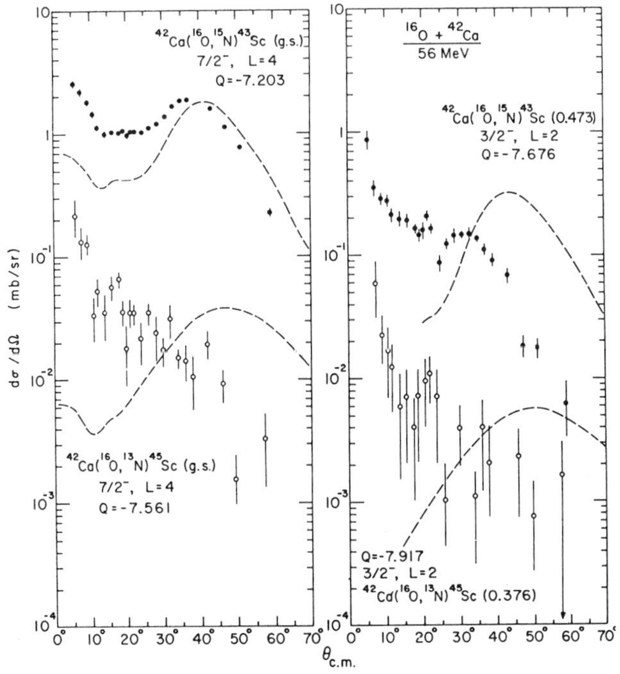

Figure 3.9

In searching for a fast dissipative mode, we are led naturally to think of giant multipole excitations. The dipole resonance has a characteristic time of 10^{-22} sec, and is one of the fastest motions known in nuclear physics. There are two characteristic times in heavy-ion collisions. The first is the time during which the nuclei are in contact, i.e.

$$\frac{\tau_{coll}}{\hbar} \approx \left[\frac{E}{A}\left(1 - \frac{V_c}{E}\right)\right]^{-\frac{1}{2}} \text{MeV}^{-1} \tag{3.3}$$

For collisions not to far above the barrier this time is of the order $0.5 \rightarrow 1\ \text{MeV}^{-1}$ ($\approx 6 \times 10^{-22}$ sec). The second one measures the degree of adiabaticity of the process, which is concerned with the time, τ_{char}, over which the form factor changes by a factor of two. For close collisions τ_{char} may be an order of magnitude shorter than τ_{coll}, i.e. $\approx 6 \times 10^{-23}$ sec. The quantum transitions, that can take place during the collision, are therefore limited adiabatically to have $\Delta E \lesssim \hbar/\tau_{char} \approx 10\text{-}20$ MeV. This region of excitation is just where the major fraction of the sum rule for inelastic excitation of giant resonances is exhausted.[267] A number of theoretical papers deal with these modes as the mechanism for deeply inelastic scattering,[268] via multistep excitation.

From the experimental point of view one might hope to get a clue about the role of these modes by looking for structure in the deeply inelastic continuum. So far this has appeared as a featureless bump, but more refined data for the Ca + Ca and Cu + Cu at approximately 300 MeV, acquired with a magnetic spectrometer, reveal quite a complex structure. (See Fig. 3.10: (a) for Ti isotopes from Ca + Ca; (b) for Zn isotopes from Cu + Cu, and (c) ^{63}Cu produced in Cu + Cu; the total excitation energies are also indicated).

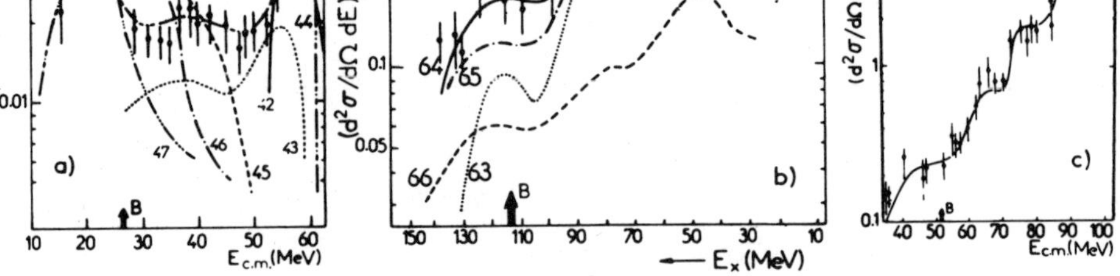

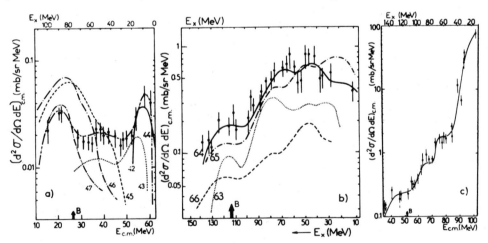

Figure 3.10

Even more pragmatically, we might look for the direct excitation of giant multipoles in inelastic heavy-ion scattering. The probability that either fragment will emerge in a single giant resonance depends, however, on the system. For heavy systems, the large energy loss implies a dominance of multiple excitation, but for lighter systems, the shorter collision times and the higher excitation, lead to stronger single excitation. The E2 mode has been observed in $^{16}O + {}^{27}Al$,[270] $^{16}O + {}^{208}Pb$,[271,272] and ^{12}C, ^{14}N + Zr, Pb.[273] For the $^{16}O + {}^{27}Al$ system, Fig. 3.11(a) shows the excitation probability for different regions of θ, together with the ratio (shaded) for excitation of the giant quadrupole resonance compared to everything else.[274] Even for this light system the probability is unexpectedly small, and it remains to be seen if the quantitative models[268] account for the strength. In (b) is shown a "Wilczynski Plot" for the inelastic scattering (compare Fig. 3.4) which also shows the ridge, between -7 and -20 MeV, characteristic of deeply-inelastic scattering and negative angle scattering.

Finally an example of E2 excitation for $^{16}O + {}^{208}Pb$ at 315 MeV is shown[272] in Fig. 3.12. Both at 140 MeV and at 315 MeV, the observed strength apparantly exhausts the energy weighted sum rule;[271] therefore the multiple step excitation of the deeply-inelastic continuum (a cross section of 400 mb at 135 MeV[275]) does not reduce the single excitation, possibly raising an element of doubt over the role of these resonances for the damping. Further comparisons at different energies are required. An interesting feature of Fig. 3.12 is the appearance of higher lying structures. The frequency of oscillation of multipole modes can be derived[275] from the liquid drop model to depend on the multipolarity as $\ell^{3/2}/A$; for quadrupole oscillations $\omega_2 \cdot \hbar \approx 0.8$ Mev. An evaluation of ω as a function of ℓ and A, tells us that the associated

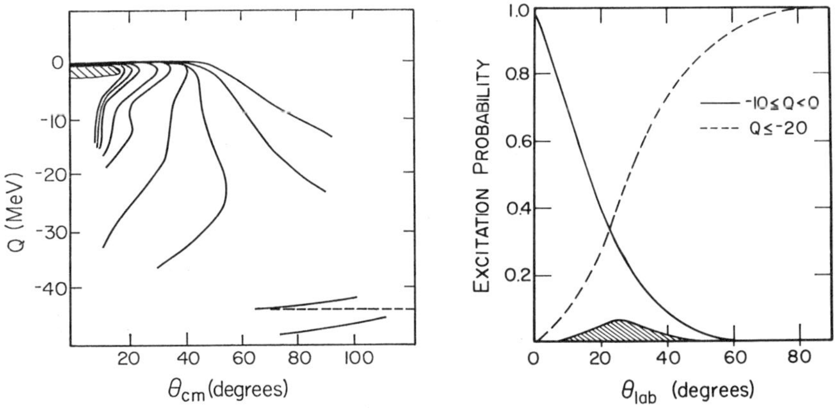

Figure 3.11

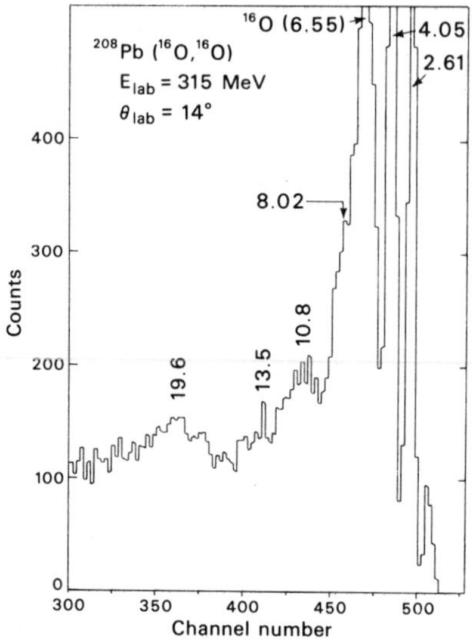

Figure 3.12

velocities, $v = \omega \cdot R$, will call for collision speeds in excess of 20 MeV/A for the excitation of higher lying multipoles, which may therefore be appearing in Fig. 3.12. (The giant quadrupole resonance corresponds to the bump at 10.8 MeV.)

Now let us turn to the alternative energy dissipation mechanism via single particle motion. In this picture, as the two nuclei rotate in close contact, an exchange of nucleons takes place through the window that opens up in the neck between them. Consider the nuclei as containers in which the nuclei have a random motion.[276] A nucleon in nucleus 1 can escape through the neck and be absorbed by nucleus 2, and vice versa. Let the area of the interface of the composite system be A(t), and the window integral in the reaction,

$$\overline{A}\Delta t = \int_{\mathrm{orbit}} A(t)\, dt \qquad (3.4)$$

The probability per second that a nucleon crosses the interface from 1 to 2 is $n_{12}A$ and similarly from 2 to 1 is $n_{21}A$. These rates depend on dynamics and are functions of time. This dependence will be weak if the number of transferred nucleons is much less than the total. So say n_{ik} is constant. Then the variance of the number transferred is:

$$\delta n = [(n_{12} + n_{21}) \int A(t)\, dt]^{\frac{1}{2}} \qquad (3.5)$$

while the flow of mass from 1 to 2 is

$$\langle n \rangle = (n_{12} - n_{21}) \int A(t)\, dt \tag{3.6}$$

and the normalized distribution of the number transferred might be expected to be a Gaussian,

$$P(n) = \frac{1}{\sqrt{2\pi}\,\delta n} \exp - \left[\frac{(n - \langle n \rangle)^2}{2\delta n^2} \right] \tag{3.7}$$

Now a good guess for the transfer rate is:

$$n_{12} \approx n_{21} \approx \frac{1}{6} \rho v \tag{3.8}$$

where ρ is the nuclear matter density, 0.17 nucleons/fm^3, and $v \approx 9 \times 10^{22}$ is the typical speed of a nucleon inside the nucleus. With an interface area of $\bar{A}$ = 10 fm^2 and a typical direct reaction time of $t \approx 5 \times 10^{-22}$ sec for the collision of ^{40}Ar on ^{50}Ti at 236 MeV,[177] we get $\delta n \approx 5$. The Z and A distribution of fragments in this reaction are illustrated in Fig. 3.13 (which were obtained by combining the Z and A information of Fig. 2.16) and we see that the spread in A values is indeed *the order* of δn. (It is difficult to see the Gaussian profiles in the 2-D plot, but such indeed are the observed shapes.)

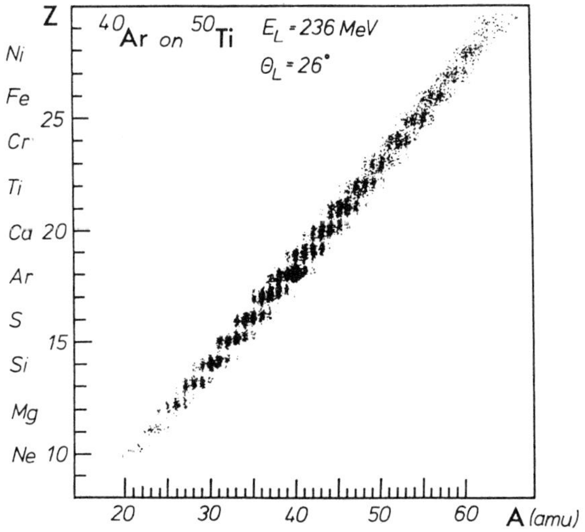

Figure 3.13

3.3 More Formal Theory

The theory presented here will be only slightly more formal, with an emphasis on the extraction of physical quantities from the data. Rigorous approaches are described in other Lectures of this School. The generalization of the discussion in the previous section to diffusion processes in the rotating dinuclear system leads to the Focker-Planck equation[251,253,277,278] for the population distribution of a macroscopic variable x as a function of time, P(x,t):

$$\frac{\partial P(x,t)}{\partial t} = -v\,\frac{\partial P(x,t)}{\partial x} + D\,\frac{\partial^2 P(x,t)}{\partial x^2} \qquad (3.9)$$

the solution of which is:

$$P(x,t) = \frac{1}{\sqrt{4\pi Dt}}\,\exp\left[-\frac{(x-vt)^2}{4Dt}\right] \qquad (3.10)$$

The mean value of the distribution x moves with time at constant velocity, and the variance $\sigma^2 = \langle x - \langle x\rangle\rangle^2 = 2Dt$ increases linearly with time (see Fig. 3.14). The transport coefficients v and D are known as the drift and diffusion coefficients. The FWHM of the curve is given from $\Gamma^2 = 16\,\ell n\,2(Dt)$. Amongst the macroscopic variables which have been measured are kinetic energy, the N/Z degree of freedom and the mass asymmetry degree of freedom $A_1 - A_2/A_1 + A_2$.

As an example of how these methods work,[279,280] consider the charge distribution as a function of angle. This can be derived

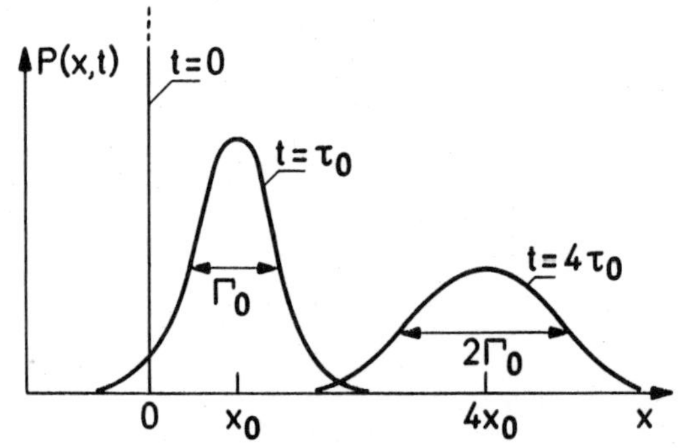

Figure 3.14

from an analysis of distributions of cross sections such as Fig. 3.3 for each Z. They would be expected to have Gaussian distributions,

$$P(z,t) = \frac{1}{\sqrt{4\pi D_z t}} \exp\left[-\frac{(z - z_o - v_z t)^2}{4D_z t}\right] \tag{3.11}$$

where $z - z_o$ stands for the number of protons transferred during the interaction time t. The quantities v_z and D_z represent average proton drift and diffusion coefficients. In order to relate *angle* information to *time* information we write,

$$\tau_{int} = \frac{1}{\bar{\omega}}(\theta_{gr} - \theta) \tag{3.12}$$

where τ_{int} is the interaction time for the rotating dinuclear system, rotating with mean rotational frequency $\bar{\omega}$. (The rotation is measured from the grazing angle.) Now,

$$\bar{\omega} \approx \frac{\hbar \ell}{\mathcal{J}} \tag{3.13}$$

where $\mathcal{J}$ is the moment of inertia of the system, and

$$\bar{\ell} \doteqdot \frac{2}{3}\,\frac{\ell_g^3 - \ell_{crit}^3}{\ell_g^2 - \ell_{crit}^2} \tag{3.14}$$

where we attribute deeply-inelastic collisions to the band of partial

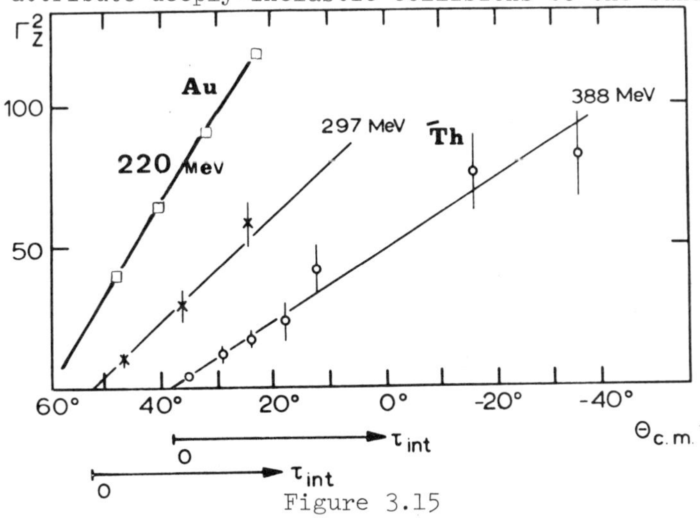

Figure 3.15

waves from ℓ_{crit} (inside of which fusion takes place) to ℓ_g (see Fig. 3.1).[281]

For the reaction Ar + Th depicted in Fig. 3.3 at 388 MeV, ℓ_g and ℓ_{crit} have been determined as 222 and 94 respectively.[243] For $\mathcal{J}$ we can assume rigid body rotation of the dinuclear complex:

$$\mathcal{J} = \frac{2}{5} M_1 R_1^2 + \frac{2}{5} M_2 R_2^2 + \mu R^2 \quad . \tag{3.15}$$

The plot of Γ^2 versus θ in Fig. 3.15 can then be regarded as a plot of Γ^2 versus $\tau_{int} = t$, and the slope $\Gamma^2/t \propto D_Z$. In fact, the same value of D_Z is derived for the different reactions studied at different energies (on the figures, the t-scale is different for the different reactions, since this is transformed by $1/\overline{\ell}$). The derived value was $D_Z \approx 10^{22}$ (charge units)^2/sec. Other quantities can be determined by similar analysis. One finds typically:[251,277]

Energy drift coefficient $v_E \approx 4 \times 10^{23}$ MeV/sec

Energy diffusion coefficient $D_E \approx 4 \times 10^{24}$ (MeV)^2/sec

Charge drift coefficient $v_Z \approx 10^{21}$ (charge units)/sec

Charge diffusion coefficient $D_Z \approx 10^{22}$ (charge units)^2/sec

These values are not expected to be very accurate due to the crude method of estimating the interaction time. In a more refined approach[282] a better relation between impact parameter ($\equiv \ell$) and scattering angle is derived by constructing a proper deflection function. Energy and angular momentum dissipation are taken into account. Interaction times calculated in this way can vary by a factor of 3 from the simple estimate.

A characteristic of the deeply-inelastic collision is the large energy damping. This energy loss also appears to take place rapidly while the two ions are in contact. On a microscopic picture the energy loss could be mediated by particle-hole excitation and also by transfer of nucleons between the colliding ions. Such a nucleon, with mass m, deposits a momentum $\Delta p = m|\dot{r}|$, where $\dot{r}$ is determined from the energy of the system prior to the transfer, and the *resultant energy loss is therefore proportional to the energy available* ($\delta E \propto (\Delta p)^2$). This argument justifies the introduction of a frictional damping force proportional to the velocity[279,283-285]

$$F_t = -kv \tag{3.16}$$

Then we can write for the rate of energy loss:

$$\frac{dE}{dt} = \mu v \frac{dv}{dt} = v \cdot F = -kv^2 = -2 \frac{k}{\mu} E \tag{3.17}$$

Integrating the expression,

$$\ell n \left(\frac{E_o}{E} \right) = 2 \frac{kt}{\mu} \tag{3.18}$$

Now we have just shown that a time scale is established by the relation $t = \Gamma_Z^2/2D_Z$, and therefore we expect that there should be a linear relation between $\ell n(E_o/E)$ and Γ_Z^2; the gradient yields a value for $k/\mu D_Z$. As Fig. 3.16(a) dramatically demonstrates,[280] there certainly is a clear correlation between the width of the charge distribution and kinetic energy loss, which is shown on this figure for successive 50 MeV wide bins in the reaction of Bi + Xe.

In Fig. 3.16(b), the values of σ_Z^2 from Fig. 3.16(a) are plotted as a function of the interaction time $\tau(\ell)$ in units of 10^{-22} sec, and appear to increase linearly, i.e., $\sigma_Z^2(\ell) = 2D_Z(\ell)\,\tau(\ell)$. The time scale on the figure was derived from the deflection function. This deflection function was constructed by assuming a sharp cut-off model, where the cross section up to ℓ_j is given by $\sigma_j = \pi\lambda\!\!\!^{-2}(\ell_j+1)^2$. Then using the experimental results on the cross section as a function of kinetic energy loss, the angular momentum can be related[286] to the energy loss by:

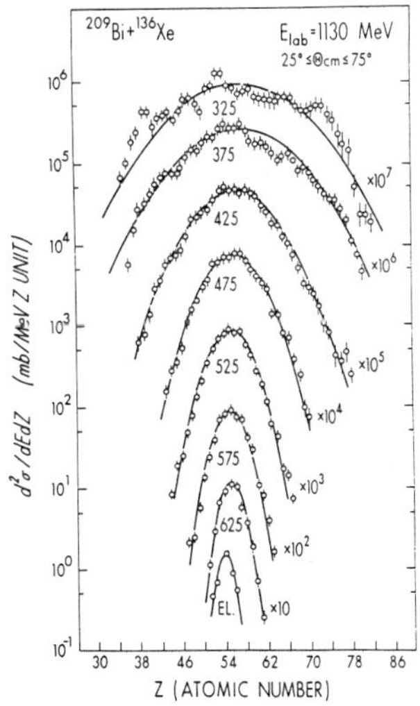

Figure 3.16(a)

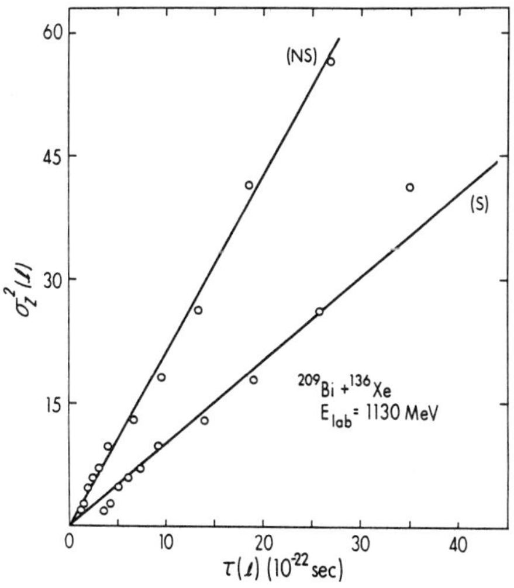

Figure 3.16(b)

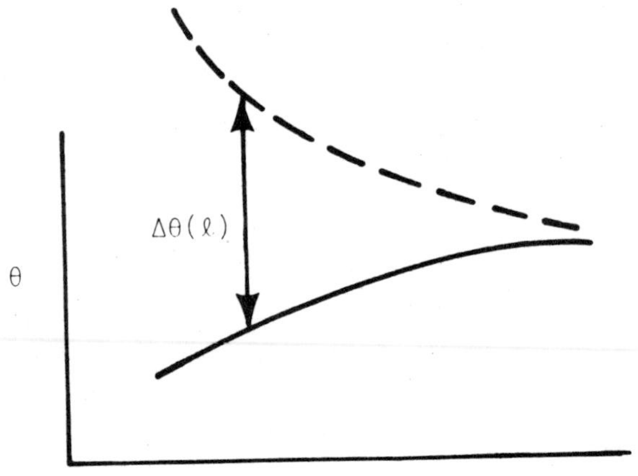

Figure 3.16(c)

$$\ell_i = \left[(\ell_j+1)^2 - \frac{\Delta\sigma_{ij}}{\pi\lambda\!\!\!^{-}{}^2}\right]^{\frac{1}{2}} - 1 \tag{3.19}$$

where $\Delta\sigma_{ij} = \sigma_j - \sigma_i$ is the cross section in an energy window between E_i and E_j. The average scattering angle for a particular energy loss is also an experimental quantity (see Fig. 3.4), so the curve of θ versus ℓ can be deduced as in Fig. 3.16(c). The angular momentum dependent interaction time is then calculated from the expression[287,288]

$$\tau(\ell) = \frac{\Delta\theta(\ell)\,\mathcal{J}(\ell)}{\hbar\ell} \tag{3.20}$$

where $\Delta\theta$ is the difference between the Coulomb deflection angle (dashed) and the actual reaction angle. From these results we extract the values of Γ_z^2 (the FWHM of the Gaussian functions in Fig. 3.16(a)) as a function of E and construct the plot shown in Fig. 3.17, which is indeed remarkably linear. Since we previously deduced a value of D_z we can now use these results to calculate the coefficient of friction $k = 0.6 \times 10^{-21}$ MeV sec fm^{-2}. (A much more sophisticated treatment involving deformation is given in Ref. 289.)

It is instructive to see how the large value of k can be understood,[276] using the simple model of matter transfer discussed earlier in section 3.2. Suppose that the speed of nucleus 1 relative to 2 is tangential and equal to v_t. The rate of nucleon "hits" from 2 to 1 through the window is:

$$\frac{dn}{dt} = \frac{1}{2}\rho v A \cos\theta\, p(v) \tag{3.21}$$

where θ is the inclination of the *nucleon* speed v, of distribution p(v). Each nucleon of mass m deposits the excess momentum $-mv_t$, and therefore the average force acting in the tangential direction is:

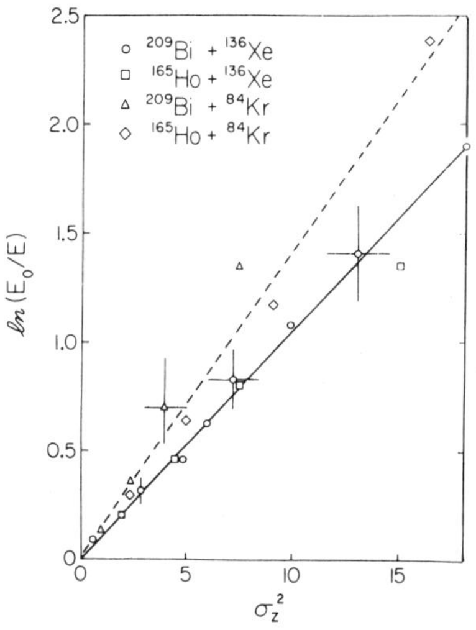

Figure 3.17

$$F_t = -\frac{1}{2}\, m\rho A v_t \int_0^{\pi/2} v\, p(v)\, \cos\theta\, \frac{d\Omega}{2\pi}\, dv \tag{3.22}$$

$$\approx -\frac{1}{4}\, m\rho A v_t\, \overline{v}$$

By identifying this expression with the fruction force -kv, we derive that

$$k \approx \frac{1}{4}\, m\rho A\overline{v} \tag{3.23}$$

Assume, as in Equ. 3.5, a window area of $A \approx 10\ \mathrm{fm}^2$, and the average nucleon speed $\overline{v} = 3/4\ v_F \approx 3/16\ c$ and the nucleon density of nuclear matter, $0.17\ \mathrm{fm}^{-3}$. Then:

$$k \approx 200\ \mathrm{MeV/fm\cdot c} \tag{3.24}$$

i.e., 0.7×10^{-21} MeV sec fm^{-2}, in good agreement with the value extracted from experiment! In fairness, however, we must note that comparable agreement can be reached[290] using the relation,

$$\frac{dE}{dt} = \frac{dE}{dn} \cdot \frac{dn}{dt} = \langle \Delta E \rangle \frac{2}{\hbar} W \tag{3.25}$$

where $\langle \Delta E \rangle$ is the average loss per collision, taken as a typical giant resonance excitation and W is the imaginary optical potential, deduced from direct reactions (Lecture 1).

A more careful examination suggests that the agreement with

the one body dissipation mechanism may be less than perfect.[291,292] Remember that the basic tenet of this model is expressed via the relation:[293,294]

$$\delta E = \frac{m}{\mu} E \qquad (3.26)$$

where δE is the loss of kinetic energy per nucleon exchange and E is the available energy at that time. (This equation is quite consistent with our earlier equations. Thus in equ. 3.17 we can write $dE/dt = \delta E\, dn/dt$ where dn/dt is the nuclear flux, and by the analysis leading to equ. 3.22 this is just $2k/m$; hence the above result for δE. The validity of the equation relies on weak coupling of intrinsic and collective degrees of freedom, an assumption that has been challenged.[295]) Now δE must be deduced from the experimental data (Fig. 3.17) which essentially gives energy loss as a function of σ_Z^2. Regarding the nucleon exchange process as a random walk process, the number of protons exchanged is just $N_Z = \sigma_Z^2$. The experimental observation of the fast equilibration of the mass to charge asymmetry degree of freedom indicates that neutron and proton exchange rates must be very similar[242] and therefore the total number of nucleons exchanged is $N = (A/Z)\,\sigma_Z^2$. Differentiation of the curve of E v σ_Z^2 with respect to $(A/Z)\,\sigma_Z^2$ leads to $\delta E = dE/dN$, which is plotted versus $\frac{m}{\mu}E$ in Fig. 3.18. The dashed line represents the one body dissipation of Equ. 3.26 and it appears that this mechanism accounts for only 30% of the energy loss. Before attributing the additional loss to other mechanisms such as the fast collective dissipation, discussed in section 3.1, the whole validity of the analysis must be examined. It has been pointed out, for example, that the relation between angular momentum and energy implied by eqn. 3.19 is oversimplified,[296] and a more rigorous treatment may remove the discrepancy with the one body dissipation model.

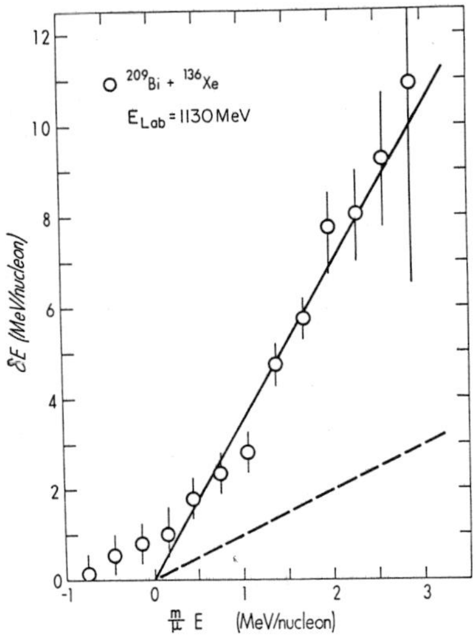

Figure 3.18

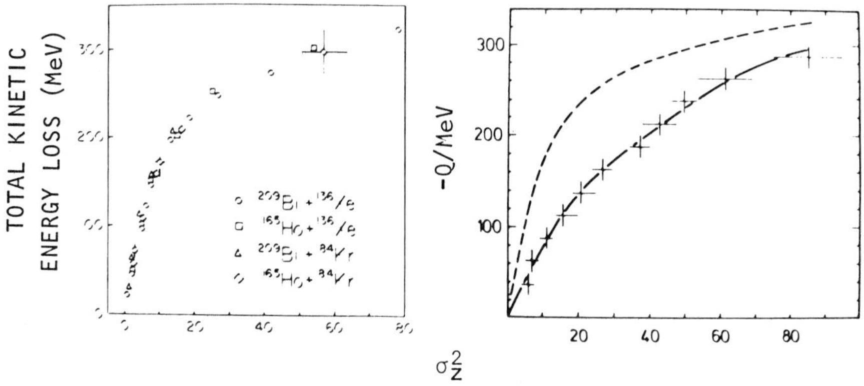

Figure 3.19

The simple approaches have nonetheless given great encouragement to the researchers on superheavy elements, as we mentioned briefly at the end of Lecture 2. It has been found that the curve of energy loss v. σ_Z^2 (represented in Fig. 3.19, with a different ordinate from Fig. 3.17) is not universal. For U + U, as shown in the right hand portion, a much wider charge distribution is found.[237] This observation has important repercussions for making superheavy elements, where the problem is to keep the excitation energy low enough for survival against fission. Consider[230] as an example $^{238}U(^{238}U, ^{181}Yb^*)^{295}SH_{114}^*$. For a relative fission width $\Gamma_f/(\Gamma_f+\Gamma_n)$ of 50% the excitation energy of the superheavy must be about 30MeV. Assuming partition of the energy according to the mass (as we justified in Section 3.1) the Yb nucleus then carries 18 MeV and the total excitation energy is 48 MeV. The Q-value for the reaction is -55 MeV, so we can tolerate a total energy loss of 103 MeV and still have reasonable survival probability. From Fig. 3.19, the associated charge variance is $\sigma_Z^2=14$. The cross section can then be calculated from $\sigma(114) = \sigma_0(92)\ \exp(-(\Delta Z)^2/2\sigma_Z^2)$ for a total kinetic energy window of ± 10 MeV, $\sigma_0(92)$ is 4 mb and with $\Delta z = 22$, we obtain $\sigma(114) = 10^{-34}\ cm^2$.

The hope of reaching the Holy Grail of superheavy elements will no doubt stimulate more accurate calculations of the production cross sections. There is much to be done. The mechanisms of dissipation we have discussed may be adequate for the early stages of deeply-inelastic reactions, where the window is open, i.e., whenever there is solid contact between the ions. There is also "two body" friction, analogous to viscosity in liquids.[297] More generally, a friction force of the type we have been discussing can be represented[298] as:

$$F = -k \int d^3 r \, \rho_1 \rho_2 \, |\dot{\bar{r}}| \tag{3.27}$$

where ρ_1 and ρ_2 are the density distributions of the two nuclei and the integral is taken over the overlap region. The rate of dissipation has also been calculated using a proximity formalism (rather similar to our discussion of proximity potentials in Lecture 1), with the result[294,299,300]

$$\frac{dE}{dt} = 4\pi \frac{n_o}{\mu} \frac{R_T R_P}{R_T + R_P} b\chi(\xi_o)E \tag{3.28}$$

where $n_o = 2.5 \times 10^{-23}$ MeV•sec•fm^{-4} is the transfer flux density, R and b are the nuclear half-density radius and diffuseness, and $\chi(\xi_o)$ is a universal flux function. An application of this formalism to the above reactions for Kr and Xe on heavy targets yields[242]

$$\frac{1}{E} \frac{dE}{dt} \approx 10^{21} \chi(\xi_o) \approx 0.7 - 2.1 \times 10^{21} \text{ sec}^{-1} \tag{3.29}$$

which is actually in very good agreement with the value of $2k/\mu \approx 2 \times 10^{21}$ sec which follows from Fig. 3.17.

3.4 Dynamical Aspects

The previous section was intended to give the flavor of the approaches to understanding the diffusion processes in deeply-inelastic scattering. The evidence strongly suggests the idea of an intermediate complex consisting of two well defined fragments in contact, undergoing equilibration, and the time constants of these relaxation processes have been determined. Now we consider the transfer of orbital angular momentum into the rotation of the two fragments constituting the complex. The angular momentum transfer induced by the frictional forces passes through several stages.[298] Initially, a sliding friction term makes the two bodies start to roll on each other, and then a rolling friction term causes the two bodies to get stuck in rigid rotation.

At the onset of sliding the moment of inertia characterising the system is simply

$$\mathcal{J}_{NS} = \mu R^2 \tag{3.30}$$

where μ is the reduced mass and R the distance between the centers of the fragments. For the sticking configuration (using the theorem of parallel axes) the moment of inertia is

$$\mathcal{J}_S = \mu R^2 + \mathcal{J}_1 + \mathcal{J}_2 \qquad 3.31$$

where $\mathcal{J}_{1,2}$ are the moments of inertia of the fragments, $2/5\, M_i R_i^2$. The *maximum* value $\Delta\ell$ of orbital angular momentum transformed into intrinsic spin can then be calculated from $\ell_i \mathcal{J}_{NS} = \ell_f \mathcal{J}_S$ as

$$(\ell_i - \ell_f) = \Delta\ell = \frac{(\mathcal{J}_S - \mathcal{J}_{NS})}{\mathcal{J}_S} \ell_i \qquad 3.32$$

which appears as *intrinsic spin* of the fragments. For equal mass nuclei we obtain $\Delta\ell = 2/7\, \ell_i$, and the fraction varies depending on the mass asymmetry, as shown below:

$\alpha = \frac{M_1}{M_1+M_2}$	$\frac{\mathcal{J}_S}{\mathcal{J}_{NS}}$	$\frac{\Delta\ell}{\ell_i}$
0.1	2.87	0.65
0.2	1.83	0.45
0.3	1.54	0.35
0.4	1.43	0.30
0.5	1.40 = 7/5	0.29 = 2/7

In the case of *rolling* friction, however, the friction $\Delta\ell/\ell_i = 2/7$ *independent of the masses of the two nuclei.*[298]

For certain cases, it is possible to show that the nuclei must have reached the sticking configuration from an analysis[301,302] of the final channel kinetic energies. The total kinetic energy of a rotating system at scission is given by:

$$E_f = V_c(R) + V_N(R) + \frac{L_f(L_{f+1})\hbar^2}{2\mu R^2} \qquad 3.33$$

In classical friction models it is usual to rewrite the last term as $f^2\, Li(Li+1)\hbar^2/\, 2\mu R^2$, where f is a numerical factor depending on the relevant type of friction. For sticking $f = \mu R^2/(\mu R^2 + \mathcal{J}_1 + \mathcal{J}_2)$, and the value of f often leads[303] to the experimental E_f values, using a value of $R \simeq R_{crit}$ as discussed in Lecture 2.

A better test is to measure $\Delta\ell$ from the γ-ray multiplicity associated with different fragments arising from the decay of the complex.[304-306] As discussed in Lecture 2 it is reasonable to assume that the intrinsic angular momentum is just twice the

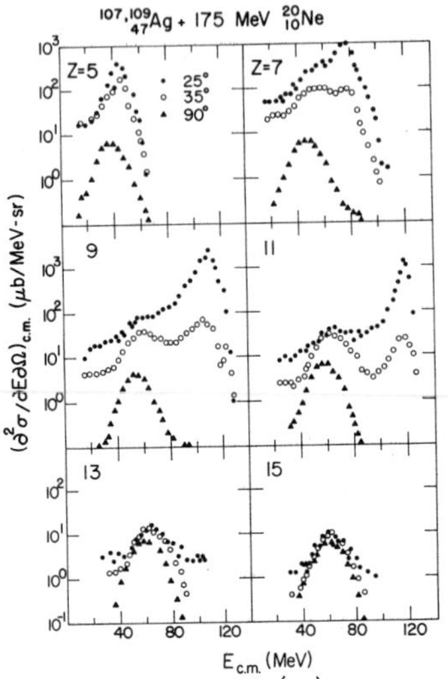

Figure 3.20(a)

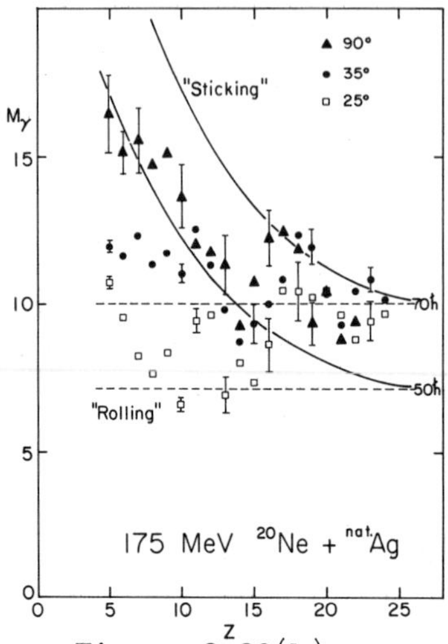

Figure 3.20(b)

multiplicity (assuming that the angular momentum is carried off mainly by the E2 yrast cascade). An example is the ^{20}Ne + Ag system at 175 MeV for which energy spectra are shown in Fig. 3.20(a) at three different angles. We see that in proceeding to more backward angles the quasi-elastic component disappears and the deeply-inelastic dominates, just as in Fig. 3.3. The multiplicities as a function of the Z of the detected fragments are shown in Fig. 3.20(b) for the deeply-inelastic component. For comparison the predicted values for the cases of rolling and sticking are drawn for two values of entrance channel angular momenta (50ħ and 70ħ). The value 70ħ is expected from the sum of the known evaporation residue cross section of 900 mb (corresponding to $\ell_{crit} = 57\hbar$) and the deeply-inelastic cross section of 400 mb, using our customary formulae. (The line for 50ħ corresponds to the limit for compound nuclear formation.) Then the rolling limit is given by

$$\Delta\ell = \frac{2}{7}\,\ell_i = 20\hbar \doteqdot 2M_\gamma \qquad 3.34$$

At 90°, where the rotating dinuclear complex has remained in contact for a long time, the sticking limit appears to be reached, with ℓ_i between 50 and 70ħ. At more forward angles the fragments appear to be still rolling on each other. These data furnish strong evidence that the intermediate complex approaches rigid rotation in a time comparable to the rotation period.

A similar experiment has been conducted[307] on the much heavier systems ^{86}Kr+^{165}Ho and ^{86}Kr+^{197}Au. (See Fig. 3.21). On the left hand side (quasi-elastic transfer) the multiplicities reflect simple transfer reactions where the angular momentum is transferred by particles without the formation of the dinuclear complex.

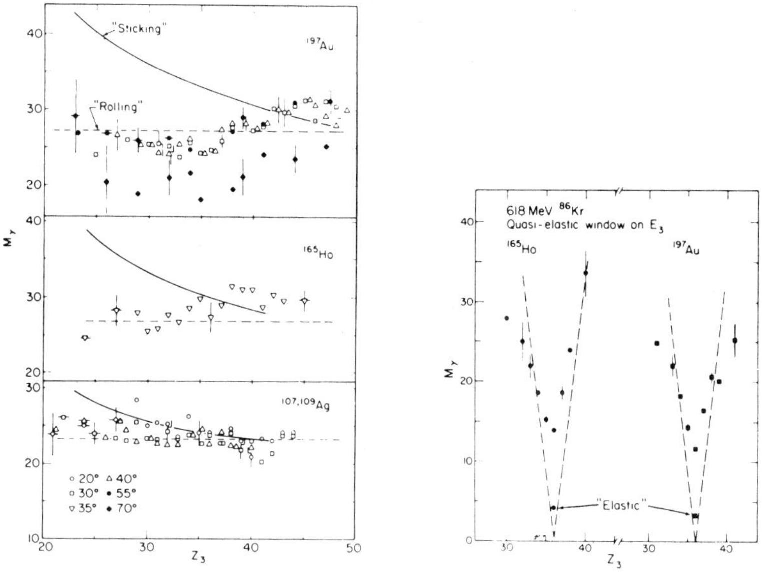

Figure 3.21

In that case we expect $\Delta\ell \approx \Delta M/M\ \ell_i$ where ΔM is the transferred mass and M is the incident mass. This formula leads to the characteristic V-shape in the figure. In contrast to our above example, the deep-inelastic components seem to be closer to the rolling limit (calculated as 2/7 $\langle \ell \rangle$, with $\langle \ell \rangle$ taken to be 2/3 ℓ_{MAX}, i.e, a triangular ℓ-distribution). This result is paradoxical since the energy is completely relaxed. The plausible escape from the dilemma is to assume that the low Z fragments are preferentially populated by low ℓ-waves. This explanation is supported by inspection of the curves of potential energy versus the Z of the fragment for a

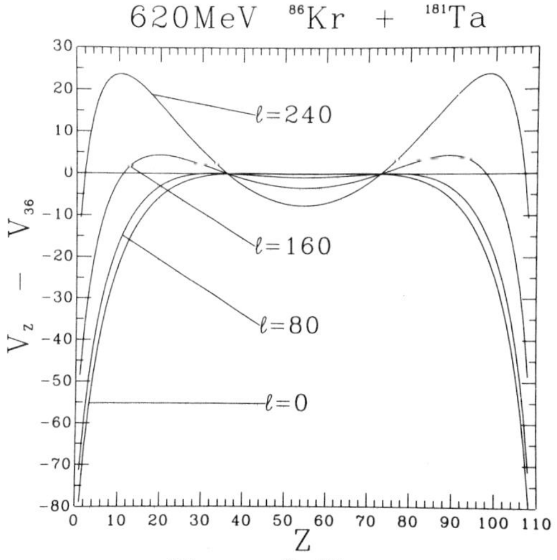

Figure 3.22

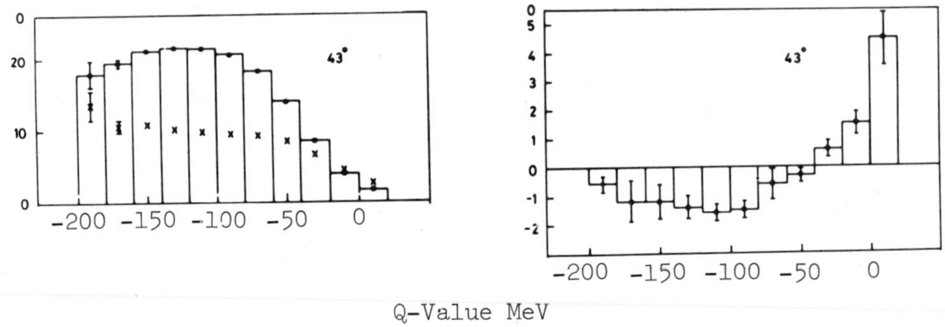

Figure 3.23

similar system in Fig. 3.22. At the Z of entrance channel (where the potential is scaled to be zero), the potential slopes towards symmetry for small angular momentum, becoming progressively steeper for higher ℓ. Therefore only the lowest ℓ-waves contribute to the population of fragments much lighter than the projectile, a so-called "fractionation of the angular momentum distribution."

Clearly a better test of the theories will come from measuring higher order quantities in the experiments. For example, a recent experiment[308] with ^{86}Kr on ^{144}Sm at 490 MeV, in addition to measuring the mean multiplicity $\langle M \rangle$ of γ-rays in coincidence with quasi- and deeply-inelastic scattering, also measured the distribution of multiplicity by using an array of γ-detectors (as we described in Lecture 2). Then quantities such as the standard deviation $\nu = (\langle M^2 \rangle - \langle M \rangle^2)^{\frac{1}{2}}$ and the skewness $\langle (M - \langle M \rangle)^3 \rangle / \nu^3$ are accessible, examples of which are plotted in Fig. 3.23. The left part shows $\langle M \rangle$ and ν as a function of reaction Q-value. The right part shows the skewness. For Q-values close to zero, the skewness is positive indicating a preponderance of low M events, with the reverse in the deeply inelastic region. On a sticking model is it not possible to get the correct values of $\langle M \rangle$, ν and the skewness simultaneously. Another piece of experimental fine tuning comes from measurement of γ-rays to discrete final states. These determine the degree of alignment of the final fragments, which can be compared with the predictions of the sticking model.[309]

Another classic experiment has capitalized on the fission decay mode (rather than γ-decay) which is dominant in heavy systems. The experimental arrangement[310] in which ^{209}Bi was bombarded with 610 MeV ^{86}Kr ions is shown in Fig. 3.24(a). The angular correlation of one of the fission fragments, in coincidence with a projectile-like fragment, was measured both in-plane and out-of-plane. Classical arguments tell us that the fission fragments should be most intense in the plane, if the target-like fragment has a large

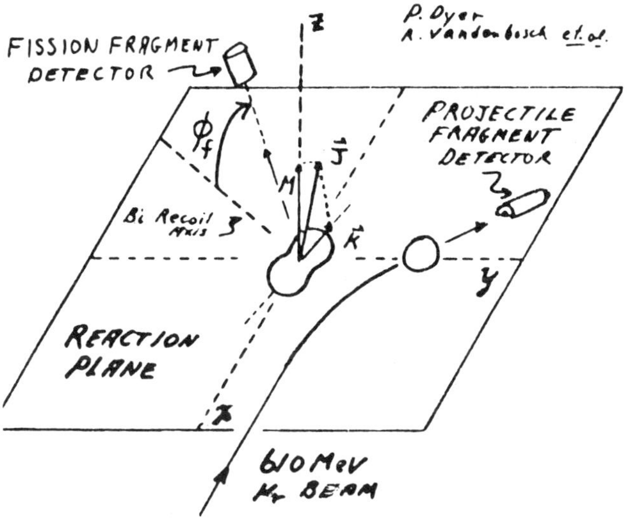

Figure 3.24(a)

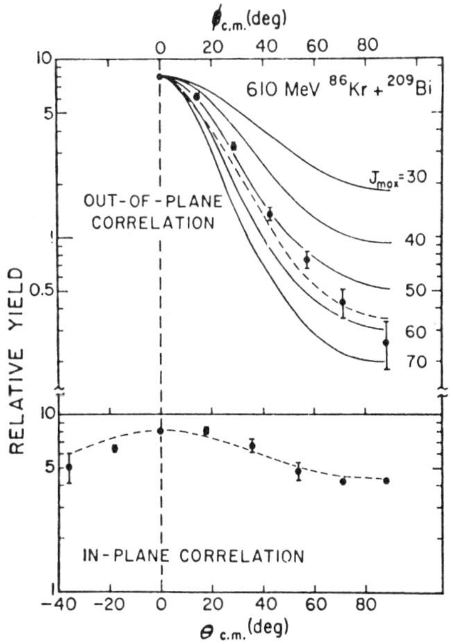

Figure 3.24(b)

angular momentum perpendicular to the reaction plane. The out-of-plane correlation for the fission fragments depends on the quantum number K, the projection of the total angular momentum on the symmetry axis of the fissioning nucleus. Then,

$$\text{Yield} \propto \sum_{JMK} P(J)\; P(M)\; P(K)\; W^{J}_{MK}(\phi) \qquad 3.35$$

where

$$W^{J}_{MK}(\phi) \propto (2J+1)\left|d^{J}_{MK}(\phi)\right|^{2} \qquad 3.36$$

The distributions P(K), P(M), and P(J) represent the probability for finding the system with these quantum numbers. P(K) can be determined from independent fission experiments. As a first estimate we can also assume complete alignment, so P(M) = P(J) with M=J. To determine P(J), the probability that a target-like fragment has angular momentum J, is the goal of the experiment. Assuming that the amount of angular momentum transferred, J, is proportional to the initial orbital momentum ℓ,

$$P(J) \propto (2J+1)$$

(because the partial deeply-inelastic cross section $\sigma_{DI}(\ell) \propto (2\ell+1)$). The distribution has an upper limit J_{max} to be determined.

The results are shown in Fig. 3.24(b) and indicate that $J_{max} = 58\hbar$, from a simultaneous fit to the in-plane and out-of-plane correlations. (Note that a recent study of sequential fission in a similar reaction attributes the out-of-plane distribution to the deeply-inelastic process itself by the excitation of collective bending oscillations.[311]) For the $^{86}Kr+^{209}Bi$ system, the *fraction* of the initial orbital angular momentum transferred is 0.29 ℓ_i for sticking. The value of ℓ_i in this reaction is 235$\hbar$ and therefore the measured value of J = 58$\hbar$ is close to the sticking limit of 68$\hbar$. This experiment is a refinement on the previously described γ-ray experiment, because in principle it could determine *the angular momentum associated with one of the fragments*. Now the angular momentum is divided between the fragments as follows:[298]

$$\text{For sticking:} \quad \left(\frac{J_1}{J_2}\right) = \left(\frac{M_1}{M_2}\right)^{5/3} \qquad 3.37$$

$$\text{For rolling:} \quad \left(\frac{J_1}{J_2}\right) = \left(\frac{M_1}{M_2}\right)^{1/3} \qquad 3.38$$

As the asymmetry becomes larger, this becomes a highly sensitive method for distinguishing between rolling and sticking.

The separation of γ-ray multiplicities between light and heavy fragments is possible in principle by measuring[312] the energy as well as the multiplicity. Then we can write:

$$\langle M_\gamma\rangle_H \langle E_\gamma\rangle_H + \langle M_\gamma\rangle_L \langle E_\gamma\rangle_L = \langle M_\gamma\rangle\langle E_\gamma\rangle$$

$$\langle M_\gamma\rangle_H \langle M_\gamma\rangle_L = \langle M_\gamma\rangle \qquad 3.39$$

and extract $\langle M_\gamma\rangle_H$ and $\langle M_\gamma\rangle_L$. The results for 237 MeV $^{40}Ar + {}^{89}Y$ give a ratio of $\langle M_\gamma\rangle_L/\langle M_\gamma\rangle_H$ in the region of 12 for fragments far removed from the initial channel. By the above equation this result implies an approach to the sticking limit.

Ultimately it will be necessary to make a full solution of the dynamical equations of motion with conservative and dissipative forces for comparison with the experiments.[298,313] For the Kr + Bi case discussed above these equations have been solved using a tangential friction component which was weak compared to the radial component[314] and resulted in a total angular momentum transfer to both fragments of only 38ħ, considerably below the experimental value.

3.5 The Limits of Space and Time

We have seen that in deeply-inelastic scattering, macroscopic concepts such as viscosity and friction, are of great current interest. On the other hand, in conventional nuclear physics, the statistical model, which assumes thermodynamical equilibrium, has been generalized to include pre-equilibrium behavior.[315] Since energy dissipation includes not only viscosity but also heat conductivity, it may be possible to make a link between the two approaches.[316,317] A new generation of experiments is aimed at studying the formation of "hot-spots" in nuclear matter. This concept is very old. To quote from an historical paper,[318] "If a nuclear particle of energy E, comparable with the nuclear interaction energy, strikes a nucleus, it will lose practically all its energy in the 'surface layer' of the nucleus. This process will cause intense local heating of the part of the nucleus struck. The 'heat' will then gradually spread over the whole nucleus." A calculation[319] of the heat conductivity, specific heat etc. of nuclear matter from a Fermi gas model was already completed in 1938.

First consider some typical time scales of deeply-inelastic reactions.[197] For the rotational motion, we have an angular velocity ω and an angle of rotation θ through which the fragments remain in contact. Therefore:

$$\tau_{DI} \approx \theta/\omega \tag{3.40}$$

Values of E_{rot}, ℓ and $\mathcal{J}$ can be estimated, so we can use

$$\omega = \frac{2E_{rot}}{\hbar\ell} \quad \text{or} \quad \omega = \frac{\hbar\ell}{\mathcal{J}} \tag{3.41}$$

to obtain ω. For example, a reasonable estimate of ℓ is $5/7\ \ell_i$, corresponding to rolling fragments, and $E_{rot} = E_{CM} - E_{Coul} + Q$. For the reaction $^{40}Ar + ^{232}Yh$ at 379 MeV (Fig. 3.3) $E_{rot} \approx 150$ MeV, $\ell \approx 150$ (see discussion of Equ. 3.15) so $\omega \approx 3 \times 10^{21}$ sec^{-1} and $\tau \approx 3 \times 10^{-22}$ sec for a typical rotation angle of 1 radian.

We can also estimate the time it takes an equilibrated excited nucleus to emit a particle. An empirical fit to the measured widths of compound nuclei for A = 20-100 yields:[111]

$$\Gamma(\text{MeV}) = 14 \exp(-4.69\sqrt{A/E^*}) \tag{3.42}$$

Relating the temperature T to the excitation energy by $E = aT^2$, where $a \approx A/8$, we have

$$\tau_{particle} \approx 0.5 \exp(13/T) \tag{3.43}$$

where T is in MeV and τ in units of 10^{-22} sec. An excitation energy of 3.25 MeV/A yields a temperature of 5 MeV and a lifetime of 7×10^{-22} sec. If local temperatures of this magnitude should be produced in heavy-ion collisions, then the lifetime for particle emission is so short that the rotating dinuclear complex will emit particles before it scissions. We say *local* temperatures because *total* center of mass energies in deeply-inelastic experiments are < 10 MeV per *projectile* nucleon, and therefore the achievement of, say, 3 MeV/nucleon in some region requires a concentration of energy into a "hot-spot."[316-318]

Delving slightly deeper we can write the relaxation time for dissipating the initial energy deposition as:[316]

$$\tau_R = \frac{R^2}{\chi} = \frac{R^2}{v_F \Lambda}, \quad \chi = \frac{K}{\rho c_p}. \tag{3.43}$$

Here v_F is the Fermi velocity, Λ is the mean free path for nucleon-nucleon scattering, K is the thermal conductivity, ρ is the density and c_p the specific heat of nuclear matter. Expressions for K and

c_P can be derived from the Fermi gas model.[319,320] Thus,

$$K = \frac{7}{48\pi\sqrt{2}} \frac{\varepsilon_F^{3/2}}{m^{\frac{1}{2}}TQ}, \quad c_P = \frac{1}{2}\frac{\pi^2 T}{\varepsilon_F} \tag{3.45}$$

where ε_F is the Fermi energy, T is the temperature and Q is the effective nucleon-nucleon cross section. (≈ 27 mb). For a temperature of ≈ 1 MeV, τ_R is 4×10^{-22} sec. From the above equations, τ_R varies as T^2 (essentially because the mean free path decreases as more nucleons are excited above the Fermi level), and, at high enough temperatures, becomes longer than the time for particle emission. These trends are illustrated in Fig. 3.25 from an old calculation[321] (left hand side) and a recent calculation.[322] In both calculations as the incident energy (temperature) increases we reach a point where the compound nuclear lifetime is less than the relaxation time, just the condition for the formation of a hot-spot. (Also shown on the right are the passing times for two A = 50 nuclei, the nucleon-nucleon collision time.) The critical temperature appears to be around 8 MeV[323]. (We shall return to this temperature in Lecture 4.)

Several coincidence experiments have recently been performed, with the general philosophy directed at observing hot-spots.[324] All these experiments have studied the angular correlation of light particles (e.g., alphas) in coincidence with the projectile-like heavy fragment emitted in quasi- or deeply-inelastic scattering at a fixed angle. A typical example[325] is shown in Fig. 3.26 for

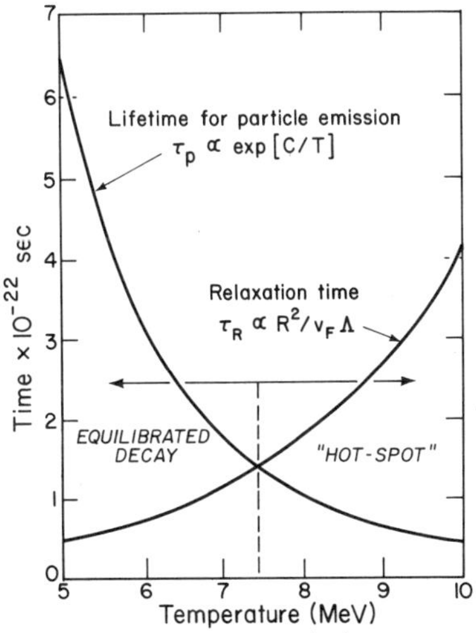

Figure 3.25(a)

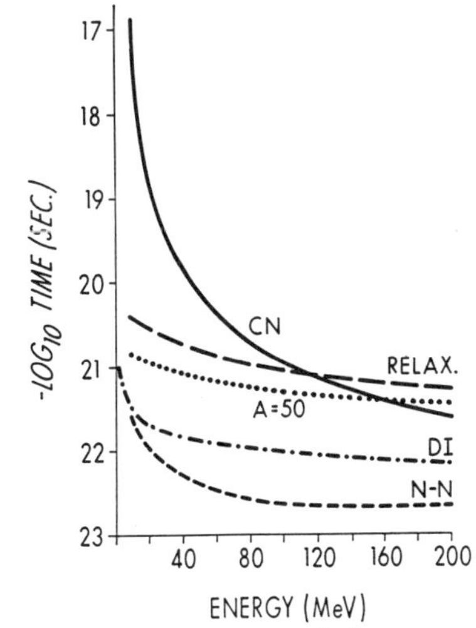

Figure 3.25(b)

reactions of $^{16}O + ^{208}Pb$ at 140 MeV and 315 MeV. For a variety of projectile fragments, the correlations are very narrow and peak roughly in the direction of the fragment (marked with an arrow) or between this direction and the beam axis. Note that the channel attainable by *pure* projectile fragmentation ($^{12}C + \alpha$) has a double peak, as expected, but the other channels (e.g. $^{14}N + \alpha$) give very similar overall distributions. The fact that all these patterns are reminiscent of the decay of an excited projectile-like fragment is also confirmed by a kinematic contour plot. This is shown in Fig. 3.27 for a similar reaction,[326] $^{14}N + ^{93}Nb$ at 90 MeV leading to ^{10}B and α fragments. The two islands are consistent with decay of a prefragment $^{14}N^*$ at an excitation of ≈ 1 MeV (denoted by the dotted kinematic constraint) traveling with a kinetic energy of ≈ 55 MeV (dashed lines).

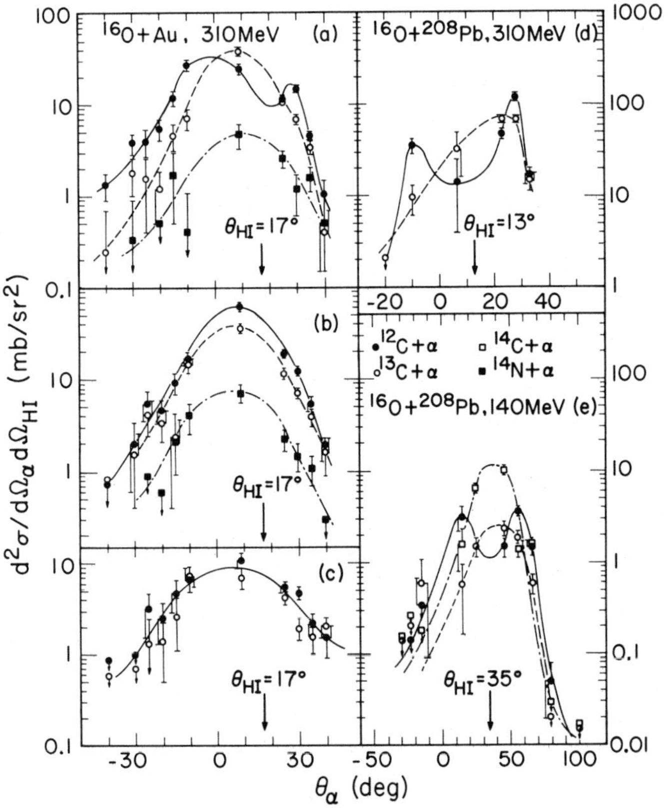

Figure 3.26

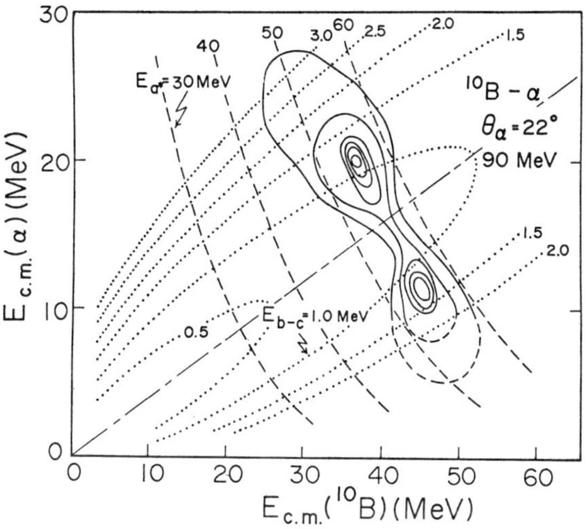

Figure 3.27

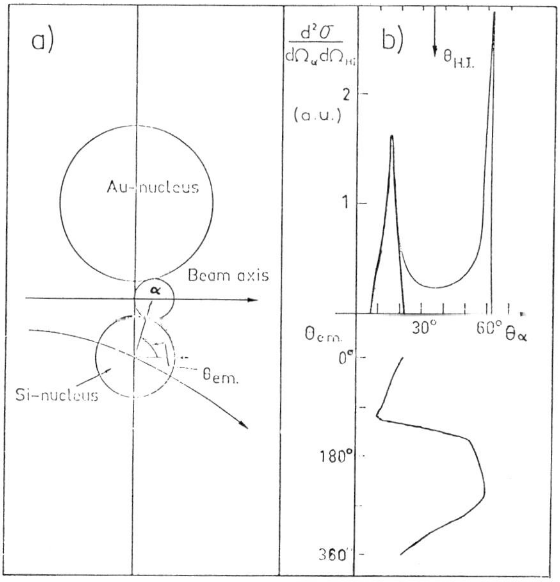

Figure 3.28

A possible interpretation of similar correlations of α-particles observed in reactions of $^{32}S + ^{197}Au$ at 12 MeV/nucleon[327] is given in Fig. 3.28. The ^{32}S moves along the Rutherford trajectory up to the distance of closest approach. Then it emits an alpha from the surface in any possible direction. The subsequent motion of the α, ^{28}Si and ^{197}Au nuclei in the Coulomb field is calculated numerically, generating two peaks in the correlation. Only the left hand peak appears in the data, which is associated with the region of the projectile between the projectile and target (i.e. a localized region). The first experiment[328] to reveal such a phenomenon (actually emitted from a "hot-spot" on the target) was the reaction $^{16}O + ^{58}Ni$ at 92 MeV. The confusing effects of projectile break-ups were eliminated by searching for α-particles in coincidence with ^{16}O scattering. The rather detailed analysis[329] of this experiment assumes that a hot-spot is created on the surface of the target, the α-emission from which has a high temperature component emitted outwards from the pole, and a low temperature component from the diffusion of the α-particles through the nuclear matter in the opposite direction. The final solution is complicated by Coulomb and nuclear deflections and by angular momentum, which makes the hot spot rotate. Nevertheless, some idea of the results is conveyed in Fig. 3.29. The top part shows the α-correlation measured from an origin in the direction of the projectile. Both the fast and the slow modes lead to the narrow angular correlations, characteristic of all the experiments we have been discussing. The bottom middle section displays contour plots of the cross section in an $E\alpha$-$\theta\alpha$ diagram, the projections of which onto the $E\alpha$ axes (left and right) show the expected α-particle spectra. The high temperature component (≈ 7 MeV) is close to the temperatures required for the observation of a hot-spot (see Fig. 3.25) whereas the low temperatures are characteristic of greater equilibration. The experiment[328] yielded temperatures of 3-4 MeV in the forward direction. Using the expressions $Ex = aT^2$ and the value of Ex = 28 MeV extracted from the experiment, the value of $a = N/8$ given $N \approx 18$ particles. For a fully equilibrated system $N \approx 70$ and the temperature would have been only 1.8 MeV. Such experiments can lead to a determination of the thermal conductivity and specific heat of nuclear matter, and are an alternative to preequilibrium theories.[316,317]

There are several other experiments on the production of fast non-equilibrium light particles,[227,228,330-332] with interpretations ranging over emission from the neck between the colliding nuclei[330] (like ternary fission and maybe even like a hot-spot) to backward splashes of α particles accompanying fusion.[228] The fun is just beginning. The theoretical possibilities are also diverse. A possible mechanism[333] for the production of fast, non-equilibrium α-particles is the strong radial friction damping force, which ejects a particle on the opposite side of the nucleus from where the projectile and target first make contact (see Fig. 3.30). This

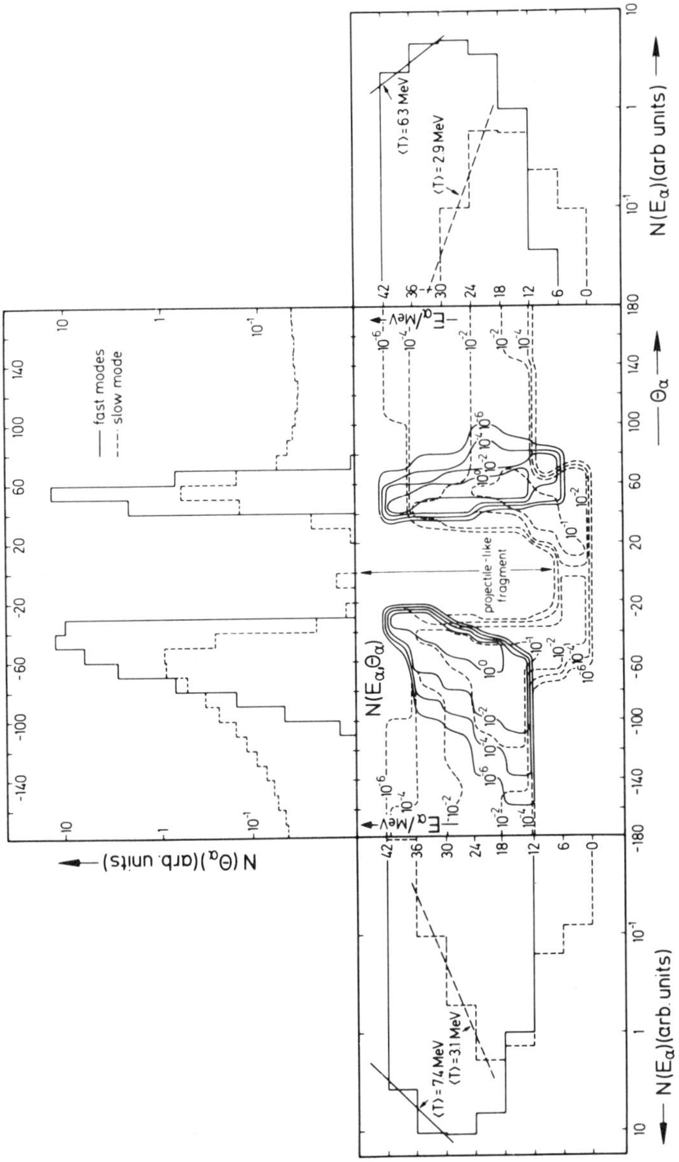
N(Θα)(arb. units)
fast modes
slow mode
N(Eα,Θα)
projectile-like fragment
Eα/MeV
Θα
N(Eα)(arb. units)
⟨T⟩=7.4 MeV
⟨T⟩=3.1 MeV
⟨T⟩=6.3 MeV
⟨T⟩=2.9 MeV

Figure 3.29

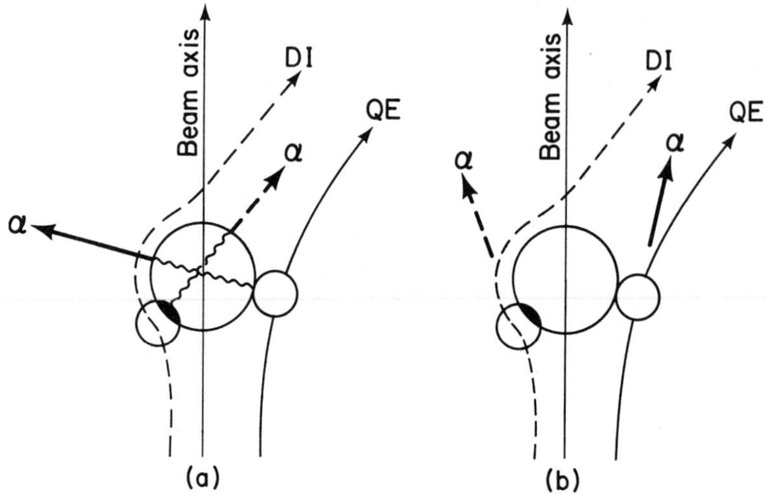

Figure 3.30

leads to a correlation with the α and the heavy fragment on the *same* side of the nucleus which would not be consistent with many of the above experiments. Another possibility is illustrated in part (b) of the figure,[325] which by similar arguments would attribute the α-production to strong *tangential* friction, certainly essential as we have seen to account for the results of γ-ray multiplicity and the fission fragment experiments. This picture can explain how in Fig. 3.26 alpha particles are observed in coincidence with heavy fragments that could not arise from simple projectile fragmentation, but which nevertheless bore close resemblances. This picture has also been said to represent a "sparking process,"[334] and is consistent with our discussion of "hot-spots" in this section, i.e., a zone of slightly higher complexity and concentration than occurs in simple projectile excitation. We note in Fig. 3.26, however, that at the higher energy the relative importance of these more complicated channels diminishes and the pure fragmentation channel becomes dominant. This simplification sets our path towards Asymptotia, the subject of the last lecture.

4. ASYMPTOTIA

In this lecture we leave behind the familiar territory of *Microscopia*, and even the still recognizable landmarks of *Macroscopia*, to venture into the New World of *Asymptotia*. Before setting out it is just as well to have a navigation chart,[335] which appears in Fig. 4.1. The abscissa is the projectile energy in MeV/nucleon and the ordinate is the projectile mass plotted as $A^{1/3}$. The shaded bands define regions of fundamental parameters such that when we cross a band, we can be confident that the underlying physics will change. The three characteristic center of mass energies of 20 MeV, 140 MeV and 930 MeV are estimates of where the subsonic, mesonic and relativistic domains merge. Macroscopic phenomena come into prominence when $A^{1/3} \gg 1$. The band at $Z \approx \frac{1}{2}$ (170) is a reminder of the changes that may occur when (2Zx fine structure constant) becomes large compared to unity. Most of this space is unexplored apart from the two axes, the left-hand side with the low energy heavy-ion machines, and the horizontal axes with high energy, hadron accelerators. Although some possibility for exploring the remaining space (where most of the crossing bands lie) has existed with Nature's own accelerators, the Cosmic radiation,[336,337] it is the development of high energy, heavy-ion accelerators, such as the Berkeley Bevalac, that has sharpened and focussed these studies. Combined with parallel developments on increasing the energy of existing Cyclotrons (at Berkeley and Texas A and M) up to 35 MeV/ nucleon, it is now possible to trace the evolution of heavy-ion reaction mechanisms across some of the critical boundaries of Fig. 4.1 We begin with a discussion of this evolution in peripheral collisions, then deal with the more dramatic (possibly) central collisions and end with a few words on exotic phenomena.

4.1 Evolution of Peripheral Collisions

In order to make a conceptual link with the last lecture, let us consider how deeply-inelastic scattering might evolve with energy.[338] Imagine two nuclei with radii R colliding with relative velocity u. The collective kinetic energy is

$$E \approx \left(\frac{4}{3}\pi R^3 \rho\right) u^2 \qquad (4.1)$$

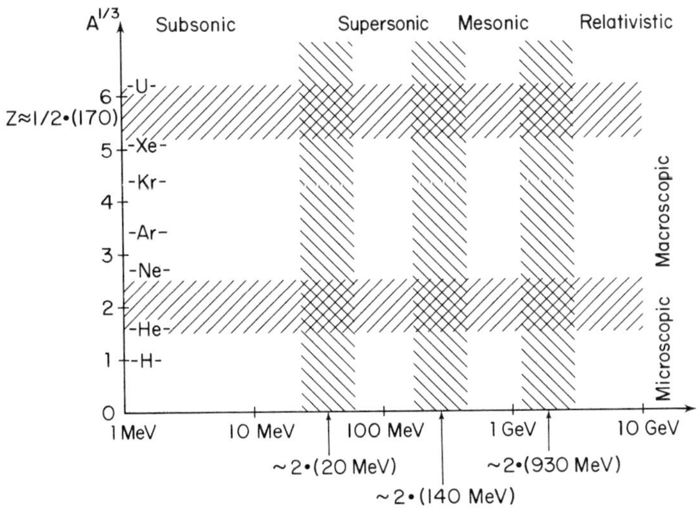

Figure 4.1

(We are dropping factors of order unity.) If the nuclei are in communication through a window of area πa^2 (as discussed in Lecture 3, equ. 3.21, etc.), we have

$$\frac{dE}{dt} \approx \tfrac{1}{4}\, \rho\, \bar{v}\, (\pi a^2)\, u^2 \tag{4.2}$$

where $\bar{v}$ is the average intrinsic nucleon speed. Therefore the characteristic damping or stopping time is of order:

$$t_{stop} \approx R^3 \rho\, u^2/f\, \bar{v}\, a^2 \approx \left(\frac{R}{a}\right)^2 \left(\frac{R}{\bar{v}}\right) \tag{4.3}$$

We compare this time with the collision time, $t_{coll} = R/u$ to give:

$$\frac{t_{stop}}{t_{coll}} \approx \left(\frac{R}{a}\right)^2 \left(\frac{u}{\bar{v}}\right)^2 \left(\frac{R}{a}\right)^2 \sqrt{\frac{\text{Energy/nucleon}}{\text{Fermi Energy}}} \tag{4.4}$$

Therefore if "a" is not too small, as the incident energy approaches the Fermi energy, complete damping plays less of a role. We must then ask the question, what process takes over the large deeply-inelastic cross section?

It appears that multibody fragmentation phenomena replace the essentially two-body processes of deeply-inelastic scattering.[339] Below 10 MeV/nucleon, the collision time is longer than the transit time of a nucleon at the Fermi level; consequently the whole nucleus can respond coherently to the collision, and the dominant phenomena are characteristic of the mean field.[340] At relativistic energies of GeV/nucleon, on the other hand, the reaction processes are dominated by independent collisions of individual nucleons.[341] The transition region might be set by requiring the complete disjunction of the two colliding nuclei in momentum space, i.e., at a few tens of MeV/nucleon. This transition, which could be labelled[323] "from nuclei to nucleons," has been observed in peripheral collisions.

The approach is to measure the production cross sections and energy spectra of projectile-like fragments from ^{16}O induced reactions on targets such at Pb, Au as a function of incident energy.[342,343] Some typical spectra for outgoing ^{12}C products at incident energies of 140, 218, 250 and 315 MeV are shown[344] in Fig. 4.2. The spectra all have a characteristic Gaussian form, peaked at an energy (labelled Ep) corresponding to the fragment travelling with a velocity close to that of the incident beam. At low energies, if two-body deeply-inelastic scattering is the relevant mechanism, this behavior implies a high excitation of the residual fragments (compare the energy, labelled g.s. in Fig. 4.2, associated with the production of the nuclei in the ground states). The

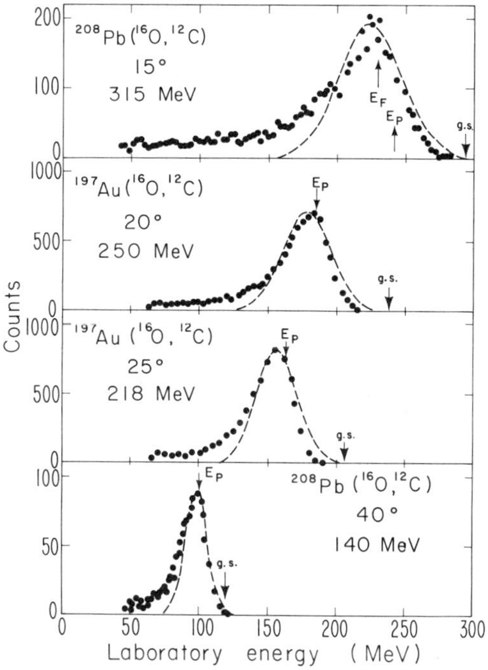

Figure 4.2

continuum could also correspond to transfer reactions to a high density of states[345,346] in the continuum, with an optimum Q-value.[347]

The continuum is also characteristic of multibody fragmentation at high energies. An example of similar spectra at 2.1 GeV/nucleon is shown[348] in Fig. 4.3. Here the spectrum is plotted in the projectile rest frame, so that a fragment emerging with beam velocity would correspond to $P_{11} = 0$, where P_{11} is the longitudinal momentum in the projectile frame. In fact, just as in Fig. 4.2, the Gaussian shaped distributions are shifted slightly below this point. Both at 2.1 GeV/A and 20 MeV/A this shift (ΔP_{11}) is well accounted for by the separation energy of the projectile into the observed fragment together with residual nucleons and alpha particles[349,350] (e.g. the arrow labelled E_F in the top part of Fig. 4.2). In Fig. 4.2 we observe that the widths of the spectra increase rapidly with energy, which is a manifestation of the transition in the nature of the reaction mechanism.

First we use the concept of temperature to find systematic trends in the data. At low energies (< 10 MeV/A) the production cross sections of isotopes, in reactions of the type reported here, have an exponential dependence,[351,352] $\sigma \propto \exp(Qgg/T)$, where Qgg is the two-body, transfer ground state Q-value. A good example is

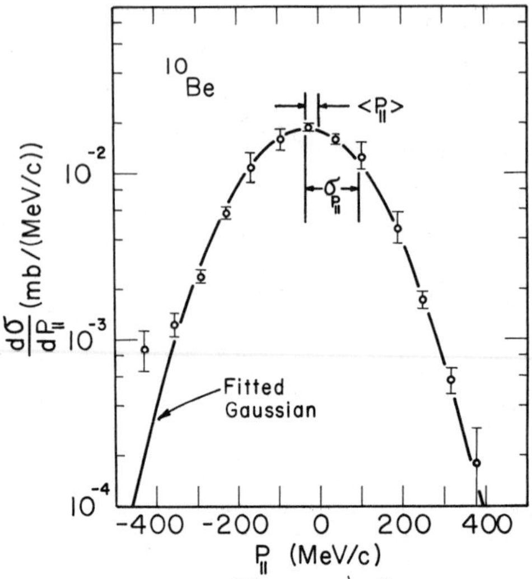

Figure 4.3

shown in Fig. 4.4 for the system $^{16}O + ^{232}Th$ (similar to ^{16}O + Au, Pb), in which the cross sections were obtained by integrating spectra similar to Fig. 4.2. The exponential dependence on Qgg over *five orders of magnitude* would not be expected from a simple direct reaction model,[352] relating the cross section to the Q-value at the *peak* of the distribution, which might be 50 to 100 MeV more negative. The systematics do however have a natural explanation in terms of a

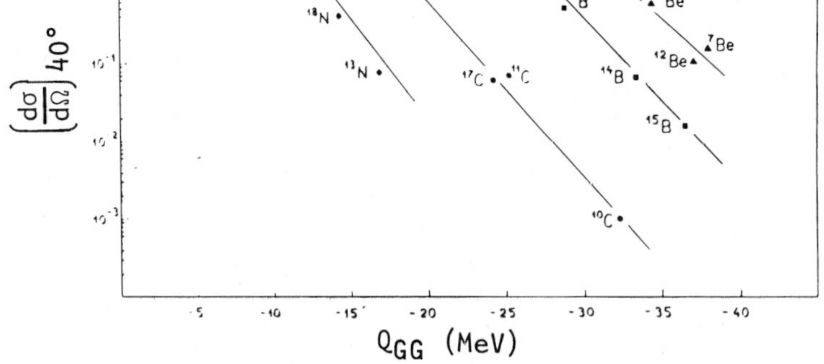

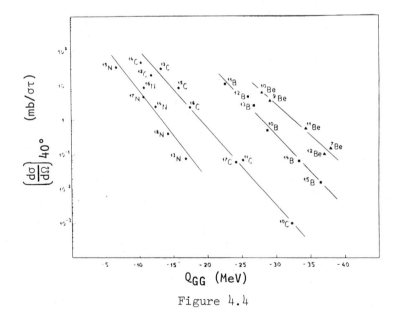

Figure 4.4

rotating dinuclear system undergoing partial statistical equlibrium at temperature T.[351,352] In a statistical reaction, the cross section is given by:[352]

$$\sigma \propto f_f(E^*) \propto \exp \frac{E^*}{T} \qquad (4.5)$$

proportional to the level density of states at excitation E*, which can be written E* = Qgg-Q, and the Q-value is made up of the changes of Coulomb energy, rotational energy and other excitation processes. Therefore,

$$\sigma \propto \exp \frac{Qgg-\Delta Vc}{T} \qquad (4.6)$$

where we have included only the Coulomb term in Q, since some of the others are not strongly coupled to the degrees of freedom participating in the equilibration.[352]

The temperatures derived from this approach for a variety of data (including those of Fig. 4.2, and of the extensive analysis of ^{16}O, ^{15}N + ^{232}Th reactions[351]) are shown in Fig. 4.5 by the filled circles, plotted as a function of the incident energy above the barrier (top scale). The variation initially follows the trend of the Fermi gas equation of state, $E^* \approx (E_c-V) = aT^2$, where E_c is the center of mass energy, V the Coulomb barrier in the incident channel, and "a" is the level density parameter, equal[354] to A/8, with A the mass number of the intermediate complex. Hence T is proportional to $\sqrt{E_c-V}$, the variable used on the bottom scale.

At relativistic energies the concept of temperature has also been useful in explaining isotope production cross sections, where

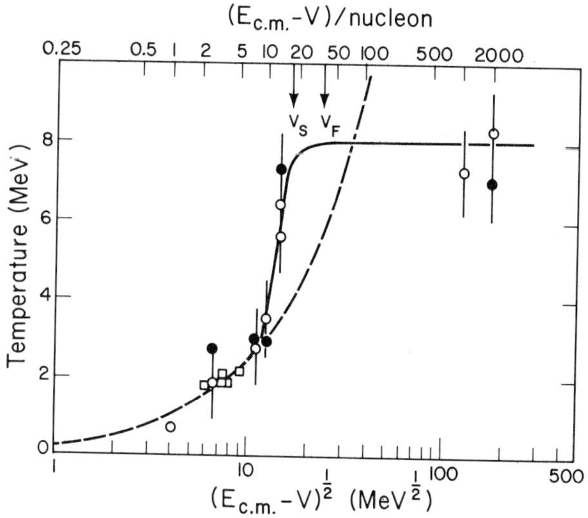

Figure 4.5

the "emitter" is the projectile rather than the dinuclear complex.[342,355-357] Then $\sigma \propto \exp(Q_F/T)$, with Q_F equal to the fragmentation Q-value, and T is the projectile temperature. This approach has been applied to the data in Fig. 4.5 at 315 MeV (≈20 MeV/A)[343] and at 2.1 GeV/A;[348] the values of T are also displayed in Fig. 4.5. Following the initial trend of the Fermi gas equation, a rapid rise sets in between 10 and 20 MeV/A, after which the temperature appears to saturate at approximately 8 MeV. Above 15 MeV/A, where the curve departs from the prediction of the Fermi gas for heating the entire complex, only a part of the total system can be heated (compare our discussion of hot-spots at the end of the last Lecture). The saturation at 8 MeV could be interpreted by assuming that A' (<<A) nucleons participate and carry less than BA' of excitation energy, where B is the binding energy of a nucleon (≈8 MeV), for the system to survive to emit a complex fragment. If this subsystem is excited like a Fermi gas, the result T ≈ 8 MeV follows immediately from the equation $8A' = A'/8T^2$. Since higher temperatures would result in a disintegration of the fragment,[339] it is natural to refer to this temperature as the "boiling point of nuclear matter" (It is interesting to make an analogy with Fig. 1.4, where a limiting temperature is also observed for hadronic matter; this has also been referred to as a boiling point of hadronic matter.[358]

Although temperature is a useful concept for organizing the data, and for understating the limiting behavior in the high energy region, an alternative interpretation comes from the abrasion model[359,360] in which the primary fragments emerge by the sudden shearing of the projectile without prior excitation. The dependence $\sigma \propto (Q_F/T)$ can also be derived analytically with this model.[344] The basic idea of this model is illustrated in Fig. 4.6 (top part).[361]

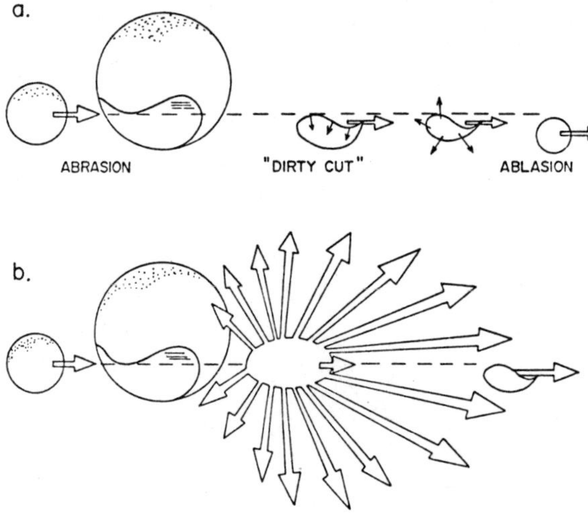

Figure 4.6

The incident projectile in the region of overlap with the target has a part sliced out.[362] The cross section for this process can be calculated using Glauber theory[363] or from geometrical considerations. The cut is not clean but creates a hot region which causes the remaining fragments to be highly excited, so that they proceed to evaporate additional particles (ablation). In the Glauber model at high energies the nucleus-nucleus cross section for an event in which n projectile nucleons are scattered out of the projectile A is:

$$\sigma_n = \binom{A}{n} \int d^2\underline{b}\ (1-P(b))^n\ P(b)^{A-n} \qquad 4.7$$

where

$$P(b) = \int dz\ d^2\underline{s}\ \rho_A(s-bz)\ \exp[-A_T\sigma_{NN}\int dz'\ \rho_T(s,z')] \qquad 4.8$$

Here (1-P(b)) is the probability of finding a projectile nucleon in the overlap zone when b is the impact parameter. Equation 4.7 is then the cross section for n projectile nucleons to be in the overlap and (A-n) outside. It turns out that σ_n changes very little between 20 MeV/A and 2 GeV/A in spite of a large change in σ_{NN}. However, at high energies the momentum transfer is sufficient to knock nucleons out, but at low energies they appear to stay in the prefragment and deposit their energy. The subsequent fate of the projectile fragment (the ablation stage) is rather different in the two cases. This model[364] appears to account both for the isotope differences and the element similarities observed in O^{16} induced reactions at 20 MeV/A and 2.1 GeV/A.

For the primary distribution of fragments, eq. 4.7, 4.8 lead to a distribution in mass and mass and isospin, we use the formulation of the abrasion model in Ref. 365 :

$$\sigma \propto \exp - \left[\frac{(a-a_o)^2}{2\sigma_a^2} - \frac{(t_3-t_{30})^2}{2\sigma_{t_3}^2} \right] \qquad 4.9$$

where a = N+Z, the number of nucleons abraded, $t_3 = (N-Z)/2$ and σ_a, σ_{t_3} are the dispersions around the mean values a_0, t_{30}. Transforming to the variables N,Z yields the distribution of isotopes about the mean:

$$\sigma \propto \exp \left[-(N-N_o)^2 \left(\frac{1}{2\sigma_a^2} + \frac{1}{8\sigma_{t_3}^2} \right) \right] = \exp \left[-\frac{(N-N_o)^2}{\alpha} \right] . \qquad 4.10$$

Values of σ_a, σ_{t_3} are derived from a model with correlations built into the nuclear ground state, viz. $\sigma_{t_3} \approx 0.24\ A^{1/3}$, $\sigma_a \approx 4.9\ \sigma_{t_3}$ (see later).

In the production of a series of isotopes the changes in Q_F are determined primarily by the N-dependent terms in the liquid drop mass formula. For a fragment of mass A_F this term can be written;

$$\frac{a_s(A_F-2N)^2}{A_F} - \frac{a_{ss}(A_F-2N)^2}{A_F^{4/3}} \qquad 4.11$$

where a_s and a_{ss} are the symmetry and surface symmetry coefficients respectively. It is then simple to derive a quadratic dependence of Q_F on $(N-N_o)^2$, viz.

$$Q_F = 4\left(\frac{a_s}{A} - \frac{a_{ss}}{A^{4/3}}\right)(N-N_o)^2 = \beta(N-N_o)^2 \qquad 4.12$$

From Eqs. 4.10 and 4.12 we get,

$$\sigma \propto \exp\left(\frac{Q_F}{\alpha\beta}\right) \qquad 4.13$$

which is equivalent to the result of the thermal model, with T replaced by $\alpha\beta$. By inserting the values[365] of σ_a, σ_{t_3} and of the mass formula coefficients,[366] we deduce that T = 9MeV (or 5 MeV with values of σ neglecting[365] correlations). This derivation of isotope distributions ignores the subsequent redistribution by nucleon cpature and evaporation,[364] but the value of 9 MeV is close to the required saturation value of 8 MeV in Fig. 4.5. This parameter in the exponential dependence of σ on Q_F is, however, identified with *the onset of the fast abrasion mechanism,* rather than with the saturation of nuclear temperature in the slower, equilibrating process.

In the saturation region above 20 MeV/nucleon, the abrasion model also accounts consistently for the momentum distribution of fragments in the projectile rest frame,[356]

$$\frac{d^3\sigma}{dp^3} \approx \exp\left[-\frac{(p-p_o)^2}{2\sigma^2}\right] \qquad 4.14$$

where p_o is the momentum corresponding to the peak of the distribution, of width:

$$\sigma^2 = \sigma_o^2 \frac{F(A - F)}{(A - 1)} \qquad 4.15$$

F, A are the masses of the observed fragment and the projectile respectively. This value of σ^2 is just related to the mean square momentum of F nucleons in the projectile suddenly going off as a single fragment. Not surprisingly, therefore, it is also closely related to the Fermi momentum by $p_F = \sigma_o\sqrt{5}$ which has been measured[367] as 235 MeV/c for ^{16}). The analysis of the heavy-ion spectra yields $\sigma_o \approx 86$ MeV/c or $p_F = 192$ MeV/c. The Gaussian distribution shown in Fig. 4.3 is calculated with the above equations. For the energy distributions in the laboratory frame at angle θ, transformation of Eq. 4.14 yields:[343]

$$\frac{d^2\sigma}{dEd\Omega} \propto \sqrt{2A_F E} \;\exp\left[-\frac{A_F}{\sigma^2}\left(E - 2aE^{1/2}\cos\theta + a^2\right)\right] \qquad 4.16$$

where $a^2 = 1/2\, M_F v_P^2$, v_P is the velocity corresponding to the peak of the energy distribution. This formula is used to generate the theoretical curve in Fig. 4.2 for the top set of data at 20 MeV/A, again using $\sigma_o \approx 86$ MeV/c in the expression for σ^2.

The energy distribution in Eq. 4.16 is also expected from a statistical model of fragment emission.[356] Therefore, the formula can equally well be applied to the lower energy spectra in Fig. 4.2, where we have already shown that equilibration processes at temperature T are relevant. By conservation of energy and momentum, T and σ_o are related[356] by

$$\sigma_o^2 = T\, m\, \frac{A_P - 1}{A_P} \quad ,$$

where m is the nucleon mass in MeV. (For $\sigma_o = 86$ MeV/c, $T \approx 8$ MeV, consistent with the two interpretations of the isotope distributions in the high energy region). The values of T required to fit the data at all energies are shown in Fig. 4.2 by the open circles. Also included are data for oxygen on nickel at 315 MeV and on tantalum[368] at 96 MeV. Although only results for ^{12}C fragments are presented, similar trends were observed in the energy spectra of other particles.[343] At low energies (<10 MeV/nucleon) the temperatures extracted from the momentum and isotope distributions are in agreement, supporting the temperature model. At high energies (>20 MeV/nucleon) the saturation of the widths of the momentum and isotope distributions at 8 MeV is consistent with a fast abrasion mechanism, although the alternative interpretation of a localized thermal excitation is not excluded.

If we adopt the abrasion model for the description of the high energy data, then the sudden transition from equilibration to fragmentation must contain information on characteristic properties of nuclear matter, such as the relaxation time for spreading the localized deposition of energy, or "hot-spot",[316,317] over the nucleus. The initial excitation may be in the form of uncorrelated particle-hole excitations, in which case this relaxation time is related to the Fermi velocity. On the other hand, if the initial excitation is carried by coherent, collective compressional modes, then this time is related to the frequency of these modes, which in turn depends on the speed of sound in nuclear matter.[369] Recent experiments,[370] determining the frequency of the monopole mode, lead[371] to a value of the compressibility coefficient $K \approx 300$ MeV, and an implied velocity of sound $v_s = \sqrt{K/9m}$ of 0.19c (m is the nucleon rest mass). This velocity and the Fermi velocity in nuclear matter (equivalent to 36 MeV/nucleon) are marked in Fig. 4.2. Although it would be premature to specify which (if either) defines the change of mechanism without a detailed model, the velocity of sound is certainly close to the transition region.

A formal approach to the break-up of nuclear matter was given recently,[372] by writing for the stress, S:

$$S = P = \frac{\partial E}{\partial V} = \rho^2 \frac{\partial (E/A)}{\partial \rho} \tag{4.17}$$

with

$$\frac{E}{A} = \frac{\hbar^2}{2m} k^2 + A\rho + B\rho^2 \tag{4.18}$$

In this equation the three terms represent the kinetic energy and the effects of the ordinary and velocity dependent nucleon nucleon potentials. Then the stress becomes:

$$\frac{P}{\rho} = \frac{2}{5} \hbar^2 \frac{k_F^2}{2m} \left(\frac{\rho}{\rho_o} \right)^2 + A\rho + 3B\rho^3 \tag{4.19}$$

from which information on the tensile strength of nuclear matter is obtained in the condition of maximum stress $dP/d\rho = 0$, which is equivalent to the classical condition of the sound velocity going to zero. In central collisions the energy per particle comes out at a few MeV/A. This approach, if extended to the type of peripheral collisions we have discussed in above, could be a fruitful way of studying continuum properties of nuclear matter.

The equivalence of two extreme models for the ^{16}O-induced reactions is an intriguing problem. One model assumes thermal equilibration whereas the other is a fast abrasion process from the nuclear ground state. The degeneracy might be removed by using heavier projectiles such as ^{40}Ar, with which the deeply-inelastic scattering processes at low energies are better developed (as we discussed in Lecture 3). A new series of experiments to study the isotope production cross sections as a function of energy has been initiated. An example of the first experiment[373] with 213 MeV/A Argon on Thorium and Carbon is shown in Fig. 4.7. The identification of isotopes was achieved by multiple ΔE-E identification in a 9 element detector telescope, and imposing a χ^2-criterion that the identification be similar in all detectors.[374] All isotopes up to Argon were resolved although this is difficult to see in the illustration.)

The momentum spectra for ^{16}O and ^{34}S are shown in Fig. 4.8. These are representative of all the isotopes are chosen as examples close to and far removed from the projectile. The theoretical curves come from Equ. 4.14 and 4.16, with values of $\sigma_o \approx 90$ MeV/c (see Equ. 4.15). (The associated temperature is 8.9 MeV.) In the framework of Fig. 4.5 this result fits into the pattern of ^{16}O, and we take it as confirmatory evidence for the fast abrasion mechanism.

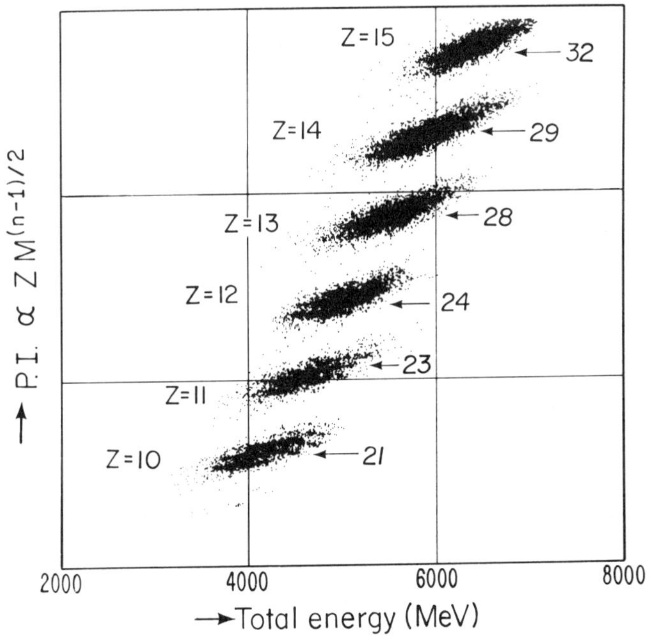

Figure 4.7

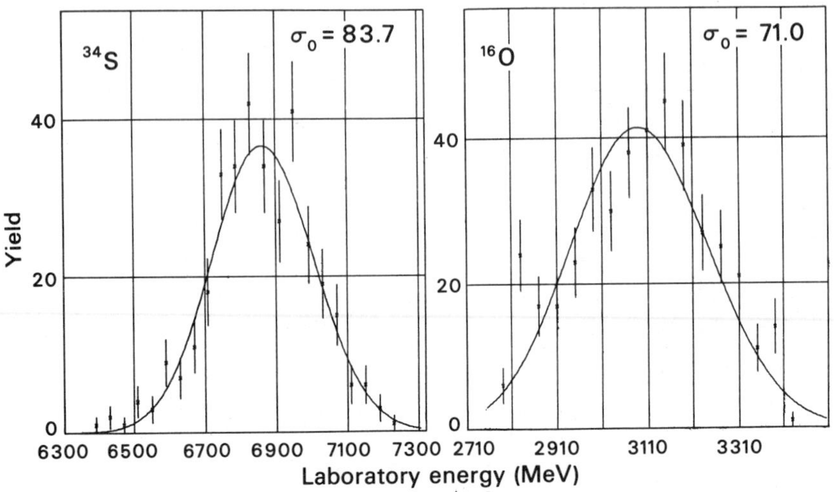

Figure 4.8

On the thermal equilibrium model we might conjecture that the temperature would have come out lower than for ^{16}O, as the initial localized deposition is cooled more rapidly by the larger thermal capacity of the heavy projectile.[337] The crucial test will come from the equivalent study of the *isotope* distributions, since the parameter $\alpha\beta$ in Equ. 4.13 *is* A-dependent, whereas the Fermi momentum parameter σ_o which characterizes the momentum distribution in the abrasion model is not.

Although the analytical comparison for the Argon reactions has not been completed, the preliminary results do indeed indicate that the "T" or "$\alpha\beta$" parameter is quite different from ^{16}O, although it appears[375] to *increase* to approximately 12 MeV, rather than decrease as predicted by the (oversimplified) analyses of Equ. 4.19-4.13. A value of 14 MeV in the expression $\sigma \propto \exp(Q_F/T)$ has been deduced in a similar experiment[376] with 250 MeV/A ^{12}C on Ca in which the *target* fragmentation yields were measured by γ-ray counting (this is effectively the inverse experiment). The predicted curve, using only the leading Q_F value of Q_{min} is shown in Fig. 4.9(a). The likely success of the abrasion-ablation approach is also encouraging from the predictions[377] for the magnesium isotope distribution[373] (hatched curve) in Fig. 4.9(b) compared to the data (solid points); the calculation reproduces the width of the distribution fairly well, although the peak is shifted from the experimental maximum.

The widths of the isotope distributions in the abrasion model is of considerable interest in view of recent attempts to account for them by building correlations into the nuclear ground state.[365,378] In the absence of correlations the abrasion model just calculates the dispersions (e.g., σ_a and σ_{t_3} in Equ. 4.9) in the number of protons and neutrons removed as equivalent to the relative number of ways of distributing neutrons and protons in an assembly of "a" nucleons

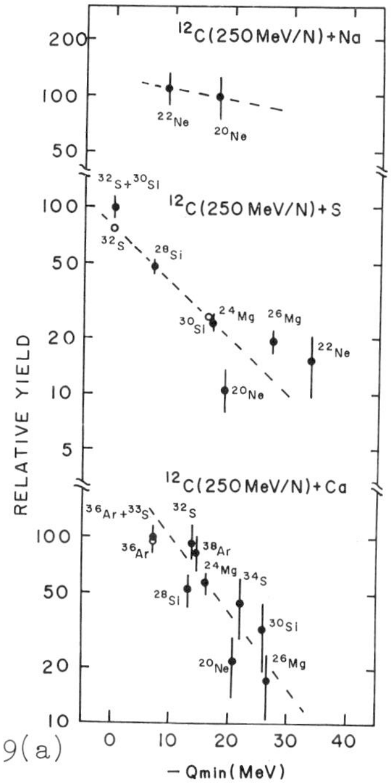

Figure 4.9(a)

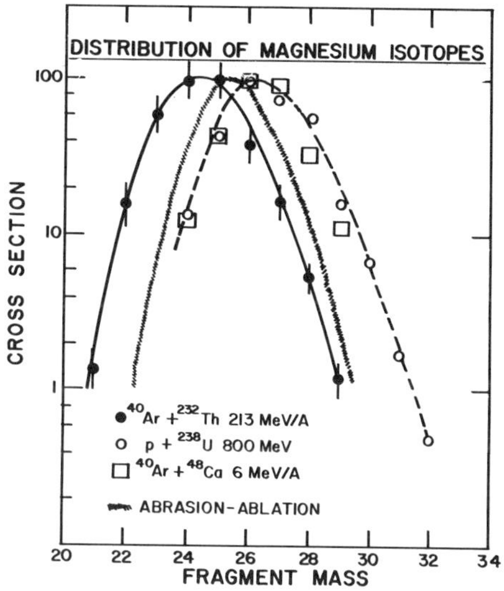

Figure 4.9(b)

(see also Equ. 4.7). Fig. 4.10 shows some representative primary product change distributions for ^{12}C + ^{238}U at 2.1 GeV/A.[378] (The data were acquired by the radiochemical method, as in Fig. 3.2.) An alternative model for the dispersions assumes that fluctuations in the number of swept-out protons (see Fig. 4.6) arise from zero-point vibrations of the giant dipole resonance, which is an out-of-phase vibration of protons and neutrons.[379] The predictions with (GDR) give a narrower width in better agreement with the experimental

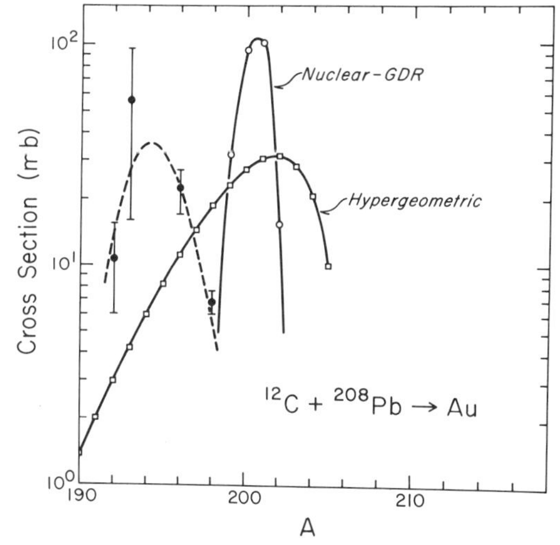

Figure 4.10

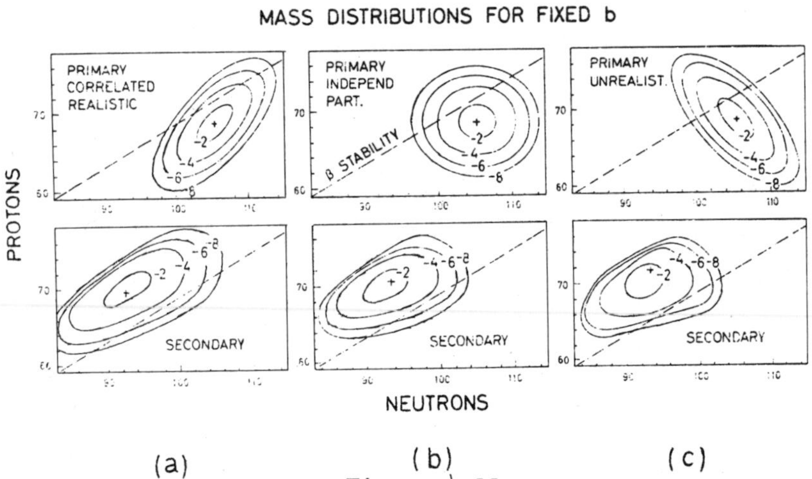

Figure 4.11

data. The uncorrelated calculation (hypergeometric) gives too large a width, essentially because it allows for unphysical possibilities such as removing *all* "a" nucleons as neutrons and protons alone. (The shift of the theory from the data is due to the neglect of the ablation stage.) Very similar considerations entered into the evaluation of the correlated widths σ_a, σ_{t_3} in Equ. 4.9, 4.10.

The subsequent ablation stage, in drifting the primary distribution back to the valley of stability, tends to erase the memory of the primary. The effect is illustrated in Fig. 4.11; the top sections display the primary abrasion distributions for (a) correlated, (b) uncorrelated and (c) unrealistic ground state motion. After the ablation stage (bottom) the distributions begin to look similar, but some influence of the primary persists.[365] Returning to experimental data in Fig. 4.9(b), it is clear that very careful measurements will be called for, since the completely different deeply-inelastic reaction ^{40}Ar + ^{48}Ca *at 6 MeV/A*[177] and the P + ^{238}U reaction at 800 MeV[380] give very similar distributions. (The points for both reactions were deduced from adding up counts from the published spectra and are thereby not very accurate.) Remember that the deeply-inelastic cross sections *also* arise from an equation like 4.9 (see 3.7), but the physics in the *primary* dispersions is quite different. What is clear however is the radical difference in the position of the peaks of the distributions. A more graphic demonstration appears in Fig. 4.12 which shows that the $\overline{N}/Z$ value for the deeply-inelastic reactions reflects more the value of the composite dinuclear system (due to the rapid equilibration of this degree of freedom, see Lecture 3) whereas at high energy the faster abrasion mechanism reflects the N/Z of the projectile, and the target acts as a "spectator." It is also clear that abrasion reactions such as ^{40}Ar + ^{232}Th, or better ^{48}Ca + ^{232}Th, at energies in the region of 200 MeV/A could be a powerful means of producing nuclei far from

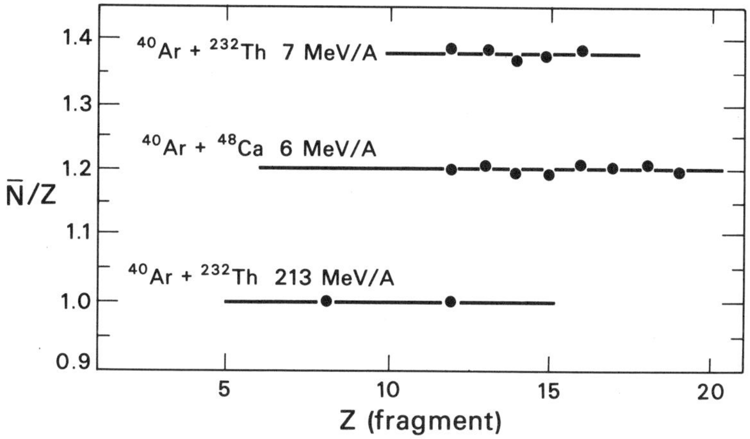

Figure 4.12

stability,[365,373] where the detection problems are simplified by the high emerging velocity of the fragments.

More detailed measurements as a function of energy for many systems must be made before a clear picture will emerge. Already departures from the skeletal framework of Fig. 4.5 may be cropping up in recent studies of ^{16}O + ^{197}Au reactions at 90 MeV/A.[381] One piece of evidence appears in Fig. 4.13, where the momentum widths of the fragments are compared with the parabolic dependence inherent in Equ. 4.15, evaluated with $\sigma_0 \approx 86$ MeV/c typical of the other data in Fig. 4.5. The systematics are obviously grossly violated. The data at 20 MeV/A may not therefore reside in Asymptotia as suggested by our earlier discussion, and implied by some other features. One characteristic of asymptotic behavior is factorization of the cross

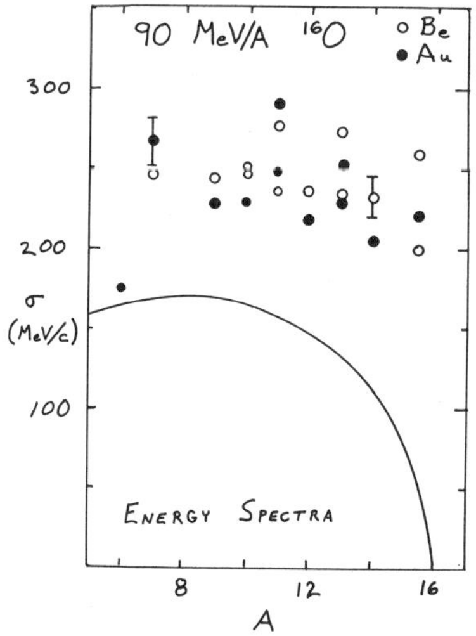

Figure 4.13

sections into a projectile and target term.[382-384] For the reaction A + T → F + Anything:

$$\sigma^F_{AT} = \sigma^F_A \ \gamma_T \tag{4.20}$$

This behavior is a logical consequence of the dependence $\sigma \ \alpha \ \exp(Q_F/T)$ but not of the deeply-inelastic dependence

$$\exp \ \frac{Qgg-\Delta Vc}{T}$$

of Equ. 4.6, since the substantial differences of Q-value for different targets would lead typically to an order of magnitude change between Pb and Au targets. The factorization appeared to hold at both 20 MeV/A and 2.1 GeV/A but not at 8 MeV/A.[382] A direct reaction model of peripheral fragmentation also leads to the observed factorization.[385] The phenomenon is also reminiscent of the Bohr independence hypothesis.[357,386] A dramatic illustration of the factorization and limiting fragmentation hypothesis (i.e., yields independent of energy[388]) is given in Fig. 4.14, which compares the yields of target fragments produced by protons at 3.9 GeV/A ^{14}N ions (upper curve) and 3.9 GeV protons (lower curve). (The data are displaced by a factor of 10 for display.) Other experiments also indicate that the distributions become similar for protons, of equivalent total energy as the heavy-ion, rather than of similar velocity.[389]

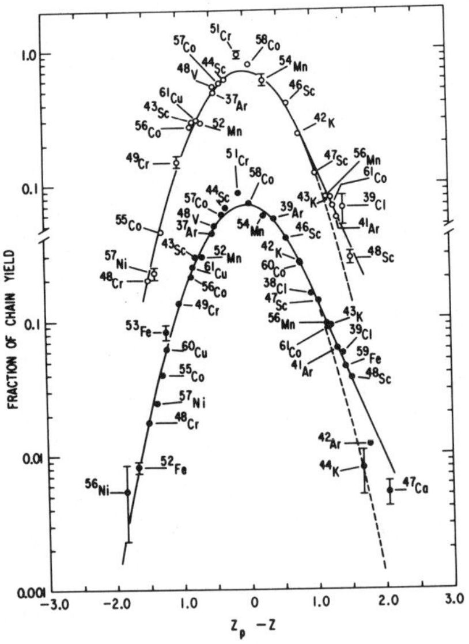

Figure 4.14

4.2 Central Collisions

Relativistic energies mark a change in the ability of a nucleon to pass through the nucleus. Above 1 GeV the longitudinal momentum decay length appears to grow to over 4 fm and begins to approximate nuclear dimensions; the colliding nuclei could then pass right through each other.[390] The consequences of the collision will vary depending on whether the collision is peripheral or central. Fig. 4.15(a) and (b) illustrates examples of the two types. In (a) the peripheral collision[391] of 870 MeV/A ^{12}C results in a small number of particles, continuing in the projectile direction. For the central collision in (b), there is a star explosion[392] of Ar + Pb at 1.8 GeV/A; the total multiplicity of charged particles ranges up to 130 suggesting that the initial system is completely disintegrated

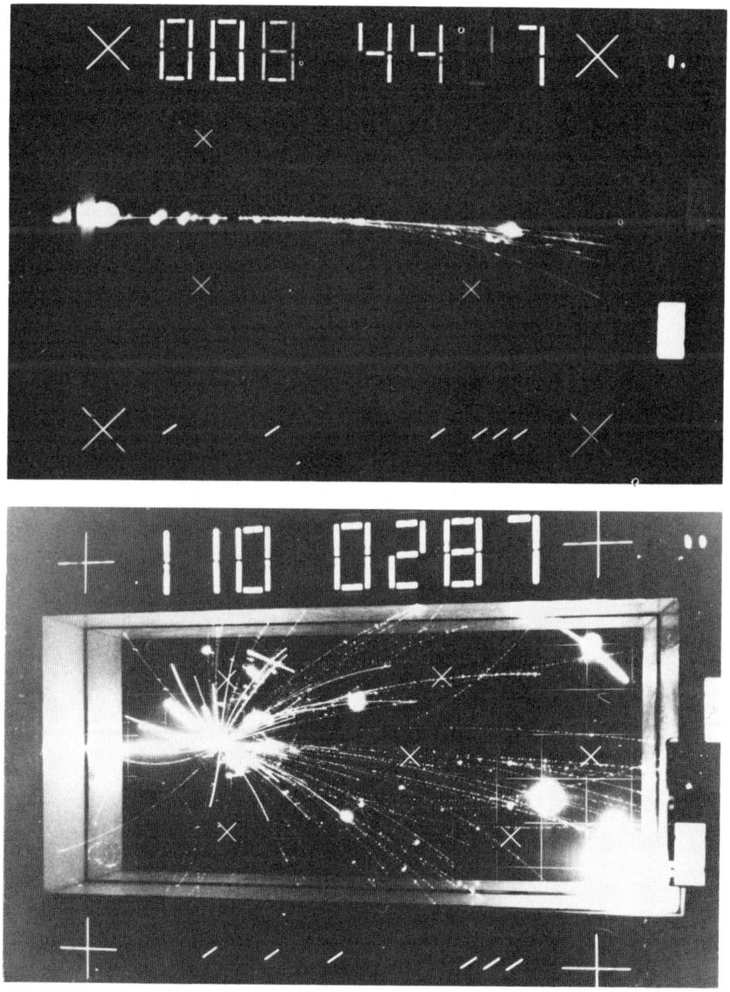

Figure 4.15

(far from passing through each other!). At lower energies, we have seen that central collisions lead to fusion or fission. Although the nature of the central collision is very different in the two regimes, it appears that the onset of these more catastrophic processes takes place at roughly the same overlap of nuclear matter densities.[49] To see this[342,350] we write the reaction cross section as the sum of peripheral and central cross sections:

$$\sigma_R = \sigma_P + \sigma_c \tag{4.21}$$

and compare values of σ_c deduced from this equation by subtracting the summed peripheral cross sections of all reaction products in $^{16}O + ^{208}Pb$ at 20 MeV/A and 2.1 GeV/A (last section) from the reaction cross section, which has been measured directly at 2.1 GeV/A and was deduced from the optical model analysis of elastic scattering at 20 MeV/A.

Energy	Reaction	Peripheral σ (mb)	Total reaction σ (mb)	Central σ (mb)
20 MeV/A	$^{16}O+^{208}Pb$	1295	3460	2160
2.1 GeV/A	$^{16}O+^{208}Pb$	930	3100	2260

The reaction cross section has also been determined from ^{16}O reactions in emulsions in the energy range 75-150 MeV/A and appears to give similar values.[393] Such an energy independence would not be expected from the known (large) variation of the nucleon-nucleon cross section over the same energy region.[394]

In the central collisions of the type in Fig. 4.15(b), the most exotic features of high-energy heavy-ion collisions will be hidden--one says hidden because they must be separated from the large background of (possibly) trivial effects which are the outcome of the superposition of all the free nucleon-nucleon cross sections, properly folded with the particle distributions of position and momentum. The basic layout of a system designed to make quantitative studies of central collisions is shown in Fig. 4.16, which combines a particle identification telescope to identify a particular particle, with an array of plastic scintillators to determine the multiplicity of charged particles associated with each event.[395] A large multiplicity is used as a signature of a central collision.

Proton energy spectra from Ne and He bombardments of U are shown in Fig. 4.16 for angles of 30°, 60°, 90°, 120° and 150°

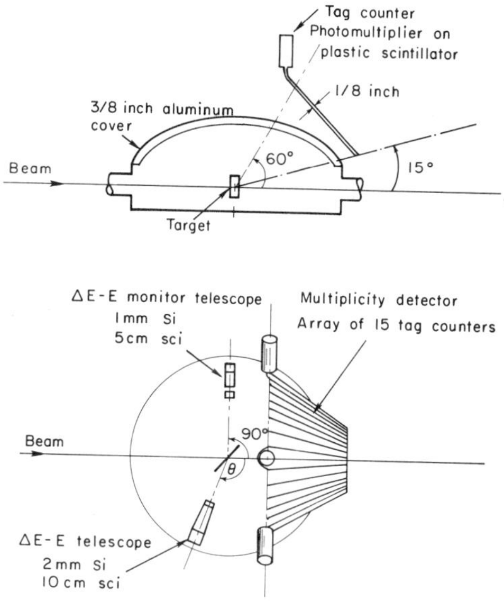

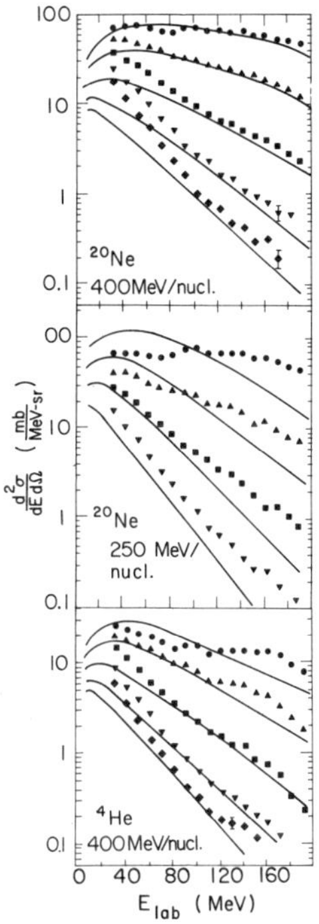

Figure 4.16

(except for Ne). The spectra have Maxwellian shapes corresponding to high temperature. These spectra have been elegantly explained with a fireball model,[395,396] illustrated schematically in Fig. 4.6(b). The model is an extension of the abrasion-ablation picture used previously for peripheral reactions. In the more central collision, nucleons swept out from the target and projectile form a quasi-equilibrated fireball at high temperature, equal to the available energy per nucleon. The velocity of the fireball is assumed to be that of the center of mass system of the nucleons swept out. The fireball expands isotropically in its center of mass system with a Maxwellian distribution in energy.

Assuming spherical nuclei and straight-line trajcctories, the participating volume of each nucleus is easily calculated as a function of impact parameter. The number of participating protons as well as the division between projectile and target are shown in Fig. 4.17 for Ne on U. At the bottom is the effective weight, $2\pi b N_{proton}$, given to each impact parameter. The velocity of the center of mass of the fireball is then given by,

$$\beta_{cm} = \frac{P_{lab}}{E_{lab}} = \frac{N_p[t_i(t_i+2m)]^{\frac{1}{2}}}{(N_p+N_t)m+N_p t_i} \tag{4.22}$$

where P_{lab} is the lab momentum, E_{lab} the total energy, t_i the

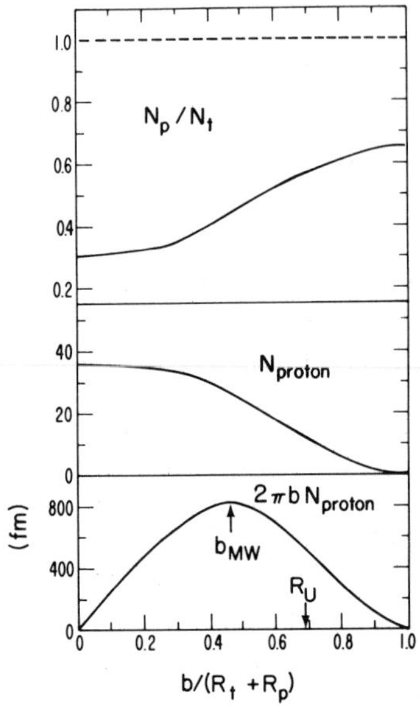

Figure 4.17

projectile incident energy/nucleon, and m the nuclear mass. The total energy in the center of mass of the fireball is

$$E_{cm} = \left[E_{lab}^2 - P_{lab}^2 \right]^{\frac{1}{2}} \tag{4.23}$$

If one assumes there are sufficient degrees of freedom in the fireball, and that there is a mechanism to randomize the available energy, one can define a temperature T, which can be expressed (non-relativistically) by:

$$\varepsilon = 3/2T \tag{4.24}$$

where ε is the available kinetic energy per nucleon in the center of mass, i.e., $E_{cm}/N_t + N_p$). The quantities β and ε (calculated relativistically) are given in Fig. 4.18 as a function of impact parameter. The momentum distribution of the fireball nucleons in the center of mass is then:

$$\frac{d^2N}{p^2 dp d\Omega} \propto (2\pi mT)^{-3/2} \; e^{-p^2/2mT} \tag{4.25}$$

where p is the momentum of a nucleon in the center of mass. Using the earlier expressions this distribution can be transformed to an energy distribution in the laboratory, which must then be integrated over impact parameter weighted appropriately (Fig. 4.17). The resultant distributions are shown in Fig. 4.16 (typical values of β and T can be derived from Fig. 4.18 at the point of maximum weight ($\beta \approx 0.25$ and $T \approx 50$ MeV)). Fairly satisfactory agreement with the data is obtained. (Note: the data shown in Fig. 4.16 have an error of absolute normalization, and the authors of Ref. 395 should be consulted for corrections.) Recently more advanced versions of the model such as the diffuse firestreak[397] have been developed, but its success is less obvious in view of the data errors. For a review of the various approaches, see Ref. 398.

It is possible to advance further and explain the distributions of other fragments heavier than the proton with a *coalescence model*.[397] If any number of protons and nucleons corresponding to a bound nucleus are emitted in the reaction with momenta differing by less than a "coalescence radius" p_0 (a parameter to be adjusted which comes out at 130 MeV/c typical of Fermi momenta), they are assumed to coalesce. The cross sections for these heavier nuclei are then trivially related to those for the proton. However, there are also thermodynamic models which extend the fireball concept to the emission of complex fragments.[400]

Fragments from central collisions may originate from several qualitatively different subsystems, such as the fireball, the target spectators, or even an explosion of the fused target

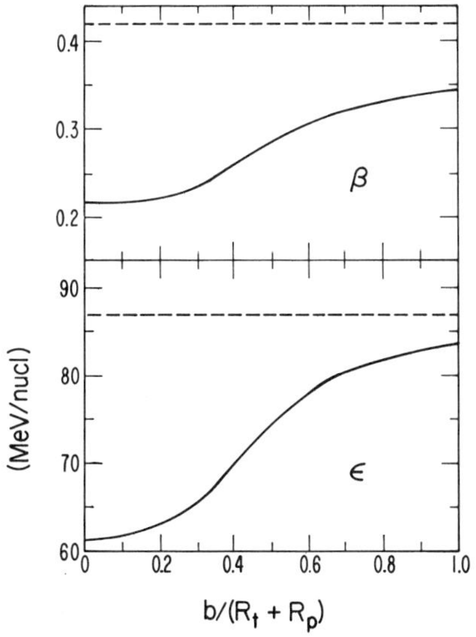

Figure 4.18

projectile system. The detailed distribution of the longitudinal and transverse momenta of all the fragments give information on these subsystems. For this purpose it is convenient to characterize the distribution of longitudinal momentum by the *rapidity variable*:

$$y = \tfrac{1}{2} \ln \frac{(E + p_{\parallel})}{(E - p_{\parallel})} \qquad (4.26)$$

where E and $p_{\parallel}$ are the total energy and longitudinal momentum of the particle. (This variable is convenient in relativistic systems because it transforms in Galilean fashion in changing frames.) Contour plots of invariant cross sections, which are measured as a function of angle, are transformed to these variables in Fig. 4.19 for inclusive proton spectra for the reactions[401] 800 MeV/A ^{20}Ne + Pb → p + x. These data were taken with a target centered rotating magnetic spectrometer to obtain data at high $p_{\perp}$ for production angles $15° \leqslant \theta_c \leqslant 145°$ and proton momenta in the interval $0.4 \leqslant p \leqslant 2.4$ GeV/c. The half rapidity line that corresponds to the velocity of the nucleon-nucleon center of mass frame is marked. The mountain top of the cross section is found for $p_{\perp} \lesssim 300$ MeV/c, $y \simeq \beta \lesssim 0.1$. Most of the protons have small transverse momentum and come from a source that moves slowly in the laboratory (target spectator decays). Towards high $p_{\perp}$ the contour lines move up in y but always bend round at a y smaller than $(y_T + y_p)/2$. The apparent proton source moves slower than the nucleon-nucleon center of mass. Over a wide range of $p_{\perp}$ the apparent source rapidity coincides with the fireball, which by equ. 4.22, is around 0.4 for this system. Similar studies for Ne + NaF (i.e., an almost equal mass target and projectile) which should have $y = (y_T + y_p)/2$, do not entirely support the elemental concept of the fireball but, at the least, call for refinements that allow a continuum of source-velocities.

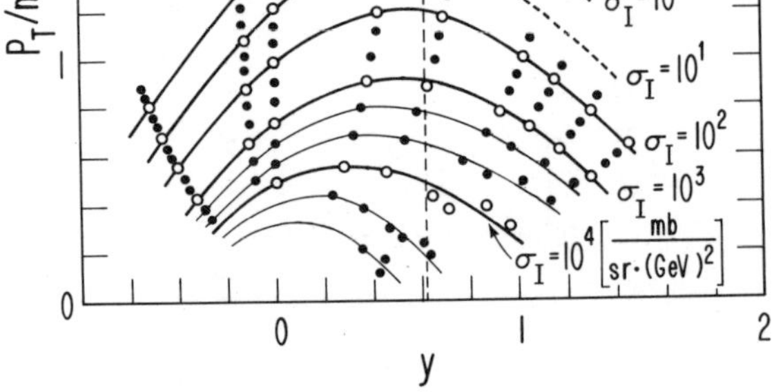

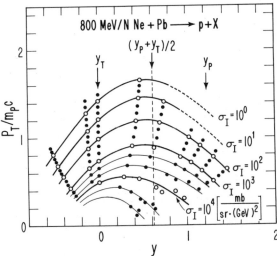

Figure 4.19

Data obtained with the very different techniques of stacked Lexan foil detectors give evidence for emission of complex fragments from a source moving with low velocity *and* high temperature,[402] which cannot be accommodated in the framework of a fireball. These fragments appear to originate from non-equilibrium emission from a system like the entire target, where the internal energy does not have to reach the value of $\frac{3}{2}$ T per nucleon. The radial emission velocity in the source frame is strongly correlated with the source velocity, independent of the mass of the fragment observed. This behavior is uncharacteristic of a thermalized source.[403] Various cooperative, non-thermal processes can be imagined, amongst which are compressional wave phenomena or the release of preexisting clusters. These ideas will be the topic of the last section, 4.3.

The fireball model was introduced in relativistic hadron and heavy-ion collisions and led to great early insight into the complex processes that take place when heavy ions collide. The model no longer enjoys unqualified success in its own territory, but it is now applied to shed light on reactions at much lower energy. Beggars (the nuclear physics community) do not mind picking up the crumbs that fall from the rich man's table (high energy physics). We can consider the logical limit of the fireball approach as the incident energy is decreased. It works at 250 MeV/A and it might work at 100 MeV/A. At still lower energies there cannot be a fireball, clearly separated from the intersecting nuclei, but can we imagine that the process degenerates into a local heated region? The possibility of the process depends on the reaction time compared to the time for transporting the local excitation outwards into the surrounding nuclear media. As we have seen (Fig. 3.25) this time increases at high energies. Presumably this concept of the "fireball" merges with the "hot-spot" discussed in Section 3.5 of Lecture 3. Some justification for the validity of this approach at least down to 20 MeV/A comes from the successful application of the Glauber model to describe complex fragment yields at 315 MeV (See Eq. 4.7, 4.8). To establish another link between the asymptotic and low energy regimes, let us look again at the fireball data of Fig. 4.16 compared with a cascade calculation [404] in Fig. 4.20. The simultaneous evolution of all projectile and target cascade particles is followed. Pion production and absorption are included via $NN \rightleftarrows N\Delta$, and experimental cross sections are used to determine the outcome of two-body collisions. Diffuse nuclear surfaces, Fermi motion, the exclusion principle and binding energy effects are also included. The inner workings of these very expensive and complicated computer calculations are beyond the comprehension of non-technicians, but they clearly do a good job in describing the data. This success does not signal a defeat for the fireball model; because the cascade model shows that complete thermalization is achieved for central collisions (but not for larger impacts parameters).

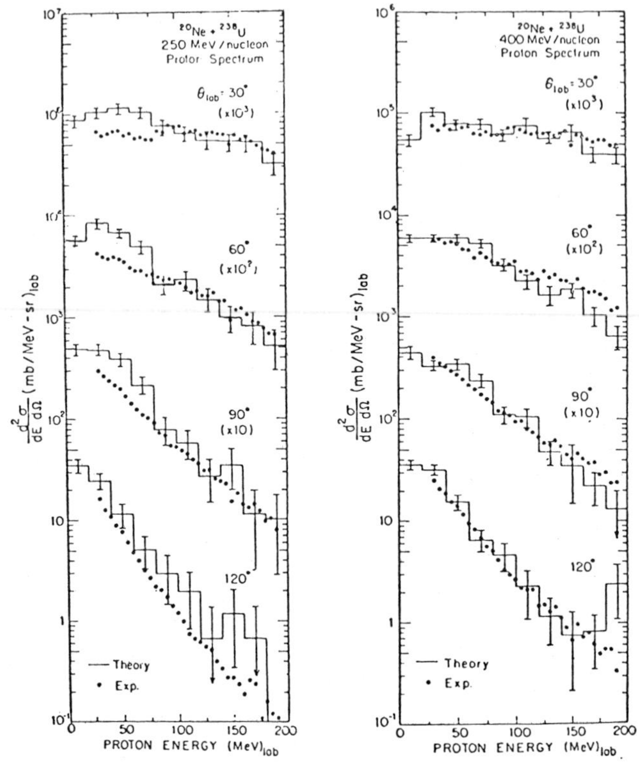

Figure 4.20

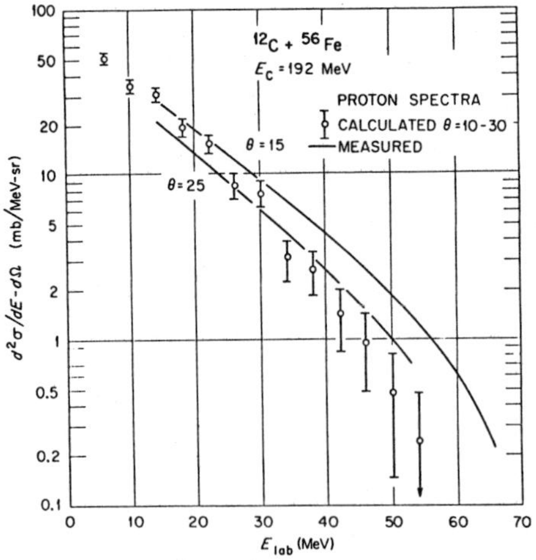

Figure 4.21(a)

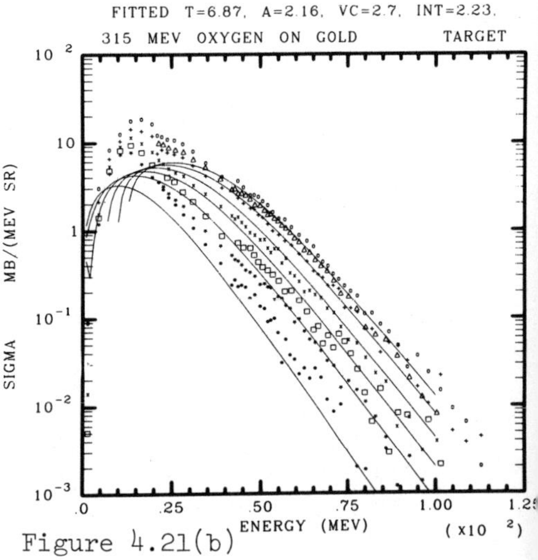

Figure 4.21(b)

Compare now the proton energy spectra[405] from the collision of ^{12}C with ^{56}Fe at a *total* energy of 192 MeV (i.e. only 16 MeV/A!) in Fig. 4.21(a). The trend of the data is indicated by the solid lines, again the spectra are statistical in appearance, but by extending with substantial cross sections up to 70 MeV, requires a temperature far in excess of the compound nucleus. (The center of mass energy of 130 MeV above the barrier gives rise to T = 3.9 from the expression $E^* = \frac{A}{8} T^2$, and a resultant decrease of 10^5 in in cross section between 10 to 60 MeV compared to the observed factor of 10^2). These data are also fitted by a cascade calculation[406] open circles). An analysis of the output suggests that the protons are evaporated by the *projectile*, which is excited in the collision and sequentially decays.[407] The high energy protons are produced by the vector addition of the low velocity decay in the projectile frame and the high projectile velocity.

Closer investigation suggests that this explanation may have a flaw. The data[408] in Fig. 4.21(b) for ^{16}O + ^{288}Pb at 315 MeV cover a wider range of angles from 20° to 80°. Over this region the spectra do not fall off sufficiently rapidly to be attributed to projectile decay. On the other hand they fall off too quickly to originate from the compound nucleus. Rather the data call for an intermediate number of nucleons moving with an intermediate velocity, *just as in the fireball*. The solid lines are in fact fits to the high energy parts of the spectra using eqs. 4.22-4.25 but replacing the ideal gas (Eq. 4.24) by the equivalent expression for a degenerate Fermi gas. The fits result in a temperature of 6.9 MeV (compared to the strict fireball prediction of 5.9 MeV) from a source of approximately 30 nucleons moving with half the projectile velocity. The temperature of 6.9 MeV is almost the same as the value deduced for the emission of complex fragments at the same incident energy (see the discussion of Fig. 4.5).

Similar descriptions of proton spectra have been reported in α-particle induced reactions at energies of 25 MeV/A[409] and 180 MeV/A.[410] The formation of a localized hot spot has also been

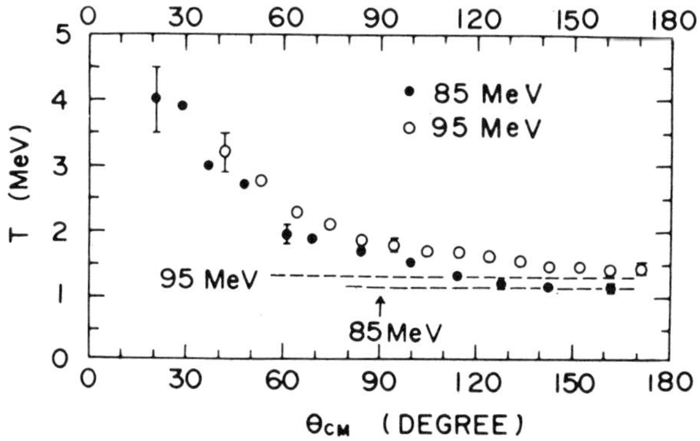

Figure 4.22

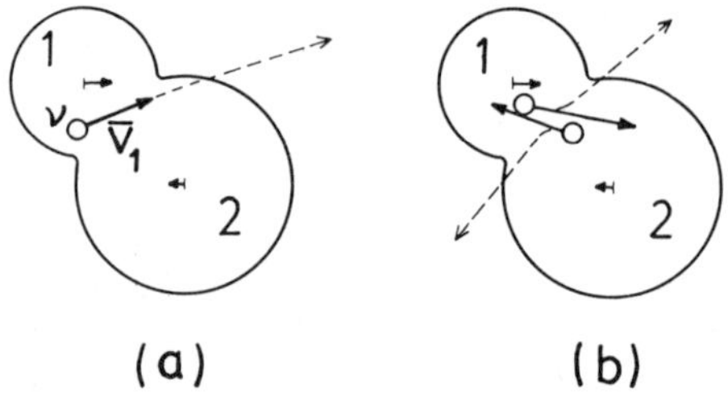

Figure 4.23

discussed in the analysis of a preequilibrium component in neutron spectra of $^{20}Ne + ^{150}Nd$, leading to a temperature of 6 MeV and 25 participating nucleons.[411] Yet another approach[412] is to describe the energy spectra with an angle dependent temperature in reactions with ^{14}N on ^{209}Bi. Local heating takes place at the contact point, due to strong frictional forces, and alpha particles are emitted from the rotating surface (compare our discussion of hot-spots in Section 3.5). We have already seen that the rotation angle is intimately related to reaction time, in deeply-inelastic phenomena. As the system rotates the temperature drops according to the conductivity and specific heat of nuclear matter. Figure 4.22 shows the temperature and number of participating nucleons as a function of angle. The values for a completely equilibrated compound nucleus are given by the dashed lines, which are approached after 3/4 of a revolution.

There are other explanations in vogue for the explanation of energetic light particle emission in heavy-ion reactions. For example, Fig. 4.23 shows[339] a heavy-ion reaction at relative speed V of nucleus 1 at the ion-ion barrier. A nucleon ν moving from 1 to 2 has on arrival a velocity $\underline{v}_2 = \underline{v}_1 + \underline{V}$ where v is its velocity in nucleus 1, with a maximum of $v_F + V$. The maximum kinetic energy is:

$$E(\max) = E_F + E_{rel} + 2\sqrt{E_F E_{rel}} \qquad 4.27$$

For a 20 MeV/nucleon with E_F= 35 MeV, E reaches 108 MeV. An extension of the model to "Fermi-Jets" has recently been developed[413] and studied experimentally.[414] The emission of fast light particles is also encountered in time-dependent-Hartree-Fock calculations[415] and in hydrodynamic calculations.[416] A standing wave is set up and the nucleus fractures at the weakest point, which is a node of the standing wave located at a distance π/k_F from the surface. The two types of calculations are compared in Fig. 4.24 for a collision

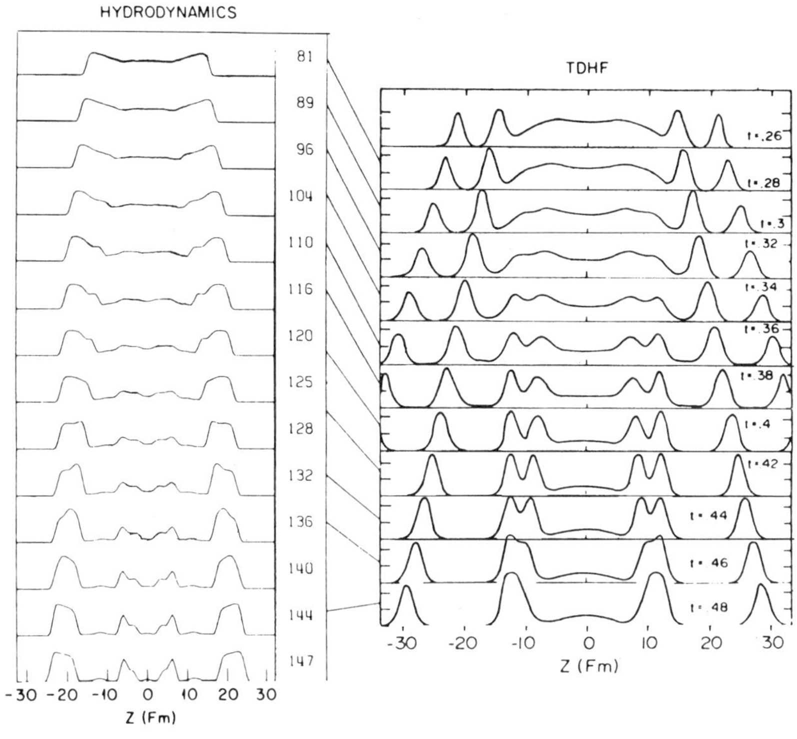

Figure 4.24

energy of E/A = 100 MeV/nucleon. The numbers at the right give the time expressed in units of fm/c in the hydrodynamical calculations, and in units of 10^{-21} sec for the TDHF. In both calculations a small piece of nuclear matter is ejected with higher than beam velocity.

In low energy light-ion reactions there are well developed preequilibrium theories for fast particle emission (see refs. 315-322 & 417). A critical question in these theories is the correct initial exciton number to use. For α-particle induced reactions there is evidence that the correct number is four, for two protons and two neutrons.[417] In heavy-ion reactions one might assume that the heavy ion, eg. ^{12}C, breaks up into 6p + 6n, and the number of excitons would be 12. Calculations[417] based on this hypothesis for

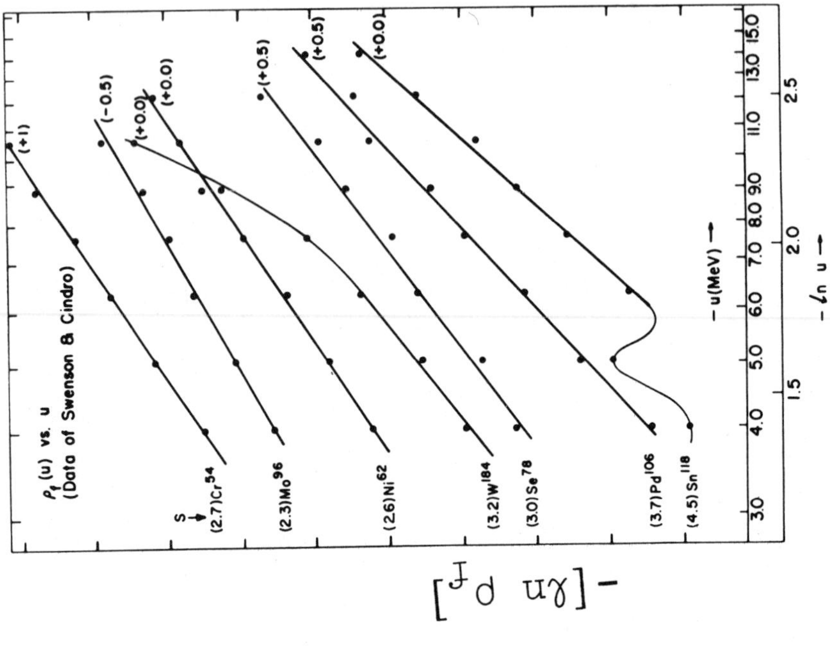

$\rho_f(u)$ vs. u
(Data of Swenson & Cindro)
(2.7)Cr^{54}
(2.3)Mo^{96}
(2.6)Ni^{62}
(3.2)W^{184}
(3.0)Se^{78}
(3.7)Pd^{106}
(4.5)Sn^{118}
— u(MeV) →
— ℓn u →
−[ℓn ρ$_f$]

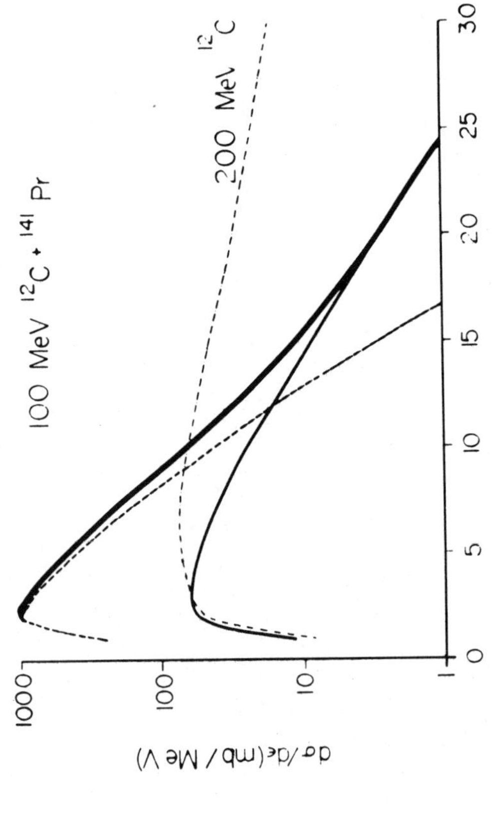

100 MeV ^{12}C + ^{141}Pr
200 MeV ^{12}C
E$_n$ (MeV)
dσ/dε(mb/MeV)

Figure 4.25

the $^{141}Pr(^{12}C,n)$ reaction are shown in Fig. 4.25(a). The dashed line for 100 MeV represents essentially compound nuclear evaporation (dashed line) with a small preequilibrium component. Now note the dramatic change at 200 MeV, where the huge increase of the preequilibrium emission leads to a cross section extending out to very high energies, just as in the ^{12}C and O^{16} induced reactions of Fig. 4.21. The preequilibrium emission becomes important when the excitation energy of the compound system becomes comparable with the particle binding energy/exciton. A method of finding out the number of excitons is to plot[418] the log of the differential cross section versus the log of the residual excitation and the slope gives the (number of excitons-2). An example for α-induced reaction is given in Fig. 4.25(b) on a variety of targets; the slopes, marked on the left hand side, are typically about 3 (a similar plot for the O^{16} induced reactions of Fig. 4.21(b) yields 25, very close to the number of particles in the fireball calculation!)

All the above lengthy discussions, (which are a considerable digression from our description of central, relativistic heavy-ion collisions) are meant to emphasize that the questions of localization, hot spots, high temperatures and the like are not unique to the province of Asymptotia. These phenomena are firmly rooted throughout the whole physics of light and heavy-ion collisions and their interpretation will call for all the tools of nuclear dynamics, whether microscopic or macroscopic, at high or low energies. We have to understand how the central collisions at low energies evolve from fusion, fission and deeply-inelastic processes to the more catastrophic event of Fig. 4.15. There are already intimations on how to treat these problems.[419,420]

As a final illustration[317] look at the two spectra in Fig. 4.26, which compares p+p collisions at 100 GeV/c with $^{89}Y(p,n)$ at low

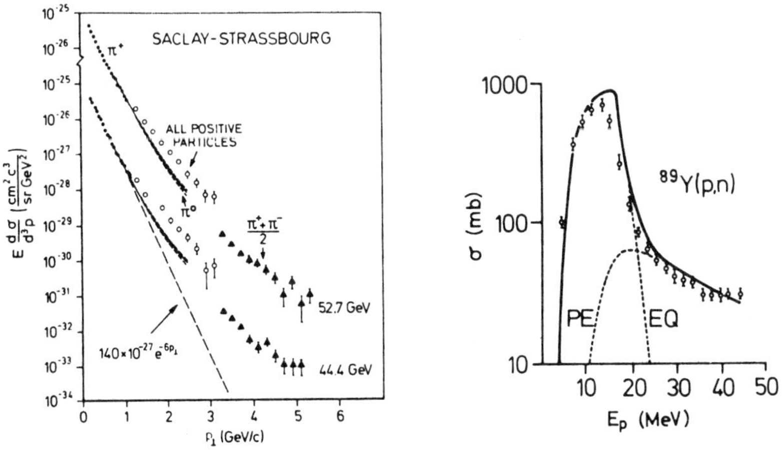

Figure 4.26

energies of 30 MeV. Both spectra have a "low temperature" component (in the p+p case $T \approx m_\pi$ the limiting temperature discussed in the introduction to Lecture 1) and a "preequilibrium tail". Both can be considered as local, instantaneous equilibrium in a hot spot.[317] All the experiments proposed to measure[421] the size and lifetime of the fireball from the small angle correlations[422] by adapting the Hanbury-Brown and Twiss technique to measure the size of stellar objects, must also be applied to the lower energy region. The method has already been used with pions to determine[423] the size of the *nucleonic* fireball. This unity between low and high energy regimes is aesthetically pleasing, but it is to be hoped that relativistic heavy-ion collisions will also reveal totally new phenomena on the nature of nuclear matter under extreme conditions, which will never be realized at lower energies or with lighter particles. To this topic we devote the last section of our Lectures.

4.3 Six Impossible Things

My title is taken from "Alice in Wonderland", when Alice finds herself in the Garden of Live Flowers (remember the search for exotic phenomena is often compared to searching for flowers among weeds!) confronted by the White Queen. "You, my dear, must learn to think of impossible things," said the Queen. "But that is so difficult," said Alice. "And besides, what's the use? One can't *believe* impossible things." "I daresay you haven't had much practice," said the Queen. "Why, when I was your age, I always did it for six hours per day. Sometimes I've even thought of as many as six impossible things before breakfast!" By and large I have not talked to you of impossible things. Rather, as quoted from James Joyce in the introduction to the Annotated Version of "Alice in Wonderland", I have "wiped my glasses with what I know"! For a detailed discussion of impossible things, I refer you to many excellent review articles.[337,340,341,424-426]

An important basic question in complex nucleus-nucleus interactions is to what extent they can be traced back to quasi-free hadron-hadron collisions. In the total energy available in the system, viz. $E_{inc}\, A_1A_2/A_1+A_2$ GeV, the important quantity or is it just $\approx (\sqrt{2(E_{inc} + 2)}-2$ GeV that is available $\leqslant A_1$ nucleon-nucleon reactions? The difference between these pictures is important. If we find pion production at 0.1 GeV/A, the former expression must be relevant, and *collective* phenomena are important, and have already been claimed to be observed,[427] but more recent experiments yield contradictory evidence.[428-429] Many experiments are in progress, and it is clear that the great majority of events can be easily explained in an independent nucleon-nucleon model.[404] There are also some indications in pion multiplicities for production via strong nucleon correlation effects,, which hopefully may be a signature for shock waves.[361]

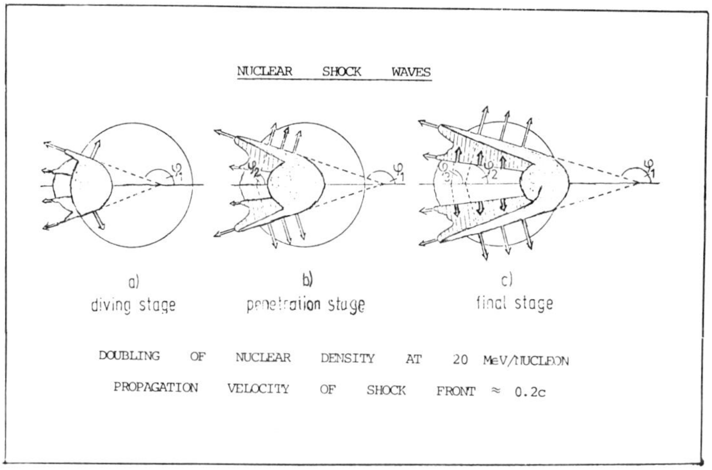

Figure 4.27

It has been suggested that a compressed zone of high energy density may be formed in a central collision, which propagates as a shock wave and could lead to the emission of energetic fragments upon impinging at the nuclear surface.[430-431] Such a propagation of high compression ($\rho > \rho_o$) and with velocities $v_s > 0.2c$ has been called a "shock wave." The progress of this wave is illustrated in Fig. 4.27. In the initial phase a "splashing tidal wave" is expected at a backward angle $\sin\phi_1 = v_t/v_i$, where v_t is the expansion velocity of the shock compression zone. In the second stage a strong compression shock is created accompanied by a Mach cone traveling outwards in the direction $\phi_2, \cos\phi_2 = v_s/v_i$, where v_s is the shock expansion velocity. In the final stage, matter is emitted in the directions ϕ_1(splashing) and ϕ_2 (Mach).

In reality the projectile would slow down considerably and the simple Mach cone picture is distorted. The emission is then spread out over a wider angular region, which actually appears to be a feature of hydrodynamical calculations of collisions of nuclear matter, treated as a classical compressible fluid.[434] The criterion for compressibility is whether flow velocities are comparable to the speed of sound. For nuclear matter with an incompressibility K(MeV) the speed of sound is[435]

$$v_s = (K/9mo)^{\frac{1}{2}} \qquad 4.28$$

and the projectile energy/nucleon above the Coulomb barrier required to reach such a velocity is:

$$E/A = K/18 \qquad 4.29$$

For typical values of K between 150 and 300 MeV, v_s is derived to be 0.13 and 0.19c, for E/A of 8 and 17 MeV. Apparently compressibility will be important at the relativistic energies we have been discussing. For a hydrodynamic description to be valid, the mean free path of the microscopic particles should be small compared to the macroscopic dimensions. From the known nucleon-nucleon cross section of 40 mb at 2 GeV, we can estimate the mean free path $\lambda \approx 1/\rho\sigma \approx 2$ fm. So the criterion is only marginally fulfilled. The hydrodynamical equations have been solved[4,434] for collisions of ^{20}Ne on U (the reaction used for the fireball discussion) at 250 MeV/A. Figure 1.3 showed the time development of the density as represented by the distributions of particles, for different impact parameters. For the nearly central collision (labeled 0.1) the neon penetrates into the uranium nucleus and sets off a strong shock wave (clearly visible at 5.1×10^{-23} sec). Subsequently most of the energy of the projectile is thermalized and the nucleus expands. The other two sections illustrate an intermediate impact parameter (which should come close to the fireball description), and a peripheral collision in which we see a part of the projectile sheared off (just as in the abrasion picture). When the angular distributions for central collisions are computed from the distribution of nucleons in the final state they lead to rather featureless exponential forms, with no sharp shock wave peak.

Another way of treating the density problem is by introducing statistical microscopic calculations.[436] These make Monte Carlo simulations of colliding samples of almost free point nucleons. The nucleon-nucleon scattering follows the known cross sections, conservation of energy, momentum, and angular momentum. The position and velocity of each nucleon is known (in principle) at each time. These calculations indicate that the transparency effects are too large to give high enough compression to produce shock waves.

Nevertheless, they have been searched for,[437] and the first experiments made extensive studies of high multiplicity events in track detectors using AgCl crystals and emulsions. The distributions of $d\sigma/d\theta$ were measured for events with more than 15 prongs, and a typical example[437] appears in Fig. 4.28(a). The sharp peak seemed to shift its position in a way characteristic of Mach shocks with a propagation velocity,

$$v_s = v_i \cos\theta(\text{peak}) \qquad 4.30$$

and the peak moves *backwards* with increasing energy. These peaks have not been found in other emulsion experiments, nor are they present in the differential cross sections obtained with the live counter techniques.[438] It seems that the peaks are due to

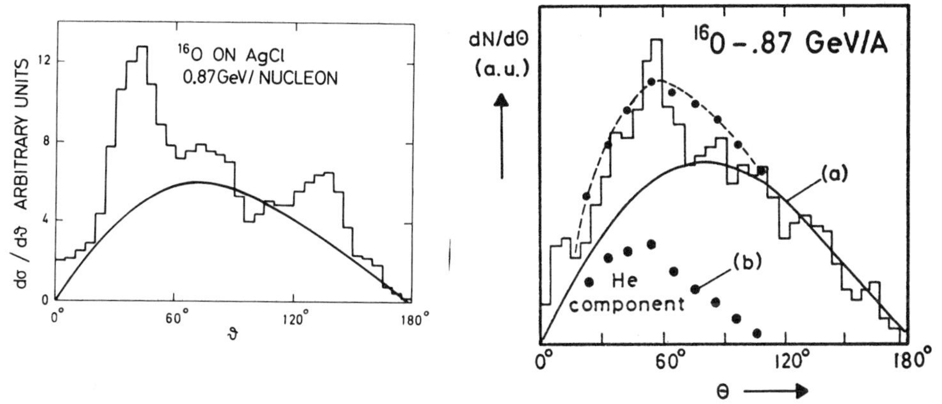

Figure 4.28

combinations of different particle types, such as protons and alphas which were selected by the experimental technique at different energies.[439] (Fig. 4.28(b) shows both components and the sum.) Other experimental searches for shock waves have not yielded positive results (see Ref. 337, p. 38 for a summary) and it must be concluded that there is no proof of their existence. Equally, though it is not clear these experiments were capable of establishing the existence of such effects, in that they were predominantly single particle inclusive measurements, lacking essential information on multiplicities. This criterion cannot be levelled at a recent study of the $^{40}Ar + {}^{9}Be$ reaction at 1.8 GeV/A.[440] A test was made of the possible correlations between a particular multiplicity M and the inclusive cross section W, by the ratio $r = W_M(\theta_L, Y_L)/W(\theta_L, Y_L)$ as a function of the laboratory

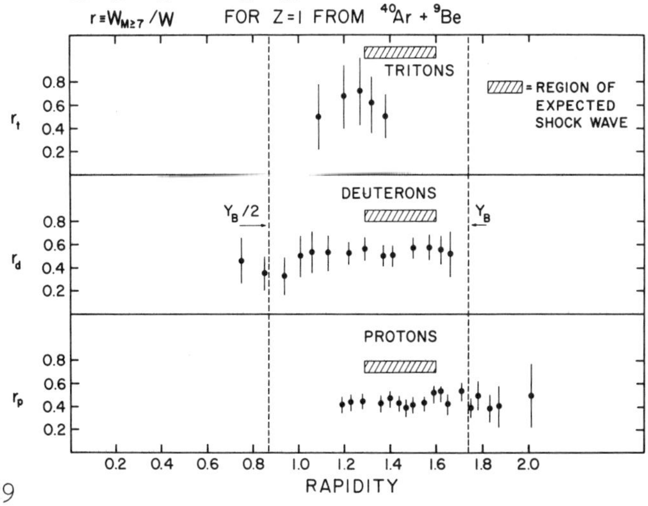

Figure 4.29

angle θ_L and the rapidity YL. This ratio is shown for p, t, d in Fig. 4.29. The multiplicity requirement was that at least seven fragments are detected by an Array of Cerenkov detectors. According to a shock wave model,[441] the fragments from a shock wave in the projectile would peak at rapidities indicated by the shaded region. The evidence is negative.

Only the first generation of experiments have been completed, which have primarily looked at single particle inclusive spectra. There are many refinements in progress to search for collective effects of nuclear matter at extreme density and pressure — conditions which are also probably realized in the interior of

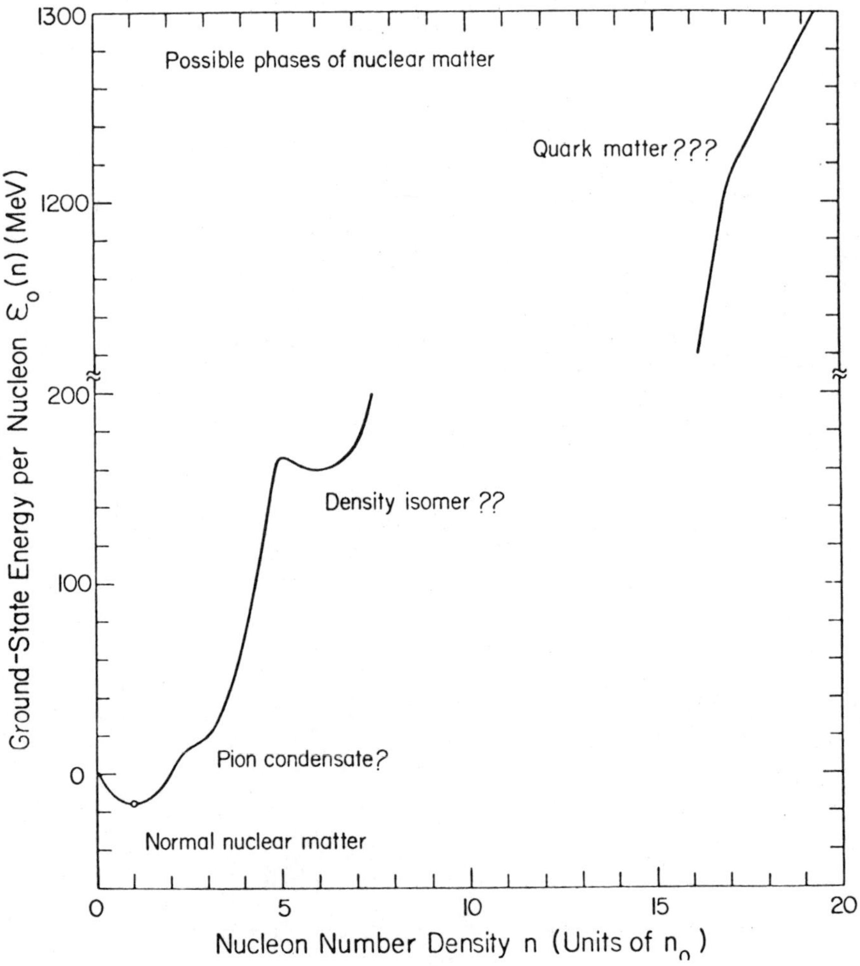

Figure 4.30

neutron stars. As an indication of some of the exciting possibilities ahead, Fig. 4.30 shows the anticipated equation of state. This equation, at densities above twice normal, can be affected by collective phase transitions to Lee-Wick abnormal matter,[442] density isomers or higher order transitions to a pion condensate,[337,341,443] the experimental signatures of which have recently been discussed.[444] In the absence of these effects the energy would simply increase monotonically with density. Since pressure in a hydrodynamic models is proportional to $dE/d\rho$, a change to negative slope above twice normal density would imply negative pressure, e.g. condensation to abnormal matter. The most favored possibility now is a transition to quark matter, in which these hypothetical constituents of strongly interacting particles (hadrons) would not be confined to individual nucleons but instead could move separately through the nucleus.[445] A possible signal for these new states of matter would be some unusual thermodynamic property of matter at high baryon density. One proposal (discussed by Glendenning in this Study) extends the speculations about hadron structure[358] to the heavy-ion domain, raising the possibility that dense matter might exhibit a limiting temperature $T \sim m_\pi \simeq 140$ MeV, as we discussed[7] at the beginning of Lecture 1, and which many have been observed in hadron collisions. It has been said[446] that "U+U collisions in the region of 4 GeV/A might produce important new phenomena, perhaps even practical applications. It should be noted that unlike hadron collisions these effects are not duplicated in accessible astronomical processes. They would be unlikely to occur except in gravitationally collapsing *objects*, or in the inverse process to the Big Bang. The lack of astronomical information means that we must depend on theoretical estimates to deduce the consequences of the stability of matter with supernormal density. Evidently this could be a potential energy source, since it could swallow up nucleons and disgorge energy, but equally evident is the possibility that the swallowing forces would be hard to control."

Whatever the theoretical speculations, the ultimate test will come from the experiments conducted on the present heavy-ion accelerators (see Fig. 4.31) although some of these studies call for yet another generation of accelerators, reaching energies of 20 to 100 GeV/nucleon, beyond even the range of the upgraded Bevalac. With the last statement, I must surely have covered at least Six Impossible Things and I shall stop!

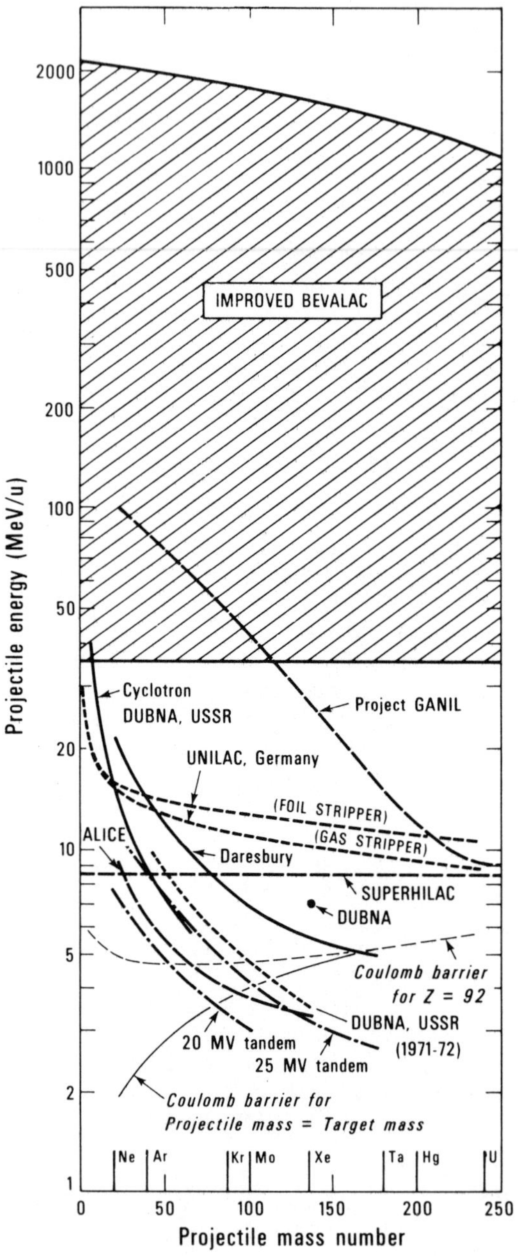
IMPROVED BEVALAC
Projectile energy (MeV/u)
2000
1000
500
200
100
50
20
10
5
2
1
Cyclotron
DUBNA, USSR
Project GANIL
UNILAC, Germany
(FOIL STRIPPER)
(GAS STRIPPER)
ALICE
Daresbury
SUPERHILAC
DUBNA
Coulomb barrier
for Z = 92
DUBNA, USSR
(1971-72)
20 MV tandem
25 MV tandem
Coulomb barrier for
Projectile mass = Target mass
Ne
Ar
Kr
Mo
Xe
Ta
Hg
U
0
50
100
150
200
250
Projectile mass number

Figure 4.31

4.4 Envoi

In these lectures I have attempted to given an overview of current activities in the different areas of nuclear reactions with heavy ions. My selection of material was guided to some extent by an attempt to show that the subjects of Microscopia, Macroscopia and Asymtotia are not separate and distinct. The rate of exploration and development on all three continents is truly remarkable, and dispenses with the criticisms of many "doubting Thomases" in the early days of heavy-ion research, who insisted that the processes would be so complicated as to defy even a qualitative understanding. Nor should we be deterred by the critics who insist that all the same phenomena can be studied more easily in hadron reactions. The fact is that they *were not so studied* until stimulated by heavy-ion research, and this is true of locating high spin states in nuclei or of forming nuclear fireballs. We have only to look at the quality of heavy-ion *data* and the sophistication of our present microscopic theories of multistep processes in deformed rare-earth nuclei, to wonder whether our tools would be of poorer quality without the advent of heavy ions.

My lectures must seem a little like a helicopter tour over the Continental Jungles. We have not flown very high (this is the task of other lecturers) but neither was your pilot skillful or knowledgable enough of the terrain to set down in the dense undergrowth. The metaphor of the Jungle is apt, because that is what experimental physics is like. Since this School is mainly a Theoretical Study we do well to recall Max Born's words[447] on the relationship of Experimental Theory in Physics. "I believe there is no philosophical high road in Science, with epistemological signposts. No, we are in a jungle and find our way by trial and by error, building our road behind us as we go. We do not find signposts at the crossroads, but our own scouts erect them to guide the rest. Theoretical ideas may be such signposts. The difficulty is that they often point in opposite directions: two theories each claiming to be built on "a priori" principles, but widely different and contradictory."

At the moment it is not clear where the many paths will lead in heavy-ion physics, but wherever, we can be assured that we have embarked on one of the voyages of the Century. The analogy is often made that research in heavy-ions is like looking for flowers among the weeds, and if any sign of flowers are evident in the weeds of these lectures, then it is only because I have merely "made up a bunch of other men's flowers and provided little of my own but the string to bind them".[448] Therefore as a tribute to the many people whose research I have used, without proper interpretation or acknowledgement, let me end with a description[449] of how the Jungle will look one day, as neat bordered lawns and flower beds.

"...Such gardens are not made,
By singing: 'Oh, how beautiful!' and
sitting in the shade,
While better men than we go out and,
start their working lives.
At grubbing weeds from gravel paths
with broken kitchen knives.
Oh, Adam was a gardener and God
who made him sees
That half a proper gardner's work is spent
upon his knees,
So when your work is finished you can
wash your hands and pray
For the Glory of the Garden, that it may
not pass away!
And the Glory of the Garden it shall never
pass away!

ACKNOWLEDGMENTS

I wish to thank the many workers in the field of heavy-ion physics who have written excellent reviews which helped me to prepare these lectures, and the many individuals mentioned in the text who sent me preprints on recent developments. Also, I owe an enormous debt to my immediate colleagues in Berkeley, M. Buenerd, M. Bini, P. Doll, C.K. Gelbke, D.L. Hendrie, J.L. Laville, J. Mahoney, G. Mantzouranis, A. Menchaca-Rocha, M.C. Mermax, C. Olmer, T.J.M. Symons, Y.P. Viyogi, K. Van Bibber, H. Wieman, P.J. Siemens, H. Faraggi, B.G. Harvey, F. Beiser, H. Crawford, D.E. Greiner, H.H. Heckman, P. Lindstrom, G.D. Westfall, C. McParland, S.A. Chessin, J.V. Geaga, J.Y. Grossiard and L.S. Schroder. In addition I have absorbed many ideas from the wider milieu of research and research workers at Berkeley. Finally it is a pleasure to thank: Shirley Ashley, Antoinette Czerwinski, Elaine Thayer, and Mike Baublitz of the Word Processing Dept; John Flambard and the staff of the Illustration Dept.; Marty Casazza, George Kagawa, Armuda Ludtke, Alma Pedersen, Steve Yarborough, Robert Patterson, Church Dees and Steve Adams of the Photographic Dept. They all performed superhuman efforts cheerfully to get these lectures prepared.

This work was supported by the Department of Energy.

References

The list of references is not scholarly either in its completeness, or in its attention to historical development. The references are illustrative, and were readily accessible or well known to the author at the time of writing the Lectures. A careful reading of them will nevertheless provide an excellent introduction to heavy-ion experiments!

1. Halton Arp, Atlas of Peculiar Galaxies, published by California Institute of Technology, 1966.

2. A. Toomre and J. Toomre, Astrophys. Journal 178, 623 (1972).

3. J.P. Vary, Second High Energy Heavy-Ion Summer Study (Berkeley, 1974), Lawrence Berkeley Laboratory Report LBL-3675, p. 371.

4. A.A. Amsden, F.H. Harlow and J.R. Nix, Phys. Rev. C15, 2059 (1977).

5. F. Seidl and A. Cameron, Astrophys. and Space Sci. 15, 44 (1972).

6. W. Scheid, J. Hofman and W. Greiner, Ref. 3, p. 1.

7. N.K. Glendenning and Y. Karant, Phys. Rev. Lett. 40, 374 (1978).

8. A.T. Laasanen, C. Ezell, L.J. Gutery, N.W. Schreiner, P. Schubelin, L. von Linden and F. Turkot, Phys. Rev. Lett. 38, 1 (1977).

9. S. Weinberg, The First Three Minutes, A Modern View of the Origin of the Universe (Basic Books, N.Y. 1977).

10. Proceedings of the International Conference on Nuclear Reactions Induced by Heavy Ions, editors R. Bock and W.R. Hering (North-Holland, Amsterdam, 1970).

11. Proceedings of the International Conference on Heavy-Ion Physics (Dubna, 1971), JINR Report D7-5769.

12. Symposium on Heavy-Ion Reactions and Many-Particle Excitations (Saclay, 1971), Colloque du Journal de Physique 32, C6 (1971).

13. Proceedings of the Symposium on Heavy-Ion Scattering (Argonne, 1971), Argonne Report ANL-7837.

14. European Conference on Nuclear Physics (Aix-en-Provence, 1972), Colloque du Journal de Physique 33, C5 (1972).

15. Proceedings of Heavy-Ion Summer Study (Oak Ridge, 1972), editor S.T. Thornton, USAEC Report CONF-720669.

16. Proceedings of the Heavy-Ion Transfer Reaction Symposium (Argonne, 1973), Argonne Report PHY-1973B.

17. Proceedings of the International Conference on Nuclear Physics, editors J. de Boer and J.H. Mang (North-Holland, Amsterdam, 1973).

18. Proceedings of the International Conference on Reactions between Complex Nuclei, editors R.L. Robinson, F.K. McGowan, J.B. Ball and J.H. Hamilton (North-Holland, Amsterdam, 1974).

19. Second High Energy Heavy-Ion Summer Study (Berkeley, 1974), Lawrence Berkeley Laboratory Report LBL-3675.

20. Proceedings of the Symposium on Classical and Quantum Mechanical Aspects of Heavy-Ion Collisions, editors H.L. Harney, P. Braun-Munzinger and C.K. Gelbke, Lecture Notes in Physics 33 (Springer, Berlin/Heidelberg/New York, 1975).

21. Proceedings of the INS-IPCR Symposium on Cluster Structure of Nuclei and Transfer Reactions Induced by Heavy Ions (eds. H. Kamitsubo, I. Kohno and T. Marumori) Tokyo, 1975, IPCR Progress Report Supplement 4.

22. Proceedings of the Symposium on Macroscopic Features of Heavy-Ion Collisions (Argonne, 1976), Argonne Report ANL/PHY-76-2.

23. Third Summer Study of High Energy Heavy Ions (Berkeley, 1976), Lawrence Berkeley Laboratory Report.

24. European Conference on Nuclear Physics with Heavy Ions (Caen, 1976), Colloque du Journal de Physique 37, C5 (1976).

25. Theoretical Aspects of Heavy-Ion Collisions (Falls Creek, Tennessee, 1977), ORNL Report 770602.

26. Proceedings of the International Symposium on Nuclear Collisions and Their Microscopic Description (Bled 1977), Fizika 9, Supplement 3 and 4 (1977).

27. Proceedings of the IPCR Symposium on Macroscopic Features of Heavy-Ion Collisions and Pre-Equilibrium Processes (Hakone, 1977), eds. H. Kamitsubo and M. Ishihara, IPCR Progress Report, Supplement 6.

28. Proceedings of the International Conference on Nuclear Structure, ed. T. Marumori (Tokyo, 1977). Supplement-6, J. Phys. Soc. Japan 44 (1978).

29. Heavy-Ion Collisions, ed. R. Bock (North Holland), to be published 1978.

30. D.K. Scott, Heavy Ion Experiments, Lectures presented in part at the Scottish Universities Summer School in Physics (St. Andrews, 1977) and the Latin American Summer School in Physics (Mexico City, 1977), Lawrence Berkeley Laboratory Preprint 7111 (1977) and to be published.

31. D.A. Bromley, Ref. 17, p. 22.

32. A.S. Goldhaber, Ref. 19, p. 382.

33. D.A. Goldberg, Symposium on Heavy-Ion Elastic Scattering (Rochester, N.Y., 1977).

34. L.C. Vaz, J.M. Alexander and E.H. Auerbach; L.C. Vaz and J.M. Alexander, to be published in Phys. Rev. C, 1978.

35. T. Koeling and R.A. Malfliet, Phys. Reports 22C, 181 (1975).

36. J. Knoll and R. Schaeffer, Ann. of Phys. 97, 307 (1976).

37. D.M. Brink, Lectures on Heavy-Ion Reactions (Orsay, March, 1972).

38. W.E. Frahn, Ref. 20, p. 102 and refs. therein; see also the series of papers I, II and III on the Generalized Fresnel Model for Heavy-Ion Scattering by W.E. Frahn, to be published.

39. J.S. Blair, Phys. Rev. 95, 1218 (1954).

40. J.B. Ball, C.B. Fulmer, E.E. Gross, M.L. Halbert, D.C. Hernley, C.A. Rudemann, M.J. Saltmarsh and G.R. Satchler, Nucl. Phys. A252, 208 (1975).

41. G.R. Satchler, Ref. 18, p. 171.

42. P.R. Christensen and A. Winther, Phys. Lett. 65B, 19 (1976).

43. C. Olmer, M. Mermaz, M. Buenerd, C.K. Gelbke, D.L. Hendrie, J. Mahoney, D.K. Scott, M.H. Macfarlane and S.C. Pieper, Lawrence Berkeley Laboratory Preprint LBL-6553, to be published in Phys. Rev. C.

44. D.A. Goldberg and S.M. Smith, Phys. Rev. Lett. 29, 500 (1972).

45. J.G. Cramer, R.M. DeVries, D.A. Goldberg, M.S. Zisman and C.F. Maguire, Phys. Rev. C14, 2158 (1976).

46. D.F. Jackson and R.C. Johnson, Phys. Lett. 49B, 249 (1974).

47. R.M. DeVries, D.A. Goldberg, J.W. Watson, M.S. Zisman and J.G. Cramer, Phys. Rev. Lett. 39, 450 (1977).

48. J. Wilczynski and K. Siwek-Wilczynska, Phys. Lett. 55B, 270 (1975); J. Wilczynski, Nucl. Phys. A216, 386 (1973).

49. W.D. Myers, Nucl. Phys. A204, 465 (1973).

50. J. Blocki, J. Randrup, W.J. Swiatecki and C.F. Tsang, Ann. Phys. 105, 427 (1977).

51. D.M. Brink, Ref. 24, p. C-47.

52. J.R. Birkelund and J.R. Huizenga, Phys. Rev. C17, 126 (1978).

53. R. Bass, Phys. Rev. Lett. 39, 265 (1977).

54. J.P. Vary and C.B. Dover, Phys. Rev. Lett. 31, 1510 (1973).

55. G.R. Satchler, Ref. 18, p. 171.

56. W.G. Love, Phys. Lett. 72B, 4 (1977).

57. F. Beck, K-H. Müller and H.S. Köhler, Phys. Rev. Lett. 40, 837 (1978) and refs. therein.

58. D.M. Brink and Fl. Stancu, Oxford University Preprint, 84/77.

59. M.C. Lemaire, Phys. Reports 7C, 280 (1973).

60. W.R. Phillips, Rep. Prog. in Phys. 40, 345 (1977).

61. A.J. Baltz and S. Kahana, Adv. Phys. 9, 1 (1977).

62. M.C. Lemaire, Proc. of Int. Conf. on Clustering Aspects of Nuclear Structure and Nuclear Reactions (Winnipeg, Manitoba, 1978).

63. H.T. Fortune, Ref. 28, p. 99.

64. D.H. Wilkinson, The Investigation of Nuclear Structure Problems at High Energy. Proceedings of the International School of Nuclear Physics (Erice, 1974), ed. H. Schopper (North Holland, 1975), p. 1.

65. A. Mechaca-Rocha, D. Phil. Thesis, Oxford University (1974).

66. D.M. Brink, Phys. Lett. 40B, 37 (1972).

67. N. Anyas-Weiss, J.C. Cornell, P.S. Fisher, P.N. Hudson, A. Mechaca-Rocha, D.J. Millener, A.D. Panagiotou, D.K. Scott, D. Strottman, D.M. Brink, B. Buck, P.J. Ellis and T. Engeland, Phys. Reports 12C, 201 (1974).

68. K.P. Artemov, V.Z. Goldberg, I.P. Petrov, V.P. Rudakov, I.N. Serikov and V.A. Timofeev, Sov. J. Nucl. Phys. 24, 1 (1976).

69. A. Menchaca-Rocha, M. Buenerd, A. Dacal, D.L. Hendrie, J. Mahoney, C. Olmer, M.E. Ortiz and D.K. Scott, to be published.

70. H. Hamm and K. Nagatani, Phys. Rev. C17, 586 (1978) and refs. therein.

71. B. Buck and A.A. Pilt, Nucl. Phys. A295, 1 (1978) and refs. therein.

72. R.M. DeVries, D. Shapiro, W.G. Davies, G.C. Ball, J.S. Forster and W. McLatchie, Phys. Rev. Lett. 35, 835 (1975).

73. W.G. Davies, R.M. DeVries, G.C. Ball, J.S. Forster, W. McLatchie, D. Shapira, J. Toke and R.E. Warner, Nucl. Phys. A279, 477 (1976).

74. R.M. DeVries, J.S. Lilley and M.A. Franey, Phys. Rev. Lett. 37, 481 (1976).

75. D.F. Jackson and M. Rhoades-Brown, Nucl. Phys. A286, 354 (1977).

76. R.M. DeVries, D. Shapira, M.R. Clover, H.E. Gove, J.D. Garrett and G. Sorensen, Phys. Lett. 67B, 19 (1977).

77. R.A. Broglia and P.F. Bortignon, Phys. Lett. 65B, 221 (1976).

78. D.K. Scott, B.G. Harvey, D.L. Hendrie, L. Kraus, C.F. Maguire, J. Mahoney, Y. Terrien and K. Yagi, Phys. Rev. Lett 33, 1343 (1974).

79. G.T. Hickey, D.C. Weisser, J. Cerny, G.M. Crawley, A.F. Zeller, T.R. Ophel and D.F. Hebbard, Phys. Rev. Lett. 37, 130 (1976).

80. G.C. Ball, W.G. Davies, J.S. Forster and H.R. Andrews, Phys. Lett. 60B, 265 (1976).

81. I. Paschopoulos, E. Müller, H.J. Körner, I.C. Gebrich, K.E. Rehm and J. Scheerer, Munich Preprint (1978), to be published in Phys. Rev. C.

82. X. Campi, H. Flocard, A.K. Kerman and S. Koonin, Nucl. Phys. A251, 193 (1975).

83. K.R. Greider, Ref. 10, p. 217.

84. One of the most recent techniques for evaluating this integral is given by D.H. Glockner, M.H. MacFarlane and S.C. Pieper, Argonne National Laboratory Report ANL-76-11 (1976).

85. P.D. Bond, C. Chasman, J.D. Garrett, C.K. Gelbke, O. Hansen, M.J. Levine, A.Z. Schwartzchild and C.E. Thorn, Phys. Rev. Lett. 36, 300 (1976).

86. R.C. Fuller, Phys. Lett. 69B, 267 (1977).

87. E.A. Seglie and R.J. Asciutto, Phys. Rev. Lett. 39, 688 (1977).

88. G. Delic, K. Pruess, L.A. Charleton and N.K. Glendenning, Phys. Rev. Lett. 69B, 20 (1977).

89. For a recent summary, see. K.S. Low, Ref. 24, pC5-15.

90. K.S. Low, T. Tamura and T. Udagawa, Phys. Lett. 67B, 59 (1977).

91. P.D. Bond, M.J. Levine and C.E. Thorn, Phys Lett. 68B, 327 (1977).

92. K.E. Rehm, W. Henning, J.R. Erskine and D.G. Kovar, to be published in Phys. Rev. Lett. 40, 1479 (1978).

93. N.K. Glendenning, Rev. Mod. Phys. 47, 659 (1975).

94. D.K. Scott, B.G. Harvey, D.L. Hendrie, U. Jahnke, L. Kraus, C.F. Maguire, J. Mahoney, Y. Terrien, K. Yagi and N.K. Glendenning, Phys. Rev. Lett. 34, 895 (1975).

95. P.D. Bond, H.J. Körner, M.C. Lemaire, D.J. Pisano and C.E. Thorn, Phys. Rev. C16, 177 (1977).

96. J.C. Peng, M.C. Mermaz, A. Greiner, N. Lisbona and K.S. Low, Phys. Rev. C15 1331 (1977).

97. M.C. Lemaire and K.S. Low, Phys. Rev. C16, 183 (1977).

98. K.A. Erb, D.L. Hanson, R.J. Ascuitto, B. Sorensen, J.S. Vaagen and J.J. Kolata, Phys. Rev. Lett. 33, 1102 (1974).

99. C.F. Maguire, D.L. Hendrie, U. Jahnke, J. Mahoney, D.K. Scott, J.S. Vaagen, R.J. Ascuitto and K. Kumar, Phys. Rev. Lett. 40, 358 (1978).

100. B. Sorensen, Phys. Lett. 66B, 119 (1977).

101. K.W. McVoy, Proceedings of the Oaxtepec Symposium on Nuclear Physics (Mexico 1978); B.V. Carlson and K.W. McVoy, Nucl. Phys. A292, 310 (1977).

102. A.J. Baltz and S. Kahana, Phys. Rev. C17, 555 (1978).

103. T. Tamura, Phys. Reports 14C, 61 (1974).

104. For a review of experimental techniques, see P. Armbruster, Ref. 24, p. 161; the forthcoming issue of Nuc. Inst. and Meth. on Spectrometer Detection Systems, ed. by D.A. Bromley; and B. Engeland, Lectures at Scottish Universities Summer School in Physics (St. Andrews, 1978).

105. C. Olmer, M.C.Mermaz, M. Buenerd, C.K. Gelbke, D.L. Hendrie, J. Mahoney, A. Menchaca-Rocha, D.K. Scott, M.H. Macfarlane and S.C. Pieper, Phys. Rev. Lett. 38, 476 (1977) and refs. therein.

106. T.A. Belote, N. Anyas-Weiss, J.A. Becker, J.C. Cornell, P.S. Fisher, P.N. Hudson, A. Menchaca-Rocha, A.D. Panagiotou and D.K. Scott, Phys. Rev. Lett. 30, 450 (1973).

107. H.V. Klapdor, H. Reiss and G. Rosner, Eighth Summer School on Nuclear Physics (Mikolajki, Poland, 1975), Nukleonika 21, 763 (1976).

108. H.V. Klapdor, G. Rosner and H. Willmes, International Workshop V on Gross Properties of Nuclei and Nuclear Excitations (Hirschegg, Austria, 1977).

109. N. Marquardt, J. L'Ecuyer, C. Cardinal, R. Volders, and M.W. Greene, Ref. 17, Vol. 1, p. 476.

110. H.V. Klapdor, H. Reiss, G. Rosner and M. Schrader, Phys. Lett. 49B, 431 (1974).

111. M. Böhning, Ref. 10, p. 633; R.G. Stokstad, Ref. 18, p. 327.

112. H.V. Klapdor, H. Reiss and G. Rosner, Phys. Lett. 58B, 279 (1975).

113. D.A. Bromley, Second International Conference on Clustering Phenomena in Nuclei (Maryland, 1975); International Conference on the Resonances in Heavy Ion Reactions (Hvar, Yugoslavia, 1977).

114. H. Feshbach, Ref. 24 p. C5-177; Europhysics Conference on Medium Mass Nuclei (September, 1977).

115. R.H. Siemssen, Ref. 21, p. 233.

116. M.L. Halbert, C.B. Fulmer, S. Raman, M.J. Saltmarsh, A.H. Snell and P.H. Stelson, Phys. Lett 51B, 341 (1974).

117. A. Gobbi, R. Wieland, L. Chua, D. Shapira and D.A. Bromley, Phys. Rev. C7, 30 (1973).

118. R. Vandenbosch, M.P. Webb and M.S. Zisman, Phys. Rev. Lett. 33, 842 (1974).

119. K.S. Low and T. Tamura, Phys. Lett. 40B, 32 (1972).

120. E.R. Cosman, T.M. Cormier, K. van Bibber, A. Sperduto, G. Young, J. Erskine, L.R. Greenwood and O. Hansen, Phys. Rev. Lett. 35, 265 (1975).

121. A.J. Lazzarini, E.R. Cosman, A. Sperduto, S.G. Steadman, W. Thoms and G.R. Young, to be published in Phys. Rev. Lett. (1978); A.H. Lumpkin, G.R. Morgan, J.D. Fox and K.W. Kemper, Phys. Rev. Lett. 40, 104 (1978).

122. A. Arima, G. Scharff-Goldhaber and K.W. McVoy, Phys. Lett. 40B, 8 (1972).

123. K. Ikeda, Ref. 21, p. 23.

124. N. Cindro, F. Coçu, J. Uzureau, Z. Basrak, M. Cates, J.M. Fieni, E. Holub, Y. Patin and S. Plattard, Phys. Rev. Lett. 39, 1135 (1977).

125. B. Imanishi, Nucl. Phys. A125, 33 (1969).

126. H.J. Fink, W. Scheid and W. Greiner, Nucl. Phys. A188, 259 (1972).

127. J. Eisenberg and W. Greiner, Nuclear Theory (North-Holland, 1975) Vol. 1, p. 149.

128. T.M. Cormier, J. Applegate, G.M. Berkowitz, P. Braun-Munzinger, P. Cormier, J.W. Harris, C.M. Jachcinski, L.L. Lee, J. Barrette and H.E. Wegner, Phys. Rev. Lett. 38, 940 (1977).

129. T.M. Cormier, C.M. Jachcinski, G.M. Berkowitz, P. Braun-Munzinger, P.M. Cormier, M. Gai, J.W. Harris, J. Barrette and H.E. Wegner, Phys. Rev. Lett. 40, 924 (1978).

130. C.M. Jachcinski, T.M. Cormier, P. Braun-Munzinger, G.M. Berkowitz, P.M. Cormier, M. Gai and J.W. Harris, Phys. Rev. C17, 1263 (1978).

131. D. Shapira, R.M. DeVries, M.R. Clover, R.N. Boyd and R.N. Cherry, Phys. Rev. Lett. 40, 371 (1978).

132. M. Golin, Phys. Lett. 74B, 23 (1978).

133. P. Braun-Munzinger, G.M. Berkowitz, T.M. Cormier, C.M. Jachcinski, J.W. Harris, J. Barrette and M.J. Levine, Phys. Rev. Lett. 38,944 (1977).

134. K.W. McVoy, Ref. 20, p. 127; Phys. Rev. C3,1104 (1971).

135. J. Barrette, M.J. Levine, P. Braun-Munzinger, G.M. Berkowitz, M. Gai, J.W. Harris and C.M. Jachcinski, Phys. Rev. Lett. 40, 445 (1978).

136. M.R. Clover, R.M. DeVries, R. Ost, N.J.A. Rust, R.N. Cherry and H.E. Gove, Phys. Rev. Lett. 40, 1008 (1978).

137. K. McVoy, Proceedings of the Symposium on Heavy Ion Elastic Scattering (Rochester, 1977).

137(a). D.M. Brink and N. Takigawa, Nucl. Phys. A279, 159 (1977); S.Y. Lee, N. Takigawa and C. Marty, IPNO/TH 77-19 Preprint (1977).

137(b). D. Dennard, V. Schkolnik and M-A Franey, Phys. Rev. Lett. 40, 1549 (1978).

138. M. Paul, S.J. Sanders, J. Cseh, D.F. Geesaman, W. Henning, D.G. Kovar, C. Olmer and J.P. Schiffer, Phys. Rev. Lett. 40, 1310 (1978).

139. For some recent discussions of this field, see J.L.C. Ford, First Oaxtepec Symposium on Nuclear Physics (Mexico, 1978).

140. D.L. Hillis, E.E. Gross, D.C. Hensley, C.R. Bingham, F.T. Baker and A. Scott, Phys. Rev. C16, 1467 (1977).

141. M.W. Guidry, P.A. Butler, R. Donangelo, E. Grosse, Y. El Masri, I.Y. Lee, F.S. Stephens, R.M. Diamond, L.L. Riedinger, C.R. Bingham, A.C. Kahler, J.A. Vrba, E.L. Robinson and N.R. Johnson, Phys. Rev. Lett. 40, 1016 (1978).

142. R. Donangelo, L.F. Oliveira, J.O. Rasmussen and M.W. Guidry, Lawrence Berkeley Laboratory Preprint, LBL-7720 (1978); M.W. Guidry et. al, Nucl. Phys. A295, 482 (1978).

143. J.R. Birkelund, J.R. Huizenga, H. Freiesleben, K.L. Wolf, J.P. Unik and V.E. Viola, Phys. Rev. C13, 133 (1976).

144. I.Y. Lee, D. Cline, P.A. Butler, R.M. Diamond, J.O. Newton, R.S. Simon, and F.S. Stephens, Phys. Rev. Lett. 39, 684 (1977).

145. K. Kumar and M. Baranger, Nucl. Phys. A122, 273 (1968).

146. W.G. Love, T. Teresawa and G.R. Satchler, Nucl. Phys. A291, 183 (1977); A.J. Baltz, S.K. Kaufmann, N.K. Glendenning and K. Pruess, Phys. Rev. Lett. 40, 20 (1978).

147. P. Doll, M. Bini, D.L. Hendrie, S.K. Kauffmann, J. Mahoney, A. Menchaca-Rocha, D.K. Scott, T.J.M. Symons, K. Van Bibber, Y.P. Viyogi, H. Wieman and A.J. Baltz, Lawrence Berkeley Laboratory Preprint LBL-7195 (1978).

148. A. Johnson, H. Ryde and J. Sztarkier, Phys. Lett. 34B, 605 (1971).

149. For a recent review, see A. Faessler, Ref. 18, p. 437; D. Ward, Ref. 18, p. 417.

150. F.S. Stephens, Comments on Nuclei and Particles VI, 173 (1976); International School of Physics (Varenna, Italy, 1976).

151. M.A.J. Mariscotti, G. Scharff-Goldhaber and B. Buck, Phys. Rev. 178, 1864 (1969).

152. B.R. Mottelson and J.G. Valatin, Phys. Rev. Lett. 5, 511 (1960).

153. P. Thieberger, Phys. Lett. 45B, 417 (1973).

154. F.S. Stephens, Rev. Mod. Phys. 47, 43 (1975).

155. I.Y. Lee, M.M. Aleonard, M.A. Deleplanque, Y. El Masri, J.O. Newton, R.S. Simon, R.M. Diamond and F.S. Stephens, Phys. Rev. Lett. 38, 1454 (1977).

156. S.M. Harris and P.J. Evans, Phys. Rev. Lett. 39, 1186 (1977).

157. L.K. Peker and J.H. Hamilton, Phys. Rev. Lett. 40, 744 (1978).

158. I.Y. Lee, D. Cline, R.S. Simon, P.A. Butler, P. Colombani, M.W. Guidry, F.S. Stepehns, R.M. Diamond, N.R. Johnson and E. Eichler, Phys. Rev. Lett. 37, 420 (1976).

159. N.R. Johnson, D. Cline, S.W. Yates, F.S. Stephens, L.L. Riedinger and R.M. Ronningen, Phys. Rev. Lett. 40, 151 (1978).

160. A. Bohr and B.R. Mottelson, Phys. Scripta 10A, 13 (1974).

161. R.M. Diamond, Australian J. of Phys., to be published.

162. A. Bohr and B.R. Mottelson, Nuc. Structure (Benjamin N.Y.). Vol. II.

163. B.R. Mottelson, Proceedings of the Nuclear Structure Symposium of the Thousand Lakes (Joutsa, Finland, 1970), Vol. II, p. 148.

164. F.S. Stephens, Conference on Highly Excited States of Nuclei (Jülich, Germany 1975).

165. R.S. Simon, M.V. Banaschik, R.M. Diamond, J.O. Newton and F.S. Stephens, Nucl. Phys. A290, 253 (1977).

166. P.J. Tjøm, F.S. Stephens, R.M. Diamond, J. de Boer and W.E. Meyerhof, Phys. Rev. Lett. 33, 593 (1974).

167. R.S. Simon, M.V. Banaschik, P. Colombani, D.P. Soroka, F.S. Stephens and R.M. Diamond, Phys. Rev. Lett. 33, 596 (1974).

168. J.O. Newton, S.H. Sie and G.D. Dracoulis, Phys. Rev. Lett. 40, 625 (1978).

169. M.A. Deleplanque, I.Y. Lee, F.S. Stephens, R.M. Diamond and M.M. Aleonard, Phys. Rev. Lett. 40, 629 (1978).

170. P.O. Tjøm, I. Espe, G.B. Hagemann, B. Herskind and D.L. Hillis, Phys. Lett. 72B, 439 (1978).

171. G. Anderson, S.E. Larsson, G. Leander, P. Möller, S.G. Nilsson, I. Ragnarsson, S. Aberg, R. Bengtson, J. Dudek, B. Nerlo-Pomarska, K. Pomorski and Z. Szymanski, Nucl. Phys. A268, 205 (1976).

172. J. Pedersen, B.B. Back, F.M. Bernthal, S. Bjørnholm. J. Borggreen, O. Christensen, F. Folkmann, B. Herskind, T.L. Khoo, M. Neiman, F. Pühlhoffer and G. Sletten, Phys. Rev. Lett. 39, 990 (1977).

173. T. Døssing, K. Neergaard, K. Matsuyanagi and Hsi-Chen Chang, Phys. Rev. Lett. 39, 1395 (1977).

174. A. Faessler and M. Ploszjcak, Phys. Rev. C16, 2032 (1977).

175. M. Hillman and Y. Eyal, Ref. 24, p. 109.

176. F. Pühlhofer, Nucl. Phys. A280, 267 (1977).

177. J. Barrette, P. Braun-Munzinger, C.K. Gelbke, H.L. Harney, H.E. Wegner, B. Zeidman, K.D. Hildenbrand, and U. Lynen, to be published in Nucl. Phys. A (1978).

178. V.S. Ramamurthy, S.S. Kapoor and S.K. Kataria, Phys. Rev. Lett. 25, 386 (1970).

179. F. Pühlhofer, W.F.W. Schneider, F. Busch, J. Barrette, P. Braun-Munzinger, C.K. Gelbke and H.E. Wegner, Phys. Rev. C16, 1010 (1977).

180. T.M. Cormier, E.R. Cosnan, A.J. Lazzarini, H.E. Wegner, J.D. Garrett and F. Pühlhofer, Phys. Rev. C15, 654 (1977).

181. G. Sauer, H. Chandra and U. Mosel, Nucl. Phys. A264, 221 (1976).

182. S. Cohen, F. Plasil and W.J. S atecki, Ann. of Phys. 82, 557 (1974).

183. J. Gomez del Campo, M.E. Ortiz, A. Dacal, J.L.C. Ford, R.L. Robinson, P.H. Stelson and S.T. Thornton, Nucl. Phys. A262, 125 (1976).

184. J. Gomez del Campo, J.L.C. Ford, R.L. Robinson, M.E. Ortiz, A. Dacal and E. Andrade, Nucl. Phys. A297, 125 (1978).

185. U. Mosel and D. Glas, to be published.

186. W.J. Swiatecki, Ref. 14, p. C4-45.

187. W.J. Swiatecki and S. Björnholm, Phys. Reports 4C, 325 (1972).

188. F. Plasil, Phys. Rev. C17, 823 (1978).

189. F. Plasil, R.L. Ferguson, H.C. Britt, B.H. Erkkila, P.D. Goldstone, R.H. Stokes and H.H. Gutbrod, Oak Ridge Preprint (1978).

190. M. Lefort, Ref. 20, p. 274; Reports on Prog. in Phys. 39, 129 (1976).

191. M. Lefort, Ref. 24, p. C5-57.

192. M. Lefort, J. Phys. A7, 107 (1974).

193. J. Galin, D. Guerreau, M. Lefort and X. Tarrago, Phys. Rev. C9, 1018 (1974).

194. D. Glas and U. Mosel, Nucl. Phys. A237, 429 (1975).

195. W.D. Myers, Nucl. Phys. A 204, 465 (1973); Ref. 18, p. 1.

196. S. Vigdor, Ref. 22, p. 95.

197. R.G. Stokstad, Ref. 25.

198. R.G. Stokstad, R.A. Dayras, J. Gomez del Campo, P.H. Stelson, C. Olmer and M.S. Zisman, Phys. Lett. 70B, 289 (1977).

199. B. Kohlmeyer, W. Pfeffer and F. Pühlhoffer, Nucl. Phys. A292, 288 (1977).

200. J. Natowitz, Ref. 25.

201. E. Seglie, D. Sperber and A. Sherman, Phys. Rev. C11, 1227 (1975).

202. R. Bass, Phys. Lett. 47B, 139 (1973); Nucl. Phys. A231, 45 (1974).

203. R. Beck and D.H.E. Gross, Phys. Lett. 47B, 143 (1973).

204. D.H.E. Gross and H. Kalinowski, Phys. Lett. 48B, 302 (1974).

205. J.P. Bondorf, M.I. Sobel and D. Sperber, Phys. Reports C15, 83 (1974).

206. K. Siwek-Wilczynska and J. Wilczynski, Nucl. Phys. A264, 115 (1976); Nukleonika 21, 517 (1976).

207. J.R. Birkelund, J.R. Hu J.N. De and D. Sperber, Phys. Rev. Lett. 40, 1123 (1978).

208. B.B. Back, R.R. Betts, C. Gaarde, J.S. Larsen, E. Michelsen and Tai Kuang-Hsi, Nucl. Phys. A285, 317 (1977).

209. H. Doubre, A. Gamp, J.C. Jacmart, N. Poffé, J.C. Roynette and J. Wilczynski, Phys. Lett. 73B, 135 (1978).

210. D. Glas and U. Mosel, Phys. Lett. 49B, 301 (1974); Nucl. Phys. A264, 268 (1976).

211. D.J. Morrissey, W. Loveland, R.J. Otto and G.T. Seaborg, Phys. Lett. 74B, 35 (1978).

212. D. Kovar, Ref. 27, p. 18.

213. S. Harar, Colloque Franco-Japonais de Spectroscopie Nucléaire, Dogashima, Japan (Univ. of Tokyo, 1977). p. 191.

214. P. Sperr, T.H. Braid, Y. Eisen, D.G. Kovar, F.W. Prosser, J.P. Schiffer, S.L. Tabor and S. Vigdor, Phys. Rev. Lett. 37, 321 (1976).

215. R.M. Freeman and F. Hass, Phys. Rev. Lett. 40, 927 (1978).

216. R.A. Broglia, C.H. Dasso, G. Pollarolo and A. Winther, Phys. Rev. Lett. 40, 707 (1978).

217. P.A.M. Dirac, Proc. Camb. Phil. Soc. 26, 376 (1930).

218. S.E. Koonin, K.T.R. Davies, V. Maruhn-Rezwani, H. Feldmeier, S.J. Krieger and J.W. Negele, Phys. Rev. C15, 1359 (1977).

219. See also detailed reviews by A.K. Kerman, Proceedings of the Enrico Fermi Summer School (Varenna, Italy, 1977) and Ref. 28, p. 711; P. Bonche, Ref. 24, p. C5-213; J.W. Negele, Ref. 25, and this school.

220. G.F. Bertsch and S.F. Tsai, Phys. Reports 18C (1975).

221. P. Bonche, B. Grammaticos and S. Koonin, Saclay Preprint DPh-T/DOC/77/128, to be published in Phys. Rev. C (1978).

222. H. Gauvin, R.L. Hahn, Y. Le Beyec and M. Lefort, Phys. Rev. C10, 722 (1974).

223. C. Cabot, H. Gauvin, Y. Le Beyec and M. Lefort. J. de. Phys. L36, 289 (1976).

224. S. Della-Negra, H. Gauvin, H. Jungclas, Y. Le Beyec and M. Lefort, Z. Phys. A282, 65 (1977).

225. H.C. Britt, B.H. Erkkila, P.D. Goldstone, R.H. Stokes, B.B. Back, F. Folkmann, O. Christensen, B. Fernandez, J.D. Garrett, G.B. Hagemann, B. Herskind, D.L. Hillis, F. Plasil, R.L. Ferguson, M. Blann and H.H. Gutbrod, Phys. Rev. Lett. 39, 1458 (1977).

226. L. Kowalski, J.M. Alexander, D. Logan, M. Rajagopalan, M. Kaplan, M.S. Zisman and T.W. Debiak, Stony Brook Preprint, 1978.

227. T. Inamura, M. Ishihara, T. Fukuda, T. Shimoda and H. Huruta, Phys. Lett. 68B, 51 (1977).

228. D. Horn, H.A. Enge, A. Sperduto and A. Grane, Phys. Rev. C17, 118 (1978).

229. J.P. Schiffer, Ref. 17, p. 813.

230. J.M. Nitschke, Symposium on Superheavy Elements (Lubbock, Texas, 1978), Lawrence Berkeley Laboratory Preprint LBL-7705 (1978).

231. W. Reisdorf and P. Armbruster, International Meeting on Reactions of Heavy Ions with Nuclei (Dubna, 1977). GSI Preprint-Bericht-M-2-78.

232. R.V. Gentry, T.A. Cahill, N.R. Fletcher, H.C. Kaufmann, L.R. Medsker, J.W. Nelson and R. Flocchini, Phys. Rev. Lett. 38, 479 (1977).

233. C.J. Sparks, S. Raman, E. Ricci, R.V. Gentry and M.O. Krause, Phys. Rev. Lett. 40, 507 (1978).

234. W.J. Swiatecki and C.F. Tsang, Lawrence Berkeley Lab Preprint LBL-666 (1971); R. Kalpakchieva, Yu. Ts. Oganessian, Yu. E. Penionzhkevich, H. Sodan and B.A. Gvozdev, Phys. Lett. 69B, 287 (1977).

235. E.K. Hulet, R.W. Lougheed, J.F. Wild, J.H. Landrum, P.G. Stevenson, A. Ghiorso, J.M. Nitschke, R.J. Otto, D.J. Morrissey, P.A. Baisden, B.F. Gavin, D. Lee, R.J. Silva, M.M. Fowler, and G.T. Seaborg, Phys. Rev. Lett. 39, 385 (1977).

236. Yu. Ts. Oganessian, H. Bruchertseifer, G.V. Buklanov, V.I. Chepigin, Choi Val Sak, B. Eichler, K.A. Gavrilov, H. Gaeggeler, Yu. S. Korotin, O.A. Orlova, T. Reetz, W. Seidl, G.M. Ter-Akopian, S.P. Tretyakova, and I. Zvara, Nucl. Phys. A294, 213 (1978).

237. H. Sann, A. Olni, Y. Cirelekoglu, D. Pelte, U. Lynen, H. Stelzer, A. Gobbi, Y. Eyal, W. Kohl, R. Renfordt, I. Rode, G. Rudlof, D. Schwalm and R. Bock, Ref. 25, p. 281; K.D. Hildenbrand, H. Freiesleben, F. Pühlhofer, W.F.W. Schneider, R. Bock, D.V. Harrach and H.J. Specht, Phys. Rev. Lett. 39, 1065 (1977).

238. G. Wolschin and W. Norenberg, Z. Phys. A284, 209 (1978).

239. V.F. Weisskopf, Phys. Today 14, 18 (1960).

240. H. Feshbach, Rev. Mod. Phy. 46, 1 (1974) and refs. therein.

241. One of the earliest experiments related to this phenomenon is that of K. Kaufman and W. Wolfgang, Phys. Rev. 121, 192 (1961). Other early works include: G.F. Gridnev, V.V. Volkov and J. Wilczynski, Nucl. Phys. A142, 385 (1970); J. Galin, D. Guerreau, M. Lefort, J. Péter and X. Tarrago, Nucl. Phys. A159, 461 (1970).

242. For a recent review, see W.U. Schröder and J.R. Huizenga, Ann. Rev. Nucl. Sci. 27, 465 (1977).

243. A.G. Artukh, G.F. Gridnev, V.L. Mikheev, V.V. Volkov and J. Wilczynski, Nucl. Phys. A211, 299 (1973); A215, 91 (1973).

244. J. Wilczynski, Phys. Lett. B47, 484 (1973).

245. H.H. Deubler and K. Dietrich, Phys. Lett. 56B, 241 (1975).

246. M. Berlanger, P. Grangé, H. Hofmann, C. Ngo and J. Richert, Phys. Rev. C17, 1495 (1978).

247. J. Galin, Ref. 24, p. C5-83.

248. G.J. Mathews, G.J. Wozniak, R.P. Schmitt and L.G. Moretto, Z. Phys. A283, 247 (1977).

249. V.V. Volkov, Ref. 18, p. 363 and V.V. Volkov, Sov. J. Nucl. Phys. 6, 420 (1976).

250. V.V. Volkov, Ref. 20, p. 253.

251. L.G. Moretto and R. Schmitt, Ref. 24, p. C5-109.

252. J. Galin, Ref. 24, p. C5-83.

253. W. Norenberg; Ref. 24, p. C5-141.

254. H.A. Weidenmuller, Ref. 25.

255. R. Babinet, B. Cauvin, J. Girard, H. Nifenecker, B. Gatty, D. Guerreau, M. Lefort and X. Tarrago, Nucl. Phys. A296, 160 (1978).

256. P.D. Bond, Phys. Rev. Lett. 40, 501 (1978).

257. K. Sugimoto, N. Takahashi, A. Mizobuchi, Y. Nojiri, T. Minamisono, M. Ishihara, K. Tanaka and H. Kamitsubo, Phys. Rev. Lett. 39, 323 (1977).

258. M. Ishihara, K. Tanaka, T. Kammuri, K. Matsuoka and M. Sano, Phys. Lett. 73B, 281 (1978).

259. W. Trautmann. J. de Boer, W. Dünnweber, G. Graw, R. Kopp, C. Lauterbach, H. Puchta and U. Lynen, Phys. Rev. Lett. 39, 1062 (1977).

260. F. Plasil, R.L. Ferguson, H.C. Britt, R.H. Stokes, B.H. Erkkila, P.D. Goldstone, M. Blann and H.H. Gutbrod, Phys. Rev. Lett. 40, 1164 (1978).

261. H. Feshbach, First Oaxtepec Symposium on Nuclear Physics (Oaxtepec, Mexico, 1978).

262. H. Feshbach, "Statistical Multistep Reactions," Proc. of Conf. on Nuc. React. Mechanisms (Varenna, Italy, 1977).

263. H. Oeschlev, J.P. Coffin, P. Engelstein, A. Gallmann, K.S. Sim and P. Wagner, Phys. Lett. 71B, 63 (1977).

264. T. Tamura and T. Udagawa, Phys. Lett. 71B, 273 (1977).

265. S. Tsai and G.F. Bertsch, Phys. Lett. 73B, 247 (1978).

266. T. Udagawa, Ref. 28, p. 667.

267. G. Bertsch and S.F. Tsai, Phys. Reports 18C, 126 (1975).

268. R.A. Broglia, C.H. Dasso and A. Winther, Phys. Lett. 61B, 113 (1976); R.A. Broglia, O. Civitarese, C.H. Dasso and A. Winther, Phys. Lett. 73B, 405 (1978). N. Takigawa, University Of Münster Preprint, 1978.

269. K.F. Liu and G.E. Brown, Nucl. Phys. A265, 385 (1976).

270. R.R. Betts, S.B. DiCenzo, M.H. Mortensen and R.L. White, Phys. Rev. Lett. 39, 1183 (1977).

271. D. Ashery, M.S. Zisman, R.B. Weisenmiller, A. Guterman, D.K. Scott and C. Maguire, Lawrence Berkeley Laboratory Annual Report (1975), p. 97; A. Guterman, D. Ashery, J. Alster, D.K. Scott, M.S. Zisman, C.K. Gelbke, H.H. Wieman and D.L. Hendrie, to be published.

272. P. Doll, D.L. Hendrie, J. Mahoney, A. Menchaca-Rocha, D.K. Scott, T.J.M. Symons, K. Van Bibber, Y.P. Viyogi and H.H. Wieman, to be published.

273. M. Buenerd, D. Lebrun, J. Chauvin, Y. Gaillard, J.M. Loiseaux, P. Martin, G. Perrin and P. de Saintignon, Phys. Rev. Lett. 40, 1482 (1978).

274. R.R. Betts, Private communication, 1978.

275. J.P. Bondorf, Lectures at International School of Physics (Enrico Fermi), 1974.

276. J.P. Bondorf, Ref. 18, p. 383.

277. W. Nörenberg, Phys. Lett. 52B, 289 (1974).

278. G. Wolschin and W. Nörenberg, Z. Phys. A284, 209 (1978) and refs. therein.

279. J.R. Huizenga, Ref. 25.

280. J.R. Huizenga, J.R. Birkelund, W.U. Schröder, K.L. Wolf and V.E. Viola, Phys. Rev. Lett. 37, 885 (1976).

281. For a fuller discussion, see M. Lefort, Symposium on New Avenues in Nuclear Physics (Rehovot, Israel, 1976).

282. G. Wolschin and W. Nörenberg, Z. Phys. A284, 209 (1978) and refs. therein.

283. R. Beck and D.H.E. Gross, Phys. Lett. B47, 143 (1973).

284. J.P. Bondorf, M.I. Sobel, and D. Sperber, Phys. Rev. C15, 83 (1974).

285. D.H.E. Gross, Nucl. Phys. A240, 472 (1975).

286. W.U. Schröder, J.R. Birkelund, J.R. Huizenga, K.L. Wolf and V.E. Viola, Phys. Rev. C16, 623 (1977).

287. J.P. Bondorf, J.R. Huizenga, M.I. Sobel and D. Sperber, Phys. Rev. C11, 1265 (1975).

288. J.R. Huizenga, Nukleonika 20, 291 (1975).

289. J.N. De and D. Sperber, Phys. Lett 72B, 293 (1978).

290. B. Sinha, Phys. Lett. 71B, 243 (1977).

291. W.U. Schröder, J.R. Huizenga, J.R. Birkelund, K.L. Wolf and V.E. Viola, Phys. Lett 71B, 283 (1977).

292. W.U. Schroder, J.R. Birkelund, J.R. Huizenga, K.L. Wolf and V.E. Viola, University of Rochester Preprint UR-NSRL-171 (1978), to be published in Physics Reports.

293. W.J. Swiatecki, J. de Phys. 33, 451 (1972).

294. J. Randrup, NORDITA Preprint. 77-45 (1977) to be published.

295. D. Agassi, C.M. Ko and H.A. Weidenmuller, Reports MPIH-1976-V25 and V5 (Heidelberg, 1976, 1977); Ann. of Phys. 107, 140 (1977).

296. J.S. Sventek and L.G. Moretto, Phys. Rev. Lett. 40, 697 (1978).

297. K. Albrecht and W. Stocker, Nucl. Phys. A278, 95 (1977).

298. C.F. Tsang, Phys. Scripta 10A, 90 (1974).

299. J. Blocki, J. Randrup, W.J. Swiatecki and C.F. Tsang, Ann. Phys. 105, 427 (1977).

300. J. Randrup, Lawrence Berkeley Laboratory Preprint LBL-4317 (1975) and NORDITA Preprint-78/9.

301. R. Eggers, M.N. Namboodiri, P. Gonthier, K. Geoffroy and J.B. Natowitz, Phys. Rev. Lett. 37, 324 (1976).

302. P. Braun-Munzinger, T.M. Cormier and C.K. Gelbke, Phys. Rev. Lett. 37, 1582 (1976).

303. R.R. Betts and S.B. DiCenzo, Yale University Preprint-3074-390 (1978).

304. M. Ishihara, T. Numao, T. Fukada, K. Tanaka and T. Inamura, Ref. 22, p. 617.

305. P. Glassel, R.S. Simon, R.M. Diamond, R.C. Jared, I.Y. Lee, L.G. Moretto, J.O. Newton, R. Schmitt and F.S. Stephens, Phys. Rev. Lett. 38, 331 (1977).

306. M. Berlanger, M.A. Deleplanque, C. Gerschel, F. Hanappe, M. Leblanc, J.F. Mayault, C. Ngô, D. Paya, N. Perrin, J. Peter, B. Tamain and L. Valentin, J. Phys. Lettres 37, L323 (1976).

307. M.M. Aleonard, G.J. Wozniak, P. Glassel, M.A. Deleplanque, R.M. Diamond, L.G. Moretto, R.P. Schmitt and F.S. Stephens, Phys. Rev. Lett. 40, 622 (1978).

308. P.R. Christensen, F. Folkmann, O. Hansen, O. Nathan, N. Trautner, F. Videbaek, S.Y. van der Werf, H.C. Britt, R.P. Chestnut, H. Freiesleben and F. Pühlhofer, Phys. Rev. Lett. 1245 (1978).

309. K. Van Bibber, R. Ledoux, S.G. Steadman, F. Videbaek, G. Young and C. Flaum, Phys. Rev. Lett. 38, 334 (1977).

310. P. Dyer, R.J. Puigh, R. Vandenbosch, T.D. Thomas and M.S. Zisman, Phys. Rev. Lett. 39, 392 (1977).

311. G.J. Wozniak, R.P. Schmitt, P. Glässel, R.C. Jared, G. Bizard and L.G. Moretto, Phys. Rev. Lett. 40, 1436 (1978).

312. J.B. Natowitz, M.N. Namboodiri, P. Kasiraj, R. Eggers, L. Adler, P. Gonthier, C. Cerruti and T. Alleman, Phys. Rev. Lett. 40, 751 (1978).

313. K. Siwek-Wilczynska and J. Wilczynski, Nucl. Phys. A264, 115 (1976); Nukleonika 21, 517 (1976).

314. D.H.E. Gross and H. Kalinowski, Ref. 20, p. 194.

315. M. Blann, Ann. Rev. Nucl. Sci. 25 (1975).

316. R. Weiner and M. Weström, Nucl. Phys. A286, 282 (1977); Phys. Rev. Lett. 34, 1523 (1975).

317. R. Weiner, Phys. Rev. Lett. 32, 630 (1974); Phys. Rev. D13, 1363 (1976).

318. H. Bethe, Phys. Rev. 53, 675 (1938).

319. S. Tomonaga, Z. Phys. 110, 573 (1938).

320. R.K. Pathria, Statistical Mechanics (Academic Press, NY, 1972) Chapter 8; A. Isihara, Statistical Physics (Academic Press, NY, 1971) Chapter 13.

321. A. Kind and G. Paternagnani, Nuovo Cimento 10, 1375 (1953).

322. M. Blann, A. Mignerey and W. Scobel, Nukleonika 21, 335 (1976).

323. D.K. Scott, From Nuclei to Nucleons, 1st Oaxtepec Symposium on Nuclear Physics (Mexico, 1978), Lawrence Berkeley Laboratory Preprint LBL-7703 (1978).

324. For a review, see J. Galin, Ref. 28, p. 683.

325. C.K. Gelbke, M. Bini, C. Olmer, D.L. Hendrie, J.L. Laville, J. Mahoney, M.C. Mermaz, D.K. Scott and H.H. Wieman, Phys. Lett. 71B, 83 (1977).

326. T. Shimada, M. Ishihara, H. Kamitsubo, T. Motobayashi and T. Fukada, IPCR - Cyclotron Preprint 46 (1978).

327. A. Gamp, J.C. Jacmart, N. Poffé, H. Doubre, J.C. Roynette and J. Wilczynski, Phys. Lett. 74B, 215 (1978).

328. H. Ho, R. Albrecht, W. Dunnweber, G. Graw, S.G. Steadman, J.P. Wurm, D. Disdier, V. Rauch and F. Schiebling, Z. Phys. A283, 235 (1977).

329. P.A. Gottschalk and M. Weström, Phys. Rev. Lett. 39, 1250 (1977).

330. J.M. Miller, G.L. Catchen, D. Logan, M. Rajagopalan, J.M. Alexander, M. Kaplan and M.S. Zisman, Phys. Rev. Lett. 40, 100 (1978).

331. J.M. Miller, D. Logan, G.L. Catchen, M. Rajagopalan, J.M. Alexander, M. Kaplan, J.W. Ball, M.S. Zisman and L. Kowalski, Phys. Rev. Lett. 40, 1074 (1978).

332. L. Kowalski, J. Alexander, D. Logan, M. Rajagopalan, M. Kaplan, M.S. Zisman and T.W. Debiak, Preprint (1978).

333. D.H.E. Gross and J. Wilczynski, Phys. Lett. 67B, 1(1977).

334. J.P. Wurm, Ref. 27.

335. W.J. Swiatecki, Ref. 19, p. 349.

336. For some of the earliest studies, see, P. Freier, E.J. Lofgren, E.P. Ney, F.Oppenheimer, H.L. Bradt and B. Peters, Phys. Rev. 74, 213 (1948) and Phys. Rev. 74, 1818 (1948); H. Alfvén, Nature 143, 435 (1939).

337. A.S. Goldhaber and H.H. Heckman, Lawrence Berkeley Laboratory Preprint LBL-6570 (1978), to be published in Ann. Rev. Nucl. Sci.

338. J. Blocki, Y. Boneh, J.F. Nix, J. Randrup, M. Robel, A.J. Sierkand, W.J. Swiatecki, Lawrence Berkeley Laboratory Preprint LBL-6536, to be published in Ann. of Phys. (1978).

339. J.P. Bondorf, Ref. 24, p. 65-195.

340. G. Bertsch, Lecture Notes for the Les Houches Summer School (1977).

341. For a recent review, see J.R. Nix, Los Alamos Preprint LA-UR-77-2952.

342. M. Buenerd, C.K. Gelbke, B.G. Harvey, D.L. Hendrie, J. Mahoney, A. Menchaca-Rocha, C. Olmer and D.K. Scott, Phys. Rev. Lett. 37, 1191 (1976).

343. C.K. Gelbke, D.K. Scott, M. Bini, D.L. Hendrie, J.L. Laville, J. Mahoney, M.C. Mermaz and C. Olmer, Phys. Lett. 70B, 415 (1977).

344. D.K. Scott, M. Bini, P. Doll, C.K. Gelbke, D.L. Hendrie, J.L. Laville, J. Mahoney, A. Menchaca-Rocha, M.C. Mermaz, C. Olmer, T.J.M. Symons, Y.P. Viyogi, K. VanBibber, H. Wieman and P.J. Siemens, Lawrence Berkeley Laboratory Preprint LBL-7729 (1978).

345. H. Kamitsubo, Ref. 22, p. 177; Ref. 21, p. 263.

346. H. Kamitsubo, M. Yoshie, I. Kohno, S. Nakajima, I. Yamane and T. Mikumo, Ref. 16, p. 540.

347. T. Mikumo, I. Kohno, K. Katori, T. Motobayashi, S. Nakajima, M. Yoshie and H. Kamitsubo, Phys. Rev. C14, 1458 (1976).

348. D.E. Greiner, P.J. Lindstrom, H.H. Heckman, B. Cork and F. Bieser, Phys. Rev. Lett. 35, 152 (1975).

349. N. Masuda and F. Uchiyama, Phys. Rev. C15, 1598 (1977).

350. C.K. Gelbke, C. Olmer, M. Buenerd, D.L. Hendrie, J. Mahoney, M.C. Mermaz and D.K. Scott, Lawrence Berkeley Laboratory Preprint LBL-5826 (1977), to be published in Phys. Reports (1978).

351. V.V. Volkov, Sov. J. of Nucl. Phys. 6, 420 (1976).

352. J.P. Bondorf, F. Dickman, D.H.E. Gross and P.J. Siemens, Ref. 12, p. C6-145; R. Billerey, C. Cerruti, A. Chevarier, N. Chevarier and A. Deneyer, Z. Phys. A284, 389 (1978).

353. A.Y. Abul-Magd and K.I. El-Abed, Prog. Th. Phys. (Japan) 53, 480 (1975).

354. A. Bohr and B. Mottelson in Nuclear Structure (Benjamin, 1969), Vol. 1, p. 187.

355. V.K. Lukyanov and A.I. Titov, Phys. Lett. 57B, 10 (1975).

356. A.S. Goldhaber, Phys. Lett. 53B, 306 (1974).

357. H. Feshback and K. Huang, Phys. Lett. 47B, 300 (1973).

358. R. Hagedorn, Cargèse Lectures in Physics, Vol. 6, ed. E. Schatzman (Gordon Breach, N.Y., 1973).

359. J.D. Bowman, W.J. Swiatecki and C.F. Tsang, LBL Preprint-2908 (1973).

360. J. Hufner, K. Schafer and B. Schurmann, Phys. Rev. C12, 1888 (1975).

361. B. Jacobson, Heavy-Ion Interactions at High Energies, Publication CRN/PN-77-3, Strasbourg (1977).

362. G.D. Westfall, J. Gosset, P.J. Johansen, A.M. Poskanzer, W.G. Meyer, H.H. Gutbrod, A. Sandoval and R. Stock, Phys. Rev. Lett. 37, 1202 (1976).

363. R.J. Glauber, Lectures in Theoretical Physics, eds. W.E. Britten and L.G. Dunham (Interscience, NY, 1959) Vol. 1.

364. J. Hufner, C. Sander and G. Wolschin, Phys. Lett. 73B, 289 (1978).

365. J.P. Bondorf, G. Fai and O.B. Nielsen, NBI Preprint 78-6.

366. M.A. Preston and R.K. Bhaduri, Structure of the Nucleus (Addison Wesley, 1975) p. 202.

367. E.J. Moniz, I. Sick, R.R. Whitney, J.R. Ficenec, R.D. Kephart and W.P. Trower, Phys. Rev. Lett. 26, 445 (1971).

368. F. Videbaek, R.B. Goldstein, L. Grodzins, S.G. Steadman, T.A. Belote and J.D. Garrett, Phys. Rev. C15, 954 (1977).

369. P.J. Johansen, P.J. Siemens, A.S. Jensen and H. Hofmann, Nucl. Phys. A288, 152 (1977).

370. D.H. Youngblood, Bull. Am. Phys. Soc. 23, 42 (1977).

371. V.R. Pandharpande, Phys. Lett. 31B, 635 (1970).

372. G. Bertsch and D. Mundinger, Phys. Rev. C17, 1646 (1978).

373. F. Beiser, H. Crawford, P. Doll, D.E. Greiner, C.K. Gelbke, H.H. Heckman, D.L. Hendrie, P. Lindstrom, J. Mahoney, D.K. Scott, T.J.M. Symons, K. Van Bibber, Y.P. Viyogi, G.D. Westfall and H. Wieman, to be published (1978).

374. D.E. Greiner, Nucl. Inst. and Meth. 103, 291 (1972).

375. Y.P. Viyogi and H. Faraggi, private communication.

376. J. Rasmussen, R. Donangelo and L. Oliveira, Ref. 27, p. 440 and to be published in Nucl. Phys. (1978).

377. L. Oliveira, private communication (1978).

378. D.J. Morrissey, W.R. Marsh, R.J. Otto, W. Loveland, and G.T. Seaborg, LBL Preprint 6579 (1978).

379. W.D. Myers, W.J. Swiatecki, T. Kodama, L.J. El-Jaick and E.R. Hilf, Phys. Rev. C15, 2032 (1977).

380. G.W. Butler, D.G. Perry, L.P. Remsberg, A.M. Poskanzer, J.B. Natowitz and F. Plasil, Phys. Rev. Lett. 38, 1380 (1977).

381. K. Van Bibber, S.A. Chessin, J.V. Geaga, J.Y. Grossiord, D.L. Hendrie, L.S. Schroeder, D.K. Scott and H.H. Wieman, to be published.

382. C.K. Gelbke, M. Buenerd, D.L. Hendrie, J. Mahoney, M.C. Mermaz, C. Olmer and D.K. Scott, Phys. Lett. 65B, 227 (1976).

383. H. Bøggilo and T. Ferbel, Ann. Rev. of Nucl. Sc. 24, 451 (1974).

384. W.R. Frazer, I. Ingber, C.H. Mehta, C.H. Poon, D. Silverman, K. Stowe, P.D. Ting and H.Y. Yesian, Rev. Mod. Phys. 44, 284 (1972).

385. D.F. Jackson, Phys. Lett. 71B, 57 (1977).

386. H. Feshbach, Lectures at École d'Été de Physique Theorique, Les Houches (1977).

387. J.B. Cumming, P.E. Haustein, R.W. Stoenner, L. Mausner and R.A. Naumann, Phys. Rev. C10, 739 (1974); Phys. Rev. C14, 1554 (1976) and Brookhaven National Laboratory Preprint BNL-25378 (1977).

388. P.J. Lindstrom, D.E. Greiner, H.H. Heckman, B. Cork and F.S. Bieser, LBL-Preprint 3650 (1975).

389. D.J. Morrissey, W. Loveland, W.R. Marsh and G.T. Seaborg, LBL Preprint 7718 (1978).

390. M.I. Sobel, P.J. Siemens, J.P. Bondorf and H.A. Bethe, Nucl. Phys. A 251, 502 (1975).

391. See, for example, An Atlas of Heavy-Ion Topology, H.H. Heckman, D.E. Greiner, P.J. Lindstrom and D.D. Tuttle (unpublished).

392. L.S. Schroeder, Proc. of Topical Meeting on Multiparticle Production of Nuclei at Very High Energy, ICTP (Trieste, 1976).

393. R. Kullberg, K. Kristiansson, B. Lindkvist, and I. Otterlund, Nucl. Phys. A280, 491 (1977).

394. P.J. Karol, Phys. Rev. C14, 1203 (1975).

395. J. Gosset, H.H. Gutbrod, W.G. Meyer, A.M. Poskanzer, A. Sandoval, R. Stock and G.D. Westfall, Phys. Rev. C16, 629 (1977).

396. G.D. Westfall, J. Gosset, P.J. Johansen, A.M. Poskanzer, W.G. Meyer, H.H. Gutbrod, A. Sandoval and R. Stock, Phys. Rev. Lett. 37, 1202 (1976).

397. W.D. Myers, Nucl. Phys. A296, 177 (1978).

398. J. Gosset, J.I. Kapusta and G.D. Westfall, LBL Preprint 7139 (1978).

399. H.H. Gutbrod, A. Sandoval, P.J. Johansen, A.M. Poskanzer, J. Gosset, W.G. Meyer, G.D. Westfall and R. Stock, Phys. Rev. Lett. 37, 667 (1976).

400. A. Mekjian, Phys. Rev. Lett. 38, 640 (1977); A. Mekjian, R. Bond, P.J. Johansen, S.E. Koonin and S.I.A. Garpman, Phys. Lett. 71B, 43 (1977).

401. S. Nagamiya, I. Tanihata, S. Schnetzer, L. Anderson, W. Brückner, O. Chamberlain, G. Shapiro and H. Steiner, J. Phys. Soc. (Japan) 44, 378 (1978).

402. P.B. Price, J. Stevenson and K. Fraenkel, Phys. Rev. Lett. 39, 177 (1977).

403. P.B. Price and J. Stevenson, LBL Preprint 7702 (1978).

404. R.K. Smith and M. Danos, Ref. 25.

405. J.B. Ball, C.B. Fulmer, M.L. Mallory and R.L. Robinson, ORNL Preprint (1978).

406. H.W. Bertini, T.A. Gabriel, R.T. Santoro, O.W. Hermann, N.W. Larson and J.M. Hunt, ORNL Report TM-4134 (1974).

407. J.P. Bondorf and W. Nörenberg, Phys. Lett. 44B, 487 (1973).

408. M. Bini, P. Doll, C.K. Gelbke, D.L. Hendrie, J. Mahoney, G. Mantzouranis, D.K. Scott, T.J.M. Symons, K. Van Bibber, Y.P. Viyogi and H. Wieman, to be published (1978).

409. H. Löhner, H. Eickhoff, D. Frekers, G. Gaul, K. Poppensieker and R. Santo, Munster Preprint IKP-2-78 (1978).

410. R.R. Doering, T.C. Schweizer, S.T. Thornton, L.C. Dennis, K.R. Cardell, K.O.H. Ziock and J.C. Comiso, Phys. Rev. Lett. 40, 1433 (1978).

411. L. Westerberg, D.G. Sarantites, D.C. Hensley, R.A. Dayras, M.L. Halbert and J.H. Barker, ORNL Preprint (1978).

412. T. Nomura, H. Utsunomiya, T. Motobayashi, T. Inamura and M. Yanokura, Phys. Rev. Lett. 40, 694 (1978).

413. M.C. Robel and W.J. Swiatecki, Private communication, 1978.

414. L.G. Moretto, Gordon Conference in Nuclear Chemistry (June, 1978).

415. P. Bonche, S. Koonin and J.W. Negele, Phys. Rev. C13, 1226 (1976).

416. C.Y. Wong, J.A. Maruhn and T.A. Welton, Phys. Lett. 66B, 19 (1977).

417. M. Blann, Nucl. Phys. A235, 211 (1974); D. Agassi, H.A. Weidenmuller and G. Mantzouranis, Phys. Rep. 22, 145 (1975).

418. J.J. Griffin, Phys. Lett. 24B, 5 (1967).

419. V.E. Viola, R.G. Clark, W.G. Meyer, A.M. Zebelman and R.G. Sextro, Nucl. Phys. A261, 174 (1976); V.E. Viola, C.T. Roche, W.G. Meyer and R.G. Clark, Phys. Rev. C10, 2416 (1974).

420. G. Bertsch and A.A. Amsden, Los Alamos Preprint LA-78-838 (1978).

421. G.J. Kopylov, Phys. Lett. 50B, 472 (1974).

422. S.E. Koonin, Phys. Lett. 70B, 43 (1977).

423. A.T. Laasanen, C. Ezell, L.J. Gutay, W.N. Schreiner, P. Schübelin, L. von Lindern and F. Turkot, Phys. Rev. Lett 38, 1 (1977).

424. R. Stock, Heavy-Ion Collisions, Vol. 1, ed. R. Bock (North Holland), in press; Phys. Reports, in press.

425. L.S. Schroeder, Acta Phys. Polonica, B8, 355 (1977).

426. H. Steiner, Proc. of VII Int. Conf. on High Energy Physics and Nuclear Structure (Zurich, 1977) ed. M. Locher, p. 261.

426a. M. Gyulassy, Ref. 26, p. 623; Phys. Rep. (1978), in press.

427. P.J. McNulty, G.E. Farrell, R.C. Filz, W. Schimmerling and K.G. Vosburgh, Phys. Ref. Lett. 38, 1519 (1977).

428. P. J. Lindstrom, H.G. Crawford, D.E. Greiner, R. Hagstrom and H.H. Heckman, Phys. Rev. Lett. 40, 93 (1978).

429. R. Kullberg, A. Oskarsson and I. Otterlund, Phys. Rev. Lett. 40, 289 (1978).

430. A. E. Glassgold, W. Hechrotte and K.M. Watson, Ann. Phys. 6, 1 (1959).

431. W. Scheid, H. Muller and W. Greiner, Phys. Rev. Lett. 32, 741 (1974); 36, 88 (1976).

432. A.A. Amsden, G.F. Bertsch, F.H. Harlow and J.R. Nix, Phys. Rev. Lett. 35, 905 (1975).

433. M. I. Sobel, P.J. Siemens, J.P. Bondorf and H.A. Bethe, Nucl. Phys. A251, 502 (1975).

434. For a review, see J.R. Nix, Symposium on Relativistic Heavy Ion Research GSI, (1978), Los Alamos Preprint LA-UR-78-571.

435. J.A. Maruhn, Ref. 25.

436. For a review see A. Bodmer, Ref. 25; A. Bodmer and C.M. Panos, Phys. Rev. C15, 1342 (1977).

437. H.G. Baumgart, J.U. Schott, Y. Sakamoto, E. Schopper, H. Stocker, J. Hofman, W. Schied and W. Greiner, Z. Phys. A273, 359 (1975).

438. A.M. Poskanzer, R.G. Sextro, A.M. Zebelman, H.H. Gutbrod, A. Sandoval and R. Stock, Phys. Rev. Lett. 35, 1701 (1975).

439. For a discussion, see H.H. Heckman, Ref. 26, and E. Schopper, Ref. 25.

440. M.M. Gazzaly, J.B. Carroll, J.V. Geaga, G. Igo, J.B. McClelland, M.A. Nasser, H. Spinka, A.L. Sagle, V. Perez-Mendez, R. Talaga, E.T.B. Whipple and F. Zarbakash, LBL Preprint 7278 (1978) to be published in Phys. Rev. Lett. (1978).

441. H. Stocker, J. Hoffman, W. Scheid and W. Greiner, Proc. Int. Workshop on Gross Properties of Nuclei, Austria (1976).

442. T.D. Lee, Rev. Mod. Phys. 47, 267 (1975).

443. V. Ruck, M. Gyulassy and W. Greiner, Z. Phys. A277, 391 (1976).

444. M. Gyulassy and S.K. Kauffman, Phys. Rev. Lett. 40, 298 (1978).

445. J.C. Collins and M.J. Perry, Phys. Rev. Lett. 34, 1353 (1975); L. Susskind, Phys. Rev. D (SLAC-2070, 1978); G. Chapline and M. Nayenberg, Phys. Rev. D16, 450 (1977).

446. A.S. Goldhaber, LBL Preprint, 6595.

447. Max Born, Experiment and Theory in Physics, (Dover Publications, 1953).

448. A.P. Wavell, Other Men's Flowers (Cape London, 1944); D.R. Bates, Phys. Reports 35, 306 (1978).

449. Rudyard Kipling, The Glory of the Garden.

450. H. Friedrich, K. Langanke, A. Weiguny and R. Santo, University of Münster Preprint, 1978; A. Weiguny, Ref. 21, p. 203 and refs. therein.

451. V. Shkolnik, D. Dehnhard, S. Kubono, M.A. Franey, and S. Tripp, Phys. Lett. 74B, 195 (1978).

SEMI-CLASSICAL APPROXIMATION FOR HEAVY IONS

Richard Schaeffer

Service de Physique Théorique, C.E.A.-Saclay, BP n°2
91190 Gif-sur-Yvette, France

INTRODUCTION

There has always been a strong interest for describing nuclear reactions by means of classical, or nearly classical concepts (Alder et al 1956, Ford 1959, Broglia 1972). This allows some insight into the reaction mechanism, that is to say what happened during the collision, what made the angular patterns look as they are. The aim of any cross-section measurement is not only to obtain the angular distribution, but mainly to use the latter and find out the properties of the scattering nuclei, their shape or whether they are opaque or transparent (this is already a classical concept). One can understand this is much easier in case one is able to describe the reaction in terms of classical trajectories. One can in this case follow the interacting ions during the whole reaction time.

Classical concepts can be used when the wavelength is sufficiently short as compared to a characteristic length of the system. In case of potential scattering, it has to be small as compared to the distance over which the potential V varies appreciably. One has roughly (this is not the exact condition)

$$\frac{1}{\lambda} >> \left|\frac{\vec{\nabla} V}{V}\right| \qquad \text{or} \qquad \frac{p}{\hbar} >> 2\pi\left|\frac{\vec{\nabla} V}{V}\right| \tag{1}$$

this is the meaning of the usual statement of $\hbar$ being in some sense "small". This condition is usually well fulfilled for Coulomb scattering since

$$\left|\frac{\vec{\nabla} V_{Coul}}{V}\right| \sim \frac{1}{R} \tag{2}$$

where R is the nuclear radius. The product pR is always large as compared to $\hbar$. On the other hand, the condition (1) is never met for nuclear scattering since the sharp nuclear surface leads to very large values of $|\vec{\nabla}V|/V$:

$$\frac{|\vec{\nabla}V|}{V} \sim \frac{1}{a} \qquad (3)$$

where a is the surface thickness. So, one has

$$\frac{p}{\hbar} \approx 2\pi \left|\frac{\vec{\nabla}V}{V}\right| \quad . \qquad (4)$$

The condition for diffraction to occur is precisely that the wavelength is comparable to the size of the system. So, one gets diffraction which is a quantum mechanical process.

A theory which aims to explain heavy ion cross-sections in terms of classical concepts has thus to be powerful enough to handle the nuclear potential correctly. This is done by the semi-classical description of the scattering process (and its generalizations), the word "semi" meaning one incorporates some of the quantum mechanical effects into the theory. The first requirement for such a theory is of course that it reproduces reasonably well the cross-sections calculated by solving exactly the Schrödinger with the optical potential. It has thus to be capable of handling the absorption as well as the diffraction which may be produced by the sharp edge of either the real or the imaginary parts of this potential. One would also, if possible, be able to understand the angular pattern in terms of trajectories. Indeed, in case one can associate to each scattering angle one or several trajectories scattering through that angle, by following this trajectory, it will be possible to determine which part of the scattering potential produces the angular pattern in the vicinity of that angle. One can thus hope to get a simple picture of how the reaction proceeds.

This can be achieved by using, in addition to the classical trajectories due to the optical potential, complex trajectories (Balian 1974, Knoll 1974, 1976 and 1977) that are complex solutions of the same classical equations of motion. Even diffractive processes are described by such trajectories and the semi-classical approximation can be extended to cases where the potential has sizeable variations within a wavelength. The use of the complex semi-classical approximation is of tremendous importance in a great variety of domains : chemistry (Miller 1970, 1974), phase transitions and renormalization in solid state and elementary particle physics (Balian 1978), nuclear physics (Balian 1972, Knoll 1974, Malfliet 1975, Knoll 1976, Avishi 1976, Knoll 1977, Brink 1977), vibration phenomena (Balian 1972).

Before describing in Chapters 2 and 3 how the theory can be built, let us give an example of what kind of answers can be expected from a complex semi-classical description of elastic heavy ion scattering. All cross-sections (Fig.1) show at low energies a flat part for small angles, and then an exponentially damped part (Fig.1).

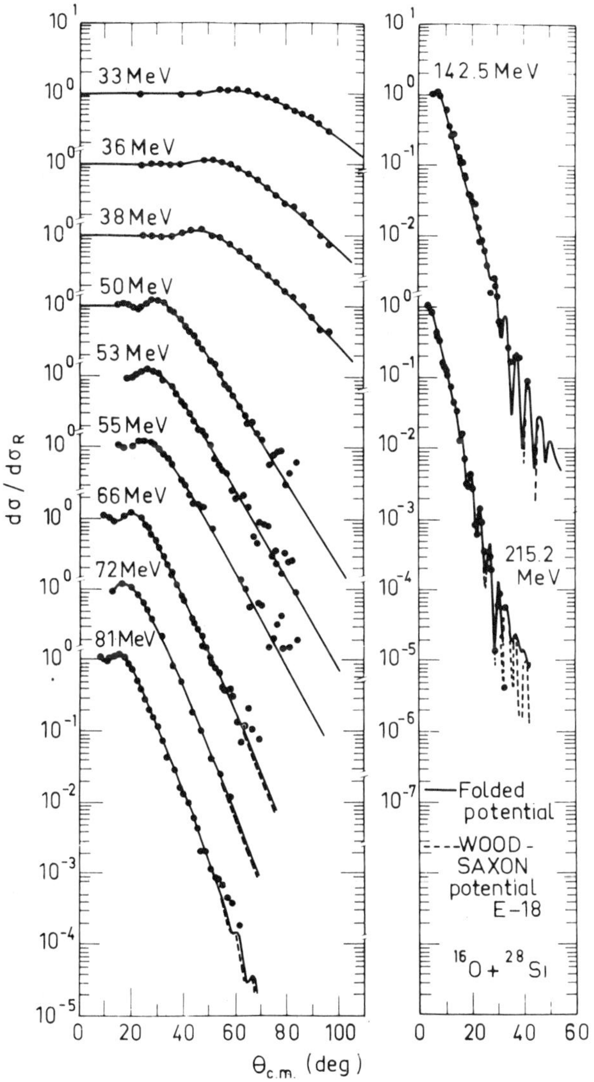

Fig.1 - From (Satchler 1977). Measured (Cramer 1974) cross-sections for $^{16}O + ^{28}Si$ *as a function of energy and fits using various optical potentials. See discussion in (Lemaire 1977).*

The transition angle θ_R is roughly the scattering angle of a Coulomb trajectory reaching the nuclear surface.

$$\theta_R = 2 \text{ Arctg } \frac{\eta}{L_R}$$

with

$$\eta = \frac{Z_1 Z_2 e^2}{v} \quad , \qquad L_R = pR \sqrt{1 - \frac{2\pi}{pR}} \quad .$$

How can this change of slope be explained ? Oscillations appear at the higher energies for large angles. Which feature of the optical potential is producing these wiggles ? Several explanations have been given (Ford 1959, Frahn 1973, Da Silveira 1973, Malfliet 1973, Knoll 1976). Some invoque refraction, some other absorption and diffraction. This will be discussed in Chapter 4 for model as well as realistic cases.

2. ONE DIMENSIONAL WKB APPROXIMATION

The semi-classical approximation to the solution of the Schrödinger equation has received the name WKB (Wentzel, Kramers, Brillouin 1926) of its pionneers. The aim is to find an approximate solution of the Schrödinger equation for smooth potentials, or "small" $\hbar$. Most of the material in this chapter can be found in standard textbooks (Messiah, Landau- Lifshitz). We present it here in a way that will introduce the complex WKB approximation in a natural way.

2.1. WKB Wave Function

Let us consider one-dimensional potential scattering, and the wave function defined by

$$-\frac{\hbar^2}{2m}\frac{d^2}{dx^2}\psi(x) + V(x)\,\psi(x) = E\,\psi(x) \tag{1}$$

For a constant potential, the wave function would be of the form

$$\psi(x) \sim e^{i(npx/\hbar)} \tag{2}$$

$$p = \sqrt{2mE} \qquad\qquad n = \sqrt{1 - V/E}$$

(n is the usual refraction index of geometrical optics).

So, for smooth potentials, it is tempting to look for wave functions that behave as

$$\psi(x) \sim A(x)\, e^{i\, S(x)/\hbar} \tag{3}$$

where A and S are smooth functions of x, and independent of $\hbar$.

From (1) it is possible to obtain the equation

$$S'^2 - i\hbar\left(S'' + \frac{2A'S'}{A}\right) - \hbar^2 \frac{A''}{A} = p^2 n^2(x) \qquad (4)$$

$$n(x) = \sqrt{1 - \frac{V(x)}{E}}$$

Neglecting the term proportional to $\hbar^2$, one gets from the requirement of A and S being independent of $\hbar$

$$S'^2 = p^2 n^2(x)$$
$$(A^2 S')' = 0 \qquad (5)$$

which leads to the solution

$$S'(x) = \pm \int^x p\ n(\xi)\ d\xi$$
$$A(x) = \frac{C}{\sqrt{n(x)}} \quad , \qquad (6)$$

where C is an arbitrary constant which depends on the boundary conditions of the problem. There are thus two solutions of the form (3)

$$\psi_+ = \frac{C_+}{\sqrt{n(x)}} \exp\left\{\frac{i}{\hbar} \int^x p\ n(x)\ dx\right\} \qquad (7a)$$

and

$$\psi_- = \frac{C_-}{\sqrt{n(x)}} \exp\left\{-\frac{i}{\hbar} \int^x p\ n(x)\ dx\right\} \quad . \qquad (7b)$$

The velocity associated to these two solutions

$$v_\pm = -\frac{i\hbar}{2m}\left(\psi_\pm^* \nabla\psi_\pm - \psi_\pm \nabla\psi_\pm^*\right) / \psi\psi^*$$

is equal to

$$v_\pm = \pm \frac{p}{m} n(x) \quad .$$

The two solutions correspond thus to particles propagating in opposite directions. The number of particles propagating to the right is in the interval dx

$$\frac{dN_+}{dx} = |\psi_+|^2 = \frac{|C_+|^2}{n(x)} \qquad (8)$$

and the number of particles propagating to the left

$$\frac{dN_-}{dx} = \frac{|C_-|^2}{n(x)} \quad . \qquad (9)$$

The classical density of particles at point x is thus

$$\frac{dN_{cl}}{dx} = \frac{|C_+|^2 + |C_-|^2}{n(x)} \quad . \tag{10}$$

One can also construct the wave function at point x . Using the original Schrödinger equation, one knows from the <u>superposition principle</u> that the general solution is a combination

$$\psi(x) = \frac{C_+}{\sqrt{n}} e^{\frac{i}{h}\int pn\,dx} + \frac{C_-}{\sqrt{n}} e^{-\frac{i}{h}\int pn\,dx} \tag{11}$$

It is very important to note that it is necessary to go back to the original, quantal, equation (1) in order to get (11). The density is thus

$$\frac{dN}{dx} = |\psi(x)|^2 = \frac{dN_{cl}}{dx} + \frac{2\,\mathrm{Re}\,C_+C_-^* \exp\left\{\frac{2i}{\hbar}\int^x pn\,dx\right\}}{n} \quad . \tag{12}$$

It is usually called the "semi-classical density", and contains an interference term, which oscillates with a period $\Delta x \approx \hbar\pi/pn = \lambda(x)/2$. If dN/dx is averaged over an interval large as compared to the wavelength, the oscillating term vanishes. To know the density in an interval smaller than the wavelength, the classical density (10) has to be replaced by (12). It is important to note this additional term has been obtained by using the superposition principle that is only contained in the quantum mechanical equation (1). The <u>interference term in (12) is a quantum mechanical correction to the classical density (10)</u>.

2.2. Classical Turning Points

Let us now discuss the corrections to (3). They can be evaluated (Messiah 1959) by adding a small part $\hbar^2\Sigma(x)$ to S . Up to second order in $\hbar$ we thus get from (4)

$$2S'\Sigma' = \frac{A''}{A} \quad . \tag{13}$$

This gives for Σ the expression

$$2p\,\Sigma(x) = -\frac{1}{2}\frac{n'}{n^2} - \frac{1}{4}\int^x \frac{n'^2}{n^3}\,d\xi \quad . \tag{14}$$

The error is small whenever $e^{i\hbar\Sigma}$ is nearly equal to 1. This implies

$$\hbar\,\frac{1}{4p}\,\frac{n'}{n} << 1 \quad , \tag{15}$$

that is the local wavelength at point x , $\lambda(x) = \hbar/pn(x)$ has to be small as compared to the variations of the potential.

The approximation (3)-(6) breaks down at the points x_o where $n(x_o) = 0$, that is $V(x_o) = E$. These points are called classical

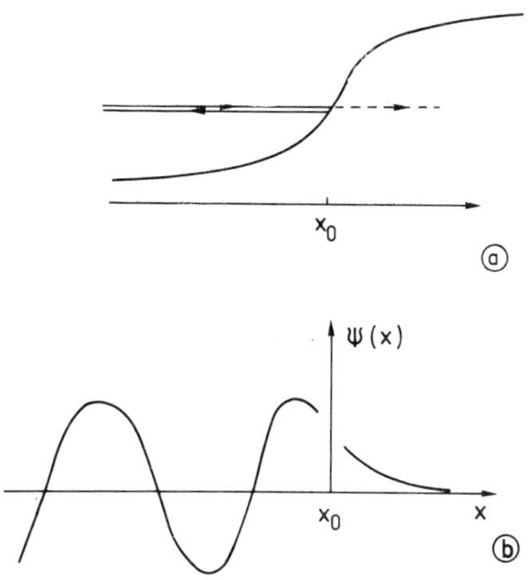

Fig 2. Above: Reflection from a potential barrier. Below: Wave function of a particle hitting a potential barrier.

turning points. The simplest example is the case of a particle hitting a potential wall (Fig.2a). For $x < x_o$, one has $E > V(x)$, so n is a real number, but for $x > x_o$, since $E < V(x)$, n is complex. So, in one region the two solutions (7a) and (7b) are oscillating functions behaving as $\exp\{\pm ipn\ \delta x\}$, in the other regions they are exponentials $\exp\{\pm p|n|\delta x\}$. Up to now, we have no way of finding the connection between the wave functions in the two regions. This is beyond our approximation since the solution of (1) cannot be written in the form (3) near x_o . By using (1) and making the expansion $V(x) \sim V(x_o) + (x-x_o)\ V'(x_o)$ it is possible to find the quantum mechanical solution near x_o , (it is a Bessel function of order 1/3 or Airy function), and to match the semi-classical solutions in the two regions (Messiah 1959, Landau 1967, Berry 1972). It can for instance be shown that, among other solutions, the one that decays exponentially for $x > x_o$ corresponds to $C_- = -iC_+$ for $x < x_o$. More precisely, the correspondance is

$$\frac{1}{\sqrt{n(x)}} \exp\left\{\frac{i}{\hbar}\int_{x_o}^{x} pn\ dx\right\} - i\frac{1}{\sqrt{n(x)}} \exp\left\{-\frac{i}{\hbar}\int_{x_o}^{x} pn\ dx\right\}$$

$$\longleftrightarrow \quad e^{-i\pi/4}\ \frac{1}{\sqrt{n(x)}} \exp\left\{-\frac{i}{\hbar}\int_{x_o}^{x} p\ \nu(x)\ dx\right\} \tag{16}$$

$$n(x) = \sqrt{1-\frac{V(x)}{E}} \qquad \nu(x) = \sqrt{\frac{V(x)}{E}-1}$$

This solution oscillates for $x < x_o$ and decays exponentially for

$x > x_o$ (Fig.2b). The quantum mechanical wave has of course the same qualitative behaviour. The classical turning points correspond to places where the quantum mechanical wave has important qualitative changes. The reflected wave exists because there is a turning point. Even far away from x_o this wave is of the same magnitude than the incoming wave. The presence of a turning point has thus effects at large distances since it changes the value of the constants $C_\pm$. This is also true (Pokrovskii 1967, Balian 1974) for complex turning points, i.e. complex solutions x_o of $n(x) = 0$. They will modify the wave even on the real axis. Precise rules have been given in (Knoll 1976) for finding which complex turning points are to be retained for constructing the wave. In more than one dimension, other singularities of the semi-classical wave (focal surfaces or points) will play the role the turning points play at one dimension.

Let us consider some simple examples. The semi-classical wave associated to a particle coming from the point at $x = -\infty$ can be obtained from (16) by multiplying both sides by $C \exp\{i \int_{-\infty}^{x} pn\, dx\}$

$$\psi(x) = \frac{C}{\sqrt{n(x)}} \exp\left\{\frac{i}{\hbar}\int_{-\infty}^{x} pn\, dx\right\} - i\frac{C}{\sqrt{n(x)}} \exp\left\{-\frac{i}{\hbar}\int_{-\infty}^{x_o} pn\, dx - \frac{i}{\hbar}\int_{x_o}^{x} pn\, dx\right\} \quad \text{for } x < x_o$$

$$\psi(x) = e^{-i\pi/4} \frac{C}{\sqrt{\nu(x)}} \exp\left\{\frac{i}{\hbar}\int_{-\infty}^{x_o} pn\, dx - \frac{1}{\hbar}\int_{x_o}^{x} p\nu\, dx\right\} \quad \text{for } x > x_o \tag{17}$$

Considering the trajectories 1 and 2 in Fig.3a, or 2 in Fig.3b, the wave $\psi(x)$ can be expressed as

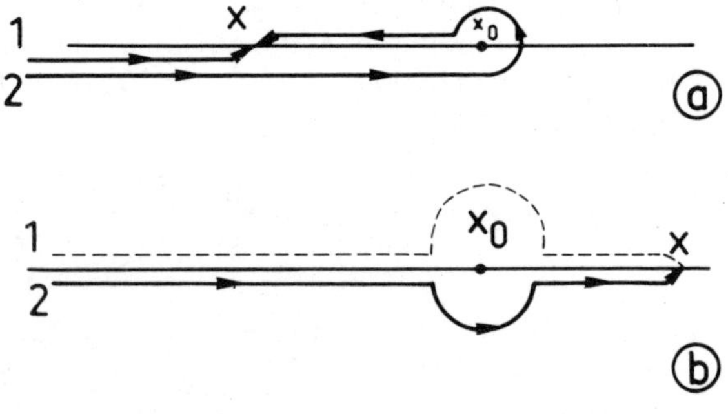

Fig.3 - Complex integration contour for the classical action at point x . *The point* x_o *is a classical turning point. Above :* x *is in the classically allowed region. Below :* x *is in the classically forbidden region.*

$$\psi(x) = \sum_\lambda \psi_\lambda(x) = \sum_\lambda \frac{C}{\sqrt{n(x)}} \exp\left\{\frac{i}{\hbar}\int_\lambda p\,n(x)\,dx\right\} \tag{18}$$

that is as a sum over all trajectories in the complex x plane "allowed" by the continuation rules. Here, trajectories 1 and 2 of Fig.3a are allowed for $x < x_o$, but only 2 of Fig.3b is allowed for $x > x_o$. For instance, trajectory 1 of Fig.3b is not allowed for $x > x_o$.

Let us check that (18) is, indeed, identical to (17). For $x < x_o$, the contribution associated to 1 is

$$\psi_1(x) = \frac{C}{\sqrt{n(x)}} \exp\left\{\frac{i}{\hbar}\int_{-\infty}^{x} p\,n(x)\,dx\right\} \tag{19}$$

the contribution associated to 2 can be evaluated by noticing that the phase of $E - V(x)$ changes by $+2i\pi$ when one turns around x_o. Thus n changes by a factor $e^{i\pi} = -1$ and $\sqrt{n}$ by a factor $e^{i\pi/2} = i^o$. We get

$$\psi_2(x) = -i\frac{C}{\sqrt{n}} \exp\left\{\frac{i}{\hbar}\int_{-\infty}^{x_o} pn\,dx - \frac{i}{\hbar}\int_{x_o}^{x} pn\,dx\right\} \tag{20}$$

For $x > x_o$, on the contour 1 the phase of $E - V(x)$ changes by $-i\pi$, hence $n(x) = e^{-i\pi/2}\,\nu(x)$, so

$$\psi_1(x) = e^{i\pi/4}\frac{C}{\sqrt{\nu(x)}} \exp\left\{\frac{i}{\hbar}\int_{-\infty}^{x_o} pn\,dx + \frac{1}{\hbar}\int_{x_o}^{x} p\nu\,dx\right\} \tag{21}$$

It increases exponentially. On the other hand

$$\psi_2(x) = e^{-i\pi/4}\frac{C}{\sqrt{\nu(x)}} \exp\left\{\frac{i}{\hbar}\int_{-\infty}^{x_o} pn\,dx - \frac{1}{\hbar}\int_{x_o}^{x} p\nu\,dx\right\} \tag{22}$$

decays. We have

$$\begin{aligned} \psi(x) &= \psi_1(x) + \psi_2(x) && \text{for } x < x_o \\ \psi(x) &= \psi_2(x) && \text{for } x > x_o \end{aligned} \tag{23}$$

which is identical to (17).

This is the proper way to find the coefficients C_+ and C_- which enter the expression (12) of the density at point x. The density for $x < x_o$ is thus

$$\frac{dN}{dx} = \frac{2|C|^2}{n}\left[1 + \cos\left(\frac{2}{\hbar}\int_{x_o}^{x} pn\,dx + \frac{\pi}{4}\right)\right] \quad . \tag{24}$$

The interference term is a genuine quantum mechanical correction as already discussed earlier. This is not only because of the $1/\hbar$ factor in the cosine which makes this term vanish whenever the density is averaged over an interval larger than a wavelength. The occurrence of two contributions associated to the two trajectories is intimately related to the existence of a classical singularity at x_o. Otherwise trajectory 2 would not exist for $x < x_o$. So even when x_o is far from x, the flux that comes back is due to that place x_o where the classical approximation is not valid.

For $x > x_o$, the density is

$$\frac{dN}{dx} = \frac{|C|^2}{\nu} \exp\left\{-\frac{2}{\hbar}\int_{x_o}^{x} p\nu \, dx\right\} \quad . \tag{25}$$

It decays by a factor 1/e over d distance $d = \lambda/4\pi$ and represents the quantum mechanical penetration of the wave into this classically forbidden region. Nevertheless, the phase of $\psi_2(x)$ as given by (22) is obtained by the classical action integral calculated along trajectory 2 of Fig.3b. This trajectory is still a classical trajectory in the sense momentum and position variables are related by the classical equation of motion

$$\frac{p^2(x)}{2m} + V(x) = E \quad . \tag{26}$$

The only difference with usual trajectories is that the momentum is now complex. So (22) or (25) are obtained using only classical (but complex) quantities. They nevertheless represent quantum mechanical effects.

2.3. Above Barrier Reflection

Let us consider a nice example of a quantum mechanical correction that can be obtained from the complex WKB (CWKB) approximation.

A quantum mechanical particle hitting a barrier V(x), but with an energy larger than the top of the barrier is partially reflected (Fig.4a). In the classical approximation, it simply crosses the barrier, and the corresponding WKB wave function is

$$\psi_1(x) = \frac{C}{\sqrt{n(x)}} \exp\left\{i\int_{-\infty}^{x} p\, n(x)\, dx\right\} \tag{27}$$

for all values of x. But although there are no real points x_o such as $E = V(x_o)$, there are in general complex points where this relation is satisfied. When the barrier is real, they occur in pairs x_o and $\bar{x}_o$ which are complex conjugate (Fig.4). The WKB wave $\psi_1(x)$, when calculated for complex values of x is still given by (27), but $\psi_1(x_o)$ is singular. As in the previous example, there is no WKB wave at x_o. But a connection rule similar to (16) can be derived by exactly the same technique, i.e. using (1) with V(x) expanded around $x = x_o$. One can then show the point x_o lying in the upper half plane which is the closest of all turning points near the real axis leads to an allowed trajectory (Fig.4b). There are then two contributions to $\psi(x)$ for $x < x_1$ and one for $x > x_1$. The new term is

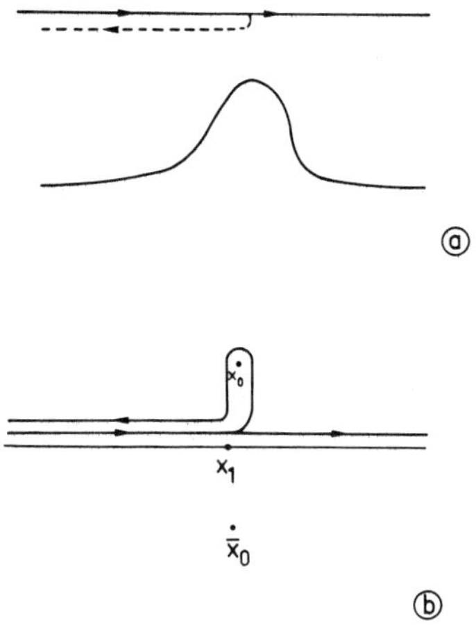

Fig. 4. Reflection of a wave due to a potential barrier. Above: Schematic sketch of a particle going over a potential barrier. Below: complex integration contour for the classical action. The points x_0 *and* $\bar{x}_0$ *are complex turning points.*

$$\psi_2(x) = -i\,\frac{C}{\sqrt{n}}\,\exp\left\{i\int_\lambda \text{pn}\,dx\right\} \tag{28}$$

$$= -i\,\frac{C}{\sqrt{n}}\,\exp\left\{\frac{i}{\hbar}\left[\int_{-\infty}^{x_1}\text{pn}\,dx + \int_{x_1}^{x_0}\text{pn}\,dx - \int_{x_0}^{x_1}\text{pn}\,dx - \int_{x_1}^{x}\text{pn}\,dx\right]\right\}$$

where x_1 is chosen such as the integral $\int_{x_1}^{x_0}\text{pn}\,dx = i\sigma_0$ is purely imaginary (it is usually not very far from $\text{Re}\,x_0$) and we have

$$\begin{aligned} \psi(x) &= \psi_1(x) + \psi_2(x) \qquad \text{for } x < x_1 \\ \psi(x) &= \psi_1(x) \qquad \text{for } x > x_1 \quad . \end{aligned} \tag{29}$$

The reflection coefficient corresponding to ψ_2 is

$$R = -i\,e^{-2\sigma_0/\hbar} \tag{30}$$

and the transmission coefficient for ψ_1 in the region $x > x_1$

$$T = 1 \quad . \tag{31}$$

In case x_0 and $\bar{x}_0$ are far from each other ($R \ll 1$), the turning point $\bar{x}_0$ has no effect on the incident wave. More correct formulae that are also valid when x_0 and $\bar{x}_0$ are close to each other and obey the unitarity relation

$$|R|^2 + |T|^2 = 1 \tag{32}$$

are given by (Froman 1965, 1970 ; Berry 1972).

In the classical limit, ψ_2 vanishes. This is also a quantum mechanical correction that can be obtained following complex classical trajectories. This correction exists, despite the WKB wave function is nowhere singular on the real axis.

2.4. Discussion

The CWKB techniques are extremely powerful and have been used for vastly different problems. They have been extensively studied by Miller (Miller 1970, 1974) in order to describe classically forbidden processes during chemical reactions. Also Balian and Bloch (Balian 1972, 1974) use complex trajectories for a variety of different vibration phenomena. There is an excellent review article by Berry and Mount (Berry and Mount 1972) that contains all was known at that time. The application to heavy ion scattering will be discussed in the next chapter.

Although it may sometimes be complicated to find all allowed trajectories λ , the CWKB wave function

$$\psi(x) = \sum_{\lambda} \psi_\lambda(x) \tag{33}$$

is usually fairly close to its quantum mechanical analogue, except of course at the points one of the ψ_λ is singular. There are different regions in space where the number of allowed trajectories is not the same. The transition from one to another describes a qualitative change of the quantum mechanical wave. A complete discussion of how to find all allowed trajectories that contribute to (33) for given boundary conditions has been given in (Knoll 1976).

3. SEMI-CLASSICAL SCATTERING

Let us now consider elastic scattering of two nuclei. Many features of the observed cross-section can be related to the refractive properties of the real part of the optical potential. The pioneering work of Ford and Wheeler (Ford 1959) who introduced the concepts of rainbow and glory for α particle scattering became much more popular in atomic and molecular physics than in nuclear physics. The main reason is that one was fearing the strong imaginary part, through the absorption and diffraction it produces, will invalidate semi-classical approaches (Frahn 1973, Gelbke 1974, Glendenning 1975) So, one has developed the Fresnel-Fraunhofer diffraction theory on one side (Frahn 1964, 1973, 1978), and the semi-classical theory on the other (Broglia 1972, Da Silveira 1973, Malfliet 1973). One has only recently (Knoll 1974, 1976, 1977, Koeling 1975) learned how to unify both approaches, using the complex WKB method. As we shall see in this chapter, this allows to treat the refractive, absorptive and diffractive properties of the optical potential at the same level. One can relate the cross-section at some angle to the three-dimensional complex trajectories that scatter through this angle. Other approaches (Koeling 1975, Brink 1977) using the partial wave expansion will also be discussed whenever necessary.

Let us say from the beginning that the scattering by a central potential is actually a two-dimensional problem, since the interacting particles stay in a plane usually called the scattering plane. Scattering by a deformed potential can, as in quantum mechanics, be solved by a set of coupled equations (Broglia 1972). We do not consider this case, but the method we use here can be extended in a straightforward manner (Knoll 1977) although this has not yet been done in a practical case.

3.1. The Classical Approximation

3.1.1. The deflection function. The classical equation of motion for a particle of mass m

$$F = m\gamma \tag{34}$$

where F is the force and γ the acceleration can be partially integrated since for a central potential the energy E and the angular momentum M are conserved

$$\begin{aligned} \frac{1}{2} mv^2 + V(r) &= E \\ m \vec{r} \wedge \vec{v} &= M \end{aligned} \tag{35}$$

From $\vec{v} = \left(\frac{dr}{dt}, \frac{1}{r}\frac{d\varphi}{dt}\right)$ where r and φ are the radial and angular coordinates of the particle, we get

$$\left(\frac{d\varphi}{dr}\right)^2 = \frac{M^2/r^2}{2m(E-V) - M^2/r^2} \tag{36}$$

This gives the equation relating r and φ on a trajectory of angular momentum M . With the choice of coordinates of Fig.5, φ decreases along the trajectory whereas r first decreases until it reaches the point r_o given by

$$2m\left[E - V(r_o)\right] - M^2/r_o^2 = 0 \tag{37}$$

Afterwards it increases. Thus initially

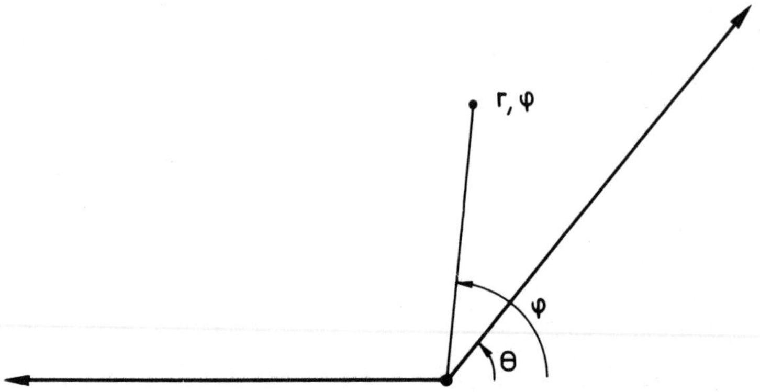

Fig.5 - Choice of coordinates in the case of the scattering by a central potential. From (Schaeffer 1977).

$$\frac{d\varphi}{dr} = \frac{M/r}{\sqrt{2m(E-V) - M^2/r^2}}$$

and after the turning point

$$\frac{d\varphi}{dr} = -\frac{M/r}{\sqrt{2m(E-V) - M^2/r^2}} \quad .$$

Since the initial value of φ is π , the scattering angle is given by

$$\theta = \pi + \int_{\infty}^{r_o} \frac{M/r}{\sqrt{2m(E-V) - M^2/r^2}}\, dr - \int_{r_o}^{\infty} \frac{M/r}{\sqrt{2m(E-V) - M^2/r^2}} \quad .$$

Or, as a function of the momentum

$$p = \sqrt{2mE}$$

and the impact parameter

$$b = M/p$$

the scattering angle is

$$\theta = \pi - 2\int_{r_o}^{\infty} \frac{b}{r^2\sqrt{n^2(r) - b^2/r^2}}\, dr$$

$$b = r_o\, n(r_o) \qquad (38)$$

$$n(r) = \sqrt{1 - \frac{V(r)}{E}}$$

An example of such a deflection function is shown for the $^{16}O + ^{16}O$ system (Fig.6). For large values of b , θ is positive due to the Coulomb repulsion. For smaller values of b , the nuclear attraction deflects the projectile to negative angles. The Coulomb potential was taken as $Z_1 Z_2 e^2/r$, so at very small impact parameters one gets again repulsion.

The relation (38) gives the scattering angle θ for each value of the impact parameter b . It can be written in the form

$$\begin{aligned} \theta &= 2\,\frac{d\delta}{p\,db} \\ \delta &= p\int_{r_o}^{\infty}\left(\sqrt{n^2(r) - b^2/r^2} - 1\right)dr - p\,r_o + \frac{\pi}{2}\,pb \quad . \end{aligned} \tag{39}$$

The quantity δ (which has the dimension of an action, i.e. of $\hbar$) is the WKB approximation for the quantum mechanical phase-shift of the partial wave $\ell = M/\hbar = pb/\hbar$.

The general form of (39) that we shall need later on is

$$f(\theta,b) = 0$$

with

$$f(\theta,b) \equiv \theta - 2\,\frac{d\delta}{p\,db} \quad . \tag{40}$$

3.1.2. <u>Classical cross-section</u>. The number of particles having trajectories with impact parameters between b and b+db is

$$dN = 2\pi\, b\, db$$

for an incoming flux equal to 1 .

Due to their motion along the trajectory, they scatter into the solid angle $d\Omega = 2\pi \sin\theta\, d\theta$, so the classical cross-section is

$$\frac{d\sigma}{d\Omega} = \left|\frac{2\pi\, b\, db}{2\pi\, \sin\theta\, d0}\right| = \left|\frac{b\ db/d\theta}{\sin\theta}\right| \tag{41}$$

In case there are several trajectories λ with different impact parameters b_λ corresponding to the same scattering angle, the total flux in the solid angle $d\Omega$ is the sum of all contributions, and we have

$$\frac{d\sigma}{d\Omega} = \sum_\lambda \frac{d\sigma_\lambda}{d\Omega} = \sum_\lambda \left|\frac{(b\ db/d\theta)_\lambda}{\sin\theta}\right| \tag{42}$$

The impact parameters b_λ are all solutions of the equation

$$f(\theta,b_\lambda) = 0 \tag{43}$$

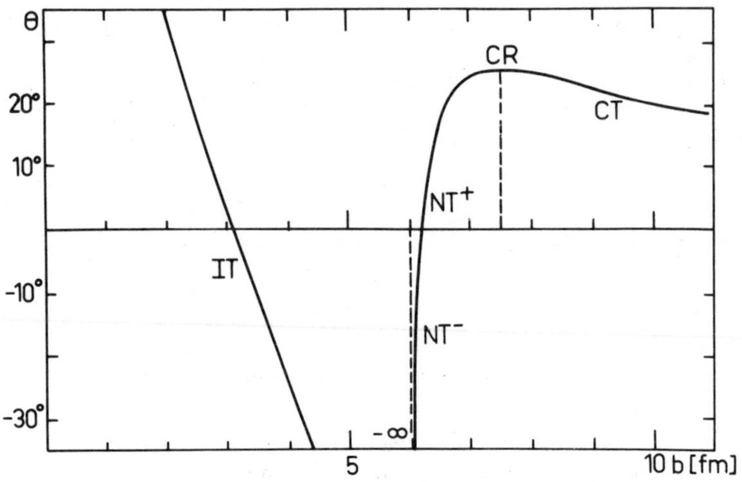

Fig.6 – *From (Knoll 1976). Deflection function given by the real part of the standard* $^{16}O + ^{16}O$ *potential at* $E_{LAB} = 50$ MeV *(*V = 50 MeV, ρ = 6.05 fm, a = 0.6 fm*)*.

for a given angle θ . Formula (42) is the analogue of the classical density (10) considered in the previous chapter. From (43), we get

$$\left(\frac{db}{d\theta}\right)_\lambda = - \frac{f'_\theta}{f'_b}\Bigg|_{b=b_\lambda} \quad . \tag{44}$$

3.1.3. Caustics. Note that the density (10) was infinite at the classical turning points $n(r_o) = 0$. Here the cross-section is infinite at the places where

$$f'_b\Big|_{b=b_\lambda} = 0$$

that is when the two equations

$$f(\theta,b) = 0$$
$$f'_b(\theta,b) = 0 \tag{45}$$

are fulfilled simultaneously. Equation (45) defines couples of points θ_R , b_R around which f can be expanded. The relation between θ and b near such points is given by

$$(\theta-\theta_R) - \frac{1}{2}(b-b_R)^2 \frac{d^2\theta}{db^2}\Bigg|_{b=b_R} = 0 \qquad \theta < \theta_R \tag{46}$$

So, for a given scattering angle, there are (assuming $d^2\theta/db^2 < 0$)

two impact parameters

$$b = b_R \pm \sqrt{\frac{2(\theta-\theta_R)}{d^2\theta/db^2}} \tag{47}$$

leading to the same angle θ . For $\theta > \theta_R$, there is no (real) solution b to (47). The places where the cross-section is infinite are thus places where the number of trajectories contributing to the cross-section (42) changes. These "critical" values θ_R and b_R play the role the turning points were playing in the one-dimensional case. One may thus expect qualitative changes in the cross-section at these angles θ_c . The regions $\theta > \theta_R$ are classically forbidden regions, and one might expect as in the one-dimensional case one of the complex impact parameter

$$b = b_R \pm i\sqrt{\frac{2(\theta_R-\theta)}{d^2\theta/db^2}} \tag{48}$$

to contribute in this region. In the one dimensional case, the coordinate x was real, here it is θ , but its conjugate coordinate p was complex as b is here.

An example of such a critical angle is shown in Fig.6 ($\theta_R = 25°$, $b_R = 7.5$ fm)

3.2. The Complex WKB (CWKB) Approximation (Knoll 1974, 1975, 1976)

We have chosen here to derive the three-dimensional WKB approximation from the partial wave series. This derivation relies on a whole series of approximations. The latter are not necessary for the formulation in terms of classical trajectories and in this case partly compensate each other. The partial wave expansion has however its own interest since it cast some light on what a complex trajectory really means. Also, it provides a more transparent procedure for choosing the allowed trajectories to be used in (51). Finally, it makes the comparison with other approaches (Koeling 1975, Brink 1977) easier.

3.2.1. The partial wave expansion. Let us consider the solution of the Schrödinger equation

$$-\frac{\hbar^2}{2m}\Delta\psi(\vec{r}) + V(r)\,\psi(\vec{r}) = E\,\psi(\vec{r}) \tag{49}$$

and its partial wave expansion

$$\psi(\vec{r}) = \sum i^\ell (2\ell+1)\,\psi_\ell(r)\,P_\ell(\cos\varphi) \tag{50}$$

From the outgoing wave

$$\psi_\ell(r) \underset{r\to\infty}{\sim} \frac{e^{ipr/\hbar}}{r}\; e^{2i\delta_\ell}$$

we get the scattering amplitude

$$f(\theta) \;=\; \frac{i\hbar}{2p} \sum (2\ell+1)\, e^{2i\delta_\ell}\, P_\ell(\cos\theta) \qquad 0 < \theta < \pi \quad . \tag{51}$$

The WKB approximation can then be derived in three steps :

1) Replace the quantal wave function $\psi_\ell(r)$ - or the phase shift - and P_ℓ by their WKB approximations.
2) Replace the discrete sum over ℓ by an integration over the impact parameter $b = \hbar\,(\ell + 1/2)/p$.
3) Perform the integration over b using the saddle point method.

Let us examine these steps in some detail.

3.2.2. CWKB approximation to the phase shift. The equation for the radial wave $\psi_\ell(r)$ is given by

$$\frac{1}{r^2}\frac{d}{dr}\, r^2 \frac{d}{dr}\, \psi_\ell(r) + \left[\frac{p^2 n^2}{\hbar^2} - \frac{\ell(\ell+1)}{r^2}\right] \psi_\ell(r) \;=\; 0 \tag{52}$$

Using $\ell(\ell+1) = \dfrac{p^2 b^2}{\hbar^2}$, we get an effective refraction index

$$n^2_{eff}(r) \;=\; n^2(r) - \frac{b^2}{r^2} \tag{53}$$

that is writing

$$n^2_{eff}(r) \;=\; 1 - \frac{V_{eff}(r)}{E} \quad , \tag{54}$$

an effective potential

$$V_{eff}(r) \;=\; V(r) + E\,\frac{b^2}{r^2} \;=\; V(r) + \frac{L^2}{2m\hbar^2 r^2} \quad . \tag{55}$$

For large r , one has $n_{eff} \sim 1$, and ψ_ℓ has the form (17)

$$\psi_\ell(r) \;=\; C \exp\left\{\frac{i}{\hbar}\int_{+\infty}^{r} p\, n_{eff}(p)\, dp\right\}$$

$$- i\, C \sum_{\mu} \exp\left\{\frac{i}{\hbar}\left[\int_{\infty}^{r_\mu} p\, n_{eff}\, dp - \int_{r_\mu}^{r} p\, n_{eff}\, dp\right]\right\} \quad . \tag{56}$$

The points r_μ are all turning points (real or complex) that lead to a reflected wave. The first term on the right hand side of (56) is the incoming wave.

From

$$\psi_\ell(r) \sim \sin \frac{pr - \frac{1}{2}\, p\, b\pi + \delta_\ell}{\hbar} \quad ,$$

the phase shift is given by

The sign ± refers to the sign of θ . The condition $-\pi<\theta<\pi$ can also be dropped. It is possible to show (Berry 1972) that the replacement of the discrete sum over the partial waves by a continuous integral is exact, provided one includes in (70) the contribution of all scattering angles differing from θ by $2M\pi$. These equations are the usual, classical equations of motion written for real scattering angles, but for real as well as complex trajectories since the turning points r_o and the impact parameters b might be complex.

3.3. Corrections to the CWKB Approximation

The corrections to the phase shift evaluation (57) as well as the saddle point integration (67) can be estimated by an expansion in powers of $\hbar$.

3.3.1. Correction to the phase shift. The error on the phase shift due to the fact there is a turning point on the trajectory can be evaluated using (4) and writing

$$S'^2 = p^2 n^2(x) + \hbar^2 \frac{A''}{A} \tag{71}$$

instead of (5). The $\hbar^2$ term gets large near a turning point. Its contribution to the integral of S' leads to a correction to the phase shift $\delta^{(\mu)}$ of the radial equation

$$\varepsilon \sim \hbar \frac{\kappa}{\eta} \tag{72}$$

where κ is some constant (which may contain additional powers of $\hbar$) and

$$\eta = [n^2_{eff}(r)]'_r \Big|_{r=r_\mu} \quad . \tag{73}$$

Since r_μ is usually a simple zero of $n^2_{eff}(r)$, η is usually finite. But in case r_μ is a double zero of $n^2_b(r)$, i.e. a double turning point, the error ε is infinite. So, if near r_μ there is a second turning point $r_{\mu'}$, the error gets large. This is for instance the case (Fig.7) for the (allowed) turning point r_1 which lies near the (forbidden) turning point r_3 . In this case, the approximation (57) has to be refined. There are standard techniques ("uniform" approximations) to treat the reflection from two close turning points (Froman 1965, 1970, 1972, Avishai 1976, Brink 1977). They have to be employed whenever the phase shifts (57) are directly used in the partial wave summation (51).

3.3.2. Correction to the saddle point integration. The double turning point correction has not to be applied to (57) whenever this formula is used for the saddle point evaluation (67). This is because $db/d\theta$ gets very small, and is proportional to η^2 , so $\varepsilon(db/d\theta)^{1/2}$ is no longer very large. It is indeed cancelled by the

$$e^{\frac{2i}{\hbar}\delta_\ell(\hbar)} = \sum_\mu e^{\frac{2i}{\hbar}\delta_\ell^{(\mu)}}$$
$$\delta_\ell^{(\mu)} = p\,b\frac{\pi}{2} + \int_{r_\mu}^{\infty}[p\,n_{eff}(r) - p]\,dr - p\,r_\mu \tag{57}$$

In (57), $\delta_\ell^{(\mu)}$ has been slightly modified in order to cure the divergence at $r=\infty$. This amounts simply to multiply C by an infinite phase $\exp\{ipR/\hbar\}$ as can be seen by replacing the upper limit of the integration by some large but finite value R . The $\delta_\ell^{(\mu)}$ are the CWKB phase shifts. They are several and correspond to each allowed trajectory of the kind shown in Figs.3 and 4. The turning points r_μ are the solutions of

$$V_{eff}(r) = E \tag{58}$$

In case the optical potential has an imaginary part, they are all complex. Let us for simplicity consider a real potential (Fig.7).

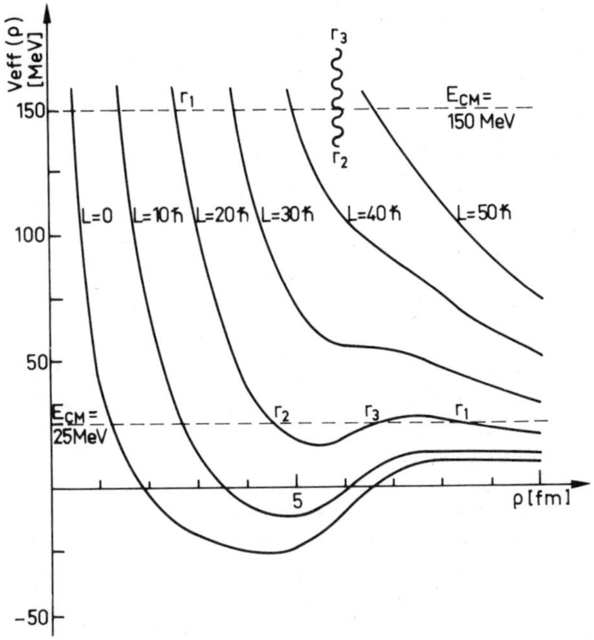

Fig.7 – From (Knoll 1976). Real part of the effective potential $V(\rho) + L^2/2m\rho^2$ *. The parameters of the optical potential are those of the* $^{16}O + {}^{16}O$ *of Table I. The turning points (*r_1 *,* r_2 *and* r_3*) are those of Figs.15 and 16, the wiggly line indicates they are complex (*r_2 *and* r_3 *at* $E_{cm} = 150$ MeV*) even for real potentials.*

At low energies, e.g. $E = 25$ MeV, for moderate values of L ($\sim 20\hbar$), there are three real turning points. The CWKB connection rules show only r_1 and r_2 should be retained in the sum (57). For r_1, the phase is real, but for r_2, it is complex since the integration between r_3 and r_1 has to be done in a classically forbidden region. For very small, or very large values of L, there is only one real turning point, the two others are complex. The real one, and one (say r_2) of the complex turning points are to be retained. The complex turning point r_2 corresponds to above barrier reflection for low values of L ($L < 17\hbar$). For large values of L ($L > 23\hbar$) it corresponds to a classically forbidden process, and is present because of the rapid variations of the nuclear potential. (For L very large, r_2 goes over to the pole of the Woods-Saxon potential, i.e. $r_2 \sim R + i\pi a$).

At high energies, e.g. $E = 150$ MeV, there is always one real (r_1) and one complex turning point which contribute. The complex one is still associated with the nuclear surface, and corresponds to above barrier reflection.

To summarize, in the model case of Fig.7 (but this is true for all Woods-Saxon plus Coulomb potentials), there are always two contributions to the sum (57)

$$\exp\left\{\frac{2i}{\hbar}\,\delta_\ell\right\} = \exp\left\{\frac{2i}{\hbar}\,\delta_\ell^{(A)}\right\} + \exp\left\{\frac{2i}{\hbar}\,\delta_\ell^{(F)}\right\} \tag{58}$$

One of the phase shifts is ($\delta^{(A)}$) real, the other one ($\delta^{(F)}$) complex. In most cases, the latter is associated with the nuclear surface and is due to the rapid variations of the nuclear potential. In the usual WKB approximation, only the real phase shift $\delta^{(A)}$ would be present since it corresponds to a classically allowed process, so we get

$$\delta_\ell^{WKB} \sim \delta_\ell^{(A)} \tag{59}$$

Note that the second contribution to (58), whose modulus is proportional to

$$\exp\left\{-\frac{2\,\mathrm{Im}\,\delta_\ell^{(F)}}{\hbar}\right\} \tag{60}$$

vanishes in the strict $\hbar \to 0$ limit. But for finite $\hbar$, it is finite. When the optical potential contains an absorptive part, the CWKB phase shift has still the form (58) in case the real and imaginary potentials have the same shape. (When the shapes are different, there may be three terms in the sum (57) since there is also a contribution from the sharp edge of the imaginary part). However, in case there is absorption both phases $\delta^{(A)}$ and $\delta^{(F)}$ are complex. For small values of ℓ, the turning point associated to $\delta^{(A)}$ is very close to the scattering center. So, if the absorption is strong enough, only the second term in (58) which is not very sensitive to the absorption contributes to the phase shift (Fig.8)

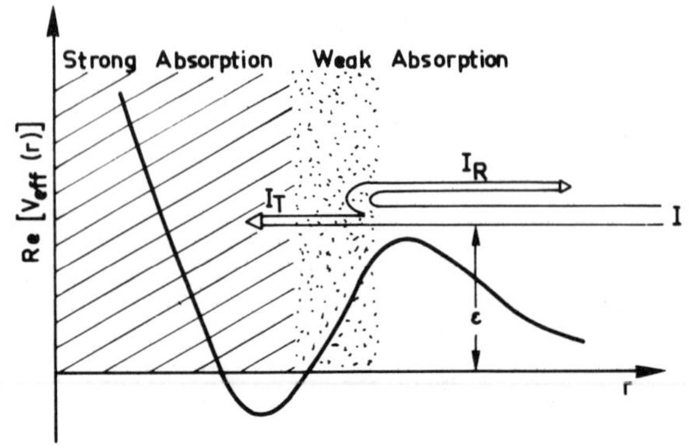

Fig.8 - From (Avishai 1976). Above barrier reflection for a given partial wave. The reflection occurs at a place where the absorption is still weak and when used in the partial wave series leads to diffraction by the edge of the real potential (if W <V*). The transmitted wave is usually absorbed.*

$$\delta_\ell^{CWKB} \sim \delta_\ell^{(F)} \tag{61}$$

whereas the usual WKB approximation would suggest to take

$$\delta_\ell^{WKB} \sim \delta_\ell^{(A)} \tag{62}$$

which predicts extremely strong damping, due to the absorption, but does not give the dominant contribution to the partial wave sum. The approximation (61) has been used by Koeling and Malflieth (Koeling 1975) and inserted into the partial wave series. This is a good approximation for scattering near the Coulomb barrier. Brink and Tagikawa (Brink 1977) have shown the contribution from the inner turning point is in some cases responsible for the large angle cross-sections.

Using (51) does not, however, provide a very precise relation between the scattering angle and the place near the scattering center the contribution comes from. For each value of θ, there is a whole set of partial waves contributing to the cross-section and thus a whole set of turning points. One can however determine whether contribution comes more from the "surface" or from the "inside".

3.2.3. <u>Saddle point integration</u> (Knoll 1974, 1976). Let us examine more precisely the relation between the scattering angle and the turning points associated to the phase shifts. In (51), the Legendre polynomial $P_\ell(\cos\theta)$ can be replaced by its semi-classical evaluation

$$P_\ell(\cos\theta) = \left(\frac{\hbar}{2\pi i\, p\, b\, \sin\theta}\right)^{1/2} \exp\{\frac{i}{\hbar}\, p\, b\, \theta\} + \left(\frac{\hbar}{-2\pi i\, p\, b\, \sin\theta}\right)^{1/2} \exp\{-\frac{i}{\hbar}\, p\, b\, \theta\} \tag{63}$$

Replacing the sum in (51) by an integral over b , we get

$$f(\theta) = f_+(\theta) + f_-(\theta) \tag{64}$$

$$f_\pm(\theta) = \frac{1}{(2\pi\, \sin\theta)^{1/2}}\, e^{\pm i\frac{\pi}{4}} \sum_\mu \int_0^\infty db \left(\frac{pb}{\hbar}\right)^{1/2} \exp\left\{\frac{2i\delta_\ell^{(\mu)} \mp ipb\theta}{\hbar}\right\}$$

Since the phase shifts δ_ℓ are complex, the usual stationary phase method has to be replaced by the saddle point method. In the limit $\hbar$ is small, the integrand oscillates very rapidly. The dominant contribution to the integral arises thus from the impact parameters for which the coefficient of $1/\hbar$ is stationary, that is the (complex) values of b given by

$$\theta = 2\,\frac{d\delta^\mu(b)}{p\, db} \quad . \tag{66}$$

This is just relation (39), that is <u>the classical equation for b</u> . But whereas (39) was written for real phase shifts and real turning points, (66) holds for complex phases, complex turning points. Even if δ^μ is real, (66) may have complex solutions b that are to be retained whenever the integration contour can be deformed so as to have the integration path go through the stationary point (Fig.9). The way this path has to be deformed is explained in (Knoll 1976). Although the method used here in order to find the "good" saddles is quite different from the one given previously for choosing the "good" turning points, both are intimately related. So, the "good" saddle points are solutions (but not <u>all</u> solutions) $b_{\lambda\mu}$ of (66). Near such a saddle point the exponent of (65) can be expanded up to second order in $b-b_{\lambda\mu}$, and the Gaussian integration then performed, leading to

$$f_\pm(\theta) = \sum_{\lambda\mu} \left(\frac{b\, db}{\sin\theta\, d\theta}\right)^{1/2} \exp\left\{\frac{iS(\theta,b)}{\hbar} \pm i\,\frac{\pi}{4}\right\}\Bigg|_{b=b_{\lambda\mu}} \tag{67}$$

$$S(\theta,b) = 2\,\delta_\ell^{(\mu)} \mp p\, b\, \theta\Big|_{b=b_{\lambda\mu}} \quad .$$

Eq.(67) gives a scattering angle which *i)* is independent of $\hbar$ and *ii)* obeys the classical equations of motion. Instead of performing the saddle point integration separately for each term μ of the sum (57), one could think performing it on the whole sum $\exp \frac{2i}{\hbar}\, \delta_\ell(\hbar)$. The result is awkward

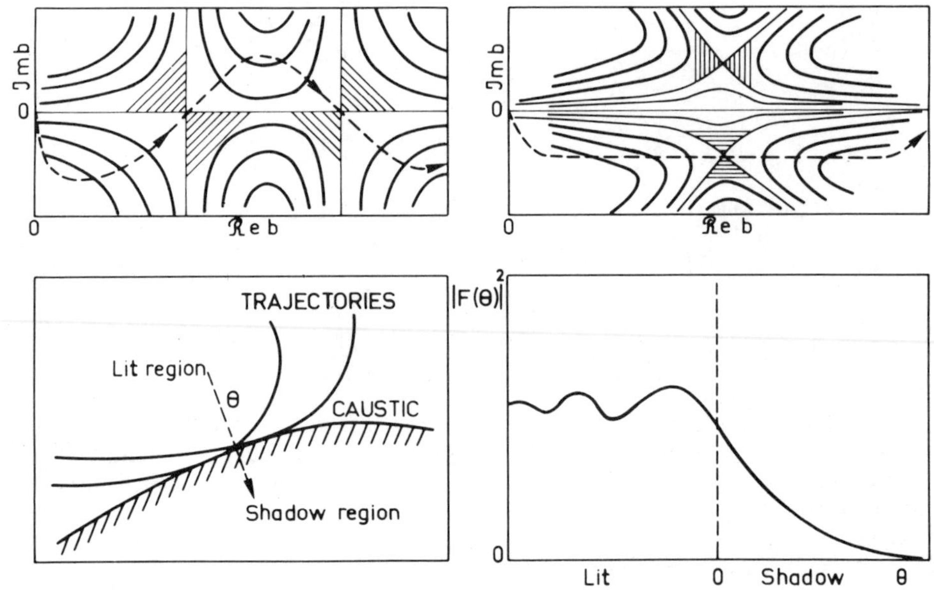

Fig.9 - From (Knoll 1976). Real caustic. Above : topography of the saddles and lines of equal hight (full lines). The mountains around the saddles are hatched, the dashed line shows the integration contour over the saddle. Case a : the classically allowed region with two real trajectories ; case b ; the classically forbidden region with two complex conjugate saddles. Only the low lying saddle can be passed by the integration contour. Below : two trajectories touching the caustic (left) and the qualitative behaviour of the scattering amplitude when passing from the allowed to the forbidden region.

$$\theta = 2\,\frac{d\delta_\ell(\hbar)}{p\;db} = 2\,\frac{\sum\limits_\mu \exp\frac{2i}{\hbar}\,\delta_\ell^{(\mu)}\,\dfrac{d\delta_\ell^{(\mu)}}{p\;db}}{\sum\limits_\mu \exp\frac{2i}{\hbar}\,\delta_\ell^{(\mu)}} \quad .$$

In particular, whenever two contributions μ are of the same order of magnitude, θ has extremely large oscillations. The derivative of $\delta_\ell(\hbar)$ cannot be considered as a classical scattering angle. This is also the case for $\theta = 2\,\mathrm{Re}\,\frac{d\delta_\ell(\hbar)}{p\,db}$ or the so-called "quantal deflection function" $\theta = 2\,\mathrm{Re}\,\frac{d\delta_\ell}{d\ell}$ where δ_ℓ is the quantal phase shift.

At first guess, it seems complicated, for each b to find the turning points $r_\mu(b)$ such as $b = r_\mu\, n(r_\mu)$ and then look for the saddles. A tremendous simplification arises if we consider (66) as an equation for the turning point r_o itself

$$\theta = \pi - 2\int_{r_o}^{\infty} \frac{b(r_o)}{r^2\sqrt{n^2(r) - b(r_o)^2/r^2}} \tag{68}$$

with

$$b = r_o\, n(r_o) \quad .$$

Solving (68) for all complex values of r_o is straightforward. One gets in general a discrete set of values. From the discussion of (Knoll (1976))one can first eliminate those values lying obviously too far from the real axis. Then by looking at the value of b given by (68) it is possible to check afterwards whether r_o was an allowed turning point for the phase shift calculation. A further simplification in the search of "good" turning point r_o arises from the fact the values of r_o are nearly energy independent as we shall see later on, and practically the same for all heavy ion cross-sections. They just depend on the shape of the potential.

This can be most easily seen by expanding (68) in powers of 1/E for fixed r_o

$$\begin{aligned} \theta &= \frac{\varepsilon_o(r_o)}{E} + \mathcal{O}(1/E^2) \\ \varepsilon_o(r_o) &= -\int_{r_o}^{\infty} \frac{r_o}{\sqrt{r^2 - r_o^2}}\, V'(r)\, dr \approx -\frac{\pi}{2}\, r_o\, V'(r_o) \quad . \end{aligned} \tag{69}$$

Since the product $E\theta$ is real, all turning points lie on the line

$$\mathrm{Re}\ \varepsilon_o(r_o) = 0$$

which is a universal curve, independent of energy. Its pattern is qualitatively the same for all Woods-Saxon (plus Coulomb) potentials.

One finally gets

$$\begin{aligned} \frac{d\sigma}{d\Omega} &= |f_+(\theta) + f_-(\theta)|^2 \\ f_\pm(\theta) &= \sum_{r_o} \left(\frac{b\ db}{\pm\sin\theta\ d\theta}\right)^{1/2} \exp\left\{\frac{iS(\theta,b)}{\hbar}\right\} \\ S &= 2\delta - p\,b\,\theta \\ \delta &= -p\int_{r_o}^{\infty}\sqrt{n^2(r) - b^2/r^2} + p\,b\frac{\pi}{2} \\ b &= r_o\, n(r_o) \\ \theta &= \pi - 2\int_{r_o}^{\infty}\frac{b}{r^2\sqrt{n^2(r) - b^2/r^2}}\, dr \\ & \qquad -\pi < \theta < +\pi \end{aligned} \tag{70}$$

higher order corrections to the saddle point result. This is why it has not been used in (Knoll 1974, 1976). It has been checked explicitly (Avishai 1976) that the saddle point result is almost the same whether or not this correction is applied to the phase shift, the better approximation being no phase shift correction at all (Fig. 10).

On the other hand, (67) gets obviously inaccurate at the caustics, that is when $d\theta/db = 0$. This is shown in Fig. 11, where at some angle θ_R this derivative vanishes. But this incorrect evaluation of the scattering amplitude is just due to the second order expansion of the phase for the saddle point integration. The quadratic term in the phase is proportional to $d\theta/db$ and in case the latter vanishes, the cubic term has to be taken into account. This also is a standard calculation, and very accurate techniques exist now (Berry 1972, Knoll 1976, 1977). Only the phase $S(\theta,b)$ is involved in this correction that has been applied in the calculations of Figs. 10 and 11.

3.4. What a Complex Trajectory Can Describe

A complex trajectory describes quantum mechanical effects. This is why it allows the WKB approximation to be extended far beyond its usual domain (smooth potentials, short wavelength) of validity. Let us give some examples.

3.4.1. Classical shadows. At a caustic, since $db/d\theta$ vanishes, b is related to the scattering angle through (46). Assuming for instance that $d^2\theta/db^2$ is real and negative, (47) can be rewritten

$$b = b_R \pm \sqrt{\frac{2(\theta_R - \theta)}{d^2\theta/db^2}} \tag{74}$$

For $\theta < \theta_R$, there are two saddles (Fig. 11) corresponding to the two classical trajectories considered in section 3.1. This angular region is a classically allowed region where the usual WKB approximation is valid. There are thus two contributions to the scattering amplitude (70) which interfere and lead to oscillations in the cross-section (Fig. 11).

For $\theta > \theta_R$, the saddle point integration allows only to go over the lowest saddle (the "mountains" of Fig. 11 are the regions where $|e^{iS/\hbar}|$ increases when one leaves the saddle point). So, there is only one contribution corresponding to a damped wave. The corresponding value of b

$$b = b_R - i\sqrt{\frac{2(\theta - \theta_R)}{d^2\theta/db^2}} \tag{75}$$

is now complex. This complex trajectory describes quantum mechanical

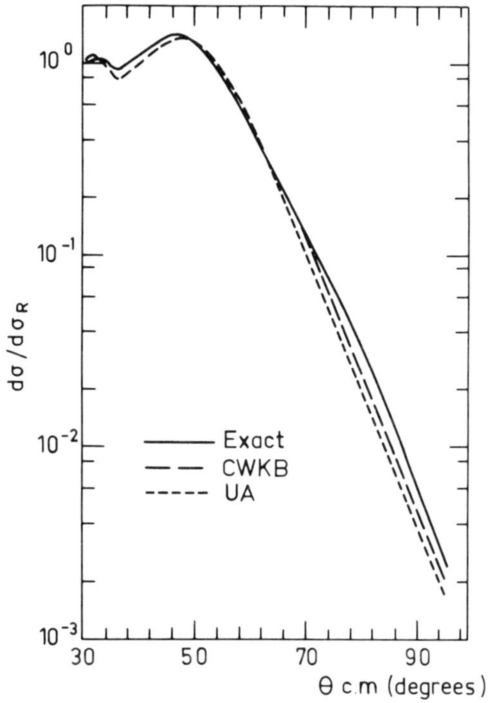

Fig.10 - *From (Avishai 1976). Exact quantal calculation (full line), complex trajectory calculation* (CWKB) *(long dashes), partial wave summation using* CWKB *phase shifts and the coalexing turning point correction* (UA).

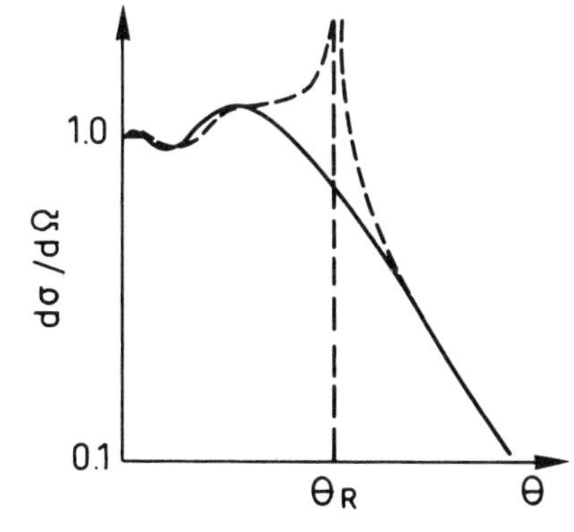

Fig.11 - *From (Knoll 1977). Behaviour of the cross-section around the rainbow angle θ_R. The primitive semi-classical result diverges at θ_R (dashed line), but the uniform approximation extended to complex values is regular. For $\theta > \theta_R$ one complex path describes the exponential damping into the classical shadow.*

penetration into the classical shadow region.

3.4.2. Slope of the cross-section. Let us assume that in some angular region, there is only one contribution to the amplitude (70). The associated cross-section is

$$\frac{d\sigma}{d\Omega} = \left|\frac{b\,db}{\sin\theta\,d\theta}\right| \exp\{-\frac{2\,\mathrm{Im}\,S}{\hbar}\}$$

that is

$$\frac{d\sigma}{d\Omega} = \left(\frac{d\sigma}{d\Omega}\right)_{cl} \exp\{-\frac{2\,\mathrm{Im}\,S}{\hbar}\} \quad . \tag{76}$$

The variations of $S[\theta,b(\theta)]$ can be obtained from

$$\frac{dS}{d\theta} = \frac{\partial S}{\partial b}\frac{\partial b}{\partial \theta} + \frac{\partial S}{\partial \theta} = \frac{\partial S}{\partial \theta} = \frac{pb}{\hbar} = \frac{L}{\hbar} \tag{77}$$

So, the slope of the cross-section is given by

$$\frac{d}{d\theta} \log \frac{d\sigma}{d\Omega} \sim - 2\,\mathrm{Im}\,\frac{L}{\hbar} \tag{78}$$

The quantal wave penetrates into the classically forbidden region over a distance $\Delta\theta$ such as

$$\Delta\theta\,\mathrm{Im}\,L \sim \hbar \quad . \tag{79}$$

So, the imaginary part of L is related to the uncertainty principle.

Whenever there is no real value of L (or b) that makes the sum over the partial wave stationary for some angle θ , the usual classical relation of b and θ does not exist. The contribution to the cross-section is then spread over several partial waves. This is the meaning of the usual statement that "the system does not behave classically". But there is a complex value of L that renders the sum stationary. The imaginary part of L describes this spreading, i.e. takes the uncertainty principle into account. It is remarkable that this genuine quantum mechanical effect can be obtained from totally classical (but complex) equation.

3.4.3. Diffraction. Diffraction occurs typically when a wave is scattered from a sharp edged object, and in case of potential scattering whenever the potential varies appreciably within a wavelength. Thus the condition for diffraction to occur is just the condition for the classical approximation to break down, a wave being in that case split into many pieces. The observation of diffraction was thus considered as an evidence, no classical approximation could be used. Let us consider the simple case of an angular momentum window

$$e^{2i\delta_\ell} \sim \exp\left\{-\frac{1}{2}\left(\frac{L-\lambda\hbar}{\Delta}\right)^2\right\} \exp\left\{i\,\frac{L}{\hbar}\,\theta_o + \frac{1}{2}\,\frac{i}{\hbar^2}\,(L-L_o)^2\,\theta'_o\right\} \tag{80}$$

The integral (65) can be evaluated by the stationary phase method whenever $\Delta >> \hbar$. Neglecting the variations of the first factor, one gets the stationary point

$$L_S = L_o + \frac{\theta-\theta_o}{\theta'_o}\hbar \tag{81}$$

and

$$f(\theta) \approx \exp\left\{-\frac{1}{2}\left(\frac{L_S-\lambda\hbar}{\Delta}\right)^2 - \frac{1}{2}i\frac{(\theta-\theta_o)^2}{\theta'_o}\right\} .$$

But in case $\Delta << \hbar$ this evaluation is incorrect. On the other hand, the saddle point method leads to

$$L_S = \frac{\theta-\theta_o+L_o\theta'_o - i\hbar^2(\lambda/\Delta^2)}{\theta'_o - i\hbar^2/\Delta^2} . \tag{82}$$

When $\Delta >> \hbar$ this reduces to the stationary phase evaluation except for small corrections of order $\hbar^2/\Delta^2$. In the case $\Delta << \hbar$ the same formula can be rewritten

$$L = \hbar\,\frac{\lambda + i\Delta^2/\hbar^2\ (\theta-\theta_o+L_o\theta'_o)}{1+i(\Delta^2/\hbar^2)\theta'_o} . \tag{83}$$

So, L has a complex part of the order of $\Delta^2/\hbar^2$ which is rather small. Its real part $\approx \lambda\hbar$ is nearly constant. The amplitude is proportional to

$$f(\theta) \sim \exp\left\{-\frac{1}{2}\frac{\Delta^2}{\hbar^2}(\theta-\theta_o+L_o\theta'_o)^2\right\} \tag{84}$$

and can by no means be obtained by calculating the corrections to (81) through an expansion in powers of $\hbar/\Delta$. Note the angular spreading is of the order of $\hbar/\Delta$ in accordance with the uncertainty principle.

While the stationary phase (WKB) method fails for $\Delta/\hbar << 1$, the saddle point method leads to the exact result since the exponent is quadratic in L . So, the CWKB approximation describes also diffractive effects.

4. HEAVY ION SCATTERING

4.1. Scattering by a Repulsive Sphere

Let us consider (Knoll 1976) a very sharp repulsive sphere (Woods-Saxon potential, with V = 100 MeV and a = 0.2 fm , E = 25 MeV with the $^{16}O + ^{16}O$ kinematics). The quantal cross-section (full line Fig.12b) displays very strong diffractive oscillations. The classical deflection function is monotonous and only one trajectory

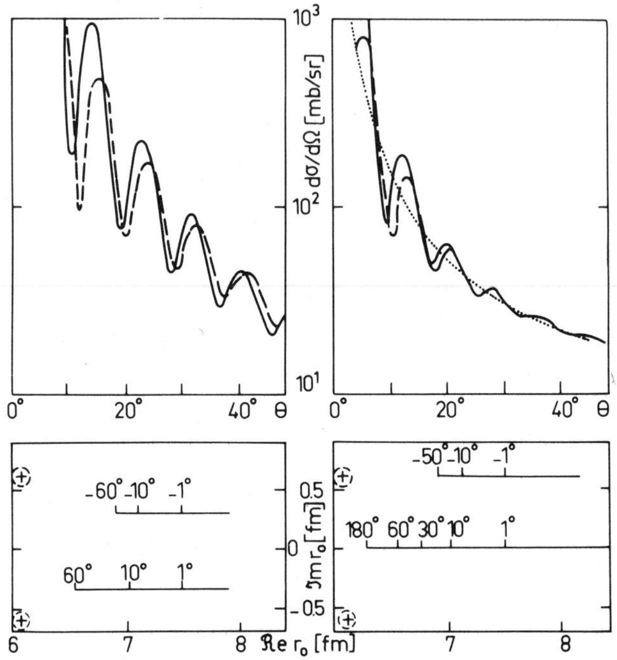

Fig.12 - From (Knoll 1976). Example of Fraunhofer scattering by an absorptive sphere (left). There are no classical trajectories, but two complex ones at $\mathrm{Im}\, r_0 \simeq \pm\pi/2\, a$ *which interfere (dashed line) in order to reproduce the quantal cross-section. At the right, we show a case of Van der Hulst diffraction by a sharp edged repulsive sphere. The dotted line is the classical result, the dashed, the* CWKB *result once the interference with the diffractive trajectory at* $\mathrm{Im}\, r_0 \sim \pi a$ *is taken into account. The full line is the quantal result.*

contributes to the scattering amplitude (70). The cross-section is thus smooth (dotted curve, Fig.12b) and represents some average of the exact quantal result. In the CWKB approximation, there is an additional, complex, contribution that can be characterized by the corresponding turning point $r_0 \approx 7\text{ fm} + i\pi a$ (Fig.12). This trajectory leads to negative angle scattering. Its contribution interferes with the classical amplitude so as to reproduce the diffractive oscillations. From the imaginary part of r_0, we get

$$\mathrm{Im}\, L = p\, a\, \pi \qquad (85)$$

that is about $2\hbar$, leading to an angular spreading of $\Delta\theta \sim \hbar/\mathrm{Im}\, L \sim 30°$ as can be seen in Fig.12b. The diffraction occurs in the surface region ($r_0 \sim 7$ fm as compared to the potential radius $R \sim 6$ fm). The imaginary part of L is associated to the surface thickness, and is rather small as can be expected from the discussion of section 3.4.3. Note that for $\mathrm{Im}\, r_o \sim i\pi a$, $V'(r_o)$ is real as predicted by the approximate formula (69).

Diffraction by a sharp edged refractive object has been studied by Van der Hulst (1957) and discussed by (Ford 1959, p.259) and (Newton 1966, p.67).

4.2. Diffraction by an Absorptive Sphere

Let us consider a strongly absorptive potential (W = 100 MeV) with a sharp edge (a = 0.2) and the same kinematics as previously. There are no classical trajectories associated with the scattering by such a potential. But (Fig.12a), there are two complex ones, for positive and negative angles respectively. The complex part of the turning points is now $\mathrm{Im}\, r_o = \pm\pi/2\ a$ that is

$$\mathrm{Im}\, L \sim p\, a \frac{\pi}{2} \quad .$$

The diffractive oscillations produced by these two complex trajectories extend over an angular range of $\Delta\theta \sim \hbar/\mathrm{Im}\, L \sim 60°$. The latter is twice as broad, as can be seen in Fig.11a despite the geometry of the potential as well as the kinematics are the same as previously. The CWKB (dashed curve) and quantal (full curve) agree quite well.

The diffraction by an absorptive object with a sharp edge is usually called Fraunhofer scattering (Fig.13).

4.3. Fresnel Diffraction

The diffraction by an absorptive screen with a sharp edge is usually referred to as Fresnel diffraction when the source is at finite distance. For heavy ion scattering, the source is (virtually)

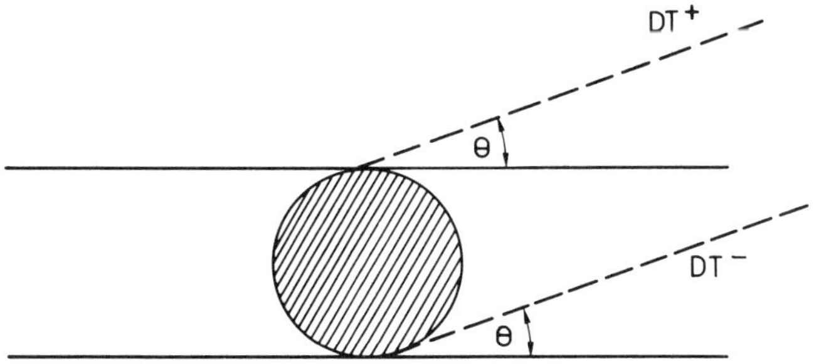

Fig.13 - From (Knoll 1976). Schematic sketch of the Fraunhofer diffractive trajectories.

brought at finite distance by the Coulomb refraction. This is schematically sketched in Fig.14, for an absorptive sphere. In region I there is classical flux from the source, but region II is a shadow region. The diffraction from the edge gives an additional contribution in region I, and provides some flux in the forbidden region II. We use the same absorptive sphere as in the previous section, but with a more realistic surface thickness $a \sim 0.6$ fm and add the Coulomb potential for $^{16}O + {}^{16}O$ scattering.

Far away from the scattering center, there is only Coulomb scattering. So, for large values of r_o, this turning point is real (Fig.15, bottom) and corresponds to small angle scattering ($\theta < 20°$) The ratio of the cross-section to the Rutherford value is nearly unity in this region. As soon as the trajectories start to feel the absorptive potential, they become complex, giving rise to the DT^+ branch of Figs.14 and 15, along which $\mathrm{Im}\, r_o \sim -i\frac{\pi}{2} a$ for large angles, as in the previous case with no Coulomb potential. The analogue of the diffractive trajectory can be seen near $r_o \sim 9 + i\frac{\pi}{2} a$ fm. It corresponds now to positive angle scattering, too, and has been called FT^+ in Figs.14 and 15. At $r = r_c$, one has $db/d\theta = 0$. So, there is a caustic which separates two regions : $\theta < \theta_c \sim 25°$ with two contributions (RT and FT) and $\theta > \theta_c$ with only one (DT^+), the FT^+ contribution being not allowed according to the saddle point rules. This leads to the cross-section of Fig.15 (top) with some oscillations at small angles, due to the $RT - FT^+$ interference, and an exponentially decreasing part (DT^+) due to the quantal penetration in the region II. The CWKB calculation (dashed line) is almost identical to the exact quantum mechanical evaluation (full line).

4.4. Strong Attraction and Rainbow

Let us now consider a more realistic case, still with the 25 MeV $^{16}O + {}^{16}O$ kinematics, but with a typical heavy ion optical potential ($V = 50$ MeV, $W = 20$ MeV, $R = 6.05$ fm, $a = 0.6$ fm). With only the real potential, the deflection function (Fig.16, top) has

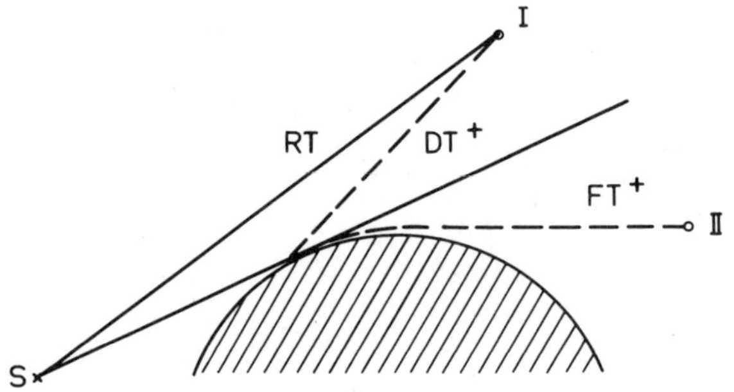

Fig.14 - From (Knoll 1976). Schematic sketch of the Fresnel diffraction case.

a maximum at $b \sim 8$ fm, $\theta \sim 25°$ corresponding to the place where the Coulomb and nuclear deflection balance each other. Then it has a sharp decrease due to the strong attraction (NT^-). There is also

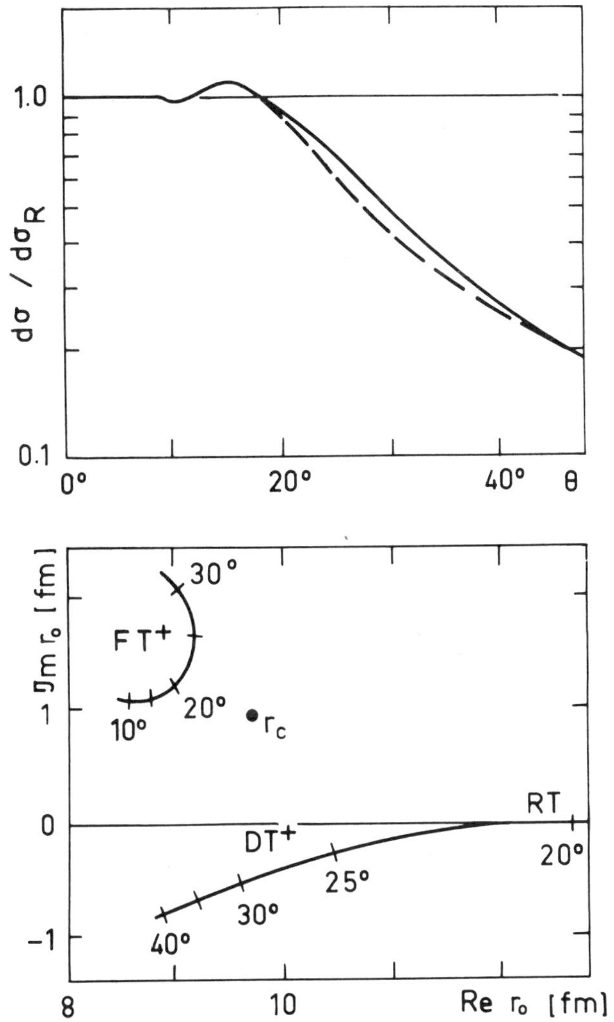

Fig. 15 – *From (Knoll 1976). Fresnel diffraction (Coulomb plus absorptive potential, no attraction). Above : cross-section divided by the Rutherford cross-section. Exact : full line,* WKB *: dashed line. Below : position of the complex turning points* r_0 *that lead to real angles. The different types of trajectories Coulomb (*CT*), diffractive (*DT^+*) and Fresnel (*FT^+*) are indicated as in Fig. 14. In the complex trajectory picture, absorption and diffraction in presence of a Coulomb field leads to a complex caustic at in* $\mathrm{Im}\, r_c \simeq \frac{\pi}{2} a$.

an internal trajectory, corresponding to deep interpenetration (IT). The Coulomb branch CT ($0 < \theta < 25°$) corresponds to real turning points ($r_o > 9$ fm), as well as the nuclear branch ($-\infty < \theta < 25°$, $7\text{fm} < r_o < 9\text{fm}$). There is a complex branch DT^+ due to the caustic. It starts at $r_o = 9$ fm and for large angles corresponds to $\mathrm{Im}\, r_o \sim -i\pi a$. So, for a purely real potential one would have the usual Coulomb rainbow at $\theta = 25°$, followed by diffraction by the edge of the (real) nuclear potential, i.e. Van der Hulst diffraction ($\mathrm{Im}\, r_o \sim 0$ on the NT^- , $\mathrm{Im}\, r_o \sim -i\pi a$ on the DT^+) .

When the imaginary part of the optical potential is included, this turning point pattern is only slightly modified. On the NT^-

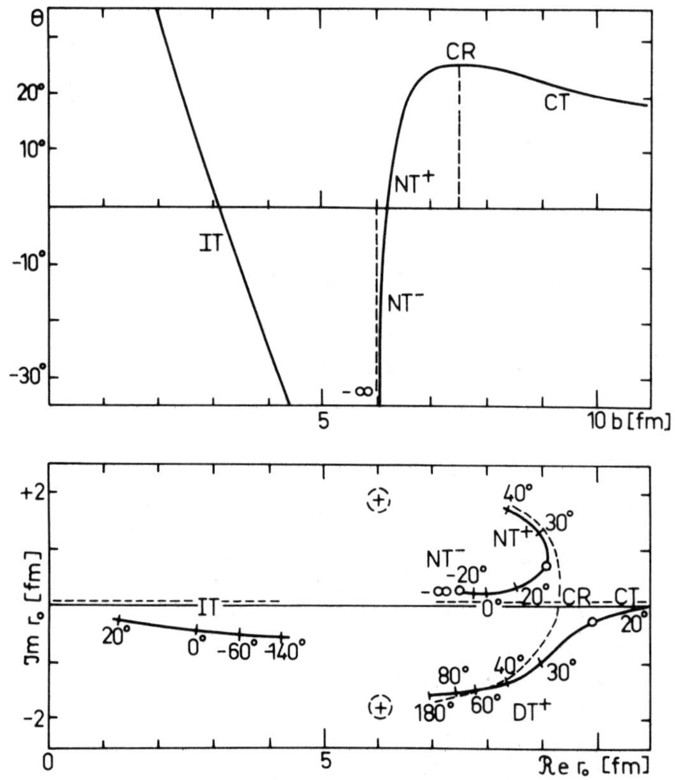

Fig.16 - From (Knoll 1976). Above : deflection function given by the real part of the $^{16}O + ^{16}O$ potential at 50 MeV. Below : position of the complex turning points leading to real angles with the full potential. The Coulomb trajectory is denoted by CT, the diffractive one by DT^+, the nuclear trajectories by $NT^\pm$ according to the sign of the scattering angle, the internal by IT. The Coulomb rainbow is indicated by CR, the orbiting by $-\infty$. The poles(crosses) of the potential give an indication of the radius and diffuseness of the Woods-Saxon form.

one has still $\mathrm{Im}\, r_o \approx 0$, and $\mathrm{Im}\, r_o \approx -\pi a$ on the DT^+ . This is very far from the Fresnel pattern which would lead to $\mathrm{Im}\, r_o \approx \pm\frac{\pi}{2} a$. It is easy to understand. At $r_o \approx 9$ fm , one is still quite far from the nuclear edge, and so the absorption is weak. So, even when the absorption is included, the angular pattern near $\theta \sim 25°$ corresponds to Van der Hulst diffraction. The diffraction is due to the sharp edge of the real potential. The calculated cross-section is shown Fig. 17 (dashed curve). The oscillations at small angles are due to the NT^+ - CT interference, the fall off to the penetration into the shadow of the Coulomb rainbow (DT^+) and the large angle oscillations correspond to DT^+ - NT^- interferences due to the Van der Hulst diffraction. On the NT^- whose turning points lie closer to the nuclear surface, the absorption starts to be important. The dotted curve in Fig. 17 corresponds to the semi-classical perturbation theory for the imaginary potential. (In the perturbation theory with classical trajectories calculated using the real part of the optical potential, there is no diffraction due to the absorption since the trajectories are modified by the absorption.) This perturbation theory is seen to be good for $\theta < 60°$, but there is not enough damping at larger angles.

The internal trajectory is totally damped by the absorption. Indeed, there is strong absorption, but very little diffraction due to the imaginary part of the optical potential. This is summarized

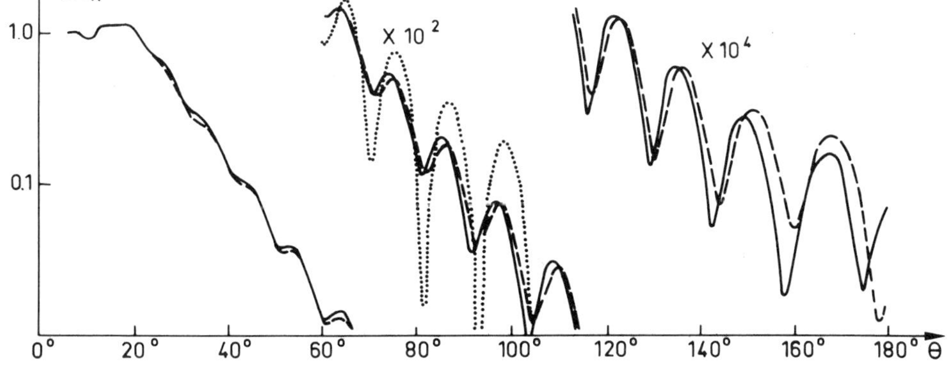

Fig. 17 - From (Knoll 1976). Cross-section (relative to the Rutherford cross-section) for the $^{16}O + {}^{16}O$ *potential at 50 MeV. The exact quantal partial wave calculation (full line) is compared to the* CWKB *result (dashed line) and to the* CWKB *result with the real potential only, adding the imaginary part by perturbation (dotted line).*

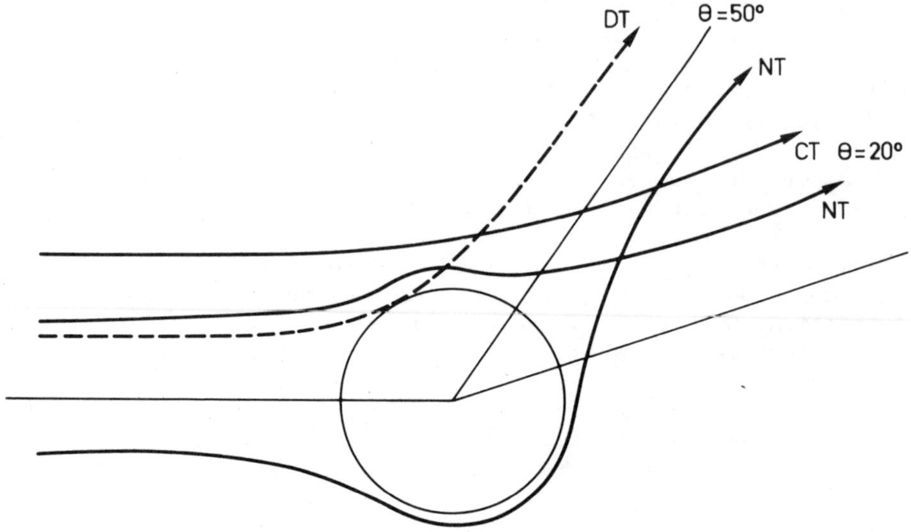

*Fig.18 - From (Knoll 1976). Most important trajectories for the scattering of two heavy ions : Coulomb trajectory (*CT*), nuclear trajectories (*NT*) which may be of diffractive nature in some cases, and diffractive trajectory (*DT*).*

in Fig.18 which shows the trajectories building up the angular distribution pattern : CT and NT at small angles (almost real), NT and DT at large angles. The latter is a new, complex trajectory describing the diffraction by the sharp edge of the real potential.

4.5. Scaling

From (69), the pattern of the turning points can be seen to be almost independent of energy. Indeed, r_o should be only a function of the product $\varepsilon = E\theta$. For the same example as in the previous section, these turning points have been calculated for $E_{CM} = 25$ MeV and $E_{CM} = 150$ MeV (Fig.19). They are nearly identical as a function of ε . This simplifies considerably the search for the allowed complex values or r_o . But the cross-sections, too, should then be universal functions of ε . This can be seen for the measured $\alpha + {}^{24}Mg$ cross-sections (Fig.20), as well as in Fig.1.

4.6. Weak Attraction

Let us now consider the ${}^{16}O + {}^{28}Si$ system (Fig.1) which has been studied over a large range of energies (Cramer 1974, Satchler 1977). These data can be fitted by an energy independent potential (E-18)

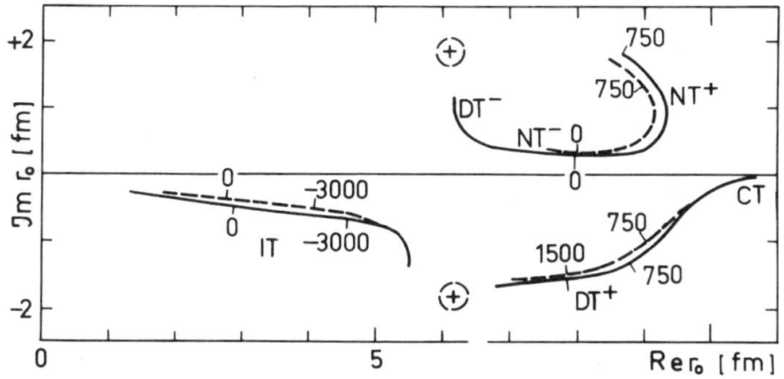

Fig. 19 - From (Knoll 1976). Universal position of turning points for trajectories leading to real angles. Dashed line, $E_{CM} = 25$ MeV ; *full line,* $E_{CM} = 150$ MeV. *The parameter on the curves is* $\varepsilon = E_{CM}\theta$ (MeV×degree).

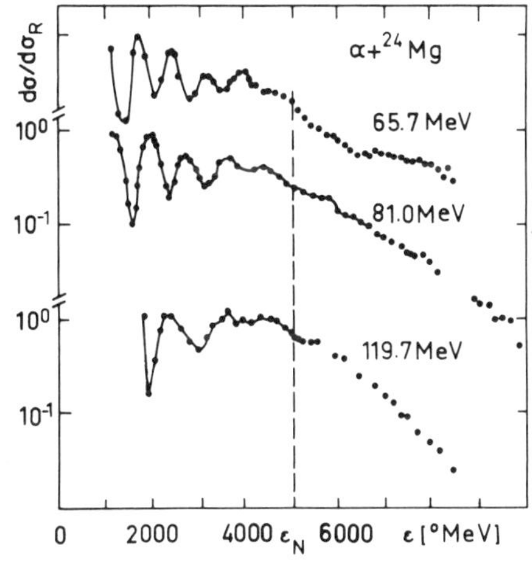

Fig. 20 - From (Knoll 1976). Experimental evidence for the scaling property of the nuclear rainbow in the elastic scattering of $\alpha + {}^{24}Mg$ *taken from (Reed 1967) and scaled here to* $\varepsilon = \theta E_{Lab}$. *The data are linked for guiding the eye.*

which has a very shallow real part ($V = 10$ MeV , $R_V = 7.5$ fm , $a_V = 0.618$ fm , $W = 23$ MeV , $R_W = 6.83$ fm , $a_W = 0.552$ fm), or by strongly attractive potentials (Satchler 1977). In the latter case this system behaves as discussed in section 4.5. Let us thus consider the weakly attractive potential.

The deflection function (Lemaire 1977, 1978) for the real potential (Fig.21) has a maximum at the Coulomb rainbow ($\theta = 40°$), but no negative branch if only real trajectories are considered. There is a minimum scattering angle ($\theta \sim 5°$) at the nuclear rainbow, and

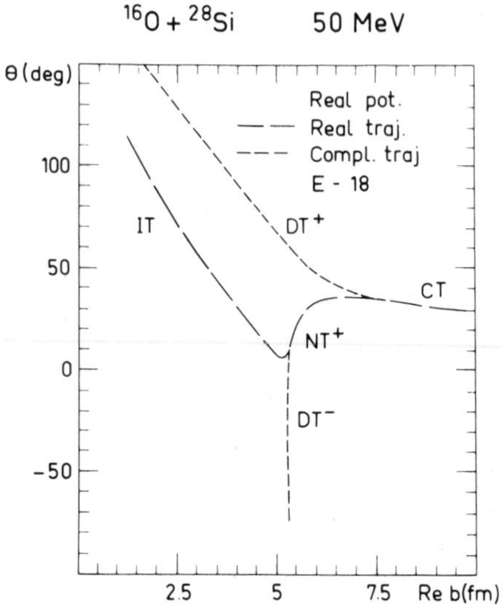

Fig.21 - From (Lemaire 1978). Deflection function for the E-18 *potential of (Cramers et al. 1974), calculated with the real part of the potential only. The complex trajectories are plotted as a function of the* Re b *although their impact parameter is complex.*

then the deflection function raises again for the smaller impact parameters. The maxima and minima correspond to caustics, and there are complex trajectories (DT^+ and DT^-) associated to them (Figs.21 and 22). As for the strongly attractive case, there is no qualitative change at large impact parameters when the absorption is included. The latter simply suppresses the internal contribution (Fig.23). The deflection function pattern is almost independent of energy. Also CWKB and quantum mechanical calculations (Fig.24) agree quite well from 33 MeV to 215 MeV. Note, at 33 MeV, that the energy available above the Coulomb barrier (15 MeV) is extremely small. So, the wavelength $\lambda \sim 5$ fm is extremely large as compared to the variations of the potential ($a \sim 0.6$ fm). The CWKB approximation is nevertheless good. The oscillations around $\theta \sim 30°$ at 215.2 MeV are due to the complex DT^- trajectory which describe the diffraction to negative angles due to the sharp edge of the potential (Fig.25). This diffraction exists already for a purely real potential. Its contribution to the cross-section for the full potential is shown in Fig.25. It can be calculated with only the real potential, and is then (Lemaire 1978) already of the correct order of magnitude (10^{-3}). So it is due, too, mostly to the diffraction by the edge of the real potential.

The forward pattern is due to the CT^+, DT^+ and NT^+ trajectories. In order to see whether the diffraction described by the DT^+ is due

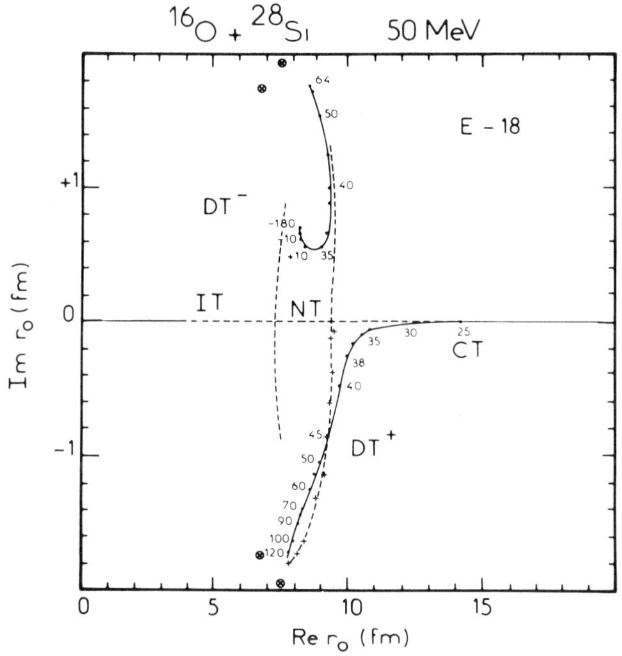

Fig. 22 - *From (Lemaire 1978). Turning point pattern in the complex plane. The dashed curve corresponds to the real part of the optical potential only. The full curve to the full potential.*

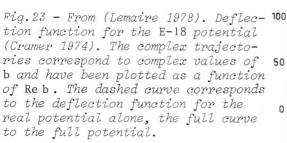

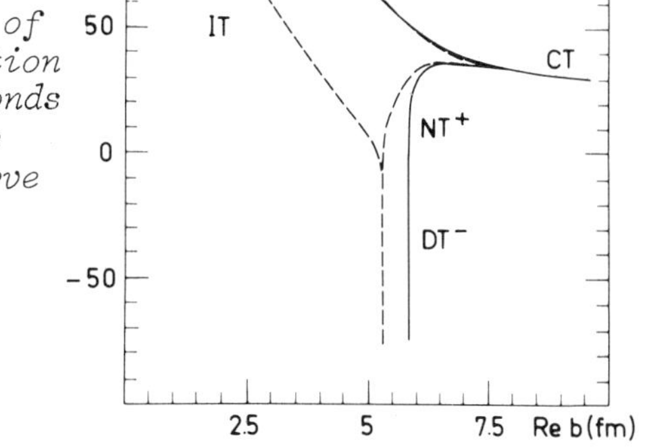

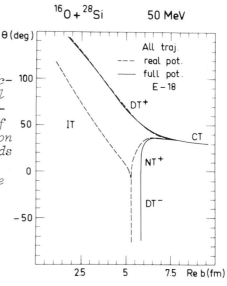

Fig. 23 - *From (Lemaire 1978). Deflection function for the* E-18 *potential (Cramer 1974). The complex trajectories correspond to complex values of* b *and have been plotted as a function of* Re b. *The dashed curve corresponds to the deflection function for the real potential alone, the full curve to the full potential.*

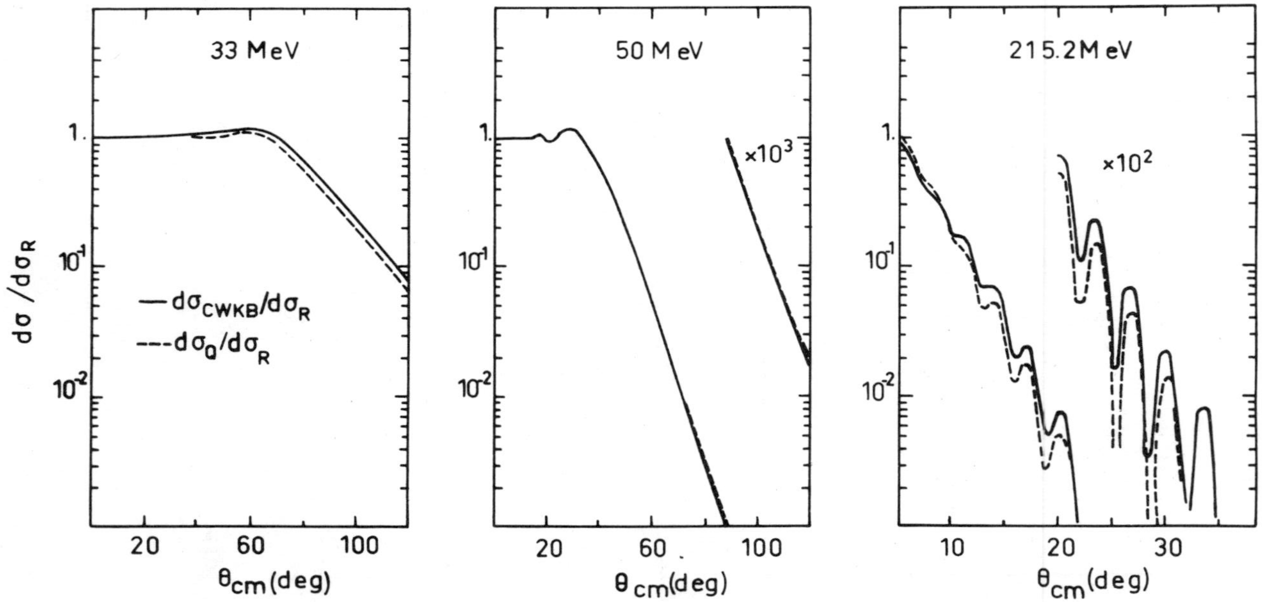

Fig. 24 – *From (Lemaire 1978). Comparison of the quantal (full line) and semi-classical (dashed line) cross-sections for* $^{16}O + {}^{28}Si$ *at various energies.*

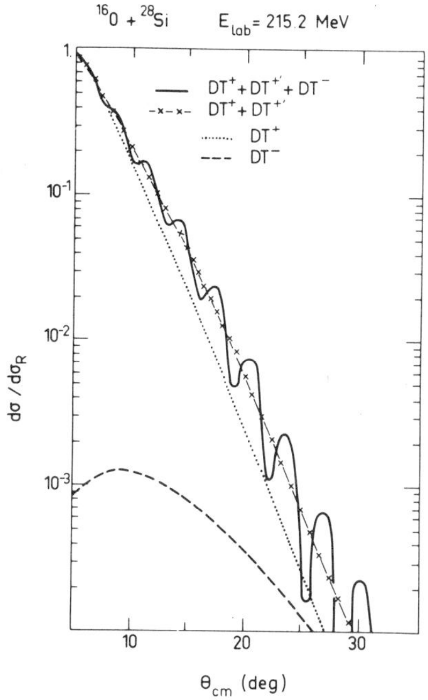

Fig.25 - From (Lemaire 1978). Cross-section for $^{16}O + ^{28}Si$ *at 215.2 MeV showing the various contributions of the different trajectories. Full line : total cross-section. Crosses : positive angle contribution ; dots positive angle contribution without the* DT'^{+} *trajectory due to the different geometry of the real and imaginary nuclear optical potential ; dashes negative angle contribution* (DT^-) *which does not exist in the real trajectory picture and which leads to the oscillations in the cross-section.*

to the real or the imaginary potential, one can consider (Lemaire 1978) the cross-section due to these trajectories only (Fig.26). The full curve is the optical model prediction calculated with the CWKB method. The dashed curve is the same calculation with the real potential suppressed. It corresponds to the picture of the reaction where there is mainly Coulomb scattering and strong absorption, the latter producing also diffraction. The dotted curve is the calculation with no absorption and corresponds to a pure Coulomb rainbow. The latter is much closer to the cross-section calculated with the full potential. So one can conclude the forward angular pattern is very close to a Coulomb rainbow.

There is of course strong absorption since the internal trajectory is completely damped. Suppressing its contribution in the $W = 0$ calculation simulates most of the features produced by the imaginary part of the optical potential. The absorption absorbs, but leads

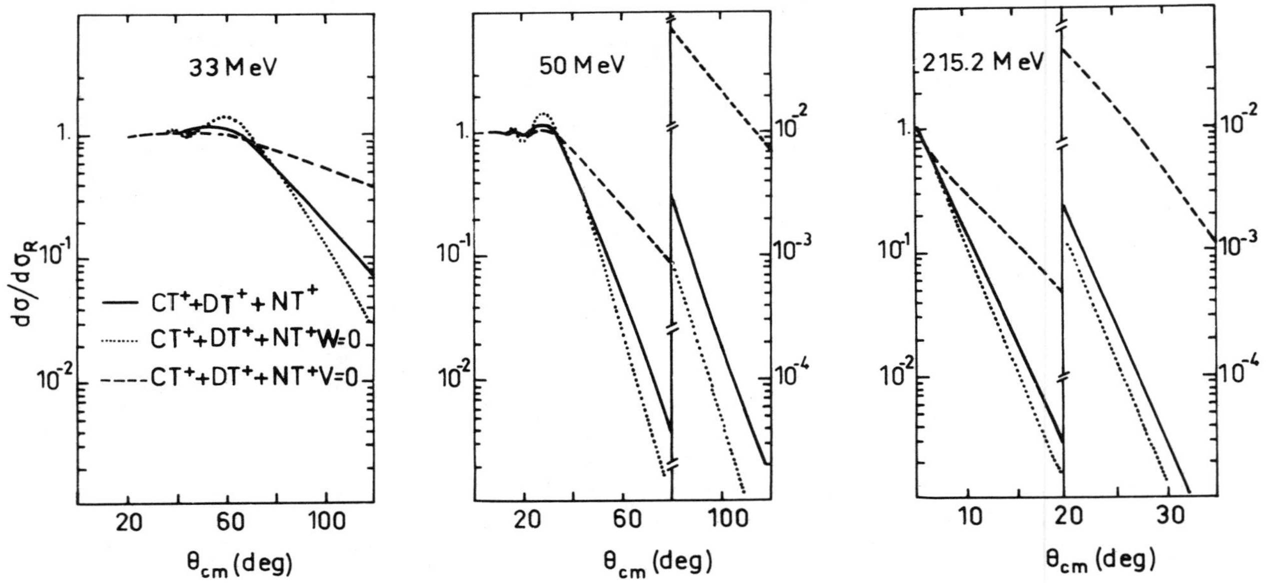

Fig. 26 – *From (Lemaire 1978). Comparison of the cross-sections due to the outer trajectories for the total, Coulomb plus absorption* (V = 0), *Coulomb plus refraction* (W = 0) *cases for the* $^{16}O + ^{28}Si$ *reaction.*

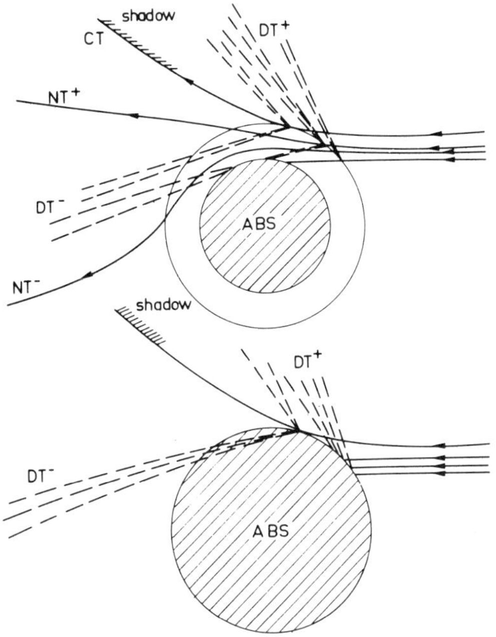

Fig.27 - The two pictures for heavy ion scattering a) *diffraction by the edge of the real potential* b) *diffraction by a "black" sphere. From (Schaeffer 1977).*

to much less diffraction than the edge of the real potential (Fig.27).

At backward angles, however, oscillations have been observed (Braun-Munziger 1977). They can be explained using complex trajectories (Lemaire 1978) as arising either from internal contributions as in α particle scattering (Brink 1977), or from surface diffraction due to the sharp edge of the imaginary potential. The correct explanation is not yet known, and depends on the optical potential used.

5. CONCLUSION

Heavy ion reactions can be described using semi-classical concepts, WKB phases or classical trajectories. The complex solutions of the classical equations of motion reproduce most of the quantum mechanical features of the cross-sections. Many objections have been made against the semi-classical approximation (Gelbke 1974, Frahn 1973, Glendenning 1975). They do not hold when complex contributions are included. In particular the WKB approximation was associated with small angle scattering. There is no restriction for the

CWKB approximation which works at any angle, essentially because it includes diffractive effects.

A semi-classical insight into the reaction process allows to understand the observed cross-sections in terms of interference patterns. Each of the trajectories contributing to the cross-section can be associated with its turning point, i.e. the "place it comes from". So, there is a close association between the optical potential shape and the form of the angular distribution.

The diffraction by the edge of the real potential (Fig.27) that was never considered seems to be one of the important features of the light ion cross-sections, at least in the energy range available presently (up to a few hundred MeV).

The complex method, which keeps the advantages of the WKB approximation, may be used for many other problems. The domain of applicability of this method is quite broad and by no means restricted to Nuclear Physics. We hope this lecture has provided some stimulation for further studies.

REFERENCES

K. Alder, A. Bohr, T. Huus and B. Mottelson, Rev. Mod. Phys. 28 (1956) 432.

Y. Avishai and J. Knoll, Zeit. für Phys. A 279 (1976) 415.

R. Balian and C. Bloch, Ann. Phys. 63 (1971) 592.

R. Balian and C. Bloch, Ann. Phys. 69 (1972) 76.

R. Balian and C. Bloch, Ann. Phys. 85 (1974) 514.

R. Balian, G. Parisi and A. Voros, to be published.

M. Berry and K. Mount, Rep. Prog. Phys. 35 (1972) 315.

P. Braun-Munziger et al., Phys. Rev. Lett. 38 (1977) 944.

D. M. Brink and N. Tagikawa, Nucl. Phys. A279 (1977) 159.

R. A. Broglia and A. Winther, Phys. Rep. 4C (1972) 153.

J. C. Cramer et al., Phys. Rev. C14 (1974) 2158.

R. Da Silveira, Phys. Lett. 45B (1973) 211.

K. W. Ford and J. A. Wheeler, Ann. Phys. 7 (1959) 259.

W. E. Frahn, Fundamentals in Nuclear Physics, IAEA Vienna 1967.

W. E. Frahn, Heavy Ion, High Spin States and Nuclear Structure, IAEA Vienna 1975.

W. E. Frahn and K. E. Rehm, Phys. Rep. 37C (1978) 1.

N. Froman and P. O. Froman, JWKB Approximation, North Holland, Amsterdam 1965.

N. Froman and P. O. Froman, Nucl. Phys. A147 (1970) 606.

C. K. Gelbke, Z. Physik 271 (1974) 399.

J. Knoll and R. Schaeffer, Phys. Lett. 52B (1974) 131.

J. Knoll and R. Schaeffer, Ann. Phys. 97 (1976) 307.

J. Knoll and R. Schaeffer, Phys. Rep. 31C (1977) 159.

T. Koeling and R. A. Malfliet, Phys. Rep. 22C (1975) 181.

M. C. Lemaire, Florence, 1977.

M. C. Lemaire and R. Schaeffer, to be published.

L. Landau and E. Lifchitz, Quantum Mechanics, MIR (1967).

R. A. Malfliet, S. Landowne and V. Rostokin, Phys. Lett. 44B (1973) 238.

A. Messiah, Mécanique Quantique, Dunod 1959.

J. Miller, J. Chem. Phys. 53 (1970) 1949, 3578.

J. Miller, Adv. in Chem. Phys. 25 (1974) 69.

R. G. Newton, Scattering Theory of Waves and Particles, McGraw-Hill 1966.

V. L. Pokrovskii and I. M. Khalatnikov, Sov. Phys. JETP 40 (1961) 1713.

M. Reed, B. H. Harvey and D. L. Hendrie, BAPS 12 (1967) 912; H.H. Duhm, Nucl. Phys. A118 (1968) 563.

R. Satchler, Nucl. Phys. A279 (1977) 493.

H. C. Van der Hulst, Scattering of Light by Small Particles, Wiley, New York 1957.

WKB: G. Wentzel, Z. Physik 38 (1926) 518
H. A. Kramers, Z. Physik 38 (1926) 828
L. Brillouin, C. R. Acad. Sci., Paris 183 (1926) 24.

Some of the material in this course is taken from a lecture given by the author at the "Les Houche" Summer School, 1977.

NUCLEAR DYNAMICS: THE TIME-DEPENDENT MEAN-FIELD APPROXIMATION AND BEYOND*

J. W. Negele

Center for Theoretical Physics
Department of Physics and Laboratory for Nuclear Science
Massachusetts Institute of Technology
Cambridge, Massachusetts 02139

PREFACE

Since physicists vary in their degree of interest in the theory of nuclear dynamics, these lectures address several levels of theoretical sophistication. Whereas one might mumble excuses concerning the broad distribution in the background of students, in fact my time and endurance simply gave out before I had produced a completely satisfactory pedagogical sequence which started at the very beginning and derived everything required to reach the end. In addition to sincere apologies to all concerned and a few homework problems designed to fill in the most obvious gaps, I also offer the following brief guide to the reader.

The Introduction is very general, is intended to motivate the entire work, and provides a basic introduction to TDHF. It should be understandable to anyone. Section II is primarily for theorists. Whereas I believe the elucidation of the general theory is crucial to the intellectual credibility of the entire approach, experimentalists can in fact skip it and still enjoy the pretty pictures in Sections III and IV. Of the examples in Section III, the one-dimensional slabs play the most crucial role in displaying the salient physical features of the mean-field theory. The exactly solvable models are directed primarily at theorists who are interested in using some technical expertise to put the general theory to real quantitative tests. However, any misguided individuals who labor under the delusion that we are in the business

*This work is supported in part through funds provided by the U.S. Department of Energy (DOE) under contract EY-76-C-02-3069.

of calculating many-body wave functions instead of expectation values of few body operators are strongly encouraged to take a peek at the results for two interacting Lipkin systems. Section IV presents an up-to-date summary of the methods and results of applying the time-dependent-mean-field approximation to realistic experimental situations. In the style of Arthur Fiedler, the low-brow offerings to the masses have been held back until adequate doses of the classics had been administered in the earlier sections. Finally, some of the open questions are briefly discussed in Section V.

I. INTRODUCTION

A fundamental problem arising in virtually every field of theoretical physics is that of dealing with systems possessing large or infinite numbers of degrees of freedom. Although these lectures are presented in the context of nuclear physics, the actual problem addressed is far more general: the formulation of a microscopic theory of the dynamics of non-relativistic self-bound, composite systems. With the overwhelming evidence that hadrons are comprised of underlying constituents, for example, a relativistic generalization of such a theory will be crucial in replacing the present phenomenology with a fundamental theory. Small drops of liquid He^3 and He^4, which are only beginning to be explored experimentally,[1] provide a particularly striking example of a system which cries out for such a theory; not only are the two-body forces and bulk liquid properties well known, but also one may explore directly the differences arising from Fermi and Bose statistics. These, and many other applications from condensed matter physics and quantum chemistry underscore the need for a general, practical microscopic theory.

Irrespective of the general significance of a microscopic theory of dynamics, there exist compelling reasons for investigating such a theory within the context of nuclear physics. Despite the underlying mesonic origin of nuclear forces, the nuclear many-body problem may be described non-relativistically as an adequate first approximation. Even in nuclear ground states, one observes strong interplay between single-particle structure and shape degrees of freedom. The phenomenology of excited states is even richer, with the full spectrum of single-particle and collective degrees of freedom arising in appropriate reactions. The accessibility of nuclei to diverse probes allows the methodical exploration of the charge, current, and magnitization densities with electromagnetic probes and the systematic excitation of states of most desired quantum numbers with mesonic and hadronic probes. Finally. the possibility of studying reactions with projectiles ranging from a single nucleon to the heaviest nuclei at energies ranging up to hundreds of MeV per nucleon offer incomparable opportunities to investigate dynamics under diverse conditions. For all these reasons, then, nuclear physics provides a natural testing ground in which to try out our first tentative ideas in formulating a general dynamical theory.

1.1. Objectives

Given the complexity of the full time-dependent many-body problem, it is desirable to formulate a general hierarchy of successive approximations. The lowest order theory should provide an intuitively motivated and physically sound approximation which yields at least a qualitative description of a wide range of dynamical processes. Furthermore, there must definitely exist a completely systematic series of corrections so that, at least in principle, the evaluation of any observable to arbitrary order is conceptually unambiguous. In view of the appreciable progress in the microscopic theory of finite nuclei, it would also be desirable that in the special case of stationary states, the general theory make contact with existing theories for which convergence properties with nuclear potentials are known.

Certainly, one great appeal of such a general hierarchy of successive approximations is the freedom in each order from any adjustable parameters or adjustable assumptions. Given the nuclear hamiltonian and appropriate initial conditions, the theory itself determines the relevant collective or single-particle degrees of freedom. If some significant feature of the force is changed, the system responds appropriately. By simply changing the initial conditions, one should be able to obtain a single unified description of such diverse phenomena as transition to excited states induced by external fields, large amplitude collective oscillations, fission, fusion, compound nucleus formation, dissipation, strongly-damped collisions, fragmentation, and high density self-sustaining spin-isospin instabilities.

Such a microscopic description is in marked contrast to the plethora of models which, guided by the phenomenology, have deliberately built into them just those features one ultimately wishes to observe in a particular application. Only certain degrees of freedom are included, and no change in the interaction or initial conditions can bring in other, suppressed degrees of freedom. When one changes from the description of one phenomemon to another, a new model with different degrees of freedom and assumptions is invoked, thereby losing all contact with a general unified description. Not only is there no systematic program of successive corrections to a hydrodynamic model, various friction models, or the assumption of random nuclear matrix elements, but often due to the multiplicity of adjustable parameters and/or assumptions, it is not even possible to perform a definitive test of the underlying theory. Indeed, it is most worthwhile to fully savor the essential shortcomings of the various models on the market so as to confront with adequate vigor and enthusiasm the fundamental challenge of formulating a viable microscopic theory.

1.2 General Considerations from Many-Body Theory

Given our aim of addressing the full time-dependent quantum many-body problem, it is useful to first review several pertinent aspects of many-body theory. Although the discussion will systematically be restricted to fermions, virtually every aspect of the theory may be carried over to bosons by assigning a fictitious hidden quantum number with N allowed values and requiring that the non-interacting wave function be totally antisymmetric in the hidden variable.[2]

1.2.1 The distinction between approximations to matrix elements of few-body operators and to the complete wave function

At the outset, it is essential to accept the fact that whereas many-body theory may under certain circumstances yield excellent approximations to expectation values of few body operators, it is inevitably inadequate to describe the full N-body wave function. Consider the familiar example of a fully-interacting N-particle ground state in which each particle has a small probability ε of being excited out of its normally occupied single-particle state. Whereas a determinant of the normally occupied states will reproduce expectation values of few body operators to order ε, the probability of all N particles simultaneously being in the occupied states is $(1-\varepsilon)^N \sim e^{-\varepsilon N}$ rendering the overlap between the determinant and the exact wave function exponentially small. An alternative physical argument may be made in coordinate space. Whereas the mean-value of a finite-range two-body operator is obviously sensitive to two-body correlations, it is unaffected by multi-particle correlations specifying what additional particles are simultaneously doing far away elsewhere in the system. In fact, the error in describing the behavior of other particles far away from a correctly correlated pair simply contributes an arbitrary normalization factor which cancels out of the numerator and denominator of $\langle \mathcal{O} \rangle = \langle \psi | \mathcal{O} | \psi \rangle / \langle \psi | \psi \rangle$. In contrast, full knowledge of N-particle correlations and hence at least N orders of perturbation theory are required to describe adequately the full wave function. For large systems we must therefore give up all pretense of calculating wave functions and deal exclusively with mean values of few-body operators.

The distinction between mean values and overlaps of N-body wave functions is particularly crucial in addressing time-dependent problems. Sloppiness in making the corresponding distinction for stationary states is seldom disastrous because virtually all experimental measurements deal with expectation values of one or two-body operators. Thus, binding energies, removal energies, and charge density distributions do not really probe the full wave function, but rather just the expectation value of the hamiltonian and one-body density operator. Similarly, despite loose talk concerning testing RPA wave functions, in practice only transition densities induced by one-body operators can be compared with experiment. Time-dependent applications, however, are far more dangerous, since experimentalists insist on confronting theorists with S-matrix elements, which are overlaps of N-body wave functions evolved through some interaction time with other appropriate N-particle wave functions describing final asymptotic states. Since we can say nothing about individual S-matrix elements, we must restrict our attention to mean values of appropriate operators. In heavy ion collisions, such operators might include the fragment mean proton number, neutron number, and c.m. momentum or higher moments like the dispersion in these quantities.

It should also be obvious from this discussion that arguments concerning the "lifetime of a determinant"[3] are as irrelevant to the validity of a theory for the evolution of mean values as the overlap between a determinantal approximation and the fully interacting wave function, since they ask the same inappropriate questions concerning the overlaps of N-body wave functions.

Because of the crucial importance of distinguishing between mean values and S-matrix elements and errors in the recent literature in this connection, a specific case of a solvable model will be elaborated in detail in Section 3. Also for this and a second solvable model, the irrelevant determinant lifetime will be contrasted with the actual period of validity of the many-body approximations.

1.2.2 Many-body theory for stationary states

Although limitations of time and space preclude a thorough discussion, it is useful to tersely summarize the present status of nuclear many-body theory for stationary states.

The simplest system with which to begin is the fictitious infinite uniform system comprised of equal numbers of neutrons and protons interacting with nuclear forces in the absence of Coulomb interactions which has come to be known as nuclear matter. After decades of effort, the two complementary approaches of variational methods and perturbation theory appear to be completely consistent.[4]

Using hypernetted chain techniques, variational calculations with Jastrow trial functions of the form

$$\psi(r_1,r_2\ldots r_n) = \prod_{i,j} f(r_i-r_j)\,\phi(r_1,r_2\ldots r_n) \qquad (1.1)$$

now yield reliable upper bounds on the binding energy with purely central forces for strongly interacting systems of either fermions or bosons. Furthermore, for bosons, comparison with exact Green's function Monte Carlo calculations[5] for liquid He^4 shows that the variational upper bound is exceedingly close to the exact energy for densities appropriate to nuclear matter.

Perturbative calculations of nuclear matter reorganize the Goldstone expansion into the form of a hole-line expansion in which diagrams are grouped according to the number of independent hole lines they contain. Each additional hole line diminishes the contribution by roughly a factor of κ, the probability of excitation of a pair of particles out of their normally occupied states in nuclear matter. Additional particle lines do not decrease the contribution of a general diagram, so one calculates a hierarchy of successive approximations. First, at the two-hole-line level, all ladder diagrams involving two nucleons rescattering any number of times into unoccupied states are summed, yielding the G-matrix or reaction matrix. At the three-hole-line level, the three-body Faddeev diagrams are summed, and, in principle, terms to any order may be summed. For all the model systems with central forces for which variational methods have been compared with perturbation theory, the two results are consistent within errors estimated by the power of κ representing the leading omitted terms. Although variational calculations are much more difficult for a realistic state-dependent nuclear force, recent variational calculations of nuclear matter with the Reid potential appear to be consistent with perturbation theory, thus lending strong support to the validity of the perturbative approach.[6] One should note, however, that the Reid potential appears to produce an equilibrium density much higher than observed in the interior of finite nuclei. That the Reid potential should be in error is no surprise, since it also underbinds the exactly solvable three-body problem, but it is important to bear in mind the shortcomings of the Reid potential when comparing microscopic calculations with experiment.

Both because of strong evidence of adequate convergence in nuclear matter and because it offers the advantage of systematic refinement, perturbation theory provides the most fundamental treatment of finite nuclei. Finite nuclei are much more difficult to calculate than nuclear matter, however, because the single particle basis is no longer specified by translation invariance and because the lack of momentum conservation introduces entire new topologies of Goldstone diagrams. Thus, except in the special case of light nuclei, one must be somewhat less ambitious in the diagrams which are summed in explicit calculations.

Consider the evaluation of some one body operator, $\mathcal{O}$, denoted by a dot in each diagram. Adding and subtracting a one-body potential to the hamiltonian

$$H = (T + U) + (v - U) \tag{1.2}$$

and denoting -U by an X and v by a dashed line, some typical low order contributions to $\langle\mathcal{O}\rangle$ are given by:

$\langle\mathcal{O}\rangle$ =

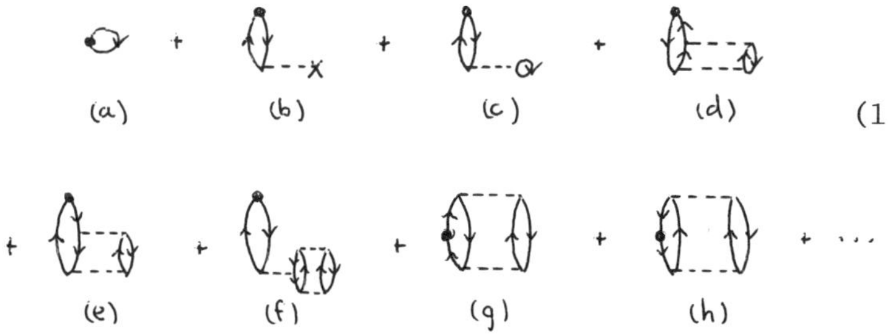

(1.3)

The first selective summation is to replace all upgoing ladders by the G-matrix, denoted by a wavy line and defined by the integral equation

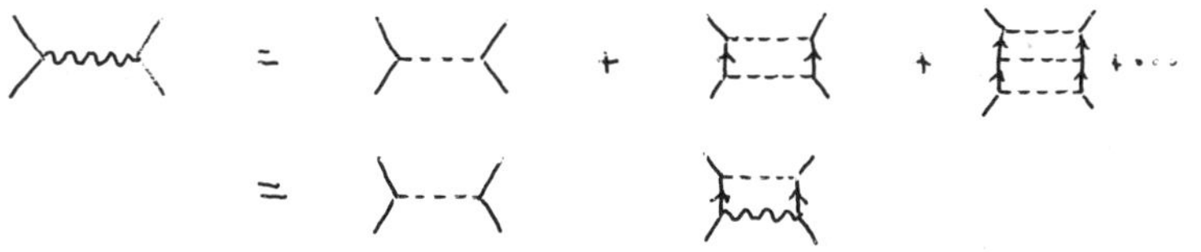

(1.4)

Then diagrams c and d of 1.3 are generated by the single diagram

(1.5)

and all the remaining diagrams in 1.3 come from diagrams of identical topology with each v replaced by G.

The next selective summation is accomplished by defining U to cancel some class of other diagrams. The simplest prescription is to define 1.3b to exactly cancel c. That is, denoting particles by ρ and holes by ν

$$\langle\rho|U|\nu\rangle = \sum_{\nu'} \langle\rho\nu'|v|\nu\nu' - \nu'\nu\rangle \quad . \tag{1.6}$$

This is just the familiar Hartree Fock (HF) approximation in which the single-particle potential is just defined to be the field generated by summing the contribution of each other occupied orbital interacting through the two-body potential. Evidently, a more physical choice would be to define 1.3b to exactly cancel 1.5. Then the complete ladder sum contributes to the potential,

$$\langle\rho|U|\nu\rangle = \sum_{\nu'} \langle\rho\,\nu'|G|\,\nu\nu' - \nu'\nu\rangle \tag{1.7}$$

where technicalities related to the specification of energy denominators have intentionally been neglected. This clearly goes beyond the HF approximation and is called a mean-field approximation. The entire sum of successive interactions of one particle with some other particle in the medium represents the mean field experienced by the first due to the second much better than the single Born term included in H.F. The terminology "mean-field approximation" is obviously not unique since one may build arbitrarily large classes of diagrams into the definition of a mean field. The class of mean-field diagrams which has been shown to be most crucial in calculating finite nuclei[7] is defined by the generalizations of terms c through f in (1.3)

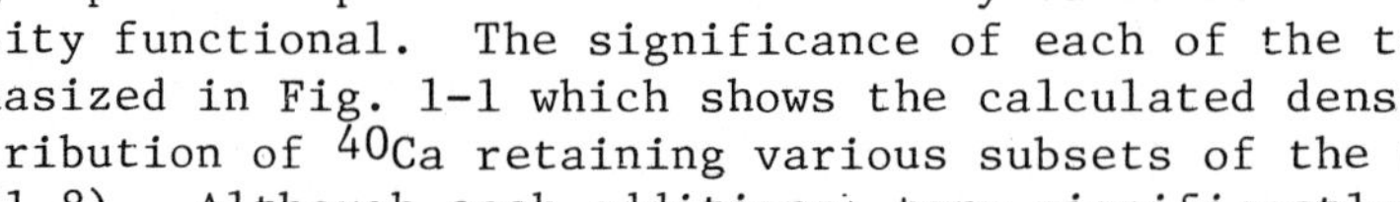

(1.8)

These are the graphs retained in the density-dependent Hartree Fock (DDHF) theory in which the density matrix dependence of the G matrix is approximated by density dependence and the single particle potential is obtained by variation of the energy density functional. The significance of each of the terms is emphasized in Fig. 1-1 which shows the calculated density distribution of ^{40}Ca retaining various subsets of the terms in (1.8). Although each additional term significantly decreases the interior density, the final result is still somewhat above the lowest curve which yields an excellent fit to electron scattering.

The coupled cluster or e^S formalism[8] is an alternative formulation of successive approximations to the time-independent

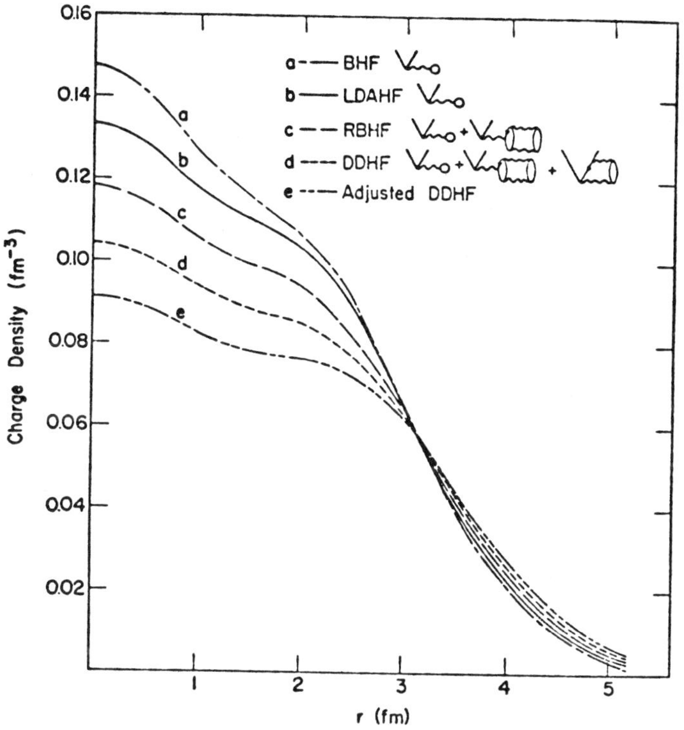

Fig. 1-1. Charge density of ^{40}Ca calculated with five different definitions of the mean field.

Schrodinger equation, and is explained in detail in Section 2. For the present discussion, it is sufficient to note that the lowest order truncation, denoted BHF, is essentially equivalent to summing ladder diagrams, the next order, FBHF(3), sums the three-body Faddeev diagrams, and the following order, FBHF(4), includes four-body Faddeev sums. This hierarchy has actually been implemented in spherical cloud shell nuclei as heavy as ^{40}Ca, without any of the additional simplifying approximations built into DDHF. The facts that comparisons of different coupled cluster calculations in ^{16}O show that diagrams 1.8 a,b and c play the same dominant role and that a full calculation of ^{40}Ca agrees quite well with the DDHF result provide strong corroboration of the DDHF theory.

Since the Reid potential predicts nuclear matter to be too dense, it is quite consistent that both DDHF and coupled cluster calculations of ^{40}Ca using the Reid potential also produce interior densities higher than determined from experiment. Until a more suitable force is discovered, it is essential to phenomenologically modify the effective interaction in finite nuclei to yield the proper binding energy and density. Although painful for the purist to accept, and a step which, in principle, can be eliminated sometime in the future when we are less ignorant, this adjustment is crucial in making the mean field represent quantitatively what actually occurs in observed nuclei rather than in a fictitious universe held together with the Reid potential. In fact, since DDHF contains only two adjustable parameters fitted once and for all to the energy and radius of a single nucleus, a number of stringent consistency checks strongly substantiate this phenomenological adjustment.

1.2.3 Successes of the mean-field theory

In order to fully appreciate the time-dependent mean-field theory, it is useful to recall the rather appreciable successes of the corresponding static approximation. Comparison with experiment is only warranted in nuclei for which crucial criteria for validity are satisfied, namely those having well-closed shells and sufficiently deep energy-of-deformation minima. For such nuclei, DDHF systematically reproduces single particle energies and the binding energy per particle at the level of several MeV and fractions of an MeV respectively. Since the original interaction was adjusted to fit ^{40}Ca, the precise agreement[9] with electron scattering cross sections, although quite pleasing, is not a definitive test of the theory. The impressive agreement in Figs. 1-2 and 1-3 of the theoretical predictions[10] for Pb and ^{238}U calculated using the same interaction with the subsequently measured cross sections constitutes an exceedingly stringent test of the mean field theory. Although less is known about neutron distributions, Fig. 1-4 shows that the predicted DDHF neutron densities are consistent with asymptotic neutron densities inferred from sub-coulomb pickup[11] and the surface density measured in α scattering.[12] In addition, in contrast to the high energy proton scattering data from Gachina, recent measurements at both Saclay and Los Alamos are also consistent with the DDHF densities.[13]

1.3 The time-dependent Hartree-Fock approximation

Just as the static HF approximation is the simplest form of the time-dependent mean-field approximation, so too time-dependent Hartree Fock (TDHF) constitutes the simplest form of the time-dependent mean-field theory. Thus, TDHF is a natural introduction to the time-dependent problem.

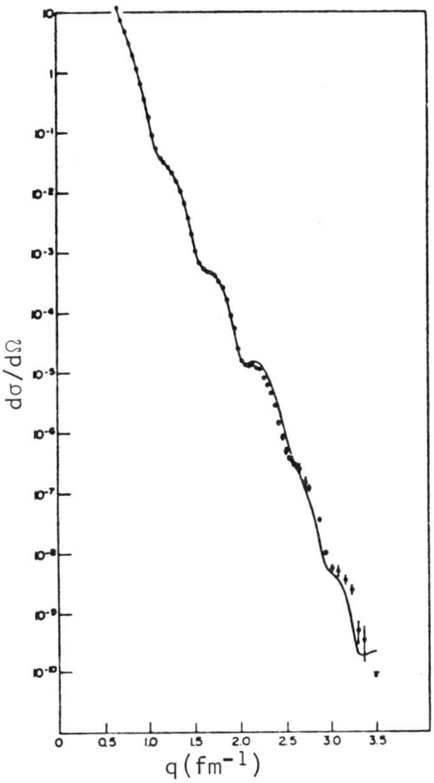

Fig. 1-2 Cross sections for elastic scattering of ^{208}Pb at 502 MeV compared with the DDHF prediction (solid line).

1.3.1 Similarity to static H.F.

There are seven features of the stationary state H.F. theory which are particularly relevant to our discussion of the time-dependent theory:

1) The intuitive argument of Hartree suggests that the motion of each particle should be governed by the mean field of the others.

2) The theory avoids arbitrary parameterization of the shell model potential by oscillator or Wood Saxon wells.

3) Instead, the system and H uniquely specify the energy, shape and radial distribution.

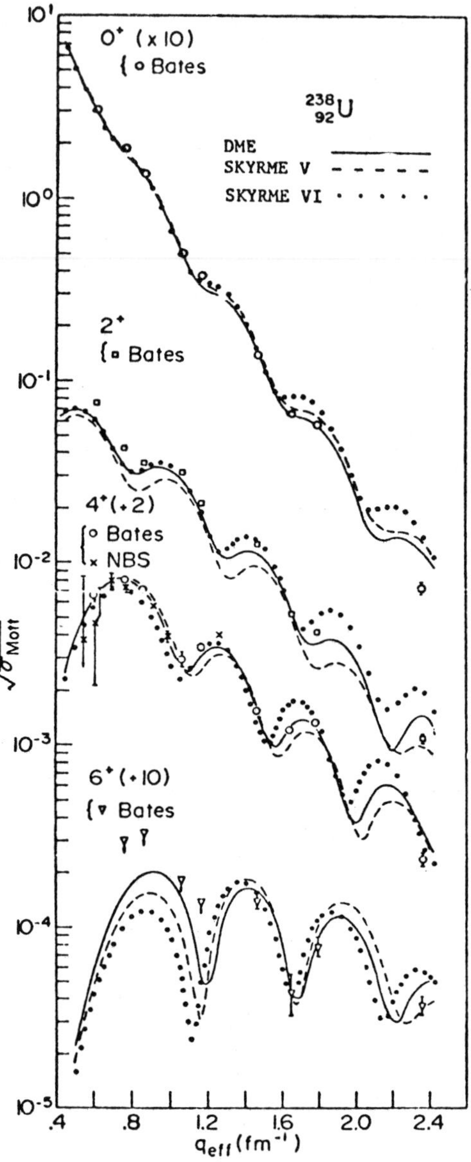

Fig. 1-3 Cross sections for elastic and inelastic electron scattering from ^{238}U compared with the DDHF prediction (solid line) and various phenomenological forces.

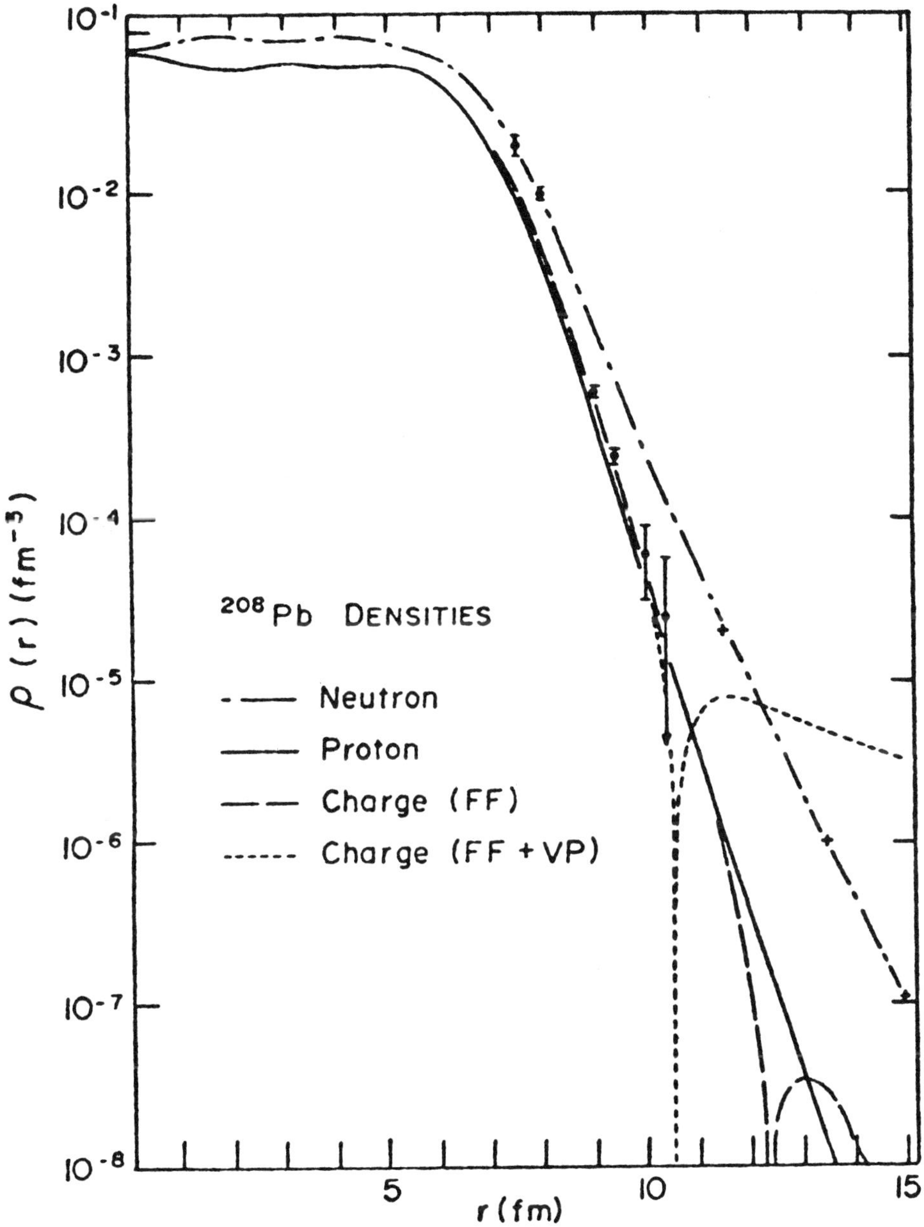

Fig. 1-4 Comparison of DDHF density predictions with results of elastic electron scattering (proton error bars) α scattering (neutron error bars) and sub-Coulomb pickup (crosses).

4) Single particle equations may be derived variationally from $\delta\langle\psi|H|\psi\rangle/\langle\psi|\psi\rangle$, with the restriction that $|\psi\rangle$ be a Slater determinant.

5) The structure of linear quantum mechanics is replaced by a system of coupled non-linear equations.

6) The single-particle wave functions obtained in (4) provide an optimal basis for systematic perturbation corrections.

7) The theory has unavoidable semiclassical aspects. The localized center of mass wave function represents a wave packet of momentum eigenstates and a deformed HF wave function must be understood as a wave packet of angular momentum eigenstates.

Each of these seven features has a precise analog in the time-dependent theory:

(1) Intuitively, the mean field is the most obvious mechanism to communicate collective information.

(2) The theory avoids arbitrary selection of collective and intrinsic variables and the need for parameterizing shapes.

(3) Instead, the initial conditions and H uniquely specify the dynamics.

(4) Single particle equations may be derived variationally from the stationarity of the action $\delta\langle\psi(t)|i\frac{\partial}{\partial T} -H|\psi(t)\rangle$ and restricting ψ to be a Slater determinant.

(5) Linear quantum mechanics has again been replaced by a system of coupled non-linear equations.

(6) The basis of time-dependent single-particle wave functions provides an optimal basis for definition of reaction channels. In contrast to the usual optical model description in terms of ground states of the scattering nuclei, no one-particle one-hole amplitudes remove probability from the entrance channel.

(7) As before, the initial conditions represent a wave packet for the impact parameter and orientation of two interacting nuclei. The final state evolved by the equations of motion approximates the final wave packet evolving from the original packet.

Thus, TDHF is the obvious generalization of the HF approximation to dynamics.

1.3.2 Variational derivation

For the case of a two-body potential and a determinantal wave function, the action may be written

$$I = \int dt \, \langle\psi(t)| i\hbar \frac{\partial}{\partial t} - H |\psi(t)\rangle$$

$$= \int dt \, \{ \sum_{\nu} \int dx \, \phi_{\nu}^{*}(x) \, (i\hbar \frac{\partial}{\partial t} + \frac{\hbar^2}{2m} \frac{\partial^2}{\partial x^2}) \, \phi_{\nu}(x)$$

$$- \frac{1}{2} \sum_{\mu\nu} \int dx_1 \, dx_2 \, \phi_{\nu}^{*}(x_1) \, \phi_{\mu}^{*}(x_2) v(x_1 - x_2) \quad \times \tag{1.9}$$

$$[\phi_{\nu}(x_1)\phi_{\mu}(x_2) - \phi_{\nu}(x_2) \, \phi_{\mu}(x_1)] \} \quad .$$

Although (1.9) should actually be varied with respect to the real and imaginary parts of an arbitrary ϕ_{ν}, as shown in Problem 1-1 the same result is obtained by variation with respect to ϕ_{ν}^{*}. Thus, stationarity of the action with respect to $\delta\phi_{\nu}^{*}(x)$ yields

$$i\hbar \, \dot{\phi}_{\nu}(x) = - \frac{\hbar^2}{2m} \frac{\partial^2}{\partial x^2} \, \phi_{\nu}(x) + \sum_{\mu} \int dx' \, v(x-x') |\phi_{\mu}(x')|^2 \, \phi_{\nu}(x)$$

$$- \sum_{\mu} \int dx' \, v(x-x') \, \phi_{\mu}^{*}(x') \, \phi_{\mu}(x) \, \phi_{\nu}(x') \quad . \tag{1.10}$$

For compactness of notation, Eq. (1.10) is usually written in a simpler, although less explicit form. One trivial step is to choose units such that $\hbar = c = 1$. Although it may seem pedantic, one should recall that any one-body operator may be written in the form $\mathcal{O}(x,x')$. With an implied integration over repeated variables,

$$\mathcal{O}\psi(x) \equiv \int dx' \, \mathcal{O}(x,x') \, \psi(x') \quad .$$

In this notation, the kinetic energy operator may be written in the form

$$T(x,x') = - \frac{\hbar^2}{2m} \, \delta(x-x') \, \frac{d^2}{dx'^2} \quad . \tag{1.12}$$

The TDHF equations (1.10) may then be rewritten

$$i\dot{\phi}_\nu = (T+W)\,\phi_\nu \equiv h\phi_\nu \tag{1.13}$$

which stands for

$$i\dot{\phi}_\nu(x) = \int dx'\,(T(x,x') + W(x,x'))\,\phi_\nu(x')$$
$$\equiv \int dx'\; h(x,x')\,\phi_\nu(x') \tag{1.14}$$

where

$$W(x,x') = \delta(x-x')\int dx''\; v(x-x'')\;\rho(x'')$$
$$-\;\rho(x,x')\;v(x-x') \quad . \tag{1.15a}$$

This is a special case of the general result

$$W(x_1,x_3) = \int dx_2 dx_4\; \langle x_1x_2|v|x_3x_4 - x_4x_3\rangle\; \rho(x_4,x_2) \tag{1.15b}$$

for a local potential

$$\langle x_1x_2|v|x_3x_4\rangle = \delta(x_1-x_3)\,\delta(x_3-x_4)\;v(x_1-x_2) \quad .$$

In Eq. (1.15), we have introduced the one-body density matrix, which is defined in general as

$$\rho(x,x';t) \equiv \langle\hat{\psi}^+(x',t)\hat{\psi}(x,t)\rangle \quad . \tag{1.16}$$

As shown in Problem 2, in the special case of evaluation in a determinantal wave function, the density matrix becomes

$$\rho(x,x';t) = \sum_\nu \phi^*_\nu(x',t)\phi_\nu(x,t) \quad . \tag{1.17}$$

The density $\rho(x)$ is defined as the diagonal part of the one-body density matrix

$$\rho(x) \equiv \rho(x,x) \quad . \tag{1.18}$$

Physically, the two terms in Eq. (1.15) are very simple to understand. The first term is local, and simply represents the

Hartree potential arising from the convolution of the two-body potential with the density. The second term is non-local and arises from the exchange contribution in Eq. (1.9) and thus represents the effect of antisymmetry. Note that the range of non-locality is limited by the off-diagonal range of $\rho(x,x')$. When $x=x'$, every term in (1.17) is positive and adds coherently. However, when $x-x' \gtrsim \frac{\pi}{k_F}$ a significant fraction of the wave functions have essentially random phases, so $\rho(x,x')$ rapidly approaches zero. In the special case of uniform nuclear matter, $\rho(x,x')$ has the familiar Slater form

$$\rho(x,x') = \rho \frac{3j_1(k_F|x-x'|)}{k_F|x-x'|} \quad . \tag{1.19}$$

Using Eq. (1.14), the equation of motion for the density matrix, Eq. (1.17), is

$$\begin{aligned} i\dot{\rho}(x,x') &= \sum_\nu \phi^*_\nu(x') \int dx'' \, h(x,x'')\phi_\nu(x'') \\ &\quad - \sum_\nu \int dx'' \, \phi^*_\nu(x'')h^*(x',x'')\phi_\nu(x) \\ &= \int dx'' \, \{h(x,x'')\, \rho(x'',x') \\ &\quad - \rho(x,x'')h(x'',x')\} \end{aligned} \tag{1.20a}$$

or, more compactly

$$i\dot{\rho} = [h, \rho] \quad , \tag{1.20b}$$

where we have used the property $h^*(x',x'') = h(x'',x')$ which follows from Eqs. (1.15) and (1.17). Even though it is often convenient to express the TDHF equations in a specific basis as in Eq. (1.13), the fact that Eq. (1.20) with h defined in (1.15) is expressed solely in terms of the density matrix emphasizes the fact that the theory is in fact representation independent. A representation independent statement of the fact that ρ is derived from a determinantal wave function is the condition derived in Problem 3

$$\rho^2 = \rho \quad , \tag{1.21a}$$

that is

$$\int dx'' \; \rho(x,x'')\rho(x'',x') = \rho(x,x') \quad . \tag{1.21b}$$

1.3.3 Basic features of TDHF

The TDHF equations imply several conservation laws. The scalar product of two single-particle wave functions remains constant in time, since

$$\begin{aligned}\frac{d}{dt}\int dx\phi_{\nu}^{*}(x)\phi_{\nu'}(x) &= -i\iint dxdx'\,[\phi_{\nu}^{*}(x)h(x,x')\phi_{\nu'}(x')-\phi_{\nu}^{*}(x')h^{*}(x,x')\phi_{\nu'}(x)] \\ &= -i\iint dxdx'\,\phi_{\nu}^{*}(x)\,[h(x,x')-h(x,x')]\phi_{\nu'}(x') \\ &= 0 \quad .\end{aligned} \tag{1.22}$$

In practice, this means that if one begins a calculation with an orthonormal basis, the wave functions remain orthonormal for all time. (The same proof clearly applies to any time-dependent hamiltonian, $h(x,x')$, not just the TDHF mean field.) Thus, the Pauli principle is built in from the start, in the sense that the wave functions in two colliding nuclei are automatically orthogonal by virtue of evolving in the same mean field. Since one often sees explicitly repulsive terms to represent the effect of the Pauli principle, it is important to think about how the same physics arises in this theory. (See Problem 4). Number conservation is immediately obtained from Eq. (1.22) by setting $\nu=\nu'$ and summing over ν.

Another important conserved quantity is the expectation value of H, which may be proved (Problem 5) by explicit substitution of Eq. (1.10) and its adjoint into $\langle\psi(t)|H|\psi(t)\rangle$. A more economical proof follows from the cyclic property of the trace, abbreviated tr, from noting

$$\begin{aligned}\langle T\rangle &= \mathrm{tr}\; T\rho \\ \frac{d}{dt}\langle T\rangle &= \mathrm{tr}\; T\dot{\rho} \\ \langle V\rangle &= \frac{1}{2}\,\mathrm{tr}\; W[\rho]\rho \\ \frac{d}{dt}\langle V\rangle &= \mathrm{tr}\; W\dot{\rho}\end{aligned} \tag{1.23}$$

and hence

$$i \frac{d}{dt} \langle H \rangle = tr(T+W)i\dot{\rho} = tr\ h[h,\rho]$$
$$= tr(h^2\rho - h^2\rho) = 0 \quad . \tag{1.24}$$

Although energy is conserved overall, the TDHF approximation certainly allows free exchange of excitation energy between collective and intrinsic degrees of freedom. Although the latter exchange is often described phenomenologically in terms of viscosity or dissipation, note that specific intrinsic or collective variables need never be defined in this theory and that TDHF is manifestly invariant under time reversal, thus ensuring microreversibility of reactions.

Several limiting forms of the TDHF equations are worthy of note. In the case of stationary states, for which $\phi_\nu(x,t) = \phi_\nu(x)e^{-i\varepsilon_\nu t}$ Eq. (1.13) reduces to the usual static H.F. equation

$$h\,\phi_\nu = i\,\dot{\phi}_\nu = \varepsilon\,\phi_\nu \quad . \tag{1.25}$$

The representation independent form of the static HF equation

$$[h,\ \rho] = 0 \tag{1.26}$$

follows immediately upon noting that the time dependent phase factors cancel out of the density matrix

$$\rho(x,x';t) = \sum_\nu \phi^*_\nu(x)e^{i\varepsilon_\nu t}\ \phi_\nu(x)e^{-i\varepsilon_\nu t} = \sum_\nu \phi^*_\nu(x)\ \phi_\nu(x) \quad . \tag{1.27}$$

An interesting and useful consequence of the connnection between the time-dependent and time-independent theories is the fact that the ground state for the time-independent equations may actually be solved by evolution of the time-dependent equations in complex time.[14] This property is made obvious for the full Schrodinger equation with a time-independent hamiltonian by expressing an arbitrary initial wave function in the basis of eigenstates of H. Let

$$\psi(0) = \sum_i a_i \psi_i \tag{1.28a}$$

where

$$H\,\psi_i = E_i\psi_i \quad . \tag{1.28b}$$

Then

$$\psi(t) = e^{-iE_o t} [\sum_i a_i e^{-t(E_i - E_o)} \psi_i] . \qquad (1.29)$$

Replacing t by $-i\tau$, one observes that all components are damped exponentially relative to the first in the limit of large τ

$$\lim_{\tau\to\infty} e^{\tau E_o} \psi(-i\tau) = \lim \sum_i a_i e^{-(E_i - E_o)\tau} \psi_i = \psi_o . \qquad (1.30)$$

The TDHF case is more complicated, since h is time-dependent by virtue of its explicit dependence on ρ. However, for $\Phi(-i\tau)$ reasonably close to Φ_o, $h(-i\tau)$ will be sufficiently close to h_o that when each infinitesimal step in complex time refines Φ, the change will be overwhelmingly toward Φ_o and will bring h closer to h_o. Although it is by no means obvious that an arbitrary apporoximation to the Schrodinger equation will be amenable to this complex time technique, it is of significant technical value in allowing one to use virtually the same computer code to solve static and time-dependent problems.

A second limiting case of TDHF is the random phase approximation (RPA). Consider a density matrix ρ which differs from the ground state density matrix ρ_o by an infinitesimal deviation ρ_1.

$$\rho \equiv \rho_o + \rho_1 . \qquad (1.31)$$

Explicitly displaying the functional dependence of W on ρ,

$$i(\dot\rho_o + \dot\rho_1) = [(T+W[\rho_o]) + W[\rho_1], \rho_o + \rho_1] , \qquad (1.32)$$

using the fact that ρ_o satisfies the static equation, and retaining only first order infinitesimals, the following equation is obtained[15]

$$i\dot\rho_1 = [W[\rho_1], \rho_o] + [T+W[\rho_o], \rho_1] \qquad (1.33)$$

Letting $\alpha,\beta,\gamma,\delta,\xi$ denote occupied or unoccupied orbits in the HF basis

$$T + W[\rho_o] |\alpha> = E_\alpha|\alpha> \qquad (1.34)$$

and using (1.15b), $<\alpha|\dot{\rho}|\beta>$ is given by

$$[i \frac{\partial}{\partial t} - (E_\alpha - E_\beta)] <\alpha|\rho|\beta> = <\alpha|[W[\rho_1], \rho_o]|\beta>$$

$$= \sum_{\gamma,\delta,\xi} \{<\alpha\gamma|v|\xi\delta \; \delta\xi><\delta|\rho_1|\gamma><\xi|\rho_o|\beta> \tag{1.35}$$

$$- <\alpha|\rho_o|\xi><\xi\gamma|V|\beta\delta-\delta\beta><\delta|\rho_1|\gamma> \} \; .$$

Noting that the most general infinitesimal change from the ground state is given by a superposition of particle-hole excitations, the most general form of infinitesimal deviation in the density matrix with a single frequency ω is

$$\rho_1(t) = \rho_1 \, e^{-i\omega t} + \rho_1^\dagger \, e^{i\omega t} \tag{1.36}$$

where ρ_1 has only particle-hole matrix elements. Thus, denoting particles and holes by ρ and ν as usual, substituting (1.30) in (1.35) and equating terms with the same time dependence yields

$$[\omega-(E_\nu-E_\rho)]<\nu|\rho_1|\rho> = - \sum_{\gamma\delta} <\nu\gamma|V|\rho\delta-\delta\rho><\delta|\rho_1|\gamma>$$

$$= \sum_{\nu',\rho'} (- \nu\nu'|v|\rho\rho'-\rho'\rho><\rho'|\rho_1|\nu'>$$

$$-<\nu\rho'|v|\rho\nu'-\nu'\rho><\nu'|\rho_1|\rho'> \tag{1.37}$$

$$[-\omega-(E_\nu-E_\rho)] <\nu|\rho_1^+|\rho> = \sum_{\nu',\rho'} (-<\nu\nu'|v|\rho\rho'-\rho'\rho><\rho'|\rho_1^+|\nu'>$$

$$<\nu\rho'|v|\rho\nu' - \nu'\rho><\nu'|\rho_1^+|\rho'> \; .$$

The standard form of the RPA equations is obtained with the definitions

$$A_{\nu\rho\nu'\rho'} \equiv (E_\rho-E_\nu) \, \delta_{\nu\nu'}\delta_{\rho\rho'} + <\nu\rho'|v|\rho\nu'-\nu'\rho>$$

$$B_{\nu\rho\nu'\rho'} \equiv <\nu\nu'|v|\rho\rho'- \rho'\rho>$$

$$X_{\nu\rho} \equiv <\nu|\rho^+|\rho> \tag{1.38}$$

$$Y_{\nu\rho} \equiv <\rho|\rho_1^+|\nu> \; ,$$

in which case, (1.37) becomes

$$\begin{bmatrix} A & B \\ B^* & A^* \end{bmatrix} \begin{bmatrix} X \\ Y \end{bmatrix} = \begin{bmatrix} X \\ -Y \end{bmatrix} . \tag{1.39}$$

The physical interpretation of the RPA is particularly clear in the context of TDHF. Since the amplitude of ρ_1 is totally undetermined (except for the condition that it is small) the wave function of the excited state has not really been calculated. Rather, we simply have a method of determining the linear response of a system to an external field. In the case of inelastic electron scattering, where the external field is the transient electromagnetic field of the scattered electron and thus both weak and very accurately known, RPA linear response theory can be spectacularly successful in describing transitions to collective states. Fig. 1-5 shows an RPA calculation [16] for the 3^- state in ^{208}PB, and one observes that both the amplitude and detailed structure of the transition density are accurately predicted. An interesting technical question which has not yet been resolved is whether simply numerically evolving a TDHF wave function in a time-dependent external field is as practical as standard techniques in linear response theory. Certainly the widespread quantitative success of the RPA provides another strong argument in

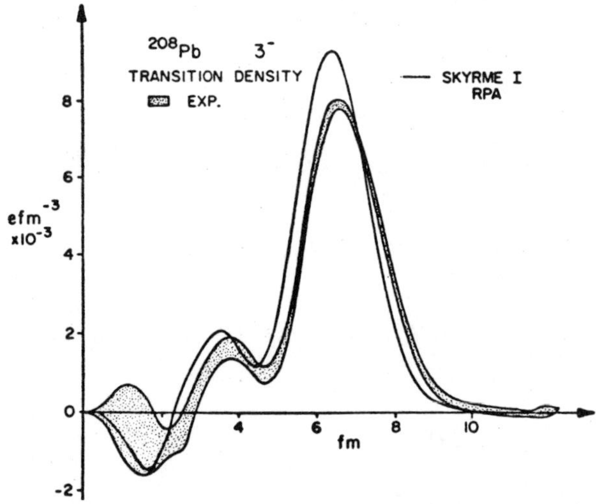

Fig. 1-5 Transition density for the first 3^- state in ^{208}Pb calculated in the random phase approximation.

addition to the success of static HF in support of the time-dependent mean-field theory as a general starting point for nuclear dynamics.

The final important limit of TDHF is the adiabatic limit. Since this limit has been discussed exhaustively in a recent tome,[17] only the basic ideas and objectives will be mentioned here. A general density matrix satisfying $\rho^2=\rho$ may be written in the form

$$\rho = e^{i\chi} \rho_o e^{-i\chi} \tag{1.40}$$

where ρ_o is time even and satisfies $\rho_o^2 = \rho_o$ and χ is time odd. If χ is sufficiently small, which because it is time-odd implies that all velocities are small, a systematic expansion may be developed in which approximate equations of motion for ρ_o and χ are determined from the TDHF equation. This formulation provides an elegant and general method for deriving a collective hamiltonian which contains not only the usual simple potential energy terms, but also the crucial mass parameters associated with each collective variable. Whereas applications of the adiabatic theory to date have been limited, it offers an appealing alternative to solution of the full TDHF theory in many important cases.

Problems

1) Verify, by writing $\phi_r \equiv \chi_r + i\pi_r$; with χ and π real that variation of the action with respect to χ and π yields the same equation as the shortcut of variation with respect to ϕ^*.

2) Derive Eq. (1.17), using $\hat{\psi}(x) = \sum_\alpha \phi_\alpha(x) a_\alpha$ and

$$|\Phi\rangle \equiv a^+_{\nu_1} a^+_{\nu_2} \dots a^+_{\nu_n} |0\rangle$$

3) Show that the one-body density matrix for a wave function is idempotent with trace N, i.e., $\rho^2 = \rho$ and $\mathrm{tr}\rho = N$ if and only if the wave function is an N particle determinant.

4) Explain the role of the Pauli principle in two simple cases:

a) Two long, one-dimensional attractive rectangular wells filled with single particle states approach each other at a relative velocity small compared to the Fermi velocity.

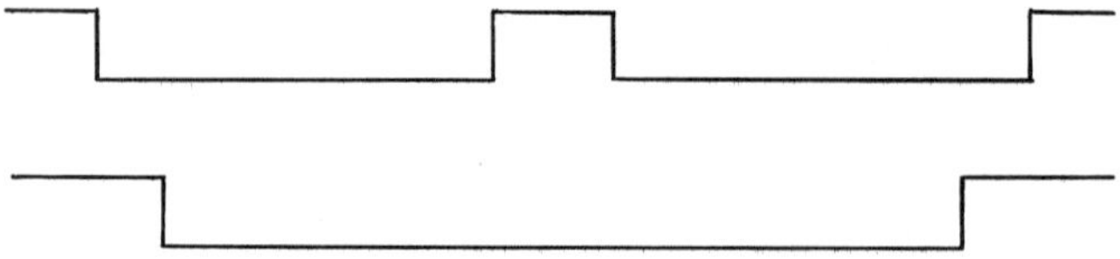

After they touch, the well is described for short times by a single rectangular well of the same depth. By decomposing the occupied state wave function into travelling waves, determine how much the density increases in the interior as the two systems try to "pass through" each other. (Compare with results in Section 3.)

b) Consider two nuclei which approach each other at velocity v. At some time τ before they hit, the two-body potential is turned off (e.g., mesonic forces suddenly cease to exist) so that h becomes T. The proof (1.22) still shows that the wave functions in the two nuclei remain orthogonal at all subsequent times. However, the equation of motion for either nucleus is the same as if that nucleus were evolving in free space. Resolve this paradox for all values of v and τ.

5) Prove $\langle H \rangle$ is conserved by the TDHF equations.

6) Since the most general variation of a determinant $|\phi\rangle$ may be written $\delta|\Phi\rangle = \sum_{\rho\nu} c_{\rho\nu} a_\rho^+ a_\nu |\Phi\rangle$, one should be able to derive the TDHF equations by requiring $\langle\Phi| a_\nu^+ a (H - i\frac{\partial}{\partial t})|\Phi\rangle = 0$ for all ν and ρ. Show that this yields the equation

$$\int \phi_\rho^* (h - i\frac{\partial}{\partial t}) \phi_\nu \, dx = 0$$

Whereas the particle-hole matrix elements are identical to (1.13), there is no statement about hole-hole matrix elements. By noting that any unitary transformation among the occupied wave functions leaves the density matrix and determinantal wave function unchanged, reconcile this apparent discrepancy.

7) Show that if ρ_o is a static solution of the HF equations, then

$$\rho(x,x';t) = \rho_o(x-vt, x'-vt)\, e^{imv(x-x')}$$

is a solution to the TDHF equations. That is, a Galilean

translation of the static HF solution is a TDHF solution.

A particularly easy way to verify $\rho(x,x',t)$ is a solution is to examine the equation of motion of a single particle wave function in the mean field $W(x,x',t)$ generated by $\rho(x,x',t)$. Write

$$\phi_\nu(x,t) = \phi_\nu^o(x-vt)\, e^{-i(\varepsilon_\nu + \frac{mv^2}{2})t}\, e^{imvx}$$

and show ρ has the form stated above and that

$$W(x,x',t) = W^o(x-vt, x'-vt)\, e^{imv(x-x')}$$

where $W^{(0)}$ is the mean field generated by the $\phi^{(0)}$'s. By substitution of ϕ and W into the TDHF equation, Eq. (1.13), verify that $\phi^{(0)}$ satisfies the HF equation, Eq. (1.25).

In addition calculate the difference in energy between a solution with velocity v and velocity 0 and thereby show that the collective mass, i.e., the coefficient M in $E(v)-E(0)=1/2\, Mv^2$ is given correctly.

REFERENCES FOR CHAPTER I

1. P. Stephens, Ph.D. dissertation, Massachusetts Institute of Technology, (1978).

2. B. Brandow, Ann. Phys. (NY) 64, 21 (1971).

3. P. C. Lichtner and J. J. Griffin, Phys. Rev. Lett. 37, 1521 (1976).

4. J. W. Negele, Nucleon Nucleon Interactions - 1977, AIP Conference Proceedings, No.41; Comments in Nuclear and Particle Physics 6, 144 (1976).

5. M. H. Kalos, Phys. Rev. A2, 250 (1970); M. H. Kalos, D. Levesque, and L. Verlet, Phys. Rev. A9, 2178 (1974).

6. V. R. Pandharipande and R. B. Wiringa, Nucl. Phys. A266, 269 (1976) and to be published.

7. J. W. Negele, Proceedings of Conference on Hartree-Fock and Self-Consistent Field Theories in Nuclei, ed. G. Ripka and M. Porneuf, North Holland (1975). Proceedings of International Conference on Effective Interactions and Operators in Nuclei, ed. B. Barrett, Springer-Verlag (1975).

8. H. Kümmel, K. H. Lührmann, and J. G. Zabolitzky, Phys. Reports 36C (1978).

9. J. W. Negele, Phys. Rev. Letters 27, 1291 (1971).

10. J. W. Negele, Proceedings of VII International Conference on High Energy Physics and Nuclear Structure, Zurich (1971); Proceedings of Conference on Radial Shapes of Nuclei, Cracow, A. Budzanowski and A. Kapuscic ed., (1976).

11. J. W. Negele, Phys. Rev. C9, 1054 (1974).

12. A. M. Bernstein and W. A. Seidler, Phys. Lett. 39B, 583 (1972).

13. G. Varma and L. Zamick, Phys. Rev. C in press.

14. Mort Weiss, private communication.

15. J. Goldstone and K. Gottfried, Nuovo Cimento [X] 13, 849 (1959).

16. G. F. Bertsch and S. F. Tsai, Phys. Reports 18, 125 (1975).

17. M. Baranger and M. Veneroni, Ann. Phys. (NY) in press.

II. FORMULATION OF THE GENERAL TIME-DEPENDENT COUPLED-CLUSTER APPROXIMATION

In spite of the intuitive appeal and physical motivation emphasized in the Introduction, the time-dependent mean-field approximation is essentially deficient until it can be understood as a natural starting point in a completely systematic and general hierarchy of successive approximations. Even if realistic numerical calculations can only be implemented at present in lowest order, their ultimate significance can only be assessed in the context of a general theory. From a careful comparison of the advantages and limitations of various alternative generalizations of the mean-field theory, the time-dependent coupled-cluster approximation emerges as the optimal theoretical framework and is described in detail.

2.1 Alternative Generalizations of the Mean-Field Theory

Since virtually all general methods in Fermion many-body theory reduce to the mean-field or Hartree-Fock (HF) approximation in lowest order or some appropriate limit, there is no unique series of systematic corrections to the mean-field theory. Thus, it is useful to motivate the coupled-cluster theory by briefly explaining why it appears preferable to alternative methods that have been considered.

2.1.1 Variational methods

The time-dependent variational principle yields the time-dependent Hartree-Fock (TDHF) approximation when the trial function is restricted to the set of all Slater determinants. The most satisfactory systematic hierarchy of successive variational approximations appears to arise from the extended Jastrow trial functions

$$\Psi^{(M)}(r_1 \ldots r_N) = \prod_i f^{(M)}(r_{i_1} r_{i_2} \ldots r_{i_M}) \prod_j f^{(M-1)}(r_{j_1} r_{j_2} \ldots r_{j_{M-1}}) \ldots \prod_k f^{(2)}(r_{k_1} r_{k_2}) \, \Phi_{SD}(r_1 r_2 \ldots r_N) \quad , \qquad (2.1)$$

where $f^{(M)}$ is an M-body correlation function and Φ_{SD} is a Slater determinant. Based on experience with the energy and radial distribution function in liquid helium and nuclear matter,[2] there is every reason to expect rapid convergence in expectation values of few-body operators in this hierarchy. The fundamental problem, however, is the difficulty in obtaining a sufficiently accurate expression for the energy in terms of the correlation functions and single-particle wave functions to use in the variational principle. In the special case of infinite matter and central forces, it is already difficult to solve the time-independent Euler-Lagrange equations in the lowest order Fermi Hypernetted chain (FHNC) approximation.[3] Given the fact that the lowest order FHNC is of marginal accuracy, state dependent nuclear potentials have never been treated adequately, and that calculation of finite nuclei requires inclusion of additional terms which vanish in the FHNC infinite-matter limit, the present variational technology is grossly inadequate even for formulation of a time-dependent variational theory containing two-body correlation functions.

2.1.2 Density matrix hierarchies

An alternative approach is the truncation of a suitable hierarchy of coupled equations for expectation values of products of n pairs of Fermion field creation and annihilation operators. Depending upon how one specifies relative time arguments and whether time-ordered products are introduced, the theory may be formulated in terms of n-particle density matrices or n-body Greens functions.

The Green's function formulation is somewhat better known and widely used, due to its close connection to diagrammatic perturbation theory, and will therefore be discussed first. The Heisenberg equation of motion for the field operator $\hat{\psi}(x,t)$ is

$$i \frac{\partial}{\partial t} \hat{\psi}(x,t) = [\hat{\psi}(x,t), \hat{H}]$$
$$= T(x)\hat{\psi}(x,t) + \int dx'' \; \hat{\psi}^{+}(x'',t)v(x,x'')\hat{\psi}(x'',t)\hat{\psi}(x,t) \;, \tag{2.2}$$

where the last line is derived in Problem 2.1.

Substitution of Eq. (2.2) into the definition of the n-particle Green's function as a time-ordered product of n creation and n annihilation operators

$$(i)^n \, G_n(x_1t_1,\ldots x_nt_n;x_1't_1',\ldots x_n't_n')$$
$$= \langle T\hat{\psi}(x_1t_1)\ldots\hat{\psi}(x_nt_n)\hat{\psi}^{+}(x_n't_n')\ldots\hat{\psi}^{+}(x_1't_1')\rangle \;, \tag{2.3}$$

where T denotes the time ordering operator, yields the Martin-Schwinger Green's function hierarchy.[4] The lowest order equation is

$$i \frac{\partial}{\partial t} G_1(xt,x't') = \frac{\partial}{\partial t}\{\theta(t-t')\langle\psi(x,t)\psi^{+}(x',t')\rangle$$
$$- \theta(t'-t)\langle\psi^{+}(x',t')\psi(x,t)\rangle\}$$

$$= \delta(t-t')\langle[\psi(x,t),\psi^{+}(x't')]_{+}\rangle + \langle T \frac{\partial}{\partial t} \psi(x,t)\psi^{+}(x't')\rangle \tag{2.4}$$
$$= \delta(t-t')\delta(x-x') + \frac{1}{i} T(x) \langle T\psi(x,t)\psi^{+}(x't')\rangle$$
$$+ \frac{1}{i} \int dx'' \langle T \hat{\psi}^{+}(x'',t)v(x,x'')\hat{\psi}(x'',t)\hat{\psi}(x,t)\hat{\psi}^{+}(x',t')\rangle.$$

Denoting a time infinitesimally greater than t by t^{+}, noting that

$$-i\int dx'' \; v(x,x'')G_2(xt,x''t;\; x't',\; x''t^{+}) \tag{2.5}$$
$$= -i \frac{1}{i^2}\int dx'' \; v(x,x'')\langle T\hat{\psi}^{+}(x'',t)\hat{\psi}(x,t)\hat{\psi}(x'',t)\hat{\psi}^{+}(x't')\rangle \;,$$

and abbreviating the space-time variables (x_nt_n) by (n), the lowest order equation becomes

$$\left(i \frac{\partial}{\partial t_1} - T(x_1)\right) G_1(1,1') = \delta(1,1') - i\int dx_2\, v(x_1-x_2)\, G_2(1,2;1'2^+)_{t_2=t_1} \quad . \tag{2.6}$$

By similar algebra in Problem 2, the general equation in n^{th} order is

$$\left(i \frac{\partial}{\partial t_1} -T(x)\right)G_n(1,2,\ldots n;\ 1',2'\ldots n') = \sum_{i=1}^{n} \delta(1,i')(-1)^{i-1}\, G_{n-1}(2\ldots n;1'\ldots(i-1)',(i+1)'\ldots n') - i\int dx_{n+1} v(x_1-x_{n+1})G_{n+1}(1,2,\ldots(n+1);1',2'\ldots(n+1)'^{+})_{t_{n+1}=t_n} \quad . \tag{2.7}$$

In general, the time evolution of G_n is coupled to G_{n+1} and G_{n-1}, so that the hierarchy may be truncated by specifying a physically motivated prescription for approximating some G_{n+1} in terms of G_m and lower order Green's function. The TDHF approximation arises in this context by assuming that two particles propagate in the system independently, (but properly antisymmetrized) so that $G_2(12,1'2') = G_1(11')G_1(22')-G_1(12')G_1(21')$. Thus, from Eq. (2.6)

$$\left(i \frac{\partial}{\partial t_1} -T(x_1)\right)G_1(1,1')=\delta(1,1')-i\int dx_2 v(x_1-x_2)\, [G_1(1,1')G_1(2,2^+) - G_1(1,2^+)G_1(2,1')]_{t_2=t_1} \tag{2.8}$$

or

$$i \frac{\partial}{\partial t_1} G_1(1,1')=\delta(1,1')+ \int dx_3 h(x_1,x_3;t_1)G_1(x_3t_1;x_1't_1') \quad , \tag{2.9}$$

where

$$
\begin{aligned}
h(x_1,x_3;t_1) &= \delta(x_1-x_3)[T(x_3)-i\int dx_2 v(x_1-x_2)G_1(2,2^+)_{t_2=t_1}] \\
&\quad + i\, v(x_1-x_3)G_1(1,3^+)_{t_3=t_1} \\
&= T(x_1,x_3)+\delta(x_1-x_3)\int dx_2\; v(x_1-x_2)\rho(x_2,x_2;t_1) \\
&\quad - v(x_1-x_3)\rho(x_1x_3;t_1) \quad .
\end{aligned}
\tag{2.10}
$$

In the last line of Eq. (2.10), Eqs. (2.3) and (1.16) were used together with the fact that the time-ordering operator interchanges $\hat{\psi}(t)$ and $\hat{\psi}^+(t)$ and introduces a negative sign. The quantity h in Eq. (2.10) is obviously the single particle hamiltonian of Eqs. (1.14) and (1.15) at time t_1, so that the one particle Green's function in this approximation does evolve in the same mean field as the TDHF single particle wave functions. Evaluating the difference between Eq.(2.9) and the adjoint equation derived in Problem 3

$$
-i\frac{\partial}{\partial t_1'}G_1(1,1')=\delta(1,1')+\int dx_3\; G(x_1t_1;x_3t_1')h(x_3,x_1;t_1')
\tag{2.11}
$$

for $t_1' = t_1^+$ yields the TDHF equation

$$
\begin{aligned}
& i\langle\dot{\hat{\psi}}^+(x',t)\hat{\psi}(x,t) + \hat{\psi}^+(x',t)\dot{\hat{\psi}}(x,t)\rangle \\
&= i\dot{\rho}(x,x',t) \\
&= \int dx_3[h(x,x_3;t)\rho(x_3,x';t)-\rho(x,x_3;t)h(x_3,x';t)]
\end{aligned}
\tag{2.12}
$$

Similarly, higher order truncations of the Martin-Schwinger hierarchy which express all Green's functions for n>m in terms of lower Green's functions can be evaluated at a single time with infinitesimals chosen to yield density matrices, producing a set of coupled equations in $\rho_1(x_1,x_1';t)\ldots\rho_m(x_1\ldots x_m,x_1'\ldots x_m';t)$. In addition to being closely related to diagrammatic perturbation theory, the Green's function hierarchy offers the additional advantage that truncation prescriptions are very easy to check for preservation of conservation laws,[5] as discussed in Problem 4. If one is willing to forgo the convenience of Green's functions, the same hierarchy of coupled equations for density matrices of arbitrary order may be obtained by using the equations of motion for field operators, Eq. (2.2) directly in the definition of an n-body density matrix:

$$\rho(x_1 \ldots x_n; x_1' \ldots x_n'; t) \equiv \langle \hat{\psi}^+(x_1',t) \ldots \hat{\psi}^+(x_n',t) \hat{\psi}(x_n,t) \ldots \hat{\psi}(x_1,t) \rangle \tag{2.13}$$

The TDHF equation then follows automatically from the truncation $\rho(x_1 x_2; y_1 y_2) = \rho(x_1 y_1)\rho(x_2 y_2) - \rho(x_1 y_2)\rho(x_2 y_1)$.

The primary disadvantage of either the Green's function or density matrix formulation relative to the coupled cluster theory described subsequently is the computational cumbersomeness of dealing with objects having as many spatial variables as n-body density matrices. Certainly the integral form of the Green's function hierarchy, obtained by multiplying Eq. (2.7) by the noninteracting one-particle Green's function and integrating over its space and time coordinates is ill-suited to time-dependent problems because it involves integrations of n-body Green's functions over relative times. However, even when the differential form of the theory has been reduced to the special case of a single time argument, one is still faced with evolving a function of 2n spatial variables, thereby even rendering treatment of two-particle correlations intractable in all but the most oversimplified applications.

2.1.3 The coupled cluster approximation

The formulation which appears most suited to the present problem is a time-dependent generalization of the exp(S) or coupled-cluster approximation pioneered by Coester and Kummel[6] and subsequently applied extensively to fermion systems.[7] In this theory, the full many-body wave function is written in the form

$$\Psi = \exp(S)\Phi \tag{2.14}$$

where

$$S = \sum_{n=1}^{N} S^{(n)} \tag{2.15}$$

and $S^{(n)}$ represents the most general n-particle, n-hole operator defined relative to the Slater determinant of occupied states Φ. The time-independent or time-dependent Schrodinger equation implies a hierarchy of non-linear, coupled equations for the static or time-dependent n-particle, n-hole amplitude, which may be truncated in the same manner as a density-matrix or Green's function hierarchy. However, the n-particle amplitude may be expressed in terms of n spatial coordinates and n occupied state labels rather than 2n coordinates since $S^{(n)}$ may be written

$$S^{(n)} = \langle x_1 x_2 \ldots x_n | S^{(n)} | \nu_1 \nu_2 \ldots \nu_n \rangle \hat{\psi}^+(x_1) \hat{\psi}^+(x_2) \ldots \hat{\psi}^+(x_n) a_{\nu_n} \ldots a_{\nu_2} a_{\nu_1} \tag{2.16}$$

where

$$\langle x_1 x_2 \ldots x_n | S^{(n)} | \nu_1 \nu_2 \ldots \nu_n \rangle$$

$$\equiv \sum_{\rho_1 \ldots \rho_n} \langle x_1 | \rho_1 \rangle \ldots \langle x_n | \rho_n \rangle \langle \rho_1 \ldots \rho_n | S_n | \nu_1 \ldots \nu_n \rangle$$

and $\psi^+(x)$ is the field creation operator at position x. Thus, for finite systems, the coupled-cluster formalism provides a particularly economical description of many-body correlations.

Whereas the basic ideas underlying the coupled-cluster hierarchy are very simple, the technical details involved in applying even the time-independent theory to the most general case are somewhat tedious and notationally cumbersome.[7] Therefore, these lectures will emphasize the essential features and general structure of the theory. As a preparation for the application to time-dependent problems, the time-independent theory will first be reviewed in Section 2.2. Then the general time-dependent theory with the simplest truncation procedure will be presented in Section 2.3. The truncation for strongly repulsive forces is not discussed, but follows straightforwardly from the time-independent treatment of Ref. 7. Finally, the application of the coupled-cluster formalism to an exactly solvable model is presented in detail in Section 3.2.2 to demonstrate how the method is carried out in practice and how well it converges to exact results.

2.2 Review of the Time-Independent Coupled-Cluster Approximation

For stationary states, equations for the n-particle, n-hole amplitudes appearing in Eqs. (2.14-2.15) are obtained by projecting the Schrodinger equation

$$e^{-S} H e^{S} | \Phi \rangle = E | \Phi \rangle \tag{2.17}$$

onto a complete set of m-particle, m-hole states. Denoting unoccupied states by ν_i, the following hierarchy of equations arises:

$$\langle\Phi|e^{-S}\ H\ e^{S}|\Phi\rangle\ = E \tag{2.18a}$$

$$\langle\Phi|a^{+}_{\nu}a_{\rho}\ e^{-S}\ H\ e^{S}|\Phi\rangle = 0 \tag{2.18b}$$

$$\vdots$$

$$\langle\Phi|a^{+}_{\nu_1}\ldots a^{+}_{\nu_m}a_{\rho_m}\ldots a_{\rho_1}\ e^{-S}\ He^{S}|\Phi\rangle = 0\ . \tag{2.18c}$$

From the identity

$$e^{-X}\hat{O}\ e^{X} = \hat{O} + [\hat{O},X] + \frac{1}{2!}\ [[\hat{O},X],X] + \ldots \tag{2.19}$$

it is evident that if H contains only one and two-body operators, repeated commutation with H can remove at most two pairs of creation and annihilation operators from the particle-hole operators $S^{(n)}$. Thus, since $e^{-S}He^{S}|\Phi\rangle$ must connect to $|\Phi\rangle$ in Eq. (2.18a) and at most two pairs of particle-hole operators have been contracted, the resulting equation can involve only S_1 and S_2. In fact, for the special case of Eq. (2.18a), the exact equation

$$\langle\Phi|H(S^{(1)} + \frac{1}{2}\ S^{(1)}S^{(1)} + S^{(2)})|\Phi\rangle = E \tag{2.20}$$

is trivially obtained by noting $\langle\Phi|S = 0$ and expanding e^{S}. Similarly, Eq. (2.18b) involves amplitudes only through $S^{(3)}$ and in general Eq. (2.18c) includes amplitudes through $S^{(m+2)}$. The explicit form of the general equation is somewhat complicated, but follows straightforwardly from substitution of Eq. (2.19), which terminates after 5 terms when $\hat{O}$ = H, in Eq. (2.18c).

Since the resulting hierarchy of equations for the $S^{(n)}$'s is equivalent to the original Schrodinger equation, physical approximations are introduced by the method of truncation. The simplest truncation prescription, which we shall use in this present work, is to specify that $S^{(n)} = 0$ for all $n > m$. This has the effect of treating particle-hole correlations of up to m particles exactly while retaining only those correlations for more than m particles which arise from products of lower order amplitudes. In terms of familiar perturbation theory, truncation at m=2 sums particle-particle and hole-hole ladders as well as RPA ring diagrams, and this approximation has been shown to be accurate for a single Lipkin model and for systems with long-range forces.[7] For potentials which are infinitely repulsive at short distances, it is inconsistent to define the higher $S^{(n)}$'s to be zero and instead one must impose a prescription which makes the wave function vanish when n particles are within a hard core radius. For finite, but strongly repulsive cores, a similar condition is physically reasonable, and this more complicated truncation procedure will be addressed subsequently in Section 2.4.

Retaining only m non-zero amplitudes yields m equations of the form (2.18c) in m unknowns which completely specify the $S^{(n)}$'s. With these amplitudes, Eq. (2.18a) yields an mth order approximation to the energy which is distinct from the expectation value of H with the wave function

$\exp(S^{(1)} + S^{(2)} + \ldots + S^{(m)})|\Phi\rangle$ since $-S \neq S^{+}$.
Truncation at m=1 yields the H.F. approximation, which is most obvious by noting that the most general determinant may be written $\exp(S^{(1)})|\phi\rangle$ by Thouless' theorem.[8] In general, one always has the freedom to specify $S^{(1)} \equiv 0$ and to solve an equation for the single-particle wave functions comprising the determinant $|\Phi\rangle$.

2.3 The Time-Dependent Coupled-Cluster Approximation

The time-dependent coupled-cluster theory is obtained analogously by projecting the time-dependent Schrodinger equation:

$$e^{-S} He^{S}|\Phi\rangle = i\, e^{-S} \frac{\partial}{\partial t} e^{S}|\Phi\rangle \tag{2.21}$$

onto n-particle, n-hole states:

$$\langle\Phi|e^{-S} He^{S}|\Phi\rangle = i\langle\Phi|e^{-S} \frac{\partial}{\partial t} e^{S}|\Phi\rangle \tag{2.22a}$$

$$\langle\Phi|a^{+}_{\nu}a_{\rho}\, e^{-S} He^{S}|\Phi\rangle = i\langle\Phi|a^{+}_{\nu}a_{\rho}\, e^{-S} \frac{\partial}{\partial t} e^{S}|\Phi\rangle \tag{2.22b}$$

$$\langle\Phi|a^{+}_{\nu_1}\ldots a^{+}_{\nu_m} a_{\rho_m}\ldots a_{\rho_1}\, e^{-S}He^{S}|\Phi\rangle = i\langle\Phi|a^{+}_{\nu_1}\ldots a^{+}_{\nu_m} a_{\rho_m}\ldots a_{\rho_1}\, e^{-S}\frac{\partial}{\partial t} e^{S}|\Phi\rangle \tag{2.22c}$$

The general structure of these equations is observed by noting that

$$e^{-S} \frac{\partial}{\partial} e^{S} = \frac{\partial}{\partial t} + \dot{S} + \frac{1}{2!}[\dot{S},S] + \frac{1}{3!}[[S,S],S] + \ldots \tag{2.23}$$

and writing $S^{(n)}$ in an arbitrary time-dependent basis,

$$S^{n}(t) = \sum_{\rho_i \nu_i} \langle\rho_1\ldots\rho_n|S^{n}(t)|\nu_1\ldots\nu_n\rangle\, a^{+}_{\rho_1}(t)\ldots a^{+}_{\rho_n}(t)a_{\nu_n}(t)\ldots a_{\nu_1}(t) \tag{2.24}$$

Since

$$\frac{\partial}{\partial t} a_\alpha^+ = \sum_\beta \langle\beta|\dot{\alpha}\rangle a_\beta^+ \quad , \tag{2.25}$$

each non-vanishing term in $S^{(m)}$ either contains an a_ρ^+ or a_ν term, for which $[\dot{S}^{(m)}, S^{(n)}] = 0$ or contains at most one a_ν^+ or a_ρ term, in which case the commutator $[\dot{S}^m, S^n]$ contains only $a_{\rho_i}^+$ and a_{ν_i} operators. Hence, in any case, the multiple commutator $[[\dot{S},S],S]$ must vanish and

$$\begin{aligned} &\langle\Phi|a_{\nu_1}^+ \ldots a_{\nu_m}^+ a_{\rho_m} \ldots a_{\rho_1} e^{-S} \frac{\partial}{\partial t} e^S|\Phi\rangle \\ &= \langle\Phi|a_{\nu_1}^+ \ldots a_{\nu_m}^+ a_{\rho_m} \ldots a_{\rho_1} (\frac{\partial}{\partial t} + \dot{S} + \frac{1}{2}[\dot{S},S])|\Phi\rangle \quad . \end{aligned} \tag{2.26}$$

Although Eq. (2.26) yields many terms, the only non-vanishing term involving the time derivative of an $S^{(n)}$ amplitude, as opposed to the derivative of a basis function $\langle\beta|\dot{\alpha}\rangle$, is $\frac{d}{dt} \langle e_1 \ldots e_m|S^m(t)|r_1 \ldots r_m\rangle$. Thus, Eq. (2.22b) specifies $\frac{d}{dt} \langle e|S^{(1)}(t)|r\rangle$ in terms of known functions at time t, providing a first order differential equation in time for the one-particle, one-hole amplitudes. In general, Eq. (2.22c) provides a first-order equation for the m-particle, m-hole amplitude, so that given initial conditions at time t=0 the amplitudes may be evolved in time by numerically integrating a system of first order equations. The first equation, Eq. (2.22a) clearly plays no role in the time evolution since no time derivatives of particle-hole amplitudes survive in the right-hand side. It could be satisfied identically by introducing an appropriate time-dependent phase in the definition of Φ, but since such an overall phase is unobservable, Eq. (2.22a) is devoid of physical content. Formally, it is satisfied identically when the order of truncation, m, equals the number of particles.

Truncation of the time-dependent hierarchy proceeds precisely as in the time-independent theory. In lowest order, setting all $S^{(n)}$'s equal to zero for $n \geq 2$ yields the TDHF approximation, which is again most readily apparent by choosing the basis in which $S^{(1)}$ is identically zero. In this case, Eq. (2.22b) becomes

$$\langle\Phi|a_\nu^+ a_\rho (H - i \frac{\partial}{\partial t})|\Phi\rangle = 0 \tag{2.27}$$

which implies the TDHF equation

$$<\rho| i \frac{\partial}{\partial t} |\nu> = <\rho|h|\nu> \qquad (2.28)$$

with h as defined in Eq. (1.14). Making the usual arbitrary choice for hole-hole matrix elements yield the more familiar form

$$i \frac{\partial}{\partial t} |\nu> = h|\nu> \qquad (2.29)$$

Truncation at n=2 yields two closed coupled equations for $\dot{S}^{(1)}$ and $\dot{S}^{(2)}$ in terms of $S^{(1)}$ and $S^{(2)}$ which describe the time evolution of two-body correlations and in n^{th} order one obtains m equations for $\dot{S}^{(1)}$ through $\dot{S}^{(m)}$.

A particularly attractive feature of the theory is the fact that solutions to the truncated time-independent equations at any order m are stationary solutions to the truncated time-dependent equations truncated at the same order by virtue of the fact that the left hand terms of Eqs. (2.18) are identical to those of Eqs. (2.22). Thus, appropriate initial conditions for time-dependent problems can be obtained by turning on interactions or arranging collisions between systems which are in initial eigenstates calculated with precisely the same approximation. A further advantage is the fact that the static equations may actually be solved by evolving an initial guess for a wave function in complex time or by beginning with a solution to a one-body hamiltonian, $H_0=T+U$, and adiabatically switching on the interaction $H'=V-U$ in the time-dependent theory.

Problems

1. Using the fermion anticommutation relations

   $[\hat{\psi}(x,t), \hat{\psi}^+(x',t)]_+ = \delta(x-x')$ and the identity

   $[A,BC]_- = [A,B]_{\mp} C - B[C,A]_{\mp}$, evaluate the commutator

   $[\hat{\psi}(x,t), \hat{H}]$, where

   $$\hat{H} = \int dx\, \hat{\psi}^+(x)T(x)\hat{\psi}(x) + \frac{1}{2}\iint dx dx' \hat{\psi}^+(x)\hat{\psi}^+(x')v(x,x')\hat{\psi}(x')\hat{\psi}(x)$$

2. Derive the equation of motion for an n-particle Green's function, Eq. (2.7).

3. Using the equation of motion for $\hat{\psi}^+$, repeat all the steps that lead to Eq. (2.9) to obtain (2.11).

4. For certain applications in many-body theory, it is crucial to enforce number, momentum, and energy conservation order by order in an approximation scheme. Show that the following two conditions[5] on G_2 ensure these conservation laws:

a) G_2 must satisfy Eq. (2.6) and the adjoint equation

$$\left(-i\frac{\partial}{\partial t_1'} - T(x_1')\right)G_1(1,1')$$
$$= \delta(1,1') - i\int dx_2 G_2(1,2;1'2^+)v(x_2-x_1')_{t_2=t_1}$$

b) G_2 must satisfy the symmetry condition:

$$G(12;1'2') = G(21;2'1').$$

To prove number conservation subtract the two equations in (a) and obtain a continuity equation, and integrate over all space. To prove momentum conservation, apply $\frac{\nabla_1-\nabla_1'}{2i}$ to each of the equations in (a) and subtract. One of the terms vanishes only upon invoking assumption (b) Energy conservation is similar to momentum conservation. Finally show that the Hartree Fock approximation,

$$G_2(1,2';1',2') = G_1(1\,1')G_1(2\,2') - G(1\,2')G_1(2\,1')$$

is conserving whereas the approximation

$$G_2(12;1'2') = G_1(11')\,G_1(22') - G_1(12')\,G_1(21')$$
$$+\, i\iiiint G_1^{(0)}(13)\,G_1^{(0)}(24)\langle 34|v|56\rangle\, G_2(56;1'2')$$

is not.

5. Prove Thouless' theorem[8] that any determinant which is not orthogonal to $|\phi\rangle$ may be expressed

$$|\phi'\rangle = \exp\{\sum_{\substack{\rho>N\\ \nu\leq N}} S_{\rho\nu}\, a_\rho^\dagger\, a_\nu\}|\phi\rangle \quad .$$

First show that if $|\psi\rangle$ is written $(\prod_{\nu=1}^{N} a_\nu^+)|0\rangle$, then a general determinant may be expressed

$$|\phi\rangle = \{\prod_{\nu'=1}^{N}(\sum_{\nu=1}^{N} f_{\nu'\nu}\, a_\nu^+ + \sum_{\rho=N+1} f_{\nu'\rho}\, a_\rho^+)\}|0\rangle$$

Using the non-orthogonality assumption, normalize $|\phi'\rangle$

such that $\langle\phi|\phi'\rangle = \det f_{\nu\nu'} = 1$ and show that the inverse of $f_{\nu'\nu}$, denoted F, must exist.

Define $S_{\rho\nu} = \sum_{\nu'=1}^{N} F_{\nu\nu'} f_{\nu'\rho}$

Explain why the determinant of the following N linear combinations of the basis wave functions of $|\phi'\rangle$ must be equal to $|\phi'\rangle$:

$$B_{\nu''} \sum_{\nu'} F_{\nu''\nu} \{ \sum_{\nu=1}^{N} f_{\nu'\nu} a^{+}_{\nu} + \sum_{\rho=N+1}^{\infty} f_{\nu'\rho} a^{+}_{\rho} \}|0\rangle$$

$$= (a^{+}_{\nu''} + \sum_{\rho=N+1}^{\infty} S_{\rho\nu''} a^{+}_{\rho})|0\rangle$$

$$= (1 + \sum_{\rho=N+1}^{\infty} S_{\rho\nu''} a^{+}_{\rho} a_{\nu''}) a^{+}_{\nu''}|0\rangle$$

Finally, using the anticommutation relations for creation and annihilation operators, evaluate the determinant constructed from these **basis** wave functions, and show

$$|\phi'\rangle = \{ \prod_{\nu''=1}^{N} (1 + \sum_{\rho=N+1}^{\infty} S_{\rho\nu''} a^{+}_{\rho} a_{\nu''}) a^{+}_{\nu''}\} |0\rangle$$

$$= \sum_{\nu''=1}^{N} \sum_{\rho=N+1}^{\infty} (1 + S_{\rho\nu''} a^{+}_{\rho} a_{\nu''})|\phi\rangle$$

$$= \exp(\sum_{\nu=1}^{N} \sum_{\rho=N+1}^{\infty} S_{\rho\nu} a^{+}_{\rho} a_{\nu})|\phi\rangle$$

6. Derive Eqs. (2.19), (2.23), (2.25), and (2.26).

REFERENCES FOR CHAPTER 2

1. A. K. Kerman and S. E. Koonin, Ann. Phys. (NY) 100, 332 (1976).

2. J. W. Negele, AIP Conference Proceedings 41: Nucleon-Nucleon Interactions, ed., D.F. Measday et al. (1977).

3. L. J. Lantto and P. J. Siemens, Phys. Lett. 68B, 308 (1977); 311 (1977).

4. P. C. Martin and J. Schwinger, Phys. Rev. 115, 1342 (1959).

5. G. Baym and L. Kadanoff, Phys. Rev. 124, 287 (1961); G. Baym, Phys. Rev. 127, 1391 (1962).

6. F. Coester, Nucl. Phys. 7, 421 (1958); F. Coester and H. Kümmel, Nucl. Phys. 17, 477 (1960).

7. H. Kümmel, K. H. Lührmann, and J. G. Zabolitzky, Phys. Reports 36C (1978)

8. D. J. Thouless, The Quantum Mechanics of Many-Body Systems, Academic Press, N.Y. (1961).

III. APPLICATION TO SIMPLE, IDEALIZED SYSTEM

Considerable insight into the time-dependent mean-field and higher order coupled-cluster approximations is obtained by studying simple idealized problems. Numerical solutions for semi-infinite slab collisions elucidate in a particularly simple form all the essential features of mean-field calculations in more realistic geometries, and are therefore presented in some detail in Section 3.1. Whereas the two exactly solvable models of Section 3.2 are far more idealized and unrealistic than the slab problem, they offer the great advantage of providing unambiguous quantitative checks on the theory. The one-dimensional δ function model[2] of Section 3.2.1 provides a self-bound composite system with analytical scattering solutions for both the exact Schrodinger and TDHF equations. Using numerical techniques, the interacting systems described by the Lipkin Hamiltonian in Section 3.2.2 may be solved exactly, in the TDHF approximation, and in the second-order coupled cluster theory.[3]

3.1 Collisions of Semi-Infinite Slabs

Collisions of semi-infinite slabs offer a number of advantages for exploration of the time-dependent mean-field approximation. One obvious advantage is the fact that decoupling of the transverse degrees of freedom of the wave functions effectively reduces the TDHF equation to coupled equations in time and one spatial variable. Thus, numerics are extremely simple and inexpensive and one need not hesitate to explore a wide variety of cases. However, this technical advantage is not nearly as important as the fact that slab geometry clearly isolates the essential physics of the theory. As will be evident subsequently, a primary feature of TDHF is its weak coupling to transverse degrees of freedom, so slab collisions illustrate the dominant longitudinal equilibration with a minimum of complications. Furthermore, the concentration of all non-trivial spatial dependence in one dimension renders it feasible to graph and study the behavior of single-particle wave functions and the one-body density matrix in detail.

3.1.1 Slab geometry

Semi-infinite slabs of spin-isospin symmetric matter are considered which are translationally invariant in the two transverse directions. The most general structure of single-particle wave functions for such geometry is

$$\psi_{n,\vec{k}_\perp}(\vec{r}) = \frac{1}{\sqrt{\Omega}} e^{i\vec{k}_\perp \cdot \vec{r}_\perp} \phi_{n,\vec{k}_\perp}(z,t) \tag{3.1}$$

where $\vec{r}_\perp \equiv (x,y)$, $\vec{k}_\perp \equiv (k_x,k_y)$ and Ω is a transverse normalization area. So far, little has been accomplished, since a general function of (x,y,z) has been traded for a function of (k_x,k_y,z). However, if one uses the effective interaction

$$v(r,r')=t_o\ \delta(r-r')+\frac{t_3}{6}\delta(r-r')\rho(r)+V_o\ \frac{e^{-|r-r'|/a}}{|r-r'|/a}\ \left(\frac{16}{15}+\frac{4}{15}P_x\right) \tag{3.2}$$

where P_x denotes the space exchange operator, it is seen in Problem 3-1 that the mean-field appearing in the single-particle equation is

$$W(z) = \frac{3}{4}\ t_o\rho(z)+\frac{3}{16}\ t_3\rho^2(z) + 2\pi a^3\ V_o \int_{-\infty}^{\infty} dz'\rho(z')e^{-|z-z'|/a} \tag{3.3}$$

and that the HF equation is separable. The interaction (3.2) is closely related to the Skyrme forces discussed in Section 4.1.1, except that one combination of the quadratically momentum-dependent terms is replaced by a more physical finite-range force, and the other combination is constrained to be zero so that the HF equation is separable. The specific combination $4+P_x$ in the finite-range force is necessary to eliminate the spatial exchange term which would otherwise destroy separability. The interaction is adjusted to yield a nuclear matter binding energy of 15.77 MeV per particle at a saturation of $\rho=0.145$ fm^{-2} corresponding to $k_F = 1.29$ fm^{-1}.

The static HF problem is solved by spatial wave functions of the form

$$\psi_{n\vec{k}_\perp}(\vec{r}) = \frac{1}{\sqrt{\Omega}} e^{i\vec{k}_\perp \cdot \vec{r}_\perp} \phi_n^{HF}(z)\ \chi_{\tau\sigma} \tag{3.4}$$

where n denotes the spatial quantum numbers and $\phi_n^{HF}(z)$ satisfies

$$\left(-\frac{1}{2m}\frac{d^2}{dz^2} + w(z)\right) \phi_n^{HF}(z) = e_n\phi_n^{HF}(z) \tag{3.5}$$

The resulting density, obtained by filling all states below the fermi energy, ε_f is

$$\rho(z)=4\sum_n \sum_{|k|\le 2m(\varepsilon_f-e_n)} |\psi_{nk}|^2 = \sum_n \frac{2m}{\pi}(\varepsilon_f-e_n)|\phi_n^{HF}(z)|^2 \equiv \sum_n A_n|\phi_n^{HF}(z)|^2 \tag{3.6}$$

From Eq. (3.6), it is evident that the transverse wave functions play a completely passive role, simply determining the relative weighting of the various longitudinal wave functions $\phi_n^{HF}(z)$.

Similarly, the TDHF problem is solved by spatial wave function of the form

$$\psi_{n\vec{k}_\perp}(\vec{r},t) = \frac{1}{\sqrt{\Omega}}\, e^{-i\frac{k_\perp^2 t}{2m}}\, e^{i\vec{k}_\perp\cdot\vec{r}_\perp}\, \phi_n(z,t) \tag{3.7}$$

where $\phi_n(z,t)$ satisfies

$$i\,\dot{\phi}_n(z,t) = \left(-\frac{1}{2m}\frac{\partial^2}{\partial z^2} + W(z,t)\right)\phi_n(z,t)) \tag{3.8}$$

The phase factor $e^{-ik_\perp^2 t/2m}$ never contributes to any observables, so, as in the static case, the transverse wave functions play a completely passive role. A different set of transverse wave functions is associated with each longitudinal wave function

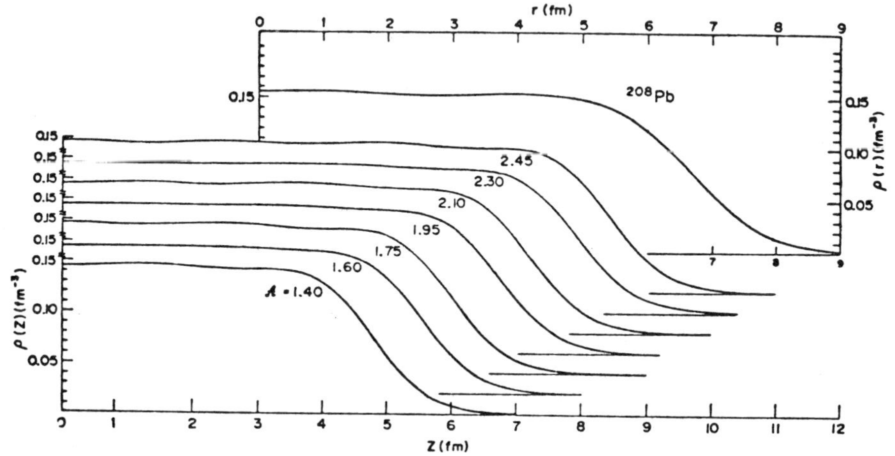

Fig. 3-1 Static slab solutions.

through the factor A_n when the initial conditions are established, and simply remain attached to that wave function for all subsequent times.

3.1.2 Initial and final states

The static HF equation, (3.5), may be solved for any specified integrated slab thickness, $\mathcal{A} = \int dz\rho(z)$ and ε_F and the number of occupied states are implicit functions of $\mathcal{A}$. Slab density profiles are shown in Fig. 3-1 for slab thicknesses corresponding to the integrated density through typical nuclei. These slabs contain from 4 to 7 different z wave functions and are qualitatively very similar in shape to the ^{208}Pb DDHF density distribution displayed in the upper right portion of the figure.

The TDHF equations (3.8) are solved numerically on a mesh of discrete values for z and t according to techniques described in Section 4.1. Since the equations are first order in time, the solution is fully specified by an initial set of complex wave functions $\phi_n(z,0)$. In order to generate appropriate initial conditions for a collision, one makes use of the result in Problem 1-7 that a static HF wave function in uniform Galilean translation satisfies the TDHF equation. Since

$$\phi_n(z,t) = \phi_n^{HF}(z-vt)\, e^{-i(e_n + \frac{mv^2}{2})t}\, e^{imvz} \tag{3.9}$$

satisfies Eq. (3.8) and translates uniformly to the right, it is clear that numerically integrating Eq. (3.8) with the initial condition $\phi_n(z,0) = \phi_n^{HF}(z)e^{imvz}$ will also yield the same result. Thus, if we wish to generate a collision between two slabs comprised of four single-particle wave functions each, the appropriate initial condition is the determinant of eight wave functions sketched in Fig. 3-2a. Each of the functions on the left is multiplied by e^{imvz} and each on right is multiplied by e^{-imvz}, so that the two slabs approach each other with relative velocity 2v. Clearly, if the slabs are sufficiently separated initially each of the eight single-particle orbitals is orthogonal to every other orbital: those within each slab by virtue of satisfying the same HF equation and those in different slabs because of the exponentially small overlap of their tails and differing phase factors.

When these single-particle orbitals are evolved in time according to the TDHF equation, they remain orthogonal as demonstrated in Section 1.3.3. If the final state is comprised of two separated fragments, in general part of each single-particle orbital is trapped in each fragment, so the final determinant is of the structure sketched in Fig. 3-2b.

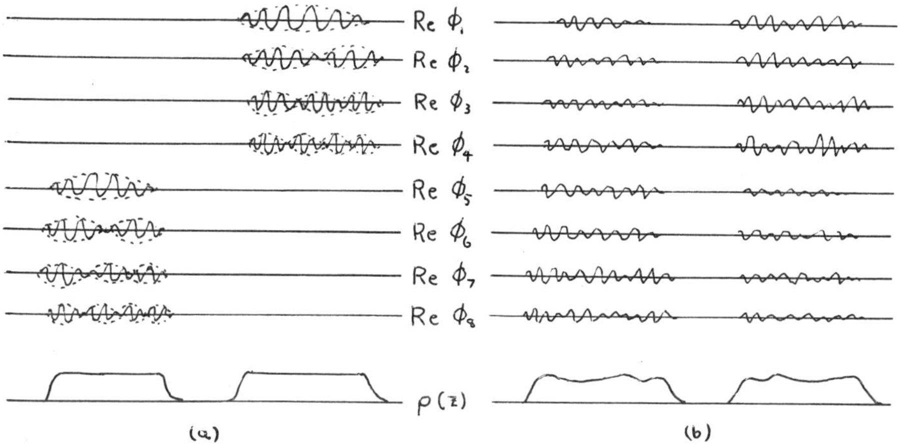

Fig. 3-2 Sketches of single-particle wavefunctions for collision initial condition (a) and final state (b).

Both the initial and final determinantal wave functions pictured in Fig. 3-2 merit some discussion, which will apply to collisions in any geometry although presented in the context of slabs. In the first place, although TDHF depends only on the one-body density matrix, we have singled out a specific and certainly non-unique representation. Any unitary transformation of the single-particle wave functions in Fig. 3-2 is physically equivalent to what has been sketched, so appearances may be misleading. By mixing the first four functions with the last four,for example, the initial state could be made to look like each orbital contributed to both the left and right fragments. Thus, in no sense can one attribute the first four functions to particles originating on the right or the last four to particles originating on the left. Indeed, throughout, one deals with approximations formulated in terms of totally anti-symmetric wave functions for indistinguishable particles. The real value of following non-unique single particle orbitals is that their evolution in the time-dependent mean-field is governed by a one-body Schrodinger equation to which one's vast intuition concerning evolution of wave packets is directly applicable.

For separated fragments such as in the initial and final states shown in Fig. 3-2, it is of interest to consider each fragment separately. Clearly, the "right half plane" density matrix

$$\rho^R(\vec{r},\vec{r}\,') \equiv \rho(\vec{r},\vec{r}\,')\theta(z)\theta(z') \qquad (3.10)$$

for the wavefunction in Fig. 3-2a is just that of a single four-orbital HF determinant translating to the left with velocity v. Thus, the left and right fragments are separately four-orbital determinants, as well as the overall wave function being an eight-orbital determinant. A different representation which mixes up the orbitals might disguise, but cannot change, this feature of separating into two determinantal subsystems.

The final fragments, however, certainly need not separately correspond to single determinants. Retaining only the right-hand portions of the orbitals in Fig. 3-2b may yield an eight-orbital determinant component, a seven-orbital component on down to a single-orbital component. Although this richness in the final state structure is interesting in the sense that it appears to naturally introduce particle transfer into the theory, it is important to recall that this lowest order theory at best describes mean values of few body operators, not the full wave-function. Hence, it is useful to address the particle number distribution in terms of the expectation value of moments of the number operator.

The number operator which counts probability in the right half plane is

$$\begin{aligned} N^R &= \iint \hat{\psi}^+(r) N^R(r,r')\psi(r')drdr' \\ &\equiv \sum_{\alpha\beta} N^R_{\alpha\beta}\, a^+_\alpha\, a_\beta \end{aligned} \tag{3.11}$$

where

$$N^R(r,r') = \delta(r-r')\theta(z)$$

and a corresponding operator can obviously be defined in any region of space containing a fragment of interest. The mean number of particles in the right fragment is then:

$$\begin{aligned} \langle\hat{N}^R\rangle & \iint N^R(r,r') \langle\psi^+(r)\psi(r')\rangle\, drdr' \\ &= \int \theta(z)\rho(r,r)\, dr = \mathrm{tr}\,\rho^R \end{aligned} \tag{3.12}$$

The fractional dispersion in particle number requires evaluation of $\langle(\hat{N}^R)^2\rangle$ in Problem 3.2, with the result

$$\frac{<(\hat{N}^R - <\hat{N}^R>)^2>^{1/2}}{<\hat{N}^R>} = \frac{[tr\{\rho^R - \rho^R\rho^R\}]^{1/2}}{tr\ \rho^R} \quad . \tag{3.13}$$

This expression for the number dispersion obviously yields zero if an individual fragment is a single determinant, as it must, from the presence of the quantity $\rho-\rho^2$ in the numerator.

To the extent to which the overall determinant appears to describe a collision between initial ground state particles with well defined relative momentum producing a superposition of alternative final states, the interpretation of collisions is straightforward. Two problems, however, arise in this lowest order mean-field description. The first is that the cm momentum of the incident HF ground states is not precisely Nmv, due to the zero-point motion of the HF solution. Whereas this initial wave function should be $e^{iNmv\ X_{cm}}\ \psi_{INTRINSIC}(x_1^{rel}\ldots x_{N-1}^{rel})$ the TDHF wave function is of the form $e^{iNmv\ X_{cm}}\ \psi^{HF}(x_1\ldots x_n)$. Instead of being a plane wave in X_{cm} over all space, since ψ^{HF} vanishes for X_{cm} outside the nucleus, the plane wave is modulated by a wave packet of the size of the nucleus. Thus, one begins with an initial cm wave packet which deviates slightly from the wave packet appropriate to a well-defined beam energy. Given this wave packet, which by construction has the correct value of $<P_{cm}>$, the TDHF equation evolves it in time to provide approximate mean values of final state observables originating from it. In higher orders of the coupled-cluster expansion, it is possible to derive successively more accurate approximations to the intrinsic state wave function, thus improving higher moments of $\hat{P}_{cm}$ and diminishing the c.m. error. For practical applications in real nuclei, the effective spreading of the beam energy due to the finite extent of the initial wave packet is not too important. Modulating the cm wave function by a wave packet the size of the nucleus introduces a spread in cm momentum $\Delta P_{cm} \sim \frac{\hbar}{2r_o A^{1/3}}$ so that the fractional error in the definition of a monochromatic beam energy is

$$\frac{\Delta E}{E} = \frac{\hbar}{\sqrt{2mE}\ r_o A^{4/3}} \tag{3.14}$$

where E is the energy per particle and m is the nucleon mass. At 2 MeV per particle, for example, this corresponds to an 8% spread in beam energy for ^{16}O and a 0.2% spread for ^{208}Pb.

A second problem concerns the interpretation of the final states. Clearly, by the non-linear nature of the theory, a number of different components of the final wave function get "trapped" in the common mean field generated by each other. One fragment may get slowed down while another gets accelerated. From the viewpoint of final states in reaction theory, this appears to be nonsense, since the various final states should be totally independent of each other. But when one recalls that the overall wave function is surely meaningless, and only mean values make sense, there is no longer an interpretation problem: the non-linear evolution in the mean field is producing just the right compromise between various components to yield accurate expectation values of few-body operators. Successive refinement by going to higher orders in the coupled cluster hierarchy should systematically improve these TDHF mean values.

3.1.3 Single particle propagation in the mean field

As explained in the last section, the essential physics of the TDHF approximation is most evident if one thinks about each single-particle wavefunction evolving independently in the time-dependent mean field. (Of course, one has to evolve all the occupied orbitals together to construct the mean field, but after one knows W(z,t) it is true that each s.p. wave function simply satisfies the one body Schrodinger equation, (3.3), in that well.) Thus, all our intuition regarding wave packets in one-dimensional potential wells is immediately applicable.

Figures 3-3a and 3-3b display the total density and the contributions to the density from the first and third orbitals of the left hand slab at successive times for an initial cm energy of 3.5 MeV per particle. (The second and fourth orbitals are also shown at the final time in 3.3b). Because of the Yukawa interaction, even when the density fluctuates, the mean field is quite smooth so the s.p. orbitals essentially propagate in a smooth 50 MeV well and are reflected from the well edges. The lump at the edge at $t=0.32\times10^{-21}$ sec, evidently arises from the reflection of the third orbital rather than from any sort of Benard instability in fluid mechanics and is chiefly responsible for beginning to move the edge of the well back to the right. At this energy, when scission occurs, most of the orbitals which began on the left are trapped on the right, but are significantly distorted from their original shape, giving rise to the high excitation energy and conspicuous density fluctuations in the emerging fragments.

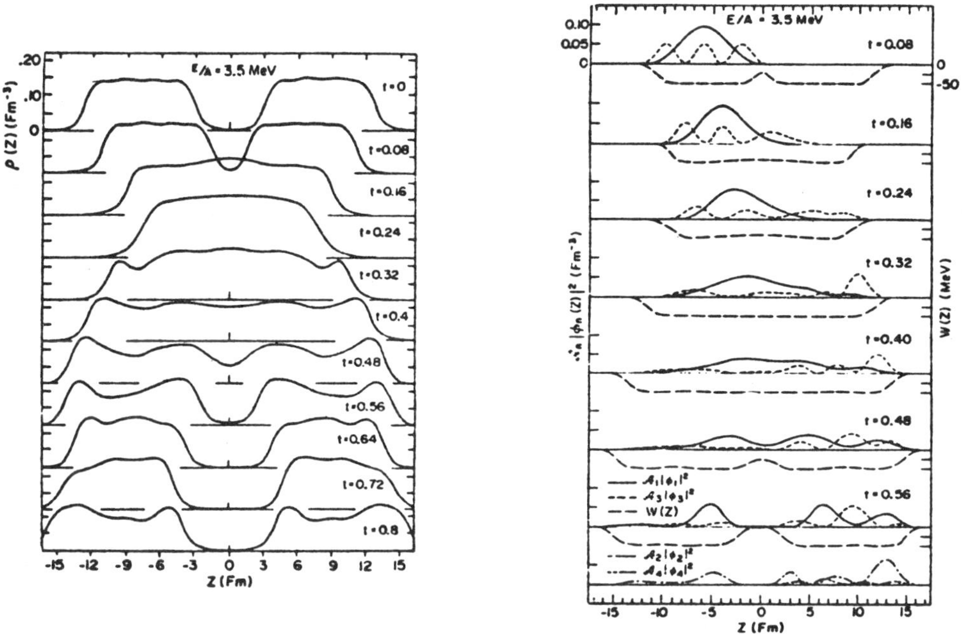

Fig. 3-3 Density distributions (a) and the contribution to total density of a single orbital (b) at sequential times for c.m. energy E/A=3.5 MeV. In the upper graphs of part b, the mean field, contribution of the lowest orbital from the left and third orbital from the left are denoted by long dashes, solid line, and short dashes respectively. The bottom graph shows the second and fourth orbitals at the final time.

At a c.m. energy of 1 MeV per particle, the corresponding density profiles are shown in Fig. 3-4a and orbitals originating from the left slab are shown in Fig. 3-4b. In this case, the first orbital was reflected from the right edge and is roughly equally distributed between the two fragments; the second, with a higher phase velocity, was almost completely reflected into the left well; the third was reflected from both the right and left well edges and began to propagate back into the right well; and the fourth and most rapid orbital made a complete cycle to the right well, back to the left, and finally all the way to the right well again.

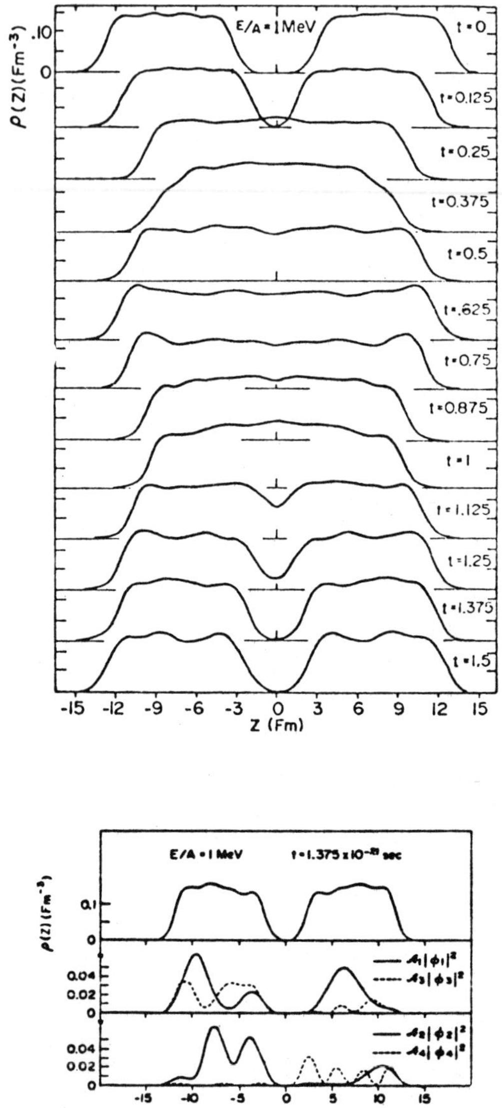

Fig. 3-4 Same as 3-3 for E/A=1 MeV.

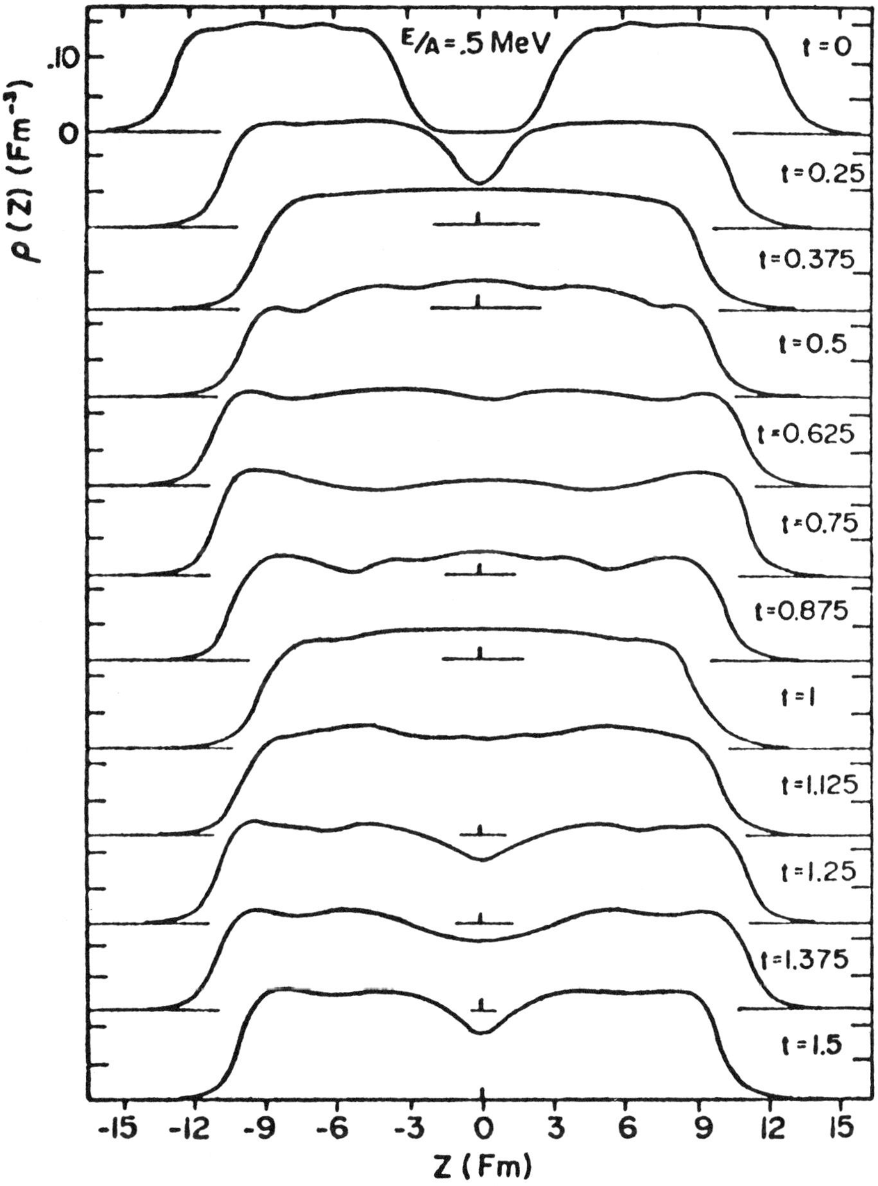

Fig. 3-5 Density distributions for E/A=0.5 MeV

At still lower energies, the repeated multiple reflections of single-particle wavefunctions from the well edges completely equilibrate the original collective translational kinetic energy into single-particle excitation, forming an oscillating compound slab as shown in Fig. 3-5 for a c.m. energy at 0.5 MeV per particle. The oscillations of this compound system have actually been calculated for a period of 2.6×10^{-21} sec. corresponding to 5 full cycles, with no hint of subsequent breakup.

If scission doesn't occur within the first few oscillations, further equilibration from subsequent reflections makes it highly unlikely later. Thus the question of whether scission occurs at a particular energy is a very delicate question depending crucially upon whether the s.p. orbitals conspire to create a relative maximum or minimum in the neck region as the compound slab reaches its point of maximum stretching. A tiny change in initial conditions modifies the arrangement of s.p. orbitals in an essential way and thus qualitatively affects the final outcome of the reaction. The resulting resonance behavior is clearly demonstrated in Fig. 3-6, which shows the ratio of final to incident translational kinetic energy for a variety of symmetric collisions. Above 2 MeV, the orbitals originating on the left primarily exit to the right, as in Fig. 3.3b, yielding excited states in which 90% or more of the collective translational energy has been converted into internal excitation energy. Below 2 MeV, complicated multiple reflections occur, giving rise to compound slabs which undergo many length oscillations without scissioning (plotted as a ratio of 0 in Fig. 3-6, as well as complicated resonances in which the slabs separate on the second or subsequent oscillation.

Figure 3-6 emphasizes two essential features of TDHF dynamics of relevance to heavy ion experiments. One feature is very strong damping in which initial translational kinetic energy is transferred into other degrees of freedom. Although the final fragment velocities in Figs. 3-3 and 3-4 are visually lower than the incident velocities, Fig. 3-6 shows that more than 90% of the original translational kinetic energy is dissipated in these low energy collisions.

The second feature is the presence of pronounced single-particle resonances, arising from the fact that one or several single-particle orbitals in the neck region can play a decisive role is determining whether or not fusion occurs. These resonances are extremely suggestive of the resonances in fusion cross sections for light ion reactions on ^{12}C. As shown in Figs. 3-7a and b, dramatic fusion cross section fluctuations are observed with ^{16}O and ^{12}C projectiles in the energy regime studied in Fig. 3.6, and structureless cross sections occur

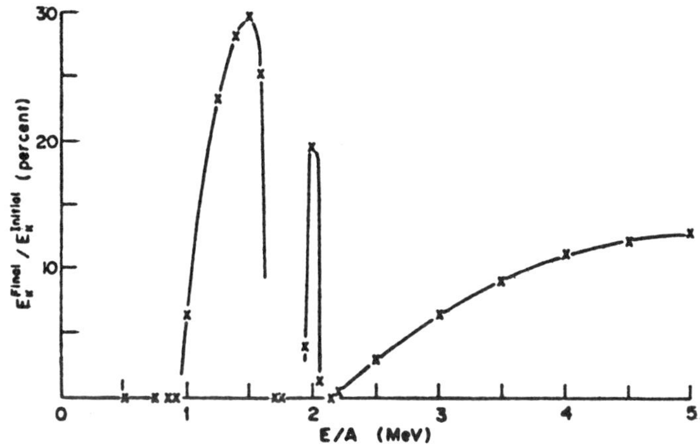

Fig. 3-6 Ratio of final to incident translational kinetic energy as a function of c.m. incident energy for symmetric slab collisions.

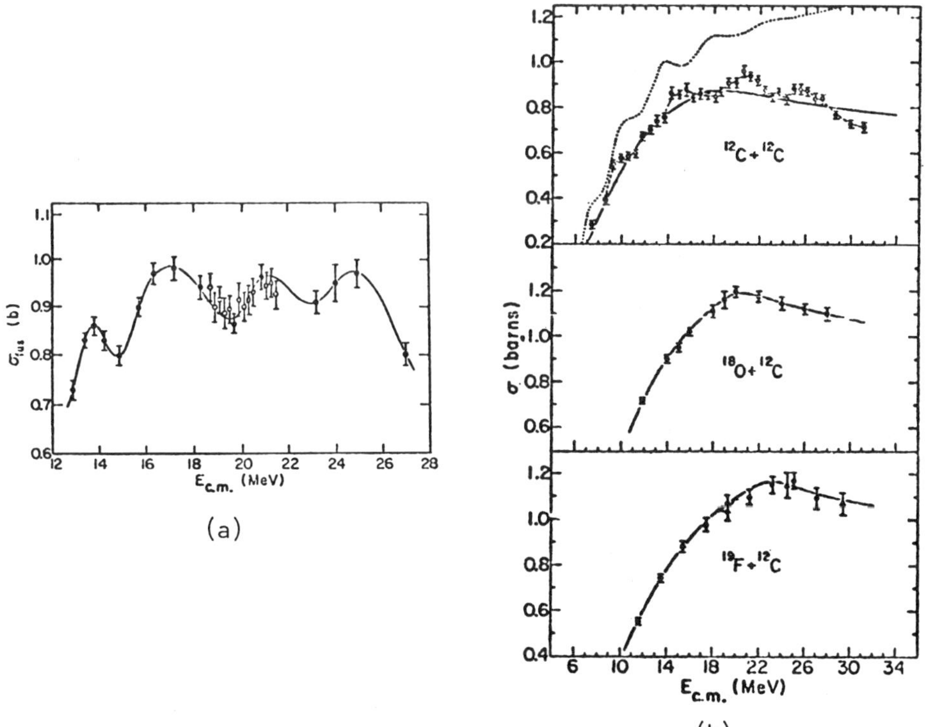

Fig. 3-7 Total fusion cross-sections as a function of c.m. energy for $^{16}O+^{12}C$ (a) and for $^{12}C+^{12}C$, $^{18}O+^{12}C$, and $^{19}F+^{12}C$ (b).

for ^{18}O and ^{19}F. If the mean-field theory is in fact successful in reproducing this phenomenology, it should give significant insight into the mechanism whereby the valence nucleons in ^{18}O and ^{19}F increase and smooth the fusion cross sections.

Another interesting feature evident in slab collisions is single - particle emission, arising from excitation of high momentum components in the single-particle orbitals of one nucleus due to the rapid passage of the potential generated by the other nucleus. Fig. 3-8 shows the density during a 2.4 MeV per particle c.m. collision on a logarithmic scale and the corresponding velocity distribution. At $t=0.32\times10^{-21}$ sec., the dashed line shows a high energy particle distribution at about one percent of the central density beginning to emerge from the compound slab. By the time scission occurs, the solid line shows the high energy tail clearly separating with particle velocities ranging well above the incident velocity of 21.5 fm/10^{-21} sec. Roughly 2/3 of the probability comes from the least bound single-particle orbital, 1/4 from the next and the remaining 1/12 from the deepest two levels. Whereas it is interesting that particle emission is already present in the lowest order mean-field approximation, in view of the crucial role of hard core collisions in particle emission, it will be particularly important to examine higher order corrections in the coupled-cluster theory.

Although the mean-field approximation with effective interactions becomes quantitatively suspect at higher energies, it is still useful to understand the qualitative behavior of high energy mean-field dynamics. Density profiles at successive time intervals are shown in Fig. 3-9 for a cm energy of 25 MeV per particle. Note that the maximum interior density almost doubles as the slabs pass through each other, and that the final slab stretches out to an extremely long, low density configuration which finally coalesces into a large number of fragments. The coalescence occurs gradually from the outer edges inward as the most rapidly moving single-particle orbitals separate from the rest, with orbitals of comparable velocities trapping each other in a common mean field. The outermost fragment has received a time advance, rather than delay, due to the acceleration it experienced from the overall attractive field of the other nucleus.

A more detailed view of the density profiles of interpenetrating slabs in high energy collisions is shown in Fig.3-10. A well-defined transition region between high and low density matter propagates at the "shock" velocity $S = \dfrac{v_i}{(\rho/\rho_o - 1)}$

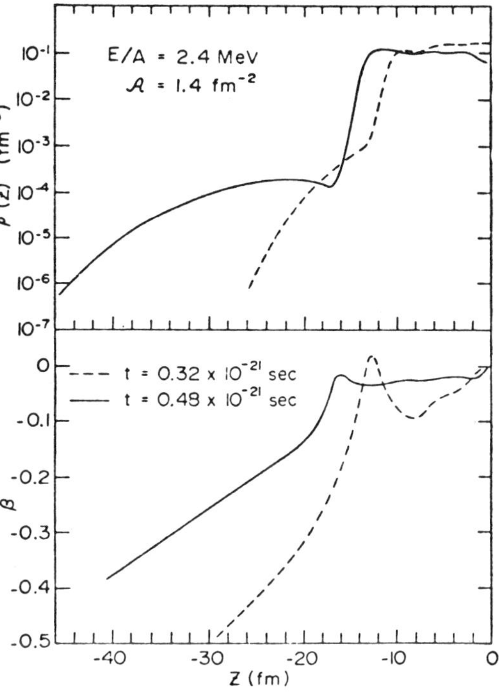

Fig. 3-8 Density profiles and velocity distributions for separating slabs showing particle emission.

obtained by applying the continuity equation between constant density regions at rest and at initial velocity v_i. Even in the mean-field approximation, non-trivial equilibration occurs across the shock front, because single particle orbitals entering the high density region see a much less attractive or even repulsive potential. Thus, they scatter from the interface just like a wave packet scattering from a finite potential barrier. Whereas for very long slabs, this scattering at the shock front would pile up density, giving rise to densities higher than twice nuclear matter density, for slabs of dimensions comparable to real nuclei, Fig. 3-11 shows that the density never exceeds the value $2\rho_o$ expected in a high energy limit in which nuclei simply pass through each other. The behavior of this density in these high energy collisions will be important for the subsequent consideration of pion condensation in Section 4.5.

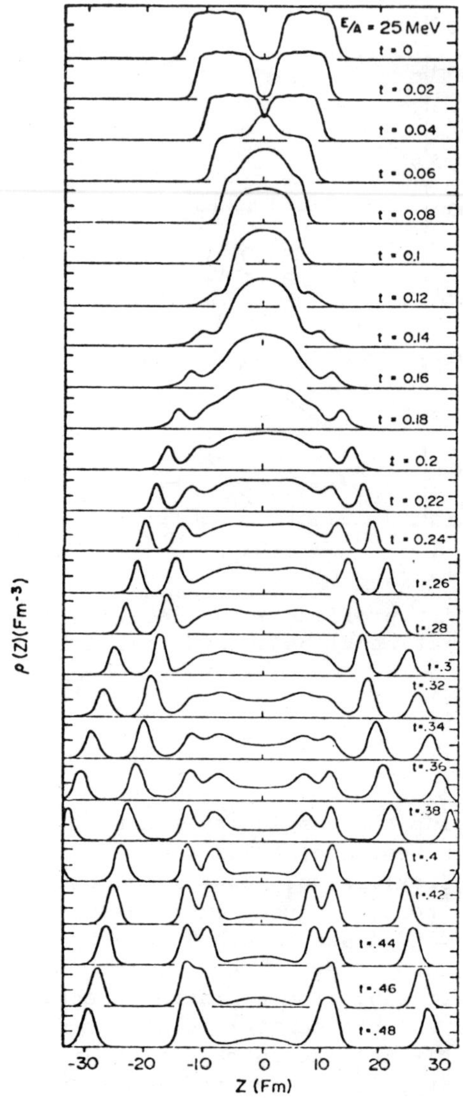

Fig. 3-9 Density profiles as a function of time for c.m. energy E/A=25 MeV.

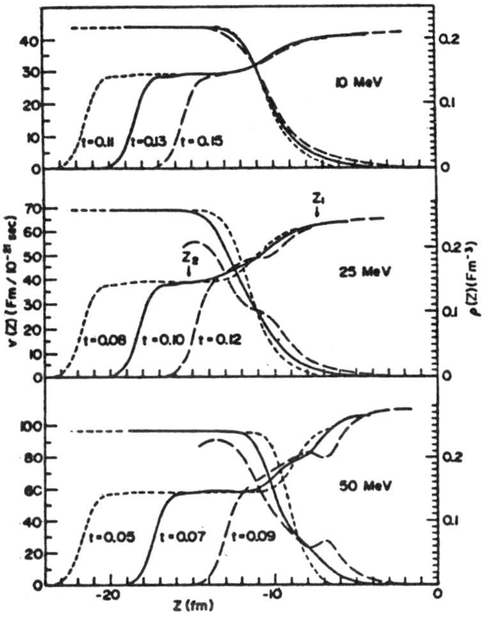

Fig. 3-10 Colliding slabs viewed in a frame moving with the shock front. The velocity profile is referred to the left scale and the density profile to the right scale.

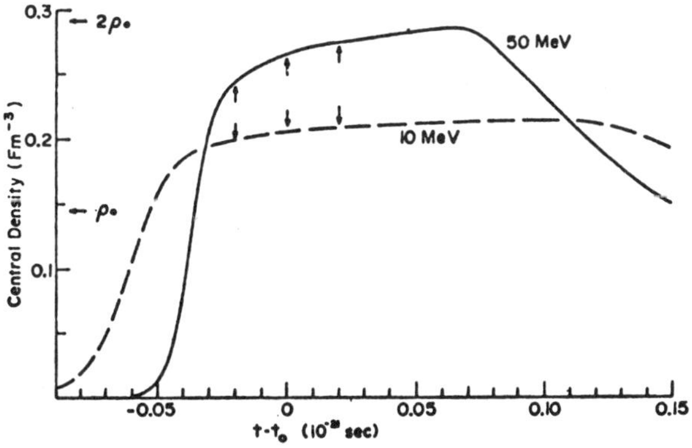

Fig. 3-11 Central density as a function of time for c.m. energy E/A=10 and 50 MeV. The arrows denote the times corresponding to Figure 3-10.

3.1.4 The one-body density matrix

Valuable insight into the role of equilibration in mean-field dynamics is provided by examination of the off-diagonal structure in the z direction of the one-body density matrix $\rho(z,z') \equiv \rho(x,y,z;x,y,z')$ in slab collisions. For convenience, it is useful to introduce relative and cm coordinates, to remove an overall complex translational phase factor, and to normalize to unity by defining

$$\tilde{\rho}(z,s) \equiv e^{-imV(z)s}\,\rho(z)^{-1}\,\rho(z+\tfrac{s}{2},\, z-\tfrac{s}{2}) \qquad (3.15)$$

where

$$V(z) = \frac{1}{m\rho(z)}\left(\frac{\partial}{\partial z''} - \frac{\partial}{\partial z'}\right)\rho(z'',z')\Big|_{z''=z'=z} .$$

Whereas from general arguments presented previously, it is clear that $\tilde{\rho}$ goes to zero for large s because of randomization of phases, the essential question relevant to equilibration is how rapidly $\tilde{\rho}$ falls off. For comparison, it is useful to refer to the corresponding form of $\tilde{\rho}$ for several different cases. For a finite-temperature three-dimensional Fermi gas at density ρ and corresponding Fermi momentum k_F,

$$\rho_T(s) = \frac{18\pi^2}{k_F^6 s}\int k\,\frac{\sin(ks)}{1+\exp(\hbar^2/2mk_BT)(k^2-\hat{k}^2)}\,dk \qquad (3.16)$$

where $\hat{k}$ is implicitly defined by $\rho_T(0)=1$. The T=0 limit is the usual Slater result,

$$\rho_{SL}(k_F s) = \frac{3j_1(k_F s)}{k_F s} . \qquad (3.17)$$

Finite temperature both narrows the range of $\rho(s)$ and damps the negative oscillation relative to the Slater function. Whereas in three dimensions, changing the density from equilibrium density ρ_o with Fermi momentum k_F^o to arbitrary density ρ scales $\tilde{\rho}(s)$ according to

$$\tilde{\rho}_{3-D}(s) = \rho_{SL}\left(k_F^o\left(\frac{\rho}{\rho^o}\right)^{1/3} s\right) , \qquad (3.18a)$$

it is shown in Problem 3.3 that with transverse wave functions frozen as they are in slabs, $\tilde{\rho}(s)$ scales as

$$\tilde{\rho}_{1-D}(s) = \rho_{SL}(k_F^o(\frac{\rho}{\rho^o})s) \qquad (3.18b)$$

With this background, reduced density matrices $\tilde{\rho}(z_1,s)$ evaluated at the position z_1 in the high density region of the 25 MeV/A collision of Fig. 3-10 are compared with various Fermi gas results in Fig. 3-12. Fig. 3-12a shows the modest narrowing of $\tilde{\rho}(s)$ arising from a finite temperature of 20 MeV at nuclear matter density ρ_o. At the more appropriate density of $2\rho_o$, however, this narrowing is rather negligible. The actual TDHF results in panel c look very different indeed from any of these finite-temperature results. Not only is the shape significantly narrower, but the negative overshoot in Re$\tilde{\rho}(s)$ at large s is much larger than in any of the calculations which assume thermal equilibration, and the imaginary part does not vanish as it should when in thermal equilibrium. The fourth panel accounts for the narrowness of the calculated $\tilde{\rho}(s)$ by demonstrating the significant difference between $\tilde{\rho}_{3-D}$ and $\tilde{\rho}_{1-D}$ of Eqs. (3.18a) and (3.18b). Physically, the reason $\tilde{\rho}(s)$ is so narrow is that because the transverse wavefunctions cannot change, all the additional high momentum components required to

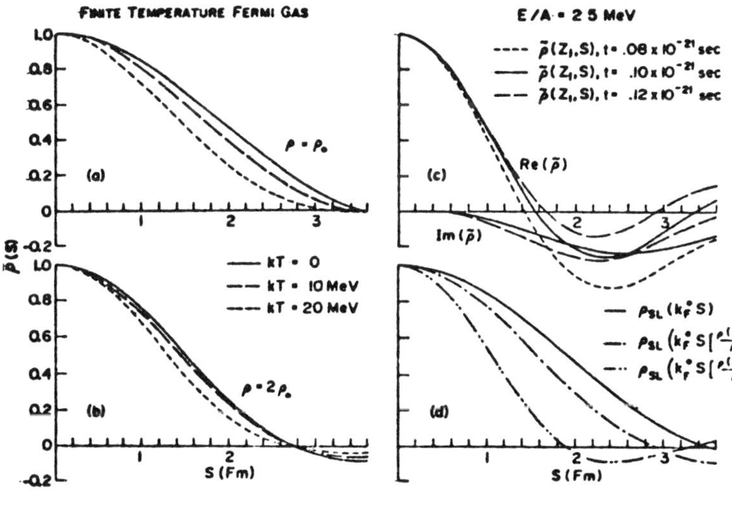

Fig. 3-12 Density matrices for finite temperature Fermi gas (a&b), for the 25 MeV collision at position z_1 of Fig. 3-10 (c), and for a zero temperature Fermi gas (d).

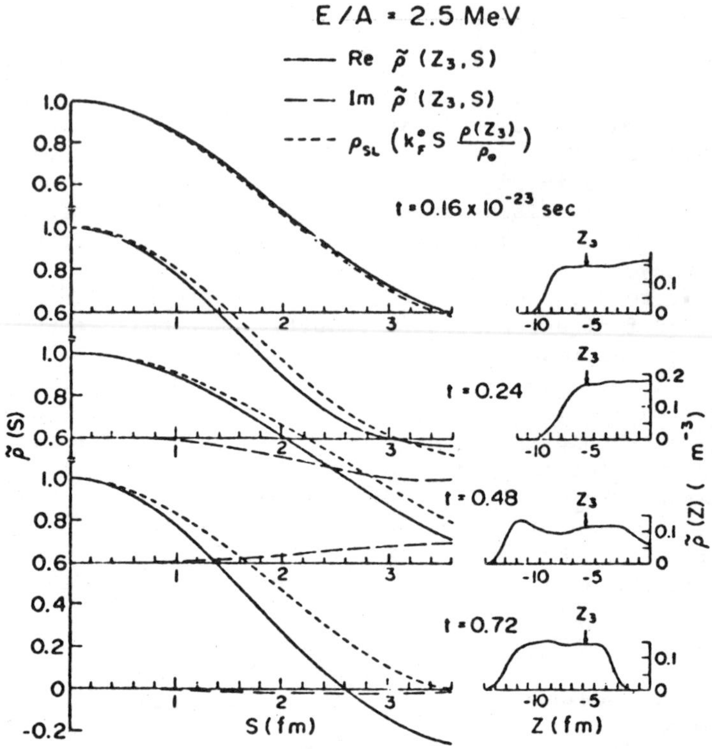

Fig. 3-13 Normalized one-body density matrices for E/A=2.5 MeV. The left-hand plots present the one-dimensional Fermi gas results (short-dashed curves) and real and imaginary parts of the calculated density matrices (solid and long-dashed curves) at the position z_3 denoted on the density distributions graphed on the right.

avoid violating the Pauli principle must occur in the z direction, and the resulting high k_z components yield phase cancellations at very small values of s.

Similar results are shown in Fig. 3-13 before and after a much lower energy collision at 2.5 MeV/A. Prior to the collision, the Slater approximation is extremely accurate, as is well known in finite nuclei.[6] Although during the overlap time and in the final fragments roughly the correct scale is set by Eq. (3.18b), it is clear from the significant overshoot of Re$\tilde{\rho}$ and non-vanishing values of Im$\tilde{\rho}$ that the one-body density matrix has in no way reached thermal equilibrium during the time scales of interest.

3.1.5 Relation to hydrodynamics

Although the time-dependent mean-field approximation and hydrodynamics obey the same mass, energy and momentum conservation equations, they differ fundamentally with respect to equilibration. In the mean field theory, equilibration arises slowly from repeated collisions with the edges of the potential well, whereas in hydrodynamics, one imposes from the outset the assumption of complete instantaneous local equilibration. The lack of equilibration during the time scales of interest in collisions have been displayed particularly dramatically in Figs. 3-12 and 3-13. Published attempts notwithstanding, it is obvious that with this fundamental difference in equilibration, there is no way to derive hydrodynamic equations from the TDHF approximation.

A particularly clear distinction between hydrodynamics and the mean field theory arises in the case of sound propagation. The complete local equilibration built into hydrodynamics yields the thermodynamic sound mode in which the Fermi sphere remains spherically symmetric and simply varies in volume periodically in space and time. In contrast, the sound mode arising in the mean-field theory is zero sound in which the Fermi sphere distorts as well as changing volume. In addition to essential differences in the role of collisions and the geometry of the Fermi surface, macroscopic features such as the sound velocity are affected, with the zero sound velocity approaching the Fermi velocity $\frac{k_F}{m}$ in the weak coupling limit, whereas the thermodynamic sound velocity approaches $\frac{k_F}{\sqrt{3}m}$ in the same limit.

A simple derivation of the spin-isospin symmetric zero sound mode is given in Problem 3-4. It is instructive, however, to look at the same result from the viewpoint of the Wigner distribution function for the special case of our slab Hamiltonian that produces a local mean field. Introducing the Wigner function

$$f(p,R) \equiv \int e^{-ip\cdot s}\, \rho(R+\frac{s}{2},\, R-\frac{s}{2}) \qquad (3.19)$$

and expanding the mean field in the TDHF equation

$$i\dot{\rho}(x,y) = [-\frac{1}{2m}(\frac{\partial^2}{\partial x^2} - \frac{\partial^2}{\partial y^2}) + W(x)-W(y)]\, \rho(x,y) \qquad (3.20)$$

yields

$$\left(i\frac{\partial}{\partial t} + \frac{1}{m}\frac{\partial}{\partial R}\frac{\partial}{\partial s} - \vec{\nabla}_s W\cdot\vec{S}\right)\rho\left(R + \frac{s}{2}, R - \frac{s}{2}\right) = \frac{\partial}{\partial s_i}\frac{\partial}{\partial s_j}\frac{\partial}{\partial s_k} W\, s_i s_j s_k\, \rho \tag{3.21}$$

so that in the long wavelength limit, f obeys the Boltzmann equation

$$\frac{\partial}{\partial t} f(p,R) + \vec{v}\cdot\vec{\nabla}_R f(p,R) - \vec{\nabla}W\cdot\vec{\nabla}_p f(p,R) = 0 \tag{3.22}$$

As shown by Landau,[7] this equation has undamped solutions for infinitesimal changes in f

$$\delta f = e^{i(\vec{k}\cdot\vec{x}-\omega t)} \frac{\cos\theta}{s-\cos\theta} \tag{3.23}$$

where s is greater than 1 and is given in Problem 3.4, and θ is the angle between k and x. When s is close to one, which is the usual case, δf is highly anisotropic, with a very large distortion in the direction of k and much smaller distortions in all other directions. For self-bound systems, like nuclei or liquid He, it is clear that the first derivative of the energy density functional with respect to ρ, which yields W, must be attractive. It is shown in Problems 3-4 that spin-isospin symmetric zero sound only propagates in nuclear matter (and similarly, spin symmetric zero sound in liquid He) if the second derivative of the energy density functional is repulsive. In the context of the slab Hamiltonian, this can occur if the contribution to W of the density-dependent repulsive t_3 term is slightly overcompensated by the attractive density independent terms. The additional differentiation to calculate the sound mode may then enhance the t_3 term relative to the others to obtain an overall repulsive result. This situation in fact occurs for the slab Hamiltonian, yielding a zero sound velocity of roughly 83 fm/10^{-21} sec. Other interesting spin modes, isospin modes, and spin-isospin modes are also possible and should be investigated when one is confident that the effective interaction is sufficiently well understood.

The transition between zero sound and thermodynamic sound, and thermodynamic sound itself, can only be understood by going to higher orders in the coupled-cluster theory. As a single-particle orbital from one nucleus propagates through the smooth mean field of another, there is nothing in the mean-field approximation to produce scattering in the transverse direction.

The "granularity" of the nuclear medium has been ignored. The two-particle, two-hole amplitudes in the second-order coupled-cluster approximation describe just these omitted effects. Since both zero sound and thermodynamic sound are observed in liquid He^3 at sufficiently low temperatures by looking at high frequencies (where collisions are negligible) and low frequencies (where collisions dominate) this system provides an ideal testing ground for exploration of higher order coupled-cluster corrections.

A final difference between hydrodynamics and mean-field dynamics concerns the macroscopic shape of compound systems created in collisions. A characteristic feature of the head-on collisions of liquid drops obeying hydrodynamics is splashing in the transverse direction. Whether or not the fluid is assumed incompressible, complete instantaneous local equilibration transfers significant amounts of longitudinal momentum into the radial direction. At low energies, if two drops fuse, they will flatten out into an oblate shape, pass back through a roughly spherical shape, and execute a series of prolate-oblate vibrations. At higher energies, fragments may actually separate off in the transverse direction. In contrast, to the extent to which a single-particle orbital propagating through a large nucleus cannot distinguish that it is not in a semi-infinite slab, there is no equilibration into transverse degrees of freedom. Thus, we shall see that in $^{16}O + {}^{16}O$ or $^{40}Ca + {}^{40}Ca$ collisions, the shape of the compound system is always prolate. The large amount of momentum concentrated in the longitudinal direction produces large forces on the longitudinal surfaces of the system when single-particle orbitals are reflected from the ends, elongating the compound system from its natural spherical shape. Only the curvature of the ends gradually transfers momentum into transverse modes, but many reflections are required to significantly populate these modes. Thus, gradually, as the compound system equilibrates, it becomes less and less prolate, but never is driven oblate. One of the great frustrations of research in this field has been the fact that in spite of such dramatic difference in the macroscopic shapes of compound systems in hydrodynamics and the mean-field theory, no experimental observables have yet been used to definitively distinguish between them.

3.2 Exactly Solvable Problems

Whereas the slab collisions of the last section elucidated many features of the mean-field theory relevant to nuclear collisions, they provided no quantitative tests of the lowest order approximation or higher order coupled-cluster corrections.

The exactly solvable models of this section provide such quantitative tests, but at the expense of being highly idealized and rather unrealistic. The first model, the one-dimensional delta function problem, provides a self-bound, composite system somewhat analogous to the preceding slab example. Its primary limitation, however, is an exceedingly limited S-matrix arising from the fact that no excited bound states exist in the model. The second model, two interacting systems described by the Lipkin Hamiltonian, provides a much richer range of non-trivial final states, but at the expense of giving up self-bound systems with non-trivial spatial dependence. Hopefully, the study of several complementary models emphasizing different aspects of the physics of the full problem somewhat compensates for the oversimplification of any individual model.

3.2.1 One-dimensional system with attractive δ-function interactions

The great advantage of the slab model of Section 3.1 was the fact that it utilized three-dimensional phase-space and a realistic effective interaction while yielding single particle equations depending only on one spatial variable. To obtain an analytically solvable problem, however, we must go to a genuinely one-dimensional model with a total unrealistic interaction.

The Hamiltonian

$$H = -\frac{1}{2}\sum_{i=1}^{N}\frac{d^2}{dx_i^2} - g\sum_{i<j=1}^{N}\delta(x_i-x_j) \qquad (3.24)$$

has spatially totally symmetric bound states solutions of the form[8]

$$\psi = \exp\left(-\frac{g}{2}\sum_{i<j=1}^{N}|x_i-x_j|\right) \quad . \qquad (3.25)$$

A wave function which is antisymmetric with respect to interchange of any two variables loses the benefit of one of the δ interactions and breaks up into two separate spatially symmetric subsystems. Thus, to obtain bound states, the wave function must be spatially totally symmetric and we must either deal with bosons or with fermions having some additional quantum number such as a spin with degeneracy 2S+1=N allowing the wave function to be antisymmetric in spin and symmetric in space. The boson case will be treated here for simplicity, but everything may be generalized to the high-spin fermion case.

3.2.1.1 Static bound state solutions: A system of N bosons interacting with hamiltonian (3.24) has a single N-particle bound state,[8,9]

$$\psi_N = N!\sqrt{(N-1)!}\, g^{N-1} \exp(-\frac{g}{2} \sum_{i<j=1} |x_i - x_j|) \tag{3.26}$$

with energy

$$E_N = -\frac{N(N^2-1)g^2}{24} \tag{3.27}$$

and density[10]

$$\rho_N(x) = \int dx_1 \ldots dx_N \; \delta(\frac{1}{N} \sum_{i=1}^{N} x_i) |\psi_N|^2 \; \delta(x_1 - x)$$

$$= g(N!)^2 \sum_{n=1}^{N-1} (-1)^{n+1} \frac{n \exp(-gnN|x|)}{(N+n-1)!(N-n-1)!} \tag{3.28}$$

In the large N limit

$$\rho_N(x) \underset{N\to\infty}{\to} \frac{N^2 g}{4\cosh^2(\frac{Ngx}{2})} \{1 + \mathcal{O}\frac{1}{N}\} \tag{3.29}$$

The mean-field approximation, which will be called the Hartree approximation, is obtained by choosing the trial wave function,

$$\psi_N^H = \sqrt{N} \prod_{i=1}^{N} \phi_o(x_i) \tag{3.30}$$

and minimizing $\langle H\rangle$ with respect to the normalized single particle wave function, $\phi_b(x)$. The resulting single particle equation

$$\{-\frac{1}{2}\frac{d^2}{dx^2} - g(N-1)|\phi_b(x)|^2 - \varepsilon\} \phi_b(x) = 0 \tag{3.31}$$

has one bound state solution[10]

$$\phi_b(x) = \frac{\sqrt{(N-1)g}}{2\cosh\{\frac{(N-1)gx}{2}\}} \tag{3.32}$$

with energy eigenvalue

$$\varepsilon_b = \frac{-(N-1)^2 g^2}{8} \qquad (3.33)$$

The resulting Hartree energy

$$E_N^H = \frac{-N(N-1)^2 g^2}{24} \qquad (3.34)$$

and Hartree density

$$\rho_N^H(x) = \frac{N(N-1)g}{4\cosh^2\{\frac{(N-1)gx}{2}\}} \qquad (3.35)$$

agree with the exact results, Eqs. (3.27) and (3.28), to leading order in N.

As usual in the mean-field theory, errors of order $\frac{1}{N}$ relative to the leading contribution arise from the spurious cm motion. In Problem 3-5, it is shown that minimization of $\langle H-H_{cm}\rangle$, where $H_{cm} = \frac{1}{2}(\sum_i P_i)^2$, yields the Hartree energy

$$\tilde{E}_N^H = -\frac{N^2(N-1)}{24} g^2 \qquad (3.36)$$

thus eliminating half of the order N^2 discrepancy between Eq. (3.27) and (3.34).

To understand why the mean-field approximation is asymptotically correct, and to gain insight into the structure of higher order corrections, it is useful to examine the N-dependence of perturbation theory corrections of arbitrary order. It is particularly convenient to apply the Goldstone expansion to this problem by assuming one is dealing with fermions having a spin degeneracy (2S+1)=N and disregarding the totally antisymmetric spin wave function. We work in the basis of wave functions generated by the Hartree potential

$$U(x) = \frac{-(N-1)^2 g^2}{4\cosh^2(\frac{(N-1)gx}{2})} \qquad (3.37)$$

which in addition to the bound state (3.32), has continuum states

$$\phi_k(x) = \frac{\sqrt{(N-1)g}}{4\pi}\, e^{ik(N-1)\frac{gx}{2}} \left[\frac{\tanh(\frac{(N-1)gx}{2})-ik}{1+ik}\right] \tag{3.38}$$

Any linked graph in the Goldstone expansion for the energy has I interactions and C closed loops. Every interaction contributes a factor of $(N-1)^2$ which arises from the normalization of the particle and hole wave functions and a factor of $(N-1)^{-1}$ which arises from the integration over spatial variables. Note that this latter factor may be extracted only because all the wave functions are functions of $(N-1)x$ and the interaction is zero-range, allowing one to change variables and remove all N-dependence from the integrand. Every closed loop contributes a factor of N arising from the sum over spin projections. For every graph, there are I-1 energy denominators, each of which contributes a factor of $(N-1)^{-2}$. Hence the overall N-dependence of any Goldstone diagram in the energy expansion is $(N-1)^{2-I}N^C \sim N^{C-I+2}$.

Similarly, the expansion for the 1-body density is derived from the energy expansion by inserting the 1-body density operator into a particle or hole line. This modification contributes a factor of N-1 arising from the normalization of the particle and hole wave functions and a factor of $(N-1)^{-2}$ arising from the addition of an energy denominator. The overall N-dependence is thus $(N-1)^{1-I}N^C \sim N^{C-I+1}$.

Thus, precisely as in the $\frac{1}{N}$ expansion in field theory, we have an unambiguous accounting of the asymptotic dependence of individual graphs on the particle number. The only diagram which contributes in order N^3 is the Hartree diagram.

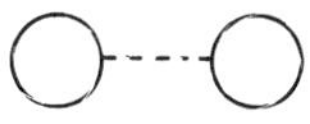

(3.39)

which has C=2 and I=1. Mean-field insertions into this (or any other diagram) are cancelled by the definition of U in Eq. (3.37), so an infinite hierarchy of diagrams formally of the same order, such as

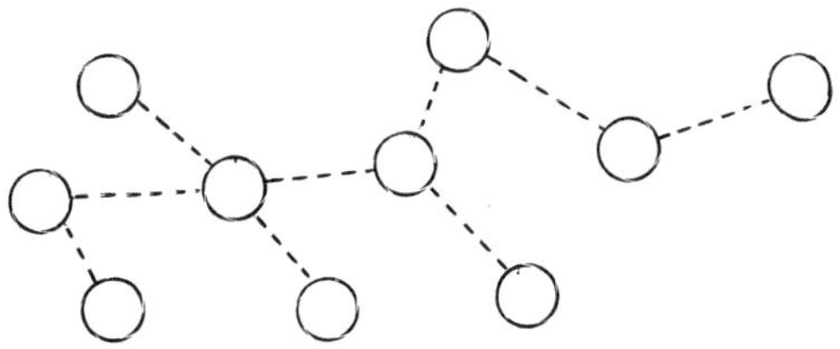

(3.40)

do not arise. The diagrams of order N^2 are the following:

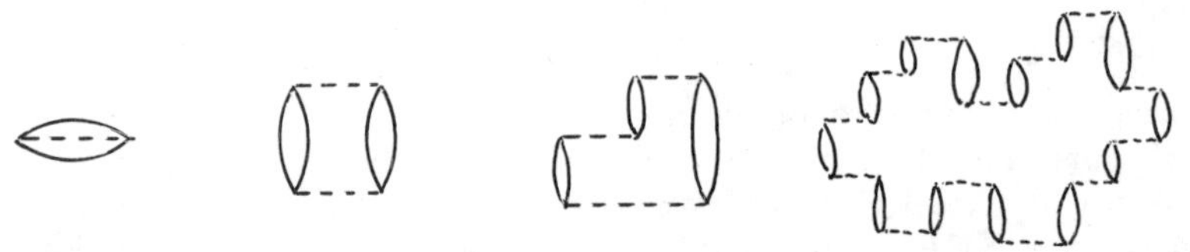

(3.41)

(a) (b) (c) (d)

Diagram a in (3.41) and diagram (3.39) are included in Eq. (3.34) and Eq. (3.37) which has also been corrected for c.m. motion. The leading remaining correction of order N^2 is (3.41b), which is shown in Problem 3-6 to yield

$$\Delta E^{(2)} \cong -0.9956\, N(N-1) \frac{g^2}{24} \tag{3.42}$$

and thus removes all but 1/2 % of the N^2 discrepancy between Eq. (3.37) and the exact result (3.27). Presumably the remaining discrepancy is removed by the higher order RPA diagrams in (3.41). Although we have not constructed an analogous time-dependent proof, it is physically plausible that a similar 1/N expansion applies to the time-dependent problem.

3.2.1.2 Exact solution for scattering of bound states: The scattering matrix for distinguishable particles interacting via δ-function potentials has been derived by Yang.[11] Applying his formulation to the scattering of N bosons by N bosons, the symmetric wave function must be of the form

$$\psi(x_1 \ldots x_{2N}) = \sum_{Q=1}^{(2N)!} \sum_{P=1}^{(2N)!} \theta_Q(x_1 \ldots x_{2N}) a_P \exp\{i \sum_{j=1}^{2N} k_{P(j)} x_{Q(j)}\} \tag{3.43}$$

where Q and P are elements of the permutation group, S_{2N}, and

$$\theta_Q(x_1 \ldots x_{2N}) = \begin{cases} 1 & \text{if } x_{Q(1)} < x_{Q(2)} < \ldots < x_{Q(2N)} \\ 0 & \text{otherwise} \end{cases} \tag{3.44}$$

Continuity of the wave function and the discontinuity of its derivative impose the following condition on the coefficients

$$a_P = \frac{i[K_{P(i)} - K_{P(i+1)}] + g}{i[K_{P(i)} - K_{P(i+1)}] - g}\, a_{P'} \tag{3.45}$$

where P and P' are related by

$$P(i) = P'(i+1)$$
$$P(i+1) = P'(i) \tag{3.46}$$

The (2N)!(2N-1) equations of the form (3.45) are internally consistent and sufficient to determine all coefficients if one of them is given.

For the case of two N-particle bound states, we choose the incoming state to be in the center of mass frame

$$\psi_i = \sum_{Q=1}^{(2N)!} \theta_Q(x_1 \ldots x_{2N}) \exp\{iK(\sum_{i=1}^{N} x_{Q(i)} - \sum_{j=N+1}^{2N} x_{Q(j)})\} \times$$
$$\times \exp\{-\frac{g}{2} \sum_{i<j=1}^{N} |x_{Q(i)} - x_{Q(j)}| \exp\{-\frac{g}{2} \sum_{k<\ell=N+1}^{2N} |x_{Q(k)} - x_{Q(\ell)}|\} \tag{3.47}$$

The final state is the outgoing wave

$$\psi_f = \sum_{Q=1}^{(2N)!} \theta_Q(x_1 \ldots x_{2N}) T_{2N} \exp\{-iK(\sum_{i=1}^{N} x_{Q(i)} - \sum_{j=N+1}^{2N} x_{Q(j)})\} \times$$
$$\times \exp\{-\frac{g}{2} \sum_{i<j=1}^{N} |x_{Q(i)} - x_{Q(j)}|\} \exp\{-\frac{g}{2} \sum_{k<\ell=N+1}^{2N} |x_{Q(k)} - x_{Q(\ell)}|\} \tag{3.48}$$

with transmission coefficient given by

$$T_{2N} = \prod_{m=1}^{N-1} \left(\frac{2K-mgi}{2K+mgi}\right)^2 \left(\frac{2K-Ngi}{2K+Ngi}\right) \tag{3.49}$$

Thus, only elastic scattering occurs in this model, and is completely specified by the phase shift, $\delta(K) = \frac{1}{2i} \ell n(T_{2N})$. For subsequent comparison, it will be useful to calculate the time delay.

$$\Delta t \equiv \frac{1}{NK} \frac{\partial}{\partial K} \delta(K) \tag{3.50}$$

which, from (3.49), has the following form

$$\Delta t = -\frac{4g}{NK}\sum_{m=1}^{N-1}\frac{m}{4K^2+m^2g^2} - \frac{2g}{K(4K^2+N^2g^2)} \tag{3.51}$$

3.2.1.3 Time-dependent mean-field solution for scattering: An initial product wave function which describes two separate bound states approaching each other with relative velocity 2K at time t_o is

$$\psi_{2N}(x_1,x_2\ldots,x_{2N};t_o) = \prod_{i=1}^{2N}\left\{\frac{1}{\sqrt{2}}\phi^+(x_i,t_o) + \frac{1}{\sqrt{2}}\phi^-(x_i,t_o)\right\} \tag{3.52}$$

where

$$\phi^{\pm}(x,t_o) = e^{\mp iKx}\,\phi_o(x\mp R) \tag{3.53}$$

and R is the initial displacement. As in the case of the TDHF slab solutions, this initial condition does not correspond exactly to two intrinsic wave functions multiplied by a cm momentum eigenfunction. However, the separated incident fragments do evolve like freely translating ground states, the expectation value of P_{cm} is correct, and the equation of motion is exactly solvable.

The time-dependent Hartree approximation for the wave function

$$\psi_{2N}(x_1\ldots x_{2N};t) = \prod_{i=1}^{2N}\phi(x_i,t) \tag{3.54}$$

is obtained by requiring that the action

$$\langle\psi_{2N}|\, i\frac{\partial}{\partial T} - H\,|\psi_{2N}\rangle \tag{3.55}$$

be stationary with respect to variations of $\phi(x,t)$. The resulting TDH equation for the single particle state ϕ is:

$$\left\{i\frac{\partial}{\partial t} + \frac{1}{2}\frac{d^2}{dx^2} + (2N-1)g|\phi(x,t)|^2\right\}\phi(x,t)=0 \tag{3.56}$$

and has the property that the norm of ϕ is a constant of the motion. This cubic Schrodinger equation, Eq. (3.56), admits a family of solitary wave solutions which have been studied extensively.[12] The solution which satisfies the initial condition (3.52) is the two soliton solution[13]

$$\phi(x,t) = \frac{\sqrt{(2N-1)v}}{2} e^{-\frac{i}{2}(K^2-a^2)t} \times$$

$$\{e^{iKx}[e^{-a(x-Kt)} + \frac{K^2}{(K-ia)^2} e^{-a(3x+Kt)}] \tag{3.57}$$

$$+ e^{iKx} [e^{-a(x+Kt)} + \frac{K^2}{(K+ia)^2} e^{-a(3x-Kt)}]\}$$

$$[1 + 2e^{-2ax} \cosh(2aKt) - 8a^2 e^{-2ax} \mathrm{Re}[\frac{e^{2iKx}}{4(K+ia)^2}] + \frac{K^4}{(K^2+a^2)} e^{-4ax}]^{-1}$$

This solution is in the center of mass frame, with relative velocity 2K, and parameter, $a = \frac{(2N-1)g}{4}$. It describes the transmission of two solitary waves through one another with time delay, Δt^H, defined by

$$\phi(x,t) \underset{t\to\infty}{\to} \frac{1}{\sqrt{2}} \{\phi^+(x,t-\Delta t^H) + \phi^-(x,t-\Delta t^H)\} \tag{3.58}$$

The time delay is given by

$$\Delta t^H = \frac{-4}{(2N-1)gK} \ln \{1 + \frac{(2N-1)^2 g^2}{16K^2}\} \tag{3.59}$$

For the collision of two many-particle wave packets moving with velocity K

$$\Delta t^H \underset{\substack{N\to\infty \\ K\ \mathrm{FIXED}}}{\to} - \frac{4}{NgK} [\ln N + \mathcal{O}(1)] \tag{3.60}$$

This result agrees with the exact time delay to order $\frac{1}{N}$. Thus we see that in the large N limit, just as in the static case, the time dependent Hartree approximation for a particular quantity of physical interest becomes asymptotically exact. One should note that the full wave function is certainly not exact, and that the "lifetime of a determinant" in this model[10] is zero. Since the time delay may be defined solely in terms of the position of the centers of mass of the two separating fragments, it is clear that we are dealing with the expectation value of a one-body operator, not an S-matrix element. In this simple model, then, we have seen a concrete example in which the mean-field theory yields an asymptotically exact first approximation to few body operators of physical interest. For any N, this lowest order approximation should be amenable to systematic higher order corrections.

3.2.2 Two interacting systems described by the Lipkin Model Hamiltonian

The Lipkin model[14] consists of a system of N identical but distinguishable fermions, each fermion having only two possible states, which are separated by an energy ε. The second quantized Hamiltonian for the system is

$$H^N = \frac{\varepsilon}{2} \sum_{p\sigma} \sigma \, a^+_{p\sigma} a_{p\sigma} + \frac{V}{2} \sum_{pp'\sigma} a^+_{p\sigma} a^+_{p'\sigma} \, a_{p'-\sigma} \, a_{p-\sigma} \tag{3.61}$$

where p labels the N particles and $\sigma = +1$ denotes the upper state and $\sigma = -1$ denotes the lower state. Operators with different particle labels commute with each other. The interaction term scatters a pair of particles from the same level to the other level. A single-particle state may be represented as a Pauli spinor, and the non-interacting ground states with V=0 is a direct product of spinors. Introducing the quasispin operators J_z, J_+ and J_-

$$\begin{aligned} J_z &= \frac{1}{2} \sum_{p\sigma} \sigma \, a^+_{p\sigma} a_{p\sigma} \\ J_+ &= \sum_p a^+_{p_+} a_{p_-} \\ J_- &= J_+^+ \quad , \end{aligned} \tag{3.62}$$

the Hamiltonian can be rewritten as

$$H^N = \varepsilon J_z + \frac{V}{2} (J_+^2 + J_-^2) \quad . \tag{3.63}$$

The operators J_z, J_+ and J_- satisfy the usual angular momentum algebra, and the operator $J^2 \equiv J_z^2 + \frac{1}{2}(J_+J_- + J_-J_+)$ commutes with each angular momentum component and therefore with the Hamiltonian. The Hamiltonian also commutes with the parity operator $\hat{\pi}$ and the number operator for the p'th fermion $\hat{n}_p$ where

$$\hat{\pi} = e^{-i\pi J_z} \quad , \quad \hat{n}_p = a^+_{p_+} a_{p_+} + a^+_{p_-} a_{p_-} \tag{3.64}$$

The eigenstates of H may thus be labelled by the eigenvalues of $\hat{\pi}$ and J. Since the interaction does not connect states of different J, the ground state may be calculated in the J=N/2 subspace, which contains the non-interacting ground state. Excitations of single particle-hole pairs are forbidden, and hence the Hartree-Fock equations are trivially satisfied in the non-interacting ground state basis.

The interacting eigenstates of the Hamiltonian in the J=N/2 subspace may be expanded in the natural basis:

$$|J\alpha\rangle = \sum \omega_{J\alpha}(m_J)|J\ m_J\rangle \quad . \tag{3.65}$$

The expansion coefficients $\omega_{J\alpha}(m_J)$ and the eigenvalues corresponding to the state $|J\alpha\rangle$ are determined from the solution of

$$\det|\langle J\alpha'|H|J\alpha\rangle - \varepsilon_\alpha\delta_{\alpha\alpha'}| = 0 \quad . \tag{3.66}$$

To obtain a solvable model problem which bears some analogy to a nuclear collision, we consider the extreme caricature of two interacting Lipkin systems. Initially, like two distantly approaching nuclei, particles in each isolated system interact only among themselves, with each system evolving in its respective ground state. At time t=0 each particle is allowed to interact with any particle in either system, like nucleons within two nuclei which are passing through each other. Finally, after some interaction time τ, the fragments are assumed to have separated by more than the range of interaction and again particles are restricted to interact only with other particles in the same system. Although questions of cm motion and particle transfer are completely eliminated from the model, one does address the essential problem of mutual excitation of two interacting systems which have an exceedingly large number of accessible states.

Distinguishing operators for particles in the two systems by subscripts and assuming the systems contain N_1 and N_2 particles respectively, the Hamiltonian for the two interacting systems is

$$H(t) = H^{N_1} + H^{N_2} + [\ \theta(t) - \theta(t-\tau)]\ H_{INT} \tag{3.67}$$

where

$$H_{INT} = V[J_{1+}J_{2+} + J_{1-}J_{2-}] \tag{3.68}$$

For times prior to t=0, the two systems are in their respective interacting ground states

$$\begin{aligned} |\Psi\rangle &= |J_1\alpha_o\rangle|J_2\beta_o\rangle \\ &= \sum_{J\gamma} c_{J\gamma}(0)|J\gamma\rangle \quad . \end{aligned} \tag{3.69}$$

By orthogonality, we find

$$c_{J\gamma}(0) = \langle J_1\alpha_o J_2\beta_o|J\gamma\rangle \quad . \tag{3.70}$$

The quantities $\langle J_1\alpha_o J_2\beta_o|J\gamma\rangle$ may be expressed in terms of the eigenvectors and the Clebsch Gordan coefficients:

$$\langle J_1\alpha_o J_2\beta_o|J\gamma\rangle = \sum_{m_{J_1} m_{J_2}} \omega_{J_1\alpha_o}(m_{J_1})\omega_{J_2\beta_o}(m_{J_2})\omega_{J\gamma}(m_{J_1}+m_{J_2})\langle J_1J_2m_{J_1}m_{J_2}|Jm_{J_1}m_{J_2}\rangle \quad . \tag{3.71}$$

During the interaction interval $0<t<\tau$, the Hamiltonian describes a "compound nucleus" with known eigenstates. Hence, application of the time evaluation operator to the initial state yields

$$|\Psi(t)\rangle = \sum_{J\gamma} c_{J\gamma}(0)\, e^{-i\varepsilon_{J\gamma}t}|J\gamma\rangle \quad . \tag{3.72}$$

It should be noted that the simple form above would have been impossible to achieve if ε or V had been chosen to be different for the two systems.

We now consider transition amplitudes to excited states of the single systems at the end of the interaction period τ. The amplitude $\tilde{c}_{\alpha\beta}(\tau)$ for finding the final state with system 1 in state α and system 2 in state β is clearly the analog of an S-matrix element for this simple model and is given by

$$|\Psi(\tau)\rangle = \sum_{\alpha\beta} \tilde{c}_{\alpha\beta}(\tau)|J_1\alpha\rangle|J_2\beta\rangle \quad . \tag{3.73}$$

Matching Eq. (3.73) at $t=\tau$ to Eq. (3.72) and projecting, we obtain

$$\tilde{c}_{\alpha\beta}(\tau) = \sum_{J\gamma} c_{J\gamma}(0)e^{-i\varepsilon_{J\gamma}t}\langle J_1\alpha J_2\beta|J\gamma\rangle \quad . \tag{3.74}$$

This expression for transition amplitudes yields all the results to be compared subsequently with approximate solutions. Since evaluation of $\tilde{\mathcal{C}}_{\alpha\beta}(\tau)$ requires knowledge of only the single system eigenvectors and the relevant Clebsh Gordan coefficients, it is feasible to calculate solutions for large numbers of particles.

3.2.2.1 Solution in the TDHF approximation: The lowest order truncation of the coupled-cluster equations, in the representation in which $S^{(1)} \equiv 0$, yields the static HF equation

$$<\phi|a_\nu^+ \, a_\rho \, H|\phi> = 0 \tag{3.75}$$

and the TDHF equation

$$<\phi|a_\nu^+ \, a_\rho(H(t) - \frac{i\partial}{\partial t})|\phi> = 0 \; . \tag{3.76}$$

A convenient parameterization for ϕ is obtained by introducing the general unitary transformation[15,16]

$$\begin{bmatrix} \gamma_{p+}^+ \\ \\ \gamma_{p-}^+ \end{bmatrix} \equiv \begin{bmatrix} \mathrm{Cos}\,\frac{\alpha}{2} & -i\mathrm{Sin}\frac{\alpha}{2}\, e^{i\psi} \\ \\ -i\mathrm{Sin}\,\frac{\alpha}{2}\, e^{-i\psi} & \mathrm{Cos}\,\frac{\alpha}{2} \end{bmatrix} \begin{bmatrix} a_{p+}^+ \\ \\ a_{p-}^+ \end{bmatrix} \tag{3.77}$$

and defining

$$|\phi> \quad \prod_p a_{p-}^+|0> \tag{3.78}$$

The non-interacting ground state, obtained by choosing $\alpha=\psi=0$ is a trivial solution to Eqs. (3.75) and (3.76), as may be noted by observing that H and $\frac{i\partial}{\partial t}$ do not connect the non-interacting ground state to any one-particle, one-hole states.

A general solution is obtained by defining the operators I_z, I_+, and I_- in analogy to the J operators.

$$\begin{aligned} I_z &= \frac{1}{2} \sum_{p\sigma} \sigma\, \gamma_{p\sigma}^+ \, \gamma_{p\sigma} \\ I_+ &= \sum_p \gamma_{p+}^+ \, \gamma_{p-} \\ I_- &= I_+^+ \end{aligned} \tag{3.79}$$

The I operators preserve the commutation relations for angular momentum operators. As before, operators defined for the two separate systems are distinguished by an integer subscript and operators not belonging to the same system commute. The transformation between the two sets of operators is

$$\begin{bmatrix} J_z \\ J_+ \\ J_- \end{bmatrix} = \begin{bmatrix} \text{Cos}\alpha & -\frac{1}{2}\text{Sin}\alpha e^{-i\psi} & \frac{i}{2}\text{Sin}\alpha e^{i\psi} \\ -i\text{Sin}\alpha e^{i\psi} & \frac{1}{2}(1+\text{Cos}\alpha) & \frac{1}{2}(1-\text{Cos}\alpha)e^{2i\psi} \\ i\text{Sin}\alpha e^{-i\psi} & \frac{1}{2}(1-\text{Cos}\alpha)e^{-2i\psi} & \frac{1}{2}(1+\text{Cos}\alpha) \end{bmatrix} \begin{bmatrix} I_z \\ I_+ \\ I_- \end{bmatrix} \tag{3.80}$$

and for $\psi=0$ each quasispin is simply rotated through the same angle α about the x axis. The Hamiltonian H(t) can now be expressed in terms of the I operators. The simplicity of the original representation is lost, however, and there are now 9 non-zero coefficients for the 9 independent linear and bilinear combinations of I_z, I_+ and I_-. We note that the Hamiltonian will contain non-diagonal one-body terms in this representation so that Eq. (3.78) will not in general satisfy the TDHF equation.

The wave function for the two systems is given by

$$|\Phi\rangle = |\phi_1\rangle|\phi_2\rangle \tag{3.81}$$

where $|\phi_1\rangle$ and $|\phi_2\rangle$ are defined by Eq. (3.78) for each system. Although the model may consist of arbitrarily many particles, by symmetry in the particle labels, each system is completely specified by the two parameters α_i and ψ_i.

The static HF solution for a single system is obtained by noting that with $|\phi\rangle$ defined in Eq. (3.78), the only projections onto particle hole states which are not trivially zero in Eq. (3.75) are

$$\langle\phi|\gamma^+_{p-}\gamma_{p+}H|\phi\rangle = 0 \ . \tag{3.82}$$

By symmetry in the particle labels, these N equations may be replaced by the single equation obtained by summing over p

$$\langle\phi|I_-H|\phi\rangle = 0 \qquad . \tag{3.83}$$

Substitution of J_z, J_+ and J_- from Eq. (3.80) and the definition of H from Eq. (3.63) in Eq. (3.83) yields the result

$$i\,\frac{N}{2}\,\mathrm{Sin}\alpha\;e^{-i\psi}[\varepsilon - (N-1)V\{i\mathrm{Sin}2\psi + \mathrm{Cos}2\psi\;\mathrm{Cos}\alpha\}] = 0 \tag{3.84}$$

The root $\alpha=0$ reproduces the non-interacting ground state as the H.F. solution. This weak-coupling solution exists for arbitrarily small values of V and has energy

$$\langle\phi|H|\phi\rangle = -\frac{N\varepsilon}{2} \quad . \tag{3.85}$$

The second root

$$\psi=0,\;\mathrm{Cos}\alpha = \frac{\varepsilon}{V(N-1)} \tag{3.86}$$

exists only for $V > \frac{\varepsilon}{N-1}$ and corresponds to the strong coupling solution. The HF energy for this root is

$$\langle\phi|H|\phi\rangle = -\frac{\varepsilon N}{4}\left\{\frac{1}{(N-1)\frac{V}{\varepsilon}} + (N-1)\,\frac{V}{\varepsilon}\right\} \quad . \tag{3.87}$$

The transition between the weak and strong coupling solutions occurs at $V_T = \frac{\varepsilon}{N-1}$ with both E(V) and $\frac{dE(V)}{dV}$ continuous at V_T.

The time evolution of the parameters α_1, α_2, and ψ_1, ψ_2 is obtained from the TDHF equation, Eq. (3.76) with H(t) given in Eq. (3.67). As in Eqs. (3.82)-(3.83), it is useful to write the non-zero $a^+_\nu a_\rho$ terms as $\gamma^+_{p-}\gamma_{p+}$ and sum over the particles in the ith system, yielding the two equations

$$\langle\Phi|I_{i-}(H(t) - i\,\frac{\partial}{\partial t})|\Phi\rangle = 0 \quad . \tag{3.88}$$

The transformation Eq. (3.77) may be differentiated with respect to time and combined with the corresponding inverse transformation and yields the following equation of motion for $\gamma^+_{p\sigma}$

$$\begin{bmatrix} \dot{\gamma}^+_{p_i+} \\ \\ \dot{\gamma}^+_{p_i-} \end{bmatrix} = \begin{bmatrix} i\dot{\psi}_i\mathrm{Sin}^2\frac{\alpha_i}{2} & -\frac{1}{2}(i\dot{\alpha}_i-\dot{\psi}_i\mathrm{Sin}\alpha_i)e^{i\psi_i} \\ \\ -\frac{1}{2}(i\dot{\alpha}_i+\dot{\psi}_i\mathrm{Sin}\alpha_i)e^{-i\psi_i} & -i\dot{\psi}_i\mathrm{Sin}^2\frac{\alpha_i}{2} \end{bmatrix} \begin{bmatrix} \gamma^+_{p_{i+}} \\ \\ \gamma^+_{p_{i-}} \end{bmatrix} \tag{3.89}$$

By equating the real and imaginary parts of Eq. (3.104) to the real and imaginary parts of

$$\langle\Phi| I_{1-}\; e^{-S^{(2)}}\; He^{S^{(2)}}\; |\Phi\rangle$$

and repeating with the labels 1 and 2 interchanged, we arrive at a 4x4 matrix equation similar in form to (3.93) but with non-zero off-diagonal blocks. The elements of the matrix may be easily worked out, and will not be displayed here.

By similar manipulations, the time derivative parts of Eqs. (3.99b) and (3.99c) are, respectively,

$$\langle\Phi| I_{1-}^{2}\; e^{-S^{(2)}}\; i\frac{\partial}{\partial t}\; e^{S^{(2)}}\; |\Phi\rangle = N_1(N_1-1)\{i\frac{\partial}{\partial t}\, \mathcal{S}_1^{(2)} - 2\mathcal{S}_1^{(2)}(1-\text{Cos}\alpha_1)\dot{\psi}_1\} \tag{3.105a}$$

$$\begin{aligned}&\langle\Phi| I_{1-}I_{2-}\; e^{-S^{(2)}}\; i\frac{\partial}{\partial t}\; e^{S^{(2)}}\, |\Phi\rangle \\ &= N_1N_2\; \{i\frac{\partial}{\partial t}\, \mathcal{S}_{12}^{(2)} - \mathcal{S}_{12}^{(2)}[(1-\text{Cos}\alpha_1)\dot{\psi}_1 + (1-\text{Cos}\alpha_2)\dot{\psi}_2\}.\end{aligned} \tag{3.105b}$$

On equating these results to $\langle\Phi| I_{1-}^{2}\; e^{-S^{(2)}}\; He^{S^{(2)}}\; |\Phi\rangle$ and $\langle\Phi| I_{1-}I_{2-}\; e^{-S^{(2)}}\; He^{S^{(2)}}\; |\Phi\rangle$ respectively, we arrive at the equations of motion for $\mathcal{S}_1^{(2)}(t)$, $\mathcal{S}_2^{(2)}(t)$, and $\mathcal{S}_{12}^{(2)}(t)$. As shown in general in Section 2, these differential equations are first order in time for $\mathcal{S}_i^{(2)}$ and $\mathcal{S}_{12}^{(2)}$ and are thus in a form directly suitable for numerical evolution.

Finally, we turn to the problem of evaluating the matrix elements of the collision Hamiltonian in Eqs. (3.99). Although one can use the straightforward means employed previously in the TDHF case, the amount of algebra is prohibitive, given the relatively complicated form of the Hamiltonian when expressed in I operators. Hence all operators have been represented by matrices of finite dimension, and the desired matrix elements obtained numerically by matrix multiplication. As an example, consider a typical matrix element which we wish to evaluate:

$$\begin{aligned}&\langle\Phi| I_{1-}^{2}\; e^{-S^{(2)}}\; H(t)\; e^{S^{(2)}}\; |\Phi\rangle \\ &= \langle\Phi| I_{1-}^{2}\; \{H(t) + [H(t),S^{(2)}] + \frac{1}{2!}\, [[H(t),S^{(2)}],S^{(2)}] + \ldots\}|\Phi\rangle.\end{aligned} \tag{3.106}$$

Since H(t) is bilinear in the operators I_{iz}, I_{i+}, I_{i-} and $S^{(2)}$

The time derivative of $|\phi_i>$ is then given by

$$\frac{i\partial}{\partial t}|\phi_i> = i\sum_j \gamma_{1-}^{+} \cdots \dot{\gamma}_{j-}^{+} \cdots \gamma_{N_i-}^{+}|0>$$
$$= \frac{1}{2}(\dot{\alpha}_i - i\dot{\psi}_i \text{Sin}\alpha_i)e^{-i\psi_i} I_{i+}|\phi_i> + \dot{\psi}_i \text{Sin}^2\frac{\alpha_i}{2}|\phi> \tag{3.90}$$

and contains one component proportional to $|\phi_i>$ and a second component of one-particle, one hole excitations. Hence

$$<\Phi|I_{i-}\frac{i\partial}{\partial t}|\Phi> = \frac{1}{2}N_i(\dot{\alpha}_i - i\dot{\psi}_i \text{Sin}\alpha_i)e^{-i\psi_i} \quad . \tag{3.91}$$

The matrix element $<\Phi|I_{i-}\ H(t)|\Phi>$ is evaluated by substituting the Hamiltonian H(t) of Eq. (3.67) after using the transformation Eq. (3.80) to express it in terms of the I operators. For the interval $0<t<\tau$, straightforward algebra yields

$$<\Phi|I_{1-}H(t)|\Phi> = -\frac{i}{2}N_1\{\varepsilon\ \text{Sin}\alpha_1\ e^{-i\psi_1} +$$
$$VN_2\text{Sin}\alpha_2(\text{Sin}^2\frac{\alpha_1}{2}e^{-i2\psi_1-i\psi_2} - \text{Cos}^2\frac{\alpha_1}{2}e^{i\psi_2}) + \tag{3.92}$$
$$V(N_1-1)\text{Sin}\alpha_1(\text{Cos}^2\frac{\alpha_1}{2}e^{i\psi_1} - \text{Sin}^2\frac{\alpha_1}{2}e^{-i3\psi_1})\} \quad .$$

Equating the real and imaginary parts of Eq. (3.88) to zero yields the equations of motion for the parameters α_i, ψ_i

$$\begin{bmatrix} N_1\text{Cos}\psi_1 & -N_1\text{Sin}\alpha_1\text{Sin}\psi_1 & 0 & 0 \\ -N_1\text{Cos}\psi_1 & -N_1\text{Sin}\alpha_1\text{Cos}\psi_1 & 0 & 0 \\ 0 & 0 & N_2\text{Cos}\psi_2 & -N_2\text{Sin}\alpha_2\text{Sin}\psi_2 \\ 0 & 0 & -N_2\text{Cos}\psi_2 & -N_2\text{Sin}\alpha_2\text{Cos}\psi_2 \end{bmatrix} \begin{bmatrix} \dot{\alpha}_1 \\ \dot{\psi}_1 \\ \dot{\alpha}_2 \\ \dot{\psi}_2 \end{bmatrix} = \begin{bmatrix} \text{Re}<\Phi|I_{1-}H|\Phi> \\ \text{Im}<\Phi|I_{1-}H|\Phi> \\ \text{Re}<\Phi|I_{2-}H|\Phi> \\ \text{Im}<\Phi|I_{2-}H|\Phi> \end{bmatrix} \tag{3.93}$$

The off-diagonal blocks are zero in the above equation, a feature which does not survive in approximations in higher order than the mean field theory. The two systems are coupled only through the appearance of both α's and ψ's in $<\Phi|I_{i-}H(t)|\Phi>$.

Inversion of the above matrix yields the final form of the TDHF equations.

$$\begin{bmatrix} \dot{\alpha}_i \\ \\ \dot{\psi}_i \end{bmatrix} = \frac{2}{N_i} \begin{bmatrix} \mathrm{Cos}\psi_i & -\mathrm{Sin}\psi_i \\ \\ \dfrac{-\mathrm{Sin}\psi_i}{\mathrm{Sin}\alpha_i} & -\dfrac{\mathrm{Cos}\psi_i}{\mathrm{Sin}\alpha_i} \end{bmatrix} \begin{bmatrix} \mathrm{Re}\langle\Phi| I_{i-} \; H(t)|\Phi\rangle \\ \\ \mathrm{Im}\langle\Phi| I_{i-} \; H(t)|\Phi\rangle \end{bmatrix} . \tag{3.94}$$

Since the non-interacting ground state trivially satisfies the TDHF equation at all times, the only interesting TDHF solution arises from the strong coupling initial condition given in Eq. (3.86). Starting with the values

$$\psi_i(0) = 0 \qquad \alpha_i(0) = \mathrm{Cos}^{-1}\left[\frac{\varepsilon}{V(N_i-1)}\right] \tag{3.95}$$

and discretizing the time variable, the set of Eqs. (3.94) may be solved numerically, thus determining the TDHF approximation to the combined state vector as a function of time. Although our TDHF equations are equivalent to those investigated by Krieger[15] the initial conditions (3.95), and thus the physical interpretation of our results differ significantly.

<u>3.2.2.2. Solution in the two-body cluster approximation</u>: The form of the most general wave function having $S_n \equiv 0$ for all $n > 2$ is especially simple for the Lipkin systems considered in this work. One-particle, one-hole amplitudes may be omitted by using $|\phi\rangle$ and $|\Phi\rangle$ of the form of Eqs. (3.78) and (3.81) and all the physical content of $S^{(1)}$ resides in the parameters α_i and ψ_i. For a single system, by symmetry in the particle labels, $S^{(2)}$ is characterized by a single parameter $S^{(2)}$

$$\begin{aligned} S^{(2)} &= \frac{1}{2}\sum \langle pp'|S^{(2)}|pp'\rangle \, \gamma^{+}_{p+}\gamma^{+}_{p'+}\gamma_{p'-}\gamma_{p-} \\ &\equiv \frac{1}{2} S^{(2)} \, I^{2}_{+} \end{aligned} \tag{3.96}$$

and for two interacting systems, the general form of $S^{(2)}$ is

$$S^{(2)}(t) = \frac{1}{2}s^{(2)}_{1}(t)\, I^{2}_{1+} + \frac{1}{2}s^{(2)}_{2}(t)\, I^{2}_{2+} + s^{(2)}_{12}(t) I_{1+}I_{2+} \; . \tag{3.97}$$

Summing over particle labels as in Eqs. (3.82) and (3.88) reduces the general static and time dependent equations (2.18) and (2.22) to the following simple form:

$$<\phi|H\, e^{S^{(2)}}|\phi> = E \qquad (3.98a)$$

$$<\phi|I_-\, e^{-S^{(2)}}\, H\, e^{S^{(2)}}|\phi> = 0 \qquad (3.98b)$$

$$<\phi|I_-^2\, e^{-S^{(2)}}\, H\, e^{S^{(2)}}|\phi> = 0 \qquad (3.98c)$$

and

$$<\Phi|I_{i-}\, e^{-S^{(2)}(t)}\,(H-i\frac{\partial}{\partial t})\, e^{S^{(2)}(t)}|\Phi> = 0 \qquad (3.99a)$$

$$<\Phi|I_{i-}^2\, e^{-S^{(2)}(t)}\,(H-i\frac{\partial}{\partial t})\, e^{S^{(2)}(t)}|\Phi> = 0 \qquad (3.99b)$$

$$<\Phi|I_{1-}I_{2-}\, e^{-S^{(2)}(t)}\,(H-i\frac{\partial}{\partial t})\, e^{S^{(2)}(t)}|\Phi> = 0 \quad . \qquad (3.99c)$$

The static equations (3.98) are solved by angular momentum algebra after expressing H in terms of the I operators, with the results

$$E = -\frac{1}{4}\varepsilon N \mathrm{Cos}\alpha\{\frac{\chi}{\mathrm{Cos}\alpha}(1-\mathcal{S}_2)-\chi\mathrm{Cos}\alpha(1+\mathcal{S}_2)+2\} \qquad (3.100a)$$

$$\mathrm{Cos}\alpha = \frac{1-(N-1)\mathcal{S}_2}{\{1-(N-3)\mathcal{S}_2\}\chi} \qquad (3.100b)$$

$$(N^2-7N+9)(1+\mathrm{Cos}^2\alpha)\chi(\mathcal{S}_2)^2 + [6\chi(N-2)(1-\mathrm{Cos}^2\alpha)+4(N-1)\mathrm{Cos}\alpha]\mathcal{S}_2 + (1+\mathrm{Cos}^2\alpha)\chi = 0 \qquad (3.100c)$$

where χ is a dimensionless parameter

$$\chi = \frac{V(N-1)}{\varepsilon} \quad . \qquad (3.101)$$

The last two equations above combine to give a single quartic equation in $\mathcal{S}^{(2)}$. The appropriate root is real, yields a real value for α, and corresponds to an energy which is continuously connected to the HF energy. In the approximation $\mathcal{S}^{(2)}=0$, Eq. (3.100) reduces to the corresponding weak and strong coupling HF results. Note that when including $\mathcal{S}^{(2)}$ the transition between weak and strong coupling solutions no longer occurs at $\chi=1$.[16]

The time-dependent equations, Eq. (3.99), involve a total of ten independent real functions of time: the four parameters $\alpha_1\alpha_2\psi_1\psi_2$ and the real and imaginary parts of the complex amplitudes $\mathcal{S}_1^{(2)}, \mathcal{S}_2^{(2)}$ and $\mathcal{S}_{12}^{(2)}$. The initial values of these quantities are fixed by requiring $\alpha_i(0)$ and $\mathcal{S}_i^{(2)}(0)$ to satisfy the static equations, and requiring the other quantities to be zero.

Evaluation of the time derivatives in Eqs. (3.99) proceeds as in Section 2 using Eq. (2.23). The time derivative of I_{i+} is

$$\frac{\partial}{\partial t} I_{i+} = i\dot{\psi}_i(1-\mathrm{Cos}\alpha_i)I_{i+} + (i\dot{\alpha}_i - \dot{\psi}_i\mathrm{Sin}\alpha_i)e^{i\psi_i} I_{iz} \tag{3.102}$$

which is established by differentiating I_{i+} and using the operator equations of motion Eq. (3.89).

To illustrate the procedure for calculating the matrix elements of the time derivative operator, we shall evaluate Eq. (3.99a) and simply quote the other results.

$$\begin{aligned} &\langle\Phi|I_{1-}\, e^{-S^{(2)}}\, i\frac{\partial}{\partial t}\, e^{S^{(2)}}|\Phi\rangle \\ &= \langle\Phi|I_{1-}\, i\frac{\partial}{\partial t}|\Phi\rangle + \frac{1}{2}\mathcal{S}_1^{(2)}\langle\Phi|I_{1-}(I_{1+}\frac{\partial}{\partial t}I_{1+} + \frac{\partial}{\partial t}I_{1+}I_{1+})|\Phi\rangle \\ &\quad + \frac{1}{2}\mathcal{S}_{12}^{(2)}\langle\Phi|I_{1-}\frac{\partial}{\partial t}I_{2+} + \frac{\partial}{\partial t}I_{1+}I_{2+})|\Phi\rangle \end{aligned} \tag{3.103}$$

Using Eq. (3.102), this simplifies to

$$\begin{aligned} \langle\Phi|I_{1-}\, e^{-S^{(2)}}\, i\frac{\partial}{\partial t}\, e^{S^{(2)}}|\Phi\rangle &= \frac{1}{2}N_1(\dot{\alpha}_1 - i\dot{\psi}_1\,\mathrm{Sin}\alpha_1)e^{-\psi_i} \\ &+ \frac{1}{2}N_1(N_1-1)\mathcal{S}_1^{(2)}(\dot{\alpha}_1+i\dot{\psi}_1\mathrm{Sin}\alpha_1)\,e^{i\psi_1} + \\ &+ \frac{1}{2}N_1N_2\,\mathcal{S}_{12}^{(2)}(\dot{\alpha}_2+i\dot{\psi}_2\mathrm{Sin}\alpha_2)e^{i\psi_2} \end{aligned} \tag{3.104}$$

is bilinear in I, and since the raising and lowering operators must occur in equal numbers for each i in order for the matrix element to be non zero, the infinite series in Eq. (3.106) terminates in finite order. In this example, the highest power of I_{i+} or I_{i-} which can occur is 4.

To represent the operators I_{iz}, I_{i+}, I_{i-} by finite dimensional matrices, it is sufficient to impose the following conditions:

a) All matrices with i=1 commute with all matrices with i=2.

b) For $n \leq 4$ and i=1,2

$$I^n_{iz}|\phi_i> = (-\frac{N_i}{2})^n|\phi_i> \tag{3.107a}$$

$$I^n_{i+}|\phi_i> = \sqrt{n!(N_i-n+1)(N_i-n+2)\ldots N_i}\,|I_i=\frac{N_i}{2}, m_{I_i}=-\frac{N_i}{2}+n> . \tag{3.107b}$$

A set of matrices which fulfill these conditions is explicitly given in Ref. 3.

3.2.2.3 Numerical results for interacting Lipkin systems: In order to quantitatively assess the validity of the lowest two orders of approximation in the coupled-cluster hierarchy, we have compared the approximations of Sections IV and V with the exact solution of Section III. Both of the interacting systems were chosen to contain 14 particles, since more than 28 particles in the combined system renders exact solution cumbersome and costly. The level spacing ε was chosen to be 1 MeV and thus crudely representative of nuclear energy scales. Interaction times up to $1.2\text{x}10^{-21}$ sec were considered, representing the time required for projectiles with energies of 10 to 40 MeV per particle to interpenetrate 5 to 10 fm respectively. Wide variations in N, V and ε make little difference in the qualitative features of the solutions, so the limited results reported here are in fact representative of the general applicability of the coupled-cluster theory for this model.

Static Solutions

Figure 3-14 shows the exact energy, E. for a single 14-particle system and the difference between truncated approximations to <H> and E as a function of interaction strength V. Since we use <H> instead of the approximate energy in Eq. (3.100a)

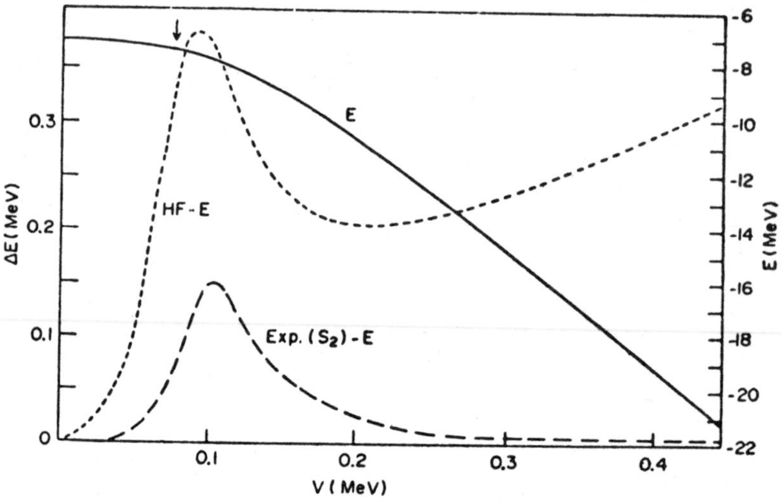

Fig. 3-14. The exact ground state energy E (solid line) deviation from E of the HF energy (short dashes) and deviation from E of the coupled-cluster energy including $S^{(2)}$ (long dashes). Note that E is referred to the scale at the right and the energy deviations use the scale at the left.

it is clear by the variational principle that the deviation is positive definite. The error in the HF approximation near the transition point, $V_T = \frac{\varepsilon}{(N-1)} \sim .077$ denoted by the arrow in Fig. 3-14 is at most 5%, and in the strong coupling limit the deviation increases with V. Including S_2 significantly decreases the deviation, not only in the weak coupling limit explored by Lührmann,[17] but also even more dramatically in the strong coupling limit. For the time-dependent calculations, we have selected NV=5 MeV, so that the potential strength $V \approx .36$ is well into the strong coupling region and yields a non-trivial TDHF solution. Other observables besides the energy for the static solutions with N=14 and NV=5 are presented in Fig. 3-15 and as the T=0 results in the time-dependent solutions.

Figure 3-15 displays the probability of observing the ground state in each state of the natural basis of Eq. (3.65)

$$P_{M_J} = |\langle J\ M_J|\psi\rangle|^2 \tag{3.108}$$

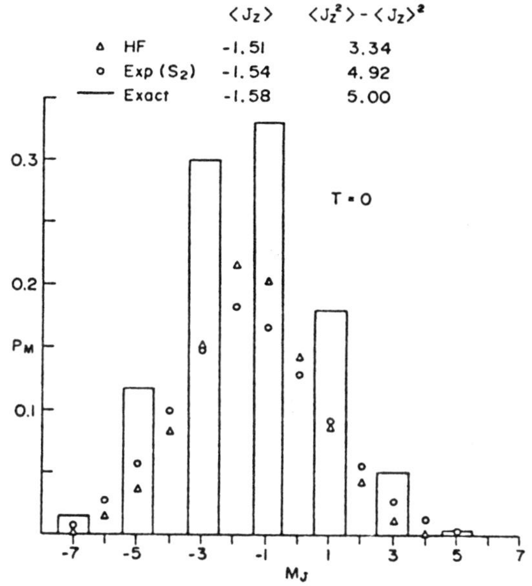

Fig. 3-15. Probability $P = |\langle \frac{N}{2} M_J|\psi\rangle|^2$ of projecting a component with M_J from the exact, HF, and second order coupled-cluster stationary-state wave functions.

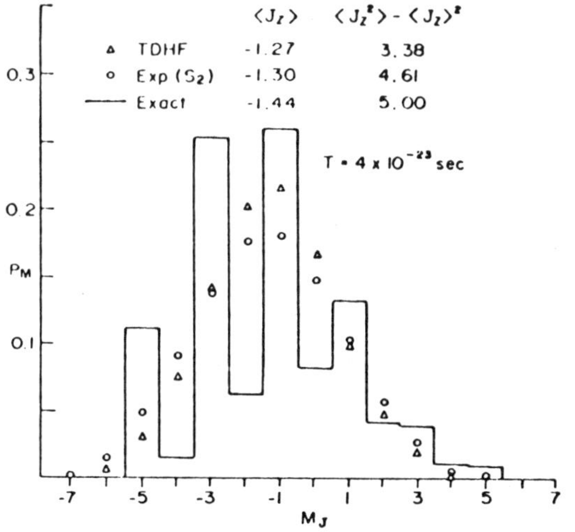

Fig. 3-16. Same as Fig. 2 for time dependent wave functions at time $T=4x10^{-23}$ sec.

Since H commutes with J^2, the only non-zero amplitudes correspond to J = N/2 and this label is suppressed for notational convenience. Because V only excites pairs of particles between levels, adiabatically evolving Ψ from the non-interacting ground state $|\frac{14}{2}, -\frac{14}{2}\rangle$ can only yield amplitudes with odd values of M_J. Thus, the exact solution yields the histogram of Fig. 3-15 with all even probabilities exactly zero and a broad, Gaussian like distribution for the odd probabilities. The most essential features of the distribution are characterized by the mean, $\langle J_z\rangle$ and width $\langle J_z^2\rangle - \langle J_z^2\rangle$ tabulated in the figure.

The probabilities $\langle J\ M_J|e^{\sum_{n=1}^{k} S_n}|\Phi\rangle$ for truncated wave functions of any order k<N differ fundamentally from the exact probabilities. Since only an N-particle correlation can enforce the condition that even numbers of particles occupy each level, there is no way of attaining the dramatic even-odd alternation characterizing the exact solution. One observes this complete lack of even-odd alternation explicitly in the numerical results for the lowest two orders denoted by triangles and circles in Fig. 3-15.

Figure 3-15 thus succinctly emphasizes the fundamental distinction between approximating the wave function and expectation values of few-body operators. In no sense do successive truncated approximations converge to the exact probabilities. In contrast, however, the mean values of $\langle J_z\rangle$ and $\langle J_z^2\rangle - \langle J_z^2\rangle$ shown in the top of the figure are approximated extremely well in very low orders.

Time Dependent Solutions

The projection of the wave function of one 14 particle fragment onto the natural basis after an interaction time of 4×10^{-23} sec. is shown in Fig. 3-16. As in Fig. 3-15, the exact solution has pronounced odd-even staggering although non-vanishing even M_J states are now possible through excitation of pairs comprised of one particle from each system. As in Fig. 3-15, although there is no convergence for the individual probabilities, good approximations are obtained for mean values of J_z and $(J_z-\langle J_z\rangle)^2$. To the degree to which $\langle JM_J|\psi(t)\rangle$ is a meaningful analog of an S-matrix element in a realistic scattering problem, this example reinforces our emphasis on approximating mean values rather than S-matrix elements.

A quantitative comparison between exact and approximate mean values of particularly relevant operators as a function of interaction time is presented in Fig. 3-17. The excitation energy, ΔE, is defined as the mean value $\frac{1}{2}\langle H^{N_1} + H^{N_2}\rangle$ for

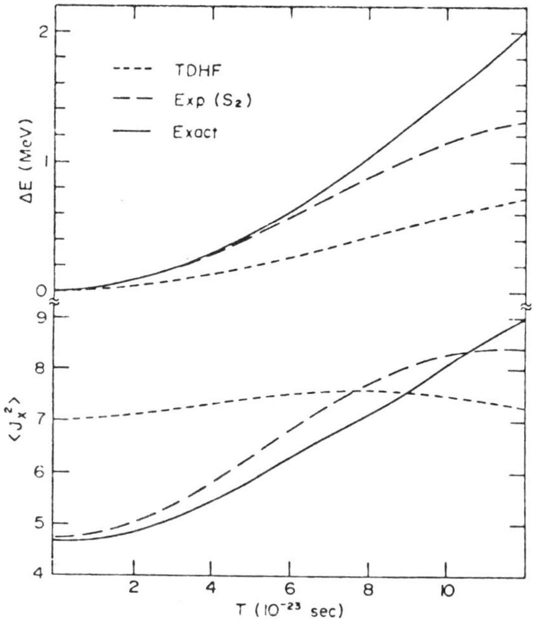

Fig. 3-17. Excitation energy E and mean value of J_x^2 as a function of interaction time for exact, TDHF, and second order coupled-cluster wave functions.

the combined system at time T minus the same quantity evaluated at time T-0 (which is just the stationary state energy for a single system in the same approximation). Roughly half of the true excitation energy is described in the mean-field approximation and for the time scales under consideration the $S^{(2)}$ truncation yields a reasonably accurate approximation to the full excitation energy. Similarly, for $\langle J_x^2 \rangle$, defined as $\frac{1}{2}\langle J_{x_1}^2 + J_{x_2}^2 \rangle$ for the combined system, when the suppressed zero of the graph is considered, the mean-field approximation yields a result of the correct order of magnitude and the inclusion of two-body clusters produces quantitative agreement with the exact solution. In both cases, the operators under consideration involved two-body components so significant contributions from $S^{(2)}$ should be expected and, in fact, do arise. Whereas $S^{(2)}$ contributes linearly to both $\langle H \rangle$ and $\langle J_x^2 \rangle$, since J_z is diagonal in a^+a, $\langle J_z^2 \rangle$ only depends quadratically on $S^{(2)}$, which explains why $\langle J_z^2 \rangle$ in Fig. 3-16 is not as accurate as $\langle J_x^2 \rangle$ in Fig. 3-17 in the e^{S2} approximation.

TABLE I

Wave Function Amplitudes Defined in Equations (6.2) and (6.3)

T (10^{-22} sec.)	C	$\hat{\mathcal{S}}^{(1)}$	$\hat{\mathcal{S}}^{(2)}$	$\mathcal{S}^{(2)}$	$\hat{\mathcal{S}}^{(2)}_{12}$	$\mathcal{S}^{(2)}_{12}$
0	0.489	0.0015	0.0156	0.0151	0	0
0.66	0.489	0.0013	0.0101	0.0097	0.0107	0.0109
1.18	0.491	0.0016	0.0024	0.0050	0.0134	0.0129

TABLE II

Comparison of Expectation Values with Various Wave Functions at Times T=0 and T=$1.18\text{x}10^{-22}$ Sec.

	$\langle J_z(0)\rangle$	$\langle J_z^2(0)\rangle$	$\langle J_z(1.18)\rangle$	$\langle J_z^2(1.18 \rangle$
ψ	-1.58	7.49	-.74	6.44
$\psi^{(2)}$	-1.54	7.28	-.11	3.55
$\hat{\psi}^{(2)}$	-1.56	7.40	-.12	3.52

One-body operators should be much more accurately reproduced in the mean-field approximation than the operators discussed above containing two-body components. Indeed, the tabulated values for $<J_z>$ in Figs. 3-15 and 3-16 bear out this expectation. Unfortunately, since $<J_x> = 0$ J_x does not provide an additional test.

One important question arising in the coupled-cluster hierarchy is whether inaccuracy in computing mean values occurs primarily from errors in the $S^{(n)}$ amplitudes arising from the fact that they are obtained from solving a truncated Schrodinger equation, or whether the inaccuracy in mean values should be attributed to the explicit contribution of higher order amplitudes to the expectation value. Although we cannot provide a general answer, the results in Tables I and II indicate that in the present model, the dominant error arises from explicit omission of higher correlation amplitudes in mean values. In the basis $|\Phi>$ defined in Eq. (3.81) with the parameters $\alpha(t)$ and $\psi(t)$ in γ^{+}_{p-} satisfying Eq. (3.99a), the n=2 coupled cluster wave function for a symmetric system has the form

$$|\Psi^{(2)}> = \exp\{\frac{1}{2}\,\mathcal{S}^{(2)}(I^2_{1+} + I^2_{2+}) + \mathcal{S}^{(2)}_{12}\; I_{1+}I_{2+}\}|\Phi> \quad (3.109)$$

In the same basis, the exact wave function, Eq. (3.72), may be written

$$|\Psi> = C\,\exp\{\hat{\mathcal{S}}^{(1)}(I_{1+}+ I_{2+}) + \frac{1}{2}\,\hat{\mathcal{S}}^{(2)}(I^2_{1+}+ I^2_{2+}) + \hat{\mathcal{S}}^{(2)}_{12}\; I_{1+}I_{2+} + \sum_{n=3}^{N} S^{(n)}\}|\Phi> \quad (3.110)$$

By projecting the exact solution onto $|\Phi>$, $I_{1+}|\Phi>$, $I^2_{1+}|\phi>$ and $I_{1+}I_{2+}|\Phi>$, the values C, $\hat{\mathcal{S}}^{(1)}$, $\hat{\mathcal{S}}^{(2)}$ and $\hat{\mathcal{S}}^{(2)}_{12}$ were obtained and are tabulated in Table I for comparison with the corresponding coupled-cluster amplitudes $\mathcal{S}^{(n)}$. Since the discrepancies $(\mathcal{S}^{(2)} - \hat{\mathcal{S}}^{(2)})$ and $(\mathcal{S}^{(1)} - 0)$ are small relative to the dominant amplitude at each time, we conclude that the $\mathcal{S}$'s evolved with the truncated Schrödinger equation are quite adequate. The normalization constant, C, indicates that only one fourth of the total wave function is comprised of one-particle, one-hole and two-particle, two-hole components.

Table II shows values of J_z and J^2_z calculated with the exact wave function ψ, the coupled-cluster wave function $\psi^{(2)}$ defined in Eq. (3.109) and a truncated wave function $\hat{\psi}^{(2)}$

obtained by using the exact $\hat{\mathcal{S}}^{(1)}$, $\hat{\mathcal{S}}^{(2)}$ and $\mathcal{S}_{12}^{(2)}$ amplitudes in Eq. (3.110) and omitting

$$\sum_{n=3}^{N} S^{(n)} \quad .$$

Since $\psi^{(2)}$ agrees much more closely with $\hat{\psi}^{(2)}$ than with ψ, clearly the dominant error arises from omission of the sum $\sum_{n=3}^{N} S^{(n)}$ rather than the small discrepancies between the $\mathcal{S}$ and $\hat{\mathcal{S}}$ amplitudes.

Application of the coupled-cluster hierarchy to the model problem of two interacting Lipkin systems thus explicitly demonstrates the theory's viability and provides valuable insight into how it actually works in practice. The lowest order mean-field approximation yields good results for one-body operators and qualitatively correct behavior for two-body operators for the interaction times investigated in this work, which were an order of magnitude longer than the "lifetime" of a determinant[18] in this model, $T_D = 1.3\times10^{-23}$ sec. Inclusion of two-particle, two-hole amplitudes yield quantitative agreement for observables containing two-body operators.

PROBLEMS

3-1. Show that with the effective interaction in Eq. (3.2)

$$\langle H - i\frac{\partial}{\partial t}\rangle = \sum_{\nu} \int \phi_{\nu}^{*}(T-i\frac{\partial}{\partial t})\phi_{\nu} + \frac{3}{8}\int d^3r(t_o\rho^2 + \frac{t_3}{6}\rho^3)$$

$$+ \frac{1}{2}\int d^3rd^3r'\ \rho(r)\ V_o \frac{e^{-|r-r'|/a}}{|r-r'|/a}\ \rho(r')$$

It is useful to carefully write out the direct and exchange matrix elements of v and to explicitly perform the spin and isospin sums. Now, vary $\langle H-i\frac{\partial}{\partial t}\rangle$ with respect to ϕ^* to obtain Eq. (3.3). Note how the density-dependent t_3 term acquires a different coefficient than if one blindly substitutes into Eq. (1.15a).

3-2. Express $<(\hat{N}^R)^2>$ in terms of the right hand density matrix of Eq. (3.10), ρ^R. First, using the anti-commutation relation for $[a^+, a]_+$ and the fact that the entire wave function (right plus left components) in a single determinant, show

$$<(\hat{N}^R)^2> = N_{\alpha\beta} N_{\gamma\delta} \{\delta_{\beta\gamma}\, \rho_{\alpha\delta} - \rho_{\alpha\delta}\, \rho_{\gamma\beta} + \rho_{\alpha\beta}\, \rho_{\gamma\delta}\} ,$$

where summation over repeated indices is implied. Then, use the fact N is idempotent and the definitions of N^R and ρ^R to obtain

$$<(\hat{N}^R)^2> = \mathrm{tr}\, \rho^R - \mathrm{tr}[\rho^R \rho^R] + [\mathrm{tr}\rho^R]^2$$

3-3. Consider a cubic box of side L with periodic boundary conditions filled with plane waves up to the equilibrium fermi momentum k_F^o corresponding to density ρ_o. If the transverse wave functions are now frozen and the length of the box in the z direction is varied to change the total density to ρ, show that the $\phi_k(z)$ wave functions become

$$\phi_k(z) \to \left(\frac{\rho}{\rho_o L}\right)^{1/2} e^{i\rho k z/\rho_o}$$

and hence the one-body density matrix becomes

$$\rho(R+\frac{s}{z}, R-\frac{s}{2}) = \frac{\rho}{\pi^2 \rho_o} \int_{-k_F^o}^{k_F^o} e^{i\frac{\rho}{\rho_o} k_z s_z} \int_0^{[(k_F^o)^2 - k_z^2]^{1/2}} e^{ik_\perp s_\perp}\, k_\perp dk_\perp dk_z$$

Derive the special cases

$$\rho(s_\perp) = \frac{3j_1(k_F^o s_\perp)}{(k_F^o s_\perp)}$$

$$\rho(s_z) = \rho\, \frac{3j_1[(\frac{\rho}{\rho_o}) k_F^o\, s_z]}{\frac{\rho}{\rho_o}\, k_F^o\, s_z}$$

3-4. With the slab hamiltonian Eq. (3.2), the H.F. approximation Eq. (2.9) yields the following equation for the one particle Green's function

$$(i \frac{\partial}{\partial t_1} + \frac{\nabla_1^2}{2m} - W(1))\ G_1(11') = \delta(1-1')$$

where W is given by

$$W(r) = \frac{3}{4} +_o \rho(r) + \frac{3}{16} t_3 \quad \rho^2(r) + V_o \int d^3r' \frac{e^{-|r-r'|/a}}{|r-r'|/a} \rho(r')$$

In the presence of an arbitrary one-body external potential, W is replaced by $U = W + U_{ext}$, and we will consider $G_1(1,1',U)$ to be the Green's function in an arbitrary total effective field U.

Show, by varying $\int d3\ G^{-1}(1,3,U)G(3,2,U) = \delta(12)$

with respect to U that $G(1,3,U)G(3,2,U) = \frac{\delta G(12)}{\delta U}$

and hence

$$\delta\rho(1) = \delta\frac{1}{i}\ G(1,1^+)$$

$$= \frac{1}{i}\int d2\ G(12)G(21)\delta U(2)$$

$$= \frac{1}{i}\int d2\ G(12)G(21)[\delta U_{ext}(2) + \int d3\ v_{eff}(23)\delta\rho(3)]$$

where

$$v_{eff}(23) = \delta(r_2-r_3)(\frac{3}{4}\ t_0 + \frac{3}{8}\ t_3\ \rho) + V_o \frac{-|r_2-r_3|/a}{|r_2-r_3|/a} \quad .$$

Fourier transform this equation to obtain

$$\delta\rho(q,\omega) = \frac{1}{i}\sum_k G(k+q)G(k)\left[\delta U_{ext}(q,\omega)+v_{eff}(q)\delta\rho(q,\omega)\right]$$

$$\frac{\frac{1}{i}\sum_r G(k+q)G(k)}{1-(\frac{1}{i}\ G(k+a)G(k))\ v_{eff}(q)} \quad \delta U_{ext}$$

Finally, find the self-sustaining, i.e., propagating, modes in the low q limit by setting the denominator to zero:

$$\frac{1}{v_{eff}(0)} = \lim_{q\to 0} \frac{4}{\Omega} \sum_k \frac{\theta(|k+q|-k_F)(1-\theta(k-k_F))(\omega_{k+q}-\omega_k)}{\omega^2 - (\omega_k - \omega_{k+q})^2}$$

$$= \lim \frac{4}{\Omega} \frac{2\Omega}{(2\pi)^3} \int_0^1 d\cos\theta \int_{k_F}^{k_F+q\cos\theta} 2\pi k^2 dk \frac{k_F q\cos\theta/m}{\omega^2 - (\frac{k_F a}{m}\cos\theta)^2}$$

$$= \frac{2mk_F}{\pi} [-1 + \frac{s}{2} \ln \frac{s+1}{s-1}]$$

where $s = \frac{\omega/a}{k_F/m}$ is the ratio of the sound speed to the fermi velocity. Note that $v_{eff}(0)$ must be repulsive for s to be real (and thus obtain a propagating mode). This means that although W is attractive to bind the system $\frac{\delta\omega}{\delta\rho} \equiv v_{eff}$ is repulsive.

3-5. Show that minimization of $\langle H-H_{cm}\rangle$ with H given in Eq. (3.24) and the ψ_N^H of Eq. (3.30) yields Eq. (3.36).

3-6. Show, by evaluating diagram 3.41b in the basis defined by Eq. (3.37) or by calculating second order Rayleigh-Schrödinger perturbation theory that

$$\Delta E^{(2)} = -\frac{1}{27} N(N-1)g^2 (1 + \frac{3}{2} \int_0^\infty \frac{(x^2+1)^{1/2} - 1}{\sinh^2 \pi\lambda} .$$

(Note, this involves somewhat non-trivial integrals) The last integral may be performed numerically or by generating an asymptotic series by exapnding $(x^2+1)^{1/2}$ and integrating term-by-term.

3-7. Derive the static solutions for a single Lipkin system in the lowest order (Eqs. (3.84)-(3.37)) and in the second order coupled-cluster approximation (Eqs. (3.100)-(3.101)).

REFERENCES FOR CHAPTER 3

1. P. Bonche, S. Koonin, and J. W. Negele, Phys. Rev. C13, 1226 (1976).

2. B. Yoon and J. W. Negele, Phys. Rev. A16, 1451 (1977).

3. P. Hoodbhoy, Ph.D. dissertation, Massachusetts Institute of Technology, 1978; P. Hoodbhoy and J. W. Negele, Center for Theoretical Physics preprint.

4. P. Sperr, S. Vigdor, Y. Eisen, W. Henning, D.G. Kovar, T. R. Ophel, and B. Zeidmann, Phys. Rev. Lett. 36, 405 (1976).

5. P. Sperr, T. H. Braid, Y. Eisen, D. G. Kovar, F. W. Prosser, J. P. Schiffer, S. L. Tabor, and S. Vigdor, Phys. Rev. Lett. 37, 321 (1976).

6. J. W. Negele and D. Vantherin, Phys. Rev. C5, 1472 (1972).

7. L. Landau, Soviet Physics JETP 5, 101 (1957).

8. H. A. Bethe, Z. Physik 71, 205 (1931).

9. J. McGuire, J. Math. Phys. 6, 432 (1965).

10. F. Calogero and A. Degasperis, Phys. Rev. A11, 265 (1975).

11. C. N. Yang, Phys. Rev. Lett. 19, 1313 (1967).

12. V. E. Zakharov and A. B. Shabot, Soviet Phys. JETP 34, 62, (1972).

13. L. Dolan, Phys. Rev. D13, 528 (1976).

14. H. J. Lipkin, N. Meshkov and A. J. Glick, Nucl. Phys. 62, 188 (1965).

15. S. J. Krieger, Nucl. Phys. A276, 12 (1977).

16. D. Agassi, H. J. Lipkin and N. Meshkov, Nucl. Phys. 86, 321 (1966).

17. K. H. Luhrmann, Ann. Phys. (NY) 103, 253 (1977).

18. P. C. Lichtner and J. J. Griffin, Phys. Rev. Lett. 37, 1521 (1972).

IV. APPLICATION TO PHYSICAL SYSTEMS

The cumulative evidence from the previous section strongly suggests that the time-dependent mean-field approximation is a viable starting point for a systematic theory of nuclear dynamics. Not only is the approximation scheme quantitatively successful in relevant solvable models, but the results in slab geometry are highly suggestive of heavy-ion reaction phenomenology.

It is clearly desirable, then, to test the general theory in finite nuclei.

Application of even the lowest order mean-field approximation to all but the lightest nuclei presents a computational problem of formidable proportions. One is required to solve coupled integro-differential equations for (A_1+A_2) complex functions of four continuous variables. Reduction of the mean-field theory to manageable proportions therefore requires the introduction of additional approximations. Aside from challenging problems dealing with computational technology, the essential physics problems arising in applying the mean-field theory to real nuclei are choosing a small number of tractable cases which provide crucial definitive tests of the theory and discriminating between shortcomings of the additional approximations and actual failures of the mean field theory itself. Given the difficulty in implementing even the mean field approximation, investigation of higher order terms of the coupled-cluster hierarchy in realistic geometries is presently impractical.

Before presenting a few representative results of calculations, it is useful to discuss the additional approximations introduced beyond the basic mean-field approximation. Therefore, Section 4.1 motivates discretization of the wave functions on a mesh in coordinate space, estimates discretization errors, discusses the choice of a simple effective interaction, and explains the basic techniques for circumventing error propagation in evolution algorithms. Alternative approximations to reduce the number of degrees of freedom from the actual number $(A_1+A_2)N_xN_yN_z$ required to describe A_1+A_2 single particle wave functions on an N_x by N_y by N_z point grid are then presented in Section 4.2. Application to collisions, fission, and pion condensation are then discussed in the subsequent sections.

4.1 Numerical Solution in Coordinate Space

Shape degrees of freedom are absolutely essential to nuclear dynamics. Therefore, it is crucial to be guided from the start by the need for dealing adequately with the wide variety of shapes which arise in large amplitude oscillations, fission, and collisions.

Two general techniques are natural candidates for static or time-dependent mean-field problems. Historically, the earliest and often the simplest to implement is a truncated expansion in some convenient set of basis functions, such as harmonic oscillator functions. The essential problem, however, is that the truncation errors are highly shape dependent. One may easily be lulled into a false sense of security by thinking that the shape of the basis is so similar to the shape of the nucleus that one overlooks the fact that even gross qualitative features of the results are highly contaminated by basis constraints. Even a 14 oscillator-shell basis is barely adequate for heavy deformed nuclei[1] and is quite unreliable for calculation of a fission barrier. Furthermore, it is exceedingly difficult to quantify the magnitude of the shape-dependent errors.

Coordinate space calculations, either directly or implicitly through the use of a particular numerical integration algorithm, assume that wave functions are defined on a discrete mesh. The fact that all points on a uniform mesh are physically equivalent immediately eliminates the problem of shape dependence. A two-fermi-thick surface in a spherical nucleus is approximated the same way as the corresponding surface of a deformed, fissioning or colliding nucleus. One only has to make available enough mesh points to keep edge effects negligible. Even in static HF calculations the implementation of coordinate space methods produced a significant breakthrough; for the range of shapes encountered in time-dependent problems, the shape independence of numerical accuracy afforded by a coordinate-space mesh is crucial and every other aspect of the approximations applied to finite nuclei should be suitably adapted to the needs of a coordinate-space formulation.

Once one is committed to thinking about the mean-field theory on a mesh in coordinate space, it is extremely useful to replace the spatial integrals in the true action, Eq. (1.9), which is a functional of A single-particle wave functions, by discrete sums with suitable weights yielding an approximate discretized action which is a function of the values of each of the single-particle wave functions on each of the mesh points.

In the limit of small mesh spacing, of course, the discretized action approaches the true action. However, for any finite mesh spacing, variation of the discretized action yields a well defined set of differential equations in time and difference equations in spatial mesh labels which approximate the TDHF equations. Such a discretized functional yields exact conservation of discrete approximations to the energy, momentum, and angular momentum, eliminates various consistency difficulties which arise in discretizing the TDHF equations directly, guarantees hermitian approximations to the single-particle hamiltonian, and provides a convenient conceptual framework for thinking about the structure of the resulting equations.

4.1.1 The Effective Interaction

Since it is impractical to implement the second-order coupled-cluster theory at present, the most sensible approximation is to construct an effective interaction which utilizes as much as is presently known about the short-range correlations in ground states of finite nuclei. The essential approximation, then, is that for the relatively low energy dynamics we wish to describe, the changes in the higher-body correlations, and thus the effective interaction, are negligible. The most appropriate effective interaction is therefore a G-matrix calculated from a realistic interaction and adjusted to produce the proper nuclear matter binding energy and saturation as described in Section 1.

One major problem, however, arises immediately in using a realistic G-matrix in coordinate space. Whereas the local Hartree contribution to the mean field, $W(r,r')$, produces a diagonal matrix and is thus very easy to treat, the exchange term generates a dense matrix and is quite intractable. Therefore, it is essential from the outset to recast the physics of the exchange potential into a form which is appropriate for coordinate space calculations. The key to reformulating the exchange term is the observation, emphasized in Section 1, that the off-diagonal range of the density matrix is given by π/k_F. Thus, a truncated expansion of the density matrix[3]

$$\rho(R+\frac{s}{2},R-\frac{s}{2}) = \rho(R)\,\frac{3j_1(k_F s)}{k_F s} + \frac{35}{2sk_F^3}\, j_3(k_F s)\left[\frac{1}{4}\nabla^2\rho(R) - \tau(R) + \frac{3}{5}k_F^2\,\rho(R)\right] \tag{4.1}$$

where

$$\tau(R) = \sum_{\nu} |\nabla\phi_{\nu}|^2 \qquad (4.2)$$

yields an extremely accurate approximation to the exchange contributions to <H> while maintaining a structure no more complicated than a Hartree theory with a position-dependent effective mass. Substitution of Eq. (4.1) into <H>, using a G-matrix derived from the Reid potential, and integration over the relative coordinate yields a Hamiltonian density functional $H(R)$ depending on $\rho(R)$ and $\tau(R)$ very similar to that obtained with a phenomenological Skyrme potential.[3] The appreciable non-locality arising from the exchange term yields an average effective mass $m^*=0.63m$ in the nuclear interior.

Once the Hamiltonian density has been reduced to such a functional of ρ and τ, it is sensible to parametrize it in a simple, convenient form, with parameters adjusted to quantitatively reproduce the appropriate properties of the original effective interaction. The effective Hamiltonian density used in most of the applications described subsequently is the following:[2,4]

$$= \int d^3r\Big(\frac{\hbar^2}{2m}(\tau_n+\tau_p) + \frac{t_o}{2}[(2+x_o)\rho_n\rho_p + \frac{(1-x_o)}{2}(\rho_n^2+\rho_p^2)$$
$$+ \frac{t_1+t_2}{4}[(\rho_n+\rho_p)(\tau_n+\tau_p) - (J_n+J_p)^2] + \frac{t_2-t_1}{8}(\rho_n\tau_n - J_n^2$$
$$+ \rho_p\tau_p - J_p^2) + \frac{t_3}{4}(\rho_n\rho_p^2 + \rho_p\rho_n^2)) + \frac{V_L}{2}[E_y(\rho_n,\rho_n)+E_y(\rho_p,\rho_p)]$$
$$+ v_u E_y(\rho_n,\rho_p) + \frac{e^2}{2}\iint d^3r d^3r' \frac{\rho_p(\vec{r})\rho_p(\vec{r}')}{|\vec{r}-\vec{r}'|} \qquad (4.3)$$

In Eq. (4.3), the Yukawa contribution is defined

$$E_y(\rho_q,\rho_{q'}) = \iint d^3r d^3r' \frac{e^{-\mu|r-r'|}}{\mu|r-r'|} \rho_q(r)\rho_{q'}(r') \qquad (4.4)$$

the current is

$$J(r) = \sum_{\nu} \mathrm{Im}\psi_{\nu}(r)\, \nabla\psi_{\nu}(r) \qquad , \qquad (4.5)$$

$\tau(r)$ is defined in Eq. (4.2), the labels p and n denote summation over only proton or neutron orbitals respectively, and the parameters are

$t_o = -104.49$ MeV fm^3	$t_3 = 9331$ MeV fm^6
$x_o = 4.01$	$V_L = -444.85$ MeV
$t_1 = 585.6$ MeV fm^5	$V_u = -355.79$ MeV
$t_2 = -27.1$ MeV fm^5	$\mu = 2.175$ fm^{-1}

The current terms arise naturally from the Galilean invariant combination $\rho\tau - J^2$. The finite range of the direct Hartree term is maintained, since reduction to the zero-range Skyrme form introduces unphysical fluctuations in the mean field, and the Yukawa parameterization is a matter of convenience since it enables evaluation of the potential by solution of a Helmholtz equation instead of the more cumbersome convolution.

4.1.2 Numerical Considerations

Numerical details are discussed at length in Refs. 2,4,5 and 6, so only a few basic concepts will be introduced here.

4.1.2.1. Discretization: An illustrative example of the discretization problem is the simple one-dimensional expression

$$I = \int dx \, \{|\nabla\phi|^2 + \phi^2(x)(u(x)-\varepsilon)\} \tag{4.7}$$

Assuming $\phi(x)$ is replaced by the discrete set of values on an evenly spaced mesh, $\phi_n \equiv \phi(x_n) = \phi(n\Delta)$, I may be approximated

$$\begin{aligned} I &\approx \sum_i \Delta |\phi'(x_i + \tfrac{\Delta}{2})|^2 + \sum_i |\phi(x_i)|^2 \, (u(x_i)-\varepsilon) \\ &\approx \sum_i \Delta \left| \frac{\phi_{i+1}-\phi_i}{\Delta} \right|^2 + \sum_i \Delta |\phi_i|^2 \, (u_i-\varepsilon) \, . \end{aligned} \tag{4.8}$$

Variation of 4.8 with respect to ϕ_n^* at some interior point n yields the difference equation

$$\frac{\phi_{n+1}+2\phi_n-\phi_{n-1}}{\Delta^2} + u_n \phi_n = \varepsilon\phi_n \tag{4.9}$$

which contains the usual 3-point approximation to ϕ'' with leading error of order Δ^2.

Several different approximations were made in arriving at Eq. (4.9). One was using a trapezoid rule for the integral. Although it might appear that being fancier by including the proper weights from Simpson's rule would increase the accuracy one order in Δ, averaging the supposedly more accurate Simpson formulas for N and N+2 mesh points yields precisely the naive trapezoidal weights in the interior with differences only arising at the mesh edges where the wave function must in any event be small. A natural way to correct for the difference between the original integral and its discrete approximation is the Euler Maclaurin summation formula

$$\int_0^{n\Delta} f(x)dx = \Delta \sum_{k=1}^{n-1} f(k\Delta) + \frac{\Delta}{2}(f(0)+f(n\Delta)) \tag{4.10}$$

$$- \frac{\Delta^2}{12}[f'(n\Delta)-f''(0)] + \frac{\Delta^4}{720}[f'''(n\Delta)-f'''(0)]+\ldots$$

Here again, however, one finds that the corrections only involve terms at the mesh edges which are negligible if the mesh is large enough for the wave functions to decay exponentially. (The absurd case of Δ so large that few points fall in the region where f is not exponentially decaying which requires many terms in the series 4.10 is, of course, excluded.)

Since the integral approximations introduce negligible error, the dominant inaccuracy arises from the approximation

$$\phi'(x_i + \frac{\Delta}{2}) = \frac{\phi_{i+1} - \phi_i}{\Delta} + \mathcal{O}(\Delta^3) \tag{4.11}$$

This leads to an error in the kinetic energy

$$-\phi''(x_i) = \frac{-\phi_{i+1} + 2\phi_i - \phi_{i-1}}{\Delta^2} + \frac{\Delta^2}{12}\phi''''(x_i) \tag{4.12}$$

For a fermi gas approximated on a three-dimensional cartesian lattice with mesh spacing Δ, the corresponding kinetic energy error per particle is

$$\frac{h^2}{m}\,\frac{3}{280}\,k_F^4\,\Delta^2 \tag{4.13}$$

corresponding to a fractional error of $\frac{1}{28}(k_F\Delta)^2$. Even for a mesh spacing as large as $\Delta=1$ fm, the discretization error produces only a 6% reduction in the kinetic energy. Since the error is highly insensitive to the position of the nucleus on the lattice, the main effect is roughly the same as increasing the nucleon mass by 6%. It will introduce quantitative discrepancies to be sure, but the qualitative behavior will not be seriously misrepresented.

Two options exist for improving the accuracy: decreasing Δ or using a more accurate differentiation formula. The next order approximation for $(\nabla\phi)^2$ is shown in Problem 4-2 to yield the usual 5-point second derivative formula, with error of order Δ^4. At least in cartesian coordinates, the savings in storage by using large Δ more than compensates the complications associated with a 5-point formula, so 3-dimensional coordinate space calculations are best done with the 5-point formula.[7]

For axially-symmetric systems, cylindrical coordinates introduce several complications. Equivalent expressions obtained from integration by parts do not yield equally accurate discretized approximations. The most accurate discretization gives rise to radially varying weights in the final difference equations. Also, since the radial variable runs from 0 to ∞, non-vanishing Euler-Maclaurin sum formula corrections arise from non-vanishing terms along the symmetry axis. Technical details concerning cylindrical coordinates may be found in Ref. 2.

In addition to the kinetic energy error estimates in Eq. (4.13), one may obtain a feeling for the significance of discretization errors by comparing the exact ^{40}Ca density calculated for the effective hamiltonian in Eq. (4.4) with those obtained on a discrete mesh in cylindrical coordinates for $\Delta=0.4$ and 0.8 fm shown in Fig. 4.1. One observes that the 0.4 fm results are extremely accurate and the 0.8 fm density is just slightly compressed due to the slight underestimation of the kinetic energy. Again, this evidence suggests that discretization errors are benign and will not affect results qualitatively.

4.1.2.2. Evolution in time: Since the numerical analysis of parabolic partial differential equations is treated extensively in the literature,[8,9] only a few fundamental aspects will be mentioned in this section. The most essential issue concerns the stability of evolution algorithms, and two alternative philosophies arise. One possible approach is to use a method which is in principle unstable but sufficiently accurate for some finite time interval through the use of very high order approximations and suppression of round-off error by the use

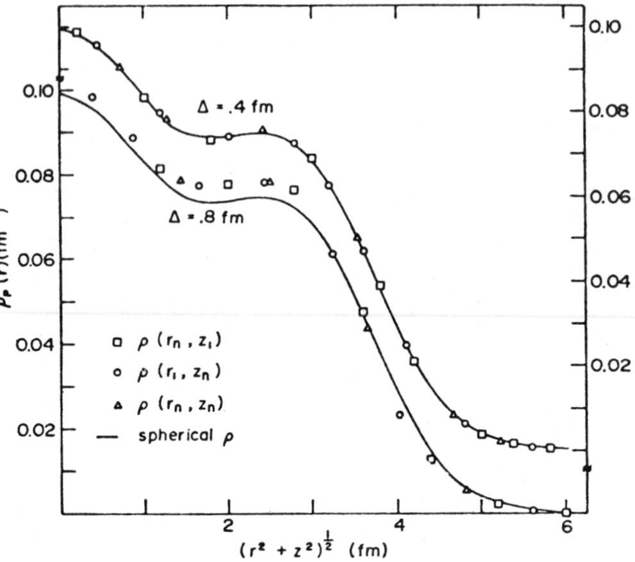

Fig. 4-1. Comparison of ^{40}Ca density distributions calculated on two meshes in cylindrical coordinates with the exact result (solid lines).

of long word length. This approach, however, contradicts ones basic instincts and sensitivities as a physicist. Already it is rather offensive to have to evolve $(A_1+A_2)N_xN_yN_z$ complex numbers in time to arrive at a viable theory of dynamics, since the number of essential physical variables in an optimally formulated theory must surely be much smaller. But to take seriously the notion that each number is meaningful to 16 decimal places is ludicrous. Hence, the alternative philosophy is to use a stable algorithm with short word length and maintain a rational balance between the level of numerical precision and the level of physical approximations already introduced.

A useful example for discussion is the evolution of single particle wave functions by a discrete single-particle hamiltonian matrix with a time independent potential, so that

$$i\dot{\phi}_j = h_{jk}\,\phi_k \tag{4.14}$$

where

$$h_{ik} = -\frac{1}{\Delta^2}(\delta_{j,k+1} + \delta_{j,k-1}) + (\frac{2}{\Delta^2} + V_k)\delta_{jk}$$

The most naive and easily evaluated approximation to Eq. (4.14) is

$$i\,\frac{\phi^{n+1}-\phi^n}{\tau} = h\phi^n \tag{4.15}$$

where the superscripts label time, $\tau \equiv t_{n+1}-t_n$, and matrix multiplication is understood in the suppressed spatial labels. Eq. (4.15) is explicit in the sense that ϕ^{n+1} is given explicitly in terms of ϕ^n

$$\phi^{n+1} = (1 - i\tau h)\phi^n \quad . \tag{4.16}$$

However, although (4.16) is simple, it gives rise to exponentially growing errors (see Problem 4-3). Since h is hermitian, the eigenvalues of $(1-i\tau h)$ have modulus greater than or equal to one and time evolution by repeated application of (4.16) results in exponential amplification of the component of ϕ associated with the eigenvalue of largest modulus. Evidently, one therefore requires a unitary approximation to the evolution operator, and two such possibilities are

$$\phi^{n+1} = e^{-i\tau h}\,\phi^n \tag{4.17a}$$

$$\sim \{1-i\tau h[1 - \frac{i\tau h}{2}(1 - \frac{i\tau h}{3}\cdots)]\}\phi^n \tag{4.17b}$$

and

$$\phi^{n+1} = \frac{1 - i\,\frac{\tau h}{2}}{1 + i\,\frac{\tau h}{2}}\,\phi^n \quad . \tag{4.18}$$

Eq. (4.17a), of course, is the exact evolution operator for the present problem, but can only be evaluated by a truncated expansion of the form (4.17b). A small number of terms, however, yields a matrix unitary to within machine precision, and since only matrix multiplications by a sparse matrix are involved, this method provides a very appealing method for three-dimensional problems.[7] The Crank-Nicholson approximation,[8,9] Eq. (4.18),is a standard method for one-dimensional problems, because of the ease with which the tri-diagonal matrix $(1+i\frac{\tau h}{2})$ may be inverted by a two-term recursion relation.

Complications arise in generalization of (4.18) to higher dimensions[9] so at some point (4.17b) becomes the more appealing method.

When the constant h of Eq. (4.14) is replaced by the self-consistent single-particle hamiltonian $h[\rho(t)]$, one additional problem arises. In the case of simple translation, it is clear that the equation

$$\phi^{n+1} = e^{-i\tau h[\rho(t_n)]}\,\phi^n \tag{4.19}$$

evolves ϕ in a potential that is always lagging behind its proper position for $t_{n+1} \geq t > t_n$. Such a constantly retarded potential gradually slows down the motion of the nucleus, giving rise to significant decrease of energy with time. Hence, in practice, it is necessary to approximate $\rho(t_n + \frac{\tau}{2})$ by first evolving ϕ from ϕ^n for time interval $\tau/2$ using $h[\rho(t_n)]$ and then evolving ϕ^n for time interval τ using $h[\rho(t_n + \frac{\tau}{2})]$ constructed from the intermediate ϕ's.

For completeness, it is useful to mention that analogous coordinate spare techniques are applicable to higher-order coupled-cluster equations (and even practical in low dimensionality or very small systems). In general, the particle-hole amplitudes S satisfy inhomogeneous discretized matrix equations:

$$i\dot{S} = AS + B \tag{4.20}$$

The exact evolution for interval t with A and B assumed to be constant (and evaluated as before at $t_n + \frac{\tau}{2}$) is given by

$$\begin{aligned} S^{n+1} &= e^{-iA\tau} S^n + (e^{-iA\tau} - 1)\, A^{-1} B \\ &= S^n + (e^{-iA\tau} - 1)\, A^{-1}\, (AS^n + B) \\ &\approx S^n - i\tau\{1 - \frac{iA\tau}{2}[1 - \frac{iA\tau}{3}(1 - \frac{iA\tau}{4}\cdots)]\}(AS^n + B) \end{aligned} \tag{4.21}$$

As before, a small number of terms in a series which involves only matrix multiplications by a sparse matrix yields stable evolution of $S^{(n)}$.

4.2 Approximations to Reduce the Number of Degrees of Freedom

Both because of the computational difficulty associated with evolution of $(A_1+A_2)N_xN_yN_z$ complex numbers to represent

A_1+A_2 particles on a three-dimensional mesh and because of a conviction that the ultimate physics is governed by far fewer variables, it is useful to consider approximations which reduce the number of degrees of freedom of the full problem.

4.2.1 Symmetries

One general technique for significantly reducing the number of degrees of freedom is specification of an initial wave function possessing symmetries which also occur in the mean field. The time-dependent mean-field equations automatically preserve any such symmetries in time. For example, if one selects an initial condition with reflection, inversion, spin, isospin and axial symmetry, these symmetries will be preserved in time. Such solutions are very special. Not only are they especially cheap to calculate because they evolve in a limited space but for the same reason they also have less freedom for randomization and thus dissipation and equilibration. The essential fact to bear in mind is that in considering a realistic ensemble of initial conditions, these special symmetric cases receive negligible weight. Thus, although those of us cursed with impoverished computer budgets may often restrict ourselves to such symmetric cases, we must at least be alert to the possible unrealistic consequences of such symmetries.

A few computational results suggest that symmetric initial conditions are not misleading. For example, three dimensional calculations of head-on collisions do not differ qualitatively from those at small but finite impact parameters. Similarly, ^{40}Ca collisions with neutron and proton orbitals assumed to evolve identically in a mean field containing half the full Coulomb potential do not differ substantially from those with neutrons and protons evolving separately.

4.2.2 Constraints

In the case of initial conditions possessing a symmetry which also occurs in the mean field, we dealt with exact solutions to the time-dependent mean-field equations. Thus, a two-dimensional axially symmetric calculation and a full three-dimensional calculation of a head-on collision should agree to within numerical errors. In contrast, when one imposes a constraint, new dynamical equations are derived which are different from the original mean field equations.

Consider, for example, the constraint that collisions at finite impact parameters must be described by a single, axially symmetric determinant in a rotating intrinsic frame. The same initial condition in a constrained two-dimensional calculation and an exact three-dimensional calculation will lead to different final states. Physically one can distinguish macroscopic and microscopic effects of an axial symmetry constraint. Macroscopically, one recognizes that the realistic shapes sketched in Fig. 4-2 must be replaced by rather different shapes if one postulates axial symmetry with respect to the axis joining the centers of two interacting nuclei. Since such deformation modes are reasonably soft, one may expect such macroscopic restrictions to be benign. Microscopically, however, as also sketched in Fig. 4-2, single particle orbitals bouncing around on non-closed random trajectories may be expected to exhibit considerably greater dissipation than the periodic longitudinal trajectories imposed by axial symmetry. At sufficiently low energies, one expects physically that the differences introduced by axial symmetry are negligible, and this expectation is borne out by explicit comparisons with three-dimensional calculations.[10]

It is a non-trivial question how best to implement the axial symmetric hypothesis in a dynamical theory. A theoretically appealing variational formulation arises from assuming the variational wave function[11]

$$\psi = e^{-iL_y} e^{i\sum_j \chi(r_j z_j \phi_j)} A \prod_j \psi_\lambda(r_j z_j) e^{im_\lambda \phi_j} \tag{4.22}$$

which is axially symmetric in an intrinsic frame that may be freely rotated by an angle η and which has a collective velocity potential χ. Using the periodicity of χ, the action is

$$S = \int d^3r \, dt\{H(\psi_\lambda^*\psi_\lambda) - i\sum\psi_\lambda^* \frac{\partial}{\partial t} \psi_\lambda + \frac{1}{2m} \rho|\nabla\chi|^2 - \dot{\eta}\rho \, \hat{e}_y \cdot r \times \nabla\chi \} \tag{4.23}$$

Variation with respect to η yields

$$\frac{d}{dt} \int d^3r \, \rho \, \hat{e}_y - r \times \nabla\chi \equiv \frac{d}{dt} \langle L_y \rangle = 0 \tag{4.24}$$

which implies $\langle L_y \rangle = L$ is a constant and variation with respect to $\langle L_y \rangle$ yields

$$\eta = \frac{\partial}{\partial \langle L_y \rangle} \frac{1}{2m} \int \rho|\nabla\chi|^2 \tag{4.25}$$

Thus, L is the dynamical variable conjugate to η and all other degrees of freedom in χ are static degrees of freedom.[12]

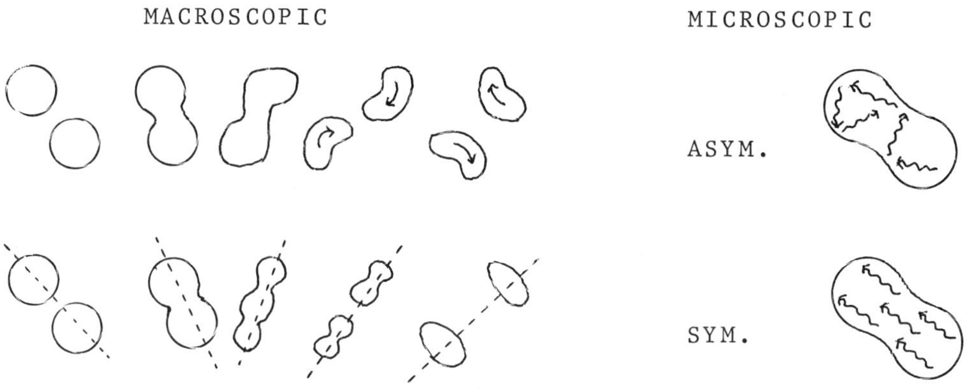

Fig. 4-2. Sketch of macroscopic and microscopic consequences of axial symmetry.

Defining

$$\theta = a\chi \tag{4.26}$$

then

$$L = \int \rho \, \hat{e}_y \circ \vec{r} x \, \frac{\vec{\nabla}\theta}{a} \tag{4.27}$$

and

$$S = \int d^3r \, dt\{H(\psi^*_\lambda \psi_\lambda) - i \sum_\lambda \psi^*_\lambda \frac{\partial}{\partial t} \psi_\lambda \tag{4.28}$$

$$+ \frac{L^2}{[\int \rho \hat{e}_y \cdot \vec{r} \times \vec{\nabla\theta}]^2} \frac{\rho}{2m} |\nabla\theta|^2 - \dot{\eta} L\}$$

Thus, the equation of motion is in canonical form with

$$\mathcal{H} = H(\psi_\lambda^* \psi_\lambda) + \frac{L^2}{2} \frac{\frac{1}{m}\int \rho|\nabla\theta|^2}{[\int \rho\, \hat{e}_y \cdot \vec{r} \times \vec{\nabla}\theta]^2} \tag{4.29}$$

The Hamiltonian equations $\dot{\eta} = \frac{\partial \mathcal{H}}{\partial L}$ and $\dot{L} = \frac{\partial \mathcal{H}}{\partial \eta}$ simply reproduce the previous equation of motion, and we define the coefficient of $\frac{L^2}{2}$ as the inverse of the moment of inertia. This moment of inertia corresponds to irrotational flow and attains the proper form in the limit of two nuclei passing by and for a spherical nucleus. Modification of the usual TDHF equations is straightforward. Variation with respect to $\partial\psi^*$ yields an additional potential for the evolution on the single particle wave functions:

$$\frac{L^2}{2m}\left\{ \frac{|\nabla^2\theta|^2}{[\int \rho\, \hat{e}_y \cdot \vec{r} \times \vec{\nabla}\theta]^2} - \frac{\int \rho|\nabla\theta|^2}{[\int \rho\, \hat{e}_y \cdot \vec{r} \times \vec{\nabla}\theta]^2}\, 2\dot{\theta}_y \cdot \vec{r} \times \vec{\nabla}\theta \right. \tag{4.30}$$

Variation with respect to $\delta\theta$ yields an equation which determines the velocity field at each time:

$$2\nabla\cdot\rho\nabla\theta = \frac{\int \rho|\nabla\theta|^2}{\int \rho\hat{e}_y \cdot \vec{r} \times \vec{\nabla}\theta}\, 2\hat{e}_y \cdot \vec{r} \times \vec{\nabla}\rho \equiv 2am\,\eta\hat{e}_y \cdot \vec{r} \times \vec{\nabla}\rho \tag{4.31}$$

As an alternative to this formulation, one may postulate a Hamiltonian functional with a physically plausible moment of inertia. For example, recent calculations have used

$$I_{rigid} = m \int d^3r\, \rho(\mathbf{r})[z^2 + \frac{1}{2} r^2] \tag{4.32a}$$

and

$$I_{point} = m \frac{A_1 A_2}{A_1 + A_2} R^2 \tag{4.32b}$$

at appropriate stages of a reaction. Note that these two prescriptions make physically different assumptions about the post-scission rotation of two fragments, as sketched in Fig. 4-3.

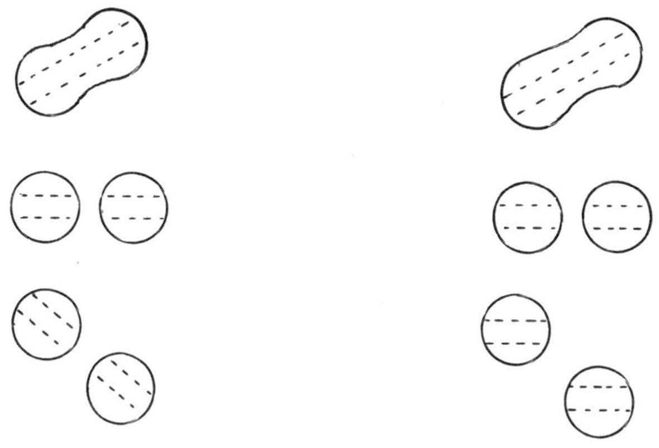

Fig. 4-3. Comparison of rigid and point post-scission moment of inertia assumptions.

Comparisons of all these alternative axially-symmetric approximations with three-dimensional calculation indicate that the most accurate results occur when the point moment of inertia is used until the density in the neck region between the approaching target and projectile reaches half nuclear matter density and the rigid moment of inertia thereafter.[10] It is not completely clear why the conceptually appealing variational theory is less satisfactory.

A second constraint, which is in some ways more appealing, is that each wave function separate into a component in the reaction plane multiplied by a normal component

$$\psi_i(r,t) = \phi_i(x,y,t)\ \chi_i(z,t) \tag{4.33}$$

Head-on axially symmetric calculations and full three-dimensional calculations support the intuitive expectation that very little equilibration occurs out of the reaction plane: it is simply very difficult for single-particle orbitals with high initial momenta in the reaction plane to acquire significant transverse momentum components from reflections at the well edge. The ansatz (4.33) capitalizes on this lack of equilibration out of the reaction plane while allowing maximal freedom for motion in the reaction plane. From the viewpoint of Fig. 4-2, the randomization arising from multiple reflections in an arbitrary shaped cavity should be more accurately reproduced by this wave function then by an axially constrained one.

As in the axially constrained case, one must derive or define equations of motion for wave functions of the form (4.33). The most theoretically appealing method is to apply the time dependent variational principle to obtain equations for ϕ_i and χ_i. Although this now appears possible,[16] significant complications arise in enforcing orthogonality.

For light nuclei, a natural approximation is suggested by the fact that harmonic oscillator functions are automatically separable and also yield reasonable approximations to HF wave functions. Thus, a sensible trial solution is[14,15]

$$\psi_i(r,t) = \phi_i(x,y,t)\chi_i^{HO}(z) \tag{4.34}$$

where the oscillator size parameter may be determined by minimizing the ground state energy with the trial function. As derived in Problem 4-5, the equations of motion for ϕ_j using an effective hamiltonian density which produces a local HF potential are

$$i\dot{\phi}_j = [-\frac{\hbar^2}{2m}(\frac{\partial^2}{\partial x^2}+\frac{\partial^2}{\partial y^2}) + \langle T_z\rangle_j + W_j(x,y)]\phi_j , \tag{4.35}$$

where W_i is related to the full local HF potential by

$$W_j(x,y) = \int dz|\chi_j^{HO}(z)|^2\, W(x,y,z) \tag{4.36}$$

and

$$\langle T_z\rangle_j = \frac{\hbar^2}{2m}\int dz\, |\frac{d\chi_j^{HO}}{dz}|^2 \tag{4.37}$$

Static and time dependent factorized solutions[15] using Eq. (4.34) are compared with three-dimensional results in Figs. 4-4 and 4-5. For the light, oscillator-like nuclei ^{16}O and ^{40}Ca, the static densities agree reasonably well, especially if one averages the shapes in the z-direction and in the x-y plane, and the binding energies and rms radii agree to within 2%. For heavier nuclei, significant discrepancies arise from the fact that oscillator functions do not produce saturating density distributions. In the Thomas Fermi limit, for example, the density distribution generated by harmonic oscillator functions is given by an inverted parabola raised to the 3/2 power. Thus, whereas the fictitious A=140 system calculated

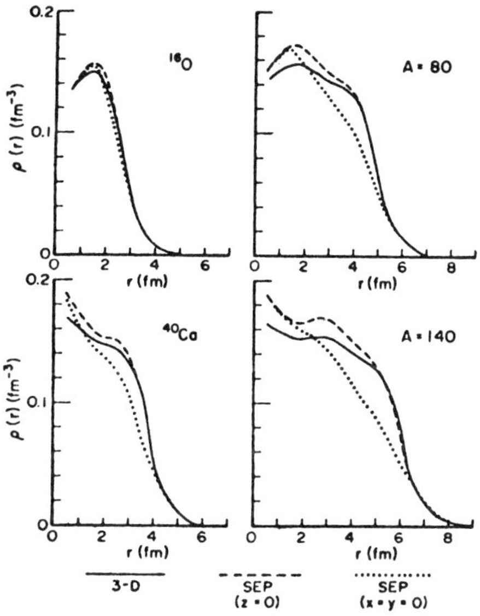

Fig. 4-4. Comparisons between three-dimensional static HF densities (solid lines) and separable results in the reaction plane (dashes) and normal to the reaction plane (dots) for several oscillator closed shell nuclei.

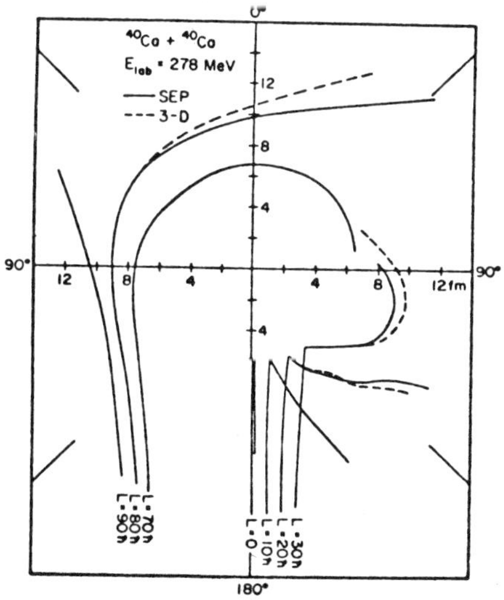

Fig. 4-5. Comparisons between three-dimensional and separable trajectories of ^{40}Ca + ^{40}Ca at c.m. energy 139 MeV. In places where a dashed curve is not drawn, the two curves coincide.

in Fig. 4-4 should have the roughly constant interior density given by the solid curve, the density profile along the z axis generated by oscillator functions, denoted by the dotted curve, follows the Thomas-Fermi estimate and is qualitatively in error. It remains an open question whether a fully variational calculation with the general wave function (4.33) is significantly better in heavy nuclei.

The agreement between separable wave functions of the form (4.34) and exact three-dimensional calculations for the scattering of $^{40}Ca + ^{40}Ca$ at 278 MeV is striking. In addition to the agreement in final scattering angle shown in Fig. 4-5, final fragment angular momenta and energies agree within 2% at all impact parameters, and the same fusion window from $L=30\hbar$ through $L=80\hbar$ occurs in both calculations.

At present, it appears that both forms of constrained calculations have their own regions of validity. The axially-constrained wave function applies to collisions between arbitrary spherical nuclei at sufficiently low energies (less than 0.5 MeV per particle above the Coulomb barrier in the c.m.). The separable approximation is best for nuclei no larger than Ca, but yields accurate results for cm energies of several MeV per particle above the Coulomb barrier.

4.3 Heavy Ion Collisions

Given the immense amount of experimental data becoming available for a wide variety of targets, projectiles, and incident energies,[17] it is neither practical nor even desirable to undertake an exhaustive program of comparing theory and experiment for all possible reactions. Rather, it is most useful to focus attention on a few special cases which are particularly amenable to calculation and provide definitive tests of specific aspects of the theory. Thus, we shall address the behavior of fusion cross sections in $^{16}O+^{16}O$ and $^{40}Ca+^{40}Ca$ collisions and the phenomena of side focussing and strongly damped collisions in Kr induced reactions on ^{20}Pb and ^{209}Bi.

4.3.1. $^{16}O+^{16}O$ and $^{40}Ca+^{40}Ca$ Collisions

The general qualitative behavior of three-dimensional calculations of symmetric medium-ion collisions has already been displayed in the polar plot in Fig. 4-5 for $^{40}Ca+^{40}Ca$ at E_{cm}=139 MeV. The separation coordinate R is defined as

$$R = \frac{2}{A} \int d^3r|z|\rho(r) \tag{4.38}$$

where the z axis is defined as the major principal axis and the origin is taken to be the c.m. The effective hamiltonian was taken to be of the form given in Eq. (4.3), but simplified to the case $m^*=m$ by setting $t_1=t_2=0$ and correspondingly adjusting the other parameters[6,7,15]

The most salient feature of these results is large dissipation. As mentioned previously, impact parameters corresponding to L from 30 through 70$\hbar$ lead to fusion. The nearly head-on collisions which do not fuse yield fragments with total cm energy of the order of 58 MeV which is just the Coulomb energy acquired in separation. Strong damping is observed in gazing collisions, with L=80 and 90$\hbar$ collisions yielding final cm energies of 88 and 120 MeV respectively, well below the total cm energy of 139 MeV. In contrast to earlier axially-constrained calculations at the same energy, which did not produce sufficient dissipation or fusion, these three-dimensional results are in qualitative agreement with observed phenomenology.

The most stringent quantitative test of the theory is posed by the systematics of fusion cross sections with energy. Using the standard $m^*=m$ interaction of Refs. 6, 7 and 15 a three-dimensional calculation[7] for $^{16}O+^{16}O$ yields the impressive agreement with experiment shown in Fig. 4-6, where the theoretical fusion cross section is defined

$$\sigma_{fus}(E_{lab}) = \frac{\pi\hbar^2}{\mu E_{lab}} [(\ell_> + 1)^2 - (\ell_< + 1)^2] \tag{4.39}$$

with $\ell_>$ and $\ell_<$ denoting the maximum and minimum angular momenta which fuse. A particularly interesting feature is the fact that even above E_{lab}= 55 where the low angular momentum transmission window sets in, agreement in shape and magnitude remains excellent. Both because of the effective spreading in the incident energy due to the initial c.m. wave packet and because of the 1$\hbar$ resolutions in ℓ, comparison of fine structure of the scale of the fluctuations in the fusion cross sections is not meaningful.

Similar fusion cross sections[7] are shown for $^{40}Ca+^{40}Ca$ in Fig. 4-7. In addition to the standard interaction, force I, alternative forces which decrease the rms radius of ^{40}Ca by 4.6% (force II) and increase the rms radius by 1.4% (force III) demonstrate the extreme sensitivity of fusion cross sections

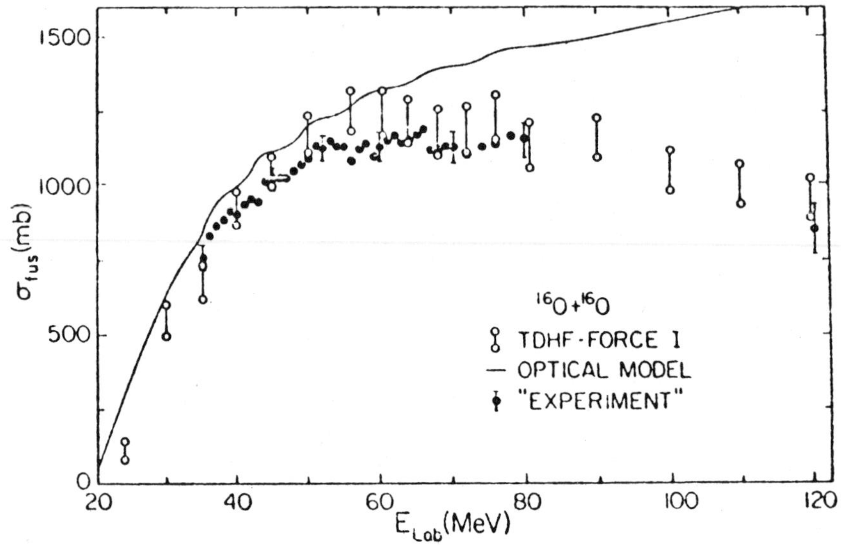

Fig. 4-6. Fusion excitation function for $^{16}O+^{16}O$.

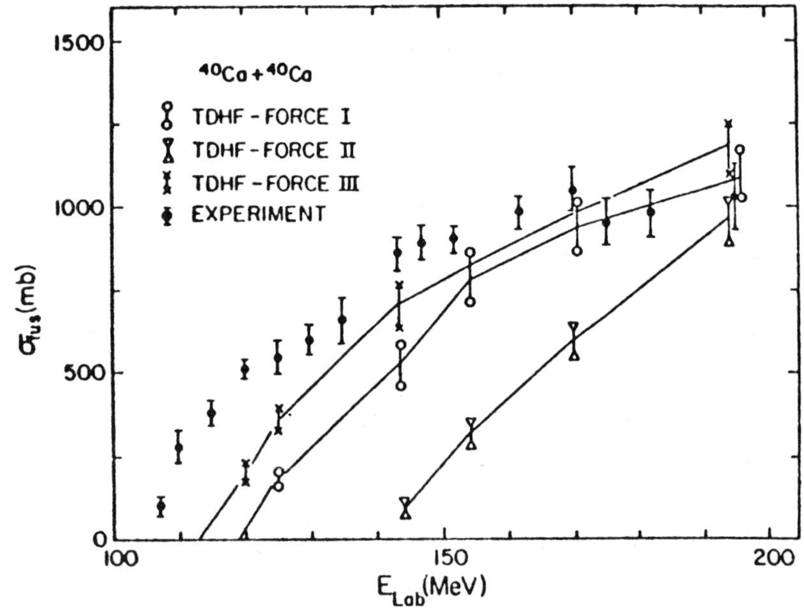

Fig. 4-7. Fusion excitation function for $^{40}Ca+^{40}Ca$.

to details of the interaction. This sensitivity alone suggests that the discrepancies do not constitute a significant breakdown of the mean-field theory. However, a much stronger argument is provided by the fact that the maximal discrepancy is at threshold, where dynamics are playing a negligible role and only the barrier height is relevant. Certainly, at least the threshold behavior will be obtained correctly when careful account is taken of the non-local mean-field, the Coulomb exchange energy is included and charge densities are required to accurately fit elastic electron scattering. From the qualitative behavior of Fig. 4-7, it is hard to imagine how the rest of the curve can go for wrong when the threshold behavior is improved.

Thus, on the basis of present evidence, the prognosis for medium-ion reactions is exceedingly optimistic. The qualitative features of strong damping and quantitative predictions of fusion cross sections appear to fall within the purview of mean-field dynamics. In addition, the provocative prediction of a transmission window for small impact parameter has emerged. Such behavior is an obvious manifestation of the low equilibration with transverse degrees of freedom which occurs when single-particle wave functions bounce off the ends, but not the sides, of the compound-nucleus cavity. Although circumstantial evidence for this window exists from fusion cross sections, unambiguous direct experimental confirmation would constitute a dramatic success of the theory.

4.3.2 Kr Induced Strongly Damped Collisions

A different, and potentially definitive test of many aspects of the mean-field theory is provided by the striking feature[17] of ^{84}Kr reactions with ^{208}Pb and ^{209}Bi. For heavy systems not too far above the Coulomb barrier, the strong side-peaking of projectile-like fragments near the grazing angle, the energy loss associated with the deep inelastic peak, and the broad fragment mass distribution centered about the initial partition place stringent demands on a microscopic theory having no free parameters.

Even with enforcing spin symmetry, the large number of orbitals in the combined Kr+Pb system necessitates constraining the form of determinantal wave function as discussed in Section 4.2.2. Hence, because the separable oscillator approximation is suspect for nuclei as large as Pb, initial calculations[19] used the axial symmetry constraint and thus considered only the low energy reactions ^{84}Kr+^{208}Pb at E_{lab}=494 MeV[20] and ^{84}Kr+^{209}Bi at E_{lab}= 600 MeV, corresponding to c.m. energies per particle above the Coulomb barrier of 0.1 and 0.4 MeV, respectively. In contrast to the ^{16}O and ^{40}Ca calculations, the effective hamiltonian of Eq. (4.3) was used, thereby incorporating a realistic effective mass. Non-closed shells were treated by assuming uniform fractional occupation of each of the states in a partially occupied shell.

The TDHF equations were integrated for 10 impact parameters in the case of $^{84}Kr+^{209}Bi$ at 600 MeV and 5 impact parameters for $^{84}Kr+^{208}Pb$ at 494 MeV and the principal results are presented in Figs. 4-8 and 4-9. The Wilczynski plots in Fig. 4-8a and b indicate that the correct overall qualitative behavior is obtained; in particular, the proper strong damping is preducted in the correct angular region. The same information is presented more quantitatively by the deflection functions and final c.m. energy as a function of angular momentum in Fig. 4-9. The deflection functions display the behavior expected from classical models[17] at large ℓ, falling below the Coulomb scattering curve just inside the grazing angular momentum. The zero slope of the Bi deflection function near 260$\hbar$ corresponds to a sharp peak in the angular distribution with small energy loss. The angle is in excellent agreement with the observed quasi-elastic peak,[20] denoted by the θ_p, so that the strong side-focussing is accurately reproduced. The Pb deflection function has a large region of nearly zero slope, which is consistent with the much broader peak observed in the angular distribution, and the position of this 40° wide peak is consistent with the region of minimum slope. At small impact parameters, the Bi deflection function is quite different from predictions of classical models, displaying fluctuations arising from single particle dynamics. Both because of limitations of the mean-field theory and the small number of calculated impact parameters, no attempt has been made to extract differential cross sections.

The deviation of the mean particle number in the Kr-like fragment from that of ^{84}Kr, $\Delta A = \bar{A}-84$, and the dispersion in particle number of the Kr-like fragment, $\Gamma_A = [\overline{A^2} - (\bar{A})^2]^{1/2}$ are shown in Fig. 4-9. For peripheral impact parameters, ΔA approaches zero as expected. In the deep inelastic region $150\hbar<\ell<260\hbar$, in which θ_{cm} fluctuates several times through 50°, $\bar{A}$ fluctuates around the experimentally observed value[15] at 50° of $\bar{A}$=84. At very low impact parameters, $\bar{A}$ increases significantly, so that A definitely increases for very large θ_{cm}. At θ_{cm}=85°, $\bar{A}$ has increased to roughly 90 which is in the right direction but of insufficient magnitude to agree with the experimental value[31] of $\bar{A}$=110. The value of Γ_A ranges from 0 to 5 and is always much less than the observed width of roughly 50.

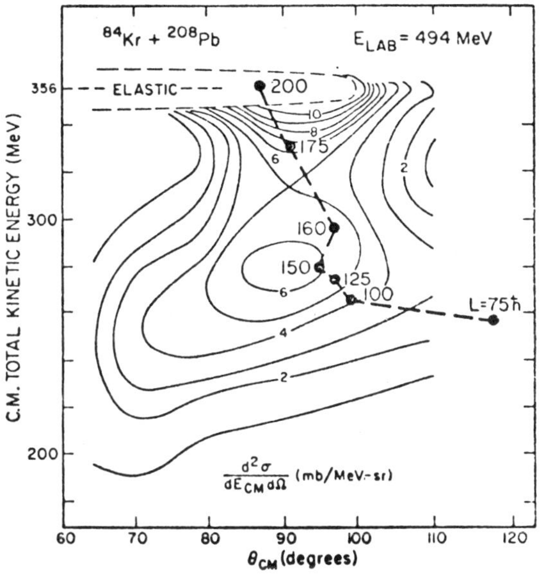

(a)

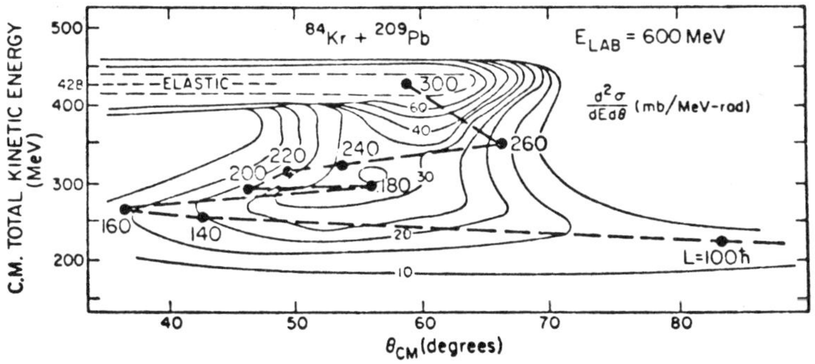

(b)

Fig. 4-8. Comparison of calculated and experimental Wilczynski plots for $^{84}Kr+^{208}Pb$ at E_{lab}=494 MeV (a) and for $^{84}Kr+^{209}Bi$ at E_{lab}=600 MeV (b). Calculated points are labeled by initial orbital angular momentum.

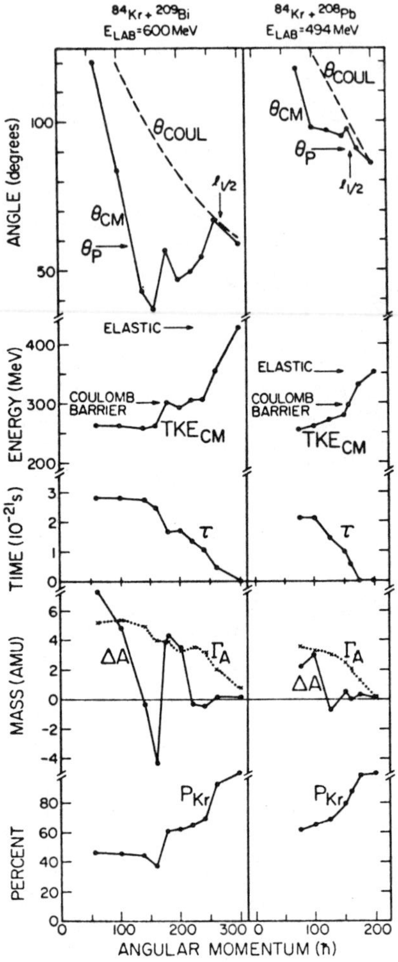

Fig. 4-9. Summary of TDHF results for Kr-induced collisions as functions of initial orbital angular momentum. The final c.m. scattering angle and total kinetic energy are denoted θ_{cm} and TKE_{cm}; τ is the contact time; A is the net mass change of the light fragment; Γ_A is the full width at half maximum of the mass distribution; and P_{Kr} is the percentage of orbitals remaining in the light fragment after the collision. Also shown on the θ_{cm} graph are θ_{coul}, the Rutherford deflection function; $\ell_{1/2}$, the angular momentum for which the optical model transmission coefficient falls to 1/2; and θ_p, the angle of the cross section peak.

The contact time, τ, is defined as the interval during which the density in the overlap region between the nuclei exceeds half nuclear matter density and is also plotted in Fig. 4.9. One observes that the energy loss is an increasing function of τ and Γ_A^2 is proportional to τ in qualitative agreement with diffusion models[17] but with much too small a coefficient. In addition, the fluctuation in angle at $\ell=180\hbar$ for the Bi collision observed in both Figs. 4.8 and 4.9 is clearly related to the unusually short contact time at this impact parameter: since the fragments separate prematurely, the Kr-like fragment is not carried around as far toward the forward direction and thus has an anomalously large scattering angle.

Two aspects of the single-particle dynamics in this mean-field theory calculation merit discussion. Although the theory is independent of the single-particle basis, it is instructive, as in Section 3.1.4., to follow the evolution of the single-particle orbitals which began in the projectile. The probability that these orbitals remain in the Kr-like fragment, P_{Kr}, is defined as the fraction of the total number of particles in the final light fragment arising from these single-particle wave functions and is graphed in Fig. 4.9. Although P_{Kr} approaches unity for peripheral collisions, the mixing of single-particle orbitals is quite dramatic at smaller impact parameters. From the viewpoint of single-particle dynamics, the compound system is comprised of two connected cavities, between which single-particle orbitals pass freely at velocities of the order of the fermi velocity. The shape of the cavities and neck region, however, is governed by the mean field which evolves much more slowly than the fermi velocity.

Given this pronounced single-particle behavior, it is not surprising that fluctuations arise in the contaft time and thus in the deflection function. Defining D as the distance along the symmetry axis between the half-density points of the outer surfaces and R as the radius of the half-density point in the minimum of the neck, one observes that the stretching and scission of the compound system have the conspicuous impact-parameter dependence shown in Fig. 4.10. The plotted portions of the L-D trajectories in Fig. 4.10b correspond only to the elongation from the most compact shape, so D is a monotonically increasing function of time. For the most compact shapes, R systematically increases with decreasing ℓ as might be expected from macroscopic considerations. At D=25 fm, an abrupt cross-over occurs, with the neck radius decreasing quite suddenly for $\ell=180\hbar$ and also less rapidly for $160\hbar$. The conspicuous contraction of the neck at $180\hbar$ is also evident from the half-density contours plotted in Fig. 4.10a at time intervals differing by $3.3x10^{-22}$ sec. The fact that the contact time in Fig. 4.9 actually decreases as ℓ is decreased from 200$\hbar$ to 180$\hbar$ seems plausibly explained by the impetus toward early scission induced by this neck behavior. In addition to emphasizing shell effects, the contour plots in Fig. 4.10a display interesting trends in the average macroscopic shape of the system at different angular momenta.

Thus, the time-dependent mean-field theory successfully accounts for three essential features of the Kr-induced reactions. Side-focussing at the appropriate angles, strong damping with the observed amount of energy dissipation, and small values of the average mass transfer are all predicted without any free parameters. Whereas single-particle effects have been demonstrated to be important, they are not quantitatively reliable in the present calculation because of the assumption of axial symmetry and the neglect of the spin-orbit force. The primary failure is in the width of the fragment mass distribution, Γ_A, which is an order of magnitude smaller than experiment. Whether this discrepancy arises from the imposition of axial symmetry or requires refinements beyond the mean field-approximation remains an important question for future investigations.

4.4 Fission

Spontaneous fission and induced fission near threshold provide interesting tests of the mean-field theory because they involve initial conditions very different from those which arise in collisions. The only viable approach to spontaneous fission appears to be the adiabatic TDHF theory described in Section I. After selecting the relevant set of collective coordinates to describe fission, one would first calculate the potential energy and mass parameters, and then solve the tunneling problem for the resulting multi-dimensional Schroedinger equation. Similarly, for induced fission, one in principle would solve the same Schroedinger equation to describe the excited nucleus rattling around statistically until it finally happens to gather most of its excitation energy in the fission degree of freedom, progresses up the fission barrier, passes slowly along a path close to the saddle point, and then begins its descent to scission. Unfortunately, both of these adiabatic TDHF calculation are beyond the purview of present capabilities.

For induced fission near threshold, however, we know that the appropriate initial condition corresponds to some wave packet of shapes which pass relatively close to the saddle configuration with low collective velocity. A first approximation would then be to release a constrained static HF solution

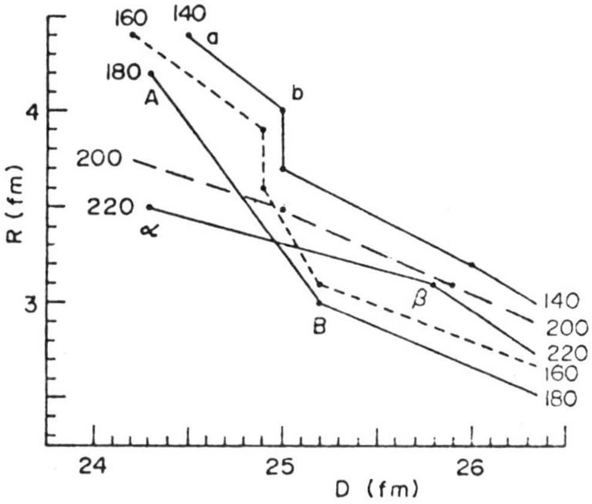

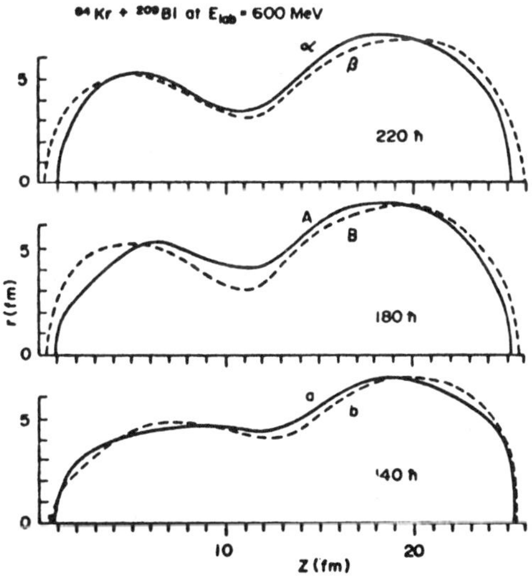

Fig. 4-10. Neck configurations for the $^{84}Kr+^{209}Bi$ compound system. The half-density contours in (a) show the density evolution in $3.3x10^{-22}$ sec. In (b), D denotes the maximum length of the half-density contour and R the minimum width in the neck region.

from slightly beyond the saddle point. Since the saddle point is rigorously independent of the form of the constraint, the initial condition is essentially unique.

Unfortunately, if one begins with an axially symmetric saddle configuration, the axial symmetry is maintained in time as discussed in Section 4.2.1. Since the hamiltonian does not connect different angular momentum projection and parity subspaces, in contrast to the case at finite axial symmetry, level crossing occurs as sketched in Fig. 4-11. Hence, an axially symmetric initial condition slightly beyond the saddle point would become trapped forever in a concave upward well. In the realistic case of a wave packet of initial configurations some arbitrarily small asymmetry would always resolve the problem . For a constrained axially symmetric problem, however, some other artifice is required.

One natural way out is to include pairing. The natural generalization of the mean-field approximation is then the Hartree-Fock-Bogoliubov (HFB) approximation. Following the notation of Valatin,[22] the dynamical equations may be expressed in terms of matrices of twice the dimension of the density matrix. In addition to the density matrix ρ and the HF potential W, the system is specified by the pairing field and pairing potential matrices

$$\chi(r,r') = \langle\psi(r)\psi(r')\rangle \tag{4.40a}$$

and

$$U(r,r') = \frac{1}{2}\iint V(r,r';r'',r''')\chi(r'',r''')dr''dr''' \tag{4.40b}$$

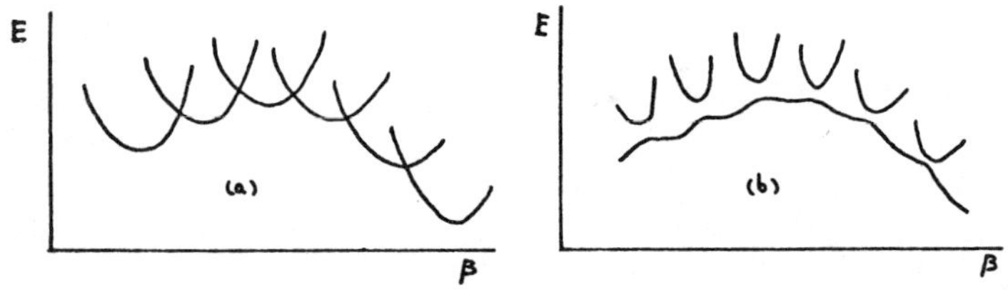

Fig. 4-11. Sketch of the energy as a function of deformation for an axially symmetric determinant (a) and with axial asymmetry and/or pairing (b).

Defining the augmented matrices

$$K = \begin{vmatrix} \rho & \chi \\ -\chi^* & 1-\rho^* \end{vmatrix} \tag{4.41a}$$

and

$$M = \begin{vmatrix} h-\lambda & u \\ -u^* & -(h-\lambda)^* \end{vmatrix} \tag{4.41b}$$

where the HF hamiltonian is given by Eq. (1.13) and the chemical potential λ specifies the mean particle number, the equation of motion is

$$i\,\dot{K} = [M, K] \tag{4.42}$$

Although $K^2=K$ in analogy with $\rho^2=\rho$, TrK is infinite, so that we cannot simply evolve a finite number of eigenfunctions of K as in the case of TDHF.

Since we turn to pairing only as an artifice to accomplish what should actually be performed by asymmetry in the mean field, it is sensible to turn to a much simpler theory than HFB, namely the time-dependent constant gap approximation.[23]

The equations of motion derived from applying the time dependent variational principle to a BCS wave function with the assumption of a constant gap are derived in Problem 3-6. The single particle wave functions evolve according to

$$i\dot{\phi}_\ell = (h - \varepsilon_\ell)\phi \tag{4.43}$$

and the u_ℓ's and v_ℓ's are determined by

$$i(v_\ell^2) = \Delta[(u_\ell v_\ell) - (u_\ell v_\ell)^*] \tag{4.44}$$

$$i(u_\ell v_\ell) = 2u_\ell v_\ell(\varepsilon_\ell - \lambda) + \Delta(2v_\ell^2 - 1) \tag{4.45}$$

where

$$\varepsilon_\ell = \int \phi_\ell^* h\ \phi_\ell \tag{4.46}$$

and the lagrange multiplier λ is defined to conserve particle number

$$\lambda = [\Delta N + \sum_{\ell>0} (\varepsilon_\ell \{u_\ell v_\ell + u_\ell v_\ell^* \} - \Delta)] \sum_{\ell>0} (u_\ell v_\ell + u_\ell v_\ell^*) \qquad (4.47)$$

The sequence of densities obtained with Δ=2.0 MeV is presented in Fig. 4-12. Particularly salient features are the fact that the time scale is rapid, essentially an order of magnitude faster than with the classical one-body dissipation formula[24] and that the neck is much more elongated than with one-body dissipation. Shapes close to the scission point for Δ=6.0 and Δ=0.7 MeV are shown in Fig. 4-13. In the space employed in this calculation, Δ=0.7,2, and 6 MeV correspond to two-body matrix elements $G \approx 0.12$, 0.17 and 0.29 MeV respectively. For comparison, 23/A yields $G \approx 0.1$ and realistic reaction matrix elements yield values ranging from 0 to 0.4 MeV in this region of the periodic table.[25]

Because of the phenomenological nature of Δ, the present calculation does not provide a definitive test of the mean-field theory. However, the results do appear both plausible and encouraging. Because of the neglect of axial asymmetry, the effective matrix element should be larger than specified by the residual interaction, and a factor of two is not unreasonable. Averaging over a wave packet of asymmetric initial conditions roughly corresponds to averaging over Δ with an appropriate weighting function. Occasionally, for almost symmetric initial conditions, the effective Δ should be very small, leading to an α particle from the neck which is observed experimentally roughly one time in 600. The particle number dispersion in the fragments has not yet been calculated, but provides another possible test of the theory. Thus, fission provides a rich opportunity for future investigations.

Unfortunately, dissipation is completely dominated by the strength of Δ. Compared to the experimental total fragment kinetic energy of 168±4 MeV, Δ = 2 MeV yields a final energy of 142 MeV, with the extra 26 MeV of dissipation arising from probability becoming trapped in highly excited orbitals. Only the very strong gap strength Δ = 6 MeV agrees with experiment, yielding final kinetic energy 166 MeV.

4.5 Pion Condensation

That the phenomenon generally referred to as "pion condensation" may be described in the time-dependent mean-field approximation is seen in two ways. One method is simply to apply mean-field approximations to a field theory containing fermions and meson fields.[26,27] The lowest order approximation immediately yields a direct Hartree term with a corresponding

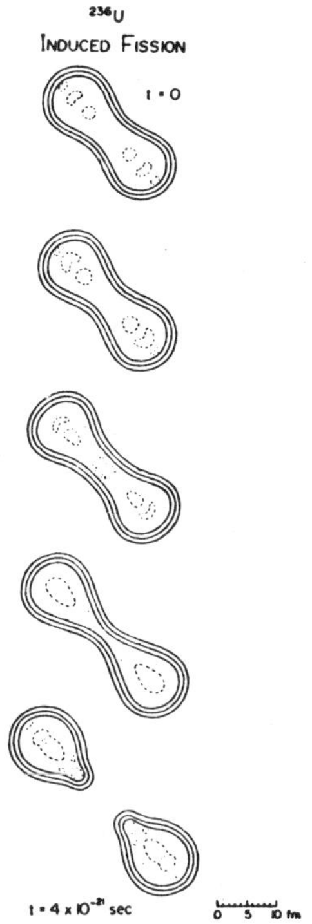

Fig. 4-12. Contour plots of the nuclear density at evenly spaced time intervals during induced fission. The outermost three solid lines denote densities of 0.02, 0.08, and 0.14 fm^{-3} and thus display the shape and extent of the surface. Interior density fluctuations are shown by dashed and solid lines which denote densities of 0.16 and 0.14 fm^{-3}, respectively.

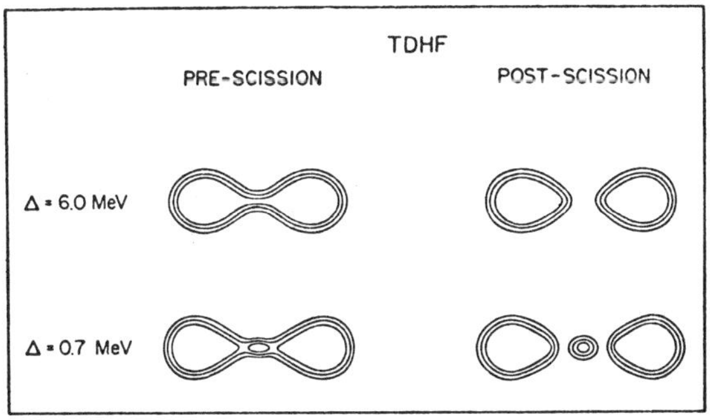

Fig. 4-13. Scission configurations for two values of gap Δ. Contours are defined as in Fig. 4-12.

Yukawa potential, and second order perturbation theory even generates the familiar exchange term.[26] A second method is to directly consider the pion propagator in the nuclear medium. Since the onset of condensation occurs at zero frequency, a static reduction of each bare pion propagator immediately yields a static one-pion exchange potential (OPEP). Given that the pion propagator is generally only calculated in the random phase approximation[28,29] and TDHF is equivalent to RPA in the infinitesimal limit, the onset of condensation should be identically described in the mean-field theory. Fig. 4-14 shows a typical set of graphs which are thereby included.

The great advantage of the mean-field approximation, of course, is that it allows one to investigate the onset of condensation in a realistic geometry. Since the same theory describes both the evolution of the nuclear density and the buildup of condensation, one may meaningfully address also questions of whether the high density overlap region lasts long enough during collisions for significant growth of a condensate, how such growth is affected by the finite nuclear size, and what the observable consequences are in the laboratory. For example, if one is interested in the pion production arising from such collisions, a simple quasi-elastic scattering calculation of the off-mass shell pions from normally occupied states in the compound nucleus yields the on-mass-shell pion spectrum and angular distribution as sketched in the box in Fig. 4-14.

Once one adopts the language of the mean-field theory, the physics of pion condensation is seen to be much less fancy than the name. One is simply exploring spin-isospin modes which become unstable at sufficiently high density, leading to growing fluctuations in the spin and isospin densities. In nuclear matter, this mode is easily visualized in terms of the alternating larger structure[27] shown in Fig. 4-15.

The driving term for this spin-isospin correlation is the direct contribution of the tensor OPEP potential

$$V_T = \frac{g^2}{12\pi} \tau_1 \cdot \tau_2 \, S_{12} \frac{e^{-r_{12}}}{r_{12}} \tag{4.48}$$

where

$$S_{12} = 3(\vec{\sigma}_i \cdot \hat{r}_{12})(\vec{\sigma}_2 \cdot \hat{r}_{12}) - (\vec{\sigma}_1 \cdot \vec{\sigma}_2) \tag{4.49}$$

Consider matrix elements between two spin aligned nuclei. Since

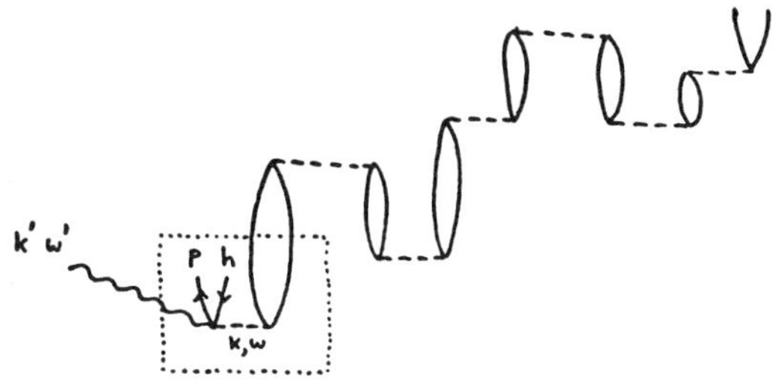

Fig. 4-14. Pion propagator in the static RPA approximation.

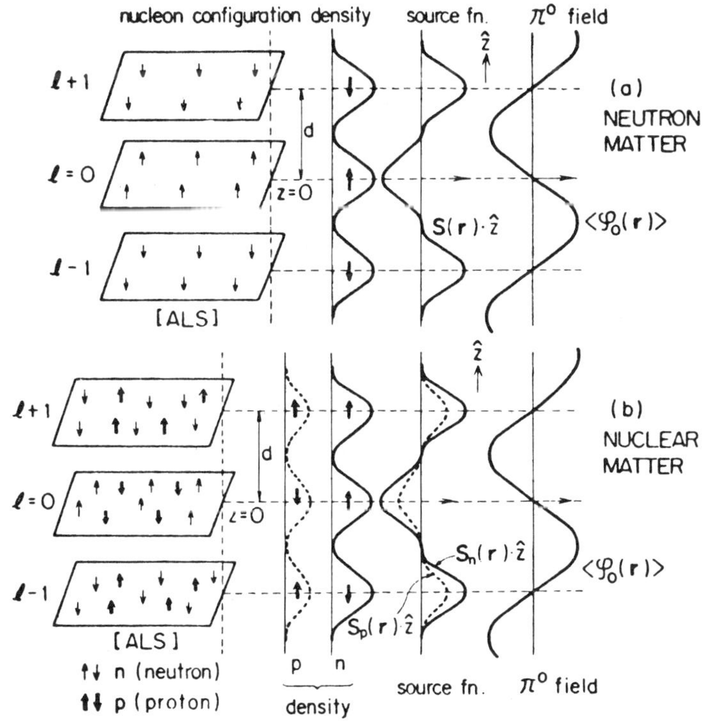

Fig. 4-15. Profile of alternating larger structure of neutron, and nuclear matter, nucleon density, $<\sigma\tau_z>$, and π^0 field.

$$\langle\chi_\uparrow(1)\chi_\uparrow(2)|S_{12}|\chi_\uparrow(1)\chi_\uparrow(2)\rangle = 3\cos^2\theta - 1 \tag{4.50}$$

where θ is the angle of r_{12} relative to the spin quantization axis and for isotriplet nucleons the coefficient of $\langle S_{12}\rangle$ in (4.48) is repulsive, the most energetically favorable orientation is $\theta \sim \frac{\pi}{2}$. Thus, like nucleons prefer to orient themselves so that all spins are aligned in a plane. For anti-aligned spins, the sign in Eq. (4.50) is reversed, so that the most favorable angle becomes $\theta=0$. Hence, successive planes of like nucleons prefer to be anti-aligned, as shown in the upper portion of Fig. 4.15 for pure neutron matter. Considering both neutrons and protons, the strong S-D component of the tensor force implies that T=0 contributions dominate T=1 terms, so that all the signs in the previous arguments are reversed. Thus, neutrons and protons within a plane want to be anti-aligned, as shown in the bottom graph in Fig. 4-15. Then because neutron and proton spins alternate between layers, opposite isospins in adjacent layers automatically have the same spin.

To undertake an explicit calculation of π_0 condensation the effective Hamiltonian density $\mathcal{H}_0$ of Eq. (4.3) must be augmented to include additional terms

$$\mathcal{H} = \mathcal{H}_0 + \frac{\beta}{2}\int d^3r\,(\vec{S}_p + \vec{S}_n)^2 + \frac{\gamma}{2}\int d^3r\,(\vec{S}_p - \vec{S}_n)^2 \cdot$$

$$+ \frac{V_0}{2}\int d^3r\int d^3r'\,[\vec{\nabla}_{(1)}(S_p(r_1)-S_n(r_1))]\frac{e^{-|r_{12}|a_0}}{r_{12}}[\vec{\nabla}_{(2)}(S_p(r_2)-S_n(r_2))] \tag{4.51}$$

where

$$\vec{S}_q = \langle\sigma\rangle = \sum_{\nu ss'} \psi_\nu^{sq}\,\vec{\sigma}_{ss'}\,\psi^{s'q}$$

The parameter β is adjusted to give a sensible $\sigma_1\cdot\sigma_2$ component in the effective interaction and γ and V_0 are adjusted to produce optimistic but plausible conditions for the onset of pion condensation, namely a critical density at $1.5\rho_0$ and critical wavelength $k_c=2\text{fm}^{-1}$. Variation of (4.51) yields additional terms in the HF potential

$$W^{HF} = W_0 + \beta\langle\sigma\rangle\cdot\vec{\sigma} + \gamma\langle\sigma\tau_3\rangle\cdot\sigma\tau_3$$

$$+ V_0\int d^3r_2\,\frac{e^{-r_{12}/a_0}}{r_{12}}\,\nabla\cdot\langle\sigma\tau_3\rangle \tag{4.52}$$

The last term is the most interesting. The source term for the creation of the pion field is $\nabla\cdot\langle\sigma\tau_3\rangle$, and the pion field is then just the convolution of this source function with a Yukawa. The quantity $\langle\sigma\tau_3\rangle$ and the Yukawa convolution of $\nabla\cdot\langle\sigma\tau_3\rangle$ are shown in Fig. 4-15 for nuclear matter.

Preliminary results from a schematic calculation of $^{15}N+^{15}N$ at 50 MeV/A are shown in Fig. 4-16. For π^+ and π^- condensation, each single-particle wave function would have four spin-isospin components, but for the simpler case of π^o condensation, proton and neutron wave functions are not mixed. Because spin symmetry is preserved by the mean field, it is essential to select a spin non-saturated initial condition, such as ^{15}N. The pion field energy, E_π, corresponding to the last term in Eq. (4.51) is shown in Fig. 4-16a. It increases by about 50% as the nuclei begin to overlap, and then begins a significant fluctuation. The source term $\nabla\cdot\langle\sigma\tau_3\rangle$ for the ground state of ^{15}N is sketched in Fig. 4-16b. The unfilled neutron single-particle orbital peaks in the z=0 plane at $r\sim 1.6$ fm with spin oriented along the z axis. Thus, the derivative with respect to z gives the simple structure shown.

The collision was arranged such that the pion source term was symmetric about the c.m. Fig. 4-16c shows $\nabla\cdot\langle\sigma\tau_3\rangle$ when the positive lobes are maximally overlapping, which obviously gives rise to the increase in E at t_1 in Fig. 4-16a. The time t_2 is selected such that the two nuclei would have completely overlapped if they had propagated freely. If target and projectile were simply superimposed with no non-linear evolution, the source term (and hence pion field) would vanish identically. As seen in Fig. 4-16d, this is far from the case. A self-sustaining mode has been set up, which even has a radial node building up corresponding to propagation of the condensate in the transverse direction.[28]

The present calculation is schematic at best. One needs to think much harder about the effective interaction, worry about including or adequately mocking up the effect of isobars, and treat much larger nuclei. But it is clear that the mean field has something useful to say about one of the most fascinating questions arising in high energy heavy ion collisions.

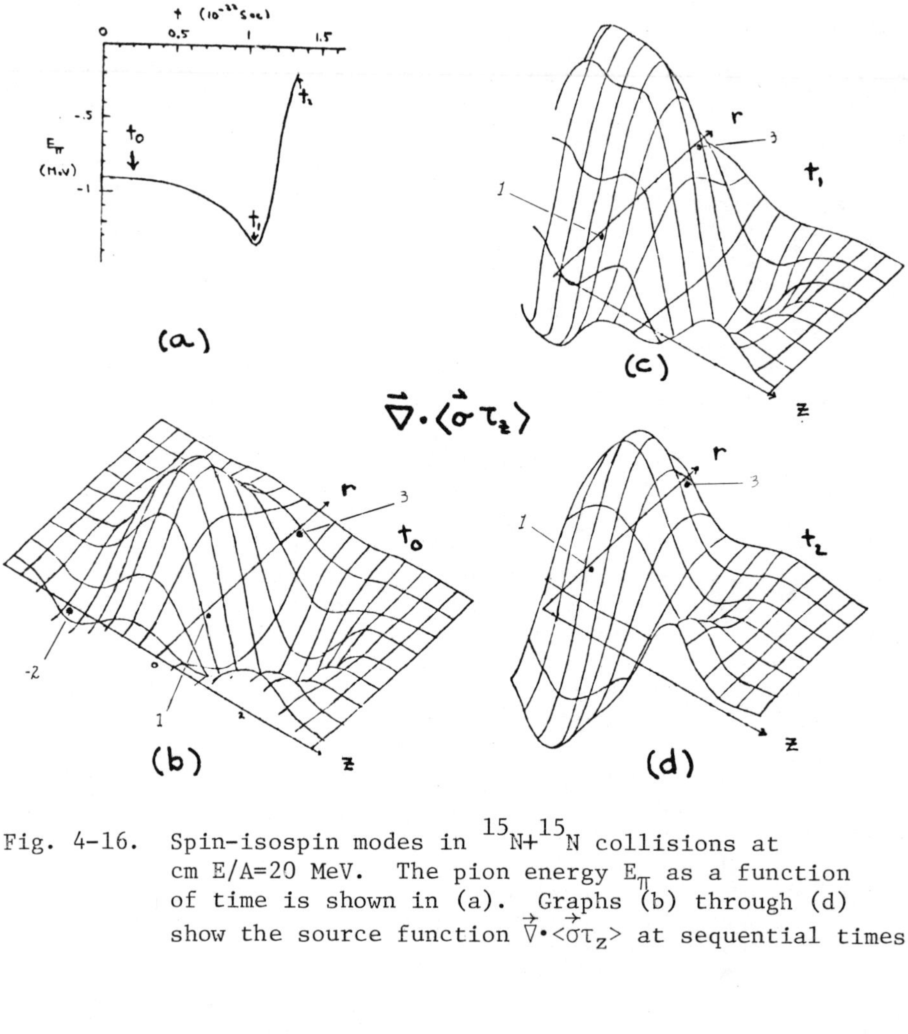

Fig. 4-16. Spin-isospin modes in $^{15}N+^{15}N$ collisions at cm E/A=20 MeV. The pion energy E_π as a function of time is shown in (a). Graphs (b) through (d) show the source function $\vec{\nabla}\cdot\langle\vec{\sigma}\tau_z\rangle$ at sequential times.

PROBLEMS

4-1. Derive Eqs. (4.12) and (4.13).

4-2. Using the differentiation formula

$$f'(x_o+p\Delta) = \frac{1}{\Delta}\{-\frac{3p^2-6p+2}{6} f_{-1} + \frac{3p^2-4p-1}{2} f_o - \frac{3p^2-2p-2}{2} f_o + \frac{3p^2-1}{6} f_2\} + \theta(\Delta^4)$$

obtain a three-point formula for $f'(x_o+\sqrt{3}\Delta)$ valid to order Δ^4. Then vary the expression

$$\sum_i \Delta|\nabla\phi(x_i+\sqrt{3}\Delta)|^2$$

to obtain the difference equation

$$\frac{1}{12\Delta^2}[-\phi_{i-2} + 16\phi_{i-1} - 30\phi_i + 16\phi_{i+1} - \phi_{1+2}] + u_i\phi_i = \varepsilon\phi_i$$

Note that the kinetic term uses the usual 5 point difference formula for ϕ'' with error of $\theta\Delta^4$.

4-3. Prove that the eigenvalues of $(1-i\Delta+h)$ are greater than or equal to 1.

Demonstrate that errors in the difference equations (4.16) grow exponentially by writing an explicit Fourier series solution to (4.16) in the case of constant (or zero) potential.

4-4. Derive Eqs. (4.23) through (4.31) in detail.

4-5. Using the time-dependent variational principle, derive the equation of motion Eqs. (4.35)-(4.37) for a separable wave function with fixed harmonic oscillator transverse components.

4-6. Derive the constant gap pairing equations, (4.43) through (4.47) from applying the time-dependent variational principle to a BCS wave function assuming constant gap.

4-7. Derive Eq. (4.50) and fill in the details of the argument for the alternating layer structure in nuclear matter.

REFERENCES FOR CHAPTER 4

1. J. W. Negele and G. A. Rinker, Phys. Rev. C15, 1499 (1977).

2. P. Hoodbhoy and J. W. Negele, Nucl. Phys. A288, 23 (1977).

3. J. W. Negele and D. Vantherin, Phys. Rev. C5, 1472 (1973).

4. S. Koonin, J. W. Negele, P. Møller, J. R. Nix, A. Sierk, and H. Flocard, Phys. Rev. C17, 1098 (1978).

5. P. Bonche, S. Koonin, and J. W. Negele, Phys. Rev. C13, 1226 (1976).

6. S. E. Koonin, K.T.R. Davies, V. Maruhn-Rezwani, H. Feldmeier, S. J. Krieger, and J. W. Negele, Phys. Rev. C15, 1359 (1977).

7. H. Flocard, S. E. Koonin, and M. S. Weiss, Phys. Rev. C17, 1682 (1978).

8. R. D. Richtmeyer and K. W. Morton, Difference Methods for Initial Value Problems, Wiley, N.Y. (1967).

9. R. S. Varga, Matrix Iterative Analysis, Prentice Hall, Englewood Cliffs (1962).

10. K. T. R. Davies, H. T. Feldmeier, M. S. Weiss, and H. Flocard, ORNL Preprint, 1978.

11. H. T. Feldmeier, Proceedings of Topical Conference on Heavy Ion Collisions, Fall Creek Falls, ORNL Rept Conf.-770602 (1977).

12. A. K. Kerman and S. E. Koonin, Ann. Phys. (N.Y.) 100, 332 (1976).

13. K. T. R. Davies, V. Maruhn-Rezwani, S. E. Koonin, and J. W. Negele, Cal. Tech. Preprint (1978).

14. K. R. Sandhya Devi and M. R. Strayer, Phys. Lett. B, in press.

15. S. Koonin, B. Flanders, H. Flocard and M. Weiss, Cal. Tech. Preprint LAP-162.

16. A. K. Kerman, private communication.

17. A thorough review of the general features and experimental data is presented by W. U. Schroder and J. R. Huizenga, Ann. Rev. Nucl. Sci. 27; 465 (1977).

18. B. Fernandez, C. Gaarde, J. Larson, S. Sorensen, and F. Videbaek (unpublished).

19. M. Conjeaud et al., Proceedings of International Conference on Nuclear Structure, Tokyo (1977).

20. R. Vandenbosch, M. P. Webb, and T. D. Thomas, Phys. Rev. C14, 143 (1976).

21. K. L. Wolf, J. P. Unik, J. R. Huizenga, J. Birkelund, H. Freieslebon, and V. F. Viola, Phys. Rev. Lett. 33 (1974) 1105.

22. J. G. Valatin, in Lectures in Theoretical Physics, Vol. IV, ed. W. E. Brittin (Interscience, N.Y.) (1962).

23. J. Blocki and H. Flocard, Nucl. Phys. A273, 45 (1976).

24. K. T. R. Davies, R. A. Managan, J. R. Nix, and A. J. Sierk, Phys. Rev. C16, 1890 (1977).

25. J. W. Negele, Nucl. Phys. A142, 225 (1970).

26. M. Bolsterli, Advances in Nuclear Physics, ed. J. W. Negele on E. Vogt, Vol. 11 in press (1978).

27. T. Takatsuka, K. Tamiya, T. Tatsumi, and R. Tamagaki, Kyoto preprint (1978).

28. Miklos Gyulassy, Proceedings of Symposium on Relativistic Heavy Ion Collisions, Darmstadt (1978); M. Gyulassy and W. Greiner, Ann. Phys. 109, 485 (1977).

29. G. E. Brown and W. Weise, Phys. Reports 27C, 1 (1976).

V. CONCLUSIONS

These lectures have by no means been complete or comprehensive. Rather they represent one personal and undoubtedly biased view of how to think about nuclear dynamics and what is interesting and important in the field. On balance, the outlook is highly encouraging. The mean-field theory appears to be a well motivated and successful first approximation for a variety of problems of interest. The coupled cluster hierarchy seems to provide a useful framework for systematic corrections.

Of the outstanding problems, the most fundamental one is the fact that we still don't really know the best way to think about systems with large numbers of degrees of freedom. Instead of evolving $(A_1+A_2)N_xN_yN_z$ numbers, it would be far more instructive to reformulate the mean-field approximation in terms of a smaller number of the most relevant variables. We need viable approximations to TDHF which clearly separate collective, statistical, and single particle effects. To the extent to which a collection of single particle orbitals bouncing around in a well is purely statistical, TDHF codes are the world's most expensive random number generators. Thus, a clean separation of statistical aspects of the problem would be most welcome. Several notes of caution, however, appear warranted. What is required is a theoretically sound, systematic reduction of the TDHF theory. Thus, unmotivated Ansätze like the time dependent Thomas-Fermi approximation are not likely to be very instructive. Furthermore, we have demonstrated that in many cases, TDHF is not statistical during the relevant time scale, so at the least, some viable criteria and restrictions for application of statistical approximations are required. One possibility under current investigation is only approximately enforcing the determinantal constraint $\rho^2=\rho$ in a formulation which deals only with quantities which depend on nearly diagonal behavior of the density matrix, like ρ, τ, J and S, since at least in large systems, they should contain the most essential physical information. In higher orders of the coupled-cluster expansion, insight into accurate, simplifying approximations is even more crucial.

One must eventually come to grips with quantitatively delineating the region of validity of the mean-field theory. Although the coupled-cluster theory in principle provides a framework for calculating or estimating errors, at present we must resort to more intuitive arguments. From the role of the Pauli principle in generating a large mean free path, it seems that the mean-field theory should be particularly good for low energy collective modes, collisions a few MeV per particle above the Coulomb barrier, and for fission. Other interesting regimes of validity, however, also seem possible. Whenever the process may be specified by the action of an external potential, the mean-field theory should provide an excellent description of linear and non-linear response. Thus, peripheral and grazing collisions at arbitrarily high energies appear to fall within the purview of the theory, with the fragmentation of peripheral relativistic heavy ion collisions being a particularly interesting possibility. Other applications might arise when one can isolate some regions of a system in which particles have low relative momentum, such as the coalescence of final fragments from an excited intermediate system.

Experimentally, the most crucial question is to think of observables corresponding to expectation values of few body operators which can definitely distinguish the mean-field theory from other approximations. It is one thing for theorists in their fairy tale world to draw pictures showing how different intermediate density shapes look in TDHF and hydrodynamics, but quite another to design an experiment which can discriminate between the two cases. The key features to consider in comparing the mean field theory with other approximations are the relatively small transverse momentum transfer, the dominance of single particle effects in appropriate regimes, and the lack of complete equilibration during the interaction time.

Particularly interesting possibilities arising from the work thus far include resonances in fusion cross-sections and the prediction of a transmission window for nearly head-on collisions.

Thus, although the mean field approximation and the coupled-cluster hierarchy show genuine promise for providing a fundamental microscopic understanding of dynamics, many exciting and challenging problems remain. These challenges provide the ultimate motivation for the presentation of these lectures.

ACKNOWLEDGEMENTS

The work described in these lectures is the outgrowth of collaborations with a number of colleagues. The coupled-cluster formulation arose from collaborations with Pervez Hoodbhoy, and the slab calculations were primarily the work of Paul Bonche. Tom Davies and collaborators at Oak Ridge bore the major computational burden of the axially symmetric collision calculations and Siegfried Krewald was responsible for the pion condensation calculation. Steve Koonin played a crucial role in virtually every phase of the development and application of the mean-field theory. A number of results with which I was not at all involved were cited, hopefully with proper credit. In addition, this work has benefitted greatly from stimulation from numerous colleagues. Conversations, suggestions and encouragement from Michel Baranger, Ernie Moniz, Felix Villars, and especially Arthur Kerman have been extremely important in all phases of this work. Also, significant interactions with colleagues outside MIT have been quite fruitful including, among others, George Bertsch, Fritz Coester, Ray Nix, Mike Strayer, and John Zabolitzky.

TRANSPORT THEORY OF DEEPLY INELASTIC HEAVY-ION REACTIONS+

Hans A. Weidenmüller*

Department of Theoretical Physics
Oxford University
Oxford, OX1 3NP, England

ABSTRACT

The term transport theory is explained, and examples of transport equations are given. The use of such equations in the analysis of compound processes and, more recently, heavy ion collisions, is described. Conditions for the applicability of transport theory are discussed in terms of the time scales involved. Nuclear transport theory is formulated as a method of describing damped large-scale collective nuclear motion at high temperatures. Various recent theoretical approaches are presented in the light of these general considerations, and their results and the open problems are discussed.

1. Introduction

Macroscopic physical systems tend towards thermodynamic equilibrium. The processes through which equilibrium is reached change various forms of energy (kinetic, electromagnetic, potential...) irreversibly into heat and lead to a monotonic, irreversible increase of entropy. They are described by equations which are not time-reversal invariant. The theory of such processes forms the fields of nonequilibrium thermodynamics and nonequilibrium statistical mechanics. The latter can be formulated both within classical mechanics, and within quantum mechanics. From a systematic point of view, the present lectures deal with a class of problems in nonequilibrium quantum statistical mechanics.

+Lectures delivered at the NATO Advanced Studies Institute on Theoretical Methods in Medium-Energy and Heavy-Ion Physics, Madison, Wisconsin, June 12-23, 1978.

*On leave from Max-Planck Institut fur Kernphysik, Heidelberg, W. Germany.

Transport theory aims at setting up equations which describe the evolution towards equilibrium, and at calculating the coefficients in these equations which describe the time irreversible flow of energy, angular momentum, mass, momentum,... Such coefficients are called transport coefficients. Examples for transport coefficients are friction coefficients, thermal and electrical conductivity, viscosity coefficients, diffusion constants, etc.

In recent years, equilibration phenomena have increasingly been observed and analyzed in the nuclear domain. Equilibration of the nuclear system through formation of a compound nucleus is the earliest example, followed historically by the discovery of preequilibrium emission of light fragments and, later, the deeply inelastic heavy ion reactions. These developments led to the formulation of phenomenological models for nuclear equilibration processes, followed by attempts to put such models on a microscopic basis. These attempts constitute an extension of quantum statistical mechanics to the nuclear domain. They pose new problems not encountered before and will challenge nuclear physicists for a considerable time to come.

The lectures consist of three parts. In Part I, we recall the evidence for transport processes in nuclei and discuss typical transport equations used in their analysis. In Part II we argue that certain conditions on time scales must be met to make a description of a system in terms of transport equations physically viable, and formulate the problem of nuclear transport theory in general terms. In Part III, we describe the approaches taken by various groups of investigators to the nuclear problem, some typical results obtained, and the open problems.

In the presentation, I attempt to explain and relate the concepts employed in the developments of various theories, rather than give a detailed account of the formal methods used. Lack of space does not permit me to go into full details of either the various theoretical approaches or the phenomenological analyses of the data.

I. EXAMPLES OF TRANSPORT EQUATIONS[1]

2. Precompound Reactions

The simplest (and oldest) example of a quantum-statistical transport equation is furnished by the master equation,

$$\frac{d}{dt} P_s(t) = -\left(\sum_e W_{s\to e}\right) P_s(t) + \sum_e W_{e\to s} P_e(t) . \qquad (2.1)$$

It was introduced by Wolfgang Pauli in 1928. In Eq. (2.1), s labels a group of dense-lying quantum mechanical states, $P_s(t)$ is the occupation probability of this group as a function of time t, and $W_{s\to e}$ is the transition probability per unit time from the group labelled s to the group labelled e. More precisely: $P_s(t)$ is the average occupation probability of any state in the group s, averaged over the members of the group, and $W_{s\to e}$ similarly the average transition probability from any member of s to any member of e.

To have some specific physical situation in mind, let us recall that Eq. (2.1) has been used to analyze precompound reactions[2,3] initiated by light projectiles. In the simplest such reaction, where the target is doubly magic and the incident particle a nucleon, the collision proceeds through a chain of n particle-(n-1) hole states, if the residual interaction is of two-body type. It is therefore natural to identify the group s with the s particle-(s-1) hole states lying in the energy interval [E, E+dE], and $W_{s\to e}$ with the associated transition probabilities. Naturally, $W_{s\to e} \neq 0$ only if $|s-e| = 1$. Eq. (2.1) is then solved with the initial condition $P_s(0) = \delta_{ss_o}$, with s_o= 2 or 3 or 4 depending on the nature of the light fragment. The transition probabilities are calculated via the formulae

$$W_{e\to s} = |M_{es}|^2 \rho_s(E) , \qquad (2.2)$$

where $\rho_s(E)$ is the density of states s at energy E,and $|M_{es}|^2 = |M_{se}|^2$ the square of the transition matrix element averaged over the groups s and e. The density $\rho_s(E)$ is found from the Fermi gas model, and $|M_{se}|^2$ from the nucleon-nucleon cross section. Such calculations yield $P_s(t)$ and, in the manner described below Eq. (2.4), reasonable agreement with the data.

Equation (2.1) is easy to interpret: The change of occupation probability of states in group s with time is determined by the balance between transitions $\sum_e W_{e\to s} P_e(t)$ feeding the group s from any other occupied group (this we refer to as the gain term), and the transitions $-(\sum_e W_{s\to e}) P_s(t)$ depleting the group s (the loss term).

Equation (2.1) has an equilibrium solution (i.e., a solution which is stationary). It has the form $P_e^o = \rho_e(E)/(\sum_m \rho_m(E))$ if $W_{e\to s}$ has the form of Eq. (2.2). This clearly is equilibrium: All states are occupied with equal probability.

Some algebra is needed to show that whatever the initial condition at time $t = 0$, the solutions of Eq. (2.1) tend exponentially towards P_e^o as $t \to \infty$. This is not done here, but the proof is indicated below. The exercise shows that Eq. (2.1) describes a system approaching equilibrium. The largest of the time scales in the exponentials is the equilibration time T.

Equation (2.1) has several features which are characteristic of transport equations in general.

(i) It describes the process in terms of probabilities, not amplitudes. All phases of quantum-mechanical wave functions have disappeared.

(ii) It is a balance equation. The net change of $P_s(t)$ is determined by the balance between gain and loss terms.

(iii) It is not invariant under the operation $t \to -t$. Therefore, it describes an irreversible approach towards equilibrium. It is easy to see that the entropy $-\sum_e P_e(t)\, \ell n(P_e(t)/P_e^o)$ increases monotonically with time.

Properties (i) to (iii) are in striking contrast to properties of the Schrödinger equation. Loss of phase memory and irreversibility are essential features of quantum statistical mechanics as opposed to quantum mechanics.

(iv) The total occupation probability $\sum_s P_s(t)$ is conserved, i.e., independent of time, if "detailed balance" holds, expressed by Eq. (2.2) and $|M_{es}|^2 = |M_{se}|^2$.

We also note that Eq. (2.1) conserves energy, since the transitions $s \to e$ occur by definition only between states of the same total energy. The coefficients $W_{s\to e}$ are transport coefficients--in this case, for the transport of probability. Eq. (2.1) is a particularly simple transport equation because it is Markoffian: The evolution in time of $P_s(t)$ is determined completely by the values of $P_e(t)$, and independent of the previous history of the system. An example of a non-Markoffian equation would be

$$\frac{d}{dt} P_s(t) = - \sum_e \int_{-\infty}^{0} W_{s\to e}(\tau)\, P_s(t+\tau)\, d\tau + \sum_e \int_{-\infty}^{0} W_{e\to s}(\tau)\, P_e(t+\tau)\, d\tau. \tag{2.3}$$

Non-Markoffian problems are much more difficult to handle. The "memory kernel" $W_{s\to e}(\tau)$ usually falls off exponentially with increasing $|\tau|$, and the difference between (2.3) and (2.1) is significant only when this falloff time is of the same order as the time Δ over which P changes significantly. The τ-dependence of W may come about, for instance, because of the finite duration of a transition $s\to e$. Since Δ increases as we decrease simultaneously the strengths of <u>all</u> the $W_{s\to e}$, Eq. (2.1) is expected to be a good approximation to Eq. (2.3) if the coupling is <u>weak</u>.

In precompound reactions, one is interested in the quantities

$$\pi_{ss_o} = \int_o^T dt\, P_s(t) \tag{2.4}$$

The index s_o is to remind us of the initial condition $P_s(0) = \delta_{ss_o}$. The product of π_{ss_o} with the decay probability of the states s into any one channel gives the precompound contribution to the cross section in that channel. This contribution is different from the Hauser-Feshbach value because the latter assumes $P_s = \rho_s/\sum_m \rho_m$. Precompound decay yields more high-energetic particles than Hauser-Feshbach because the P_s with small values of s are, for $t < T$, larger than their equilibrium values, and in configurations with a small number of particles and holes, the energy is shared among fewer constituents than assumed in the compound nucleus picture. Integration of Eq. (2.1) between 0 and T and use of $P_s(T) \cong \rho_s(E)/\sum_m \rho_m(E)$ yields for the π_{ss_o} the matrix equation

$$(-)\sum_m \{ W_{m\to s} - \delta_{ms}(\sum_e W_{s\to e})\}\ \pi_{ms_o} = (\delta_{ss_o} - \rho_s/\sum_m \rho_m) \tag{2.5}$$

This is a typical "probability balance equation": π_{ss_o} is given in terms of the inverse of the matrix in curly brackets. The diagonal elements of this matrix are the decay rates $\Gamma_s^{\downarrow}/\hbar = \sum_e W_{s\to e}$, while the nondiagonal elements are the transition rates. The approach towards equilibrium of the solutions of Eq. (2.1) can be discussed in terms of the eigenvalues of this matrix.

3. Mass Transport and the Fokker-Planck Equation

Let us write Eq. (2.1) for a situation where s is a <u>continous</u> (rather than a discrete) variable, denoted by ε, and let $\rho(\varepsilon)$ be the density of states characterized by the variable ε, so that $\sum_s \to \int d\varepsilon$. Then, we have

$$\frac{\partial}{\partial t} P(\varepsilon,t) = \int d\varepsilon'\, w(\varepsilon,\varepsilon')\ [\rho(\varepsilon)\, P(\varepsilon',t) - \rho(\varepsilon')P(\varepsilon',t) \tag{3.1}$$

where we have used Eq. (2.1) with $|M_{se}|^2 \to w(\varepsilon,\varepsilon')$. Eq. (3.1) can also be written in the form

$$\frac{\partial}{\partial t} P(\varepsilon,t) = \int d\varepsilon' \; w(\varepsilon,\varepsilon')\rho(\varepsilon)\rho(\varepsilon') \left[\frac{P(\varepsilon',t)}{\rho(\varepsilon')} - \frac{P(\varepsilon,t)}{\rho(\varepsilon)}\right] . \tag{3.2}$$

For large $|t|$, we expect $P(\varepsilon,t)$ to become proportional to $\rho(\varepsilon)$. For times t for which $P(\varepsilon,t)$ is already sufficiently close to such a distribution, the function P/ρ does not differ much from unity, and this suggests introducing the function

$$\pi(\varepsilon,t) = P(\varepsilon,t)/\rho(\varepsilon) . \tag{3.3}$$

We expect the kernel $w(\varepsilon,\varepsilon')\rho(\varepsilon)\rho(\varepsilon')$ to have a peak at or near $\varepsilon' = \varepsilon$, and to fall off with increasing $|\varepsilon-\varepsilon'|$. If this falloff is fast compared to the rate of change of $\pi(\varepsilon,t)$, it is permissible to expand $\pi(\varepsilon',t)$ in a Taylor series at ε. Keeping only terms up to second order, we find that $\pi(\varepsilon,t)$ obeys the Fokker-Planck equation

$$\frac{\partial}{\partial t} \pi(\varepsilon,t) = -\alpha_1(\varepsilon) \frac{\partial}{\partial \varepsilon} \pi(\varepsilon,t) + \frac{1}{2} \alpha_2(\varepsilon) \frac{\partial^2}{\partial \varepsilon^2} \pi(\varepsilon,t), \tag{3.4}$$

where for n = 1,2

$$\alpha_n(\varepsilon) = \int d\varepsilon' \; w(\varepsilon,\varepsilon')\rho(\varepsilon')(\varepsilon-\varepsilon')^n . \tag{3.5}$$

We discuss Eq. (3.4) under the assumption that α_1 and α_2 are independent of ε. (In the nuclear case and in situations were ε is related to excitation energy, growth of $\rho(\varepsilon')$ with ε' may invalidate this assumption as well as the earlier one pertaining to a strong peaking of the kernel. This point often requires some extra care.) We define the moments of $\pi(\varepsilon,t)$ by

$$\langle\varepsilon_n(t)\rangle = \int d\varepsilon' \; \pi(\varepsilon',t)(\varepsilon')^n / \int d\varepsilon' \; \pi(\varepsilon',t) . \tag{3.6}$$

Multiplication of Eq. (3.4) by ε^n and integration over ε yields

$$\frac{d}{dt} \langle\varepsilon_1(t)\rangle = +\alpha_1 , \qquad \frac{d}{dt} \langle[\varepsilon_2(t) - \langle\varepsilon_1(t)\rangle^2]\rangle = \alpha_2, \tag{3.7}$$

where we have used that $\int \pi(\varepsilon',t)d\varepsilon'$ is independent of time. This shows that α_1 and α_2 determine mean value and variance of the distribution and suggests for π the ansatz

$$\pi(\varepsilon,t) = \frac{1}{\sqrt{2\pi\alpha_2 t}} \exp\{-(\varepsilon-t\alpha_1)^2/2\alpha_2 t\} \quad . \tag{3.8}$$

It can be checked that this is a solution to Eq. (3.4), if α_1 and α_2 are constant. Our simple analysis suggests that even when α_1 and α_2 change smoothly with ε, the function $\pi(\varepsilon,t)$ will approach unity (i.e., equilibrium) by attaining a form which is nearly Gaussian. The further evolution is then determined by the time-dependence of the mean value, given by α_1 (which is why α_1 is called the drift coefficient), and an ever increasing spread just as in a diffusion problem (hence the name diffusioncoefficient for α_2). Clearly, Eq. (3.4) offers a much simplified picture of the equilibration process described originally by Eq. (2.1), and its use is to be preferred wherever possible.

The transition from Eq. (2.1) to Eq. (3.4) can be studied exactly for a very simple diffusion problem--that of a random walker in 1 dimension, with ε identified with the walker's position, $\rho(\varepsilon) = 1$, and $w(\varepsilon,\varepsilon')$ the probability of the random walker's making a step from ε to ε'. A nonzero drift coefficient results if the total transition probability to go right is not equal to that to go left. One finds that the Gaussian distribution is realized with very good accuracy already after a few steps. This result suggests that the vehicle used in deriving Eq. (3.4) -- the Taylor series expansion, broken off after the second term -- is useful for large times. It can indeed be checked that terms of higher order than the second give rise to terms which vanish rapidly with $t \to \infty$ in comparison with those kept in the solution (3.8), i.e., those arising from 1st and 2nd order.

Nörnberg [4] was the first to apply a transport equation to heavy-ion reactions. He argued that the mass exchange between two heavy ions might be viewed as a diffusion process, and used a Fokker-Planck equation for the analysis. Using the fact that the width of the distribution increases with the square root of time, and identifying contact time with rotation angle of the binary system, he plotted the square of the width versus rotation angle and found straight lines, supporting the idea of a diffusion process for mass. Since then, much more sophistocated analyses of mass exchange as a diffusion problem have been carried out by Nörnberg and collaborators,[5] Moretto et al.,[6] and by many others.

4. Brownian Motion, and the Dissipation of Energy and Angular Momentum in Deeply Inelastic Heavy-Ion Collisions

A Fokker-Planck equation slightly more complex than Eq. (3.4) is used in the analysis of Brownian motion. The Brownian particle is immersed in a gas, or a fluid, called the medium. If the temperature is not zero, the molecules of the medium move about in a random fashion, characterized by a Maxwellian probability distribution. The Brownian particle undergoes collisions with the molecules and consequently exchanges momentum in a random fashion with the medium. This leads to a random walk, and to a slowing down of the Brownian particle, described by a friction force, if the particle has initially a finite velocity with respect to the medium. The slowing down occurs because "head on head" collisions involve a bigger net loss of momentum than the net gain involved in "head on tail" collisions. The Brownian particle may also be subject to an external force, like gravitation, given by the gradient of a potential $V(\vec{x})$.

Because of the stochastic character of the interaction between the Brownian particle and the medium, it is not possible to predict exactly position $\vec{x}$ and momentum $\vec{p}$ of the Brownian particle at some time t in terms of the initial values at time t_0. It is possible only to calculate the probability distribution in phase space, i.e., the probability density $P(\vec{x},\vec{p};t)$ for finding the Brownian particle at time t with momentum $\vec{p}$ at position $\vec{x}$ if $P(\vec{x},\vec{p},t')$ is known at some earlier time t'. In view of the discussion in Section 3 of the physical interpretation of the Fokker-Planck equation and especially the connection between this equation and the random-walk problem, it is perhaps not surprizing that P also obeys a Fokker-Planck equation. This equation has the following form (M is the mass of the Brownian particle)

$$\frac{\partial}{\partial t} P(\vec{x},\vec{p},t) + \frac{\vec{p}}{M}\vec{\nabla}_x P(\vec{x},\vec{p},t) - \vec{\nabla}_x V \cdot \vec{\nabla}_p P(\vec{x},\vec{p};t)$$
$$= \vec{\nabla}_p \cdot [\gamma\vec{p} P(\vec{x},\vec{p},t)] + \frac{1}{2}\Delta_p [D_p P(\vec{x},\vec{p},t)] \,. \tag{4.1}$$

In order to understand the implications of Eq. (4.1), let us proceed as in Section 3 and take the moments of Eq.(4.1) with respect to $\vec{x}$ and $\vec{p}$. We define the moments in analogy to Eq. (3.6) as

$$\langle\vec{x}(t)\rangle = \int d^3x \int d^3p \; \vec{x}\, P(\vec{x},\vec{p},t) / \int d^3x \int d^3p \; P(\vec{x},\vec{p},t)$$
$$\langle\vec{p}(t)\rangle = \int d^3x \int d^3p \; \vec{p}\, P(\vec{x},\vec{p},t) / \int d^3x \int d^3p \; P(\vec{x},\vec{p},t) \tag{4.2}$$

and so on. We notice that Eq. (4.1) implies conservation of the total probability (the integral of P over $\vec{x}$ and $\vec{p}$). We find

$$\frac{d}{dt}\langle\vec{x}(t)\rangle = \frac{1}{M}\langle\vec{p}(t)\rangle ,$$

$$\frac{d}{dt}\langle\vec{p}(t)\rangle = -\vec{\nabla}_x V(\langle x(t)\rangle) - \gamma\langle\vec{p}\rangle , \qquad (4.3)$$

$$\frac{d}{dt}\langle(\vec{p} - \langle\vec{p}\rangle)^2\rangle = D_p ,$$

and more complex equations for $\langle x_i p_x\rangle$ and $\langle x_i x_j\rangle$. In evaluating the right-hand sides of Eqs. (4.3), we have assumed that $\vec{\nabla}_x V$ changes slowly over distances within which P is essentially different from zero. This yields the term $\vec{\nabla}_x V(\langle x(t)\rangle)$.

The first two Eqs. (4.3) have the form of Newtonian equations of motion with a friction force given by $-\gamma\langle\vec{p}\rangle$. It is worthwhile to appreciate the difference in form between the first term on the right-hand side of Eq. (3.7), and of the second of Eqs. (4.3). In Eq. (3.7), a constant α_1 leads to a drift of the center of the distribution as shown in Eq. (3.8), whereas a term linear in p under the differentiation operator with respect to p leads to a friction term, i.e., to a damping of the motion of the center of the distribution.+ We mention that it is quite straightforward to generalize the friction term to a situation in which the friction force is not equally strong in all directions. This is the case, for instance, in heavy-ion reactions where radial and tangential friction may differ. One replaces $\vec{\nabla}_p(\gamma\vec{p}P)$ by $\sum_{j,k} \frac{\partial}{\partial p_j}[\gamma_{jk}\, p_k\, P]$ and chooses the friction tensor γ_{jk} accordingly.

The equation for the variance of $\vec{p}$ goes beyond classical mechanics altogether. The diffusion constant D_p expresses the randomness of the collisions between the Brownian particle and the molecules of the medium. This randomness causes a random walk of the Brownian particle in momentum space, and an ensuing widening of the momentum distribution with time. Naturally, the position of the Brownian particle must also be described by a diffusion process, and an ever widening distribution. Nonetheless, the Eq. (4.1) does not contain a term like $\frac{1}{2}\Delta_x[D_x P(\vec{x},\vec{p},t)]$. This is because the primary agent to produce

+The way in which such a term arises from a master equation is briefly described in the appendix to this section.

randomness is the exchange of momentum. The diffusion in ordinary space is a latter consequence, obtained upon a proper integration of the equations of motion (4.3), or of the Fokker-Planck equation (4.1). We do not go into this here.

The left-hand side of Eq. (4.1) is reminescent of the Boltzmann equation, and can easily be interpreted. Indeed, with classical trajectories given by $\dot{\vec{x}} = (1/M)\vec{p}$ and $\dot{\vec{p}} = -\vec{\nabla}_x V$, the left-hand side is equal to the total derivative of p with respect to time, including the flow of probability due to the motion of the particle. The right-hand side is obviously caused by the collisions, and may be thought of as being derived from a balance term through a Taylor expansion, as was done in Section 3. This shows that the original Master equation from which Eq. (4.1) is to be deduced has a form similar to Eq. (2.1) with d/dt interpreted as the total derivative with respect to time.

In his famous analysis of Brownian motion, Einstein showed in 1904 that γ and D_p are not independent of each other, but related. This is understandable because both are due to the same mechanism--the random exchange of momentum between Brownian particle and the medium. The essential point of the argument is the following. Let us consider a case where V=0. For $t\to\infty$, the distribution function tends towards the stationary distribution function of a particle in contact with a heat bath at temperature T_∞. This distribution is independent of x and has the form of a Maxwell distribution in $\vec{p}$. Hence, with k the Boltzmann constant,

$$P(\vec{x},\vec{p},t) \xrightarrow{t\to\infty} (2\pi MkT_\infty)^{-\frac{1}{2}} \exp\{-p^2/(2MkT_\infty)\} \ . \tag{4.4}$$

The requirement that the right-hand side of the expression (4.4) be a solution of Eq. (4.1) leads immediately to the Einstein relation

$$D_p = 2MkT_\infty \cdot \gamma \tag{4.5}$$

connecting friction coefficient γ and momentum diffusion constant D_p. Eq. (4.5) is the special case of a large class of relations connecting a transport coefficient for dissipation (here: the friction coefficient) with a transport coefficient for diffusion (here: D_p). There is a general theorem--the fluctuation-dissipation theorem--which in its various forms extends the Einstein Relation (4.5) to more general situations.

The lesson to be learned from Eq. (4.5) is that there is never any dissipation without fluctuation. In the early days of analysis of heavy-ion experiments, when Newtonian equations with friction constants, but without dissipative terms were used, this point was not taken into account.

The Fokker-Planck equation (4.1) is expected to hold under the following conditions:

(i) The coupling is <u>weak</u> in the sense specified below Eq. (2.3). Otherwise the process would be non-Markoffian, and would not yield the simple form of Eq. (4.1).

(ii) The time is sufficiently large so that initial deviations from the Gaussian form have been washed out.

The Einstein relation (4.5), for which the conditions (i) and (ii) must naturally be satisfied, is subject to yet another constraint. Generally speaking, we cannot exclude the possibility that D_p and γ both depend on temperature <u>and</u> momentum. The derivation given for Eq. (4.5) applies only to those values of p as argument of D_p and γ for which the Maxwellian in Eq. (4.4) is essentially different from zero, or to kinetic energies $p^2/(2m) \lesssim kT_\infty$. For kinetic energies which significantly exceed kT_∞, the Einstein relation does not hold in the form (4.5). Modifications may also arise from the binding of particles in the potential $V(\vec{x})$. However, the correct generalization of Eq. (4.5) can be derived from the fluctuation-dissipation theorems, provided conditions (i) and (ii) are fulfilled. This point is of no concern for the study of Brownian motion. In heavy ion collisions, however, where $kT_\infty \lesssim$ several MeV even for highly excited nuclei, $p^2/(2m)$ may easily exceed kT_∞ by a factor 10, and corrections to the Einstein relation (4.5) are expected, and have been found, as will be discussed in Part III below.

Equations of the form (4.3) have been used by various groups to analyze the loss of energy and angular momentum in deeply inelastic heavy-ion reactions. The variables $\vec{x}$ and $\vec{p}$ were identified with the distance between the two centers-of-mass, and the associated momentum, respectively. Such a treatment presupposes from the outset that the loss of energy and angular momentum are due entirely to a stochastic process, and that other, non-stochastic ways of exciting the fragments do not contribute. We return to this point in Part II. In the analyses, both $V(\vec{x})$ and the transport coefficients are unknown parameters. A summary of these analyses is outside the scope of the present lectures. Suffice it to say that in the first analyses, only friction was taken into account, a difference being made between radial and tangential friction. Some of the contributions to this problem are listed in the references.[7] The most recent such phenomenological analysis is that of Berlanger et al.[8] It uses a Fokker-Planck equation for the variables $\vec{x}$ (the relative distance), $\vec{p}$ (the associated momentum) and $x = \frac{Z_2-Z_1}{Z_1+Z_2}$ (the charge asymmetry of the two fragments). This Fokker-Planck equation is an extension of Eq. (4.1), describing loss of energy, of angular momentum, and transfer of charge. The charge transfer is assumed to proceed in a fashion which is statistically uncorrelated with loss of energy and angular momentum, and is assumed to be overdamped. The potential $V(\vec{x})$ used in this analysis as well as the friction forces were taken from an earlier phenomenological analysis of Wilczynska and Wilczynski.[7] In both analyses, neck formation between the separating fragments was taken into account phenomenologically by choosing the exit-channel potential to be different from the entrance channel potential. A friction constant for the mass transfer was fitted to the data. The diffusion constants were then determined via the Einstein relation (4.5). The nuclear temperature $T_\infty(t)$ at any time t was given in terms of the mean excitation energy $\varepsilon_s(t)$ of both fragments at that time via the relation

$$kT_\infty(t) = (\varepsilon_s(t)/a)^{1/2} \tag{4.6}$$

where a is the usual level-density parameter. The quantity $\varepsilon_s(t)$ is obtained through energy conservation from the energy of relative motion at time t. Cross sections are calculated by integrating equations of the type (4.3) for the first and second moments in time until the fragments have separated. Thereafter, the motion is determined by the Coulomb field. The cross section is then obtained by assuming that the distribution function is Gaussian and by integrating, at given values of energy loss, scattering angle, and charge transfer, over the contributions from all impact parameters. Fig. 1 shows experimental results for the reaction $^{40}Ar + ^{58}Ni$ at 280 MeV. The triple differential cross section $\frac{d^3\sigma}{dE\,d\theta\,dZ}$ in μb/(MeV rd Z) is plotted in contour form versus c.m. scattering angle and c.m. energy of both fragments (right-hand scale). Fig. 2 shows the results of the calculation, plotted similarly except that the contour lines now refer to the cross section in mb rather than in μb as in Fig. 1. Except near the quasielastic peak where a stochastic model is not expected to work, the qualitative and semiquantitative agreement is rather satisfactory.

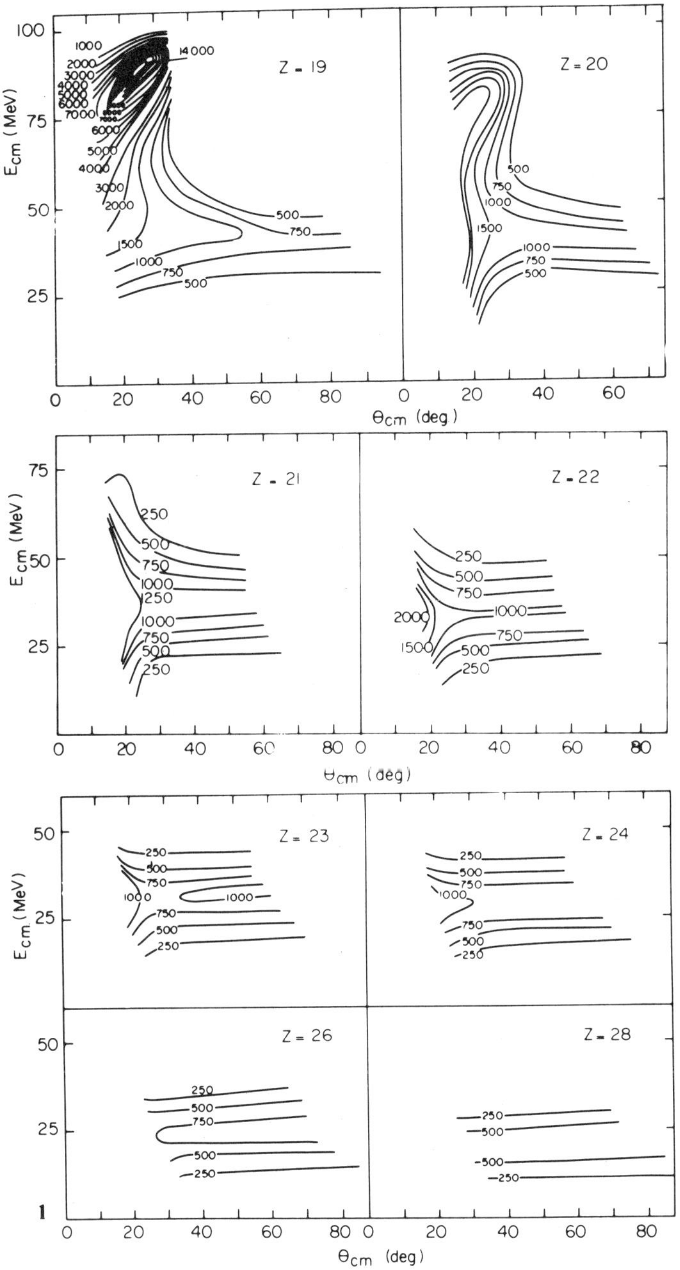

Fig. 1. Experimental contour plots of $\frac{d^3\sigma}{dE d\theta dZ}$ for ^{40}Ar + ^{58}Ni
From Ref. 21.

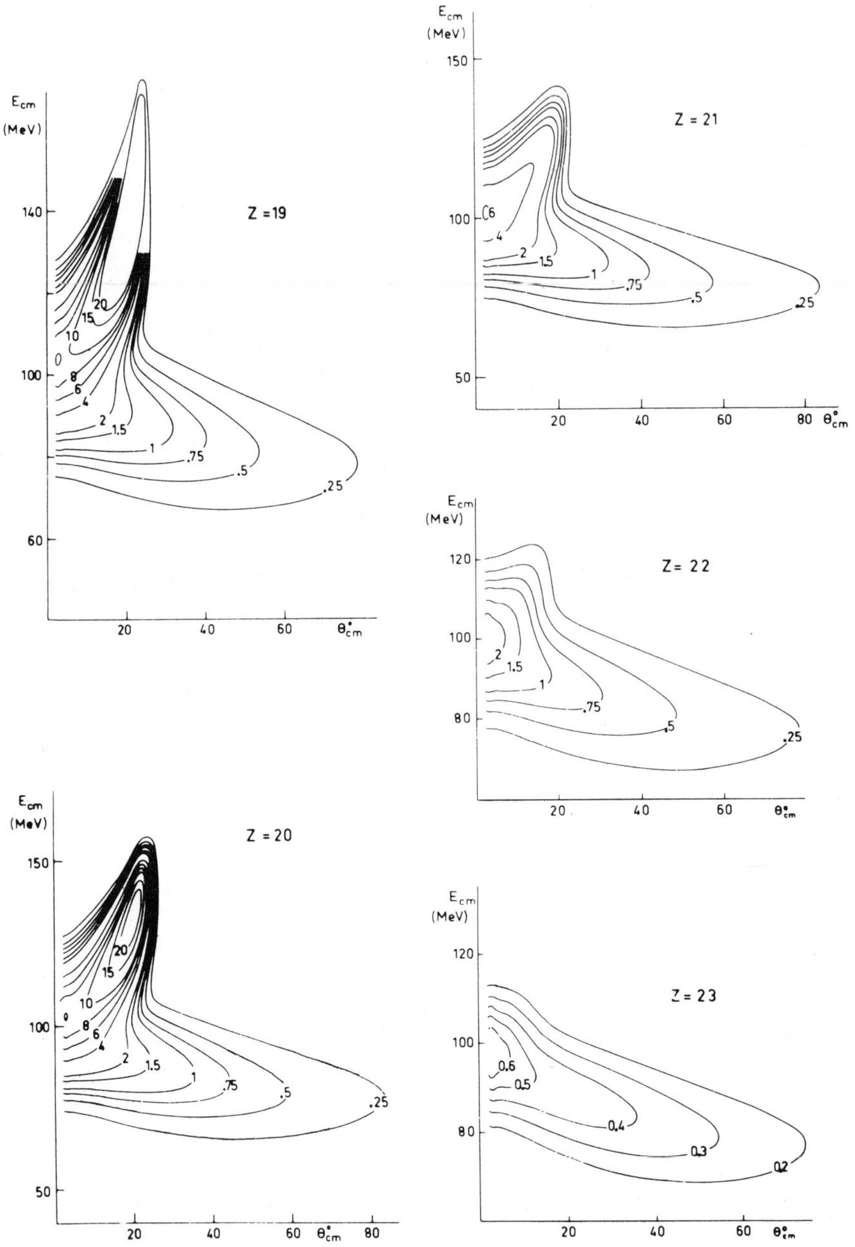

Fig. 2. Calculated contour plots of $\frac{d^3\sigma}{dEd\theta dZ}$ for $^{40}Ar + ^{58}Ni$. From Ref. 8.

APPENDIX A

FRICTION COEFFICIENT VERSUS DRIFT COEFFICIENT IN THE DERIVATION OF THE FOKKER-PLANCK EQUATION

We briefly show how terms of the form $\vec{\nabla}_p(\gamma\vec{p}\,P)$ appearing in the Fokker-Planck equation (4.1) come about in a derivation which starts from the Master equation (2.1). For simplicity, we restrict ourselves to a single one-dimensional variable which we denote by ε as in Section 3. If in Eq. (3.2) we expand $P(\varepsilon')$ and $\rho(\varepsilon')$ separately around $\varepsilon' = \varepsilon$ up to terms of second order (this is a more stringent approximation than the one used in Section 3 because of the strong dependence upon energy of $\rho(\varepsilon)$), use the symmetry of $w(\varepsilon,\varepsilon') = w(\varepsilon',\varepsilon)$, write $w(\varepsilon,\varepsilon') = W(1/2\ (\varepsilon+\varepsilon'),\varepsilon-\varepsilon')$, and expand $W(x,y)$ around $x = \varepsilon$, keeping only terms of up to second order, we find

$$\frac{\partial P}{\partial t} = \frac{\partial}{\partial \varepsilon}\left(c_1(\varepsilon)P(\varepsilon,t)\right) + \frac{\partial^2}{\partial \varepsilon^2}\left(c_2(\varepsilon)P(\varepsilon,t)\right) \qquad \text{(A.1)}$$

where

$$c_2(\varepsilon) = \frac{1}{2}\,\rho(\varepsilon)\int dy\ y^2\ W(\varepsilon,y)\ , \qquad \text{(A.2)}$$

$$c_1(\varepsilon) = \frac{1}{2}\,\rho(\varepsilon)\int dy\ y^2\,\frac{\partial}{\partial \varepsilon}\,W(\varepsilon,y) - 2\,\frac{d}{d\varepsilon}\,c_2(\varepsilon) \qquad \text{(A.3)}$$

If $c_1(\varepsilon)$ and $c_2(\varepsilon)$ are independent of energy, the solutions of Eq. (A.1) are again of the form (3.8). If $c_1(\varepsilon)$ is linear in ε, the Fokker-Planck equation takes the form of Eq. (4.1) rather than that of Eq. (3.4) with α_i = constant. This shows that the energy dependence of the terms in the master equation which give rise to the low moments appearing in the Fokker-Planck equation is of great physical significance, and must be treated carefully.

II. GENERAL ASPECTS OF NUCLEAR TRANSPORT THEORY

5. General Considerations

How can we derive the transport equations described above from a microscopical input? How can we calculate the transport coefficients? Can we find and justify transport equations for other variables like mass and charge transfer, shape deformations, density fluctuations etc.? Much about these questions can be learned from other fields of physics where transport equations have long been used. At the same time, it is necessary to keep in mind the peculiarities of the nuclear situation. Transport theory will then be seen to be one extreme of a whole spectrum of possible approaches to the nuclear problem.

As two heavy ions approach each other, multiple Coulomb excitation occurs, followed by inelastic nuclear excitation and nucleon transfer processes. The two fragments probably also incur shape deformations. For values of the impact parameter close to the grazing one, only fairly low-lying states are excited. Such states are widely spaced, have strong individual characteristics, and the use of thermodynamic or statistical concepts to describe their excitation is quite out of the question. Some sort of coupled-channels approach would be far more appropriate. The picture changes for smaller impact parameters and, consequently, stronger virtual penetration of the fragments. Contact times and, presumably, excitation energies increase, the ever more narrowly spaced levels lose their individual characteristics, and so do the excitation processes which feed them. This suggests that for the later stages of these close encounters, a statistical treatment is perhaps adequate, and is certainly to be preferred over a coupled-channels approach which becomes prohibitive. There is some evidence that the two fragments ultimately form a lump of fused hot nuclear matter rotating for a few revolutions about its center before the centrifugal and Coulomb forces cause it to break apart. In this stage of the process there appears to be a close similarity to induced fission.

The theoretical approaches used to describe these phenomena reflect the wide range of processes mentioned. Time-dependent Hartree-Fock theory[9] assigns primary importance to the idea of a self-consistent field the shape of which reflects the evolution in time of the system and determines the orbitals of the individual nucleons. The approach by Broglia, Dasso and Winther[10] emphasizes the role of surface modes in the nuclear excitation. These modes are taken to be the modes of the two independent, free fragments. In some sense, this approach extends in a simple, powerful way the essentials of a coupled-channels calculation to the domain of smaller impact parameters. Dietrich et al.[11] and, more recently, Strutinsky[12] have used a parametrization of the cross section as their starting point. Strutinsky in particular, emphasizes the difference between the coherent features dominating the near-grazing collisions, and the statistical features dominating the deeply inelastic ones. His parametrization provides a framework which must be filled in by a dynamical theory. Nuclear transport theory, finally, emphasizes the tendency towards equilibrium and is expected to work best for strong interpenetration of the fragments, long contact times, and the later stages of the process. It is the only approach we are concerned with here.

Access to nuclear transport theory is possible both from concepts familiar from nuclear reaction theory, and from concepts established in other areas of physics. Both routes have been used by various groups of authors.

Some concepts of nuclear reaction theory may be viewed as precursors of those dominating nuclear transport theory. High-lying collective states like the E1 resonance mix strongly with the numerous states in the background. This is described by a spreading width which is an average quantity and does not describe the mixing with individual states. The optical-model potential contains similarly a part which describes the coupling to more complex configurations that are populated on the way to compound nucleus formation. This potential, too, is an average quantity. The Schrodinger equation with a complex optical model potential violates time-reversal invariance. This suggests that averaging over energy is an important step in transforming a microscopic equation into transport form. However, the Schrodinger equation is still an equation for an amplitude, not for a probability, and a further element is needed to make the transition complete. There exists[3] a detailed and exact derivation of the transport equation (2.5) for pre-compound processes which can serve as a model. This derivation supplies the missing element by using the concept of random matrices.[13] This is the agent which destroys phase correlations and leads to equations for probabilities. This route to nuclear transport theory was followed by Norenberg et al.[14], and by Agassi, Ko, et al.[15]

In macroscopic systems, transport phenomena are described by considering part of the system as a heat bath. Examples are the motion of a Brownian particle in a medium of fixed temperature, and the periodic motion of a piston extending and compressing a gas of particles. Friction and diffusion of the Brownian particle can be calculated in terms of its interaction with the medium, i.e., the heat bath. The dissipative energy loss of the moving piston can be calculated from the balance of momentum exchange with the gas particles averaged over the Maxwell distribution. The density matrix of a heat bath is generally of the form $\exp\{-\beta H\}\,\mathrm{Trace}[\exp\{-H\beta\}])^{-1}$ where H is the Hamiltonian of the heat bath, $\beta = 1/(kT)$, k the Boltzmann constant, and T the temperature. This form both contains an average over a set of states, and lacks phase correlation between different states--the two ingredients characterizing transport equations. Thermodynamic averages of the interaction matrix elements transfer these properties to the macroscopic variable with which the heat bath interacts, and lead to transport equations. This route to nuclear transport theory was followed by Gross et al.,[16] Swiatecki et al.,[17] and by Hofmann and Siemens.[18]

Neither of these routes, however, can help to answer the central problem of nuclear transport theory: Which are the variables that obey transport equations? To solve this problem, we would have to be able to decompose the system into two parts, one which equilibrates quickly, and the other which does not. The variables describing the latter part might suitably be called collective variables. They obey transport equations. There are obvious candidates for collective variables--the relative distance, the mass distribution, parameters describing shape deformations, etc. Whether these really qualify in terms of the time scales involved, how they interact with the other degrees of freedom, and whether this list is complete--these are the challenging questions facing nuclear transport theory. The hope of those working in the field is, of course, that nuclear transport theory will develop into a theory of large-scale damped collective nuclear motion, describing a range of phenomena for which neither the Schrödinger equation nor the equation of macroscopic physics yield adequate formulations.

6. Time Scales in Transport Equations. Weak Coupling Versus Strong Coupling

Transport phenomena are governed by certain time scales. These are now discussed. To be specific, let us take as the collective variable the relative distance $\vec{r}$ between the two centers-of-mass, and let us work in a representation where $|s\rangle$ is the product of eigenstates of the intrinsic Hamiltonian of the two fragments. The collective variable $\vec{r}$ is meaningful only when antisymmetry of nucleons in different fragments is disregarded, and the choice of basis essentially implies neglect of neck formation and large scale deformation. The choice of variable and basis is thus suitable only for collisions in which $|\vec{r}|$ does not get smaller than something like tne sum of the half-density radii. However, it is the only case for which a microscopic derivation of the relevant time scales has so far been given.[19] The various time scales introduced below will feature prominently for <u>any</u> choice of collective variable and basis. Their numerical values may be quite different, however, for different choices.

Let $V(\vec{r}, \vec{\xi})$ denote the interaction between collective variable and intrinsic degrees of freedom $\vec{\xi}$. This interaction must be thought of as the single-particle potential of one nucleus penetrating into the mass distribution of the other, giving rise thereby to inelastic excitation and nucleon transfer reactions. (There is a contribution to V also from collisions between nucleons in different fragments but we disregard it here.) The matrix elements of V between intrinsic states

$|s\rangle$ and $|m\rangle$ are denoted by $\langle s|V|m\rangle = V_{sm}(\vec{r})$. They still depend upon r and are therefore called form factors in References 15 and 19. To simplify things as much as possible, we suppress the dependence of $V_{sm}(\vec{r})$ on angular momentum in the sequel, and we therefore regard $V_{sm}(r)$ as depending on the scalar quantity $r = |\vec{r}|$.

The biggest of the time scales involved is the total duration time τ_{coll} of the collision. That is the time during which the collective motion is more or less totally damped. Analyses of the type mentioned in Section 4 suggest that for deeply inelastic collisions,

$$\tau_{coll} = \text{a few} * 10^{-21} \text{ sec.} \qquad (6.1)$$

Let τ_{equ} denote the time during which part of the system equilibrates. For a transport description to be viable, this time must be shorter than any of the other time scales introduced in this section. For the specific choice introduced above, τ_{equ} can be estimated as follows. In the course of repeated action of $V(r, \xi)$, an ever increasing number of particle-hole states is formed in either fragment. Equilibration between particle-hole states with the same number of particles and holes is a very rapid process--about 10 times faster than transitions in which the number of particles and holes changes by one unit each, because of bigger overlap. Using estimates from the precompound model,[3] or using directly the formula $\tau_{equ} = \hbar/\Gamma\!\downarrow$ with $\Gamma\!\downarrow = 2\pi V^2 \rho(E)$ with $V \cong 200$ keV and $(\rho(E))^{-1} \gtrsim 5$ to 10 keV for the states with fixed particle-hole number, we find

$$\tau_{equ} \lesssim 3\ldots 7 \times 10^{-23} \text{ sec} \qquad (6.2)$$

The process of forming even more particle-hole states during the collision always favors states with the most probable number of particles and holes at a given excitation energy. This is because the level density of such states is highest. Transitions within each fragment between states of different numbers of particles and holes are therefore not important to reach internal equilibrium in each fragment. Thus, Eq. (6.2) provides an estimate of the total equilibrium time within each fragment. This situation is quite different from precompound reaction where initially a 2p 1h state is formed at high excitation energies. The estimate (6.2) then still applies to the equilibration within each group of sp, (s-1)h states, and provides the reason why a transport equation can be used to describe precompound reactions. However, the total equilibration involves several or many transitions between states of different particle-hole numbers, and thus times long in comparison with (6.2).

The three remaining time scales relate to the exchange of energy and momentum between the collective degree of freedom, and the intrinsic states $|s\rangle$, and are therefore determined by the properties of the form factors $V_{sm}(r)$. In a thermodynamic or statistical treatment, it is always the average of $V_{sm}(r)\, V^*_{sm}(r')$ over a sufficiently large number of levels in the vicinity of s, and of m, which appears in the formulation. We denote this average by a bar. It is characterized by three parameters which specify the strength, the dependence upon s and m, and the dependence on r and r'. In Reference 19 it was shown that

$$\overline{V_{sm}(r)\, V_{sm}(r')} = D_s^{1/2}\, D_m^{1/2} * W_o\, f\left(\frac{r+r'}{2}\right) * \exp\left(-\frac{(\varepsilon_s-\varepsilon_m)^2}{2\Delta^2}\right) * \exp(-(r-r')^2/(2\sigma^2)) \quad . \tag{6.3}$$

The three parameters are W_o, Δ, and σ. To interpret Eq. (6.3), we remark that with D_s the mean spacing of levels around the state s, the factor $(D_s^{1/2}\, D_m^{1/2})$ takes account of the increasing complexity of nuclear levels with increasing excitation energy, which reduces the overlap between levels s and m. (To understand this factor better, we should remember that the spreading width of any one level τ, a quantity depending very smoothly on energy, is given by $2\pi \overline{V_{\tau s}^2}\rho(\varepsilon_s)$.) The factor $W_o f((r+r')/2)$ is the strength of the interaction, with $f((r+r')/2)$ the density overlap of the two fragments, and W_o typically 10 MeV. The first exponential in Eq. (6.3) restricts energy transfer to values $\lesssim \Delta$. (The energy of states s is ε_s.) Since V is essentially a single-particle operator, it decreases or increases the number of particle-hole pairs by one. Level-density arguments show that this yields a significant contribution only if $\Delta \leq 7$ MeV or so. The last exponential implies that the mean value (6.3) tends to zero as $|r-r'|$ increases. The interactions then take place at different parts of the nucleus, create a different superposition of particle-hole states, and get out of phase. The correlation length σ was found to have a typical value of 3.5 fm in the nuclear surface.[19] Equivalently, we may think of $\hbar/\sigma$ as the maximum momentum transferred in a single action of $V_{sm}(r)V_{sm}(r')$. This we expect to be something like the Fermi momentum. Taking account of the reduction of Fermi momentum in the nuclear surface, we see that the value

$\sigma \cong 3.5$ fm is reasonable.

The time $\hbar/\Delta$ is the time needed to transfer the energy ΔE. We may think of $\hbar/\Delta$ as the duration time τ_Δ of a single action of V_{sm}. With $\Delta E = 7$ MeV, it is given by

$$\tau_\Delta \cong 10^{-22} \text{ sec.} \tag{6.4}$$

The correlation length σ is related to a correlation time τ_σ through

$$\tau_\sigma = \sigma/\dot{r} \cong 3 * 10^{-22} \text{ sec.} \tag{6.5}$$

where $\dot{r}$ is the velocity. In Eq. (6.5) we have used $\dot{r} = 0.05$ c. Obviously, σ is the more fundamental quantity here. The strength W_o is related to the mean free path λ, the distance travelled in r between the onset of two subsequent single actions of $V_{sm}(r)$. Since the typical energy loss in the deeply inelastic reactions is about 200 MeV, that in a single action $\Delta \cong 7$ MeV, we find that a total of about 25 actions must take place. With a total distance travelled of about 10-15 fm, this yields $\lambda \cong 0.4$ or 0.5 fm, and a corresponding time between subsequent single actions τ_λ given by

$$\tau_\lambda = \lambda/\dot{r} \cong 4 \cdot 10^{-23} \text{ sec.} \tag{6.6}$$

The numbers given in Eqs. (6.1), (6.2), and (6.4) to (6.6) are clearly only rough estimates. It was mentioned above that these estimates may turn out to be quite different for another choice of collective variable, and/or of basis. Nevertheless, they deserve some comments.

The near equality of τ_λ and of τ_{equ} shows that a description employing an equilibrated subsystem is somewhat of an idealization. There is another point which is much more worrisome. Both τ_Δ and τ_σ are considerably larger than τ_λ, and scarcely an order of magnitude smaller than τ_{coll}. During the effective duration time of a single action of V--which we take somewhere between τ_Δ and τ_σ -- 5 or 6 actions of V take place altogether, and the energy located in the interaction during this time is $5...6 * \Delta \cong 40$ MeV. This is a very substantial fraction of the total energy available for dissipation (about 200 MeV). This fact has grave consequences. It means that we are in a regime of strong coupling. The time over which a substantial fraction of energy is lost from collective motion, is comparable with the time scales τ_Δ and τ_σ. We therefore expect the transport equation for the collective variable to be non-Markoffian in time (see Section 2). The energy of collective motion and of intrinsic excitation need not add up to the total energy available, but may differ from it by 40 MeV or so. This in turn may drastically affect the calculation of transport coefficients. Since the process is non-Markoffian, the Einstein relation need not hold. Some of these points have been corroborated already by the numerical study of a one-dimensional model.[15]

These remarks illustrate the fact that this "strong coupling regime" is quite different from the familiar "weak coupling regime" characterized by the inequality

$$\mathrm{Max}(\tau_\Delta, \tau_\sigma) \ll \tau_\lambda \quad . \tag{6.7}$$

Taking W_o as a free parameter of the theory, and reducing its size, we can always formally attain the limit in which the inequality (6.7) is fulfilled. (We expect $\lambda \propto W_o^{-1}$.) In this limit, the fraction of the total energy located in the interaction at any one time is small, and, for this reason, the time over which a substantial fraction of energy is lost from the collective motion, is very large compared to τ_Δ and τ_σ, so that a Markoffian approximation is expected to hold, with the Einstein relation fulfilled (subject to the further conditions mentioned in Section 4). This is also the situation implied when a macroscopic system is in contact with a heat bath: The exchange of energy between system and heat bath is assumed to proceed in negligible units of energy.

We should thus be wary of using in a straightforward manner concepts and methods established in the thermodynamics of macroscopic bodies. Here as in other instances nuclear physics is beset with the problem that the characteristic parameters all tend to have similar values (within an order of magnitude), and that a small parameter useful for a series expansion is difficult to find. Our discussion suggests that in deriving nuclear transport equations, a critical examination of the time scales involved is of prime importance. The time scales dictate which approximations are permissible, and it is in this way--by paying proper attention to the peculiarities of the nuclear situation--that nuclear transport theory can establish a place of its own in the general framework of nonequilibrium statistical mechanics.

III. A SURVEY OF THEORETICAL APPROACHES

The existing approaches are listed under "weak coupling" and "strong coupling". I try only to describe the essential points, omitting all mathematical detail. Attention is focussed on the derivation of transport coefficients. The choice of a collective variable always implies certain values for the effective mass, and the conservative potential. Little attention has, however, been paid to this problem, save for Ref. 18.

7. Weak Coupling Approaches

Gross, Kalinowski and De[20] were apparently the first authors to advocate the use of perturbation theory for the calculation of the friction coefficient. From a general point of view, this proposal suffers from the difficulties mentioned in Section 6 since the authors certainly had in mind using the distance between heavy ions as a collective variable. In a more recent contribution,[21] Gross and Kalinowski argue that the loss of energy is a very rapid process, taking place in about 10^{-22} sec, and that the rise of statistical methods is therefore altogether inappropriate. The evidence brought forward for this claim rests, however, on the use of a specific model and is not universally accepted as compelling. The authors advocate the use of straightforward perturbation theory. They fail to explain how this could ever lead to equations of transport form. No calculations of friction coefficients have been reported.

The approach of Hofmann and Siemens[18] is based on two assumptions. The equilibration of the internal excitation happens very quickly in comparison to the time in which the collective variable, or the collective energy, change significantly,

$$\tau_{equ} << \tau_{coll} .$$

(The collective variable is treated classically.) The second assumption consists in the use of linear response theory to calculate the transport coefficients. It consists of the following steps. The intrinsic excitations are described by a temperature T. The change of T with time is eventually calculated via energy conservation from the equations of motion for the collective variable. The transport coefficients (as well as the effective mass) are calculated from expressions of the form

$$< \sum_{m} \int_{-\infty}^{t} ds \, V_{nm}(r(t)) \exp\{i(E_m - E_n)(t-s)/h\} V_{mn}(r(s)) >_T \tag{7.1}$$

where s, t are time variables, E_m and E_n the energies of intrinsic states, and the bracket denotes a thermodynamic average over the states n. The form of Eq. (7.1), especially the use of the free propagator $\exp\{iE_m t/h\}$ for the intrinsic excitations, shows that this is a weak coupling approximation. Indeed, the range of integration over s in (7.1) is limited by $|t-s| \lesssim \mathrm{Max}(\tau_\Delta, \tau_\sigma)$. This can be seen immediately if one uses the parametrization (6.3). During this time interval, V is accounted for only to lowest non-vanishing order. This is justified only if the condition (6.7) is met which is the condition for weak coupling. A further approximation consists in expanding V(r(s)) in powers of r(s) - r(t) about V(r(t)) and keeping only the linear terms. This is justified if r(t) changes very slowly with time. More precisely, it requires the condition $\tau_\Delta << \tau_\sigma$ to hold. This can again be seen by using the expression (6.3). The conditions of validity of the Hofmann-Siemens theory are thus seen to be

$$\tau_{eq} < \tau_\Delta << \tau_\sigma << \tau_\lambda \quad . \tag{7.2}$$

Because of the thermodynamic average--a summation over n of the expression (7.1), multiplied by the Boltzmann factor $\exp(-E_n/kT)$ and a proper normalization factor--interference terms of form factors with different indices n,n' never arise, and this is the necessary ingredient for a transport theory. Because of the weak coupling limit (7.2), it is not surprising that a Fokker-Planck equation of the general form of Eq. (4.1) results, with drift and friction coefficients given by terms deduced from (7.1). Such terms obviously describe the transfer of energy and angular momentum to a heat bath.

Calculations of transport coefficients within the framework of this scheme have not been reported so far. The scheme has, however, been extensively explored in its theoretical ramifications. The Einstein relation results under the restrictions mentioned in Section 4. The form of the Fokker-Planck equation, together with the Einstein relation, was used in the phenomenological analyses described in Section 4.

In the last of References 18, particular attention has been devoted to the following problem. The two fragments form a closed system for which energy conservation is of vital importance. Does this invalidate the description of a subsystem in terms of a heat bath at temperature T? Or, in other words: How does the theory change as we pass from a canonical ensemble to a microcanonical ensemble? In the framework of linear response theory, i.e., of the system of inequalities (7.2), it was found that the theory, in its essential parts, remains unaltered.

In 1975 Gross[16] proposed the use of a "piston model" to calculate the friction coefficient. A similar idea is followed in the work of Swiatecki, Randrup, and others.[17] The gas of particles inside the nucleus is characterized by a long mean free path (about 8 fm at neutron threshold). As the two heavy ions approach each other, their single-particle potentials start overlapping, and particles from one nucleus move freely into the potential well of the other. Since the two potentials are in motion relative to each other, the flow of momentum associated with the exchange of particles transforms kinetic energy of relative motion into intrinsic excitation, and vice versa. This transformation of energy becomes irreversible and leads to a transport phenomenon if particles from one nucleus, having reached the other, equilibrate there before they return. In Reference 17, this equilibration is attributed to random reflections of the particles on the wall of the potential.

This dissipation mechanism has been referred to a "one-body dissipation". It is quite different from the normal, "two-body dissipation" of ordinary fluids which is caused by the two-body collisions in the fluid. Two-body dissipation would be prevalent if the mean free path of nucleons were short compared with the linear dimension of nuclei. This would entail a local heating of the volume where the two mass distributions overlap: A "hot spot". One-body dissipation, on the other hand, implies that the two fragments have a uniform intrinsic temperature.

For the formulation of the theory, it is not important that equilibration takes place by random scatterings on the walls of the potential, although this is clearly the mechanism the authors have in mind. Once the nucleons from one fragment have travelled a good distance in the other, two-body collisions could do as well. The only important point is that two-body collisions are not the primary agent of momentum and energy transfer.

The friction mechanism due to "one-body dissipation" operates not only when two heavy ion collide, but also when the potential well deforms, as is the case in fission. Then, one has to consider the exchange of energy between a moving wall and the gas. This leads to the "wall formula". The relevant picture for two colliding ions is the opening of a "window" in the potential wall of either. This gives rise to the "window" formula.

Both formulas are obtained using the classical kinetic theory of gases, and calculating the flux of momentum (in the case of the window formula) through the window. Under the assumption that typical velocities of the gas particles are large in comparison to the velocities of the potential wells, one finds in lowest nonvanishing order for the force of nucleus B on nucleus A the window formula

$$\vec{F}_{BonA} = \Delta\sigma \cdot n_o \cdot (2u_{\parallel} + u_{\perp}) \ . \qquad (7.3)$$

Here, $\Delta\sigma$ is the size of the window, $n_o = \rho\bar{v}/4$ (with ρ the particle density, $\bar{v}$ the average speed) is the static one-sided flux of particles in the gas, and $u_{\parallel}$ and $u_{\perp}$ are the components of relative velocity between B and A along and at right angles to the normal through the window, pointing from A to B. Such a velocity-dependent force obviously implies friction. Taking into account the stochastic nature of the flow of particles through the window, F. Randrup has recently found an expression for the diffusion constant based on similar arguments (W. Swiatecki, private communication). Using the window formula, F. Beck et al.[17] have recently calculated the trajectories, energy losses, and angular momentum losses of deeply-inelastic heavy ion scattering, and obtained reasonable results. Energy losses presumably due to deformation could, of course, not be reproduced.

Although Eq. (7.3) does not show it, the window formula is also based on a weak-coupling approximation. Indeed, the derivation of (7.3) is carried out taking u(t) and the momentum distribution of the gas particles to be stationary. The formula is also based on classical physics, although the Pauli principle is taken into account in some approximation.

It is interesting to observe that the last two schemes starting as they do from an approach well-founded in macroscopic thermodynamics, are weak-coupling schemes.

8. Strong-Coupling Approaches[14,15]

Both these approaches use as an input not the contact of the collective variable with a heat bath, but rather an ansatz which has played a major role in other areas of nuclear physics. Both in the statistical theory of spectra[13] and in the statistical theory of nuclear reactions[3] one uses the concept of a random matrix ensemble to describe average nuclear properties, and the statistical fluctuations around them. The basic idea of this approach goes back to Wigner. In a situation where many valence nucleons are present, or where we consider a highly excited nucleus, the details of the nuclear Hamiltonian defy detailed theoretical description, because of the complexity and large number of states involved. Therefore, one replaces the Hamiltonian by a matrix ensemble of Hamiltonians with a random distribution of matrix elements. The distribution law must, of course, be chosen in accord with whatever little we can say about properties of nuclear matrix elements. Average properties of the ensembles are then compared with average properties of actual nuclear levels. In the present context, the matrix elements $V_{sm}(\vec{r})$ are taken to be randomly distributed objects, with mean value zero for $s \neq m$, and with a Gaussian distribution. This can be justified[19] if it is assumed that the states $|s\rangle$ and $|m\rangle$ be sufficiently high to be amenable to a statistical description, and are described as eigenvectors of a random matrix. The second moment of $V_{sm}(r)$ then has the form given in Eq. (6.3), and this second moment determines completely the Gaussian distribution for $V_{sm}(r)$. This, then, is the input which replaces the assumption of a heat bath. It contains the assumption that all states in an infinitesimal energy interval δE have equal a priori occupation probabilities, and this can only be so if the modes excited during the collision equilibrate rapidly with the rest of the system. Hence, this assumption implies $\tau_{equ} < \mathrm{Min}(\tau_\lambda, \tau_\sigma, \tau_s)$, as does the thermodynamic approach. It does not, however, involve a priori, a weak-coupling picture, or the use of a temperature to describe the intrinsic excitations.

In the basic physical mechanism used, this picture is quite similar to that of "one body dissipation" mentioned in Section 7. The parametrization (6.3) is based on the idea of the mean field of one fragment creating particle-hole excitations in the other, and leading to nucleon transfer. Two-body collisions in the overlap region are not considered. The randomness hypothesis may be thought of as resulting from two-body collisions happening elsewhere in either fragment. Just as in the one-body dissipation mechanism nothing definite need be stated about the origin of thermal equilibration once the gas particles have left the overlap region, nothing definite need be said here about the physical origin of the related assumption of randomness of the matrix elements $V_{sm}(\vec{r})$.

The approaches[14,15] both use essentially Eq. (6.3) and a random-matrix hypothesis as starting points. They differ however, in emphasis and detail. The procedure of Ref. 15 consists in calculating cross sections from the statistical input with as few additional assumptions as possible. The aim is to test the underlying picture without introducing further fit parameters. The procedure of Ref. 14 has been semiphenomenological, with much more emphasis on obtaining a semiquantitative understanding of the data. Only quite recently have Norenberg et al. addressed themselves to the problem of a more rigorous approach. We describe some aspects and results of both approaches.

In Ref. 15, a transport equation is derived exactly, by taking the average over energy (or over the ensemble of V's) of the cross section. The technique of statistical averaging is that developed in Ref. 3. A transport equation emerges because of the presence of a further time scale not discussed so far--the Poincare recurrence time τ_p. This is the time after which the system returns to its original position in phase space. For the nuclear systems considered here, τ_p can be estimated by $\tau_p \cong \hbar/D$ where D is the mean level spacing. With $D \approx 10$ eV at neutron threshold and decreasing exponentially with increasing excitation energy, we see that $\tau_p >>> \tau_\lambda, \tau_\Delta, \tau_\sigma$. Therefore it is meaningful to use the ratios τ_λ/τ_p, τ_Δ/τ_p and τ_σ/τ_p as expansion parameters, and to keep only terms of lowest nonvanishing order in any of them.

This procedure yields a transport equation for $F_s(\vec{R},\vec{k})$. Here, $\vec{R}$ is the distance between the two ions, $\hbar\vec{k}$ their relative momentum, and s denotes the state of intrinsic excitation of the two fragments. In the weak-coupling limit, $F_s(\vec{R},\vec{k})$ becomes proportional to a delta function with argument $E - \varepsilon_s - \hbar^2k^2/(2\mu) - V(R)$ where E is the total energy, V(R) the sum of Coulomb and nuclear potentials, and μ the reduced mass. In this limit, $F_s(\vec{R},\vec{k})$ can be interpreted as the probability density for finding the two heavy ions at a distance $\vec{R}$ with momentum $\hbar\vec{k}$. For strong coupling, the variables ε_s and $\hbar\vec{k}$ become unrelated, in the interaction region.

The function F is independent of time. This corresponds to a scattering problem where a continuous beam of particles, described by a plane wave, hits the target and leads to an equally continuous beam of reaction products. The function F describes this stationary flow in phase space.

The transport equation derived in Ref. 15 has the form

$$\left(\frac{\hbar}{\mu}\vec{k}\vec{\nabla}_R - \vec{\nabla}_R V\vec{\nabla}_k\right)F_s(\vec{R},\vec{k}) = \sum_t \int d^3R' \int d^3k' \{ G_{st}(\vec{R},\vec{R}';\vec{k},\vec{k}')$$

$$* F_t(\vec{R}',\vec{k}') - L_{st}(\vec{R},\vec{R}';\vec{k},\vec{k}')F_s(\vec{R}',\vec{k}') \} . \quad (8.1)$$

Here, G and L stand for "gain" and "loss", respectively. These terms are given in terms of the second moment, see Eq. (6.3), and an optical-model Green's function G_s^{opt}. This function describes the loss of flux from "channel" s through collisions induced by V_{sm}. Replacement of G_s^{opt} by the free Green's function on the right-hand side of Eq. (8.1) leads to the weak-coupling limit. It was shown by Ko, Pirner and myself[15] that in this limit, a master equation governs the intrinsic excitations, and a simple expression for the friction coefficient emerges. In the strong-coupling limit, the solution is much more difficult to obtain. Existing calculations use the following scheme. Numerical studies in one dimension suggest that strong-coupling results can be simulated by a weak-coupling calculation with a renormalized strength parameter W_o. This procedure was also applied in three dimensions. As a result one obtains equations of the form (4.3). These are solved, and cross sections are calculated from the assumption that the distribution is Gaussian. The real nuclear potential was taken to be of the proximity form, and mass/charge exchange was treated as a random walk problem, with the probability for single nucleon exchange per single action of $V_{sm}(\vec{R})$ something like 0.5 or 0.7. Except for this number, the calculations are free of any adjustable fit parameter.

The derivation of the transport equation (8.1) contains a simplification. To carry out the summations over intrinsic spins $\mathfrak{J}$, the nuclear level density $\rho(\mathfrak{J}, \varepsilon)$ was written as $\rho(0, \varepsilon)(2J+1)$, the spin cutoff factor being suppressed. This is all right as long as the loss of angular momentum out of relative motion is so small that it does not populate levels near the Yrast line. The calculations show, however, that fairly large losses of angular momentum are involved, and this point has to be improved.

Some results of the calculations are shown together with data points in Figs. 3 to 5. The dashed lines are the results of the calculations. From bottom to top they are in one-to-one correspondence with the data.

It is seen that the agreement between calculations and data, while satisfactory for energy losses of up to 200 MeV or so, gets increasingly worse with increasing energy loss. This is attributed to the fact that the model used does not make allowance for shape deformations or for formation of a neck. In fact, the Coulomb barrier is at approximately 430 MeV!

The approach of Refs. 14 has consisted mainly in deriving equations of the Master type, like Eq. (2.1), and to use these together with estimates of time scales obtained semiphenomenologically to discuss the transport of energy, mass, and angular momentum. I find it very difficult to judge the stringency of the arguments used in these derivations, and therefore cannot

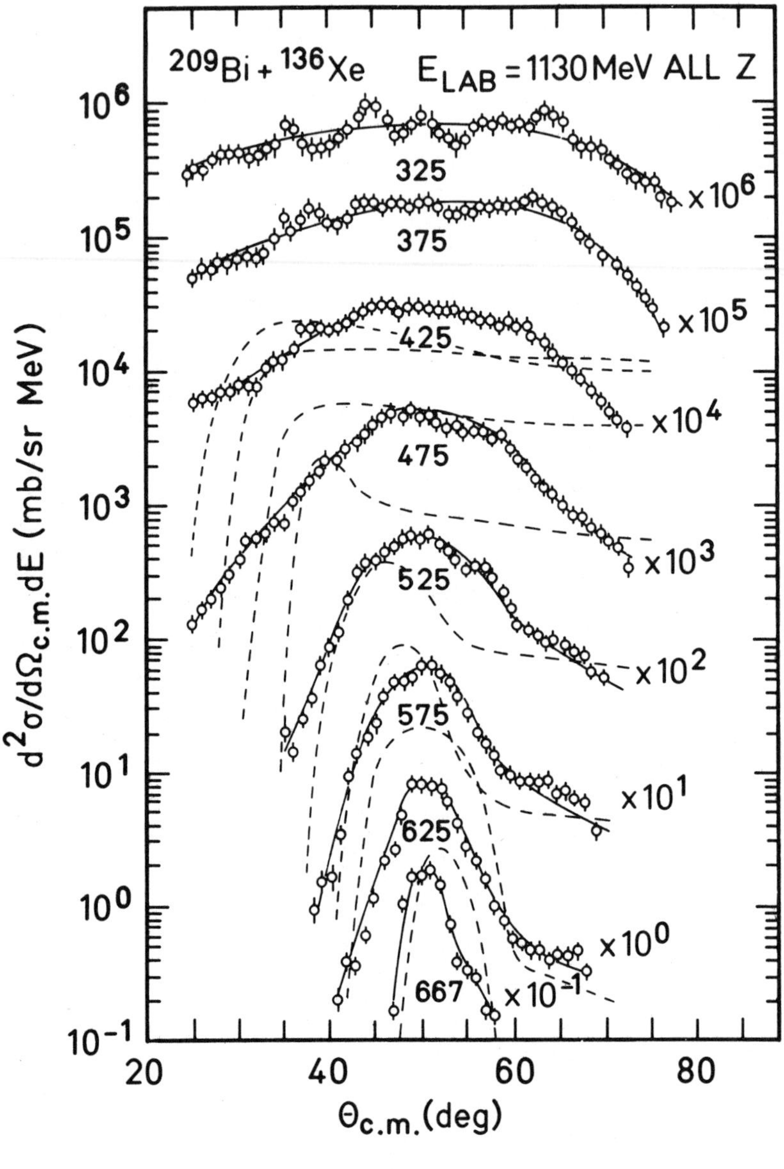

Figure 3

Distribution in angle and energy of light reaction products integrated over all masses and charges. Taken from the fourth of Refs. 15.

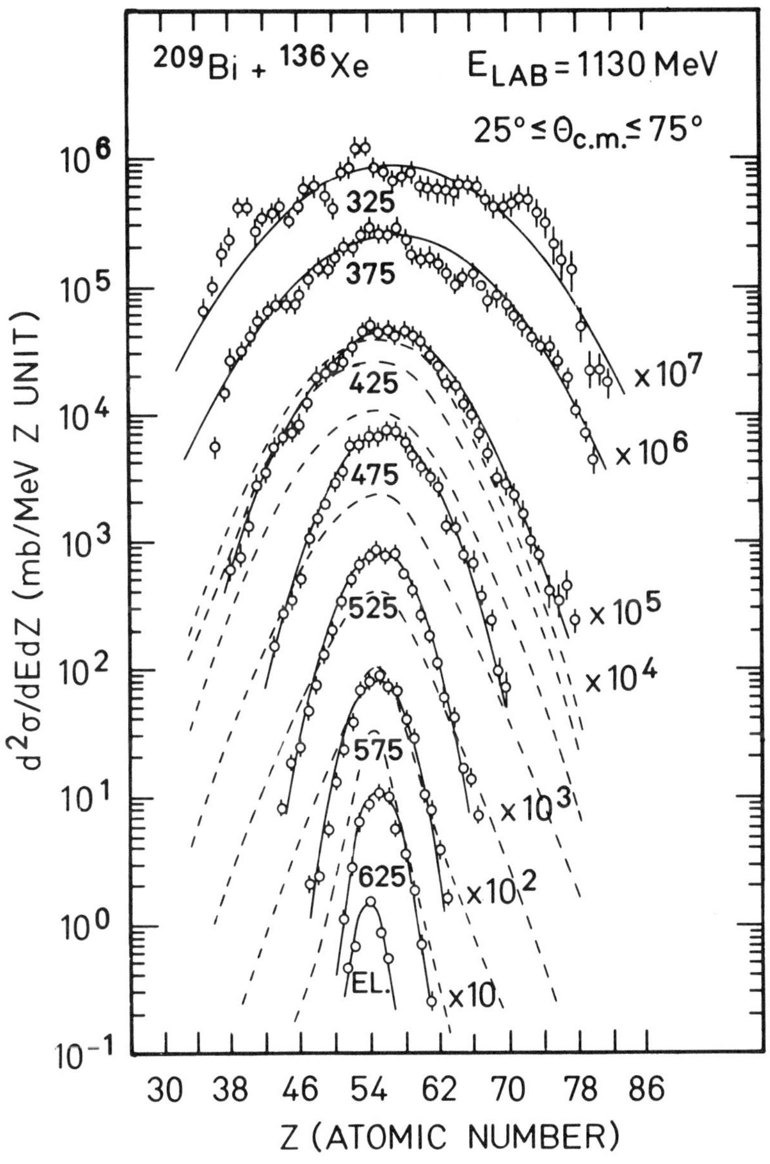

Figure 4

Distribution in charge and energy of reaction products integrated over scattering angle as indicated. Taken from the fourth of Refs. 15.

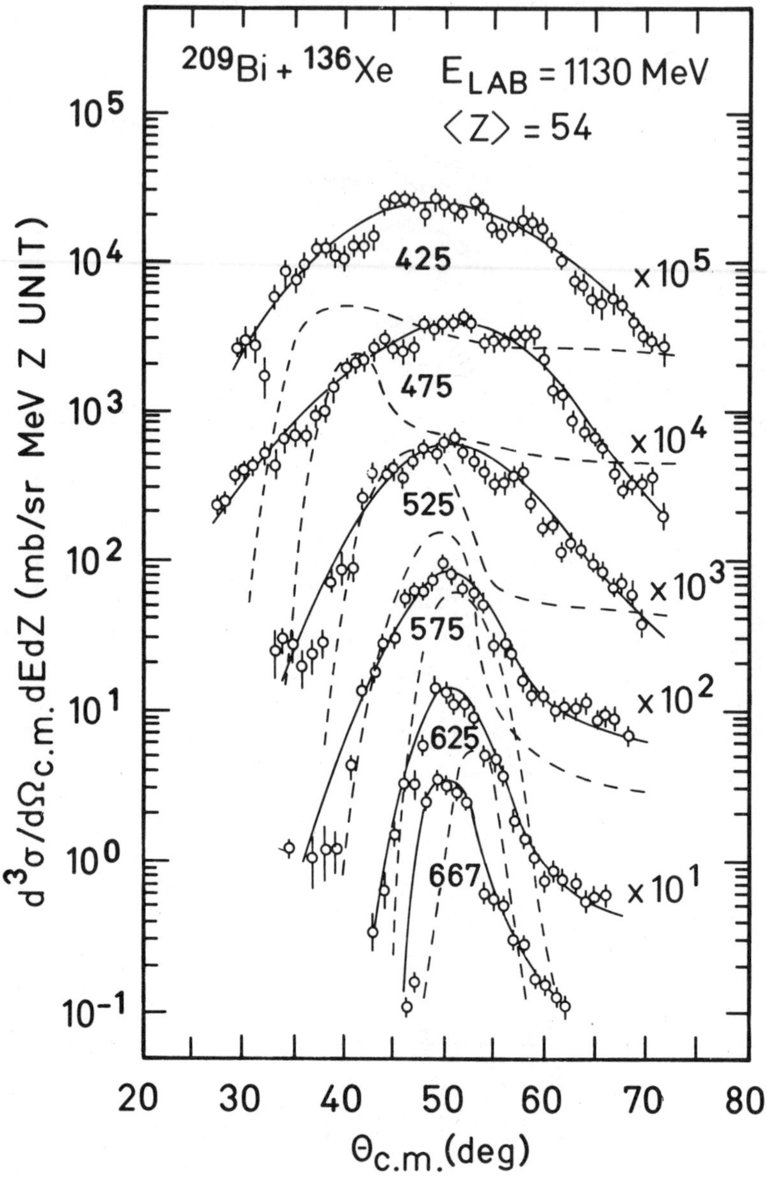

Figure 5

Triple differential cross section for reaction products with Z = 54. Taken from the fourth of Ref. 15.

comment on them. The procedure yields transport equations with well-defined transport coefficients. The times are estimated from an analysis of Coulomb trajectories leading to a given scattering angle. Two recent results obtained in this fashion are shown in Figs. 6 and 7. In Fig. 6, the dashed curve shows the result of the calculations, the solid curve the result of fitting the shift coefficient to the data. In Fig. 7, the total angular momentum $<|I_1|> + <|I_2|>$ of both fragments, and the mean value M_γ of the γ multiplicity deduced from it, are plotted versus the charge of the light fragment. The dots are data points. The dashed curve results if fluctuations around the mean value are neglected, the solid one, if they are taken into account. This is a very nice example to demonstrate the importance of fluctuations.

9. Summary. Open Problems

In these lectures, I have tried to show how nuclear transport theory works, and why it works. Time scales provide the essential part of the answer to the last question, and indicate the prevalence of a strong-coupling situation in nuclear physics. Calculations based on a strong-coupling approach yield--within the frame of the models used--good agreement with the data, and this suggests that transport phenomena play an important role in deeply inelastic scattering.

There remains, however, a number of open problems. Existing approaches are restricted to relative distance and, to a lesser extent, mass asymmetry as the two collective variables. Within this frame, a consistent framework for solving the transport equation in the strong-coupling regime is still lacking. It is necessary to extend the formulation to shape deformations and neck formation, and to include induced fission at 10 MeV or more of excitation energy into the description. Fusion cross sections should also be amenable to a theoretical treatment. Such an extension requires a careful investigation of the relevant collective variables, of their inertia parameters and conservative potentials. A discussion of the coupling between these and the other, non-collective degrees of freedom in terms of time scales is especially important for the construction of a transport theory. It remains to be seen whether the strong-coupling limit features as prominently in these cases as it does in the case of relative motion. The concepts developed in this way may hopefully be useful also at higher bombarding energies, where hydrodynamic descriptions may become more adequate for the reaction.

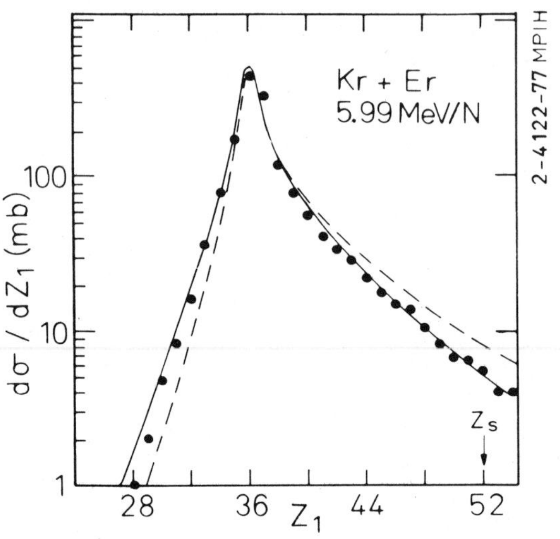

Fig. 6. Charge distribution of fragments produced in the reaction kr+Er. From G. Wolschin and W. Nörenberg, preprint, 1978.

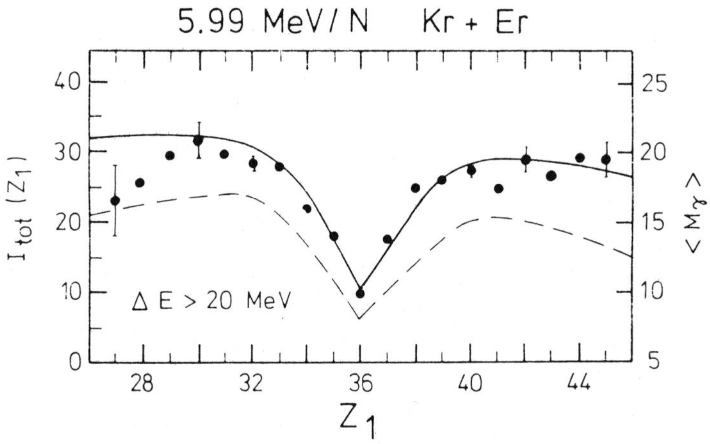

Fig. 7. Angular momentum distribution versus charge of residual fragment. From G. $olschin and W. Norenberg, preprint 1978.

The beginning stages of a deeply inelastic reaction are certainly not governed by transport equations. It must be understood how important these stages are, where they merge into a transport description, and to what extent the initial conditions that they provide influence the later development of the transport process.

Transport theory is successful because the observed smooth distributions in angle, mass, and energy of the reaction products as well as the gamma multiplicities do not suggest that the processes are dominated by single modes. Improved experimental resolution may reveal such modes through structures in the cross sections, and may thus point to the existence of long-lived highly excited nuclear states. This would be an exciting possibility which would require a considerable extension of existing transport theory to include the treatment of such modes. For theorists, a challenging problem consists in understanding better the connection and interdependence of the various approaches to the problem--TDHF, surface vibrations, transport theory.

I am sure that the investigation of deeply inelastic heavy-ion scattering and its extension to higher energies will continue to pose interesting problems for a long while, in the domains of increased experimental resolution, of phenomenological analysis, and of fairly fundamental theory.

APPENDIX B

FEATURES OF THE TRANSPORT EQUATION: CALCULATION OF MEAN VALUES

In this appendix, I indicate the essential steps involved in deriving a transport equation from a random-matrix model. To save space, I do not give the full derivation which can be found in the second of Refs. 15. I just emphasize the main steps. The reader should be able to fill in the gaps. I also indicate typical properties of the transport equation which prevail in the limits of weak and of strong coupling.

1. Second-Order Perturbation Expansion

It is useful to evaluate averages of wave functions, and of density matrices, using the perturbation expansion. In this way, one sees clearly the physical background of the more general and more formal procedure of section 2 below.

Let H be the Hamiltonian, with

$$H = H(r) + H_o(\xi) + V(r,\xi) = H_1(r,\xi) + V(r,\xi). \tag{B.1}$$

Here, $H(r)$ is the Hamiltonian of relative motion, $H_o(\xi)$ the Hamiltonian of intrinsic excitation, with solutions

$$H_o|s\rangle = \varepsilon_s|s\rangle, \quad \langle s|m\rangle = \delta_{sm} \tag{B.2}$$

and $V(r,\xi)$ the coupling between the two, with matrix elements

$$V_{sm}(r) = \langle s|V(r,\xi)|m\rangle \quad . \tag{B.3}$$

The eigenfunctions of $H_1(r,\xi)$ have the form

$$H_1(r,\xi) \mid s\rangle \, |X(E-\varepsilon_s)\rangle = E|s\rangle|X(E-\varepsilon_s)\rangle \tag{B.4}$$

where $\{X(\varepsilon)\}$ with

$$\begin{aligned} H(r)|X(\varepsilon)\rangle &= \varepsilon|X(\varepsilon)\rangle \quad \text{and} \\ \langle X(\varepsilon)|X(\varepsilon')\rangle &= \delta(\varepsilon-\varepsilon') \end{aligned} \tag{B.5}$$

are the scattering solutions of $H(r)$. Let us now suppose that $V_{sm}(r)$ have a Gaussian distribution with mean value zero, and a second moment

$$\overline{V_{sm}(r)V_{s'm'}(r')} = (\delta_{ss'}\delta_{mm'} + \delta_{sm'}\delta_{ms'}) \; \overline{V_{sm}(r)V_{sm}(r')} \qquad (B.6)$$

as given by Eq. (6.3). As explained above Eq. (6.3), averages always refer to the indices s, m, s', m', and the quantity of physical interest is the energy-averaged cross section. The average over energy implies a summation over intermediate states, and thus over the labels s, m, s', m', and leads naturally to the introduction of averages like (B.6). Formally it is much simpler, however, to replace the matrix $V_{sm}(r)$ by an <u>ensemble</u> of matrices $V^{(\lambda)}_{sm}(r,)$, where λ runs over the ensemble, and to consider averages such as (B.6) and the ones to follow as ensemble averages. This is what we shall do throughout, omitting the label λ as we continue. It is, of course, necessary to prove the identity of energy average and of ensemble average. This has been done.[22]

Let us calculate the scattering eigenfunction $\psi(E)$ of the Hamiltonian (B.1) to second order in V. We assume that for V=0, we have $\psi(E) = |i\rangle|X(E-\varepsilon_i)\rangle$. Projecting $\psi(E)$ onto an arbitrary state $|f\rangle$, we find

$$\begin{aligned}\langle f|\psi(E)\rangle &= \delta_{if}\, X(E-\varepsilon_i) + (E+i\varepsilon - H(r)-\varepsilon_f)^{-1}\, V_{fi}\, X(E-\varepsilon_i) \\ &+ \sum_m (E+i\varepsilon-H(r)-\varepsilon_f)^{-1}\, V_{fm}\, (E+i\varepsilon-H(r)-\varepsilon_m)^{-1}\, V_{mi}\, X(E-\varepsilon_i)\ .\end{aligned} \qquad (B.7)$$

We now average (B.7) over the ensemble and see that $\bar{V}=0$, and that the second moment has the form (B.6). This yields

$$\overline{\langle f|\psi(E)\rangle} = \delta_{if}\, \{1+ \sum_m (E+i\varepsilon-H(r)-\varepsilon_i)^{-1}\, \overline{V_{im}(E+i\varepsilon-H(r)-\varepsilon_m)^{-1}V_{mi}}\} \times X(E-\varepsilon_i)\ . \qquad (B.8)$$

We see that the average has a nonvanishing component only in the state $|i\rangle$. This is a direct consequence of our statistical assumptions. In second order, the scattering function $X(E-\varepsilon_i)$ is modified. The modification may be viewed as originating from an optical-model potential given by

$$V_i^{opt} = \sum_m V_{im}(E+i\varepsilon-H(r)-\varepsilon_m)^{-1}\, V_{mi}\ . \qquad (B.9)$$

This potential is nonlocal in r, and complex. The imaginary part is negative definite and thus leads to flux absorption. We conclude: Averaging the amplitude (the wave function) leads to flux absorption. This result is, of course, well known and forms the basis of the theory of the optical model.

Let us now consider the average density matrix $\overline{\langle f|\psi_E\rangle\langle\psi_E|f'\rangle}$. To simplify the notation, we introduce the Green function

$$G_m(r,r') = G_m = (E+i\varepsilon - \varepsilon_m - H(r))^{-1}. \tag{B.10}$$

A straightforward calculation yields

$$\begin{aligned}\overline{\langle f|\psi(E)\rangle\langle\psi(E)|f'\rangle} &= \delta_{ff'}\{\overline{\langle f|\psi(E)\rangle}\cdot\overline{\langle\psi(E)|f\rangle} + \\ &+ G_f\, \overline{V_{fi}|X(E-\varepsilon i)\rangle\langle X(E-\varepsilon_i)|V_{if}}\, G_f^*\}\,.\end{aligned} \tag{B.11}$$

We note: The average density matrix is diagonal. It is different from zero in all channels f for which $\overline{V_{fi}^2} \neq 0$. This is in striking contrast to the average amplitude (B.8)! The "gain term" which describes the population of states $f \neq i$ is the last term in Eq. (B.11). It does not arise in the average over the amplitude, but only in the average of the density matrix. We leave it as an exercise to the reader to show that the trace of (B.11) equals the trace of the zeroth-order term, to second order in V. This shows that flux is conserved: The loss term--i.e., the optical-model potential term--describes flux lost from channel $|i\rangle$ to populate other channels $|f\rangle$. The population of these channels is described by the gain term. As we go to higher orders in V, we shall see that the situation, although more complex, remains essentially the same. The optical model term, or the loss term, comes from averaging the amplitude (or, as we shall see, parts of the amplitude). The gain term comes from averaging pairs of V's, one in one amplitude, the other in the other, of the density matrix.

2. The General Procedure of Calculating Averages

The scattering matrix has the form

$$\begin{aligned}S_{fi} &= \langle X(E-\varepsilon_f)|\{V_{fi} + \sum_m V_{fm}G_mV_{mi} + \sum_{m,n} V_{fm}G_mV_{mn}G_nV_{ni}+\ldots\} \\ &\quad\times |X(E-\varepsilon_i)\rangle\,.\end{aligned} \tag{B.12}$$

(For f=i, there is an additional term which is without interest for our purpose.) We abbreviate our notation further and write (B.12) in the rather symbolic form

$$S_{fi} = \langle X_f | V + VGV + VGVGV + \ldots | X_i \rangle . \tag{B.13}$$

Our problem consists in calculating the mean value of the cross section, or of $|S_{fi}|^2$. Because of the random nature of V, S_{fi} itself is a fluctuating quantity of energy, and so is $|S_{fi}|^2$. The average of this quantity will be seen to show irreversible behavior. This is in line with the fluctuation-dissipation theorem mentioned below Eq. (4.5). Taking the product of a term in the Born series (B.13) with a term in the Born series of S^*_{fi}, we see that our problem consists in calculating the mean value of a product of (m+n) factors V, m in S_{fi}, n in S^*_{fi}, for arbitrary m and n. Since V has a Gaussian distribution with zero mean, we can use the following:

<u>Theorem</u>. Let $z_1, z_2, z_3, \ldots, z_j$ be j Gaussian distributed random variables, each with zero mean. (Several or all of the z's may be identical.) Then,

$$\overline{z_1\ z_2\ z_3\ \ldots\ z_j} = 0 \quad \text{if j is odd, and}$$

$$\overline{z_1\ z_2\ z_3\ \ldots\ z_j} = \sum \quad \overline{z_{\alpha_1} z_{\alpha_2}} \cdot \overline{z_{\alpha_3} z_{\alpha_4}} \ldots \overline{z_{\alpha_{j-1}} z_{\alpha_j}}$$

if j is even. The sum extends over all possible ways to arrange the z's in pairs.

This theorem shows that the mean value of (m+n) factors V vanishes for m+n=odd, and is given by a sum (over all possible ways to arrange the V's in pairs) of products of mean values of pairs of V's for (m+n)=even. We now have to find out which contributions to this sum must be carried along, and which are negligible. We indicate this by a simple example. The mean value of a pair of V's is indicated by a line connecting the two V's: $\overline{VV}$. By the rule just given, the mean value of a product of 4 V's has the following value:

$$\overline{VVVV} = \overline{VV}\,\overline{VV} + \overline{V\,\overline{VV}\,V} + VVVV \ . \tag{B.14}$$

Let us evaluate (B.14) explicitly, using Eq. (B.6), for the following case.

$$\sum_{m\ell k} \overline{V_{jm}G_m V_{m\ell} G_\ell V_{\ell k} G_k V_{ki}} = \delta_{ji}\Big(\sum_m \overline{V_{im}^2}\, G_m\Big) G_i \Big(\sum_k \overline{V_{ik}^2}\, G_k\Big)$$

$$+ \delta_{ji} \sum_m \Big(\overline{V_{im}^2}\, G_m^2 \Big(\sum_\ell \overline{V_{m\ell}^2}\, G_\ell\Big)\Big) \tag{B.15}$$

$$+ \delta_{ji} \Big(\sum_m [\overline{V_{im}^2}\, G_m]^2\Big) G_i .$$

It is important that the reader who wishes to understand this section verifies (B.15) himself. The three terms on the r.h.s. correspond to the three terms on the r.h.s. of (B.14). The first two terms on the r.h.s. of Eq. (B.15) contain two summations over intermediate states (m and k or m and ℓ), the last, only one. For sufficiently high excitation energies, each intermediate-state summation is replaced by an integration over energies, weighted by the level density:

$$\sum_m \to \int d\varepsilon_m \; \rho(\varepsilon_m). \tag{B.16}$$

The value of such an integral is typically $\rho(\varepsilon_m^o) \cdot \Delta E$, where ε_m^o is a typical excitation energy, and $\hbar/(\Delta E)$ one of the times τ_Δ, τ_σ or τ_λ introduced in Section 6. This can be verified explicitly by using Eq. (B.15) and the parameterization (6.3). We introduce the time

$$\tau_\rho = \hbar \cdot \rho(\varepsilon_m^o) . \tag{B.17}$$

It is the Poincaré recurrence time of the system. The high excitation energies and high level densities (typical level spacings are 10 eV at neutron threshold and decrease exponentially with excitation energy), mean that τ_ρ is many orders of magnitude larger than τ_λ, τ_Δ, or τ_σ. Since the last term on the r.h.s. of Eq. (B.15) contains one sum less over intermediate states than the first two, its contribution is, by the argument just given, a factor $\mathrm{Max}(\tau_\lambda, \tau_\Delta, \tau_r)/\tau_\rho$ smaller than that of the first two. This factor is extremely small in comparison to one, and the last term on the r.h.s. of Eq. (B.15) is therefore negligible. (The incredulous reader is asked to go through the algebra himself.)

A little patient playing with mean values involving products of more than four V's will convince the reader of the following rule: We call a line connecting a pair of V's a contraction line. The rule is: All patterns with intersecting contraction lines are negligible.

Let us return to the calculation of $\overline{|S_{fi}|^2}$. If we take the average of a pair of V's, one in S_{fi}, the other in S^*_{fi}, then we say that the two V's are cross-contracted. Let us consider two neighboring cross-contracted V's appearing in S_{fi} (there is no other cross-contracted V between them). In the Born series, the possible insertions between two such V's have the form

$$G + \overline{GVGVG} + \overline{GVGVGVGVG} + \ldots \tag{B.18}$$

By the rules and definitions just given, all the V's appearing in (B.18) must be contracted with each other. This exercise can be carried out, if one follows the rule just established. As a result, one finds that the object (B.18), which I denote by G^{opt}, obeys the integral equation

$$G^{opt} = G + G\overbrace{V\,G^{opt}\,V}\,G^{opt} . \tag{B.19}$$

The reader is asked to verify this statement. The possible insertions between initial state X_i and the first cross contracted V can be similarly summed up and define an optical-model wave function

$$X_i^{opt} = X_i + G\overbrace{V\,G^{opt}\,V}\,X_i^{opt} \tag{B.20}$$

The Eqs. (B.19) and (B.20) show that the optical-model potential has the form

$$V^{opt} = \overbrace{V\,G^{opt}\,V} . \tag{B.21}$$

This is a generalization of Eq. (B.9) in which G^{opt} is replaced by G. The optical-model potential (B.9) contains only a single contraction pattern, namely $\overline{VGV}$. By contrast, the optical-model potential (B.21) contains an infinite number of contraction patterns. Among them, we find the low-order terms

$$VGV,\ VGVGVGV,\ VGVGVGVGVGV,\ VGVGVGVGVGV,\ \text{etc.}$$

This can easily be verified by iterating Eq. (B.19). The difference between the optical-model potentials $\overline{VGV}$ and $\overline{VG^{opt}V}$ is that between weak and strong coupling. Indeed, all the higher-order contraction terms just listed contribute only if the duration time of the first collision (starting with the right most V) is comparable with or larger than the time between two subsequent collisions. This should be intuitively obvious from the patterns listed. The reader is urged to verify it analytically, at least for the first few low-order contraction patterns given above, using the definitions given in Section 6 of the times involved.

After the introduction of X^{opt} and G^{opt} into $\overline{|S_{fi}|^2}$, there remain by definition only the cross contracted V's. Hence,

$$\overline{|S_{if}|^2} = \langle X_f^{opt}| V \{1 + G^{opt} V + G^{opt} V G^{opt} V + \cdots\} | X_i^{opt}\rangle \langle X_i^{opt}| \{\cdots + V G^{opt*} V G^{opt*} + V G^{opt*} + 1\} V | X_f^{opt}\rangle \tag{B.22}$$

All the terms on the r.h.s. of this expression save for the two optical-model wave functions X_f^{opt} and the last two V's define an average density matrix ρ which depends upon r and r'. It obviously obeys the integral equation

$$\rho = |X_i^{opt}\rangle\langle X_i^{opt}| + G^{opt}\, \overline{V \rho V}\, G^{opt*} \quad . \tag{B.23}$$

In terms of ρ, the average cross section is proportional to

$$\overline{|S_{fi}|^2} = \langle X_f^{opt}| \overline{V \rho V} | X_f^{opt}\rangle \ . \tag{B.24}$$

This shows that the calculation of the mean value $\overline{|S_{fi}|^2}$ has been reduced to solving the two integral equations (B.19) and (B.23). Instead of calculating the expression (B.24), one may directly find $\overline{|S_{fi}|^2}$ from the asymptotic behavior of ρ for large values of r, r'. The reader is asked to verify this statement. The only approximation that went into the derivation of our result is the assumption that $\tau_\rho \gg\!> \mathrm{Max}(\tau_\lambda, \tau_\Delta, \tau_\sigma)$. In the domain of high excitation energies or small level spacings, we have thus been able to convert our statistical assumptions on V directly into the integral equation (B.19) and (B.23), without further approximations.

Equation (B.23) can be cast into the form of a transport equation by taking the Fourier transform of this equation with respect to r-r', and by commuting both sides of the equation with the Hamiltonian. These steps are quite straightforward, and are left as an exercise for the reader who in this way should find the explicit form of the gain and loss terms appearing in Eq. (B.1). Just as in the case of lowest-order perturbation theory, the gain term is seen to originate from a cross-contracted pair of V's, while the loss term is due to G^{opt}.

By summing all order of the perturbation expansion, we have thus derived a transport equation which is valid for any strength of the coupling V.

3. Properties of the Transport Equation

The derivation just sketched leads to a transport equation for the Fourier transform of the density matrix. This transform depends both on relative distance and on intrinsic excitation. It describes the whole process as a quantum-mechanical transport phenomenon. For a heavy-ion induced reaction, one would expect that a classical description of relative motion should be adequate for most processes save for those in which all energy of relative motion is converted into intrinsic excitation. Intrinsic excitation must, of course, always be formulated quantum-mechanically. This "semiclassical" description can be carried out.[23] We cannot sketch here the details of the derivation. Suffice it to say that as a result, one finds a separate description of relative motion, and of intrinsic excitation. Relative motion is described in terms of quasi-Newtonian equations of motion with a friction force and a diffusion constant such as Eq. (4.3). Intrinsic excitation is described by an averaged density matrix $\rho_{fi}(t,t')$ which describes the probability of finding the system in state f if we started out in state i. Just as the ρ in Eq. (B.23) depends on two variables, r and r', the new ρ relating to the semiclassical approximation depends on two times, t and t'. One introduces the Fourier transform of ρ with respect to (t-t'). This defines a function $F_{fi}(T,\varepsilon)$ with $T = 1/2\ (t+t')$ and ε the variable replacing (t-t') in the Fourier transformation. $F_{fi}(T,\varepsilon)$ obeys a transport equation, the kernel of which depends on the relative motion. The friction and diffusion constants are, in turn, given by proper mean values over the distribution function F_{fi} at time T. That is the time that enters into the Newtonian equations.

With the help of F_{fi}, two different time-dependent mean values of energy can be defined. One is

$$\langle\varepsilon(T)\rangle = \sum_f \int d\varepsilon\, F_{fi}(T,\varepsilon) \cdot \varepsilon , \tag{B.25}$$

the other

$$\langle\varepsilon_f(T)\rangle = \sum_f \int d\varepsilon\, F_{fi}(T,\varepsilon) \cdot \varepsilon_f . \tag{B.26}$$

Using the definition of F_{fi} and some simple algebra, one can show that

$$\left.\begin{aligned} \langle\varepsilon(T)\rangle &= \text{Trace}\,\{(H_o + V)\,\rho_{fi}(t,t)\}\ , \\ \langle\varepsilon_f(T)\rangle &= \text{Trace}\,\{H_o\,\rho_{fi}(t,t)\} \end{aligned}\right\} \qquad (B.27)$$

Hence, $\langle\varepsilon(T)\rangle$ and $\langle\varepsilon_f(T)\rangle$ differ by the mean value of the interaction energy. Energy conservation is expressed by

$$E = E_{coll} + \langle\varepsilon(T)\rangle \qquad (B.28)$$

where E_{coll} is the total energy residing in collective motion.

In the weak-coupling approximation, we can show that $\langle\varepsilon(T)\rangle = \langle\varepsilon_f(T)\rangle$, as is to be expected. Friction and diffusion constants take simple forms, and the validity of the Einstein relation can be explicitly verified in the high-temperature limit.

In the strong-coupling limit, $\langle\varepsilon(T)\rangle \neq \langle\varepsilon_f(T)\rangle$ since the interaction energy is not negligible. As a consequence, the distribution function is not proportional to a delta function in $(\varepsilon-\varepsilon_f)$ as for weak coupling, but depends independently on ε_f and ε. Friction and diffusion constants violate the Einstein relation: For $w_o \to \infty$, the friction coefficient behaves as $(w_o)^{-1/2}$ while the diffusion coefficient is proportional to $(w_o)^{+1/2}$. The calculation of the transport coefficients require a precise knowledge of $F_{fi}(T,\varepsilon)$.

ACKNOWLEDGEMENTS

The author thanks Dr. D. M. Brink and Dr. J. Richert for numerous discussions relating to the topics here discussed, and Dr. H. M. Hofmann for extensive correspondence. He acknowledges partial support by the Science Research Council.

Much of what I have said is the fruit of many discussions over the years with my collaborators, D. Agassi and C. M. Ko.

REFERENCES

1. For a general review, see S. Chandrasekhar, Rev. Mod. Phys. 15 (1943) 1.

2. M. Blann, Ann. Rev. Nucl. Science 25 (1975) 123.

3. D. Agassi, H. A. Weidenmuller, and G. Mantzouranis, Phys. Lett. 22 C (1975) 145.

4. W. Nörenberg, Phys. Lett. 53 B (1974) 289.

5. S. Ayik, B. Schürmann, W. Nörenberg, Z. Physik A 279 (1976) 145.

6. L.G. Moretto and J. S. Sventek, Phys. Lett. 58 B (1975) 26. J. S. Sventek and L. G. Moretto, Lawrence Berkeley Report LBL-5012 (1976).

7. R. Beck and D.H.E. Gross, Phys. Lett. 47 B (1973) 143. D.H.E. Gross and H. Kalinowski, Phys. Lett. 48 B (1974) 302. H. H. Deubler and K. Dietrich, Phys. Lett. 56 B (1975) 241 and Nucl. Phys. A 277 (1977) 493. J. P. Bondorf, M. I. Sobel, and D. Sperber, Phys. Lett. 15 C (1975) 83; J. P. Bondorf, J. R. Huizenga, M.I. Sobel, and D. Sperber, Phys. Rev. C 11 (1975) 1265; A. Sherman, D. Sperber, M. I. Sobel, J. P. Bondorf, Z. Physik A 286 (1978) 11; K. Siewek-Wilczynska and F. Wilczynski, Nucl. Phys. A 264 (1976) 115.

8. M. Berlanger, P. Grange, H. Hofmann, C. Ngo, J. Richert Z. Physik A 286 (1978) 207 and references therein.

9. J. W. Negele, Lectures given at this school.

10. A. Broglia, C. Dasso, and Aa. Winther, Phys. Lett. 53 B (1974) 301; ibid. 61 B (1976) 113; R. Broglia, C. Dasso, G. Pollarolo, and Aa. Winther, IPCR Symposium on Heavy-Ion Reactions and Preequilibrium Processes, Hakone, Japan, September 1977.

11. K. Dietrich and C. Leclerq-Willain, Ann.Phys. (N.Y.) 109 (1977) 41.

12. V. M. Strutinsky, Z. Phys. A 286 (1978) 77.

13. C. E. Porter, Statistical Theory of Spectra: Fluctuations, Academic Press, New York, 1965.

14. S. Ayik, B. Schürmann, W. Nörenberg, Z. Physik A 277 (1976) 299; W. Nörenberg, Proc. European Conf. on Nucl. Phys. with Heavy Ions, Caen 1976, J. de Physique 37, C5-141 (1976). G. Wolschin, W. Nörenberg, Z. Physik A 284 (1977) 2; S. Ayik, G. Wolschin, and W. Nörenberg, Z. Physik A (in press); B. Schürmann, W. Nörenberg, and M. Simbel, Z. Physik A (in press).

15. C. M. Ko, M. Pirner, and H. A. Weidenmüller, Phys. Lett. 62B (1976) 248; D. Agassi, C. M. Ko, and H. A. Weidenmuller, Ann.Phys. (N.Y.) 107 (1977) 140; Ann.Phys. (N.Y.) in press; Phys. Rev. C, in press; Phys. Lett. 73 B (1978) 284; C. M. Ko, D. Agassi and H. A. Weidenmuller, Ann.Phys. in press; H. A. Weidenmüller, Proc. Int. Conf. Nucl. Structure, Tokyo 1977, Journ. Phys. Soc. Japan Suppl. Vol. 44 (1978) 701.

16. D. H. M. Gross, Nucl. Phys. A 240 (1975) 472.

17. F. Blocki, Y. Boneh, J. R. Nix, J. Randrup, M. Robel, A. J. Sierk, and W. J. Swiatecki, Ann.Phys. (submitted F. Beck, J. Blocki, M. Dworzecka and G. Wolschin, preprint TH Darmstadt 1977.

18. H. Hofmann and J. P. Siemens, Nucl. Phys. A 257 (1976) 165, ibid. A. 275 (1977) 464; Phys. Lett. 58 B (1975) 417; H. Hofmann, Phys. Lett. 61 B (1976) 243; XIIIth Int. Winter meeting on Nuclear Physics, Bormio, Italy, 1975; Int. Symp. on Nuclear Collisions and their Microscopic Description, Bled, Yugoslavia, 1977.

19. B. R. Barrett, S. Shlomo and H. A. Weidenmüller, Phys. Rev. C 17 (1978) 544.

20. D. H. E. Gross, H. Kalinowski, and J. N. De, Symp. Classical and Quantum Mechanical Aspects of Heavy-Ion Collisions, Heidelberg 1974, Lecture Notes in Physics Vol. 33, p. 194, Springer-Verlag, Berlin-Heidelberg-New York, 1974, and J. N. De, D.H. E. Gross and K. Kalinowski, Z. Physik A. 277 (1976) 385.

21. J. Galin, B. Gatty, D. Gurreau, M. Lefort, X. Tarrago, R. Babinet, B. Gauvin, J. Girard and H. Nifeneker, Z. Phys. A. 278 (1976) 347.

22. J. Richert and H. A. Weidenmuller, Phys. Rev. C16 (1977) 1309. J. B. French, P. A. Mello and A. Pandey, preprint 1978.

23. D. M. Brink, J. Neto, and H. A. Weidenmuller, to be published.

REACTIONS BETWEEN HEAVY IONS AT HIGH ENERGY

Jörg Hüfner

Inst. f. Theoret. Physik der Universität
and MPI für Kernphysik, Heidelberg (F.R.G)

1. SOME THEORETICAL TOOLS AND CONCEPTS.

Why do I work on heavy ion reactions at high energies? A few years ago I had finished a review on pion-nucleus physics and I felt like starting something new. Heavy ion research at the Bevalac had just begun and rumors about fantastic results were running high. They made me curious and I started to look into the first experiments which had just been finished. The fantastic results which had attracted me, found their down-to-earth explanation. Yet my excitement and fascination stay. I feel like an adventurer, I am among the first to enter a new wilderness where there is no street nor path. We all keep our eyes and ears open, always prepared for the fantastique (which so rarely happens) but still happy if a new piece of nature is discovered and understood. As Jensen once said about some of his own work:"I was digging for a treasure and found a worm."

Lectures at a summerschool must not be fair reviews of a field but may express personal and biased views. I would like to take advantage of this freedom. I will present a theory of heavy ion reactions at high energies as derived from multiple scattering theory. This connects the work to methods in intermediate energy physics (pion-nucleus scattering, for example) but, as you will see, also concepts of low-energy heavy ion work, like diffusion and thermalization, are recovered. Most of the work which I shall speak about comes from Heidelberg and is performed in collaboration with A. Abul-Magd, M. Bleszynski, A. Bouyssy, J. Knoll, C. Sander, K. Schäfer, B. Schürmann and G. Wolschin.

My approach to the field of high energy heavy ions is rather oriented to the conservative side. I am familiar with various methods in multiple scattering theory (for hadron-nucleus interactions) and I believe these methods adequate also for nucleus-nucleus collisions, although other approximation schemes may be necessary. I am particularly fascinated by the Glauber approach. It may not always be realistic, but the answers to most problems can be found analytically. Therefore it always gives a direct and clear insight into the physics underlying a problem. For this reason I also choose Glauber theory to introduce you to the main ideas which underly our theory of high-energy heavy ion reactions.

1.1 Recalling Glauber's multiple scattering formalism.

Glauber himself presented his multiple scattering formalism so well in the Boulder lectures [Gl 59] , that I can limit myself to recalling a few equations and definitions. Consider a reaction, where two nuclei in their respective ground states Φ_0 and Ψ_0 scatter into states Φ_α and Ψ_β (not necessarily bound)

$$\Phi_0 + \Psi_0 \longrightarrow \Phi_\alpha + \Psi_\beta \qquad (1.1)$$

and momentum $\underset{\sim}{q}$ is transfered between the respective centers-of-mass. In Glauber theory, the scattering amplitude for this reaction is given by

$$F_{\alpha\beta}(\underset{\sim}{q}) = \frac{k}{2\pi i}\int d^2b\, e^{-i\underset{\sim}{q}\underset{\sim}{b}} \langle \Phi_\alpha(\underset{\sim}{x}_j)\, \Psi_\beta(\underset{\sim}{x}_\ell) | \prod_{\ell,j} (1-\Gamma(\underset{\sim}{x}_\ell - \underset{\sim}{x}_j + \underset{\sim}{b})) - 1 | \Phi_0(\underset{\sim}{x}_j) \cdot \Psi_0(\underset{\sim}{x}_\ell) \rangle . \qquad (1.2)$$

Here $\underset{\sim}{k}$ is the momentum per nucleon of the projectile and $\underset{\sim}{b}$ the impact parameter. The operator, which connects the initial states to the final ones contains the profile functions Γ . They are related to the nucleon-nucleon (NN) scattering amplitude f_{NN} via

$$f_{NN}(\underset{\sim}{q}) = \frac{ik}{2\pi}\int d^2b\, e^{-i\underset{\sim}{q}\underset{\sim}{b}}\, \Gamma(\underset{\sim}{b}) \quad \text{or} \quad \Gamma(\underset{\sim}{b}) = \frac{1}{2\pi i k}\int d^2q\, e^{i\underset{\sim}{q}\underset{\sim}{b}}\, f_{NN}(\underset{\sim}{q}) . \tag{1.3}$$

Eq. (1.2) generalizes the expression for nucleon-nucleus collisions given in [Gl 59] to the case of heavy ion reactions. The basic physics is simple: Two nuclei approach each other on straight lines and interpenetrate. Each nucleon j from the projectile nucleus may collicle with each target nucleon ℓ (provided they meet on their straight line) via the free NN scattering amplitude f_{NN} . There is no interaction among projectile nor among target nucleons. This is a consequence of the fixed-scatterer approximation, which also implies the neglect of excitation energies of the states Φ_α and Ψ_β . The basic problem in Glauber theory is how to evaluate eq. (1.2) for the physical situation of relevance.

Experiments in high- and low-energy heavy ion physics are usually inclusive: Only part of all the observables which could be measured are actually measured. In a typical experiment the quantum states of the nuclei after the reaction are not fully identified (as was assumed in eq. (1.1)) but only one aspect is measured. For instance the momentum $\underset{\sim}{p}$ of one nucleon or one pion is determined, the remainder is unobserved for which one writes

$$\Phi_o + \Psi_o \longrightarrow N(\underset{\sim}{p}) + X , \tag{1.4}$$

where N stands for "nucleon" and X for the unobserved. The calculation of inclusive cross sections requires special methods which will be dealt in this lecture.

Two simple but very important aspects of inclusive reactions will be shown. They are already present in nucleon-nucleus scattering. Since this case is formally easier, we stick to it in Lecture 1 . We also forget correlations in the nucleus and the inelasticity of NN collisions (e.g. processes like $NN \to N\Delta$), without losing the essential physics. This is not evident but is borne out by more accurate calculations. To be specific, we study the reaction

$$N(\underset{\sim}{k}) \quad + \quad \Psi_0 \quad \longrightarrow \quad N(\underset{\sim}{k}+\underset{\sim}{q}) + X^{(n)}, \qquad (1.5)$$

where an incoming nucleon with momentum $\underset{\sim}{k}$ hits the target nucleus Ψ_0, knocks n nucleons out of the target and receives transverse momentum $\underset{\sim}{q}$. The inclusive cross section for this reaction is

$$\frac{d^2\sigma^{(n)}}{d\underset{\sim}{q}} = \sum_{\alpha^{(n)}} \left| \int \frac{d^2b}{2\pi}\, e^{-i\underset{\sim}{q}\underset{\sim}{b}} \langle \Psi_{\alpha^{(n)}} | \prod_j (1-\Gamma_j) | \Psi_0 \rangle \right|^2 . \qquad (1.6)$$

The sum over all final states $\Psi_{\alpha^{(n)}}$ (which have n particles in excited states) is the mathematical expression of the inclusive nature of the experiment: All events of type (1.5) are accepted, irrespective of the quantum states of the n nucleons. For the very simple form of the nuclear wave functions

$$\langle \underset{\sim}{x}_1, \ldots \underset{\sim}{x}_A | \Psi_0 \rangle = \varphi_0(\underset{\sim}{x}_1) \cdots \varphi_0(\underset{\sim}{x}_A),$$

$$\langle \underset{\sim}{x}_1, \ldots \underset{\sim}{x}_A | \Psi_{\alpha^{(n)}} \rangle = \varphi_{\alpha_1}(\underset{\sim}{x}_1) \cdots \varphi_{\alpha_n}(\underset{\sim}{x}_n)\, \varphi_0(\underset{\sim}{x}_{n+1}) \cdots \varphi_A(\underset{\sim}{x}_A) \qquad (1.7)$$

eq. (1.6) simplifies to

$$\frac{d^2\sigma^{(n)}}{d\underset{\sim}{q}} = \binom{A}{n} \sum_{\substack{\alpha_i\,(\neq 0)\\ i=1,\ldots,n}} \left| \int \frac{d^2b}{2\pi}\, e^{-i\underset{\sim}{q}\underset{\sim}{b}} \prod_{j=1}^{n} \langle \varphi_{\alpha_i} | \Gamma | \varphi_0 \rangle \left(1 - \langle \varphi_0 | \Gamma | \varphi_0 \rangle\right)^{A-n} \right|^2 . \qquad (1.8)$$

The factor $\binom{A}{n}$ reflects the set-up that we do not care which n nucleons are ejected from the target. We use

(a) closure for the φ_α (note that Glauber theory has no restrictions due to energy conservation)

(b) the short range nature of $\Gamma(\underset{\sim}{b})$ (the profile function falls off over a distance of 1 fm which is small compared with typical nuclear diameters)

(c) the following relations for the profile function, which follow from eq. (1.3)

$$\int d^2b\,\Gamma(\underset{\sim}{b}) = \frac{2\pi}{ik} f_{NN}(0) = \frac{i}{2}\,\sigma_{NN}^{tot}\left(1 - i\,\frac{\mathrm{Re}\,f_{NN}(0)}{\mathrm{Im}\,f_{NN}(0)}\right), \tag{1.9}$$

$$\int d^2s\,\Gamma(\underset{\sim}{b}-\underset{\sim}{s})\,\Gamma^*(\underset{\sim}{s}) = \int\frac{d^2q}{k^2}\,e^{i\underset{\sim}{q}\underset{\sim}{b}}\,\frac{d\sigma_{NN}}{d\Omega}(\underset{\sim}{q}) \equiv \tilde{\sigma}_{NN}(\underset{\sim}{b})\,;\ \tilde{\sigma}_{NN}(0) = \sigma_{NN}^{tot}\,. \tag{1.10}$$

Then the cross section eq. (1.8) can be evaluated.

1.2 Geometry: Overlap volume.

First we discuss the integrated cross section for the knock-out of n nucleons:

$$\boxed{\sigma^{(n)} \equiv \int d^2q\,\frac{d^2\sigma^{(n)}}{d\underset{\sim}{q}} = \int d^2b \binom{A}{n} P^n(\underset{\sim}{b})\,(1-P(\underset{\sim}{b}))^{A-n}.} \tag{1.11}$$

This formula will be used to discuss the geometrical aspect of inclusive reactions at high energy. The cross section $\sigma^{(n)}$ is an integral in the impact parameter over a binominal distribution, where $P(\underset{\sim}{b})$ plays the role of a probability. The derivation of eq. (1.11) leads to

$$P(\underset{\sim}{b}) = T(\underset{\sim}{b})\,\sigma_{NN}^{tot}\left(1 - \left(\frac{2\pi}{k}\right)^2 \frac{|f_{NN}(0)|^2}{\sigma_{NN}^{tot}}\,T(\underset{\sim}{b})\right) \tag{1.12}$$

with the "thickness" function

$$T(\underset{\sim}{b}) = \int_{-\infty}^{+\infty} dz\,\rho(\underset{\sim}{b},z)\ ;\quad \int d^2b\,T(\underset{\sim}{b}) = 1 \tag{1.13}$$

Here $\rho(\underset{\sim}{x})$ is the single particle density of the ground state wave function Ψ_0. The second term in eq. (1.12) is small (few percent) and will be neglected in the following.

The geometrical nature of $P(\underset{\sim}{b})$ is most clearly exhibited if we use a uniform density distribution with radius R for the target nucleus.

$$\rho(\underset{\sim}{x}) = \Theta(R^2 - \underset{\sim}{x}^2) \Big/ \tfrac{4\pi}{3} R^3 \tag{1.14}$$

Then

$$P(\underset{\sim}{b}) \approx \sigma_{NN}^{tot} T(\underset{\sim}{b}) = \frac{2\sqrt{R^2 - b^2}\;\sigma_{NN}^{tot}}{\frac{4\pi}{3} R^3}\,\Theta(R - b) \tag{1.15}$$

The volume in the numerator corresponds to a cylinder of length $2\sqrt{R^2 - b^2}$ and σ_{NN}^{tot} as the cross section area. The projectile nucleon with impact parameter $\underset{\sim}{b}$ "cuts" exactly this volume out of the target nucleus, i.e. it scatters by all nucleons in this "overlap volume". The total volume of the target nucleus is in the denominator of eq. (1.15).

Now we can understand eq. (1.11): The cross section for the knock-out of n nucleons is proportional to the probability $P(\underset{\sim}{b})$ that n nucleons can be found in the overlap volume. This probability depends on the length of the cylinder and hence on $\underset{\sim}{b}$. Indeed for a given n the integrand is maximal for a value b_n which satisfies

$$2\sqrt{R^2 - b_n^2} = n\lambda , \tag{1.16}$$

where λ is the mean free path

$$\lambda^{-1} = \sigma_{NN}^{tot}\,\rho_0 \quad ; \quad \rho_0 = \frac{A}{\frac{4\pi}{3} R^3} \tag{1.17}$$

Therefore n nucleons are preferentially ejected if the length of the cylinder corresponds exactly to n mean free paths. Therefore values of $n > 2R/\lambda$ have rather small cross sections. Since typical values are $\lambda = 2\,fm$ (corresponding to $\sigma_{NN}^{tot} = 40\,mb$, $\rho_0 = 0.17\,fm^{-3}$), only few nucleons can be ejected by a nucleon when it ploughs through a nucleus.

1.3 Dynamics: Boltzmann Equation.

Here we study the cross section eq. (1.5), but only the outgoing nucleon $N(\underset{\sim}{q})$ is measured without looking at the target at all (sum over n). We start from eq. (1.8), use the approximations already stated and add two new ones

(d) since $A \gg n$ (see above), $\binom{A}{n} \approx \frac{A^n}{n!}$

(e) instead of the variables $\underset{\sim}{b}$ and $\underset{\sim}{b}'$, relative ($\underset{\sim}{\beta} = \underset{\sim}{b} - \underset{\sim}{b}'$) and c.m. coordinates ($\underset{\sim}{B} = (\underset{\sim}{b} + \underset{\sim}{b}')/2$) are introduced, then

$$T\left(\underset{\sim}{B} \pm \frac{\underset{\sim}{\beta}}{2}\right) \approx T(\underset{\sim}{B}) \pm \frac{1}{2}\underset{\sim}{\beta}\cdot\underset{\sim}{\nabla}_B T(\underset{\sim}{B}), \tag{1.18}$$

since the integration over $\underset{\sim}{\beta}$ is limited by the extension of $\tilde{\sigma}_{NN}(\underset{\sim}{\beta})$, which is of order 1 fm or less.

Then the one-nucleon inclusive cross section can be written [Hü 78] in the form ($\underset{\sim}{x} = (\underset{\sim}{B}, z)$)

$$\frac{d^2\sigma^{incl}}{d\underset{\sim}{q}} \equiv \sum_{n=1}^{A} \frac{d^2\sigma^{(n)}}{d\underset{\sim}{q}} = \int d^2\underset{\sim}{B}\left\{W(\underset{\sim}{x},\underset{\sim}{q}) - W_0(\underset{\sim}{x},\underset{\sim}{q})\right\}_{z=\infty} \tag{1.19}$$

with the definitions

$$W(\underset{\sim}{x},\underset{\sim}{q}) = \int \frac{d^2\beta}{(2\pi)^2} \exp\Big\{-i\underset{\sim}{\beta}\Big(\underset{\sim}{q} - \frac{2\pi}{k} A \,\mathrm{Re}\, f_{NN}(0)\, \underset{\sim}{\nabla}_B T(\underset{\sim}{x})\Big) - $$
$$- A\,[\sigma_{NN}^{tot} - \tilde{\sigma}_{NN}(\underset{\sim}{\beta})]\,T(\underset{\sim}{x})\Big\} ,$$
$$T(\underset{\sim}{x}) = \int_{-\infty}^{z} \rho(\underset{\sim}{B}, z')\, dz' \qquad (1.20)$$

and W_0 is obtained from (1.20) by setting $\tilde{\sigma}_{NN}(\beta) = 0$. Although eqs. (1.19) and (1.20) solve the problem of inclusive cross sections (they are already in [Gl 59] and [GM 70]), they do not provide insight into the physics of the problem. But we are after the physics. Therefore we derive W with respect to z and obtain a differential equation

$$\boxed{v \frac{d}{dz} W(\underset{\sim}{x},\underset{\sim}{q}) + \underset{\sim}{F}\cdot\underset{\sim}{\nabla}_q W(\underset{\sim}{x},\underset{\sim}{q}) = v\rho(\underset{\sim}{x}) \int \frac{d^2q'}{k^2} \frac{d\sigma_{NN}}{d\Omega}(\underset{\sim}{q}-\underset{\sim}{q}')\cdot\{W(\underset{\sim}{x},\underset{\sim}{q}') - W(\underset{\sim}{x},\underset{\sim}{q})\}} \qquad (1.21)$$

$$v \frac{d}{dz} W_0(\underset{\sim}{x},\underset{\sim}{q}) + \underset{\sim}{F}\cdot\underset{\sim}{\nabla}_q W_0(\underset{\sim}{x},\underset{\sim}{q}) = -v\rho(\underset{\sim}{x})\,\sigma_{NN}^{tot}\, W_0(\underset{\sim}{x},\underset{\sim}{q}) . \qquad (1.22)$$

<u>Eq. (1.21) is the Boltzmann equation for a particle of high energy which passes through a medium</u>. Since v is the velocity, z/v plays the role of the time along the straight line trajectory (defined by the impact parameter $\underset{\sim}{B}$). At times $t \to -\infty$, i.e. $z \to -\infty$, the initial condition, calculated from eq. (1.20) is

$$W(\underset{\sim}{x},\underset{\sim}{q}) = W_0(\underset{\sim}{x},\underset{\sim}{q}) = \delta^{(2)}(\underset{\sim}{q}) . \qquad (1.23)$$

i.e. no transverse momentum. While the projectile passes through the nucleus, the momentum distribution changes for two reasons

(a) the projectile particle is deflected by the force $\underset{\sim}{F}(\underset{\sim}{x})$

$$\underset{\sim}{F} = \frac{4\pi}{2M} \operatorname{Re} f_{NN}(0) \underset{\sim}{\nabla}_B \rho(\underset{\sim}{x}) = -\underset{\sim}{\nabla}_B \operatorname{Re} \mathcal{U}_{opt}(\underset{\sim}{x}), \tag{1.24}$$

where the expression for the optical potential (neglecting correlations) is used. $\underset{\sim}{F}$ is the classical force derived from a potential.

(b) Individual NN collisions change the distribution $W(\underset{\sim}{x},\underset{\sim}{q})$. This process is contained in the collision term on the r.h.s. of eq. (1.21). The new distribution after collision is obtained from the old one, by folding $W(\underset{\sim}{x},q')$ with the NN cross section $d\sigma_{NN}(q-q')/d\Omega$. The second term in the curly bracket, usually called the "loss term", helps to conserve probability, so that

$$\int d^2q \int \frac{d^2q'}{k^2} \frac{d\sigma_{NN}}{d\Omega}(\underset{\sim}{q}-\underset{\sim}{q}') \left\{ W(\underset{\sim}{x},\underset{\sim}{q}') - W(\underset{\sim}{x},\underset{\sim}{q}) \right\} = 0 \tag{1.25}$$

The same equation eq. (1.21) is also used in low energy heavy ion reactions (cf. the lecture by Weidenmüller), to describe inclusive cross sections. Therefore we are not surprised that expressions like "diffusion" or "thermalization" ("nuclear fire ball") turn up also at high energies. Indeed, if the distribution function $W(\underset{\sim}{x},\underset{\sim}{q})$ has broadened sufficiently so that $W(\underset{\sim}{x},\underset{\sim}{q}')$ varies slowly over the range where $d\sigma_{NN}/d\Omega$ falls off, a Taylor expansion in $(\underset{\sim}{q}'-\underset{\sim}{q})$ of the curly bracket in eq. (1.21) leads to

$$v\frac{d}{dz} W(\underset{\sim}{x},\underset{\sim}{q}) = -\underset{\sim}{F}\cdot\underset{\sim}{\nabla}_q W(\underset{\sim}{x},\underset{\sim}{q}) + D(\underset{\sim}{x})\, \Delta_q W(\underset{\sim}{x},\underset{\sim}{q}),$$

$$D(\underset{\sim}{x}) = v\rho(\underset{\sim}{x}) \frac{1}{4} \int \frac{d^2q}{k^2}\, \underset{\sim}{q}^2 \frac{d\sigma_{NN}}{d\Omega}, \tag{1.26}$$

which is a diffusion equation (Fokker-Planck-equation) in momentum space. The classical force $\underset{\sim}{F}$ plays the role of the drift coefficient and the factor in front of the Laplacian $\Delta_q W$ is the diffusion coefficient. Irrespective of how the elementary cross section $d\sigma_{NN}/d\Omega$ looks like, the distribution function ($W(\underset{\sim}{x},\underset{\sim}{q})$) will become a Gaussian, centered around a mean momentum determined by the classical force. When the Gaussian is reached, we may say this degree of freedom is "thermalized".

We come back to the inclusive cross section, eq. (1.19). It exhibits the momentum distribution of a projectile nucleon after it has passed through nuclear matter. The term $W_0(\underset{\sim}{x}, \underset{\sim}{q})$ describes all events, where the projectile did not collide, i.e. it describes elastic scattering. Since we wanted to exclude it, the term W_0 is subtracted out.

2. PERIPHERAL (FRAGMENTATION) REACTIONS.

A fast projectile nucleus ${}^{A_P}Z_P$ hits a target nucleus ${}^{A_T}Z_T$, they interact and one fast fragment ${}^{A_F}Z_F$ is observed in the counter, while everything else remains undetected .

$$ {}^{A_P}Z_P + {}^{A_T}Z_T \rightarrow {}^{A_F}Z_F + X , \qquad (2.1) $$

Fragmentation reactions of the type eq. (2.1) have been studied at Berkeley for several targets and projectiles at 1 and 2 GeV/nucleon [LG 75] . The same reactions have been measured at 20 MeV/nucleon and some striking similarities are found [B 76] . First a word to the identification. The fragment ${}^{A_F}Z_F$ in reaction (2.1)is that nucleus which comes out with roughly the projectile velocity, but a few nucleons are lost ($A_F < A_P$, $Z_F \leq Z_P$). At 20 MeV/nucleon the reaction (2.1) is called quasi-elastic to distinguish it from the deep-inelastic ones, where a major part of the incoming energy per nucleon is dissipated. The name "peripheral" is rather a theoretical classification and will become clear later. We want to understand the reactions eq. (2.1) at high and low energies, and begin at the high energy side. The "abrasion-ablation" model proposed by Bowman et al. [BS 75] contains the essential physics: While the projectile passes by the target nucleus, the overlapping nuclear matter is scraped away (abrasion). The remaining prefragments are highly excited and decay (ablation).

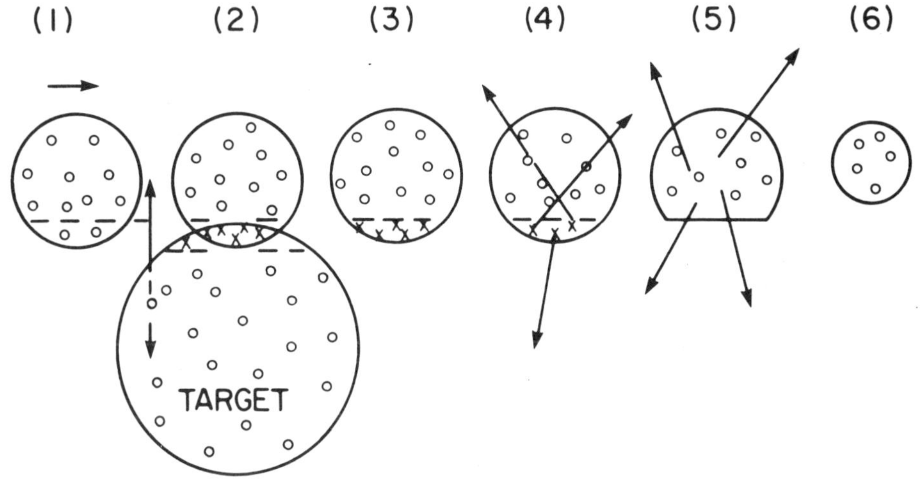

Fig. 2.1 A schematic representation of a fragmentation reaction. (1) Projectile before the reaction. (2) Nucleons of projectile and target in the overlap volume are knocked-on. (3) End of knock-on: Projectile and target nuclei have separated. Knocked-on nucleons are still at their positions but have momenta. (4) They move out of the nucleus, exciting the remaining prefragment. (5) The prefragment evaporates a few nucleons. (6) A stable nucleus (final fragment) proceeds to the counter.

To put this model into a quantitative form, we divide the fragmentation reaction into three steps: knock-on, excitation and decay.

(a) Knock-on: Projectile and target nuclei overlap. Nucleons in the overlap region are "knocked-on", i.e. they receive energy and momentum in each nucleon-nucleon (NN) collision (on the average $\langle q^2\rangle^{1/2}$=400 MeV/c, $\epsilon = \langle q^2\rangle / 2m \simeq 80$ MeV at 2 GeV/nucl. incident energy). The knock-on stage is a high-energy phenomenon since particles at relativistic energies collide. The methods of high energy multiple scattering, in particular Glauber theory, should apply here. Compared to the incident momentum k of several GeV/nucl., the momentum transfer q in each collision is not large. Therefore the motion due to the knock-on momentum q is slow. We may roughly say: While a target nucleon is knocked-on during the overlap phase, it starts to move only when projectile and target have separated. The same holds true for a knocked-on projectile nucleon in the rest--system of the projectile.

(b) Excitation of the prefragment: The energy transfer $\epsilon \simeq 80$ MeV during the knock-on stage is large compared to typical binding energies in a nucleus. Therefore the knocked-on nucleons leave the nucleus which they belonged to, and a "prefragment" is left behind. It is excited for two reasons: Holes are created in the knock-on stage, and, on its way out, the knocked-on nucleon collides with other nucleons and looses some of the initial energy ϵ. The description of the energy loss is difficult since the energy $\epsilon \approx 80$ MeV is intermediate. Neither high-energy concepts like Glauber theory nor those of low energy (precompound models) apply.

(c) Decay: The prefragment is left in an excited state, the excitation probability of which is computed in stage two. The prefragment decays by emitting nucleons and/or heavier clusters. Presumably this is a low-energy phenomenon. Therefore we teat it by compound-nucleus evaporation theory. The approximately isotropic distribution of "black tracks" (corresponding to proton energies less than 30 MeV) in emulsion experiments supports the picture.

The theory, which I shall present in the next sections is worked out in collaboration with A. Abul--Magd, K. Schäfer and B. Schürmann [HS 75], [AH 78] .

2.1 The knock-on cross section.

This is a high-energy phenomenon and Glauber theory should be applicable. If the target is a proton, and we want to calculate the knock-on of n nucleons in the projectile, the formulae from section 1.2 can be readily taken over (except that there the proton was the projectile and the target was the heavy nucleus). The cross section for the knock-on of n nucleons in the projectile nucleus is derived in eq. (1.11) to be

$$\sigma^{(n)} = \int d^2b \binom{A_P}{n} P(\underset{\sim}{b})^n \left(1 - P(\underset{\sim}{b})\right)^{A-n} \qquad (2.2)$$

where the probability $P(\underset{\sim}{b})$

$$P(\underset{\sim}{b}) = \sigma_{NN}^{tot} \int_{-\infty}^{+\infty} dz \, \rho_P(\underset{\sim}{b}, z) \qquad (2.3a)$$

(proton as target)

is identified as the overlap volume between the projectile nucleus and the tube cut by the target protons. For nucleus-nucleus collisions the same formula eq. (2.2) was derived [HS 75] with

$$P(\underset{\sim}{b}) = \int dz \, d^2s \, \rho_P(\underset{\sim}{b}+\underset{\sim}{s}, z) \left\{ 1 - \exp\left(-A_T \sigma_{NN}^{tot} \int_{-\infty}^{+\infty} dz' \rho_T(\underset{\sim}{s}, z')\right)\right\}. \qquad (2.3b)$$

(nucleus-nucleus)

It has the same geometrical meaning: The overlap volume in the projectile nucleus is cut by the target density $\rho_T(\underset{\sim}{x})$. Thus knock-on or abrasion cross sections are geometric in nature. The more nucleons are knocked-on in the projectile the stronger the two nuclei overlap. This can be seen in Fig. 2.2 . Here the integrand of eq. (2.2) is shown for n=2 (^{14}C) up to n=7 (^{9}C). The impact parameter b_n where the distribution peaks, shrinks by $1/2$ fm for $\Delta n = -1$. The width Δb of the distribution in b space is remarkable big: Δb is of order of 2 fm, rather independent of n . This width may be related to the diffuseness of projectile and target densities.

The result of a calculation with eq. (2.2), with realistic densities for target and projectile and σ_{NN}^{tot} = 40mb is given in Tab. 2.1 together with

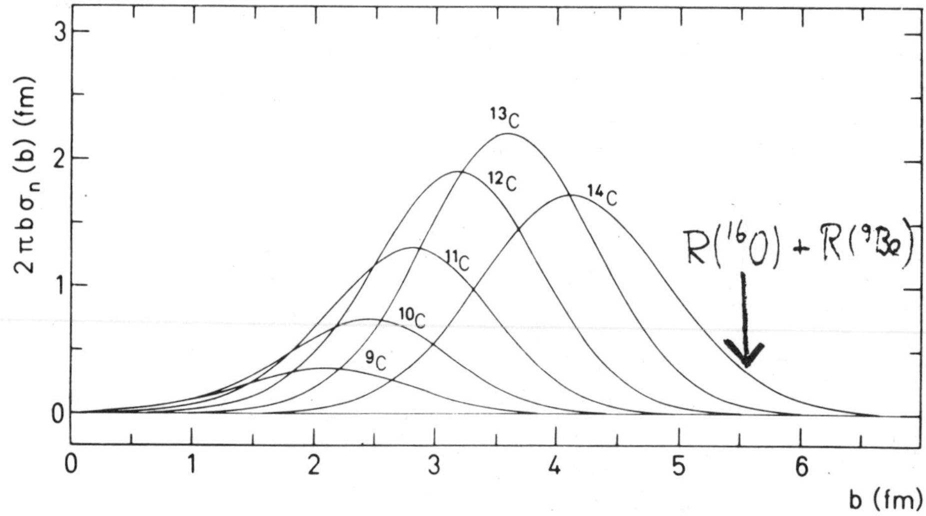

Fig. 2.2 The abrasion cross section $\sigma^{(n)}(\underset{\sim}{b})$ as a function of the impact parameter for the production of carbon isotopes in the reaction $^{16}O + {}^{9}Be \rightarrow C + X$ [HS 75].

experiment. The order of magnitude of the experimental cross sections is hit (without free parameter!) except for ^{14}O. Also the target number dependence ($\propto A_T^{0.2}$) is reproduced (c.f. the more detailed investigation by Bleszynski and Sander [BS 78]). But one cannot deny significant discrepancies. The values for one-nucleon removal ($A_F = 15$) carry the cleanest signature: The abrasion cross section is too large by a factor of two. It was this factor which lead us to propose the following excitation mechanism (not present in ref. [BS 75]).

2.2 Excitation of the prefragment.

What happens to a knocked-on nucleon? It has received a momentum transfer q ($\langle q^2 \rangle^{1/2} \approx 400$ MeV/c) and a mean energy $\langle \epsilon \rangle \approx \langle \tilde{q}^2 \rangle / 2M = 80$ MeV. $\underset{\sim}{q}$ is essentially transverse (in the plane of the impact parameter, Fig. 2.3).

$^{A_F}Z_F$	$^{16}O+p \rightarrow {}^{A_F}Z_F+X$ exp.	abrasion only	$^{16}O+{}^{9}Be \rightarrow {}^{A_F}Z_F+X$ exp.	abrasion only
^{15}O	27.3 ± 2.6	63	43.0 ± 2.1	130
^{15}N	34.2 ± 3.3	63	54.1 ± 2.7	130
^{14}O	0.75 ± 0.12	11	1.6 ± 0.1	33
^{14}N	31.0 ± 3.2	25	49.5 ± 4.0	75

Table 2.1: A few selected cross sections [mb] for fragmentation reactions. The experimental ones are compared with those calculated on the basis of "abrasion alone", i.e. the prefragment is not excited. Note:

(a) fragmentation cross sections are not small compared to the total reaction cross sections ($\sigma_{react} \approx 300$ mb and 1000mb for p and ^{9}Be as targets, respectively).

(b) the experimental cross sections increase roughly all by the same factor when going from p to Be as target. This dependence can be parametrized as $A_T^{0.22}$.

(c) calculated and experimental cross sections differ by roughly a factor 2 for one--nucleon removal ($A_F = 15$). This discrepancy points to important physics.

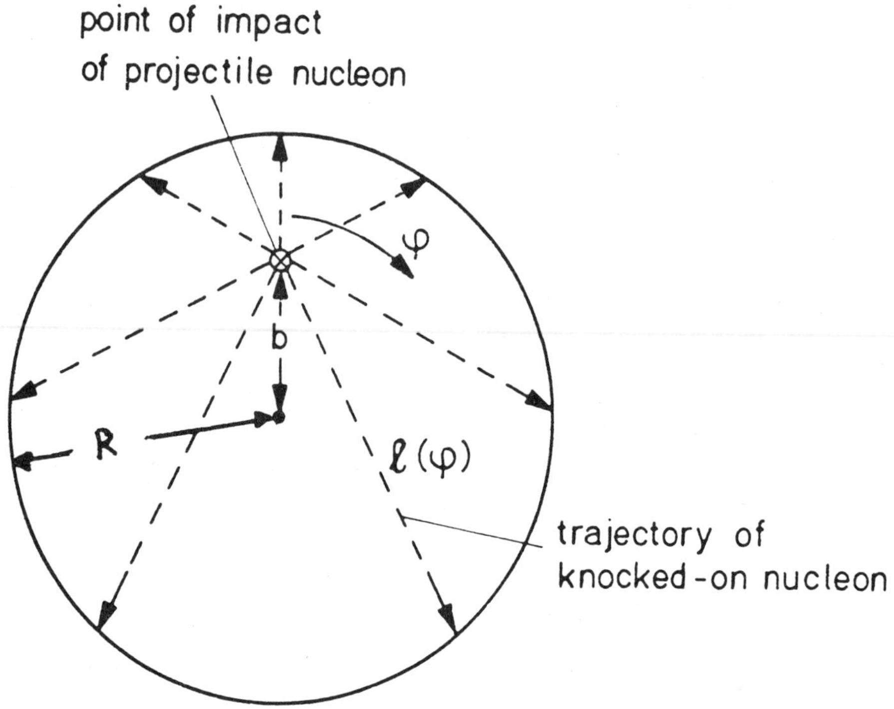

Fig. 2.3 The motion of a knocked-on nucleon after it has been hit at the impact parameter b. The motion is assumed to proceed in the impact parameter plane on a straight line characterized by the angle φ .

For simplicity we assume that the knocked-on nucleon moves in the impact parameter plane. We approximate its trajectory by a straight line, Fig. 2.3 . Its length ℓ depends on the azimuthal scattering angle φ via

$$\ell(\varphi) = +\sqrt{R^2 - b^2 \sin^2\varphi} - R\cos\varphi . \tag{2.4}$$

Of course, the straight line is only true for the mean trajectory, i.e. when the deflections due to individual NN collisions (similar to Brownian motion) are averaged out.

Since the knock-on usually happens close to the surface (see Fig. 2.2), there are "short ways" out of the nucleus ($\ell(\varphi) \ll R$ for $|\varphi| \lesssim \pi/2$) and "long ways" (for $3/2\,\pi > \varphi > \pi/2$). The probability for each case is one half. If the knocked-on nucleon leaves on the short way, a stable prefragment remains. However, a lot of energy is deposited if the knocked-out nucleon leaves on the long way, and the prefragment will lose further nucleons by decay. This is the intuitive picture for the factor of two discrepancy for $A_F=15$ in Tab. 2.1 .

A quantitative calculation proceeds as follows: Along the trajectory the nucleon collides with other nucleons and looses energy according to

$$\frac{d\epsilon}{d\ell} = -\frac{\alpha}{\lambda}\,\epsilon \tag{2.5}$$

where ϵ is the actual energy of the knocked-on nucleon and λ is the mean free path between two collisions. The dimensionless quantity α describes the fraction $d\epsilon$ of the energy ϵ which is lost in one collision. It is related to the elementary NN collision by [HK 77]

$$\alpha = \int_0^1 d\cos\theta \;\sin^2\frac{\theta}{2}\,\frac{d\sigma^{NN}}{d\Omega_{CM}}(\epsilon) \Big/ \int_0^1 d\cos\theta\,\frac{d\sigma^{NN}}{d\Omega_{CM}}(\epsilon) \tag{2.6}$$

The differential NN cross section in eq. (2.6) is taken for the energy of the knocked-on nucleon. Since this energy is relatively low ($\epsilon <$ 100 MeV), $d\sigma^{NN}/d\Omega_{CM}$ is close to isotropy and $\alpha \doteq 1/4$.

The probability distribution of excitation energies is then calculated as follows. In the knock-on stage the nucleon receives momentum $\underset{\sim}{q}$ whose direction defines the angle φ in the impact parameter plane and whose magnitude defines the initial energy $\epsilon = q^2/2M$ of the knocked-on nucleon. Since in a high-energy collision the cross section has the form

$$\frac{d\sigma}{d\Omega} \propto \exp\left(-\underline{q}^2/\langle q^2\rangle\right) , \tag{2.7}$$

the distribution of azimuthal angles φ is isotropic and the probability for initial energy ϵ is

$$W(\epsilon) = \frac{1}{\langle\epsilon\rangle} \exp\left(-\frac{\epsilon}{\langle\epsilon\rangle}\right) ; \quad \langle\epsilon\rangle = \frac{\langle q^2\rangle}{2M} . \tag{2.8}$$

The nucleon travels a distance $\ell(\varphi)$ before it leaves the nucleus with an energy

$$E_f = \epsilon\, e^{-\alpha\, \ell(\varphi)/\lambda} \tag{2.9}$$

The difference $E = (\epsilon - E_f)$ is the excitation energy of the prefragment. With the distribution W of initial energies and the isotropy in φ the probability distribution S(E) of excitation energies is calculated as

$$S(E) = \int_0^\infty d\epsilon\, W(\epsilon) \int_0^{2\pi} \frac{d\varphi}{2\pi}\, \delta\left(E - \epsilon\left(1 - e^{-\alpha\, \ell(\varphi)/\lambda}\right)\right) . \tag{2.10}$$

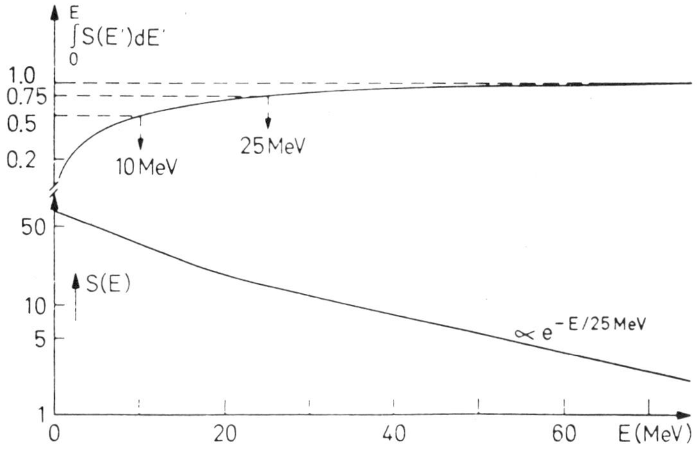

Fig. 2.4 The distribution S(E) of excitation energies in the prefragment after one knocked-on nucleon has traversed and left the target, eq. (2.10). The parameters are given in the text. Below, the distribution function S(E), above $\int_0^E dE' S(E')$ both as a function of the excitation energy E.

A typical example is shown in Fig. 2.4 . The parameters are $\langle\epsilon\rangle$ =80 MeV, α =0,25 , b=3 fm and R =3.5 fm. The distribution extends to high energies yet most probability is concentrated in the lower end. In fact, about 1/2 of the probability is concentrated below an excitation energy of 10 MeV, it corresponds to the short ways out. The distribution of excitation energies S(E), eq. (2.10) is calculated for one knocked-on nucleon. For n particle knock-on, the prefragment has $A_p - n$ nucleons and a distribution of excitation energies is obtained by folding S(E) n-times.

2.3 Decay of the prefragment.

In the third and last stage, the excited prefragment decays. We have no detailed theory for this decay. We assume that the excitation energy is distributed over many degrees of freedom and a compound nucleus is formed whose energy is lost by evaporation of nucleons and clusters. The hypothesis of evaporation finds a partial support in the observed isotropy of black tracks in fragmentation reactions and in the success of our calculation. We use a simplified version of an evaporation cascade [BN 73] . A prefragment at a given excitation energy emits protons, neutrons and α's. Typical results are shown in Tables 2.2 and 2.3 .

Fragment	Cross section [mb] Calculation	Cross section [mb] Experiment
^{15}O	29 ± 4	27.3 ± 2.6
^{15}N	31 ± 5	34.2 ± 3.3
^{14}O	0.4 ± 0.3	0.75 ± 0.12
^{14}N	31 ± 5	31.0 ± 3.2
^{14}C	5 ± 4	3.69 ± 0.38
^{13}C	13 ± 10	17.8 ± 1.7
^{13}B	0.3 ± 0.2	0.31 ± 0.05
^{12}C	13 ± 7	32.3 ± 4.8
^{12}B	1.0 ± 0.8	1.45 ± 0.17

Table 2.2 Fragmentation cross sections for the reaction $^{16}O + p \rightarrow {}^{A_F}Z_F + X$ at 2 GeV/nucl.. The experimental values are from ref. [LG 75] . The calculated values are obtained as described in the text, the error bars reflect estimated uncertainties of the approach.

The calculated values are given error bars. They reflect uncertainties of the calculation which are not fully controlled, like the production of a Δ in the knock-on stage or the relative importance of α decay and nucleon decay. Otherwise there is no free parameter in the calculation.

Two years ago, at the summer study in Berkeley, Scott presented data for the reaction

$$^{16}O + {}^{208}Pb \rightarrow {}^{A_F}Z_F + X \qquad (2.11)$$

measured at 20 MeV/nucleon and compared them with the results from the Bevalac at 2.1 GeV/nucleon [B 76] .

	20 MeV/A		2100 MeV/A	
$^{A_F}Z_F$	Exp.	Calc.	Exp.	Calc.
^{15}O	38 ± 19	54 ± 13	86 ± 22 a)	124 ± 25
^{15}N	211 ± 53	143 ± 36	110 ± 26 a)	128 ± 26
^{14}O	-	-	2.8 ± 1.5	11 ± 2
^{14}N	140 ± 42	86 ± 21	71.2 ± 23	83 ± 16
^{14}C	43^{+7}_{-22}	20 ± 5	12.3 ± 2	22 ± 5
^{12}C	198 ± 30	247 ± 41	126 ± 25	87 ± 13 b)

Table 2.3 Experimental and calculated fragmentation cross sections for the reaction eq. (2.11) [HS 78] .

a) these experimental values have been corrected for Coulomb removal of a nucleon following ref. [HL 76] .

b) Direct α -knock-off not considered.

At both energies, the energy spectrum of the emitted fragments shows a strong "quasielastic" peak, (i.e. projectile and outgoing fragment have nearly the same energy per nucleon). The angle integrated cross section in the quasielastic peak is rather similar for many fragment isotopes observed in the two experiments. But there are also some characteristic differences. For example, the production of low-Z isobars is preferred at the lower energy. The similarities in the cross sections at energies different by a factor of 100 present a big puzzle and were a kind of sensation at the Berkeley meeting. We [HS 78] attempted a model, very similar to the one presented above. The motivation for our model is as follows. Regarding the similarities we realized that other cases are known where particular nuclear cross sections do not change dramatically when the energy is increased by a factor 10 or 100. Total nucleon-nucleus or α-nucleus cross sections quickly reach the geometric value $2\pi R^2$ and then remain constant as the energy is increased. This suggests that the similarities observed in the fragmentation reactions can also be traced to a geometric origin. As for the differences, we noted that at 20 MeV/A the kinematic distinction between the projectile (fast) and target (slow) nucleons disappears. Consequently, we hypothesize that at 20 MeV/A projectile and target can exchange nucleons, Pauli exclusion significantly affects the scattering and nuclear shell effects are important. Thus, as explained in detail in HS 78 , we take the knock-on stage to be similar at the two energies. The mechanism for the excitation of the prefragment is completely different (two colliding Fermi-gases at 20 MeV/nucleon). The decay is calculated as described for the 2 GeV data. The results are shown in Table 2.3 .

3. CENTRAL COLLISIONS.

They fascinate people most. One nucleus penetrates into another nucleus. Matter from the two system mixes, a lot of energy is brought in, the density may increase.

The number of exciting phenomena which might happen is fantastique - do we see them ?

We are still far from a decent understanding. Cascade and hydrodynamic calculations are performed [BF 76] [AG 77], classical equations of motion are solved [BW 77], shock waves are predicted [BSc 75], plain thermodynamics is assumed [CJ 73]. Amazingly enough, most of these approaches fit experiment (on a logarithmic plot, though). Since these approaches differ widely in their assumptions, their fit to the data does not mean that we understand the physics. Let me explain the situation. I choose the nuclear fire ball model [WG 76] since it is the most simple and most successful at present. The fireball model rests on the following three assumptions:
(a) Straight-line geometry. The projectile nucleus moves on a straight line even while interacting with the target nucleus. In this way, an overlap or interaction zone is defined for each impact parameter (Fig. 3.1). Nucleons in this zone are called the "participants", the remaining nucleons of target and projectile are called "spectators". The number of participants $n_T(\underset{\sim}{b})$ from the target ($n_P(b)$ from the projectile) follow from geometry.

$$n_T(\underset{\sim}{b}) = \int d^2b \; \rho_0 \; 2\sqrt{R_T^2 - b^2} \; \Theta(R_P^2 - (\underset{\sim}{b} - \underset{\sim}{s})^2) \tag{3.1}$$

The basic assumption of the fireball model is that only the participants receive sufficient kinetic energy to be seen in the counter. Thus, the total energy and angle integrated double differential cross section for the observation of one nucleon is simply the mean number of participants:

$$\sigma = \int d^2b \left(n_T(\underset{\sim}{b}) + n_P(\underset{\sim}{b}) \right) = A_T \pi R_P^2 + A_P \pi R_T^2 , \tag{3.2}$$

where A_P, A_T, R_P and R_T are nucleon numbers and radii for projectile and target, respectively. We conclude that the one-nucleon inclusive cross section and therefore the normalization of the double differential cross section is a geometrical property and void of any dynamics.

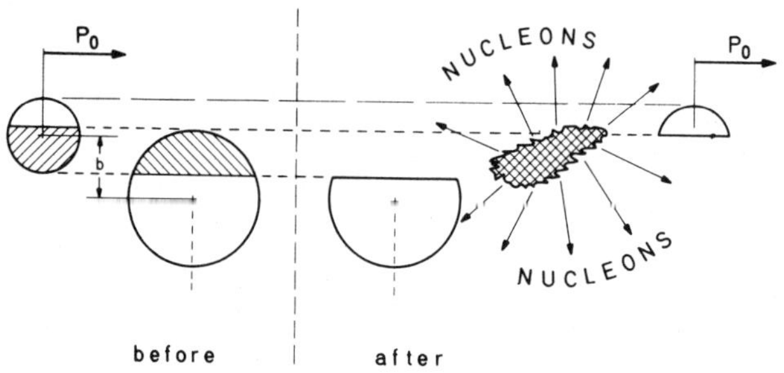

Fig. 3.1 An illustration of the straight-line geometry in collisions between nuclei at high energy. An incoming projectile with momentum p_0 per nucleon ploughs through the target. The nucleons from the overlap zone (shaded area) are called "participants". Only they collide and contribute to the one-nucleon inclusive cross section. The other nucleons (unhatched pieces) remain "spectators".

(b) Thermalization. The participant nucleons in projectile and target thermalize: The total available c.m. energy $E_{c.m.}$ in the overlap zone is converted into random motion. For non-relativistic energies

$$E_{c.m.}(\underset{\sim}{b}) = \frac{n_T(\underset{\sim}{b}) \cdot n_P(\underset{\sim}{b})}{n_T(\underset{\sim}{b}) + n_P(\underset{\sim}{b})} \cdot E_0 \quad , \tag{3.3}$$

where E_0 is the energy per nucleon of the projectile. If, in addition, the nucleons are considered as an ideal (Boltzmann) gas, each participant nucleon acquires a Maxwell distribution in momentum space with a mean linear momentum $\langle \underset{\sim}{p}(\underset{\sim}{b}) \rangle$ and a temperature $\tau(\underset{\sim}{b})$. The differential inclusive one-nucleon cross section is derived to be

$$\frac{d^3\sigma}{dp^3} = \int d^2b \left(n_T(\underset{\sim}{b}) + n_P(\underset{\sim}{b}) \right) W(\underset{\sim}{p}, \underset{\sim}{b}) \, ,$$

$$W(\underset{\sim}{p}, \underset{\sim}{b}) = \exp\left\{ -\left(\underset{\sim}{p} - \langle \underset{\sim}{p}(\underset{\sim}{b}) \rangle\right)^2 / 2m_0 \tau(\underset{\sim}{b}) \right\} / \left[2 m_0 \pi \tau(\underset{\sim}{b}) \right]^{3/2} \tag{3.4}$$

where m_0 is the nucleon mass.

(c) Energy and momentum conservation. The mean momentum $\langle \underset{\sim}{p} \rangle$ and the temperature τ (in units of energy) are the only parameters in the fireball model. They are completely determined from energy and momentum conservation in the mean

$$\langle \underset{\sim}{p}(\underset{\sim}{b}) \rangle = \frac{n_P(\underset{\sim}{b})}{n_P(\underset{\sim}{b}) + n_T(\underset{\sim}{b})} \, \underset{\sim}{p}_0 \quad ,$$

$$\frac{3}{2} \tau(\underset{\sim}{b}) = \frac{E_{c.m.}(\underset{\sim}{b})}{n_T(\underset{\sim}{b}) + n_P(\underset{\sim}{b})} \, . \tag{3.5}$$

Here p_o is the momentum per nucleon of the projectile.

Not all assumptions of the fireball model are basic to the same degree. Energy and momentum conservation (even if only on the average) must be in each theory. The geometrical separation into participants and spectators is a consequence of the straight line trajectory, which should be good at sufficiently high energies. On the other hand, the hypothesis of thermalization is a daring bypass of the complicated many-body problem of two interpenetrating nuclei. We understand this hypothesis rather as a question to nature: Does nuclear matter thermalize in reactions between fast nuclei? This is one of the most important questions of present-day investigations with heavy ions at high energy. The answer, which I shall present has been worked out with J. Knoll [HK 77] .

3.1 The geometry of "rows on rows".

When I want to gain insight into a difficult physical problem, I make a model. That means: I try to drop as much of the unessential (what I consider to be so) until I reach a stage, where the problem becomes "tractable", i.e. soluble by analytical methods. After I have the insight, more realistic calculations may be started.

An important aspect is geometry. This is clear from the formal analysis by Glauber theory: the study of peripheral collisions gives experimental support and geometry again appears in the concept of "participants" in the "fireball". Our model has this geometrical aspect also. To simplify as far as possible, we restrict ourselves to one-dimensional geometry. Only nucleons, which lie on the same straight line interact. This assumption follows directly from Glauber theory but may not be too realistic at those energies, where we want to apply it also (250 MeV/nucl.).

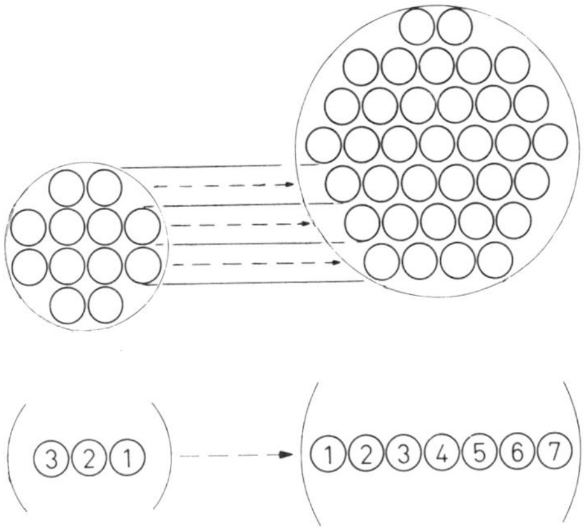

Fig. 3.2 "Rows on rows". An illustration of the one--dimensional cascade model. Projectile and target are decomposed into rows of nucleons (in beam direction). Only corresponding rows on the same straight line scatter by each other. For instance, projectile nucleon 1 scatters by target nucleons 1 to 7, then projectile nucleon 2 scatters by target nucleons 1 to 7, etc. Interactions among projectile and among target nucleons are excluded.

3.2 The dynamics of the one-dimensional cascade.

The equation for the dynamics is already derived (sect. 1.3). It is the Boltzmann equation for the distribution function in phase space, eq. (1.21). In order to get some insight, let us take one nucleon from the projectile row and have it pass through the target. This corresponds exactly to the situation for which eq. (1.21) is derived. We also neglect the mean force $\underset{\sim}{F}$. Thus the equation reads

$$v \frac{d}{dz} W(\underset{\sim}{B}, z; \underset{\sim}{q}) = -\frac{v}{\lambda(\underset{\sim}{x})} W(\underset{\sim}{B}, z; \underset{\sim}{q}) + \frac{v}{\lambda(\underset{\sim}{x})} \int d^2q' \frac{1}{k^2} \frac{1}{\sigma_{tot}^{NN}} \frac{d\sigma^{NN}}{d\Omega}(\underset{\sim}{q}-\underset{\sim}{q}') W(\underset{\sim}{B}, z; \underset{\sim}{q}'). \tag{3.6}$$

Along the trajectory, defined by the impact parameter $\underset{\sim}{B}$, the distribution function changes after each mean free path $\lambda(\underset{\sim}{x})$, in that the old distribution is folded with the NN cross section. This is a very intuitive result, which shall be generalized in the following way: We denote by $W^P_{mn}(\underset{\sim}{q})$ the momentum distribution of the m'th nucleon in a projectile row after it has scattered with the first n nucleons of the target row, and correspondingly by $W^T_{mn}(\underset{\sim}{q})$ the distribution of the n'th target nucleon after it has been struck by the first m projectile nucleons. After the collision m on n+1, the distribution for the projectile nucleon (and similarly for the target nucleon) is

$$W^P_{m\,n+1}(\underset{\sim}{q}) = \int d^3p_1\, d^3p_2\, d^3p_1'\; M(\underset{\sim}{p}_1', q \leftarrow \underset{\sim}{p}_1, \underset{\sim}{p}_2) \cdot W^P_{m\,n}(\underset{\sim}{p}_1)\, W^T_{m-1\,n}(p_2). \tag{3.7}$$

This relation is true if the process is Markovian, i.e. if it only depends on $W^P_{m,n}$ and W^T_{mn} and not on how they have been prepared in previous processes. The transition probability for the elastic scattering of two particles with masses m_1 and m_2 has the form

$$M(\underset{\sim}{p}_1', \underset{\sim}{p}_2' \leftarrow \underset{\sim}{p}_1, \underset{\sim}{p}_2) = \frac{1}{N} \delta^3(\underset{\sim}{P} - \underset{\sim}{P}')\, \delta(|\underset{\sim}{p}_{rel}| - |\underset{\sim}{p}_{rel}'|)\, \frac{d\sigma_{NN}}{d\Omega_{cm}}(\underset{\sim}{p}_{rel}), \tag{3.8}$$

where $\underset{\sim}{P}$ and P_{rel} denote total and relative momenta and N is a normalization constant.

Together with the initial conditions (Fermi-motion before the rows collide) eq. (3.7) can be solved to give the momentum distribution. The inclusive cross section is then obtained from the assumption: After two corresponding rows have collided, their nucleons proceed to the counter. The cross section formula is in [HK 77] . The solution of eq. (3.7) is straightforward but complicated, and since we are after the physics we propose an approximate solution, which, however, is transparent. We solve the equations by a moment expansion. We derive recursion relations for the moments of $W(p)$ (the linear moment is denoted by $\langle p \rangle$ for instance) and truncate after the second moment. The distributions W are then reconstructed by the assumption of a Gaussian.

For instance we find for linear moments

$$\langle \underset{\sim}{p} \rangle^{P}_{m\,n} = (1-\alpha)\, \langle \underset{\sim}{p} \rangle^{P}_{m\,n-1} + \alpha\, \langle \underset{\sim}{p} \rangle^{T}_{m-1\,n}$$

$$\langle \underset{\sim}{p} \rangle^{T}_{m\,n} = \alpha\, \langle \underset{\sim}{p} \rangle^{P}_{m\,n-1} + (1-\alpha) \langle \underset{\sim}{p} \rangle^{T}_{m-1\,n} \qquad (3.9)$$

which contains the conservation of momentum (on the mean). The constant α is an average of

$$\alpha = \langle \sin^2 \vartheta/_2 \rangle_{NN} = \frac{\int d\Omega\, \sin^2\vartheta/_2\, \frac{d\sigma_{NN}}{d\Omega_{CM}}}{\int d\Omega\, \frac{d\sigma_{NN}}{d\Omega_{CM}}} \qquad (3.10)$$

over the NN cross section. For s-wave NN scattering $\alpha = 1/4$, i.e. 25% of the momentum is exchanged in one collision, the faster nucleon becomes slower, the slower one is accelerated.

For reasons of symmetry, there are only two widths, the one in transverse direction $\langle \sigma_\perp^2 \rangle$ and the other in longitudinal direction $\langle \sigma_\parallel^2 \rangle$. For instance

$$\langle \sigma_\perp^2 \rangle^P_{mn} = \left\{ (1-\alpha) \langle \sigma_\perp^2 \rangle^P_{m,n-1} + \alpha \langle \sigma_\perp^2 \rangle^T_{m-1,n} \right\}$$
$$+ \frac{1}{2}\beta \left\{ \langle \sigma_\parallel^2 \rangle^P_{m,n-1} - \langle \sigma_\perp^2 \rangle^P_{m,n-1} + (P \leftrightarrow T) \right\}$$
$$+ \frac{1}{2}\beta \left\{ \langle \underset{\sim}{p}_\parallel \rangle^P_{m,n-1} - \langle \underset{\sim}{p}_\parallel \rangle^T_{m-1,n} \right\}^2 , \qquad (3.11)$$

where β is another moment of the NN cross section. Eq. (3.11) has an interesting structure: The first curly bracket leads to an "equilibration" between the

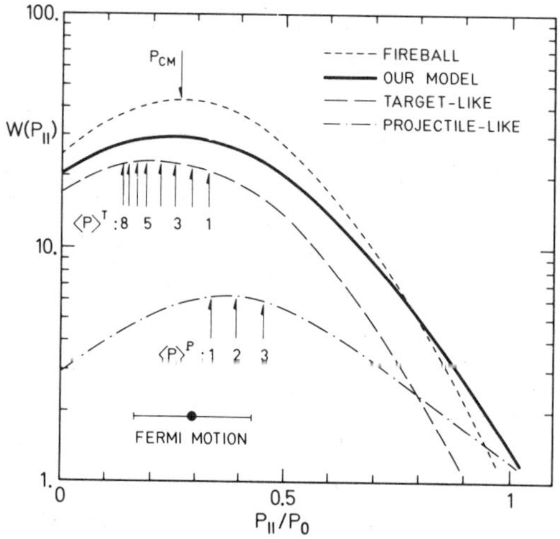

Fig. 3.3 The distributions of longitudinal momentum for a scattering of a row of 3 nucleons by a row of 8 nucleons at 400 MeV/nucleon. The position of the mean momenta of projectile and target nucleons are indicated by arrows. Projectile and target distributions are shown separately. The total distribution resulting from our cascade calculation is compared with the thermal (Maxwell) distribution ("fireball").

widths of target and projectile nucleons. The second curly bracket tries to make widths in transverse and longitudinal direction similar, and the third curly bracket converts longitudinal motion into width. Since we identify the width of a distribution with a temperature in the thermal limit, eq. (3.11) can be well understood as an approach to the thermal limit. And indeed, the thermal limit seems to be reached rather quickly (after a few collisions (Fig. 3.3)). The parameters which govern the thermalization are α and β . For a rough estimate: The number of collisions n_{th} after which a high energy particle has thermalized is of order

$$n_{th} \approx \alpha^{-1} \tag{3.12}$$

where α is defined in eq. (3.10). For isotropic NN scattering in the c.m. system $\alpha = 0.25$, $n_{th} \approx 4$. The same number should be relevant for a classical gas with hard-sphere scattering (for which the cross section is also isotropic). Indeed model calculations show thermalization after 2 to 4 collision times. But other degrees of freedom also seem to thermalize quickly. Allowing the N to convert into a Δ (1236) via

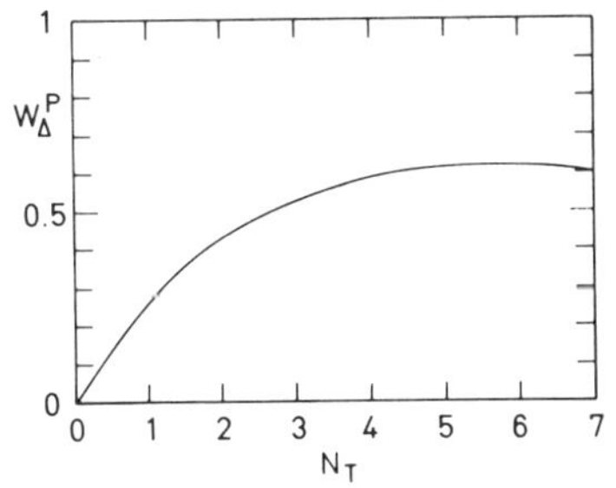

Fig. 3.4 The probability W_Δ^P of finding the first projectile nucleon in the Δ(1236) state after it has scattered by N_T target nucleons. Note the rapid saturation.

the reactions $NN \rightarrow \Delta N$, while it collides with the nucleons of its row, leads to the pattern shown in Fig. 3.4 . The Δ -degree of freedom saturates after 2 to 4 collisions.

3.3 One-particle inclusive cross sections.

After the rows have interacted in the overlap zone, a Gaussian momentum distribution is reconstructed. We assume that the nucleons from the interaction zone reach the counter without any disturbances from the spectator pieces and that spectator nucleons remain unobserved. Thus the absolute value of the inclusive nucleon or pion cross sections is determined by the overlap geometry, while the angular and energy distributions reflect the momentum distribution calculated via the cascade equation. Figs. 3.5 and 3.6 show two typical examples.

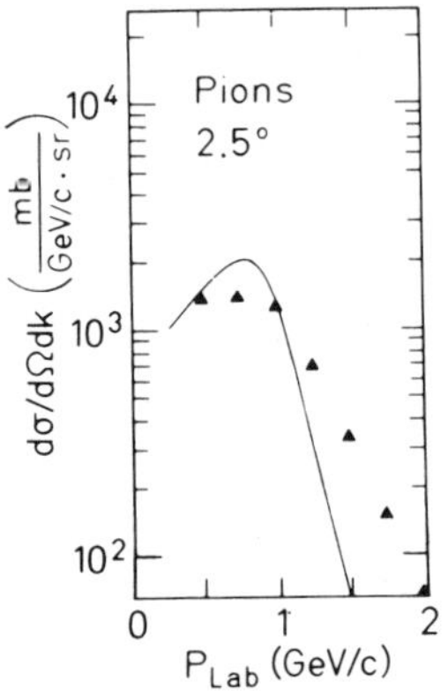

Fig. 3.5 The double differential cross section for charged pions from the reaction ^4He on ^{208}Pb at 2.1 GeV/nucleon. Experiment by Papp et al. [PJ 75] , the solid line is the result of the calculation described in the text.

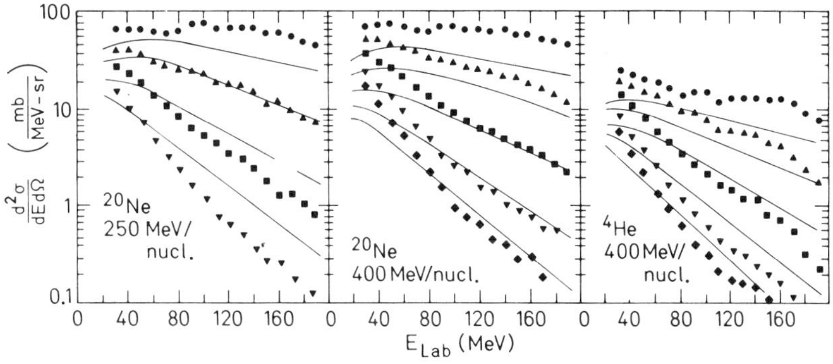

Fig. 3.6 The experiment [WG 76] and our calculation for the double differential cross section of protons from the reactions indicated. The experimental and theoretical curves correspond to various angles, Θ = 30°, 60°, 90° and 120° in decreasing order.

In view of the very crude models (and in view of the logarithmic scale !) the agreement is satisfactory. In the mean time better fits are reached with phenomenological models which assume thermalization ("fire streak"). Our model was built to give an insight into why thermalization is not so far even at high energy. But details of the process (especially thermalization of composite particle production) are rather unclear to me. There is still a long way to go.

Acknowledgements: The results presented here would not have been obtained without the intensive collaborations, which I have indicated. I thank Ms. S. Jordan for the careful preparation of the typescript of these lectures.

References.

AG 77 A. Amsden, J.N. Ginocchio, F.H. Harlow, J.R. Nix, M. Danos, E.C. Halbert and R.K. Smith, Phys. Rev. Lett. 38 (1977) 1055

AH 78 A. Abul-Magd and Jörg Hüfner, submitted to Nucl. Phys. A

B 76 M. Buenerd, et al. Phys. Rev. Lett. 37 (1976) 1191

BF 76 J.P. Bondorf, H.T. Feldmeier, S. Garpman and E.C. Halbert, Phys. Lett. 65B (1976) 217;

J.P. Bondorf, P.J. Siemens, S. Garpman and E.C. Halbert, Z. Phys. 279 (1976) 385

BN 73 J. Bondorf and W. Nörenberg, Phys. Lett. 44B (1973) 487
K.Kikuchi and M. Kawai, Nuclear Matter and Nuclear Reactions (North-Holland, Amsterdam, 1968), chapter 5

BS 75 J.D. Bowman, W.J. Swiatecki and C.F. Tsang, Berkeley 1975, unpublished

BSc 75 H.G. Baumgardt, J.U. Schott, Y. Sakamoto, E. Schopper, H. Stöcker, J. Hofmann, W. Scheid and W. Greiner, Z. Phys. A273 (1975) 359

BS 78 M. Bleszynski and C. Sander, preprint, Heidelberg (1978)

BW 77 A.R. Bodmer and C.N. Panos, Phys. Rev. C15 (1977) 1342

L. Wilets, F.M.Henley, M. Kraft and A.D. MacKellar, Nucl. Phys. A282 (1977) 341

CJ 73 G.F. Chapline, M.H. Johnson, E. Teller and M.S. Weiss,
Phys. Rev. D8 (1973) 4302

M.I. Sobel, P. J. Siemens, J.P. Bondorf and H.A. Bethe,
Nucl. Phys. A251 (1975) 502

Gl 59 R.J. Glauber,
in Lectures in theoretical physics, vol. I, ed. W.E. Brittin et al. (Interscience Publishers, NY, 1959) pp 315 - 414

GM 70 R.J. Glauber and G. Matthiae,
Nucl. Phys. B21 (1970) 135

HK 77 J. Hüfner and J. Knoll,
Nucl. Phys. A290 (1977) 460

HL 76 H.H. Heckman and P.J. Lindstrom,
Phys. Rev. Lett. 37 (1976) 56

HS 75 J. Hüfner, K. Schäfer and B. Schürmann,
Phys. Rev. C12 (1975) 1888

HS 78 J. Hüfner, C. Sander and G. Wolschin,
Phys. Lett. 73B (1978) 289

Hü 78 J. Hüfner,
From quantal multiple scattering theory to a classical transport equation,
Ann. Phys. to be published (1978)

LG 75 P.J. Lindstrom, D.E. Greiner, H.H. Heckman, B. Cork and F.S. Bieser,
report LBL-3650, Feb. 75

PJ 75 J. Papp, J. Jaros, L. Schroeder, J. Staples, H. Steiner, A. Wagner and J. Wiss,
Phys. Rev. Lett. 34 (1975) 601, 991

J. Papp,
Thesis, LBL report

WG 76 G.D. Westfall, J. Gosset, P.J. Johansen,
A.M. Poskanzer, W.G. Meyer, H.H. Gutbrod,
A. Sandoval and R. Stock,
Phys. Rev. Lett. 37 (1976) 1202

MODELS OF HIGH ENERGY NUCLEAR COLLISIONS

Norman K. Glendenning

Lawrence Berkeley Laboratory
Berkeley, California 94720

A. NUCLEAR COLLISIONS AT RELATIVISTIC ENERGIES

Introduction, Equation of State, Energy Thresholds

What do we know about nuclei? The literature of the last 20 or 30 years contains a wealth of fascinating detail about their structure, their energy levels and single particle aspects, their collective motion, and the way they interact with each other in collisions. Both the quantity and detail of the experimental data, and the sophistication of some of the theory is impressive. Yet what we know about nuclei concerns their properties at only one point on the graph of the equation of state of nuclear matter which is illustrated in Fig. 1. Aside from the trivial point at the origin, and the energy per nucleon at normal density, the curve drawn is a guess. The point where it crosses the axis at $\rho/\rho_0 \sim 2$ is based on nuclear matter calculations. We do not even know the curvature (compressibility) at normal density. Virtually everything we know about nuclei concerns their normal state!

Some interesting possibilities for the state of nuclear matter at high density are illustrated in Fig. 1. The Lee-Wick super dense state is illustrated, as is the effect of a phase transition, corresponding to a situation where a state of special correlation having the quantum numbers of the pion (pion condensate) becomes degenerate with ground state.

Perhaps the ultimate goal of research with relativistic energy nuclei is to study nuclear matter under abnormal conditions of high particle and energy density. This is a break from the past.

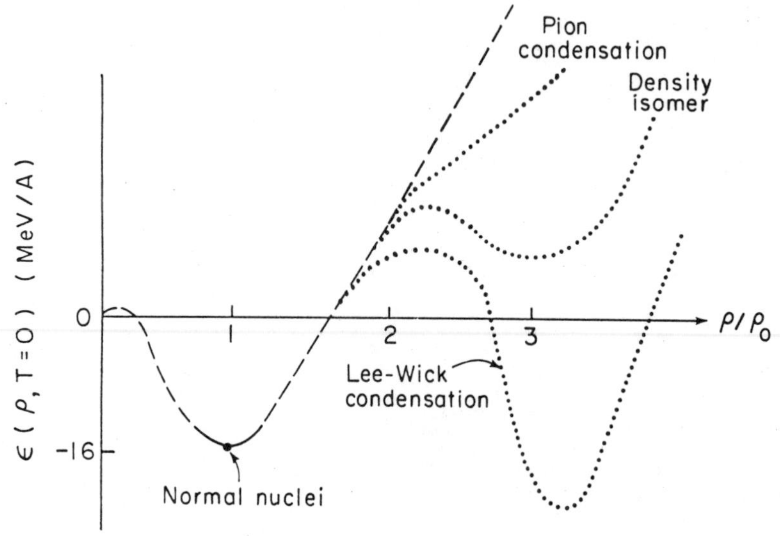

Fig. 1. Schematic of energy per nucleon versus density of nuclear matter (equation of state) showing several possible high density behaviors. (Courtesy of A. M. Poskanzer.)

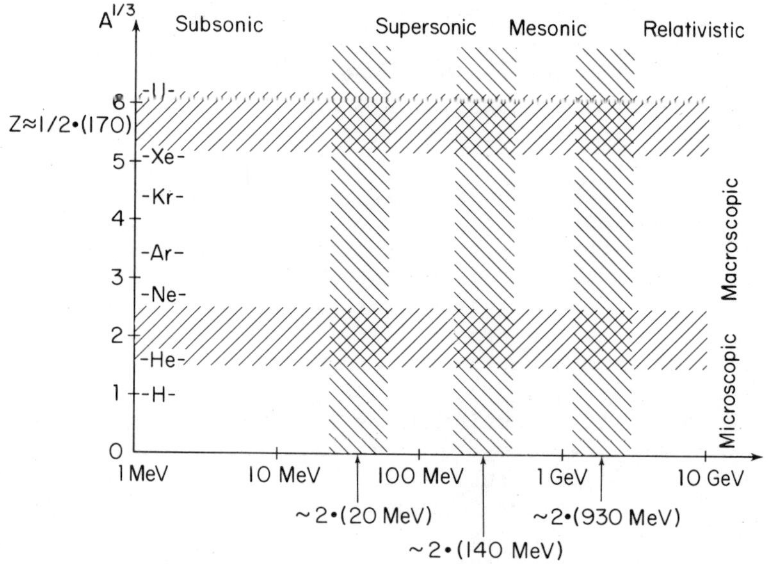

Fig. 2. Illustration by W. Swiatecki of various thresholds in the mass of projectile versus collision lab energy for symmetric collisions.

Nuclear physicists have concentrated on studying nuclei under normal conditions of low energy and temperature. High energy physicists have concentrated on putting higher and higher energy into a small volume. We do not know what surprises await us, but several possible rewards will be mentioned later.

To make it plausible why we expect to encounter new and interesting phenomena it is useful to examine Fig. 2, prepared by Swiatecki.[1] There the projectile mass for a symmetric collision is plotted on one axis, and a bombarding energy per nucleon on the other. The shaded areas indicate thresholds where qualitatively new physical features take over. The low energy region is the domain of conventional nuclear physics, and is being intensively studied at many laboratories. The region immediately adjacent to the x-axis extending to very high energies is the domain of particle physics, studied at the very large accelerators. Most of the plane is completely unknown territory. We discuss briefly the thresholds following the subsonic region of conventional nuclear physics.

Supersonic Threshold. The energy of 20 MeV per nucleon corresponds to

1) the average kinetic energy of nucleons in nuclei $\simeq \frac{3}{5} E_F$

2) minimum energy needed to compress nuclear matter to something like twice normal density (see Fig. 1).

3) estimated speed of sound in normal nuclear matter (involves the curvature at the minimum in Fig. 1).

We anticipate qualitatively different behavior as this threshold is crossed into the region of what can be labelled supersonic.

Meson Threshold. This is the threshold for particle production, starting with the pion, followed by the nucleon isobars and heavier mesons. Hot nuclear matter can be cooled by the production of these particles.

Relativistic Threshold. This corresponds to the mass of the nucleon and brings us to the full relativistic regime where not even the wave equation for $s > 1/2$ is known. Far into this region I have the hope that it may be possible to discover the general form of the hadronic mass spectrum, one of the most fundamental properties of matter.[2]

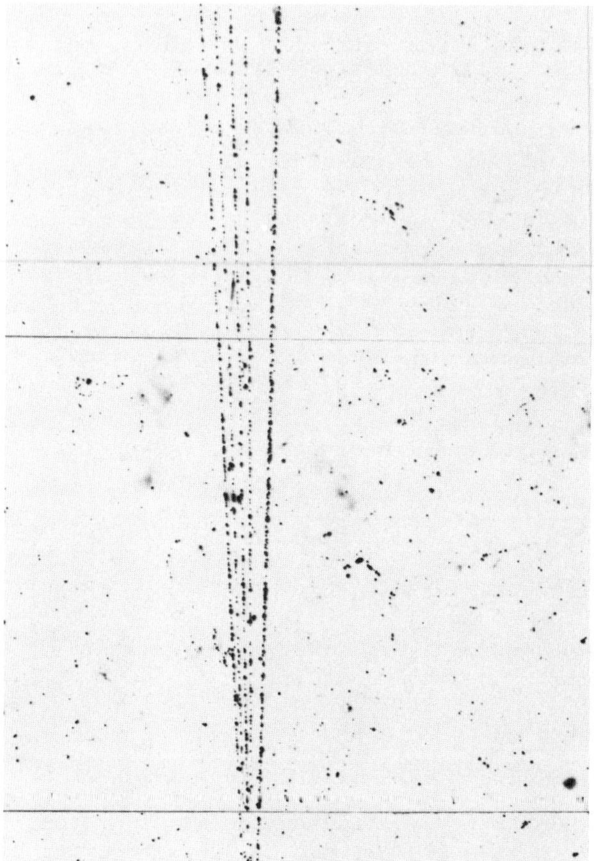

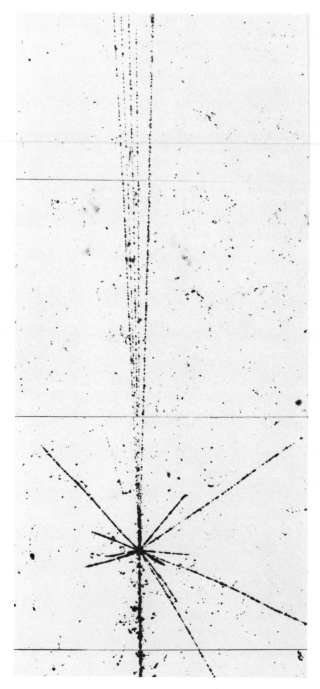

Fig. 3. A typical peripheral collision showing forward cone of (charged) projectile fragments having virtually the projectile energy.

Fig. 4. A central collision showing high multiplicity of charged particles.

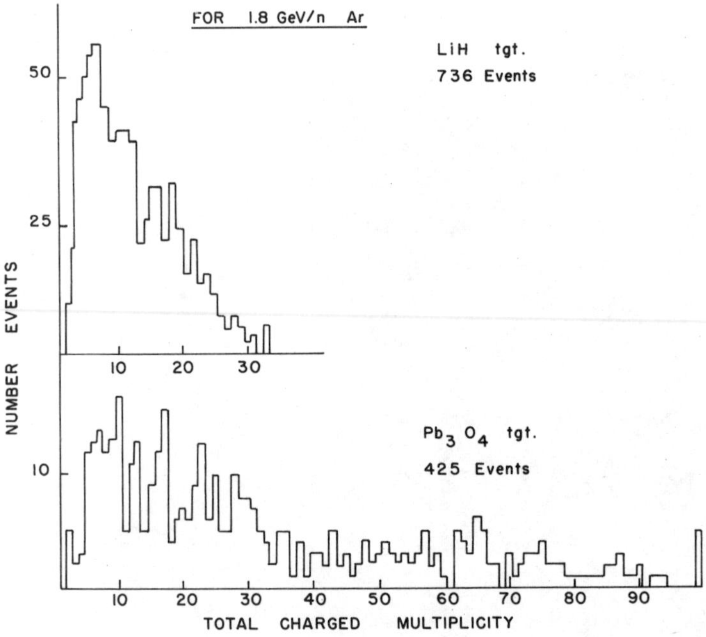

Fig. 5. Charged particle multiplicity distributions from Fung, Gorn, Kiernan, Liu, Lu, Oh, Ozawa, Poe, Van Dalen, Schroeder, and Steiner (unpublished, 1977).

Two Classes of High Energy Nucleus-Nucleus Collisions

The experimentally observed events reveal two extreme limiting types of collisions at energies in the few hundred MeV to several GeV per nucleon range.

1) Peripheral Collisions.[3] This is the most frequently observed class of collisions and is characterized by the fact that a few particles are observed in the extreme forward cone and they have almost the same speed as the projectile. They are presumably fragments of the projectile and they range in mass from one to a number of mass units, but less than the projectile. Presumably these collisions are geometrically peripheral so that a few nucleons in the overlap region are knocked out. Both the projectile and target residues are presumably excited by the sudden removal of a part of their mass, and may radiate particles after the collision (Fig. 3).

2) Central Collisions.[4] In about 10% of the collisions no fast particles having the projectile speed are observed. Instead, many particles, up to 130 charged particles are emitted even in

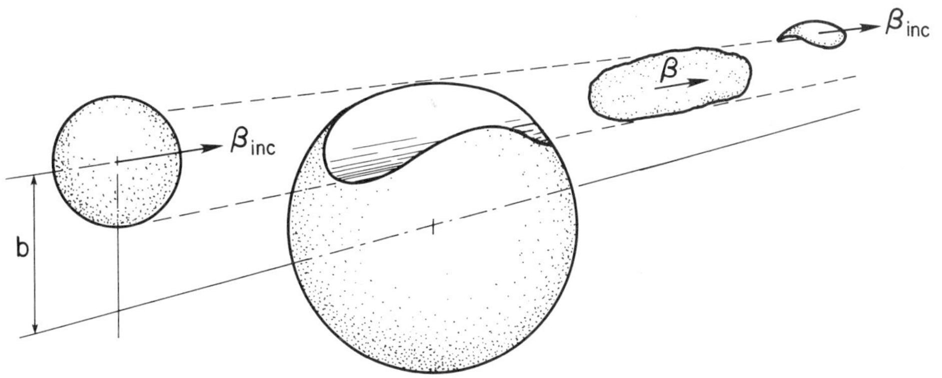

Fig. 6. Schematic showing geometrical assumptions of fireball model. The fireball is the portion swept out by the projectile having an intermediate velocity β and a temperature corresponding to an excitation energy given by application of energy and momentum conservation.

collisions with a total initial charge of 100. Many mesons are evidently produced. No remnant of the projectile in the forward cone is observed. The projectile is presumably stopped in the target and the energy shared by many particles. Fairly fast ones come out in the forward hemisphere while high Z particles (bright tracks) come out in all directions (Fig. 4,5).

What Happens in a Central Collision?

Are nuclei opaque or transparent to an incident high energy (2 GeV/nucleon) nucleus? The de Broglie wave length is so short that we argue at first on the basis of a sequence of individual nucleon-nucleon collisions. The mean free path between collisions based on the nuclear density and N-N cross section is

$$\ell \sim \frac{1}{\rho\sigma} = [(0.17/F^3)\ (40\ \text{mb})]^{-1} \sim 1.5\ \text{F} \quad .$$

The energy loss per collision is ∿100 MeV so that a 2 GeV nucleon might lose a GeV energy in traversing a lead nucleus. In other words there might be a high degree of transparency. On the other hand, in roughly 10% of collisions there is an absence of high energy particles: the projectile appears to be stopped and the composite system decays or explodes. Since we deal with systems of at most several hundred particles, deviations from the mean can be large! Clearly the opaque collisions are very interesting.

If the projectile is stopped in the target or a part thereof the resulting system is very hot. After taking account of the escape of a certain fraction of the prompt pion production, energy and momentum conservation can be used to calculate the internal excitation or temperature (Fig. 6). The temperature can be lower than this however because 1) the prompt π's that do not immediately escape can interact with nucleons to form the nucleon isobars. This lowers the number of fermions of given type which allows cooling. 2) Collisions of hot nucleons can produce secondary π's reducing thus the kinetic energy.

The high velocity (near c) of the pions and their strong interaction with nucleons provides a fast mechanism for thermalization of the composite system in addition of course to the nucleon-nucleon collisions. Indeed computer studies suggest that thermalization can occur already after only 3 or 4 collisions. Chemical equilibrium between the various species, π, N, N*, D, t takes longer but may still be fast compared to the disassembly time of the composite. Therefore a thermodynamical description may be reasonable and indeed a free ideal gas treatment of the composite, called a nuclear fireball does qualitatively account for some of the proton and composite particle spectra observed[5,6,7] (Figs. 7,8). The temperatures so determined run as high as 100 MeV, or 10^{12} °K, perhaps the highest temperature ever produced in the laboratory (however in a very small piece of matter, i.e., $\sim$nucleus) and the highest that have existed naturally since the beginning of time.

If a thermodynamic description does apply, it makes it very simple to investigate a very exciting prospect, the discovery of the assymptotic form of the hadronic mass spectrum. This subject will form the second part of these lectures.

High Density Phenomena. So far there is no convincing evidence for the propagation of shock waves in these collisions but the formation of regions of high density is expected in any case.[8] This is very interesting from several points of view. One is the possible creation of quark matter.[9] Quarks, if they exist, are believed to be the constituents of hadrons in which they are confined by the forces acting between them. However, Susskind finds that at least one model of confinement allows a transition to a plasma-like phase to occur at high temperature and energy density. The colored gluons form a plasma, which screens the color of quarks, allowing the quarks to become disassociated from their original hadrons, and to roam throughout the high density medium. Another interesting high density phenomenon is the Lee-Wick[10] density isomeric state of nuclear matter (see Fig. 1). Even another is the phase transition sometimes referred to as pion condensation.[11-14] At some critical density not so much greater than normal, it is believed that a collective state of special correlation having the quantum numbers of the pion will become degenerate with the ground state.

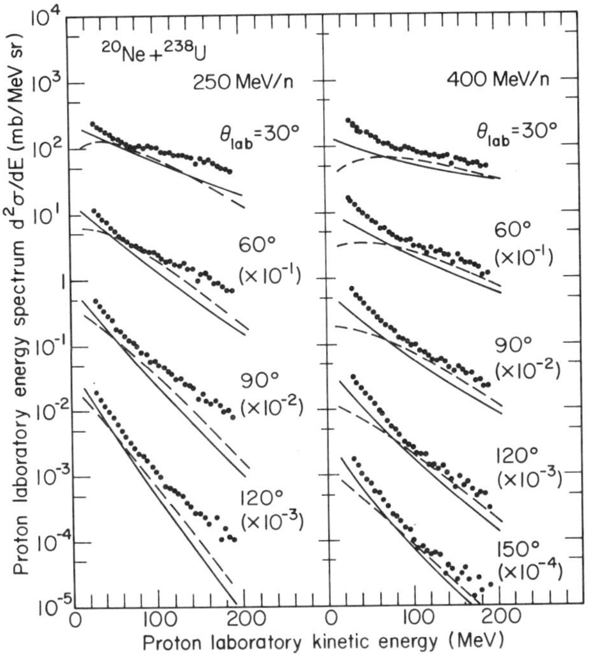

Fig. 7. Comparison of the fireball calculation[5] (dashed) and firestreak[7] (solid) with data of proton spectra.[4]

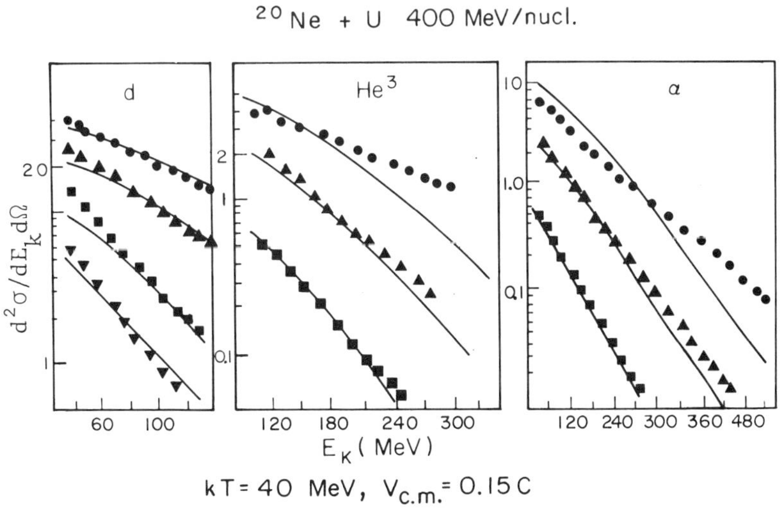

Fig. 8. Thermodynamic calculations[6] of composite particle spectra compared with data.[4]

The fact is, however, that we do not know of any way to squeeze a nucleus while keeping its temperature low. Nevertheless signals of the existence of such states, were they to exist at densities not too much greater than normal, have been sought in the spectra of ordinary nuclei.[11] The weight of recent calculations is against the existence of such pion condensates at low temperatures, and densities as low as twice normal.[12] High energy nuclear collisions may therefore give us another chance. In collisions, nuclear matter at high temperature and density is fleetingly created, offering the possibility of triggering a collective instability in the non-equilibrated matter.[13] Gyulassy has emphasized that these instabilities do not imply the participation of a large number of pions, but instead a few collective "phonons" each with the quantum numbers of a pion. The structure of the phonon is that of a spin-isospin lattice. The calculated effect of the growth of these instabilities in the nuclear medium on the scattering of nucleons passing through the medium, is to moderately increase the cross sections. As a result, if pionic instabilities occur, they will assist in rapidly thermalizing the system. However, if equilibrium is actually reached, the memory of the dynamical processes preceding it is lost! Apparently the detection of these instabilities will be very difficult.

It might be supposed that a pionic instability would lead to an enhancement in the average number of pions produced, or at the very least, change the pion-multiplicity distribution. Both these questions have been studied.[14] That the first supposition is false is a trivial consequence of the fact that at energies up to 2 GeV/nucleon most pions are produced during the non-equilibrium phase. Even at the temperature of 50 MeV, relative momenta are generally below threshold. The production rate during the pre-equilibrium phase is presumably enhanced, just as the nucleon scattering by the medium in enhanced. However the thermalization occurs sooner by a compensating factor, so that the number of pions produced is essentially unchanged by the existence of pionic instability. Even the pion multiplicity distribution is little effected.[14]

Still in a pursuit of distinguishing features of pionic instabilities, Gyulassy has made a very interesting study of correlation properties of pion produced in a medium such as colliding nuclei, which is an extension of quantum optics.[14] He studies the properties of the field equation

$$(\Box + m_\pi^{\ 2})\ \phi(\underset{\sim}{r},t) = J(\underset{\sim}{r},t)$$

where ϕ is the pion field and J is a transition current operator representing the physical processes involved in the production of pions. If J is a given C-number source, which is a reasonable assumption if the pion field is not too strong, allowing one to ignore the recoupling of the pion fields to the nucleon collisions creating them, then the solution exhibits some interesting properties. In particular for a pure coherent source, the solution has the properties of a pion laser. The two pion inclusive correlation function can be used to distinguish between a coherent source, and a chaotic one. Moreover the coherent source, which stimulates the pion laser, such as the collective pion instability, can be studied mode by mode.

This will serve to give an indication of the range of phenomena under investigation, the difficulties involved in their observation and some of the theoretical approaches employed. Other theoretical approaches include hydrodynamics, Glauber theory, cascade calculations, and classical equations of motion.[15]

B. THE ASYMPTOTIC HADRON SPECTRUM

Elementary Particles

An age old question has intrigued mankind, at the latest, since the early Greek philosophers. Is matter divisible only down to fundamental particles which are not further divisible or is matter infinitely divisible into smaller pieces? The modern experience does not come down on one side of this either-or question. Instead we find that matter can be divided again and again but not into smaller and smaller particles. Instead always some of the same particles we started with reappear together with other particles. A nucleon cannot be broken up only into smaller particles! This was not known to the Greeks, and it runs counter to our experience of all matter from the macroscopic down to and including nuclei.

Einstein's law of equivalence of mass and energy tells us that in a high energy collision between nucleons mass can be created. We see mesons and baryon anti-baryon pairs produced and the nucleons may reappear as nucleon isobars. Is this an entirely trivial consequence of Einstein's law? Does a law or physical principle underlie the indeterminancy of the outcome? Presumably so. In other areas of physics we are accustomed to considering that the state of system comprises, virtually, all possible configurations of the same symmetry. What these "configurations" are is the goal of high energy physics. Whatever the mathematical description of the modern answer to the age old question, whether it be bootstraps or quarks or whatever, it seems quite certain that we know only a few of the particles and resonances that can be produced in high energy collisions; that their number and variety is staggering.

The Known Hadrons. There are 56 named hadrons representing about 1000 hadronic states with spin, isospin, baryonic charge and strangeness quantum numbers measured (Table 1).[16] The lightest of these are the three pions $m_\pi \sim 140$ MeV. They become quite densely spaced as their mass increases to about 10 m_π. Thereafter the spectrum becomes sparse. Presumably however the cutoff is an experimental one. Figure 9 plots the number of hadronic states per pion mass interval. Already at $m = 10\ m_\pi$ there are 34 non-strange states per pion mass interval and the average width at this mass is $\Gamma \sim 100$ MeV. Production rates are expected to decrease with m. The experimental problem becomes one of intensity and resolution.

The Hadrons that Will Never be Known by Name. Theories of hadronic structure, in contrast to the known spectrum, imply that it continues indefinitely. The bootstrap hypothesis predicts a spectrum that increases exponentially.[17] The hypothesis can be stated simply as follows: From among the known particles or resonances select two (or more) and combine their quantum numbers. The multiplet so obtained are also particles or resonances (at something like the sum of the masses). Add these to the pool of known particles and continue. The spectrum thereby generated by Hamer and Frautschi[18] is also shown in Fig. 9. The implication is astonishing. The number of particles and resonances grows so fast that at only 2.5 GeV the number expected in a pion mass interval, on the basis of the bootstrap hypothesis, is $\sim 10^4$. The number of known particles is $\lesssim 10^2$ at that mass. If new particles were discovered at the rate of one a day it would require about a hundred years to verify the bootstrap prediction by a direct count, and that at only one mass!

Most high energy physicists currently favor the quark hypothesis, and there is even a theory, quantum chromodynamics (QCD), that is a candidate for the dynamical description of hadrons. The theory has not been solved in any general sense so far. The mesons are thought to consist of a quark anti-quark pair and baryons of three quarks, which in both cases are referred to as the valence quarks, and in addition there is believed to exist in each hadron, a sea of quark anti-quark pairs. Whether all the quark flavors have been identified is an open question. The modes of excitation of the quarks within a hadron, like the radial and angular momentum states in a nucleus, are also unknown.

Table I. The families of light mass multiplets, their average masses in MeV, and their baryon and strangeness quantum numbers (B, S). Total multiplicity including the unlisted multiplets is indicated in the bottom row for each family.

Family (B, S)	Π (0,0)	K (0,1)	N (1,0)	Λ (1,-1)	Σ (1,-1)	Ξ (1,-2)	Ω (1,-3)
Multiplets	π(138)	K(495)	N(940)	1116	1193	1318	1672
	η(549)	K*(892)	N*(1430)	1405	1385	1533	
	ρ(773)	K*(1421)	N*(1520)	1519	1670		
	ω(783)		N*(1515)	1670	1745		
	η'(958)		Δ(1232)	1690	1773		
	⋮		⋮	⋮	⋮		
Total Multiplicity	103	18	248	38	108	12	4 = 531

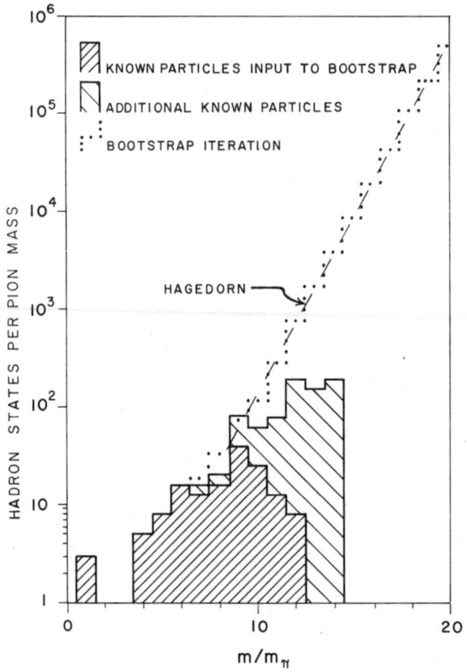

Fig. 9. The density of different hadrons, ρ, in unit interval is plotted as a function of the mass in units of the pion mass. The experimentally known particles and resonances with their multiplicities are shown in the shaded areas. The dotted histogram is a bootstrap iteration[18] on the known spectrum and the solid curve is a Hagedorn type spectrum, fitted to the bootstrap.

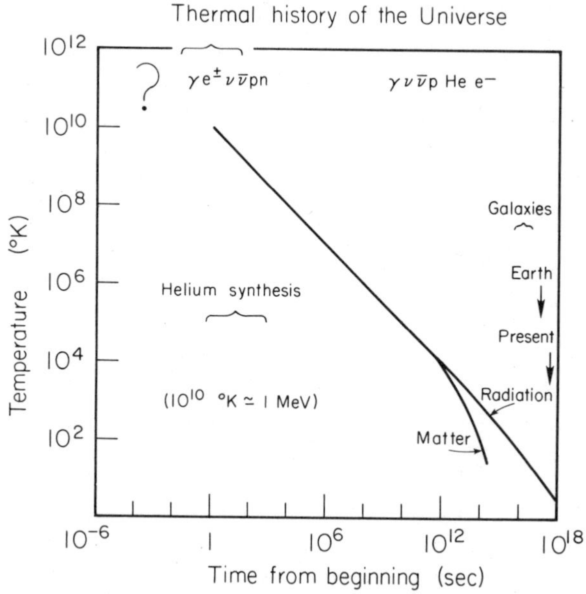

Fig. 10. Thermal history of the universe. The question mark indicating ignorance as to temperature and composition is due in part to the unknown hadronic spectrum.

QCD is not an easy theory. It is possible that the number of excitations (i.e., hadrons) are unbounded in number and mass, that the density of states grows with mass, perhaps exponentially as the bootstrap. It should be borne in mind throughout the rest of this paper, that the exponential spectrum, here associated explicitly with the bootstrap hypothesis, may also be the form of the spectrum, based on the quark hypothesis.

In any case, we have seen that it is out of the question to determine even the general form of the hadronic mass spectrum at even relatively low masses like 2 to 3 GeV, much less in the high mass region, by a direct count of individual particles and resonances. The sheer density of states is not only very large, but the widths are at least a pion mass, so of the order 10^4 or more states fall with the width of any one.

Is it Interesting? Is it important or even interesting to know the density of hadronic states in the region were they cannot be individually discovered and given a name? I think so. It is both interesting and important. Interesting because it is a fundamental property of matter on the smallest scale, and important for two reasons that I can think of. It is important in particle physics because the density of hadron states at high mass provides an asymptotic constraint on theories of hadronic structure. Let me elaborate. The properties of the low mass particles that can be individually identified provide important clues as to the group structure of the theory. Their quantum numbers (spin, isospin, strangeness) which are determined by the decay modes and so on, and their masses suggest particular classifications which any theory of hadronic structure must account for. But the symmetries are not perfect. This fact leaves a lot of room for competing theories. Certainly the quark theory is favored by many particle physicists today, but there are a number of quark models. If quarks are the fundamental building blocks, there is still no agreement as to the nature of the glue that holds them together. And because the symmetries are broken, the light particle spectroscopy cannot provide a unique way of discriminating between the theories. Yet any theory of hadronic structure, when sufficiently developed, can be made to yield a prediction of the asymptotic region (i.e., high mass). It is in this sense that the asymptotic behavior of the hadronic spectrum may become decisive in particle theory.

The asymptotic region is important also in cosmology. The thermal history of the universe can be guessed with considerable confidence back to the time of helium synthesis at temperatures of about 1 MeV. Figure 10 traces this history, backward in time from the present, through the formation of planets, the galaxies, and heavy element synthesis to the high temperature of 1 MeV (10^{10} °K). For much earlier times when the energy density was extremely high, the composition of the universe must have been very different in kind, not merely in density and temperature, from what we see today. That there were no nuclei is clear, but that there were no nucleons is likely. What there was was in fact determined by the spectrum of hadrons and leptons that could energetically exist at the energy densities prevalent. At even earlier times, at extreme particle and energy density, the hadrons may have been dissolved into a quark soup which only later condensed into hadrons. Whether there are any residual signals in the universe left over from these early instants I do not know, but doubt. Probably the temperature and composition in the earliest instants will remain forever a subject of speculation, uninformed at the present, but informed, if the form of the hadronic spectrum can be determined.

Are there contemporary astronomical events that bear on the high mass region of the hadron spectrum? I was fascinated to learn a short time ago of Hawking's work on the quantum theory of black holes.[19] I had believed that black holes are really black. That no matter or radiation can escape from them. No so in quantum theory. Black holes evaporate, at first slowly but eventually catastrophically. A process by which this evaporation can occur is illustrated in Fig. 11. The loops represent the spontaneous fluctuations of the vacuum, when a particle and anti-particle momentarily appear and then mutually annihilate. However in the vicinity of a black hole, if one member of such a pair is captured by the black hole, the other has lost the partner with which to annihilate. It appears therefore as radiation from the black hole. It is not easy to follow the reasoning that leads to the conclusion that the temperature of the black hole increases as a result. It _is_ easy to understand, as will develop later, that the energy released in the ultimate explosion of the black hole, depends crucially on the hadronic spectrum, being many orders of magnitude greater, if it increases exponentially than otherwise. The reason is simply that in the former case, energy can be stored in the benign form of mass rather than kinetic energy. In the latter case, where this possibility is limited, the explosion occurs, so to speak, prematurely.

I am sure that I have convinced you by now of two things. The general form of the hadronic spectrum is a most interesting thing to know, and it cannot be discovered by looking for the individual particles of which it is composed.

Black Holes are only grey !

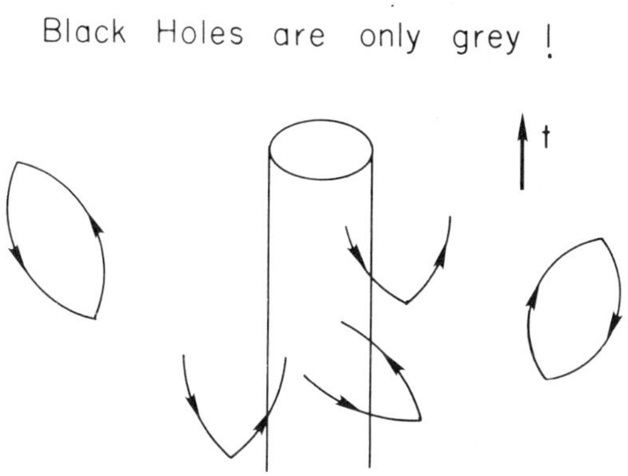

Fig. 11.   A mechanism for the evaporation of a black hole showing the capture by the hole of one partner of pair fluctuations of the vacuum.

INITIAL FIREBALL

Symmetric collision of Z = N nuclei in C.M.

assume perfect
inelasticity

Lorentz

$$\frac{1}{\gamma} = \frac{m}{E}$$

$$\rho_{initial} = 2\gamma\rho_0 = 2\frac{E}{m}\rho_0$$

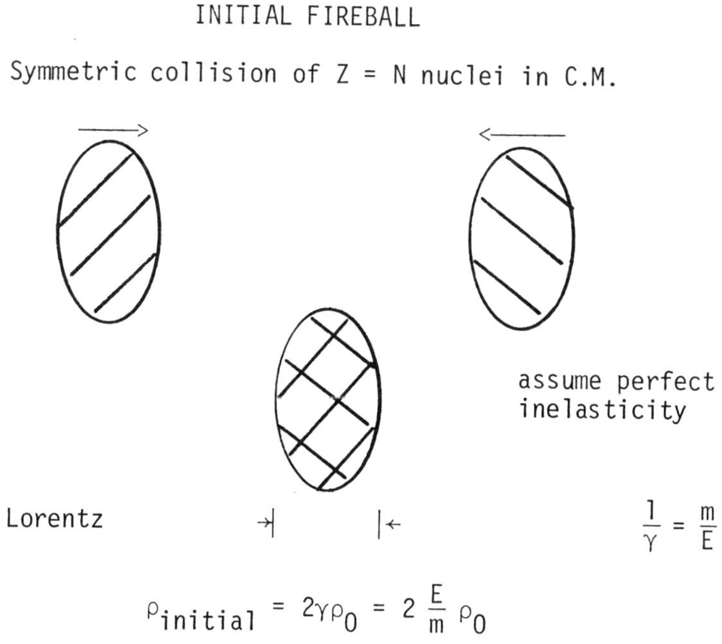

Fig. 12.   Illustrating the collision.

How Then Can it be Discovered?

Perhaps by creating as large a piece of matter as possible at high energy density and studying its properties. This is the only way I can think of. I am sure that you appreciate for example, that the specific heat of material objects depends upon their compositions. The composition of matter at high energy density depends in turn upon the number, type, and masses of hadrons that can energetically exist at that density--both those that are known, and those that are unknown, and never will be known by name!

The only means we have of producing matter at very high energy density is in collisions, by no means an ideal situation for performing calorimetric measurements. Yet it is our only hope.

So we have in mind collisions between large nuclei at high energy. Two questions come immediately to mind. 1) What is the dynamical description of the reaction? and 2) At what energy can different assumptions about the hadronic spectrum be expected to yield observable difference in the outcome of the collision?

For the dynamics I can envision two extremes. Either the collision of two nuclei at high energy 1) develops as a sequence of independent collisions, or 2) it attains thermal equilibrium and then decays.

If the first is true then for the purpose at hand, at least, there is no point in studying nuclear collisions rather than nucleon-nucleon collisions.

But I think it highly unlikely that the first is true. More likely the truth lies between the extremes.

A moment's reflection makes clear that a complete dynamical description of a collision between nuclei at very high energy involves something like the full complexity of a relativistic quantum field theory. Of course if all the ingredients of such a theory were at hand, the question raised by this paper would be moot. Since however the ultimate theory of particle structure or its solution is unlikely to emerge in the near future, it seems reasonable to attempt a model description of the dynamics of a nuclear collision.

Before attempting to explore too elaborate a model it seems prudent to me to assess whether it is worth doing so. For example, it might turn out that the energy at which sensitivity to the hadronic spectrum is achieved is so high as to be out of sight; that by no stretch of the imagination would it ever be possible to produce the required energy in the laboratory.

Therefore, Y. Karant[20] and I have assumed thermal equilibrium as a model of high energy collisions for the purpose of accessing the prospects of learning from nuclear collisions the form of the hadronic spectrum and the possibility of distinguishing between various theories of hadronic structure. If the results of such a study give an optimistic prognosis, we will feel encouraged to try harder in our treatment of the dynamics.

The attainment of a state of thermal equilibrium in a nuclear collision may seem strange at first. But at the energies in question a very large phase space is opened up by particle production. The high velocity (near c) of the pions and their strong interaction with nucleons provides a fast mechanism for thermalization in addition of course to the hadron-hadron collisions. Indeed computer studies suggest that thermalization can occur already after 3 or 4 collisions.[21] Chemical equilibrium among the various species π, N N* Δ... takes longer but may still be fast compared to the disassembly time of the composite. The extended size of the initial nuclear composite for geometrical reasons alone, slows the disassembly of the interior.[6]

There is a very extensive and beautiful literature on the thermodynamic theory of hadronic structure.[17] Also for nuclear collisions, at lower energy than we have in mind, a thermodynamic model has been introduced.[4,5,6,7,22] Inspired by the analogy of hadron thermodynamics, the hot composite system was referred to as a nuclear fireball. The model has been refined[7] and applied[23] to data on pion, proton, and composite particle spectra at various energies between 200 MeV per nucleon to 2 GeV per nucleon laboratory kinetic energy. The overall agreement with such a wide range of data is quite impressive.

Thermodynamics of Hadronic Matter

In this section we discuss the thermodynamics of the nuclear fireball in terms of an ideal relativistic gas. It may seem strange that a strongly interacting hadron system is described in such a way. However, Hagedorn[17] has argued convincingly, on the basis of statistical mechanical techniques introduced by Beth and Uhlenbeck[24] and Belenkij[25] that the hadronic spectrum _is_ the manifestation of the interactions; that by introducing the complete spectrum one has accounted for their interaction completely.

The partition function and momentum distribution for an ideal relativistic gas of Fermions or Bosons of mass m and statistical weight g = (2J+1)(2I+1) occupying a volume V at temperature T are.[17] (Units are $\hbar = c = k_{Boltzman} = 1$).

$$Z(V,T) = \frac{gV}{2\pi^2} m^2 T \sum_1^\infty \frac{(\mp)^{n+1}}{n} K_2\left(\frac{nm}{T}\right) \quad , \begin{pmatrix} F \\ B \end{pmatrix} \tag{1}$$

$$f(p,T)d^3p = \frac{gV}{2\pi^2} \frac{p^2 dp}{\exp\left(\frac{1}{T}\sqrt{p^2+m^2}\right) \pm 1} \quad , \begin{pmatrix} F \\ B \end{pmatrix} \tag{2}$$

from which the various thermodynamic quantities can be calculated. We want to describe a gas of Baryons and Mesons distributed in mass according to some unknown functions $\rho_\alpha(m)$. (α labels the families of particles, ordinary and strange baryons and mesons, some of which are shown in Table I). There are two important quantum numbers that have to be conserved, the net baryon number and strangeness. This is achieved as usual in thermodynamics, by introducing chemical potentials. If we specialize to symmetric collisions between Z = N nuclei then the conservation of baryon number and strangeness implies conservation of electric charge, since on the average $\langle Q \rangle = \frac{B+S}{2}$. We therefore make this specialization.

The average number and energy for the family of particles labelled α are

$$N_\alpha = \frac{VT}{2\pi^2} \int_{m_\alpha}^\infty dm\rho_\alpha(m)m^2 \sum_1^\infty \frac{(\mp)^{n+1}}{n} K_2\left(\frac{nm}{T}\right) \exp\frac{n\mu_\alpha}{T} \tag{3}$$

$$E_\alpha = \frac{VT}{2\pi^2} \int_{m_\alpha}^\infty dm\rho_\alpha(m)m^3 \sum_1^\infty \frac{(\mp)^{n+1}}{n} \left[K_1\left(\frac{nm}{T}\right) + \frac{3T}{nm} K_2\left(\frac{nm}{T}\right)\right] \exp\left(\frac{n\mu_\alpha}{T}\right) \tag{4}$$

Here μ_α is the chemical potential, m_α is the threshold, i.e., lowest mass particle in the family α, and K is a Kelvin function.

The baryonic charge of the system is clearly

$$B = \sum_\alpha N_\alpha B_\alpha \tag{5}$$

the sum being extended over the seven families of particles indicated in Table I(α = Π, K, N, ...) (As a family label, Π does not designate only the pions, but all the ordinary mesons, π, ρ, η, etc., and similarly for K, N, etc.) The number of antiparticles of type α and their energy are given by the above two equations (3) and (4) with $\mu \rightarrow \bar{\mu}$. We indicate by a bar, the antiparticle quantity, e.g., $\bar{N}$. Then the net baryon charge is

$$A = B - \bar{B}. \tag{6}$$

This quantity is conserved, and equal to the initial number of nucleons in the collision. The net strangeness is also conserved and is zero

$$0 = S - \bar{S} \tag{7}$$

where

$$S = \sum_{\alpha} N_\alpha S_\alpha \tag{8}$$

(The sign of the strangeness is opposite for particle and antiparticle.)

The two conditions (6) and (7) expressing baryon and strangeness conservation clearly couple all thermodynamic quantities T, μ_α. The reactions possible between the various particles dictate certain relations among the chemical potentials with the result that there are only two independent potentials, that for the nucleon and that for the kaon. The scheme we use to solve for the energy and particle populations as a function of temperature is basically the following. Choose a temperature T and find the values of the two chemical potentials that satisfy equations (6) and (7). When these are found then the populations and energies can be found from (3) and (4) and the total energy is of course

$$E = \sum_{\alpha} E_\alpha \tag{9}$$

The initial condition of the fireball is a little more complicated to solve. We consider symmetric collisions in the center of mass frame between nuclei of atomic number A/2. Each nucleus is Lorentz contracted by the factor $1/\gamma = m/E$. If the volume per nucleon in the rest frame of each nucleus is $v_\rho = \frac{4}{3}\pi\,(1.2)^3$, then in the C.M. frame it is $v_\rho\gamma$. We assume that the collision is perfectly inelastic; that each nucleus is stopped by the other. Then the largest possible volume, in which all nucleons are contained, just after the nuclei have stopped each other is the contracted volume occupied originally by one.[18] So the initial baryon density of the fireball is

$$\rho_{initial} = \frac{2\gamma}{v_0} = \frac{2E}{mv_0} = v^{-1} \qquad (10)$$

and the volume per baryon is the reciprocal (see Fig. 12). Hence the volume V multiplying all quantities (3), (4), is a function of the as yet to be determined energy. How this problem is solved can be found in Appendix B.

In case the assumption of inelasticity is questioned, Fig. 13 shows the stopping of a very high energy proton by a nucleus which is clearly highly inelastic.

Three Examples of Hadronic Spectra

The object of the rest of the paper is to show how and at what energies the thermodynamic nuclear fireball would differ under the three different assumptions for the hadronic spectrum discussed below. For brevity we shall sometimes refer to the results for different spectra as being different worlds. The ultimate object, toward which this paper is a modest start is to discover which is most like our world.

a) The Known Hadrons: As one extreme case we might suppose that all of the hadrons have already been discovered. They are listed with their properties in the Particle Data Tables[16] and their density is plotted in Fig. 9 with the exception of recent discoveries. There are 56 different multiplets known with a total particle multiplicity of 531. Together with the antiparticles these comprise the 1000 or so known hadronic states mentioned earlier. We include them all by using the average mass and width for each multiplet. For our purpose, they fall into the seven families shown in Table I.

b) Bootstrap (exponential) Spectrum: There are several mathematical formulations of the bootstrap hypothesis but the thermodynamic theory of Hagedorn[17] is most useful to us because it yields an asymptotic form for the bootstrap hadron spectrum. The bootstrap spectrum lies at the opposite extreme from the "known" spectrum since it rises exponentially and without bound. We shall test the consequences of a bootstrap theory by using the Hagedorn form of the spectrum for the non-strange mesons and baryons in the region $m > 12\ m_\pi$. Below this mass we use the discrete known particles for these two families and all known strange particles.

We normalize the Hagedorn spectrum to agree with the average density of states in five pion mass intervals around 10 m_π. Thus

$$\rho_{Bootstrap}(m) = \begin{cases} \dfrac{1.12\, e^{m/T_0}}{(m/T_0)^3}/\text{pion mass} & m > 12\, m_\pi \\ \\ \text{discrete non-strange particles} & m \leq 12\, m_\pi \end{cases} \qquad (13)$$

$$+ \text{ all strange particles}$$

$$T_0 = 0.958\, m_\pi \qquad m_\pi = 140 \text{ MeV}$$

We assume that there is an equal number of ordinary mesons and baryons in the continuous region. We might, but do not yet include continua for the families of strange particles because there is generally an insufficient number to estimate the normalization of the continua.

Recall that the quark hypothesis might also lead to an exponential spectrum, but possibly with different constants than those determined by a fit (Fig. 9) to the bootstrap iteration on the known particles.

c) Rigid Quark Bag: As an intermediate case, and so as to bring out where the sensitivity is achieved under less extreme alternatives than the first two, we consider a naive rigid quark bag. A meson is considered to be composed of 2 quarks, and a baryon of three. The walls are considered rigid and no new quark pairs are created within a hadron. Frautschi[26] finds that the density of such objects rises as m^2 and m^5 respectively. Normalizing at $m = 10\, m_\pi$ to the same value as the Hagedorn spectrum at that mass, we have for the continuous spectra for ordinary mesons (π) and baryons (N)

$$\rho_{Bag} = \begin{cases} \left.\begin{aligned} &\rho_\pi(m) = 0.154\, (m/m_\pi)^2/\text{pion mass} \\ &\rho_N(m) = 1.36 \times 10^{-4}\, (m/m_\pi)^5/\text{pion mass} \end{aligned}\right\} & m > 12\, m_\pi \\ \text{discrete non-strange particles} & m \leq 12\, m_\pi \end{cases} \qquad (14)$$

$$+ \text{ all strange particles}$$

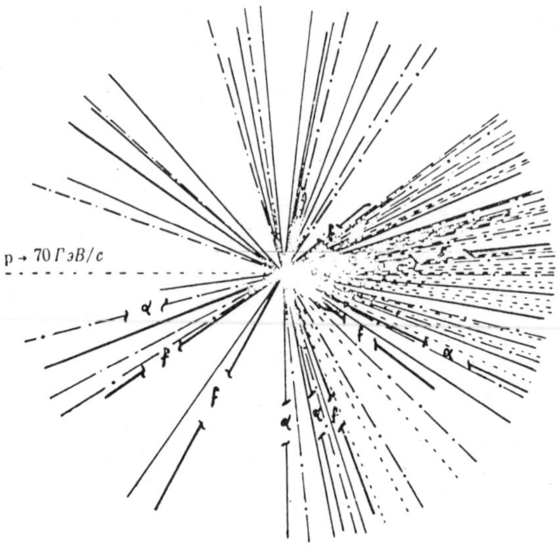

Fig. 13. Complete destruction of Ag, Br, Pb nuclei by 70 GeV/c protons and 17 GeV/c alphas (150 events analyzed). From O. Akhrorov <u>et al</u>., JINR (Dubna, 1976).

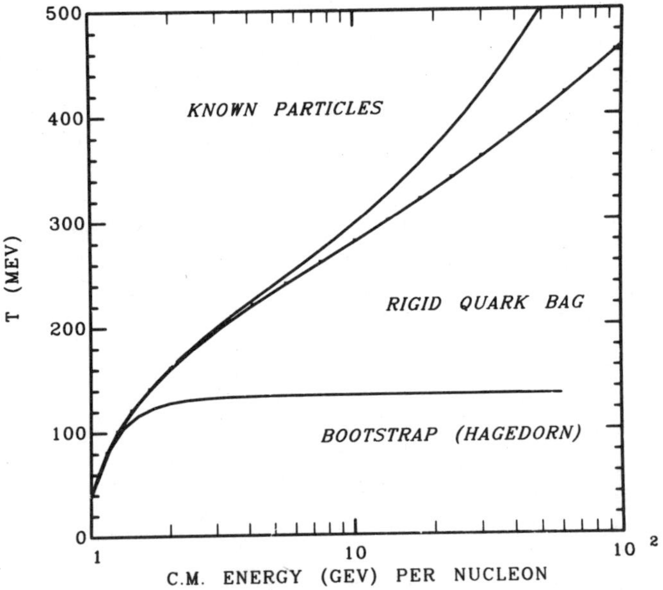

Fig. 14. For the three "worlds" considered, the temperature of hot hadronic matter assumed to be produced as a symmetric nuclear collision is plotted as a function of the C.M. total energy per nucleon of the colliding nuclei for a volume corresponding to the initial Lorentz contracted fireball.

The Temperature

The first crude indication of differences between hadronic matter constructed from the three assumed spectra is registered in the initial temperature they would be heated to for the same energy content. Since we assume a perfectly inelastic collision, the C.M. collision energy per nucleon including rest energy is the total fireball energy per nucleon. These temperatures are shown in Fig. 14. For matter composed of a hadronic spectrum limited to the known particles, the temperature is by far highest at energies greater than several GeV. Because energy goes into making additional particles in the quark bag spectrum that were not present in the known spectrum, the temperature is lower at any corresponding energy. For the exponentially rising spectrum, as first discovered and emphasized by Hagedron, the temperature is limited to a maximum value corresponding to the constant T_0 in the spectrum eq. (13).

While T_0 appears to be nearly the pion mass, its value is not determined within the theory of Hagedorn. Instead it is deduced from a comparison with data. While the data often used are $p_\perp$ measurements, we chose to fit the Frautschi[18] bootstrap iteration on the known particles.

The limiting temperature of matter, if composed of hadrons obeying the bootstrap condition (more precisely the exponential rise) is a truly remarkable property which has no analogies in other physical systems that I know of. (The boiling point of water is sometimes mentioned. This is a false analogue. The temperature of matter is limited even though the energy input is increased indefinitely! The limit to water temperature is reached because the energy is carried off by the steam. It is by comparison a trivial limit and totally different in origin.)

The mathematical nature of the limit can be seen by referring to eq. (4). For large masses, m >> T, the Kelvin functions decay exponentially like

$$K(x) \propto \frac{1}{\sqrt{x}} \; e^{-x}$$

Inserting the Hagedorn spectrum we find

$$E \propto \int_M \frac{dm}{m^{1/2}} \; \exp - \left(\frac{T_0 - T}{T_0 T}\right) m \propto \left(\frac{T\,T_0}{T_0 - T}\right)^{1/2}$$

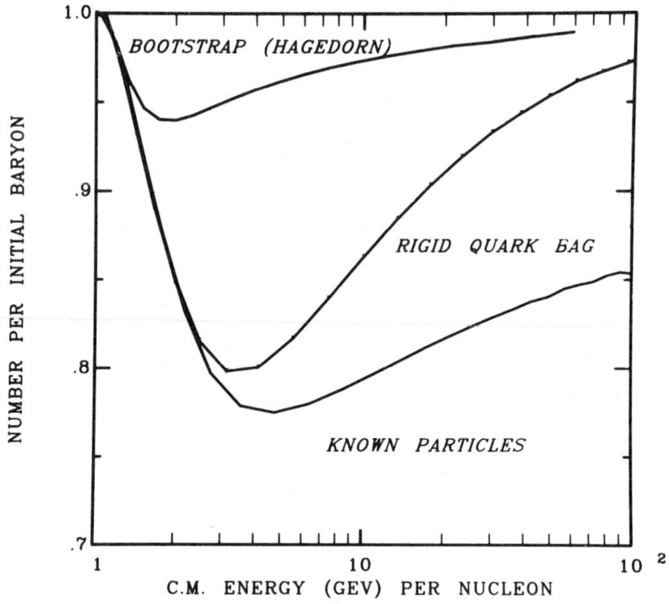

Fig. 15. The number of ordinary baryons is depleted owing to creation of strange baryons. The difference between unity and the curve corresponds to the baryon change resident in strange particles, and is also equal to the strange meson (K) population, because the net strangeness is zero.

Thus as long as $T < T_0$ the integral converges; the energy is finite. But for $T \geq T_0$ the integral diverges. It would require infinite energy to raise the temperature to T_0 or beyond!

Composition of the Initial Fireball

Neither the temperature nor composition of the initial fireball are observables because any conceivable experiment must look at the products of the collision after the fireball has disassembled. Nonetheless it is interesting to look at the calculated populations because they are the starting point of the subsequent expansion or decay of the fireball. They also give us a glimpse of what the composition of the universe might have looked like at the beginning of time for very high energy and particle density. Because of the time scales involved we do not have to consider photons and leptons in equilibrium with the hadrons, so this is an important difference from the cosmological problem.

A very immediate impression of how the three worlds differ is given by Fig. 15 which shows the degree to which the ordinary (non-strange) baryon number is depleted. Initially all of the baryonic charge resides in non-strange baryons (the original neutrons and protons). As the energy is increased, the strange particles begin to be populated. The difference between unity and the plotted curves is this strange baryon population. This is a loose statement since the strangeness quantum numbers are not limited to the value unity. The succeeding statement for kaons is exact. Since however there was initially no net strangeness, this is exactly counter balanced by kaon populations (the strange mesons). We see that in all cases, there is a sudden rise in the strange particle populations which however is quenched quickly in the bootstrap world but rises to almost 25% in the case of the "known" world. At about 5 GeV almost 25% of the baryonic charge is converted to strange particles and to corresponding kaons!

Because there are so many discrete particles, not to mention the continua, we make the following arbitrary groupings to display more detailed information. Each family of particles is broken up into light particles comprising the lightest five (when there are that many) and heavy particles comprising all the rest, including continuum particles in the case of the quark bag and bootstrap worlds. We sum the populations in each group and plot only the summed populations. Thus the ordinary (non-strange) mesons are represented by two curves, for light and heavy mesons. There are no heavy kaons, but there are anti-kaons so there are curves for both. The ordinary baryons are represented by four curves, light and heavy baryons and anti-baryons. And so on.

Figures 16-19 show truly remarkable differences of the three fireballs depending on which is the underlying hadronic spectrum. For both the known spectrum and the quark bag, the heavy baryon and anti-baryon populations eventually dominate with heavy mesons the next most populous group. In the case of the quark bag this happens at rather low energy (on a particle creation scale). The heavy mesons follow. The composition at one GeV is of course all nucleon, but the light baryon and anti-baryons become less populated than the heavy ones in the bag model at energy above 10 GeV. The Hagedorn or bootstrap world is remarkably different. The light meson population rises to 10% and then falls. The heavy baryon population rises sharply and above 3 GeV the fireball is composed of more heavy baryons than light ones. By 10 GeV about 60% of the baryons are heavy and only 40% are light.

There is another remarkable difference. In the "known" and "bag" worlds, all particle-anti-particle populations approach each other at high energy (with anti-particles slightly less numerous). In the bootstrap (exponential) world, the anti-particles and mesons have microscopic populations. It is a world dominated at high energy and density by heavy baryons. This is an inevitable consequence of the exponential rise in the bootstrap density. At high temperature the system wants to produce heavy particles.

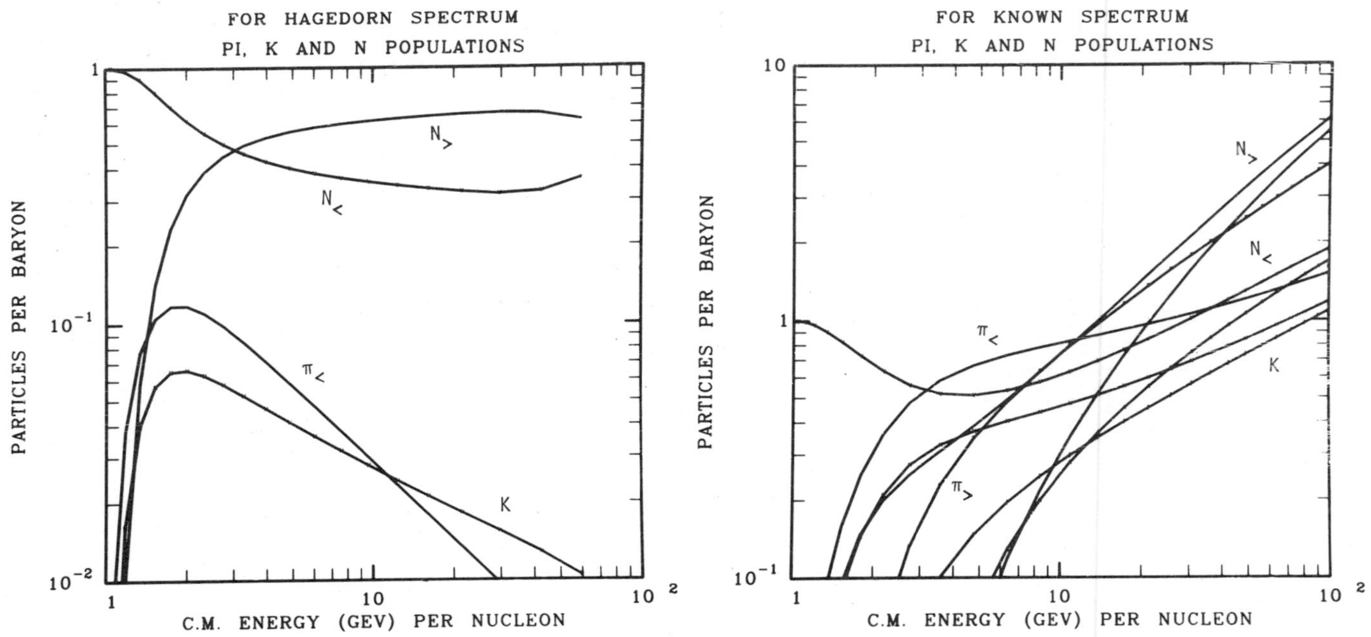

Figs. 16-17. Corresponding to the three "worlds" investigated, the populations of the light and heavy members of the family of ordinary mesons (Π), strange-mesons (K) and ordinary baryons (N), are plotted as a function of energy. Light, refers to the first five multiplets (if that many) of each family, and are denoted by < Heavy, refers to all others, including the continuum where applicable, and are denoted by >. Anti-particles, except for the bootstrap, approach the particle populations at high energy. Refer to Table I for some of the members of the families.

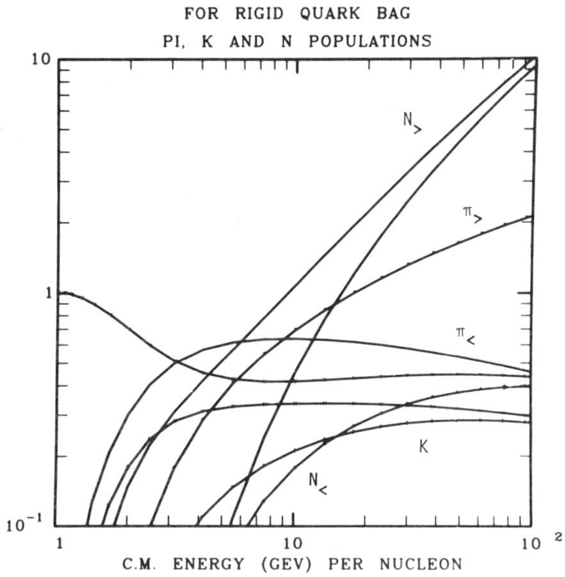

Fig. 17. See caption for Fig. 16.

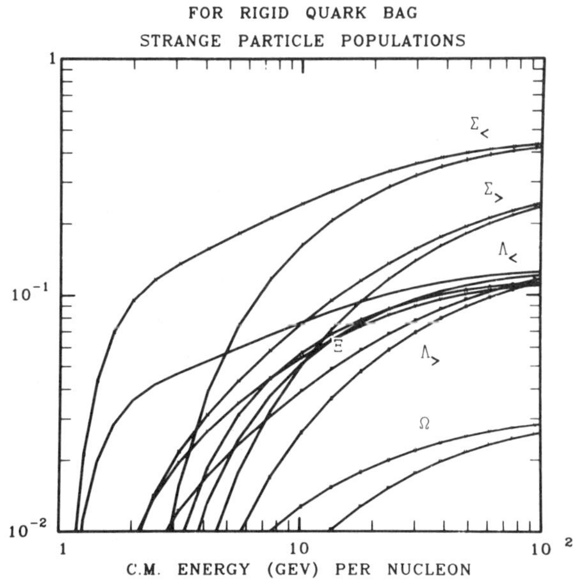

Figs. 18-19. The population of the strange baryons as a function of energy for the three "worlds". Notation similar to above.

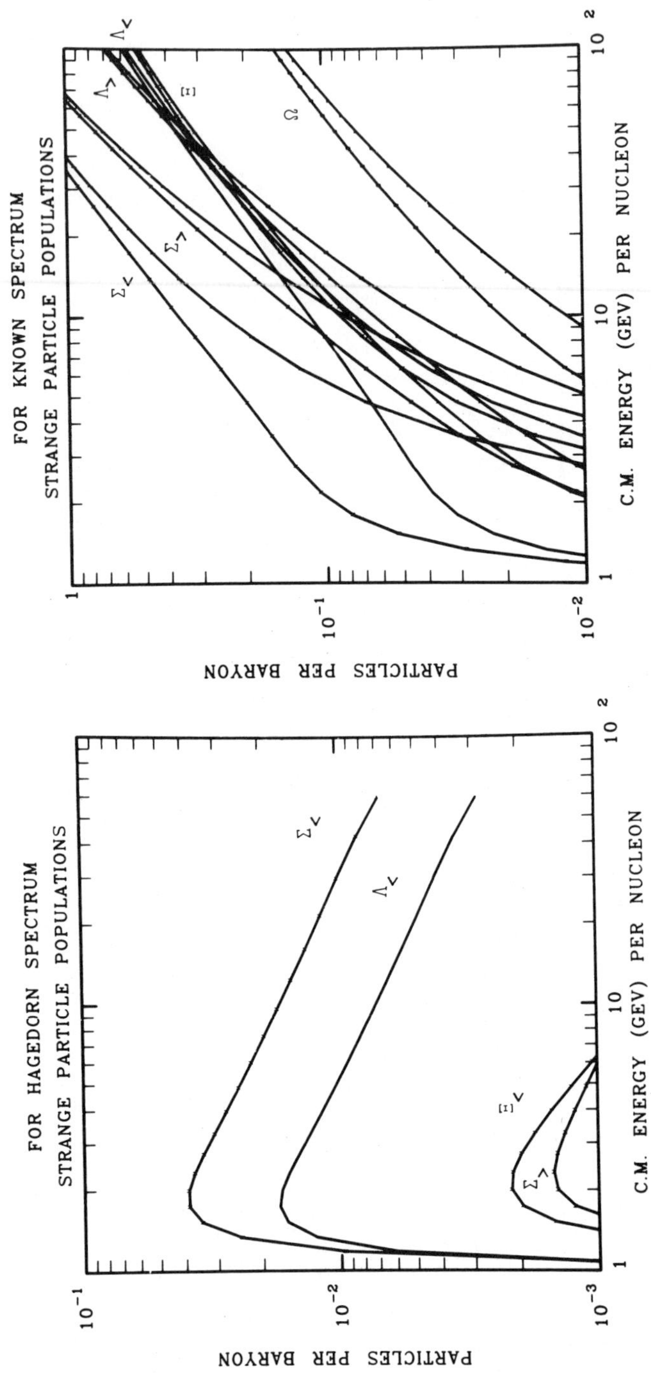

Fig. 19. See caption for Fig. 18.

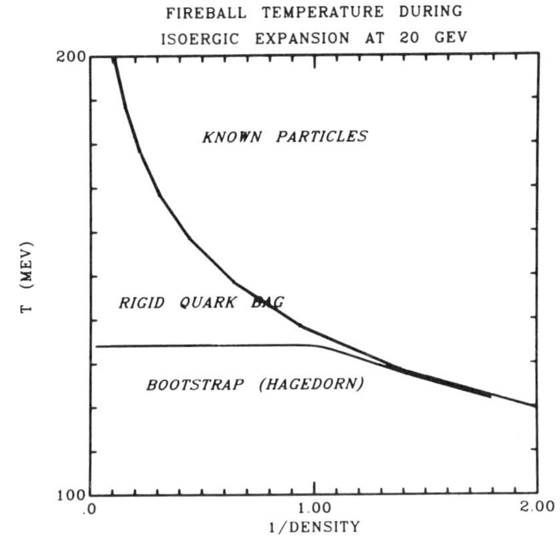

Fig. 20. The temperature of the fireball as it expands with constant energy equal to 20 GeV. The ordinate is the reciprocal of the total hadron density in units of the density of normal nuclei (0.17 Fm^{-3}). On this scale, 2 corresponds to a density such that hadron has a share of the volume corresponding to a sphere with radius equal to the pion Compton wavelength 1.4 Fm.

Since however baryon conservation is forced, the energy is committed to making heavy baryons to the exclusion of mesons.

Expansion of the Fireball

If the S-matrix of the strong interactions (for each of our model worlds) were known, then we could calculate the observables that reach the counters. It is not, and that is what led us to the statistical mechanical treatment and the reasonable assumption of an equilibrated initial state. Now we must model the subsequent expansion of the system that carries the particles to the counting apparatus. It is possibly reasonable to assume, as in cosmology, that the expansion occurs through a series of equilibrated states. At some point during the expansion, when the density falls below a critical value, thermal contact between the particles is broken. This is called the freezeout.[27] Relative populations do not change thereafter, except by the decay of isolated particles. Since however, there is no membrane surrounding the expanding fireball, fast outward moving particles can escape from the equilibrated region prior to freezeout. Therefore there are two (indistinguishable) components to the particles that reach the counters, those that escape from the fireball during its expansion and prior to freezeout, and those that remain in thermal equilibrium until the freezeout. This is the scenario that we wish to model in order to calculate the spectra of the observed stable particles.

The importance of the pre-freezeout radiation should be commented on. An equilibrated system retains no memory of earlier states of the system. As the equilibrated region expands, its temperature falls, and the distinction between the three worlds will fade. It is the pre-freezeout radiation that carries the vital information on the early high-temperature condition of the fireball. Those particles that remain in equilibrium until the end are an unwanted background.

Isoergic Expansion with no Pre-freezeout Radiation

As a first orientation to gain insight into the evolution of the composition and temperature of the fireball in the three worlds, during an expansion, we consider an isoergic expansion in which no particles leave the system prior to the freezeout density, which is the density below which thermal contact is lost.

Presumably, the freezeout density, ρ_F, is less than normal nuclear density, $\rho_N \simeq 0.17\ \mathrm{fm}^{-3}$, but is not less than the density corresponding to each particle having a sphere of radius equal to

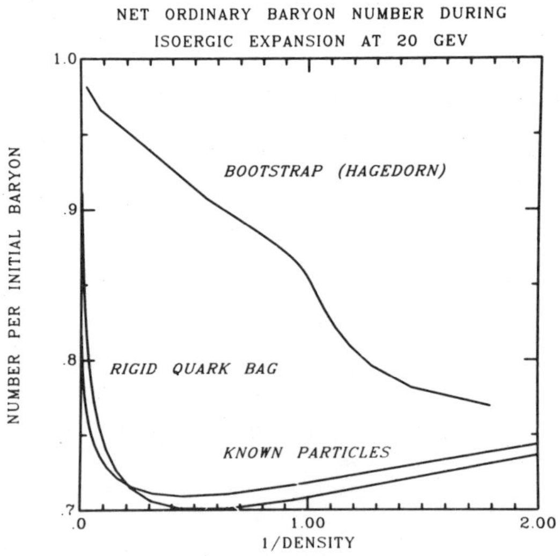

Fig. 21. Similar to Fig. 15 for the expansion at 20 GeV.

a pion wavelength, $\rho_\lambda \simeq 0.085\ \mathrm{fm}^{-3}$.

$$\rho_\lambda < \rho_F < \rho_N$$

We shall plot our results for the thermal expansion stage as a function of $1/\rho$ where the density is measured in units of the nuclear density. On such a scale, the freezeout presumably occurs between 1 and 2. (ρ is the hadron density.)

Of course the temperature falls monotomically during the expansion to the freezeout point. Therefore the temperature characterizing particles that remain in thermal contact until the freezeout is the lowest temperature the fireball possessed. In this connection it is worth remarking that the claim in the literature[28] that the ultimate temperature has been measured in hadron-hadron collisions is possibly unwarranted. Unfortunately no one possesses a thermometer that he can insert into the initially formed fireball, but instead he must wait until the fireball expands and its constituents arrive at the counters.

The fall in temperature during an isoergic expansion at 20 GeV is shown in Fig. 20. For the "known" and rigid bag worlds it drops precipitously while for the bootstrap world it remains nearly constant until a late stage. At somewhat less than normal nuclear density, it begins to fall at the more rapid rate of the other worlds. This merely corresponds to the fact that at low enough temperature all worlds look the same. This figure suggests that temperature measurements near freezeout cannot distinguish between various hadronic spectra, as remarked earlier. The possible measurement of a temperature of T = 119 MeV in hadron-hadron collisions at 28 GeV reported in the literature[28] would correspond, on our figure, to a freezeout a little to the right of the frame where all worlds have fallen to virtually the same temperature.

The way in which the ordinary baryon charge is depleted during the expansion at 20 GeV is shown in Fig. 21. The bootstrap world is again remarkably different from the others, but although there was an initial large difference between the bag and "known" world (Fig. 15) it rapidly diminishes.

The populations of the various groups during the expansion at 20 GeV are shown in Figs. 22-23. The composition of the bootstrap fireball is remarkably different from the others during the early stage. However if thermal contact were sustained for <u>all</u> time it is clear that all worlds must appear the same at low enough temperature and density. Indeed they would just return to the original neutron proton composition. The breaking of thermal contact interrupts this return however. As stated earlier, we expect freezeout to occur between $1 \lesssim \frac{1}{\rho} \lesssim 2$. It is in precisely

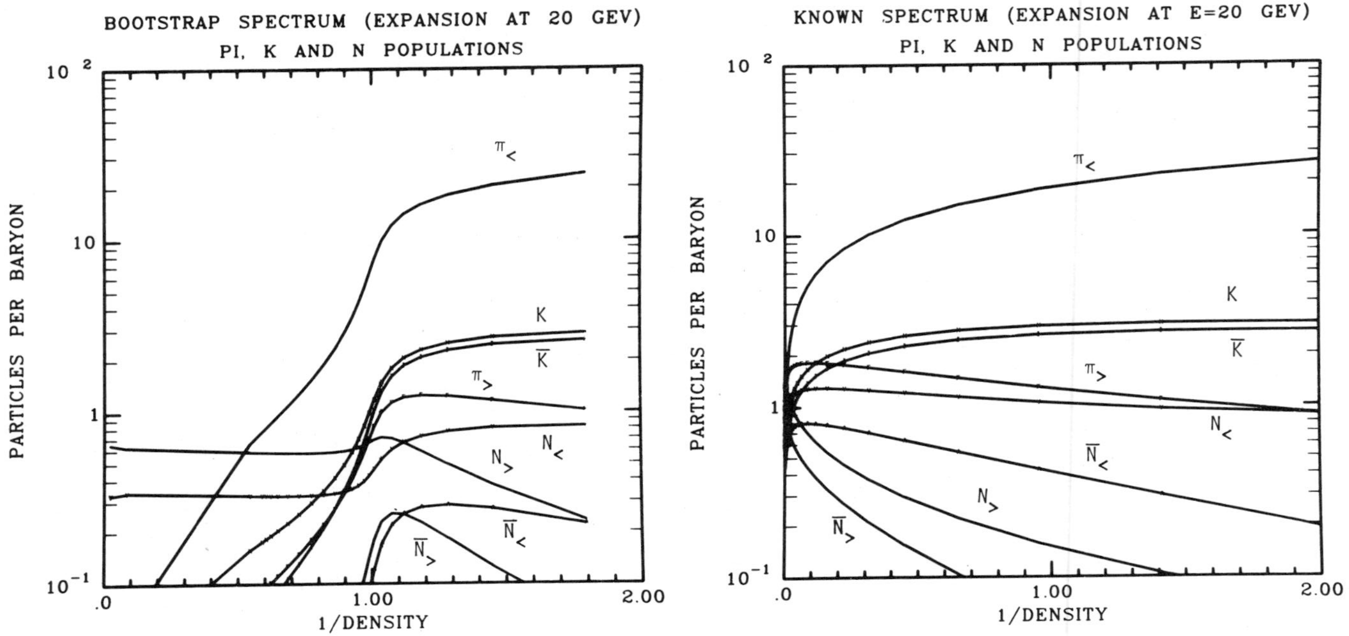

Figs. 22-23. For the expansion at 20 GeV the populations for the three "worlds".

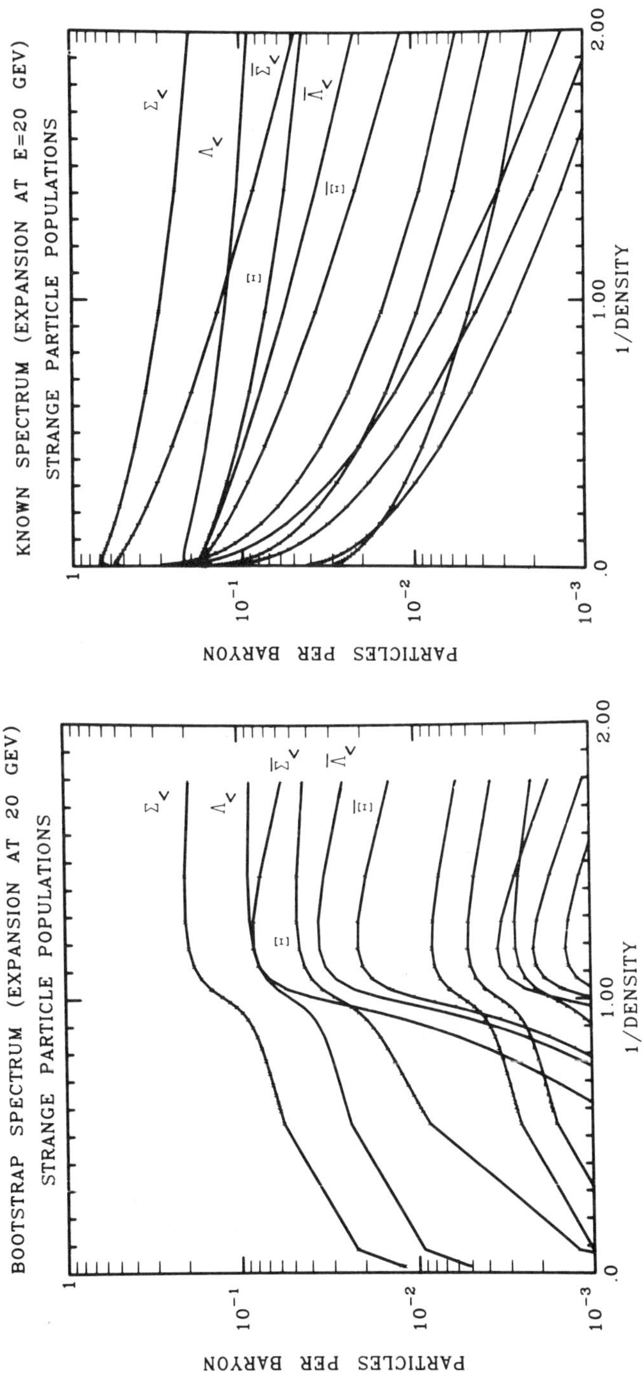

Fig. 23. See caption for Fig. 22.

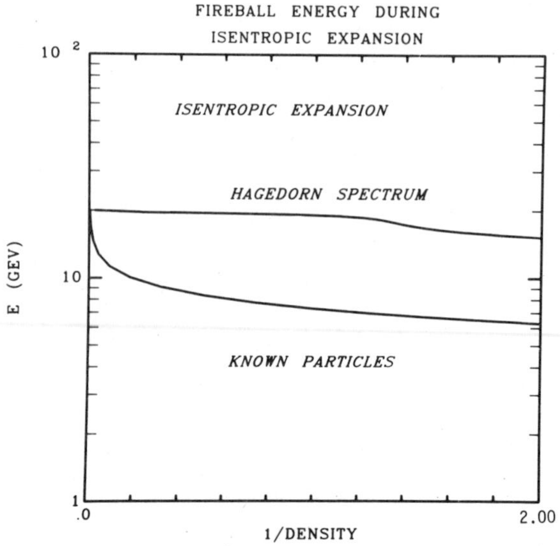

Fig. 24. The energy of the fireball decreases during an inentropic expansion. The energy lost to the thermal region is assumed to be carried off by particle radiation. No strange particles were included in this particular calculation.

this range however that all light particle populations have already become quite similar in the three worlds.

It has now become clear that if thermal contact between all constituents is sustained during an expansion to a freezeout density equal to the nuclear density or less, it is impossible to distinguish between the three worlds. On the other hand it seems most likely that some of the constituents of the fireball will escape prior to the freezeout.

Isentropic Expansion of the Fireball

In an isentropic expansion, energy is lost to that part of the fireball that remains in thermal contact. I interpret the loss to be balanced by radiation of particles from the surface during the course of the expansion. This radiation produces the pressure against which the particles remaining in the fireball work during the expansion.

Figure 24 shows the energy remaining in the fireball for an isentropic expansion starting from an energy of 20 GeV. What this picture immediately suggests is that the pre-freezeout radiation is much more copious for the "known" and bag worlds than for the

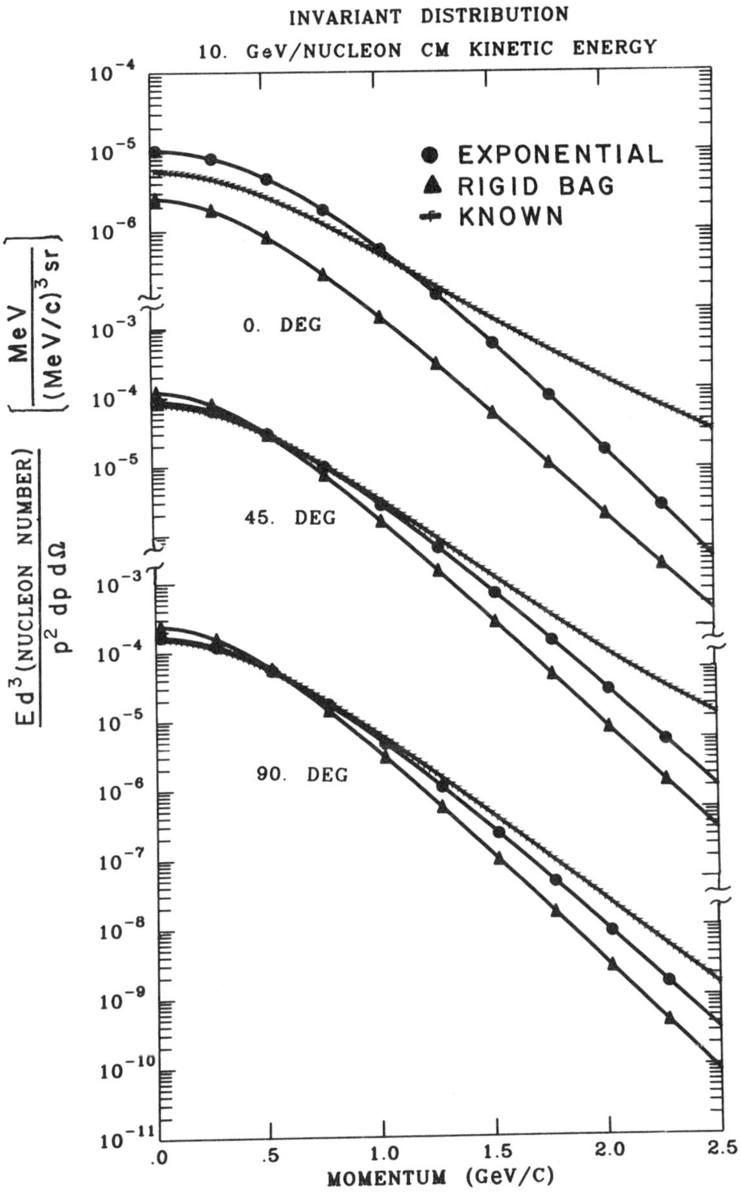

Figs. 25-28. For a symmetrical head-on collision of two A = 50 nuclei with C.M. kinetic energy of 10 GeV/nucleon (colliding beams) the number, N, of particles, nucleons, anti-nucleons, pions, and kaons) in invariant phase space, $Ed^3N/(p^2dpd\Omega)$ is plotted as a function of C.M. momentum at three angles.

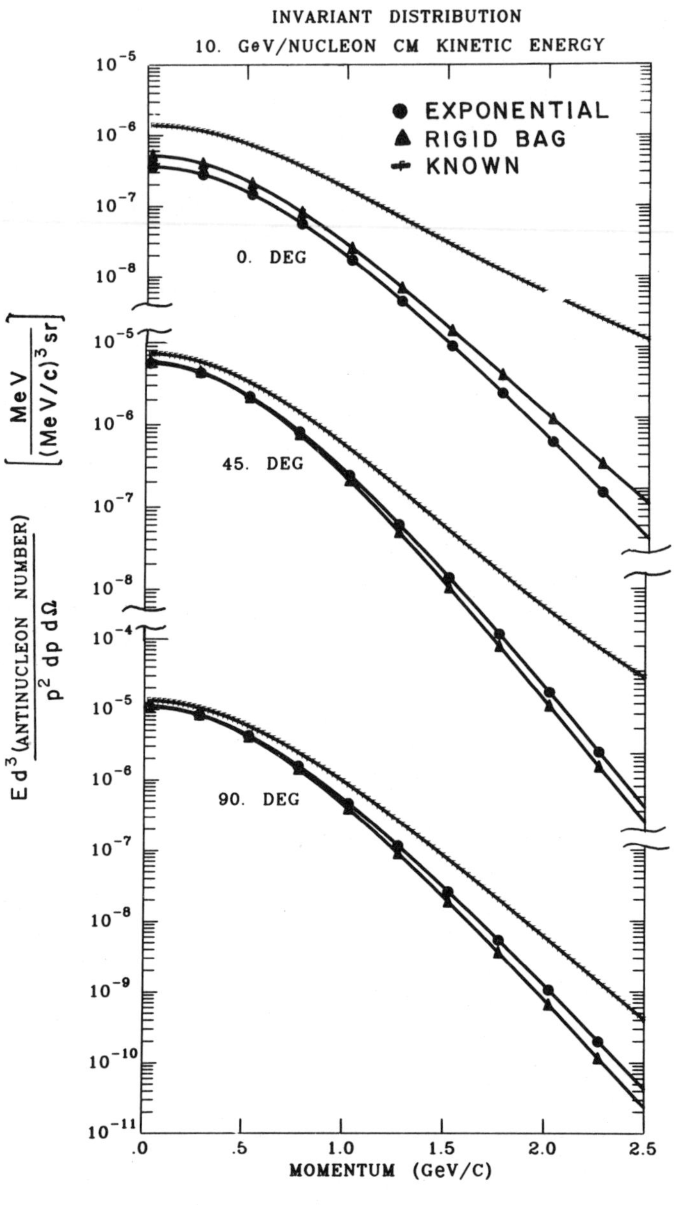

Fig. 26. See caption for Fig. 25.

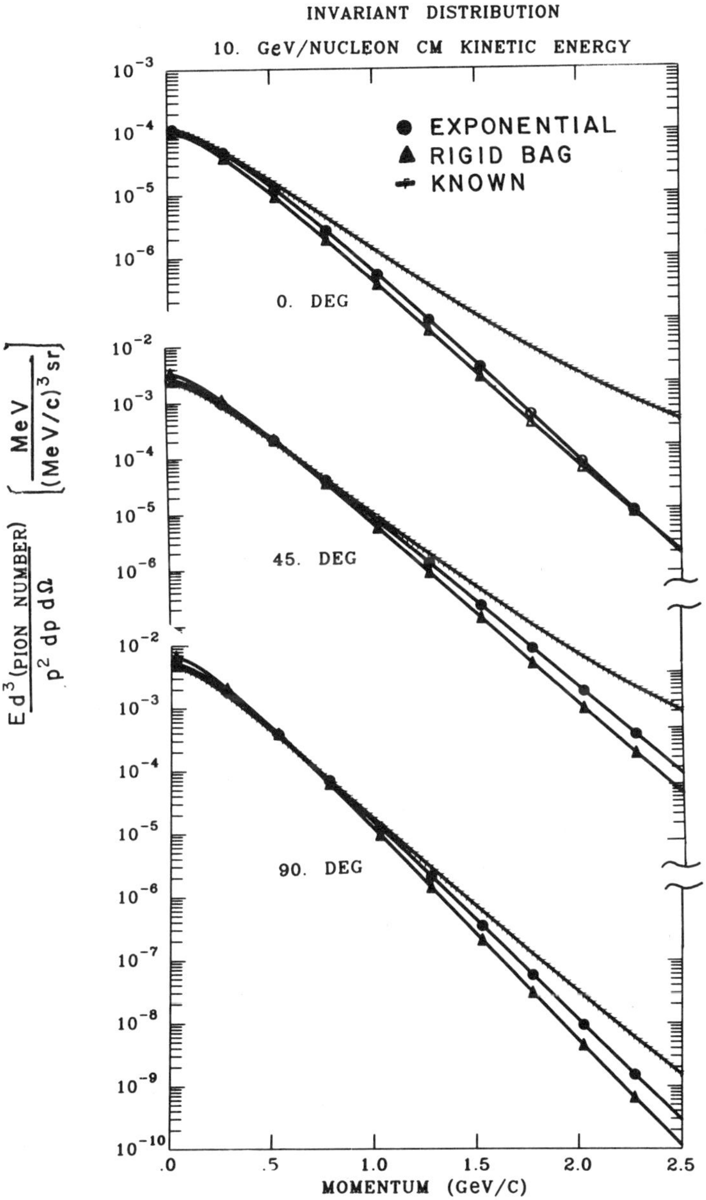

Fig. 27. See caption for Fig. 25.

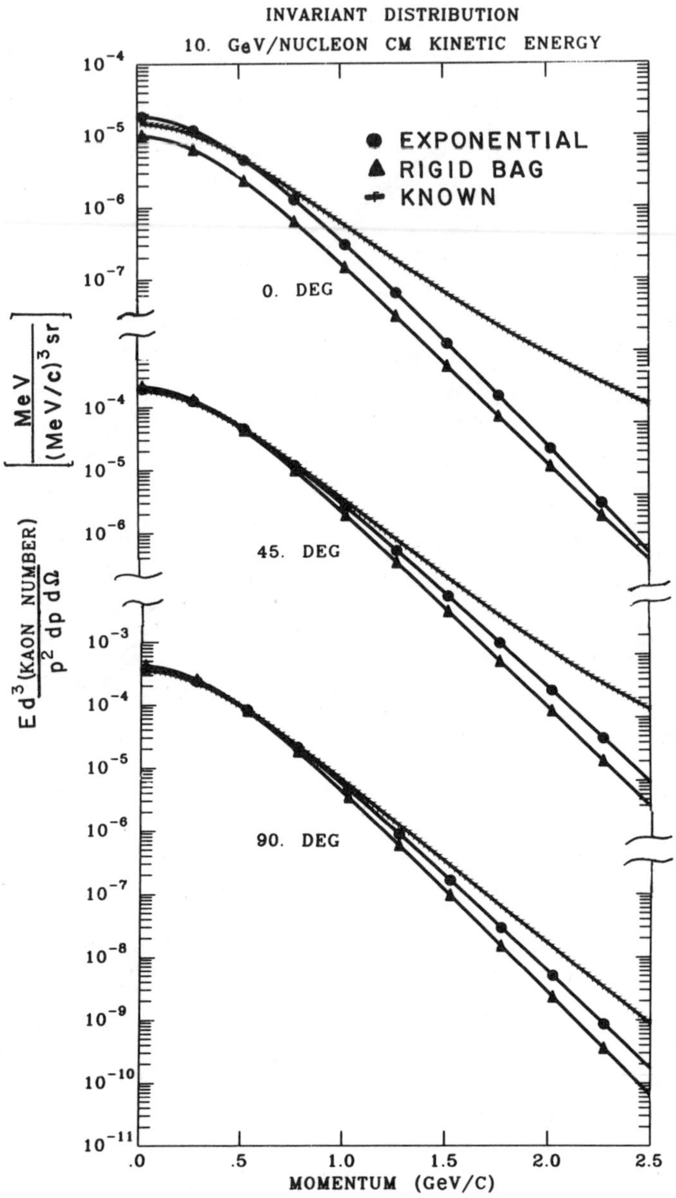

Fig. 28. See caption for Fig. 25.

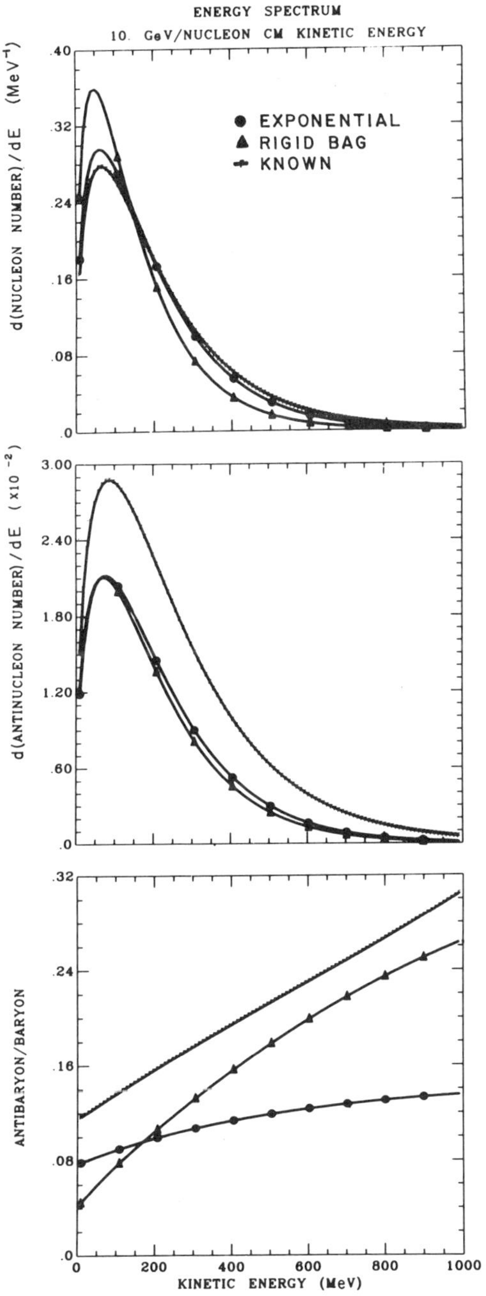

Fig. 29. For the collision described in Fig. 25 the spectrum (integrated over all angles) of nucleons, anti-nucleons and the ratio anti-baryon/baryon is plotted as a function of C.M. kinetic energy.

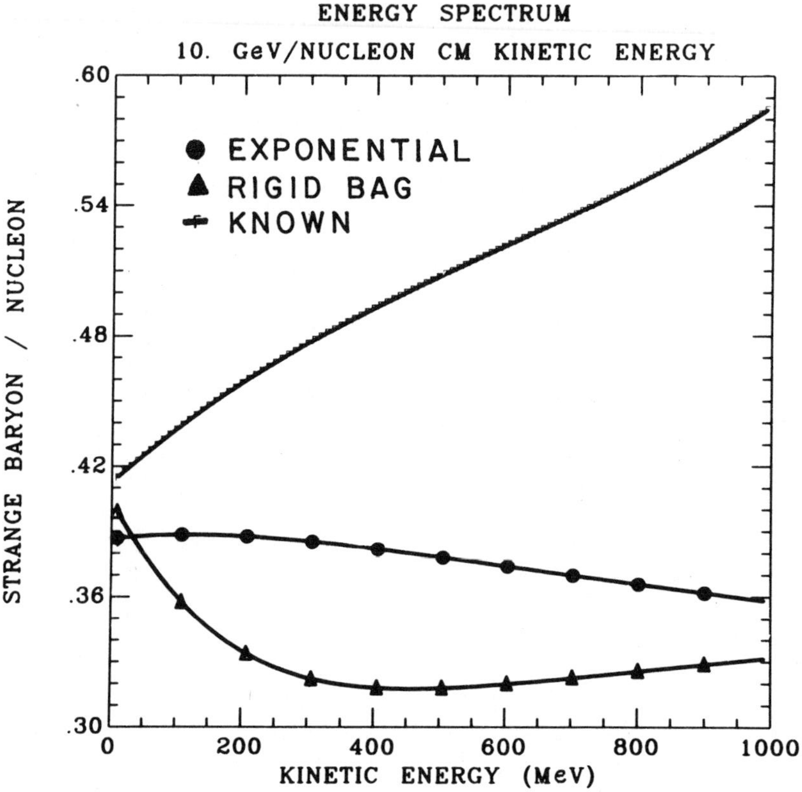

Fig. 30. Ratio of strange baryons to nucleons.

bootstrap world. The energy is trapped in massive baryons till a later stage in the expansion. Otherwise the qualitative difference between the worlds discussed in connection with the isoergic expansion appear in this expansion also.

Quasi-Dynamical Expansion

The hint gained in the preceeding study, that the pre-freezeout radiation from the three worlds will be different, because of the differing populations in the early, high-temperature phase of the expansion, provides the motive for attempting to follow the time development of the expansion.

We shall assume that at any instant the particles that lie within a mean free path of the surface of the fireball, and are directed outward will move into vacuum. Those of them that are unstable will decay, within a resonance mean life, into lighter stable and instable particles, so that in the immediate vicinity of the surface, the density remains high. Therefore we take this to define an instantaneous new surface and we assume that a new quasi-equilibrium state is established in this new volume. Meanwhile, those of the original outward moving particles that are stable, and moving faster than the unstable ones that established the position of the new surface, escape to vacuum. Their quantum numbers and energy are subtracted from those defining the state of the new quasi-equilibrated fireball. These steps are iterated until the density has dropped to the critical density or the fireball contains negligible energy and conserved quantum numbers in resonance states, whichever comes sooner. At that point the remaining particles move freely to the vacuum.

Of course the expansion does not occur isotropically in the C.M. because of the initial Lorentz contracted shape of the fireball. It is clear from the geometry that the shape of the fireball will evolve from the oblate spheroidal shape to a prolate spheroid.

We have calculated the distributions of the various stable particles and anti-particles [π, κ, η, n, Λ(1116), Σ(1193), Ξ(1318), Ω(1672)]. As anticipated, the differences between the three worlds does not register so dramatically in the final products as it did in the initial hot fireballs, since the products are emitted over the lifetime of the fireball, from its initial hot state to its cooler final state. Note that the expansion to freezeout is enormous, beginning with a Lorentz contracted nuclear volume, and ending at freezeout some 20-30 nuclear volumes.

A sample of the preliminary results, corresponding to two mass 50 nuclei impinging on each other with 10 GeV kinetic energy per nucleon (colliding beams) are shown in Fig. 25-29. At three different angles, the invariant quantity, $E\ dN/d^3p$, is plotted as a function of the momentum, for nucleons, anti-nucleons, pions and kaons. While the differences are not large, they become of the order of 10 for the high momentum particles, which are emitted predominantly during the early stage of the fireball expansion. A particularly strong signal, and easy to measure, is the ratio of anti-baryons to baryons shown in Fig. 29.

Anti-Nuclei, Hyper-Nuclei, and Quark Phase

We come now to a most remarkable difference between nuclear collisions and hadron-hadron collisions. To the extent that thermodynamics applies to each, then all I have said until now applies to nuclear fireballs as to hadron fireballs.

A reexamination of the populations reveals that the anti-baryons and also strange baryons have significant populations. As an example, for the bootstrap world during the expansion phase at densities below nuclear density, (1 on the ordinate) the population of the light Σ family is ~ 0.2 per baryon. That means that for a collision involving a hundred nucleons, 20 Σ's appear at the freezeout! There are even more light anti-baryons present, about 27. Thus although we have not yet calculated composite particle populations, we can anticipate significant production of light anti-nuclei and hyper-nuclei and possibly even strange nuclei; i.e., nuclei composed entirely of strange baryons. This appears to be a fascinating possibility. I presume little is known of the binding properties of such objects, except of course anti-nuclei, which would have to be the same as ordinary nuclei.

Moreover, we note that pre-freezeout radiation of these as with single particles would be quite different in the worlds examined.

A quark phase can also be discussed. If the whole system remains in thermal contact until the density has fallen to a freezeout density below which interactions cease, then a quark phase would be hidden. (Unless of course quarks can exist as free particles in which case some of them may not find a partner to recombine with before freezeout.) The total energy, since it is still shared by the whole system, insures that when the quarks recondense into the hadron phase, the composition will evolve with density to the freezeout in exactly the same way as if the quark phase had never existed. However, if some particles do escape from the equilibrated region before freezeout, which does seem plausible, then the numbers and types of escaping particles would depend upon whether, during part of the expansion stage, the matter was in a quark phase. During the quark phase there would be no radiation (assuming no asymptotically free quarks), or if there were a mixture of the two phases, say quark matter in the interior surrounded by a hadron halo, the radiation would likely be different than if only the hadron phase existed throughout. Assuming that quarks cannot exist as asymptotically free particles, I conclude that detection of a quark phase may be possible, but its detection would depend upon an accurate description of the disassembly stage of the fireball.

CONCLUSIONS

The calculations reported here confirm our hope that it will be possible to make a statement concerning the assymptotic form of the hadronic spectrum through a study of the products of nuclear collisions at energies from about 5 GeV per nucleon in the C.M. In particular, it should be easy to distinguish between a finite spectrum and an unbounded one. The energies required are high but they can be attained in conceivable accelerators.

We concede that the dynamics of high energy collisions is probably more complicated than thermodynamics supplemented by a quasi-equilibrium expansion. But I can see no reason to expect that a complete dynamical description, were it possible, would exhibit much less sensitivity to the outcome, than our model.

ACKNOWLEDGEMENTS

The calculations reported here were carried out in collaboration with my student, Y. Karant. This work was supported by the U.S. Department of Energy.

APPENDIX A: CHEMICAL POTENTIALS

The chemical potentials, as is usual in thermodynamics, must obey certain relations that are dictated by the possible reactions. For example

$$2p \to 2p + \pi^0 \text{ implies } \mu_\pi = 0$$

and

$$pn \to p\Delta^0$$

$$pn \to n\Delta^+$$

$$pp \to p\Delta^+$$

$$pp \to p\Delta^{++}$$

$$nn \to p\Delta^-$$

imply that

$$\mu_{\Delta^0} = \mu_n, \ \mu_{\Delta^+} = \mu_p, \ \mu_{\Delta^{++}} = 2\mu_p - \mu_n, \ \mu_{\Delta^-} = 2\mu_n - \mu_p$$

We will ignore the proton, neutron mass difference so that $\mu_n = \mu_p - \mu_B$. Then

$$\mu_\Delta = \mu_B$$

and in general all multiplets belonging to the same family have the same chemical potential.

There are also relationships between the chemical potentials of different families. The reactions

$$NN \to NK\Lambda$$

$$NN \to NK\Sigma$$

$$N\pi \to \Lambda K$$

$$NN \to N\Xi KK$$

$$NN \to N\Omega KKK$$

imply

$$\mu_B = \mu_K + \mu_\Lambda = \mu_K + \mu_\Sigma = \mu_\Lambda + \mu_\Sigma = \mu_\Xi + 2\mu_k = \mu_\Omega + 3\mu_K$$

Call $\mu_K = \mu_S$; then a solution to the equation is:

$$\mu_\Lambda = \mu_\Sigma = \mu_B - \mu_S$$

$$\mu_\Xi = \mu_B - 2\mu_S$$

$$\mu_\Omega = \mu_B - 3\mu_S$$

Now all the chemical potentials are expressed in terms of the two, μ_B and μ_S.

APPENDIX B: CONTRACTED INITIAL FIREBALL

As indicated in eq.(10) the initial volume, which multiplies all quantities, depends on the as yet undetermined energy because of the Lorentz contraction. To solve the initial fireball equations, write eqs. (3), (4), (10) as

$$N_\alpha = V\, n_\alpha(\mu,T) \qquad E_\alpha = V\, \mathcal{E}_\alpha(\mu,T) \tag{B1}$$

$$E = V\,\mathcal{E}(\mu,T) \qquad \mathcal{E} = \Sigma\, \mathcal{E}_\alpha$$

where μ stands for μ_S, μ_B, the two independent chemical potentials. Since eq. (10)

$$V = Av = A\frac{mv_0}{2E} \tag{B2}$$

we find from (B1) and (B2)

$$E = \left(\frac{v_0\; m\; \mathcal{E}}{2}\right)^{1/2}$$

$$V = A\left(\frac{mv_0}{2\mathcal{E}}\right)^{1/2} \tag{B3}$$

So the equation for baryon conservation, eq. (6), becomes

$$1 = \left(\frac{m\; v_0}{2\,\mathcal{E}(\mu,T)}\right)^{1/2} (b(\mu,T) - \bar{b}(\bar{\mu},T)) \tag{B4}$$

The equations that define the initial contracted fireball are (7) and (B4) which are to be solved for μ_B and μ_S for chosen T.

REFERENCES

1. W. Swiatecki; see also A.M. Poskanzer and R. Stock, Comments on Nucl. Part. Phys. 7, 41 (1977).

2. N.K. Glendenning, LBL-6597 and Ref. 20.

3. Heckman, Crawford, Greiner, Lindstron, and Wilson, LBL-6562, Proceedings of Topical Conference on Heavy Ion Collisions, Oak Ridge National Laboratory Report CON-770 602, p. 411 (1977).

4. J. Gosset, H.H. Gutbrod, W.G. Meyer, A.M. Poskanzer, A. Sandoval, R. Stock, and G.D. Westfall, Phys. Rev. C16, 629 (1977).

5. G.D. Westfall, J. Gosset, P.J. Johansen, A.M. Poskanzer, W.G. Meyer, H.H. Gutbrod, A. Sandoval, and R. Stock, Phys. Rev. Letters 37, 1202 (1978).

6. A. Mekjian, Phys. Rev. Letters 38, 640 (19-7); Phys. Rev. C 17, 1051 (1978).

7. W.D. Myers, Nucl. Phys. A296, 177 (1978).

8. A.S. Goldhaber, LBL-6595.

9. G. Chapline and M. Nauenberg, Phys. Rev. D 16, 450 (1977).
   L. Susskind, Phys. Rev. D (SLAC-2070, 1978).
   G.F. Chapline and A.K. Kerman, preprint.

10. T.D. Lee, Rev. Mod. Phys. 47, 267 (1975).

11. A.B. Migdal, O.A. Markin, I.I. Mishustin, Sov. Phys. JETP 39, 212 (1974).
    G.E. Brown and W. Weise, Phys. Rep. 27C, 1 (1976).

12. W. Weise, Univ. of Regensburg preprint (1978).
    J. Meyer-ter-Vehn, Inst. für Kernphysik, Jülich (1978).

13. V. Ruck, M. Gyulassy, W. Greinder, Z. Phys. A277, 391 (1976).

14. M. Gyulassy, in Symposium on Relativistic Heavy Ion Collisions, GSI, Darmstadt (1978), and LBL-7704.

15. See Reviews by J.R. Nix, Progress in Particle and Nuclear Physics (in press), Los Alamos Scientific Report 77-2952, (1977).

16. Data Particle Group, Review of Particle Properties, Rev. Mod. Phys. 48, No. 2, Part II (1976).

17. R. Hagedorn in Cargese Lectures in Physics, Vol. 6, ed. by E. Schatzman (Gordon Breach, New York, 1973). See reference to earlier literature by Hagedorn.

18. C.J. Hamer and S.C. Frautschi, Phys. Rev. D4, 2125 (1971).

19. S.W. Hawking, Communications in Math. Phys. 43, 199 (1975); Phys. Rev. D 13, 191 (1976).

20. N.K. Glendenning and Y. Karant, Bull. Am. Phys. Soc. 22, 1005 (1977), and Phys. Rev. Letters 40, 374 (1978).

21. C. Kittel, Elementary Statistical Mechanics, (John Wiley and Sons, 1958) Appendix C, p. 219.

22. G.F. Chapline, M.H. Johnson, E. Teller, and M.S. Weiss, Phys. Rev. D 8, 4302 (1973).

23. J. Gosset, J.I. Kapusta, and G.D. Westfall, LBL-7139.

24. E. Beth and G.E. Uhlenbeck, Physica 4, 915 (1937).

25. S.Z. Belenkij, Nucl. Phys. 2, 259 (1956).

26. S. Frautschi, Phys. Rev. D 3, 2821 (1971).

27. J.I. Kapusta, Phys. Rev. C 16, 1493 (1977).

28. A.T. Laasanen, C. Ezell, L.T. Gutlay, W.N. Schreiner, P. Schübelin, L. von Lindern, and F. Turkot, Phys. Rev. Letters 38, 1 (1977).

PART II

PION-NUCLEUS REACTIONS

MULTIPLE SCATTERING THEORY

W. R. Gibbs

Theoretical Division
Los Alamos Scientific Laboratory
Los Alamos, New Mexico 87545

I. INTRODUCTION TO MULTIPLE SCATTERING AND THE PION NUCLEUS

In this first section, which should take about two lectures, I hope to lay the ground work for my later lectures and those of Professors' Moniz and Kisslinger. I am starting from basics and must necessarily cover some topics rather rapidly. However, I hope that when I finish you will have at least a familiarity with most of the vocabulary. I have tried to provide a few references if you should need more information on a given subject.

TWO BODY t-MATRICES[1]

Let us start with the Schrödinger equation for the scattering from a static potential (or center of mass scattering of two particles, in which case m will be a reduced mass). We wish to solve the scattering problem, with an incident plane wave with momentum $\overline{q}$, for the properties of the wavefunction $\psi_{\overline{q}}(\overline{r})$.

$$(K + V - E)\psi_{\overline{q}}(\overline{r}) = 0 \tag{1}$$

In analogy with the plane wave expansion (Bauer's series)

$$e^{i\overline{q}\cdot\overline{r}} = 4\pi\Sigma i^{\ell} j_{\ell}(qr)Y_{\ell}^{m}(\hat{q})Y_{\ell}^{m*}(\hat{r}) \tag{2}$$

we express $\psi_{\overline{q}}$ as

$$\psi_{\overline{q}} = 4\pi\Sigma i^{\ell}\psi_{\ell}(r)Y_{\ell}^{m}(\hat{q})Y_{\ell}^{m*}(\hat{r}) \quad . \tag{3}$$

Formally we may write

$$\psi_{\overline{q}}(\overline{r}) = e^{i\overline{q}\cdot\overline{r}} + (E - K)^{-1}V(\overline{r})\psi_{\overline{q}}(\overline{r}) \tag{4}$$

where we have explicitly inserted the incident plane wave boundary condition which is permissible since

$$(E - K)e^{i\overline{q}\cdot\overline{r}} = 0 \quad . \tag{5}$$

To evaluate this equation we insert "one" in the form

$$1 = \int d\overline{r}' \, \delta(\overline{r}-\overline{r}') = \frac{1}{(2\pi)^3}\int d\overline{q}' \, d\overline{r}' \, e^{i\overline{q}'\cdot(\overline{r}-\overline{r}')} . \tag{6}$$

We have thus introduced a complete set of eigenstates of K i.e. $Ke^{i\overline{q}\cdot\overline{r}} = (q^2/2m)e^{i\overline{q}\cdot\overline{r}}$ so that

$$\psi_{\overline{q}}(\overline{r}) = e^{i\overline{q}\cdot\overline{r}} + 2m\int \frac{d\overline{q}' \, e^{i\overline{q}'\cdot\overline{r}}}{q^2-q'^2} \frac{1}{(2\pi)^3}\int e^{-i\overline{q}'\cdot\overline{r}'} V(\overline{r}')\psi_{\overline{q}}(\overline{r}')d\overline{r}' \tag{7}$$

We define

$$t(\overline{q}',\overline{q}) = \frac{1}{(2\pi)^3}\int e^{-i\overline{q}'\cdot\overline{r}'} V(\overline{r}')\psi_{\overline{q}}(\overline{r}')d\overline{r}' \tag{8}$$

as the t-matrix. Thus

$$\psi_{\overline{q}}(\overline{r}) = e^{i\overline{q}\cdot\overline{r}} + 2m\int \frac{d\overline{q}' \, e^{i\overline{q}'\cdot\overline{r}} t(\overline{q}',\overline{q})}{q^2-q'^2+i\eta} \tag{9}$$

The $+i\eta$ has been added in the Green's function so that we will have only outgoing waves at infinity.

We may write for a spherically symmetric potential

$$t(\overline{q}',\overline{q}) = \frac{(4\pi)^2}{(2\pi)^3}\sum_{\ell m} Y_\ell^m(\hat{q}')Y_\ell^{m^*}(\hat{q}) \int_0^\infty r^2 dr V(r)\psi_\ell(r)j_\ell(q'r)$$
$$= \sum_{\ell m} t_\ell(q',q)Y_\ell^m(\hat{q}')Y_\ell^{m^*}(\hat{q}) \tag{10}$$

$$t_\ell(q',q) = \frac{2}{\pi}\int_0^\infty r^2 dr V(r)\psi_\ell(r)j_\ell(q'r) \tag{11}$$

Note that

$$t_\ell(-q',q) = (-1)^\ell t_\ell(q',q) \quad . \tag{12}$$

To establish the relationship of t to the scattering amplitude let us examine the second term of Eq. 9 for large r .

$$2m\int \frac{d\overline{q}' e^{i\overline{q}'\cdot\overline{r}} t(\overline{q}',\overline{q})}{q^2-q'^2+i\eta} =$$

$$= (2m)(4\pi)\sum_{\ell m} i^{\ell} Y_{\ell}^{m}(\hat{q})Y_{\ell}^{m*}(\hat{r})\int_0^{\infty}\frac{q'^2dq'\, j_{\ell}(q'r)t_{\ell}(q',q)}{q^2-q'^2+i\eta}$$

$$= (2m)(2\pi)\sum_{\ell m} i^{\ell} Y_{\ell}^{m}(\hat{q})Y_{\ell}^{m*}(\hat{r})\int_{-\infty}^{\infty}\frac{q'^2dq'\, j_{\ell}(q'r)t_{\ell}(q',q)}{q^2-q'^2+i\eta}$$

$$= (2m)\pi\sum_{\ell m} i^{\ell} Y_{\ell}^{m}(\hat{q})Y_{\ell}^{m*}(\hat{r})\int_{-\infty}^{\infty}\frac{q'^2dq'\left(h_{\ell}^{(+)}(q'r)+h_{\ell}^{(-)}(q'r)\right)t_{\ell}(q',q)}{q^2-q'^2+i\eta}$$

$$\xrightarrow{r\to\infty}(2m)\pi\sum_{\ell m}\frac{Y_{\ell}^{m}(\hat{q})Y_{\ell}^{m*}(\hat{r})}{r}\int_{-\infty}^{\infty}\frac{q'dq'\left(i^{-1}e^{iq'r}+i^{2\ell+1}e^{-iq'r}\right)t_{\ell}(q',q)}{q^2-q'^2+i\eta}$$

Completing the contour on the first piece in the upper half-plane (where e^{iqr} dies for $q \to + i$) and the second piece in the lower half plane and using Eq. 12 we find

$$m(2\pi)^2\frac{e^{iqr}}{r}\sum_{\ell m} Y_{\ell}^{m}(\hat{q})Y_{\ell}^{m*}(\hat{r})t_{\ell}(q,q) \quad .$$

Thus

$$f(\theta) = m(2\pi)^2 t(\vec{q}',\vec{q}) \ ; \ |\vec{q}'| = |\vec{q}| \tag{13}$$

$$\frac{d\sigma}{d\Omega} = |f(\theta)|^2 \tag{14}$$

and the t-matrix is identified as the scattering amplitude (with constant factors) for $|q'| = |q|$ i.e. on shell.

Note that we may obtain an equation for t itself by multiplying Eq. 9 by $1/(2\pi)^3 e^{-i\vec{q}\cdot\vec{r}}V(\vec{r})$ and integrating

$$t(q',q) = V(\vec{q}',\vec{q}) + 2m\int\frac{d\vec{q}'\, V(\vec{q}',\vec{q}'')t(\vec{q}'',\vec{q})}{q'^2-q''^2+i\eta} \tag{15}$$

Also note that the energy in the Green's function need not be related to either momentum and we obtain

$$t(\vec{q}',\vec{q},E) = V(\vec{q}',\vec{q}) + 2m\int\frac{d\vec{q}'\, V(\vec{q}',\vec{q}'')t(\vec{q}'',\vec{q},E)}{2mE-q''^2+i\eta} \ , \tag{16}$$

an equation for the fully-off-shell t-matrix.

Now that I have written out several of the equations in detail I will begin using operator notation. To write out the full equation one must insert appropriate complete sets of states as was done above.

We define an operator $t(E)$ such that

$$t(\bar{q}',\bar{q},E) = \langle \bar{q}' | t(E) | \bar{q} \rangle \quad . \tag{17}$$

We also use

$$G \equiv \frac{1}{E - K - V} \quad , \quad G_0 \equiv \frac{1}{E - K} \quad . \tag{18}$$

Useful algebraic identities are:

$$G = G_0 + G_0 V G \tag{19a}$$

$$= G_0 + G V G_0 \quad . \tag{19b}$$

In this notation Eq. 16 may be written as (suppressing the variable E):

$$t = V + V G_0 t \quad . \tag{20}$$

Using Eqs. 19 one can show that

$$t = V + VGV \tag{21}$$

and

$$t = V + t G_0 V \tag{22}$$

which are useful alternative forms of Eq. 20.

THE MULTIPLE SCATTERING EQUATIONS

We wish to solve the $A + 1$ body Schrödinger equation given as

$$(K + \sum_{i=1}^{A} V_i + H_N - E)\Psi = 0 \tag{23}$$

where K is the pion kinetic energy, V_i is the pion interaction potential with the ith nucleon, Ψ is the total $A + 1$ - body wave function, and H_N is the nuclear Hamiltonian with eigenstates given by:

$$H_N \phi_n(\bar{r}_1 \cdots \bar{r}_A) = E_n \phi_n(\bar{r}_1 \cdots \bar{r}_A) \quad . \tag{24}$$

We will use the notations:

$$G = \frac{1}{E - K - V - H_N} \quad , \quad G_0 = \frac{1}{E - K - H_N} \quad , \quad g = \frac{1}{E - K} \quad . \tag{25}$$

Introducing the boundary condition as before

$$\Psi = \phi_0 e^{i\bar{q}\cdot\bar{r}} + G_0 \sum_{i=1}^{A} V_i \Psi \quad . \tag{26}$$

We may identify, as before, the pion-nuclear-ground-state

t-matrix as

$$T(\vec{q}',\vec{q}) = \frac{1}{(2\pi)^3}\int e^{-i\vec{q}'\cdot\vec{r}} \sum_{i=1}^{A} V_i \Psi d\vec{r} \qquad (27)$$

$$= \sum T_i(\vec{q}',\vec{q})$$

Multiplying Eq. 26 by V_i we have

$$T_i = V_i + V_i G_0 \sum_{j=1}^{A} T_j \ . \qquad (28)$$

To eliminate the pion-nucleon potential we define τ_i such that

$$\tau_i = V_i + V_i G_0 \tau_i$$

$$= V_i + \tau_i G_0 V_i \qquad (29)$$

$$= (1 + \tau_i G_0) V_i \ .$$

Thus

$$V_i = (1 + \tau_i G_0)^{-1} \tau_i \ . \qquad (30)$$

Substituting Eq. 30 in Eq. 28 we get

$$T_i = \tau_i + \tau_i G_0 \sum_{j \neq i} T_j \qquad (31)$$

which are the desired multiple scattering equations. We may get the Watson multiple scattering series by iterating these equations

$$T = \sum \tau_i + \sum_{j \neq i} \tau_i G_0 \tau_j + \sum_{\substack{i \neq j \\ j \neq k}} \tau_i G_0 \tau_j G_0 \tau_k + \cdots \ . \qquad (32)$$

These two forms of the multiple scattering equations[2] are the basis of most of the work that has been done on pion nucleus scattering.

THE τ_i OPERATOR

We would like to have an expression for pion nucleus scattering in terms of pion-nucleon t-matrices, but we do not have that since G_0 still contains the nuclear Hamiltonian. The quantity τ_i may be interpreted as the pion-nucleon t-matrix in a nuclear medium. The evaluation of this quantity has a long history and some of the later lectures will be on this subject but I will mention some of the things which have been done.

a) Ignore the nuclear Hamiltonian. This is referred to as the impulse approximation and gives $\tau_i(E) \approx t_i(E)$, the free pion-nucleon t-matrix.

b) Use closure. Inserting a complete set of nuclear states in the basic equation for τ_i we have

$$\tau_i = V_i + V_i G_0 \tau_i$$

$$= V_i + \sum_n \frac{V_i |n\rangle\langle n| \tau_i}{E - K - E_n} \quad . \tag{33}$$

Assuming that there exists some effective excitation energy $\overline{E}$ and removing the denominator from the sum we recover the free pion-nucleon t-matrix at a different energy i.e.

$$\tau_i(E) \approx t_i(E - \overline{E}) \tag{34}$$

There are several prescriptions in existence for how to chose $\overline{E}$.

c) Include true pion absorption.[3] Using the form

$$\tau_i = V_i + V_i G V_i \tag{35}$$

the complete spectrum of pion states must be included. Since in the $T = \frac{1}{2}$, $J = \frac{1}{2}$ channel the nucleon pole, corresponding to true pion absorption, is in the sum, the use of the τ_i in various optical model terms allows this effect to be included.

d) Discuss modification of the $\Delta(33)$ in nuclear matter. Since this will be discussed in detail in later lectures I will not say more.

FIRST ORDER OPTICAL POTENTIAL[4]

We may project out the scattering to the ground state by taking the expectation values of Eqs. 31 on nuclear variables only

$$\langle 0|T_i|0\rangle = \langle 0|\tau_i|0\rangle + \sum_{\substack{n \\ j\neq i}} \frac{\langle 0|\tau_i|n\rangle\langle n|T_j|0\rangle}{E - K - E_n + i\eta}$$

If we require the nucleus to remain in the ground state in intermediate scatterings (the coherence approximation) we obtain:

$$\overline{T}_i = \tau_i + \tau_i g \sum_{j\neq i} \overline{T}_j \quad . \tag{36}$$

Note that since we have assumed that everything that leaves the incident channel never comes back we expect the result to be that the resulting calculation is too absorptive. Since the wave function is antisymmetrized the τ_i are all identical and so must be the $\overline{T}_i$.

$$\overline{T} = A\overline{\tau} + A\overline{\tau}\, g \frac{A-1}{A} \overline{T} \tag{37}$$

Defining $\tilde{T} = \frac{A-1}{A} T$ (38)

$$\tilde{T} = (A-1)\bar{\tau} + (A-1)\bar{\tau} g \tilde{T} \quad (39)$$

and we may identify the first order optical potential as

$$U_0 = (A-1)\bar{\tau} \quad . \quad (40)$$

If we use a single particle model wave function for the nucleus we have:

$$\bar{\tau} = \int \phi^*(\bar{r}) e^{-i(\bar{q}'-\bar{q})\cdot\bar{r}} \tau(\bar{q}',\bar{q},\bar{p}_i,\bar{p}_f)\phi(\bar{r})d\bar{r} \quad (41)$$

where $\bar{p}_i$ is a momentum operator on the ϕ to the right and $\bar{p}_f$ operates on the ϕ to the left. This momentum dependence comes about because the τ (t) is defined in the π-nucleon center-of-mass and we are now in the π-nucleus center-of-mass. Since $\bar{p}_i$ and $\bar{p}_f$ are of the order of Fermi momenta and there is a factor of $\sim\mu/m$ reducing them compared with $\bar{q}$ and $\bar{k}$ it might be thought that they can be ignored. If we do, we have the "factorization approximation".

$$U_0 = (A-1)\tau(\bar{q}',\bar{q})S(\bar{q}'-\bar{q}) \;\; ; \;\; S(\bar{q}'-\bar{q}) = \int e^{i(\bar{q}'-\bar{q})\cdot\bar{r}}|\phi|^2 d\bar{r} \quad (42)$$

It has recently been found that it is important to correct for this approximation. I will discuss briefly this so-called "angle transformation" in a later lecture.

THE KISSLINGER POTENTIAL[5]

For heavy nuclei the form factor is strongly peaked about $\bar{q}'=\bar{q}$. If the t-matrix is slowly varying it is a reasonable approximation to evaluate it at that point. This is common practice in the nucleon-nucleus optical model. Kisslinger pointed out that, because of the strong p-wave nature of the π-nucleon interaction, the form factor does not dominate the t-matrix, especially for light nuclei. Thus he proposed a potential based on a minimum momentum dependence.

$$U_0 = (A-1)(\lambda_0 + \lambda_1 \bar{q}\cdot\bar{q}')S(\bar{q}'-\bar{q}) \quad (43)$$

This leads to a non-local potential (of a derivative type) in coordinate space.

$$U\psi = (A-1)[\lambda_0\rho + \lambda_1\nabla\cdot\rho\nabla]\psi \quad (44)$$

The idea is certainly correct but we now know that another refinement is necessary. The t-matrix used in Eq. 43 continues to become stronger the farther it is taken off-shell. This is equivalent to the statement that the pion-nucleon interaction is zero range. We know that both the pion and the nucleon are not point particles so some factor must be introduced to provide this finite range. The technique usually used for doing this is the one of separable potentials leading to separable t-matrices. This will be the subject of a later lecture.

It is perhaps worth mentioning in passing that there exists another, similar, potential called the Laplacian potential. To obtain this one notes that, on-shell,

$$\vec{q}\cdot\vec{q}' = -\tfrac{1}{2}(\vec{q}-\vec{q}')^2 + q^2$$

$$= \omega^2 - \mu^2 - \tfrac{1}{2}(\vec{q}'-\vec{q})^2 \tag{45}$$

and the t-matrix

$$\lambda_0 + \lambda_1(\omega^2 - \mu^2) - \tfrac{1}{2}\lambda_1(\vec{q}'-\vec{q})^2 \tag{46}$$

is used. Since it involves only differences of momenta it is local, reducing to a Laplacian operator on the density, hence the name.

HIGHER ORDER OPTICAL POTENTIALS

The corrections to the first order optical potential take two forms. One of these is first found in second order scattering the other in third order.

The first effect arises from correlations. If we use a simple single particle product wave function for the nucleus then the intermediate states in the second term of Eq. 32 are automatically required to be in the ground state since if the first τ excites a single particle state the second cannot deexcite it because it operates on a different single particle coordinate. Introducing a correlation operator gives a correction to the first order optical potential which depends strongly on the correlation function and the pion-nucleon range. This effect has been shown[6] to be approximately equivalent to the pionic Lorentz-Lorenz effect.[7]

The second correction is found to start in the third term and occurs because even a single particle wave function can be excited in the intermediate state when $i=k$.

II. THE FIXED NUCLEON APPROXIMATION

Since the nucleons in the nucleus are considerably more massive than the pion they will, generally speaking, be moving more slowly. The Fermi motion of the nucleons corresponds to a velocity of ~0.2c while, for example, a 50 MeV pion has a velocity of ~0.68c. For this reason let us consider the approximation of a pion scattering from a collection of fixed centers. We will later consider corrections to this picture.

To get some feeling for what this means, relative to the optical model discussed previously, I wish to refer to a derivation of the fixed scatterer approximation given by Foldy and Walecka.[8] They showed, for local potentials, that using closure in the multiple scattering equations led to fixed nucleons. Remember that we truncated the intermediate sum to the ground state to get the first order optical model while this closure approximation keeps all of the excited nuclear states with equal weight. Thus, while the first order optical model treats as "absorbed" everything which leaves the ground state in the intermediate sum, the fixed nucleon approximation brings everything back that is allowed by the type of nuclear wave function chosen. This last simply means that one still does not excite independent particle states with the double scattering term. Thus, while the first order optical potential has too much absorption, the fixed nucleon approximation tends to have too little. The use of both methods allows some estimate of upper and lower limits on absorption. I expect the true case to be closer to the fixed nucleon limit than the first order optical potential.

Unfortunately the Foldy and Walecka proof holds only for local potentials and the fixed nucleon equations are generally used for separable t-matrices so that closure is not exactly equivalent to the fixed nucleon approximation. For a discussion of the relation between the two approximations see Benayoun.[9]

By the fixed nucleon approximation I mean simply that the pion scatters from A fixed centers with the final amplitude being the average of these computed amplitudes over the positions of all of the particles. Note that this procedure is the same one used in Glauber theory. The probability distribution is given by the square of the ground state wave function.

$$\rho(\bar{r}_1 \cdots \bar{r}_A) = |\phi_0(\bar{r}_1 \cdots \bar{r}_A)|^2 \tag{47}$$

To obtain the fixed nucleon multiple scattering equations we again start with the Schrödinger equation, but this time for potentials defined realtive to fixed points, $\bar{r}_i$.

$$(K + \sum_{i=1}^{A} V_i(\bar{r}-\bar{r}_i) - E)\psi = 0 . \tag{48}$$

Since there is no nuclear Hamiltonian we have only the free Green's function, g , and

$$\psi = e^{i\bar{q}\cdot\bar{r}} + g \sum_{i=1}^{A} V_i\psi \quad . \tag{49}$$

We use the trick of operating on both sides of the equation with $1/(2\pi)^3\int d\bar{r}e^{-i\bar{q}'\cdot\bar{r}}V_i(\bar{r}-\bar{r}_i)$ to get

$$T_i = V_i + \sum_{j=1}^{A} V_i g T_j \quad . \tag{50}$$

Using $V_i = (1 + t_i g)^{-1} t_i$ we again eliminate V_i in favor of t_i to get:

$$T_i = t_i + t_i \sum_{i \neq j} g T_j \quad . \tag{51}$$

Note that the T_i are functions of all of the nuclear coordinates but the t_i are functions of only the ith one. Writing out equation 5 more fully,

$$T_i(\bar{q}',\bar{q},\bar{r}_1,\cdots\bar{r}_A) = t_i(\bar{q}',\bar{q},\bar{r}_i) + 2m\sum \frac{t_i(\bar{q}',\bar{p},\bar{r}_i)T_j(\bar{p},\bar{q},\bar{r}_1,\cdots\bar{r}_A)}{q^2 - p^2 + i\eta} . \tag{52}$$

Using the relation between the scattering amplitude and the t-matrix:

$$f_i(\bar{q}',\bar{q}) = m(2\pi)^2 t_i(\bar{q}',\bar{q}) \; ; \; F_i(\bar{q}',\bar{q}) = m(2\pi)^2 T_i(\bar{q}',\bar{q}) \tag{53}$$

We may write the equations directly in terms of the scattering amplitudes.

$$F_i(\bar{q}',\bar{q}) = f_i(\bar{q}',\bar{q}) + \sum_{j\neq i} \int \frac{d\bar{p}}{(2\pi)^3} \frac{f_i(\bar{q}',\bar{p}) 4\pi F_j(\bar{p},\bar{q})}{q^2 - p^2 + i\eta} \tag{54}$$

Since

$$\begin{aligned} V_i(\bar{q}',\bar{q}) &\equiv \frac{1}{(2\pi)^3} \int e^{i(\bar{q}-\bar{q}')\cdot\bar{r}} V^i(\bar{r}-\bar{r}_i) d\bar{r} \\ &= \frac{e^{i(\bar{q}-\bar{q}')\cdot\bar{r}_i}}{(2\pi)^3} \int e^{i(\bar{q}-\bar{q}')\cdot\bar{r}} V^i(\bar{r}) d\bar{r} \\ &= e^{i(\bar{q}-\bar{q}')\cdot\bar{r}_i} V^i(\bar{q}',\bar{q}) \end{aligned} \tag{55}$$

and

$$t_i(\bar{q}',\bar{q}) = V_i(\bar{q}',\bar{q}) + 2m \int \frac{V_i(\bar{p},\bar{q}) t_i(\bar{q}',\bar{p})}{q^2 - p^2 + i\eta} d\bar{p} \quad , \tag{56}$$

$$t_i(\bar{q}',\bar{q}) = e^{i(\bar{q}-\bar{q}')\cdot\bar{r}_i} t^i(\bar{q}',\bar{q}) \tag{57}$$

and the dependence of f_i on $\bar{r}_i$ is known.

Defining

$$G_i(\bar{q}',\bar{q}) = e^{i\bar{q}'\cdot\bar{r}_i} F_i(\bar{q}',\bar{q}) \tag{58}$$

we can rewrite Eq.(54) as

$$G_i(\bar{q}',\bar{q}) = f^i(\bar{q}',\bar{q})e^{i\bar{q}\cdot\bar{r}_i} + \frac{1}{2\pi^2}\sum_{j\neq i}\int \frac{d\bar{p}\ f^i(\bar{q}',\bar{p})e^{i\bar{p}\cdot(\bar{r}_i-\bar{r}_j)}G_j(\bar{p},\bar{q})}{q^2-p^2+i\eta} \quad . \tag{59}$$

The reason for making the replacement given in Eq.(58) is to make the partial wave decomposition of Eq.(59) feasible. Since the only appearance of $\bar{q}'$ on the right is in $f^i(\bar{q}',\bar{q})$ we need only as many partial waves to expand $G_i(\bar{q},\bar{q})$ as are needed to expand the pion-nucleon system.

Note that $f^i(\bar{q}',\bar{q})$ still carries an index even though it is independent of $\bar{r}_i$. The purpose of this index is to keep track of other properties of the nucleon, i.e. neutron or proton.

The solution of Eq.(59) requires a knowledge of the pion-nucleon off-shell amplitude. We see that we will obtain the pion-nucleus amplitudes off-shell as well. The general solution of the three dimensional coupled integral equations is too much to ask for at the present time and we must look for approximations or simplifications to help us.

One approximation which has been used is the so-called "on-shell" approximation. In this case one takes only the δ-function part of the propagator, ignoring the principal value part and all other possible singularities of the integrand. This puts both the pion-nucleon and pion-nucleus amplitudes on shell and leaves only an angular integral to be done. The name is misleading since there are other approximations which lead to only on-shell information being used.

THE BEG LIMIT

It is interesting to see how Beg's theorem[10] is related to Eq.(59). The statement of this theorem is that, for scattering from fixed, non-overlapping centers, the total amplitude can be expressed in terms of on-shell quantities (phase shifts) only.

We note that, while the integrand in Eq.(59) is allowed to have singularities other than those of the propagator, they cannot lie on the real axis, or else either the pion-nucleon or pion-nucleus wave function would have additional contributions at infinity, which is forbidden by the boundary condition. Because we shall be closing the contours as before, the contributions of these other singularities will damp out for large $r_{ij} = |\bar{r}_i - \bar{r}_j|$ (exponentially in the case of poles).

We could obtain the appropriate set of equations for the Beg conditions at this point but they will come out as a by product of the separable t-matrix approach we are about to begin.

SEPARABLE POTENTIALS

Since we wish to use separable t-matrices to represent the off-shell behavior of the pion-nucleon t-matrix let us review briefly the properties of separable potentials. For simplicity let us consider only an s-wave potential;

$$2mV(\bar{r},\bar{r}') = \lambda v(r)v(r') \quad . \tag{60}$$

We note that if we consider this to be an effective potential we may make λ energy dependent.

Taking the Fourier transform of the Schrödinger equation with this potential

$$(-\nabla^2 - k^2)\psi(\bar{r}) + \lambda v(r)\int d\bar{r}'\, v(r')\psi(\bar{r}') = 0 \tag{61}$$

we have

$$(q'^2 - k^2)\psi(\bar{q}') + \frac{\lambda v(q')}{(2\pi)^6}\int d\bar{r}'\, d\bar{p}d\bar{p}'\, e^{i(\bar{p}+\bar{p}')\cdot\bar{r}'}\, v(p)\psi(\bar{p}') = 0 \tag{62}$$

or

$$(q'^2 - k^2)\psi(\bar{q}') + \frac{\lambda v(q')\beta}{(2\pi)^3} = 0 \tag{63}$$

with

$$\beta \equiv \int d\bar{p}v(p)\psi(-\bar{p}) \quad . \tag{64}$$

The wave function, including the incident plane wave boundary condition, is given by

$$\psi(\bar{q}') = (2\pi)^3\delta(\bar{q}-\bar{q}') + \frac{\lambda v(q')\beta}{(2\pi)^3(k^2 - q'^2 + i\eta)} \quad . \tag{65}$$

Substituting Eq.(65) into (64) we obtain an equation for β

$$\beta = (2\pi)^3 v(q) + \frac{\lambda\beta}{2\pi^2}\int \frac{q'^2 dq'\, v^2(q')}{k^2 - q'^2 + i\eta} \tag{66}$$

which yields:

$$\beta = \frac{(2\pi)^3 v(q)}{1 - \frac{\lambda}{2\pi^2}\int \frac{q'^2 dq'\, v^2(q')}{k^2 - q'^2 + i\eta}} = \frac{(2\pi)^3 v(q)}{D(k)} \tag{67}$$

and

$$\begin{aligned}\psi(\bar{q}') &= (2\pi)^3\delta(\bar{q}-\bar{q}') + \frac{1}{k^2 - q'^2 + i\eta}\,\frac{\lambda v(q)v(q')}{D(k)} \\ &= (2\pi)^3\delta(\bar{q}-\bar{q}') + \frac{4\pi}{k^2 - q'^2 + i\eta}\, v(q)g(E)v(q')\end{aligned} \tag{68}$$

where

$$g(E) \equiv \frac{\lambda/4\pi}{D(k)} \quad . \tag{69}$$

Thus we see that the off-shell amplitude is given by

$$f(q',q,E) = v(q)g(E)v(q') \tag{70}$$

and is itself completely separable. If we chose λ to be energy independent then $g(E)$ is completely determined by the choice of $v(q)$, but if we allow an energy dependence of λ, g and v can be chosen independently. This last has the advantage of allowing a simple functional form to be used for $v(q)$ while still fitting the measured pion-nucleon phase shifts. If one assumes an energy independent potential the phase shifts can be used to determine the functions, v, for each partial wave. This procedure has been used to obtain these functions.[11] Pion nucleon half-off-shell t-matrices have also been obtained by dispersion theory techniques.[12]

One form commonly used for v is

$$v(q) = \frac{\alpha^2 + k^2}{\alpha^2 + q^2} \quad . \tag{71}$$

This has the advantage that $v(k) = 1$ and the on-shell amplitude is just $g(E)$. Using Eq.(71) the coordinate space wave function is

$$\psi(\overline{r}) = e^{i\overline{k}\cdot\overline{r}} + \frac{g(E)}{r}\left[e^{ikr} - e^{-\alpha r}\right] \quad . \tag{72}$$

We see that the wave function becomes asymptotic at a distance of $\sim 1/\alpha$ so this is the range of the pion-nucleon interaction.

SEPARABLE t-MATRICES IN MULTIPLE SCATTERING[13]

Our object now is to insert separable t-matrices of the form (24) into Eq.(59). I will skip details of the angular momentum algebra but I must cover the procedure used for expanding in partial waves.

We first write the free amplitudes as:

$$\begin{aligned} f^i(\overline{q}',\overline{q}) &= \frac{2\pi}{ik}\sum_{\ell m} f^i_\ell(q',q)Y^m_\ell(\hat{q})Y^{m*}_\ell(\hat{q}') \\ &= \frac{2\pi}{ik}\sum_{\ell m} \lambda^i_\ell(\omega)v_\ell(q)v_\ell(q')Y^m_\ell(\hat{q})Y^{m*}_\ell(\hat{q}') \quad , \end{aligned} \tag{73}$$

where

$$\lambda_\ell(\omega) = e^{2i\delta_\ell} - 1 \ , \quad v_\ell(q) = \left(\frac{q}{k}\right)^\ell \frac{k^2 + \alpha_\ell^2}{q^2 + \alpha_\ell^2} \quad ; \ \ell = 0 \ , \ 1 \quad . \tag{74}$$

Choosing $\overline{q}$ to be on shell and along the z-axis, we also expand the unknown function

$$G_i(\overline{q}',\overline{q}) = \frac{2\pi}{ik}\sum_{\ell m} g_i^{\ell m}(q')\sqrt{\frac{2\ell+1}{4\pi}}\, Y_\ell^m(\hat{q}') \quad . \tag{75}$$

Making these substitutions in Eq.(59) one notices that the only dependence of the right hand side on the variable q' is through a factor $v_\ell(q')$ so that we may take

$$g_i^{\ell m}(q') = g_i^{\ell m}\, v_\ell(q') \tag{76}$$

i.e. the functional dependence of $g_i^{\ell m}(q')$ is known and it is separable. Using this fact the final equations are:

$$g_i^{\ell m} = \lambda_\ell^i(\omega)\,\delta_{m,0}\, e^{i\overline{q}\cdot\overline{r}_i}$$
$$+ 2\pi\lambda_\ell^i(\omega)\sum \sqrt{\frac{2\ell'+1}{2\ell+1}}\, G^{\ell m}_{\ell' m' \lambda\mu}\, Y_\lambda^{\mu *}(\hat{r}_{ij})\, Z^\lambda_{\ell\ell'}(r_{ij})\, g_j^{\ell' m'} \tag{77}$$

where

$$G^{\ell m}_{\ell' m' \lambda\mu} \equiv \int d\Omega p\, Y_\ell^{m*}(\hat{p})\, Y_\lambda^\mu(\hat{p})\, Y_{\ell'}^{m'}(\hat{p}) \tag{78}$$

is Gaunt's integral and

$$Z^\lambda_{\ell\,\ell'}(r_{ij}) \equiv \frac{2i^{\lambda+1}}{\pi k}\int_0^\infty \frac{p^2 dp\, v_\ell(p)\, v_{\ell'}(p)\, j_\lambda(p r_{ij})}{k^2 - p^2 + i\eta} \tag{79}$$

is needed only for ℓ, ℓ', λ satisfying the triangle relations and $\ell + \ell' + \lambda$ even. It is in these "Z-functions" that the interesting physics lies. For the form 28 the integral can be done and the result is:

$$Z^\lambda_{\ell\ell'}(r_{ij}) = i^\lambda \Big\{ h_\lambda^{(+)}(k r_{ij})$$
$$- \frac{1}{(\alpha_{\ell'}^2 - \alpha_\ell^2)\, k^{\ell+\ell'+1}} \Big[(i\alpha_\ell)^{\ell+\ell'+1}(k^2+\alpha_{\ell'}^2)\, h_\lambda^{(+)}(i\alpha_\ell r_{ij})$$
$$- (i\alpha_{\ell'})^{\ell+\ell'+1}(k^2+\alpha_\ell^2)\, h_\lambda^{(+)}(i\alpha_{\ell'} r_{ij}) \Big] \Big\} \quad . \tag{80}$$

These functions have the form of regularized outgoing spherical Hankel functions. For $\alpha_\ell r_{ij} >> 1$ i.e. for the nonoverlapping region

$$Z^\lambda_{\ell\ell'}(r_{ij}) \rightarrow i^\lambda h_\lambda^{(+)}(k r_{ij}) \tag{81}$$

so these are the Beg-limit functions. Thus for a system of nonoverlapping fixed centers the solution of Eqs. (77) with the Z-functions given by(81), give the exact solution to the problem and expresses the multiple scattering in terms of the phase shifts. It is instructive to study the simple case of two real potentials taken to the concentric limit.[14] In this case only $\lambda = 0$ survives and the solution to Eqs.(77) may be written:

$$F(\theta) = \frac{1}{2ik}\sum(2\ell+1)P_\ell(\cos\theta)F_\ell \tag{82}$$

$$F_\ell = \frac{2f_\ell}{1 - \frac{1}{2} f_\ell\beta_\ell}, \quad \beta_\ell \equiv Z^0_{\ell\ell}(0) \quad . \tag{83}$$

Using

$$F_\ell = 2ie^{i\Delta_\ell}\sin\Delta_\ell \tag{84}$$

$$f_\ell = 2ie^{i\delta_\ell}\sin\delta_\ell \tag{85}$$

we see that

$$\beta_\ell = 1 + i[2\cot\Delta_\ell - \cot\delta_\ell] \tag{86}$$

$$\equiv 1 - is_\ell$$

The "1" comes from the δ-function part of the integral in Eq. 79 and s_ℓ from the principal value part. Note that this "1" guarantees that if there is no flux lost in a single scattering there is none lost from multiple scattering from fixed centers. For the v_ℓ given by Eq.(71) we have

$$s_0 = \tfrac{1}{2}(\alpha_0/k - k/\alpha_0)$$
$$s_1 = \alpha_1(3k^2 + \alpha_1^2)/(2k^3) \tag{87}$$

so that we can always choose a value of α_0 such that s_0 is given correctly, of α_1 such that s_1 is given correctly, etc. Thus using a one term separable potential it is always possible to exactly represent scattering from two concentric potentials <u>no matter what their form</u>.

Since we can be exact for $r_{ij} = 0$ and $r_{ij} \gg \frac{1}{\alpha}$ it only remains to interpolate smoothly between these two regions. A study is made of this problem relative to other scattering methods in Ref. 14.

Since most of the nucleons in the nucleus are nonoverlapping it might well be thought that one could use just the Hankel function part of Z .

In fact calculations have been made using this approximation.[15] The problem comes because, even though the nucleons are not likely to be close together, when they are the scattering is weighted with a large Hankel function when it should be weighted by a much smaller

Z-function. If short range correlations keep the nucleons apart this effect is reduced, but then one must know the correlation function and cannot try to learn it.

The clearest example of this occurs in the pion-nucleus optical model. If one uses the Ericsons' correction of the Kisslinger potential for the Lorentz-Lorenz effect one is doing the analogous thing to using the Beg functions. What is found is that the Lorentz-Lorenz effect (which corresponds to short range correlations) is very large. If one uses form factors for the pion-nucleon interaction (corresponding to using the complete Z-function) the effect is diminished greatly.

THE "ANGLE TRANSFORM"

This effect is largely the same as correcting the factorization approximation mentioned before. I wish to discuss this emphasizing the physics involved. The name comes about because the effective angle of scattering from the nucleon changes because of the motion of the nucleon in the nucleus.

Since the correction exists in the single scattering limit I will discuss it for this case only. I will also assume that we are at sufficiently low energies that we may use non-relativistic kinematics and Galilean transformations.

The basic effect comes about because the average angle of the initial and final nucleon direction depends on the pion scattering angle. We must know the form of the pion-nucleon off-shell t-matrix since (while we may take the limit of on-shell pions) the nucleon must be off-shell.

We start by considering the pion-nucleon amplitude in the pi-nucleon center of mass,

$$f = b_0 + b_1\, \bar{q} \cdot \bar{q}' \tag{88}$$

to be a Galilean invariant

$$f = b_0 + b_1\left(\frac{m}{m+\mu}\right)^2 \left(\bar{q} - \frac{\mu}{m}\bar{p}_i\right)\cdot\left(\bar{q}' - \frac{\mu}{m}\bar{p}_f\right) \tag{89}$$

and, using the conservation of momentum,

$$\bar{q} + \bar{p}_i = \bar{q}' + \bar{p}_f \quad , \tag{90}$$

we may re-express $\bar{p}_i$ and $\bar{p}_f$ in terms of their average, $\bar{p} = \frac{1}{2}\left(\bar{p}_i + \bar{p}_f\right)$.

$$\begin{aligned} \bar{p}_i &= \bar{p} - \frac{\bar{q} - \bar{q}'}{2} \\ \bar{p}_f &= \bar{p} + \frac{\bar{q} - \bar{q}'}{2} \quad . \end{aligned} \tag{91}$$

Thus

$$f = b_0 + b_1'\left[\bar{q} + \frac{\mu}{2m}(\bar{q} - \bar{q}') - \frac{\mu}{m}\bar{p}\right]\cdot\left[\bar{q}' - \frac{\mu}{2m}(\bar{q} - \bar{q}') - \frac{\mu}{m}\bar{p}\right] \tag{92}$$

If we now: 1) neglect $\left(\frac{\mu}{m}\right)^2 p^2$ because the mass ratio is very small ii) neglect linear terms in $\bar{p}$ because they will tend to average to zero and have a factor μ/m iii) keep only terms of order zero or one in the mass ratio, we find

$$f \approx b_0 - \frac{\mu}{2m}(q^2 + q'^2)b_1' + b_1'\left(1 + \frac{\mu}{m}\right)\bar{q}\cdot\bar{q}' \quad . \tag{93}$$

Putting the pion on shell

$$f \approx b_0 - \frac{\mu}{m} b_1' k^2 + \left(1 + \frac{\mu}{m}\right) b_1' k^2 \cos\theta \quad . \tag{94}$$

Note that a rather large effect remains even though there is no reference to any Fermi momentum. This is because a given change in pion momentum requires a corresponding change in Fermi momentum no matter how unlikely that value of momentum is. The probability of finding that momentum is expressed by the form factor which multiplies the pion nucleon amplitude. Thus we see that the principle effect is to add some of the p-wave strength into the s-wave. Since b_0 is negative and b_1 is positive, the magnitude of the s-wave is increased and typically it is doubled since $|b_1| >> |b_0|$.

We can start to understand the general structure of pion-nucleus differential cross sections from these considerations. If we think in terms of single scattering the first dominant feature is the p-wave nature giving a factor of $\cos^2\theta$. Multiplied by this we have the nuclear form factor giving more cross section in the forward direction than the back but leaving the zero at 90°. For the heavier nuclei the diffraction minima eventually come into this region and destroy this picture. Now we have the s-wave which shifts the zero from 90° into smaller angles. The angle transform moves it into yet smaller values. One can identify this s-p interference minimum very clearly and it is very nearly independent of energy (which used to be a mystery) and its position is independent of A . It can be seen even in Calcium where it appears as a shoulder.

MONTE CARLO

Once we have the solution of Eqs.(59) we must carry out the 3A dimensional integral over

$$F(\bar{r}_1,\cdots\bar{r}_A) = \sum_{i=1}^{A} G_i(\bar{r}_1,\cdots\bar{r}_A)e^{-i\bar{q}'\cdot\bar{r}_i} \tag{95}$$

to get the pion-nucleus scattering amplitude. The only practical technique for doing such integrals is by Monte Carlo.

The basic idea behind Monte Carlo integration is that, if one can treat the integrand as a product of an arbitrary function times a probability density function, then the integration can be considered as taking the average value of the arbitrary function over an infinite ensemble. That is:

$$\int g(x)f(x)dx = \lim_{N\to\infty} \frac{1}{N} \sum_{i=1}^{N} g(x_i) \quad , \tag{96}$$

where $\int f(x)dx = 1$ and in the sum on the right the x_i are to be chosen according to $f(x)$. This holds no matter how many dimentions are in x but it is only efficient if there are more than two.

This technique is most effective if the density can be factored or there are simple correlations thus

$$\int g(x_1,\cdots x_\mu)f_1(x_1)\cdots f_\mu(x_\mu)dx_1\cdots dx_\mu$$
$$= \lim_{N\to\infty} \frac{1}{N} \sum_{j=1}^{N} g(x_1^j,\cdots x_\mu^j) \tag{97}$$

since then the choice of one x_i^j does not depend on another. Thus we generally start from a single particle expression for the nuclear wave function. If we wish to add correlations or antisymmetrization effects they can be multiplied into $F(\overline{r}_1,\cdots\overline{r}_A)$. I will show an example of this shortly.

There is a useful algorithm for choosing random variables from a distribution. Although this is well known, in general, since this audience does not deal a great deal in random numbers, it may be worth showing.

Define

$$F(x) \equiv \int_0^x f(x')dx' \tag{98}$$

where $f(x)$ is the probability distribution of interest. Let us ask "what is the distribution of F if we choose random variables for x according to f?" To make all moments of the unknown function $G(F)$ correct we must have

$$\int_0^1 G(F)e^{iFt}dF = \int_0^\infty f(x)e^{iF(x)t}dx \tag{99}$$

or, taking the Fourier transform

$$G(F) = \int_0^\infty f(x)\delta(F-F(x))dx$$
$$= \frac{f(x)}{\frac{dF}{dx}} = 1 \qquad 0 \leqslant F \leqslant 1 \tag{100}$$
$$= 0 \qquad \text{otherwise} \quad .$$

Thus, if we invert Eq.(98) to find $x(F)$ and choose F uniformly between 0 and 1 then the $x(F)$ will be distributed according to $f(x)$.

It is perhaps worthwhile to remark that, for a Monte Carlo method to give reliable answers the distribution function should be well chosen to match the true density. The eventual answers are invariant but the errors vs. computing time are a very strong function of this matching.

ELASTIC SCATTERING ON ^{4}He

I am now going to show the results of actual calculations, made by the methods just discussed, to illustrate some of the points. Figure 1 shows π^- elastic scattering at 110 MeV. Note the large change from the long-short dashed curve to the dashed curve. This is the result of a relativistic generalization of Eq. (94). As seen, it is a critical correlation. The change to the solid curve is given by an energy correction of an analogous type. I will leave the discussion of such corrections to Professor Thomas.

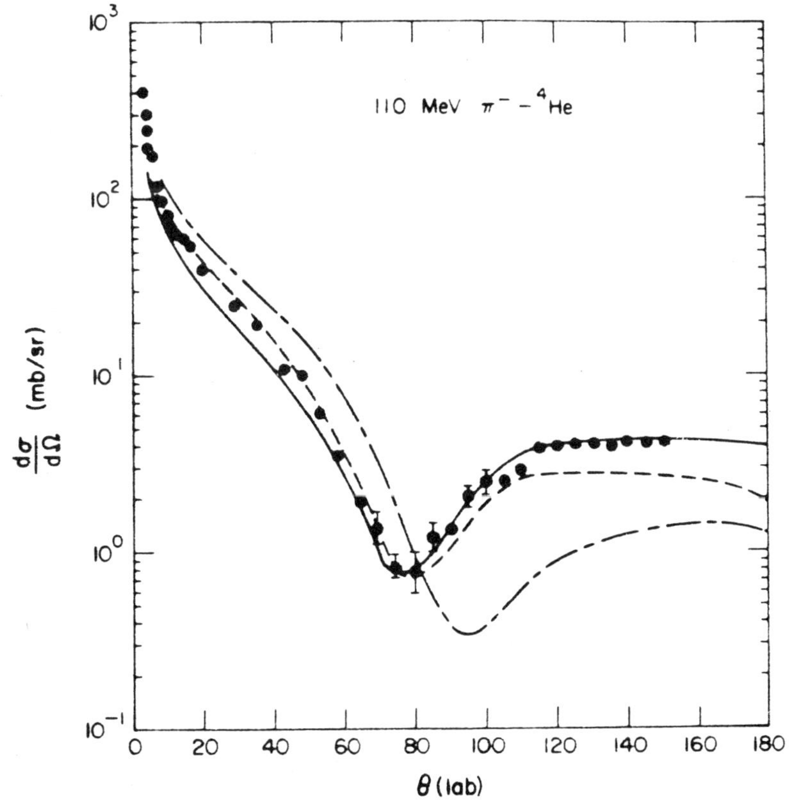

Fig. 1. Scattering of negative pions from ^{4}He at 110 MeV.

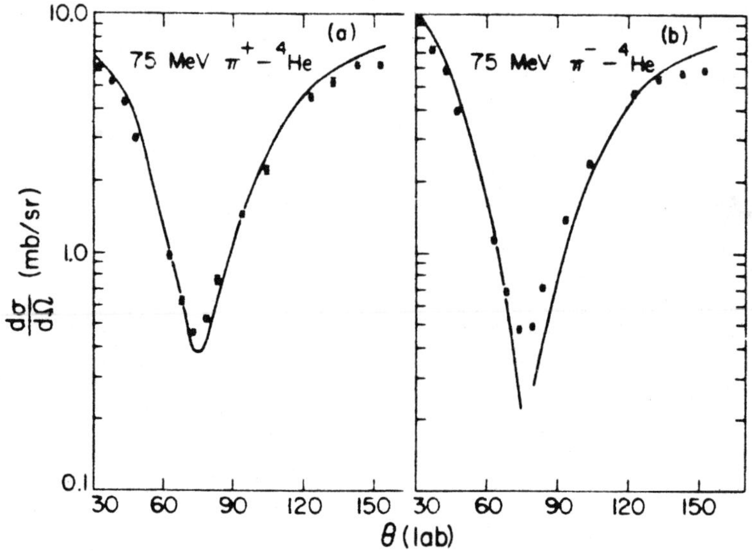

Fig. 2. Scattering of positive and negative pions from ${}^4He$ at 75 MeV.

Figure 2 shows the result for 75 MeV. The characteristics mentioned before are clear, the dominant p-wave with the shift of the zero forward of 90° by the s-p interference.

The calculation at higher energies is difficult because relativistic versions of Eq.(94) introduce d-waves (and higher) into the effective pion-nucleon amplitude. For ${}^4He$ these are not impossible to calculate and a code is being developed to do this.

CHARGE EXCHANGE

We might expect that the technique used here would be also applicable to "almost" elastic reactions as well. By "almost" elastic I mean that the final pion energy is approximately the same as the incident energy. While we may calculate inelastic scattering to low lying excited states, the principle catagory of reactions that comes to mind is single and double charge exchange. We can write down the set of operator equations using the following conventions and assuming a π^+ beam incident.

T_i , U_i , V_i are nuclear amplitudes for π^+ , π^0 , π^- ,

t_i , u_i , v_i are nucleon elastic scattering amplitudes for π^+ , π^0 , π^- ,

and the basic charge exchange amplitudes are:

$$w_i \qquad \pi^+ \to \pi^0$$

$$\tilde{w}_i \qquad \pi^0 \to \pi^+$$

$$x_i \qquad \pi^0 \to \pi^-$$

$$\tilde{x}_i \qquad \pi^- \to \pi^0 \; . \tag{101}$$

$$T_i = t_i + t_i \Sigma g T_j + \tilde{w}_i \Sigma g U_j$$

$$U_i = w_i + u_i \Sigma g U_j + w_i \Sigma g T_j + \tilde{x}_i \Sigma g V_j$$

$$V_i = \qquad v_i \Sigma g V_j + x_i \Sigma g U_j$$

These form a large system of equations and are unmanageable except for the lightest nuclei. We have found that the back-charge-exchange is a small effect (~ few percent) and we normally use the equations without the "tilde" terms.

$$T_i = t_i + t_i \Sigma g T_j$$

$$U_i = w_i + u_i \Sigma g U_j + w_i \Sigma g T_j \tag{102}$$

$$V_i = \qquad v_i \Sigma g V_j + x_i \Sigma g U_j$$

Note that we can solve for T_i and use it to drive the "π^0" equations to get U_i. Then U_i is used to drive the "π^-" equations to get V_i. Of course if one is computing single charge exchange the last step is not present.

The procedure used for solving Eq. (102) is analagous to DWBA while solving Eq. (101) directly is like doing a full coupled channel optical model.

We actually solve Eq. (102) in practice and I will discuss some details of such a solution with a realistic wave function for ^{13}C in the next section. First I wish to consider a schematic model in which the energy dependence of charge exchange is displayed in the 3-3 resonance region.

Let us start by using the coherence approximation on the first of Eqs. (102).

$$\bar{T} = A\bar{t} + (A - 1)\bar{t}g\bar{T} \tag{103}$$

where we have summed over the nucleons. We may formally solve to get:

$$\bar{T} = \frac{A\bar{t}}{1 - (A - 1)\bar{t}g} \; . \tag{104}$$

We now insert a resonant energy dependence by taking

$$\bar{t} = \frac{t_0}{E - E_0 + i\Gamma/2} , \tag{105}$$

which gives for the elastic scattering

$$\bar{T} = \frac{At_0}{E - E_0 + i\Gamma/2 \ (A-1)\ t_0 g} = \frac{At_0}{E - E_0' + i\Gamma'/2} \ . \tag{106}$$

Note that there is both a broading of the resonance and a shift in energy but, the appearance of a resonance remains. One can easily understand that the more nucleons around the more absorption and the greater the width, since there are more channels available for decay.

Performing the same operation on the "π^0" equation, assuming only one "valence" nucleon:

$$\bar{U} = \bar{w} + (A-1)\bar{u}g\bar{U} + \frac{A-1}{A}\bar{w}g\bar{T} \tag{107}$$

$$\left[1 - (A-1)\bar{u}g\right]\bar{U} = \frac{\bar{w}}{A}\left[A + (A-1)g\bar{T}\right] \ . \tag{108}$$

Using as before

$$\bar{u} = \frac{u_0}{E - E_0 + i\Gamma/2} \ , \tag{109}$$

$$\bar{w} = \frac{w_0}{E - E_0 + i\Gamma/2} \ . \tag{110}$$

$$\left[\frac{E-E_0 + i\Gamma/2 \ (A-1) \ {}_0g}{E - E_0 + i\Gamma/2}\right]\bar{U} = \frac{\bar{w}}{A}\left\{\frac{A[E-E_0 + i\Gamma/2 \ (A-1)t_0 g] + (A-1)Agt_0}{E-E_0' + i\Gamma'/2}\right\} \tag{111}$$

$$\frac{E - E_0'' + i\Gamma''/2}{E - E_0 + i\Gamma/2}\bar{U} = \frac{w_0}{E - E_0 + i\Gamma/2} \ \frac{E - E_0 + i\Gamma/2}{E - E_0' + i\Gamma'/2} \tag{112}$$

$$\bar{U} = \frac{w_0(E - E_0 + i\Gamma/2)}{(E - E_0'' + i\Gamma''/2)(E - E_0' + i\Gamma'/2)} \tag{113}$$

The energy dependence of Eq. (113) is very interesting. We see that if Γ' and Γ'' are large (so that the energy dependence of the denominator is washed out) we will see a minimum in the single charge exchange reaction at the 3-3 resonance energy. This is exactly what optical model calculations give. If the absorption is weak then the numerator is killed by one of the factors in the denominator and the free amplitude is restored. In the intermediate case one will see a more-or-less horizontal line. Because of the energy shifts one can get s-shaped (or even more complicated) curves. Of course this is only a schematic model, but its general features seem to be borne out by actual calculations.

$^{13}C(\pi^+,\pi^0)^{13}N$

In this section I will discuss our calculation of this much-looked-at analog transition.

Since the shell model of this system is very complicated it was decided to treat the nucleus in the collective model of Banerjee, Stephenson and Reed.[16] The S-state nucleons have been treated as inert as far as antisymmetrization is concerned. Of course they were allowed to scatter the pions, but only in a single-particle orbitals.

In the model of Ref.16 each p-shell nucleon is considered to move in an orbit near either the x-, y- or z-axis with a radial function multiplying it, i.e.,

$$\phi_x = xf(r), \quad \phi_y = yf(r), \quad \phi_z = z\tilde{f}(r) \quad . \tag{114}$$

It may seem strange, at first, to write down the wave function in Cartesian variables but it is only a simple change of basis from $rY_1^{\mu}(\hat{r})$.

For ^{12}C the x-y orbitals are completely filled and the z-orbital is empty. This puts all of the particles near the x-y plane and leads to the pancake shape usually ascribed to this nucleus. When the next particle is added it must go into a z-orbital, since all of the x-y orbitals are filled, and the nucleus starts to move toward a spherical shape.

The entire collection of particles in the p-shell must be antisymmetric. If we make an expansion in individual spin and isospin states (and hence go to a spin-up, spin-down, neutron, proton basis) the problem simplifies. The coordinate space wave functions which multiply each state must have a definite character.

We will conventionally take our nucleus to have spin up. Thus we have in the initial state: two spin-down neutrons, two spin-down protons, three spin-up neutrons, two spin-up protons and in the final state: two spin-down neutrons, two spin-down protons, two spin-up neutrons and three spin-up protons. Each of these groups must be spacially anti-symmetric itself. Due to the symmetry of the total isospin lowering operator the spin-isospin dependence of any one of these groups may be made equivalent to any other by relabeling. Hence we need to consider only one labeling scheme. Charge exchange will always take place on particle 7.

The general convention used is:

1 and 2	s-shell neutrons,
3 and 4	p-shell spin-down neutrons,
5 and 6	p-shell spin-up neutrons
7	p-shell spin-up neutron (initial state),
	p-shell spin-up proton (final state),
8 and 9	p-shell spin-up protons,
10 and 11	p-shell spin-down protons,
12 and 13	s-shell protons.

Thus the overlap of the initial and final wave functions (p-wave part only) is,

$$(\psi_f^*\psi_i)_p = \frac{1}{24}\begin{vmatrix} \phi_x(3) & \phi_x(4) \\ \phi_y(3) & \phi_y(4) \end{vmatrix}^2 \begin{vmatrix} \phi_x(10) & \phi_x(11) \\ \phi_y(10) & \phi_y(11) \end{vmatrix}^2 \tag{115}$$

$$\times \begin{vmatrix} \phi_x(5) & \phi_x(6) & \phi_x(7) \\ \phi_y(5) & \phi_y(6) & \phi_y(7) \\ \phi_z(5) & \phi_z(6) & \phi_z(7) \end{vmatrix} \begin{vmatrix} \phi_x(8) & \phi_x(9) \\ \phi_y(8) & \phi_y(9) \end{vmatrix} \begin{vmatrix} \phi_x(5) & \phi_x(6) \\ \phi_y(5) & \phi_y(6) \end{vmatrix} \begin{vmatrix} \phi_x(7) & \phi_x(8) & \phi_x(9) \\ \phi_y(7) & \phi_y(8) & \phi_y(9) \\ \phi_z(7) & \phi_z(8) & \phi_z(9) \end{vmatrix}$$

This expression contains many types of terms. Of course it contains (with a factor of 24) the product of single particle densities:

$$\rho_{sp} = \phi_x^2(3)\phi_y^2(4)\phi_x^2(5)\phi_y^2(6)\phi_z^2(7)\phi_x^2(8)\phi_y^2(9)\phi_x^2(10)\phi_y^2(11) \quad . \tag{116}$$

It also contains terms corresponding to two step processes (using the convention that only non-squared densities are written), e.g.,

$$x_7 \; z_7 \; x_5 \; z_5 \quad . \tag{117}$$

We may represent all possible processes contained in Eq.(115) by means of a pair of diagrams: one for spin down and one for spin up.

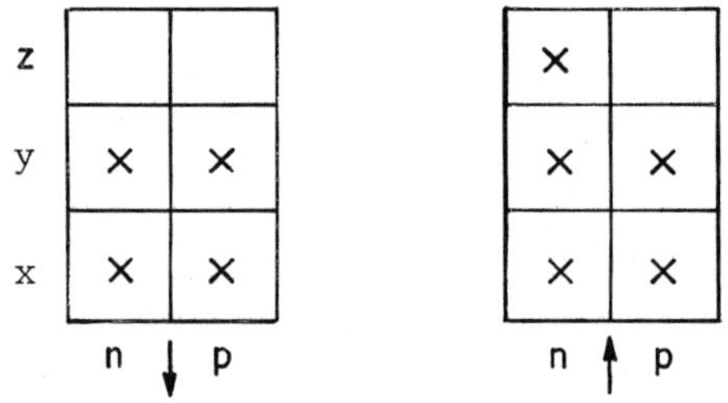

= initial state

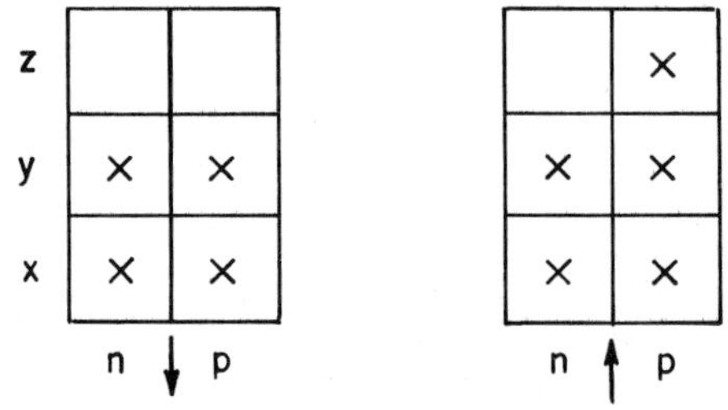

= final state

A transition may be represented as a collection of arrows always leading from the initial state to the final state. For the direct transition (Eq. (116)) we have

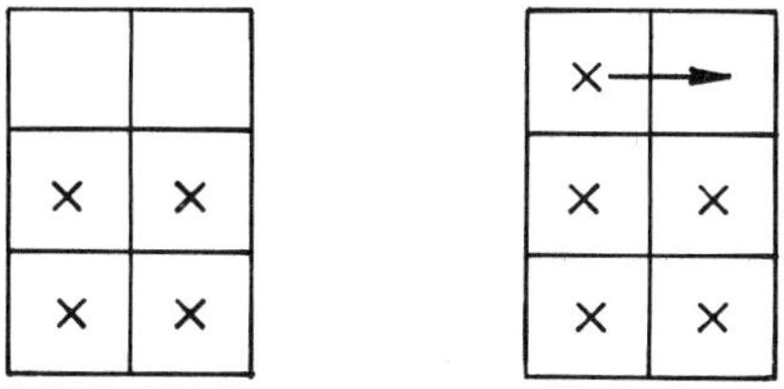

or with the x's understood:

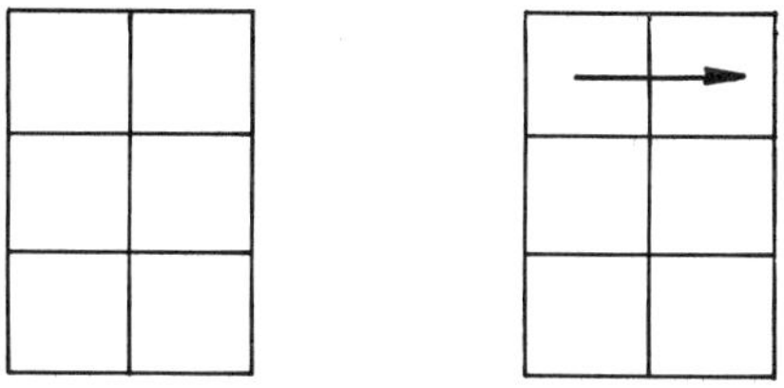

The transition in 117 would look like

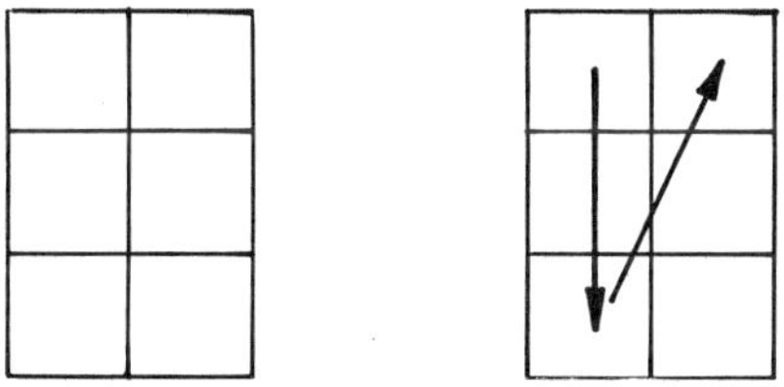

There are transitions of the type not involving the seventh particle except in the square e.g.:

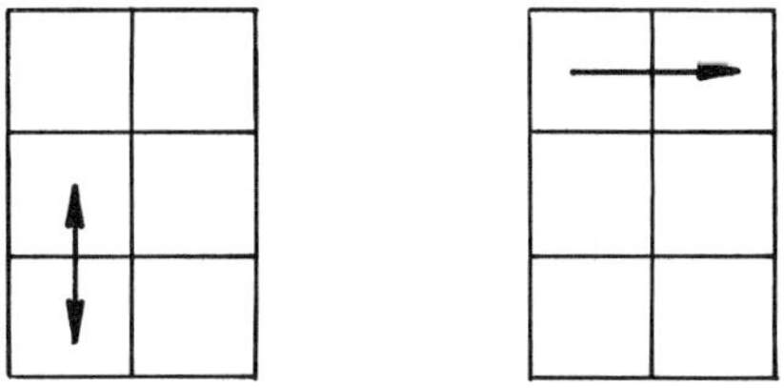

This type of transition is present in elastic scattering as well.
Since we allow only a single charge exchange, transitions of

the type

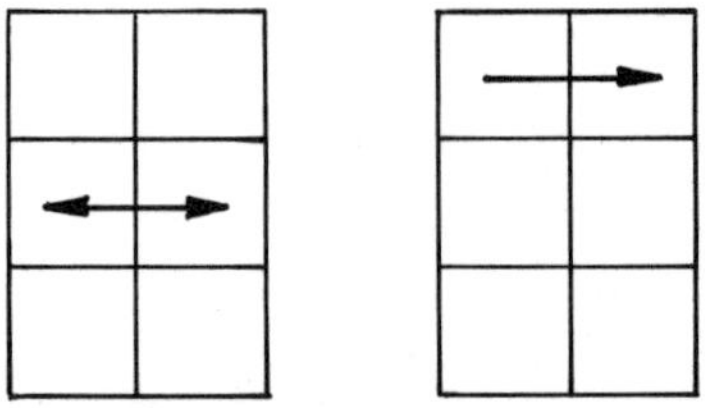

or

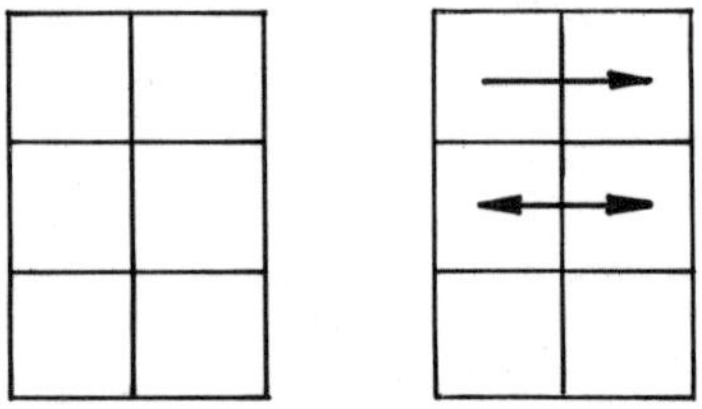

are not allowed in the present calculation. Also, transitions between the boxes are not allowed since we have not included the spin-flip operator, (we do consider spin-flip charge exchange, but not with anti-symmetrized wave functions).

While this expansion is useful for understanding the physical processes involved, the calculation of the expectation value using Eq.(115) looks rather different. Since the integral is to be done by Monte Carlo techniques, one's first thought might be to multiply and divide Eq.(115) by the product of single particle densities from Eq. 116 choose the positions according to these uncorrelated densities and multiply the operator by the ratio of 115 to 116. In fact this ratio has such a large variance as to make the computational time impractical. A more subtle approach must be used which "builds in" a large part of the correlation from the beginning. To do this we make the separation

$$(\psi_f^*\psi_i)_p = \frac{\phi_z^2(7)}{24}\begin{vmatrix}\phi_x(3) & \phi_x(4)\\ \phi_y(3) & \phi_y(4)\end{vmatrix}^2 \begin{vmatrix}\phi_x(5) & \phi_x(6)\\ \phi_y(5) & \phi_y(6)\end{vmatrix}^2 \times \begin{vmatrix}\phi_x(8) & \phi_x(9)\\ \phi_y(8) & \phi_y(9)\end{vmatrix}^2 \begin{vmatrix}\phi_x(10) & \phi_x(11)\\ \phi_y(10) & \phi_y(11)\end{vmatrix}^2 H \tag{118}$$

where

$$H = \frac{1}{\phi_z^2(7)} \frac{\begin{vmatrix}\phi_x(5) & \phi_x(6) & \phi_x(7)\\ \phi_y(5) & \phi_y(6) & \phi_y(7)\\ \phi_z(5) & \phi_z(6) & \phi_x(7)\end{vmatrix} \begin{vmatrix}\phi_x(7) & \phi_x(8) & \phi_x(9)\\ \phi_y(7) & \phi_y(8) & \phi_y(9)\\ \phi_z(7) & \phi_z(8) & \phi_z(9)\end{vmatrix}}{\begin{vmatrix}\phi_x(8) & \phi_x(9)\\ \phi_y(8) & \phi_y(9)\end{vmatrix} \begin{vmatrix}\phi_x(5) & \phi_x(6)\\ \phi_y(5) & \phi_y(6)\end{vmatrix}} \,. \tag{119}$$

The factor H will be multiplied into the fixed nucleon reac-

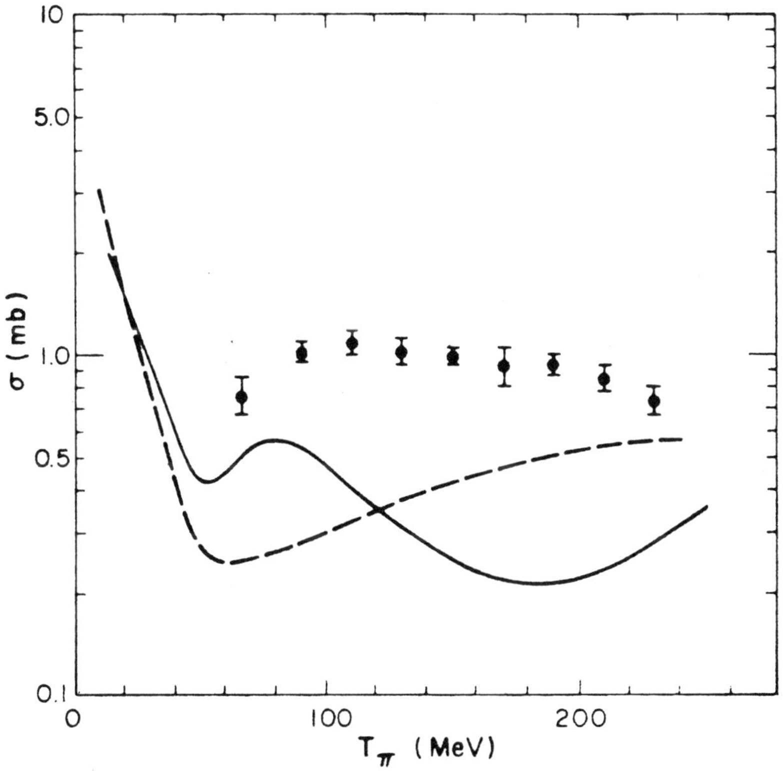

Fig. 3. Total cross section for pion charge exchange on ^{13}C.

tion amplitude as the Monte Carlo sum is being done and the particle positions will be chosen according to the multiplier of H in Eq. (119). This can be done since

$$\begin{vmatrix} \phi_x(3) & \phi_x(4) \\ \phi_y(3) & \phi_y(4) \end{vmatrix}^2 = f^2(3)f^2(4)(x_3y_4 - x_4y_3)^2$$

$$= r_3^2f^2(3)r_4^2f^2(4)(\sin\theta_3\sin\theta_4\cos\phi_3\sin\phi_4 - \sin\theta_3\sin\theta_4\cos\phi_4\sin\phi_3)^2$$

$$= r_3^2f^2(3)r_4^2f^2(4)\sin^2\theta_3\sin^2\theta_4\sin^2(\phi_4 - \phi_3) \quad .$$

Thus the r and θ variables can be chosen independently and the only correlation is in the ϕ variables. If we change variables to ϕ_3 and $\delta(\phi_4 = \phi_3 + \delta)$ then we can choose ϕ_3 uniformly in the range $-\pi \leqslant \phi_3 \leqslant \pi$ and γ according to the probability function $\sin^2\delta$. Thus a large part of the correlation is included from the beginning which improves the computational efficiency greatly.

We find that use of these anti-symmetrized wavefunctions increases the calculated total single charge exchange cross section by about 20-30%. Figure 3 shows the result of such a calculation for two different single particle wave functions for the valence nucleon.

We note that the theory still lies a factor of two below the experiment. Since we have calculated with a theory which allows the minimum of absorption and hence obtained the maximum cross section we must conclude that there is something lacking in our understanding of the basic physics.

REFERENCES

1. The material in this section is treated in several texts, e.g. "Collision Theory" by Goldberger and Watson, John Wiley & Sons, Inc., New York.

2. K. M. Watson, Phys. Rev. 105, 1388 (1957); M. M. Sternheim, Phys. Rev. B135, 912 (1964).

3. H. Garcilazo and W. R. Gibbs, Los Alamos Preprint LA-UR-78-897; G. A. Miller, Phys. Rev. C14, 361 (1976).

4. A. Kerman, H. McManus and R. Thaler, Ann. of Phys. 8, 551 (1959).

5. L. S. Kisslinger, Phys. Rev. 98, 761 (1955).

6. J. Eisenberg, J. Hufner and E. Moniz, Phys. Lett. 47B, 381 (1973); H. Garcilazo,to be published in Nucl. Phys.

7. M. Ericson and T.E.O. Ericson, Ann. Phys. (N.Y.) 36, 323 (1976).

8. L. L. Foldy and J.D. Walecka, Ann. Phys. (N.Y.) 54, 447 (1969).

9. J.J. Benayoun, Thesis, Grenoble; J.J. Benayoun, C. Cignoux and J. Gillespie, Nucl. Phys. A274, 525 (1976).

10. M.A.B. Beg, Ann. Phys. (N.Y.) 13, 110 (1961).

11. R.H. Landau and F. Tabakin, Phys. Rev. D5, 2746 (1972). J.T. Londergan, K.W. McVoy and E.J. Moniz, Ann. Phys. (N.Y.) 86, 147 (1974).

12. M. J. Reiner, Phys. Rev. Lett. 38, 1467 (1977).

13. W. R. Gibbs, A.T. Hess and W.B. Kaufman, Phys. Rev. C13, 1982 (1976).

14. W. R. Gibbs, Phys. Rev. C10, 2166 (1974).

15. Dan Agassi and Avraham Gal, Ann. of Phys. (N.Y.) 94, 184 (1975).

16. N. E. Reid, G.J. Stephenson, Jr., and M.K. Banerjee, Phys. Rev. C5, 287 (1972).

FIELD THEORY ASPECTS OF MESON-NUCLEUS PHYSICS

Leonard S. Kisslinger *

Carnegie-Mellon University
Physics Department
Pittsburgh, Pennsylvania 15213

1. INTRODUCTION

Pion-nucleus physics comprises the study of scattering and reactions involving pions and nuclei. The theoretical methods which have been developed for conventional nuclear reactions with nucleons and light ions have been and are being applied to the pionic processes. Most of the research has made use of multiple scattering theory based on the existence of a pion-nucleon potential. Inelastic scattering of pions, pion production and absorption, charge exchange scattering, and so forth are done with the distorted wave Born approximation, and related methods, often with the same computer codes used for direct nuclear reactions such as (d,p).

Yet we know in principle that the dynamics of the pion-nucleon system is very different from the nucleon-nucleon system. There are two differences which are so important that it suggests that the approach to pion-nucleus physics should be entirely different from, say, nucleon-nucleus physics. First, the pion, a

* Supported in part by the NSF Grant Phy 75-22555

boson, can disappear and reappear. The number of degrees of freedom thereby changes, and the concept of a potential is no longer directly applicable. Second, the dynamics of the medium energy region is dominated by baryon resonances. This is related to the absorptive processes. To the extent that the resonances are elementary structures the pion degrees of freedom have disappeared, and once more this is entirely different physics than that encountered in conventional nuclear physics.

One can argue that it is possible to find effective potentials for the π-N system, and that one can ignore most of this new physics. However, in recent years good data has been obtained with pion beams, and it is obvious that the simplest ideas will not work systematically. Moreover, it is also evident that the pion-nucleus processes involve strong interaction dynamics which can only be studied in the nuclear systems. The creation of a baryon resonance in a nucleus allows one to explore basic interaction mechanisms not encountered elsewhere. Therefore, my approach in these lectures is to emphasize the differences between the theory involved in pion-nucleus vs. conventional nuclear physics, and to develop some of the tools needed for carrying out relativistic calculations.

Much of the field theory which I will discuss was developed for the nucleon-nucleon interaction, which in turn grew out of quantum electrodynamics. I will devote a large part of my effort to discussing these methods for the π-nucleon system. Recall that the N-N interaction arises from meson exchange and related processes. The existence of a virtual meson has as its quantum mechanical expression the presence of an energy denominator or propagator. We shall use the notation for the momentum four-vector involved in these propagators (and other things) $k=(k_o,\vec{k})$, and $k_1 \cdot k_2 = k_{10}k_{20} - \vec{k}_1 \cdot \vec{k}_2$. The propagator has the form $1/(k_1 \cdot k_1 - m^2)$ for a spinless (scalar) meson. Since $k_o \approx o$ for a virtual meson coupled to the nucleon for the N-N force, the propagator is $\approx -1/(\vec{k}_1 \cdot \vec{k}_1 + m^2)$. Noting that the Fourier transform of this propagator is

$$\int d^3k\, e^{i\vec{k}'\vec{r}}/(\vec{k}_1 \cdot \vec{k}_1 + m^2) \simeq \frac{e^{-mr}}{r}$$

one remembers that the propagator represents a force.

In Chapter 2 we first try to derive a π-N interaction using the same bosons which can account approximately for the N-N interaction, and learn that such a boson exchange force does not work. We then review the Chew-Low Theory in perturbation theory, and see that this can lead to the Δ(1232) resonance. In the next two sections the off-shell T-matrix is discussed for nonrelativistic models. In chapter 3 we review the concept of a baryon resonance and show how this is treated as an independent quantity in the quark model. The isobar representation of a baryon resonance is discussed using projection operator techniques.

Relativistic problems and models are discussed in chapters 4 and 5. The recoil correction to the static model of the π-N-N vertex is derived starting from the relativistic pseudoscalar interaction. Those not familiar with Dirac spinors, matrices, propagators, and the Dirac equation will find a brief review in Sec. 4a. In chapter 5 the Bethe-Salpeter approach is discussed for the scattering and bound state problem. The Low equation is also introduced and briefly discussed. The off-mass-shell π-N T-matrix as derived from a model with a Bethe-Salpeter equation is given. In chapter 6, the status of the application of these concepts to the pion-nucleus optical potential is given.

The concept of "true" absorption is studied in cpt. 8 and a relativistic treatment of the πd→pp process is given. In this study various field theory and relativistic questions such as the off-mass-shell T-matrix appear. Finally, in the last chapter a brief review of phenomenological models is presented. Overall, it is seen that the theory of pion-nucleus processes is one of the most interesting areas of current scholarship in physics.

2. THE PION-NUCLEON AMPLITUDE - STATIC MODEL

a) The π-N Potential In A Nonrelativistic σ-Model

In nuclear physics we are accustomed to treating the interaction between two nucleons by a potential, usually a static local potential, which in coordinate space has the properties

$$\langle \vec{r}_1{}'\vec{r}_2{}' | v_{NN} | \vec{r}_1\vec{r}_2 \rangle = v_{NN}(\vec{r},\sigma,\tau)\ \delta(\vec{r}-\vec{r}) \tag{2.1}$$

where $\vec{r},\vec{r}'$ arc the relative coordinates, $\vec{r}_1$, $\vec{r}_2$, $\vec{r}_1'$, $\vec{r}_2'$ are the initial and final nucleon coordinates, and $\underset{\sim}{\sigma}$ and $\underset{\sim}{\tau}$ are the nucleon spin and isospin operators. In momentum space this operator has the form

$$\langle \vec{k}_1'\vec{k}_2' | v_{NN} | \vec{k}_1\vec{k}_2 \rangle = \langle \vec{k}' | v_{NN} | \vec{k} \rangle\, \delta(\vec{P}-\vec{P}') \tag{2.2}$$

where $\underset{\sim}{k}$, $\underset{\sim}{P}$ are the relative and center of mass momenta. The potential is usually considered to arise from meson exchange as illustrated in Fig. 1. In the mutliple pion transfer processes the virtual pions are interacting strongly, and these processes are ordinarily treated as the exchange of ρ,ω and other mesons.

Based on the success of the general idea of a potential derived from meson exchange for the N-N potential, an obvious line to investigate would be to try the same thing for pions. I.e., we shall derive a potential from meson exchange between a pion and a nucleon, perhaps even an "ordinary" potential such as one satisfy-

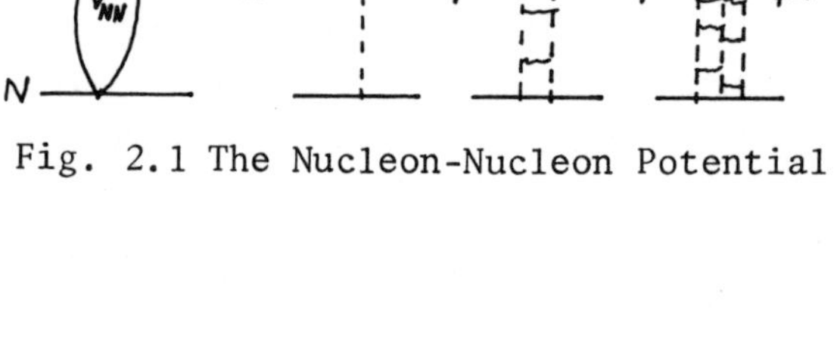

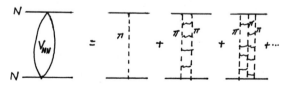

Fig. 2.1 The Nucleon-Nucleon Potential

in Eqs. 1 and 2, restricting ourselves to the mesons used in the N-N interaction. Strictly speaking, the multiple scattering approach which Dr. Gibbs has discussed implies that a nonrelativistic potential exists and can be used to obtain the pion-nucleon t-matrix in the standard (nonrelativistic) way. However, the fact that pions are bosons and can therefore be destroyed and created makes the derivation of an effective potential by meson exchange decidely less than obvious, but let us try it at least for its educational value and to define some terms and methods.

A major difference between a meson exchange potential for the π-N vs. the N-N system is that the pion cannot exchange one or an old number of pions (G-parity). Thus the most important contribution except for very short range will be (interacting) two-pion exchanges. In other words the ρ-meson (isospin one, vector $[J^P=1^-]$ and scalar mesons which have strong π-π branching ratios, which we shall call σ-mesons. The scalar mesons are not very "elementary", and are best treated as broad π-π resonances, however they are needed in one boson exchange N+N potentials. First let us consider the exchange of on isoscalar (I=0), scalar $(J^P=0^+)$ $\sigma_0$meson. In the static limit the coupling of the pion and nucleon to the σ_0 is given by the Lagrangian density

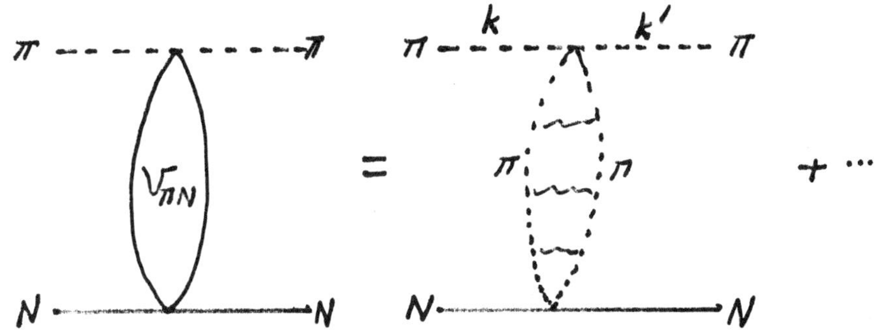

Fig. 2.2 Meson exchange processes for the N-N interaction

$$\mathcal{L}^{(\sigma_o)} = g^{\sigma_o\pi} \phi_\pi^+ \phi_{\sigma_o} + g^{\sigma_o N} \bar{\psi}_N \phi_{\sigma_1} \psi_N, \tag{2.3}$$

where the ϕ and ψ are meson and baryon field functions, and g's are coupling constants. The nonrelativistic one-meson-exchange diagram corresponding to this Lagrangian gives[(1)]

$$\frac{g^{\sigma\pi} g^{\sigma N}}{(k-k')^2 - m_\sigma^2}$$

leading to a static π-N potential

$$V^{\sigma_o}_{\pi\text{-}N} = -\frac{g^{\sigma_o\pi} g^{\sigma_o N}}{4\pi} \frac{e^{-m_{\sigma_o} r}}{r}$$

Using the nonrelativistic relation between the T-matrix and the potential

$$\langle k'|T|k\rangle = \langle k'|V|\psi(k)\rangle \quad ,$$

and the relation between the partial wave amplitude and the phase shift ($\alpha = I, \ell, J$)

$$t_\alpha = -\frac{1}{(2\pi)^2} \frac{E E_\pi}{E+E_\pi} \frac{1}{k} e^{i\delta} \sin\delta_\alpha$$

one observes that there is no isospin dependence of the phase shifts, since we have no put any into our model yet. (I shall use the notation δ_1, δ_3, δ_{11}, δ_{13}, δ_{31}, δ_{33} for the s-wave and p-wave phase shifts.) Thus in the Born Approximation ($\delta^{\sigma_o} \approx k\langle V^{\sigma_o}\rangle$) the phase shifts produced by the σ_o exchange satisfy

$$\delta_3^{\sigma_o} = \delta_1^{\sigma_o}$$

Let us review the experimental situation briefly. We know that the charge exchange amplitude at low energy satisfies[2]

$$t\ (\pi p \rightarrow \pi^o n) \sim (t_3 - t_1), \tag{2.4}$$

and that the charge exchange cross section is comparable to the elastic cross section, so that a model giving isospin independent phase shifts is clearly not acceptable. Furthermore, for low-energy elastic scattering

$$\begin{aligned} t\ (\pi^- p \to \pi^- p) &\sim 2t_1 + t_3 \\ t\ (\pi^+ p \to \pi^+ p) &\sim t_3 \end{aligned} \tag{2.5}$$

and from the low energy elastic on charge exchange scattering one has determined the low-energy S-wave phase shifts to be

$$\begin{aligned} \delta_3 &\approx -.1\ k/m_\pi \\ \delta_1 &\approx .14\ k/m_\pi , \end{aligned} \tag{2.6}$$

so that to be consistent with experiment we need a model giving an attractive interaction in the I = 1/2 and a repulsive interaction in the I = 3/2 S-wave channel. This is easy to do. First consider an isovector σ-meson, the σ_1. The Lagrandian density in the static model is

$$\mathcal{L}^{(\sigma_1)} = g^{\sigma_1 \pi}\, \phi_\pi^+ \vec{t}_\pi\, \phi_{\sigma_1} \phi_\pi + g^{\sigma_1 N}\, \bar{\psi}_N \vec{\tau}_N \cdot \phi_{\sigma_1} \psi_N , \tag{2.7}$$

where $\vec{t}_\pi$, $\vec{\tau}$ are the isospin operators for the pion, nucleon,† which leads to the nonrelativstic potential

$$V_{\pi N}^{\sigma_1} = - \frac{g^{\sigma_1 \pi} g^{\sigma_1 N}}{4\pi}\, \vec{\tau}_N \cdot \vec{t}_N\, \frac{e^{-m_{\sigma_1} r}}{r} \tag{2.8}$$

Noting that for the isospin I = 1/2, 3/2 states of the π-N system the quantity $\vec{\tau}_N \cdot \vec{t}_\pi$ is

$$\vec{\tau} \cdot \vec{t}_\pi \left\{ \begin{array}{ll} -2 & I = 1/2 \\ 1 & I = 3/2 \end{array} \right\} \tag{2.9}$$

one seems that in the Born approximation ($\delta \sim k\langle V \rangle$) that the phase shifts produced by the σ_1 exchange satisfy

$$2\delta_3^{\sigma_1} + \delta_1^{\sigma_1} \approx 0 \tag{2.10}$$

It is fairly obvious from Eqs. 2.3 and 2.10 that by taking a

---

† $\tau^\pm \psi_{P,N} = \psi_{N,P}$; $\tau^o \psi_{P,N} = \pm\psi_{P,N}$; $t^\pm \phi_\pm = \sqrt{2}\phi_o$, $t_\pi^o\, \phi_{\pm,o} = \{\pm, o\}\phi_{\pm,o}$

linear combination of σ_o and σ_1 exchange

$$V^{\sigma}_{\pi-N} = V^{\sigma_o}_{\pi-N} + V^{\sigma_1}_{\pi-N}$$

over a limited energy range one could satisfy

$$\begin{aligned} \delta_1/\delta_3 &\approx -1.4 \\ \delta_3 &\approx -.1\ k/m_\pi \end{aligned} \tag{2.11}$$

with a suitable choice if coupling constants. What we have done is to carry out a simplified, first order calculation of processes which have been considered in integral equations of the Chew-Low type (see below). The most recent calculation is by Barnerjee and Cammarata[3], who include anti-nucleon (pair) contributions as well as the σ terms.

Let us now take a look at the p-wave π-N phase shifts. The experimental results are showin in Fig. 2.3.

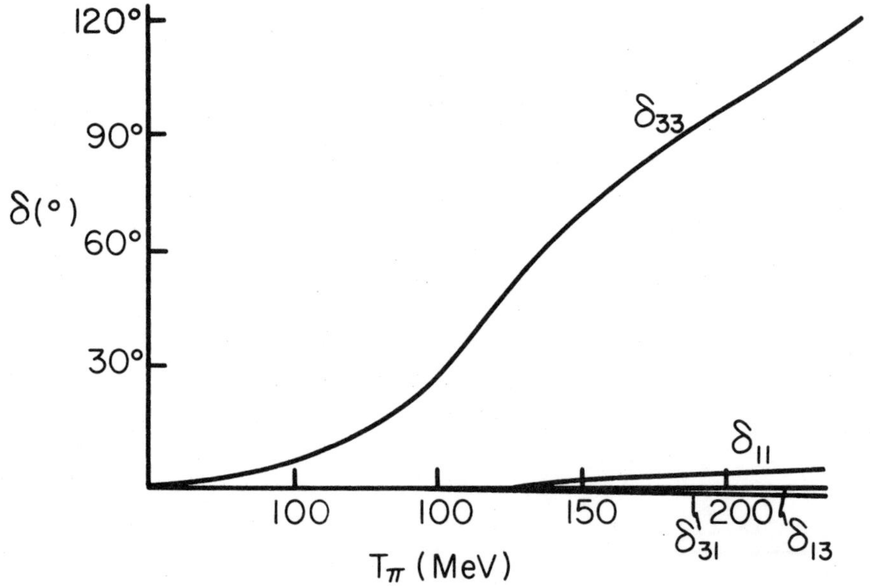

Fig. 2.3 The low-medium energy p-wave phase shifts

The most striking results is the behavior of the I = 3/2, J=3/2 δ_{33} phase shift, which passes through $\pi/2$ at about T = 200 MeV (E_{total} = 1232 MeV). This corresponds to the I = 3/2, J=3/2 $\Delta(1232)$ baryon resonance. The other p-wave phase shifts in this energy region are small and uninteresting. It is obvious that the potential V^{σ} cannot account for this behavior. In fact in that model the J=1/2 and J=3/2 p-wave phase shifts are equal.

b) The ρ-Meson Contribution To The π-N Potential

As a vector meson ($J^P=1^-$), the ρ exchange process between the π and N is a little more complicated than the σ exchange. Our simple-minded treatment including the σ and ρ mesons resembles calculations such as Ref. 4 using an effective interaction in lowest order. In the static limit the interaction Lagrangian for coupling the ρ to the π and N is

$$\mathcal{L}^{\rho} = g^{\rho\pi}\phi_{\pi}^{+}\vec{t}_{\pi}\cdot\phi_{\rho_{\mu}}\nabla_{\mu}\phi_{\pi} + g^{\rho N}\bar{\Psi}_{N}\tau_{\lambda}\sigma\times\nabla\cdot\phi_{\rho_{\lambda}}\Psi_{N}, \tag{2.12}$$

where by $\sigma\times\nabla$ we indicate the rank one tensor operator in the product of the internal nucleon and ρ-momentum spaces. The nonrelativistic potential has the form

$$V^{\rho} = -\frac{g^{\rho\pi}g^{\rho N}}{4\pi}\,\vec{\sigma}\cdot\vec{L}\;\vec{\tau}\cdot\vec{t}_{\pi}\,\frac{e^{-m_{\rho}r}}{r} \tag{2.13}$$

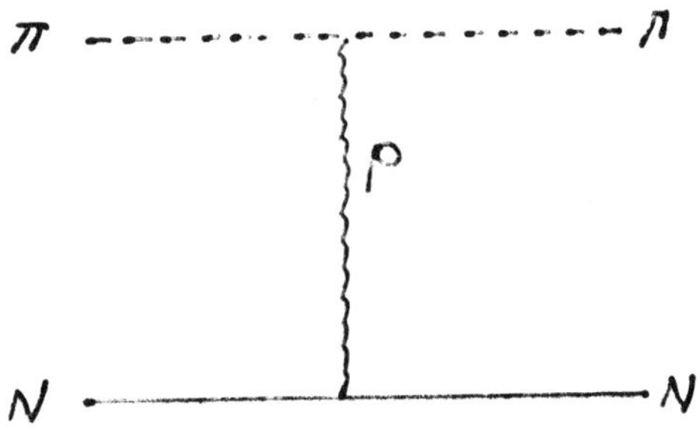

Fig. 2.4 The ρ-Exchange Contribution to the π-N Potential

Using the relationship for the p-wave

$$\vec{\sigma}\cdot\vec{L} = \begin{Bmatrix} -2 & J=1/2 \\ 1 & J=3/2 \end{Bmatrix}, \qquad (2.11)$$

analogous to Eq. 2.9, one can try to obtain the J-dependence of the phase shifts. By using $V=V^{\sigma}+V^{\rho}$ and using Eqs. 2.8, 2.9, 2.12, and 2.13, one can see how it might be possible to get a large δ_{33} phase shift, and even try for a resonance in the I=3/2, J=3/2 state. If the p-wave arises mainly from the ρ-meson, one could see how the δ_{13} and δ_{31} phase shifts could become repulsive, as V^{ρ} has opposite sign in the 3/2-3/2 vs the 3/2-1/2 and 1/2-3/2 states. However, it would be predicted that the δ_{11} phase shift is even larger than the δ_{33} phase shift and should resonate for pion energy less than 200 MeV. As one can see from Fig. 3, the δ_{11} phase shift is less than 2° in the low energy region, and is even repulsive (negative) for $T_{\pi}<150$ MeV. The potential model clearly fails. There are no other reasonable candidates for meson exchange.

c) The Chew-Low Model-Static

The theory of π-N scattering is a true field theory problem. A nonrelativistic potential model is simply not appropriate. As we shall see, this will have profound implication for the theory of π-nucleus interactions. The Chew-Low Theory of π-N scattering[5] is not only a beautiful and successful model of low-medium-energy π-N interactions, but it is one of the most important theoretical developments from the early stages of particle physics. It directly led to the development and applicaton of the theory of dispersion relations-the use of analytic properties of scattering amplitudes and vertex functions to nuclear science. Since an extensive treatment of this topic is not appropriate for the present lectures, I shall confine myself to the field theory itself and the use of the Chew-Low effective Lagrangian in perturbation theory in order to see how the $\Delta(1232)$ resonance arises. In addition to the original papers,[5] one can find detailed treatments in several texts.[1,6]

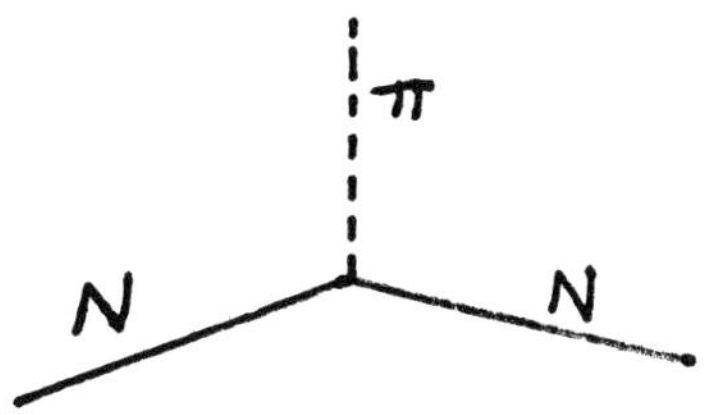

Fig. 2.5 The Basic π-N-N Vertex

The Chew-Low Lagrangian is based on pseudoscalar coupling of pseudoscalar mesons. The basic (renormalized) interaction vertex in a relativistic theory for p.s. pion coupling to a nucleon, illustrated in Fig. 2.5 is

$$\mathcal{L}^{p.s.} = -ig\, \bar{\Psi}_N \gamma_5\, \vec{\tau}\cdot\phi_\pi \Psi_N \;, \tag{2.15}$$

where γ_5 is an operator in the space of Dirac spinors which changes sign under special inversion, giving the pseudoscalar property. If you are unfamiliar with Dirac matrices, see Section 4a. The norecoil limit for the interaction Hamiltonian is

$$H_I = -g\, \Psi_N^+\, \vec{\sigma}\cdot\nabla\vec{\tau}\cdot\phi\Psi_N \tag{2.16}$$

which is the Chew Lagrangian. The relation between 2.15 and 2.16 is discussed in Sec. 4b. It is the unique form for p.s. nucleon coupling in the static limit. Since we are using a nonrelativistic form, we can use a nonrelativistic perturbation expansion for evaluating our field thoery. Let us use the standard creation and destruction operators for nucleons and mesons. For massive, static nucleons $\bar{\Psi}_N\psi_N$ just becomes a static source of mesons

$$\bar{\Psi}_N(x)\psi_N(x) = \sum_{\alpha\beta i} \delta(\vec{x}-\vec{x}_i)\, a_\alpha^+(i) a_\beta(i), \tag{2.17}$$

where α,β are spin and isospin projections, $\vec{x}_i$ is the position and a_i, a_i^+ are the destruction creaction operators for the ith nucleon. The pion field operator is

$$\phi_{\pi_t}(x) = \frac{1}{\sqrt{(2\pi)^3}}\int\frac{d^3k}{\sqrt{2\omega(k)}}\left[e^{i\vec{k}\cdot\vec{x}}\, b_{\underset{\sim}{k},t} + e^{-i\vec{k}\cdot\vec{x}}\, b^+_{\underset{\sim}{k},-t} \right], \tag{2.18}$$

where $b_{k,t}$, $b^{+}_{k,t}$ are the creation, destruction operator for a meson of three-momentum $\vec{k}$ and isospin t. We use the relativistic form for the pion energy, $\omega(k) = \sqrt{|\vec{k}|^2+m_\pi^2}$, and work in a Schroedinger representation, so the operators have no time dependence. Note, also, that the coupling factor g in Eq. (2.16) should be treated as a function of momentum

$$g = g\ (|\vec{k}|^2) \tag{2.19}$$

which has the effect of giving spacial structure to the pi-nucleon vertex. For example

$$g(|\vec{k}|^2) = \frac{g}{|\vec{k}|^2+\beta^2} \leftrightarrow \rho_N(r) \propto \frac{e^{-\beta r}}{r}, \tag{2.20}$$

so that the nucleon-pion vertex is given a spatial range of β^{-1}, which has the interpretation of an effective nucleon size. In fact, nucleons are rather large, and almost touch in a nucleus. This is very important to keep in mind, especially for various high momentum transfer processes. We shall be returning to this a number of times.

We now treat the interaction 2.16 in nonrelativistic perturbation theory. The lowest order diagrams are given in Fig. 2.6.

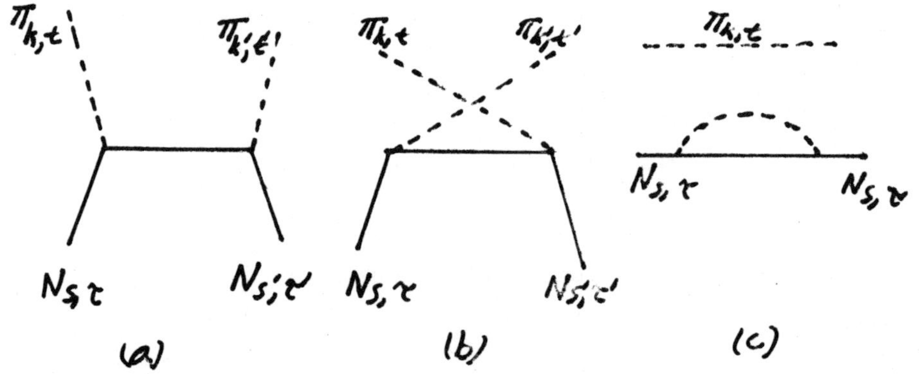

Fig. 2.6 The Second Order Diagrams in Chew-Low Theory

The third process is an unlinked one, which is already included by using physical masses, and is not to be explicitely included. The first two diagrams are nucleon pole diagrams in the s and u channels. The invariant variable s and u are defined in terms of the external 4-vectors = $s = (k+P_N)^2$ and $u = (P_N-k')^2$. For the process of Fig. 2.6 (a) $s = M_n^2$, and for that of Fig. 2.6 (b) $u = M_n^2$. These fixed points for the invariant variables indicate that they correspond to poles in the amplitude. Also, diagram b is called the crossed diagram corresponding to diagram a. To second order we find that the π-N t-matrix is

$$t_{fi}^{(2)} = \frac{\langle f|H_I|a\rangle\langle a|H_I|i\rangle}{E_i-E_a+i\varepsilon} + \frac{\langle f|H_I|b\rangle\langle b|H_I|i\rangle}{E_i-E_b+i\varepsilon} . \tag{2.21}$$

the initial state $|i\rangle = |\vec{k}t,s\tau\rangle$ and the final state $|f\rangle = |\vec{k}'t',s'\tau'\rangle$ and

$$\begin{aligned} E_i &= M_n + \omega(k) \\ E_a &= M_n \\ E_b &= M_n + \omega(k) + \omega(k') . \end{aligned} \tag{2.22}$$

The isospin and spin labels are shown in Fig. 2.6. Substituting Eq. 2.16 in 2.21, using Eqs. 2.17, 2.18 one finds

$$t_{fi}^{2} = \frac{g^2}{(2\pi)^3 2\omega(k)} \langle s'\tau'|(\vec{\sigma}\cdot\vec{k}')(\vec{\sigma}\cdot\vec{k})\tau_{t'}\tau_t - \vec{\sigma}\cdot\vec{k}\vec{\sigma}\cdot\vec{k}'\tau_t\tau_{t'}|s\tau\rangle , \tag{2.23}$$

with s,τ and s',τ' the spin and isospin of the initial and final nucleons. Note, the matrix element of Eq. 2.23 is defined only in the one-nucleon space.

The quantitiy in the nucleon matrix element of Eq. 2.23 can conveniently be expressed in terms of projection operators. For the coupling of a spin 1/2 particle (the nucleon) to a spinless particle (the pion) to a total angular momentum $\vec{J}$, when they are in a state of orbital angular momentum ℓ, the operators

$$\mathcal{P}_{J=\ell+1/2} = \frac{\ell+1+\vec{L}\cdot\vec{\sigma}}{2\ +1}$$
$$\mathcal{P}_{J=\ell-1/2} = \frac{\ell-\vec{L}\cdot\vec{\sigma}}{2\ell+1} \tag{2.24}$$

projection operators onto states of $J=\ell\pm1/2$. They satisfy the usual rules $\mathcal{P}_J\mathcal{P}_J = \mathcal{P}_J$ and $\mathcal{P}_{J^+}\mathcal{P}_{J^-} = 0$. They are operators in the direct product nucleon x meson space. Defining operators in the nucleon space (for the case $\ell=1$) by

$$\mathcal{P}^J_{\frac{3}{2}}(\vec{k}',\vec{k}) \equiv \frac{3}{4\pi k^2}\,[\vec{k}'\cdot\vec{k} - 1/3\,(\vec{\sigma}\cdot\vec{k}')(\vec{\sigma}\cdot\vec{k})]$$
$$\mathcal{P}^J_{\frac{1}{2}}(\vec{k}',\vec{k}) \equiv \frac{1}{4\pi k^2}\,\vec{\sigma}\cdot\vec{k}'\ \vec{\sigma}\cdot\vec{k}\ , \tag{2.25}$$

one can show that these are projection operators, satisfying

$$\int d\Omega_{k''}\,\mathcal{P}^J_{J_1}(\vec{k}',\vec{k}'')\mathcal{P}^J_{J_2}(\vec{k}'',\vec{k})=\delta_{J_1J_2}\mathcal{P}^J_{J_1}(\vec{k}',\vec{k}). \tag{2.27}$$

One can also readily prove the needed relationships between matrix elements in the direct products space of nucleons and mesons in a momentum representation, $|\vec{k}t,s,\tau\rangle$, and the space of nucleons, $|s\tau\rangle$:

$$k\omega\,\langle k't',s'\tau'|\mathcal{P}_{\frac{3}{2}}|\,kt,s\tau\rangle = \langle s'\tau'|\mathcal{P}^J_{\frac{3}{2}}(\vec{k}',\vec{k})|s\tau\rangle$$
$$k\omega\,\langle k't',s'\tau'|\mathcal{P}_{\frac{1}{2}}|\,kt,s\tau\rangle = \langle s'\tau'|\mathcal{P}^J_{\frac{1}{2}}(\vec{k}',\vec{k})|s\tau\rangle, \tag{2.28}$$

with the factor of $k\omega$ arising from the projection of linear to angular momentum. Thus the matrix elements in Eq. 2.23 can be written

$$\langle s'\tau'\,|\,\vec{\sigma}\cdot\vec{k}'\vec{\sigma}\cdot\vec{k}|\,s\tau\rangle = 4\pi k^3\omega\,\langle k't',s'\tau'|\mathcal{P}_{\frac{1}{2}}|kt,s\tau\rangle$$
$$\langle s'\tau'|\,\vec{\sigma}\cdot\vec{k}\,\vec{\sigma}\cdot\vec{k}'|s\tau\rangle = \frac{4\pi k^3\omega}{3}\,\langle k't',s'\tau'|2\mathcal{P}_{\frac{3}{2}}-\mathcal{P}_{\frac{1}{2}}|kt,s\tau\rangle \tag{2.29}$$

In a similar way we can relate the operators in the nuclear isospin space appearing in Eq. (2.23) to the projections isospin of the combined pion-nucleon space. Let us introduce the projection operators in isospin space for an isospin one particle (pion) coupled to an isospin 1/2 particle (nucleon)

$$\mathcal{P}^{I}_{\frac{3}{2}} = \frac{2+ \vec{t}_{\pi}\cdot\vec{\tau}}{3}$$
$$\mathcal{P}^{I}_{\frac{1}{2}} = \frac{1- \vec{t}_{\pi}\cdot\vec{\tau}}{3} , \tag{2.30}$$

which can easily be seen to satisfy $\mathcal{P}^{I}_{I_1}\mathcal{P}^{I}_{I_2} = \delta_{I_1 I_2}\mathcal{P}^{I}_{I_1}$. Proceeding as with the angular momentum projection operators, one obtains relationships between matrix elements of the nucleon isospin operators in the combined nucleon-pion space. One finds

$$\langle s'\tau'|\tau_{t'}\tau_{t}|s\tau\rangle = \langle k't',s'\tau'|3\,\mathcal{P}^{I}_{\frac{1}{2}}|kt,s\tau\rangle$$
$$\langle s'\tau'|\tau_{t}\tau_{t'}|s\tau\rangle = \langle k't',s'\tau'|2\,\mathcal{P}^{I}_{\frac{3}{2}} - \mathcal{P}^{I}_{\frac{1}{2}}|kt,s\tau\rangle , \tag{2.31}$$

From Eqs. 2.23, 2.29, and (3.31), one obtains the result for the t-matrix in second order.

$$\langle \vec{k}'t_1'\,s'\tau'|t^{(2)}|\vec{k}t_1 s\tau\rangle = \frac{g^2k^3 4\pi}{} \tag{2.32}$$
$$\times\langle k't',s'\tau'|8\,\mathcal{P}^{I}_{\frac{3}{2}}\mathcal{P}^{J}_{\frac{1}{2}} + 2\,\mathcal{P}^{I}_{\frac{1}{2}}\mathcal{P}^{J}_{\frac{3}{2}} + 2\,\mathcal{P}^{I}_{\frac{3}{2}}\mathcal{P}^{J}_{\frac{1}{2}} - 4\mathcal{P}^{I}_{\frac{3}{2}}\mathcal{P}^{J}_{\frac{3}{2}}|kt,s\tau\rangle$$

$$\begin{aligned} e^{i\delta_{33}}\sin\delta_{33} &= 4c(k) \\ e^{i\delta_{11}}\sin\delta_{11} &= -\,8\,c(k) \\ e^{i}\delta_{31}\sin\delta_{31} &= -\,2\,c(k) = e^{i\delta_{13}}\sin\delta_{13}, \end{aligned} \tag{2.33}$$

where

$$c(k) = \frac{g^2k^3 4\pi^2}{(2\pi)^3\;2\omega(k)3}$$

From this we obtain the important result that only in the J=3/2, I=3/2 state is the effective π-N interaction attractive. Of course one cannot except to calculate a resonance phonomenon in perturbation theory, but, knowing that the Born approximation underesti mates an attractive potential, it would not be unexpected that an exact solution could give rise to a resonance in the 3/2 - 3/2 channel. On the other hand, for the channels in which the phase shifts are negative one would not expect that the exact solution would be particularly interesting, as the repulsive potential tends to exclude the wave function (or its relativistic equivalent).

We shall come back to the Chew-Low Model in Sec. 5.c.

d) The π-N Interaction and the Off-Momentum-Shell T-Matrix: Structure Functions

Although we argued in Secs. 2a, b that the idea of a meson exchange potential is not appropriate to describe the π-N interaction, this by no means rules out the existence of a π-N potential. Let us consider the t-matrix $\langle\vec{k}'|t|\vec{k}\rangle$ of the previous section, in which we drop the spin and isospin indicex for convenience. The potential can be defined by the inverse problem. I.e., what is the operator V such that the matrix elements

$$\langle\vec{k}'|V|\vec{k}\rangle = \text{momentum representation of potential}$$

when inserted into the Lippman-Schwringer (Schroedinger) Eq.

$$\langle\vec{k}'|\, t(\omega)|\vec{k}\rangle = \langle\vec{k}'|V|\vec{k}\rangle + \int d^3k'\frac{\langle\vec{k}'|V|\vec{k}''\rangle\langle\vec{k}''|t(\omega)|\vec{k}\rangle}{\omega - E\,(k'') + i\varepsilon} \qquad (2.34)$$

give the t-matrix ? From experiment one can only determine the on-shell part of the t-matrix

$$\langle\vec{k}'|t(\omega)|\vec{k}\rangle_{\text{on shell}} \leftrightarrow |\vec{k}'|=|\vec{k}|= \sqrt{\omega^2-m_\pi^2}, \qquad (2.35)$$

where we continue to neglect nuclear motion. Decomposing t and V into partial waves (drop spin)

$$\langle \vec{k}'|t(\omega)|\vec{k}\rangle = \sum_{\ell} (2\ell + 1)\ t_{\ell}(k',k,\omega)\ P_{\ell}(\cos\Theta)$$

$$\langle \vec{k}'\ |V\ |\vec{k}\rangle = \sum_{\ell} (2\ell + 1)\ V_{\ell}(k',k\ )\ \ P_{\ell}(\cos\Theta),$$

using the spherical harmonic addition theorem and orthogonality of the spherical harmonics one easily derives the one-dimensional integral equation for the ℓth partial wave.

$$t_{\ell}(k'k,\omega)=V_{\ell}(k',k) + \frac{2}{\pi}\int_0^{\infty} \frac{dk''k''^2\ V_{\ell}(k',k'')\ t_{\ell}(k'',k,\omega)}{\omega - E(k'')+i\varepsilon} \quad . \tag{2.36}$$

Let us consider the special case of the separable interaction in each partial wave. I.e., for an attractive potential

$$V_{\ell}\ (k',k) = -g_{\ell}\ (k')\ g_{\ell}(k). \tag{2.37}$$

By substitution into 2.36 one can readily show that

$$t_{\ell}(k',k,\omega) = \frac{-g_{\ell}(k'^2)\ g_{\ell}(k^2)}{D_{\ell}\ (\omega+i\varepsilon)} \tag{2.38}$$

where

$$D_{\ell}(E) = 1 - \frac{2}{\pi}\int_0^{\infty} \frac{dk\ k^2\ g_{\ell}{}^2(k^2)}{E(k)\ -E,} \tag{2.39}$$

with $E(k) - E_N(k) + \omega(k)$, is a solution to the partial-wave Lippman-Schwinger equation.

Up to this point we have simply obtained the t-matrix from an arbitrary separable potential. However, under certain conditions one can easily solve the inverse problem. If the phase shifts are properly behaved one can show that the functions $g_{\ell}(k)$ of Eq. 2.37 can be determined from the on-shell t-matrix by[7]

$$g_{\ell}^2(k) = -e^{-i\delta_{\ell}(E)}\, t_{\ell}(k,k,E(k))\exp\left\{\frac{1}{\pi}P\int_{m_\pi+M}^{\infty} \frac{\delta_{\ell}(E')dE'}{E-E'}\right\} \tag{2.40}$$

Note that this is a solution, not an equation, for if the experimental phase shifts $\delta_{\ell}(E)$ are know for all energies one can obtain $g_{\ell}(k)$. This methods has been used for the πN case[7]. Landau and Tabakin assume Regge behavior for medium energies, and take

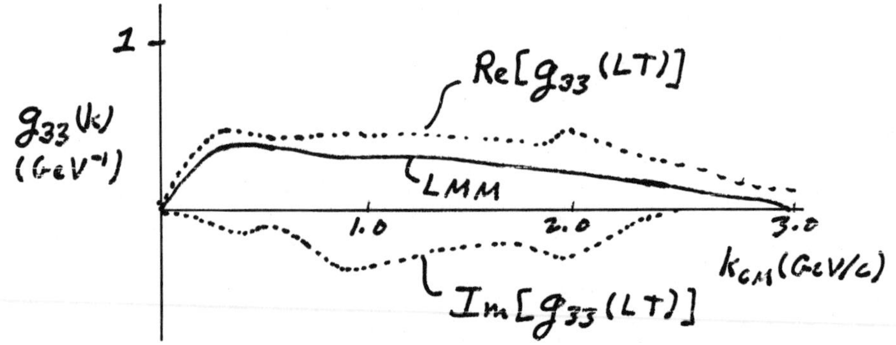

Fig. 2.7 The Structure Function in the treatments of Landau and Tabakin (LT), and Londergan, McVoy, and Moniz (LMM).

$$S_\ell(E) = e^{2i\delta_\ell(E)} \xrightarrow[E \to \infty]{} 1$$

Londergan, McVoy and Moniz use a coupled-channel, separable model to include inelasticity. Results for the 3/2-3/2 structure function are shown in Fig. 2.7.

Having determined these structure functions $g_\ell(k)$, one has an off-momentum shell representation for the T-matrix. I.e,

$$t_\ell(k',k,\omega) = \frac{g_\ell(k')\ g_\ell(k^2)}{D_\ell(\omega + i\varepsilon)} \tag{2.41}$$

As Dr. Cibbs has discussed, knowledge these off-shell forms is essential for determining the pion-nucleus potential from the π-N t-matrix. In Sec. 5 we shall discuss some relativistic approaches to this problem.

3. THE Δ(1232) ISOBAR - THE ISOBAR MODEL

The existence of baryon resonances has been the most interesting aspect of medium energy meson-nucleon physics. The Δ(1232) was the first baryon resonance discovered experimentaly, and the only one which can be studied at the meson factories which now exist. Corresponding to the resonating phase shift shown in Fig. 2.3, it shows up clearly in the π^+p cross sections as seen in Fig. 3.1. Other I = 3/2 resonances occur, and their treatment is similar except that only the Δ(1232) is almost elastic. There are also many I = 1/2 resonances. In addition to the dynamics which we have been discussing, these baryon resonances have played an important role in the development of particle spectroscopy based on the quark model. In fact, the spectroscopic classification by the quark model is so successful that we can use it as a basis for

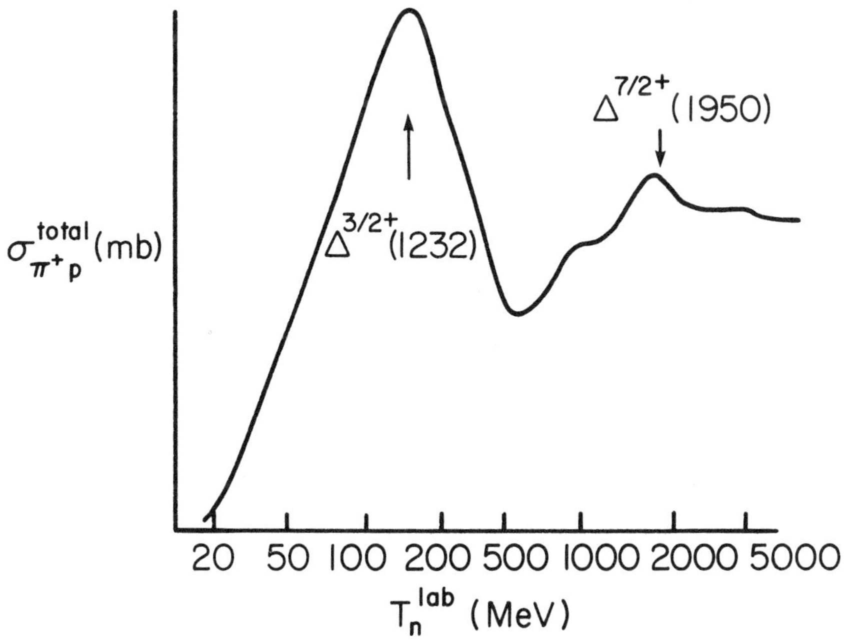

Fig. 3.1 The $^+$P Total Cross Section

our treatment of the resonances.

a) The Quark Model for the Δ(1232), the π, and the N

For the π,N, and Δ we need only the old, nonstrange quarks. The quantum numbers of the old quarks (we ignore the quarks discovered after 1970) are

	Spin	Isospin	$\frac{\text{Charge}}{e}$	Strangeness	Baryon#	Color
u	1/2	1/2	2/3	0	1/3	r,ω,b
d	1/2	1/2	-1/3	0	1/3	ω,b,r
s	1/2	0	2/3	-1	1/3	b,r,ω

Out of the u,d, and s quarks one can account for the spectroscopy of the nonstrange and strange mesons and baryons with an exceptionally simple model: the mesons are consist of quark (q) antiquark ($\bar{q}$) systems and the baryons are three quark systems. Other objects are referred to as bein "exotic". Therefore we have, e.g.,

$$|\pi^+\rangle = |[u\bar{d}]\ I=1,\ J=0\rangle$$
$$|p\rangle = |[uud]\ I=1/2,\ J=1/2\rangle \quad (3.1)$$
$$|\Delta^{++}\rangle = |[uuu]\ I=3/2,\ J=3/2\rangle$$

All three colors are present in Δ^{++}, so the wave function can be symmetric. Thus in the quark model the description of elastic π^+p scattering through the Δ^{++}

$$\pi^+ + p \rightarrow \Delta^{++} \rightarrow \pi^+ + p \quad (3.2a)$$

is

$$(u\bar{d}) + (uud) \rightarrow (uuu) \rightarrow (u\bar{d}) + (uud) \quad (3.2b)$$

This process is illustrated in Fig. 3.2.

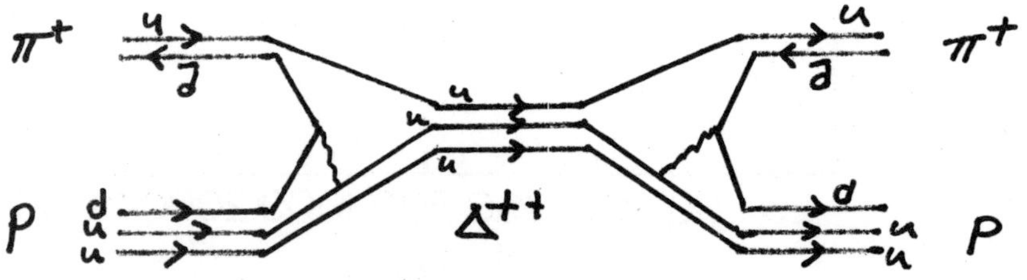

Fig. 3.2 π^+p Scattering through the Δ^{++} resonance in the quark model.

Such a picture suggests that the Δ can be treated as an elementary particle. Rather than to attempt to directly use the quark model for the theoretical treatment of the process of Eq. (3.2), which would include the detailed description of the $d\bar{d}$ annihilation and production illustrated in Fig. 3.2, we make use of the isobar picture.

b) The Isobar Model

The baryon resonances are very broad. The width of the Δ(1332) typical

$$\Gamma_\Delta \approx 110 \text{ MeV.} \tag{3.3a}$$

This corresponds to a lifetime

$$\tau_\Delta \approx \frac{\hbar}{\Gamma} \approx 10^{-23} \text{ sec} \tag{3.3b}$$

The time τ_Δ is about the time for a relativistic particle to pass by an object the size of a nucleon. For this reason nuclear physicists have traditionally not dealt with the Δ as a fundamental object. Instead we have used the π-N t-matrix in multiple scattering theory, and allowed the presence of the Δ to manifest itself in an affective π-N and π-nucleus interaction.

However, as is made evident by the quark model, as is discussed in the previous section, one can treat the Δ as an entity in itself. As will be pointed out in section 5b, and is the main topic for Dr. Moniz's lectures, there are strong motivations and many advantages in introducing the Δ explicitely into the theory of π-nuclear scattering and reactions. Here we show how one can explicitely introduce the Δ into the π-N system.

A state of total energy ≈ 1.232 GeV which at $t \to -\infty$ ("in" state) is an incoming π and N, can be considered as consisting of three parts, a nonresonating π-N, a Δ, and more complicated components.

$$|\Psi^{(\pi N)}\rangle = |\pi N, \text{nonresonant}\rangle + |\Delta + 1\rho N\rangle \tag{3.4}$$

This is illustrated in Fig. 3.3 for the π-N sector

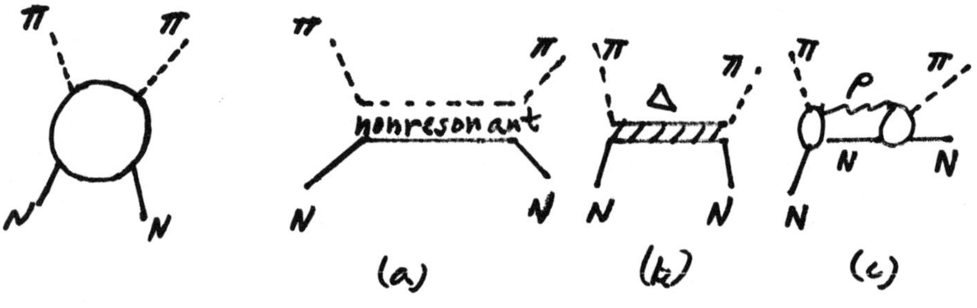

Fig. 3.3 Various Component of state with an incoming and outgoing pion and nucleon.

At the region of the Δ(1232), the contribution from inelastic states is very small, and will be neglected. It is convenient to use projection operators to define the state of Eq. 3.4. We define

$$\begin{aligned} &|\pi N, \text{nonresonant}\rangle \equiv |\pi N\rangle \equiv p|\Psi^{(\pi N)}\rangle \\ &|\Delta\rangle \equiv d\,|\Psi^{(\pi N)}\rangle \\ &p + d \approx 1 \text{ for } t_\pi < 300\text{MeV} \end{aligned} \tag{3.5}$$

We note also that asymptotically the system goes to the $|\pi N\rangle$ components

$$|\psi^{\pi N}\rangle \xrightarrow[t\to\pm\infty]{} P|\psi^{\pi N}\rangle_{\pm\infty} \tag{3.6}$$

Let us now examine the Schroedinger equation, within this projection operate formalism. This has been widely used for nuclear reactions.[8] For the time-independent treatment.

$$\begin{aligned} &(E-H)\,|\psi^{(\pi N)}\rangle = 0 \\ &|\psi^{(\pi N)}\rangle = p|\Psi^{(\pi N)}\rangle + d|\Psi^{(\pi N)}\rangle \end{aligned} \tag{3.7}$$

or

$$\begin{aligned} &(E - H_{pd})\,|\pi N\rangle = H_{pd}\,|\Delta\rangle \\ &(E - H_{dd})\,|\Delta = H_{dp}\,|\pi N \end{aligned} \tag{3.8}$$

where $H_{pp} = pHp$, $H_{dd} = dHd$, $H_{pd} = pHd$, and $H_{dp} = dHp$

Eliminating the $|\Delta\rangle$ components

$$(E-H_{pp})|\pi N\rangle = H_{pd}\,\frac{1}{E-H_{dd}+i\varepsilon}\,H_{dp}|\pi N\rangle \tag{3.9}$$

Making use of Eq. 3.6, i.e, that asymptotically one has only a pion and a nucleon, one can write down the π-N t-matrix.

$$\langle k'|\mathbf{t}_{\pi N}(\omega)|\vec{k}\rangle = \langle\vec{k}'|t_{\pi N}(\omega)^{N.R.}|\vec{k}\rangle + \langle\vec{k}'|t_{\pi N}(\omega)^{Res.}|\vec{k}\rangle \tag{3.10}$$

where the nonresonant part $t_N^{N.R.}$ arises from the interactions in H_{pp} and the resonant part is given by (define $V = H_{dp}$)

$$\langle\vec{k}'|t_{\pi N}(\omega)^{Res}|\vec{k}\rangle = \frac{\langle\pi N|V^{+}|\Delta\rangle\langle\Delta|V|\pi N\rangle}{\omega - M_\Delta + i\Gamma_\Delta/2} \tag{3.11}$$

where

$$M_\Delta + i\Gamma_\Delta/2 = \langle\Delta|H_{dd}|\Delta\rangle + \langle\Delta|V\,\frac{1}{E-H_{pp}+i\varepsilon}\,V^{+}|\Delta\rangle \tag{3.12}$$

Physically, M_Δ is the Mass and Γ_Δ is the width of the Δ in the quark model one could attempt to calculate M_Δ and Γ_Δ, say, by the interactions of the mechanism shown in Fig. 3.2. In any case we observe that for the ℓ=1, 3/2, 3/2 partial wave

$$\left(t_{\pi N}^{\text{Resonant}}(E)\right)_{\frac{3}{2}\frac{3}{2}} = \frac{\Gamma_\Delta(E)}{E-M_\Delta+i\Gamma_\Delta^{total}(E)/2}, \tag{3.11}$$

a typical Breit-Wigner form. We also note that

$$\Gamma_\Delta^{total}(E) \simeq \Gamma_\Delta(E), \tag{3.12}$$

for the Δ-region. Obviously, the partial widths Γ_Δ are closely related to the structure functions discussed in Sec. 2c.

4. RECOIL: RELATIVITY HALF-ASSUMED

The dynamic treatment of Sec. 2 was carried out in a static model. The interaction Hamiltonian is assumed to be

$$H_I = -g \, \bar{\Psi}_N^{+} \, \vec{\sigma}\cdot\nabla\vec{\tau}\cdot\phi\Psi_N \, . \tag{4.1}$$

This form is the limit neglecting nuclear motion of the pseudo-scalar interaction

$$\mathcal{L}^{P.S.} = - i \, g_{\pi N}\bar{\Psi}_N^{+} \, \gamma_5\vec{\tau}\cdot\phi_\pi\Psi_N \tag{4.2}$$

or the pseudovector interaction

$$\mathcal{L}^{P.V.} = g'_{\pi N}\bar{\Psi}_N \, \gamma_5\gamma_\mu \, \vec{\tau}\cdot\partial_\mu\phi_\pi\Psi_N \tag{4.3}$$

of a pseudoscalar pion with a nucleon. We wish to look at the limiting process to reach Eq. 4.1 from Eq. (4.2) or (4.3). In order to do this we need to use Dirac spinors. Those readers familiar with the Dirac equation can go immediately to Sec. b.

a) Dirac Equation, Dirac Matrices, and Spinors

Recall the nonrelativistic treatment of spin. E.g, if the potential has a spin-orbit force such as shown in Eq. 2.12. The Schroedinger equation is

$$(- \frac{1}{2\mu} \nabla^2 + V_{central}(r) + V_{s.o.}(r)\sigma\cdot L) \; \chi = E\chi \tag{4.4}$$

The wave function in Eq. 4.4 is a Paul spinor, a two-component quantity

$$\chi = \begin{pmatrix} \chi^+ \\ \chi^- \end{pmatrix} \, , \tag{4.5}$$

so that Eq. 4.4 decomposes into a $j = \ell \pm 1/2$ component. From Eq. 2.14 it follows that

$$(- \frac{1}{2\mu}\nabla^2 + V_{central} + V_{s.o.}) \quad \chi^+(\vec{r}) = E\chi^+(\vec{r})$$

$$(-\frac{1}{2\mu}\nabla^2 + V_{central} - 2V_{s.o.}) \quad \chi^-(\vec{r}) = E\chi^-(\vec{r})$$

More generally there are potential operators such as $\sigma \cdot B$, where B is the magnetic field, which couple the $\chi^\pm$ solutions.

The Dirac equations is a 4-component equation, and Dirac spinors are 4-component quantities. The Dirac equation for a free (noninteracting) Fermion is

$$(\gamma^\circ E - \vec{\gamma}\cdot\vec{p} - m)\psi = 0, \tag{4.6}$$

where in a frequently used representation[1)] the 4 x 4 <u>Dirac</u> matrices are

$$\gamma^\circ = \begin{pmatrix} 1 & 0 \\ 0 & -1 \end{pmatrix}, \; \gamma^i = \begin{pmatrix} 0 & \sigma_i \\ -\sigma_i & 0 \end{pmatrix}, \tag{4.7}$$

where the σ_i are the ordinary 2 x 2 Pauli spin matrices and 1 is the 2 x 2 unit matrix. The γ_5 operator is defined as

$$\gamma_5 = i\gamma^\circ\gamma'\gamma^2\gamma^3 = \begin{pmatrix} 0 & 1 \\ 1 & 0 \end{pmatrix}. \tag{4.8}$$

The wave function is often usefully treated by partitioning all of the Dirac 4 x 4 matrices in to 2 x 2 matrices. This is accomplished by writing

$$\psi = \begin{pmatrix} u \\ v \end{pmatrix}, \tag{4.9}$$

where u and v are each 2-component Pauli spinors. Then for the free Dirac particle (Eq. 4.6) one has

$$\begin{aligned} (E - m)u &= \vec{\sigma}\cdot\vec{p}\; v \\ (E + m)v &= \vec{\sigma}\cdot\vec{p}\; u \end{aligned} \tag{4.10}$$

Substituiting for $v = \vec{\sigma}\cdot\vec{p}\; u/(E + m)$ and $u = \vec{\sigma}\cdot\vec{p}\; v/(E - m)$, and using $\sigma\cdot p\; \sigma\cdot p = p^2$ one has

$$\begin{aligned} (E^2 - m^2 - p^2)u &= 0 \\ (E^2 - m^2 - p^2)v &= 0 \end{aligned} \tag{4.11}$$

as expected. From Eq. 4.11 one sees that for a free fermion of momentum p, the energy is

$$E = \pm \sqrt{p^2+m^2}$$

For a complete set of solutions the negative energy solutions must be included. These are most satisfactorily interpreted as anti-fermion solutions - predicted by the Dirac equation and confirmed for all known fermions, a triumph for theoretical physics.

In the presence of a potential the Dirac equation becomes (with V transforming as the fourth component of a 4-vector, like E)

$$(\gamma^\circ \ (E-V) \ - \vec{\gamma}\cdot\vec{p}-m)\psi = 0 \tag{4.12}$$

Using the same decomposition as before with Eq. 4.7, 4.9 one can show that

$$\begin{aligned} u &= \frac{1}{E-v-m} \vec{\sigma}\cdot\vec{p}v \\ v &= \frac{1}{E+m-V} \vec{\sigma}\cdot\vec{p}u \end{aligned} \tag{4.13}$$

For $V \ll m$ and $p^2 \ll m^2$, $E \approx m$, and it is apparent that u tends to be much larger than v. In fact, u is referred to as the large component and v as the small component of the wave function.

The adjoint spinor solution to the Dirac equation are defined by

$$\Psi = \psi^+\gamma^\circ \tag{4.14a}$$

where Ψ^+ is hermetian adjoint. They satisfies the adjoint equation

$$\Psi(\gamma^\circ E-\vec{\gamma}\cdot\vec{p}-m) = 0 \tag{4.14b}$$

Another useful quantity is the Dirac propagator. The Dirac equation with an interaction, Eq. 4.12, is

$$(\gamma^0 P_0-\vec{\gamma}\cdot\vec{P}-m)\psi = \gamma^\circ V\psi,$$

where we introduce the forth component of momentum $P_0=E$. This can be written as an integral equation, say, for the scattering problem ($\not{p}= \gamma^0 p_0 -\vec{\gamma}\cdot\vec{p}$)

$$\psi=\psi_0 + \frac{1}{\not{p} - m+i\varepsilon} \gamma^0 V\psi$$

or

$$\psi = \psi_o + G_o^{Dirac} \gamma^o V \psi , \tag{4.15a}$$

where

$$G_o^{Dirac} = \frac{1}{\not{p}-m} \equiv \frac{1}{\gamma^o p_o - \vec{\gamma}\cdot\vec{p} - m} \tag{4.15b}$$

is the free propagator for a Dirac particle (a fermion), just as $(q^2-m^2)^{-1}$ is the propagator for a boson. Comparing the Lippmann-Schwinger equation to Eq. 4.15, it is obvious that G_o^{Dirac} has the physical interpretation of a propagator.

The Dirac equation has many interesting properties, and naturally leads us into field theory, since with any interaction there is a possibility of particle-antiparticle pair creation. However, we shall only need the properties discussed in this section.

b) The Treatment of Recoil

Returning to the treatment of the π-N vertex function, let us consider the p.s. interaction of Eq. 4.2. Assuming that the nucleons are solutions to the Dirac equation, and taking matrix elements of the interaction Hamiltonian 4.2

$$\langle \psi_f(p_f)\pi(q) | \mathcal{L}^{p.s.} | \psi_i(p_i)\rangle \sim \bar{\psi}_f\gamma_5\psi_i \tag{4.16}$$

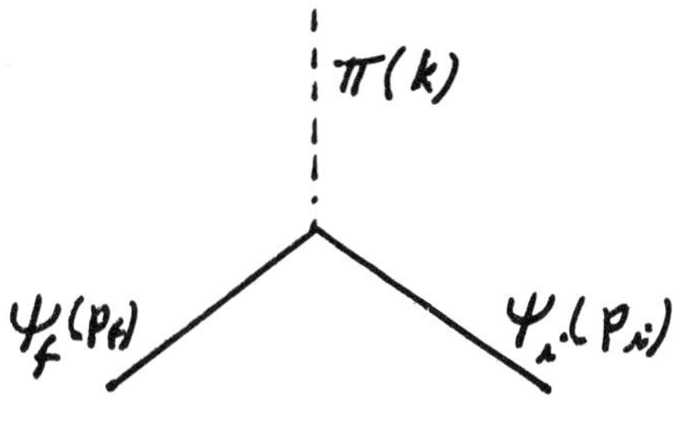

Fig. 4.1 π-N-N Vertex Function

In order to evaluate this vertex for nuclear physics, one must reduce the quantity on the right hand side of Eq. 4.16 to Pauli spinors, since only nonrelativist nuclear wave functions are available. This has been studied using a Foldy-Wouthuysen transformation.[9] The papers most closely related to my discussion are those Miller and Bolsterli et.al,[10] as well as the comment on Eisenberg et.al.[11] These references contain references to other work. Let us consider first the approximation in which the nucleons are momentum eigenstates (free particles), satisfying Eq. 4.6. Noting that

$$[\bar{\Psi}_f\gamma_5\psi_i]^{\text{free}} = (u_f^+ \ v_f^+)\,\gamma^\circ\gamma_5 \begin{pmatrix} u_i \\ v_i \end{pmatrix} = u_f^+ v_i - v_f^+ u_i, \tag{4.17}$$

and using 4.10, we find that

$$[\Psi_f\gamma_5\psi_i]^{\text{free}} = \frac{1}{E+M}\, u_f^+(\vec{\sigma}\cdot\vec{k})u_i \, 1 \, \frac{1}{2M}\, u_f^+\sigma\cdot k u_i \ , \tag{4.18}$$

which gives the reduction of the relativistic p.s. interaction of Eq. 4.2 to the Chew-Low form

$$ig_{\pi N} \ \Psi_N\gamma_5\vec{\tau}\cdot\phi_\pi\Psi_N \rightarrow \frac{g_{\pi N}}{2M}\,\psi_N^+ \ \sigma\cdot\nabla\tau\cdot\phi_\pi\psi_N \tag{4.19}$$

To find corrections to the result given by Eq. 4.18, one must consider further interactions. Of course, free nucleons cannot absorb or emit pions, so that further interactions must be considerered in any case. Returning to the general form of the matrix element of γ_5 given in Eq. 4.17 and using Eq. 4.13 one finds

$$\Psi_f\gamma_5\psi_i = u_f^+ \left[\frac{1}{E+m-v} \ \sigma\cdot\vec{p} - \vec{\sigma}\cdot\vec{p}\,\frac{1}{E-m-v}\right] u_i \tag{4.20}$$

For low momentum transfer, $\frac{|p_f-p_i|}{M} \ll 1$, and nonrelativistic motion of the nucleons

$$vu_i \simeq (E_i - m - \frac{P_i^2}{2m})u_i \ ,$$

one can expand in $v/2m$, $P_i/2m$, and $p_f/2m$ to find

$$H_I \approx \frac{g_{\pi N}}{2M}\ \psi_f^+\sigma\cdot[\vec{k}_\pi - \frac{m_\pi}{2M}(\vec{p}_i+\vec{p}_f)] \tag{4.21}$$

Eisenberg et.al. take a different approach and detain a different result in some cases and conclude that the approach of Bolsterli et.al. is unreliable. Also, Barnhill finds that the coefficient of the recoil correction, the $\sigma\cdot(\vec{p}_f+\vec{p}_i)$ term, is arbitrary.

Since, as we have just seen, the $\sigma\cdot\nabla_\pi$ form follows for a pion being created or destroyed on a free nucleon, the question of the correct operator to use for pion absorption or creation on a nucleon in a nucleus is related to the dynamics of the nucleus. We can summarize the present status of the problem by suggesting that the operator

$$H_I = \frac{g_N}{2M} \psi_f^+ \sigma\cdot[\nabla_\pi - \alpha\nabla_N] \tau\cdot\phi_\pi\psi_i , \qquad (4.22)$$

where the best guess for α is $m_\pi/2M$, is the statest form to use.

As we will se in the next sections, the proper way to ask these questions is: what is the off-mass-shell behavior of the π-N vertex function?

5. SOME RELATIVISTIC METHODS

In this section I discuss some methods which have been and are being used in pion-nucleus physics to try to account for the relativistic and field theoretical features. The method which most nearly resembles the Schroedinger Equation approach in its formal structure is the Bethe-Salpeter equation, which I discuss in secs. a and b. I also sketch the relativistic generalization of the Chew-Low Theory in sec. c. However I do not discuss one of the most interesting and important approaches,that of dispersion relations. The derivation of the dispersion relations from axiomatic field theory involves developing far too much apparatus for the time allowed. For the same reason, I shall not give the field theory derivation of the low equation in Sec. C, but rely on a heuristic approach. In sec. d I discuss the off-mass-shell pion-nucleon amplitude. Also, because of its complexity and because of our limited time, I shall not treat the PCAC method. I should also mention that for a truly satisfactory discussion of the topics in this chapter, it is essential to discuss the analytic properties of some of the structures entering into the theory, and as with the topic of dispersion relations it is not realistic to attempt this here. Therefore my treatment is incomplete, and stresses results and methods useful to pion-nuclear physics.

a) Variables for the Two-Body Problem

a.1) Nonrelativistic t-matrix. The T-matrix for two-particle scattering at first seems to be a function of 13 variables, two initial momenta, two final momenta, and the energy. Yet we know that the on-shell scattering amplitude is a function of only two variables (we are neglecting spin here).

$$\langle k_1', k_2' | T(E) | k_1, k_2 \rangle_{\text{on-shell}} = T(E, \theta), \tag{5.1}$$

the energy and scattering angle. It is the various symmetries of the scattering amplitude which reduce the number of variables. Eq. 5.1 can most easily be seen by recognizing that one can go to the center of mass system of the initial and final state simultaneously (total momentum is conserved), $\vec{k}_1+\vec{k}_2 = \vec{k}_1'+\vec{k}_2' = 0$, that the magnitudes of k_1, k_2, k_1', and k_2' are determined by the energy, and that from rotational invariance the amplitude only depends on the relative orientation of the initial and final momenta. This is usually expressed by writing T in the center of mass system as

$$\langle T \rangle^{\text{on-shell}} = \langle \vec{k}' | T(E) | \vec{k} \rangle^{\text{on-shell}} \tag{5.2a}$$

with

$$E = \sqrt{k^2+m_1^2} + \sqrt{k^2+m_2^2} = \sqrt{k'^2+m_{1'}^2} + \sqrt{k'^2+m_{2'}^2}$$

for the on-shell T-matrix, where m_1, m_2 and $m_{1'}$, $m_{2'}$ are the initial and final states, respectively.

For the scattering from a bound system one needs the off-shell amplitude. This differs from (5.2) in that energy is not conserved during the interaction - the scattering takes place in a finite time interval, so that energy is not a good quantum number if one follows the individual collisions using a nonrelativsitic prescription. This the t-matrix has the form

$$\langle T \rangle^{\text{off-shell (N.R.)}} = \langle \vec{k}' | T(E) | \vec{k} \rangle = T(E, \theta, |k|, |k'|). \quad (5.3)$$

In other words, the relative momenta are variables, and T is a function of four variables. Note that it is not a function of seven variables ($\vec{k}$, $\vec{k}'$, E), as rotational invariance eliminates the dependence on all but one angle. (Again, we are neglecting spin dependence and assuming that $V(\vec{r}) = V(r)$.)

a.2) Relativistic t-matrix. From the same arguments given in the sentences first preceeding and first following Eq. (5.1), it must also be true (neglecting spin) that the on-shell t-matrix in a relativistic treatment is a function of only two variables.

$$\langle |T| \rangle^{\text{relativistic, on-shell}} = T(E,\theta). \quad (5.4)$$

To best utilize the relativistic transformation properties, it is often very advantageous to introduce variables which are themselves relativistic invariants. Consider the scattering problem

$$a(k_1) + b(k_2) \rightarrow c(k_1') + d(k_2'),$$

where $k_1 = (k_{1_0}, \vec{k}_1)$, etc. are four vectors for the incoming and outgoing particles, satisfying

$$k_1^2 = k_{1_0}^2 - \vec{k}_1 \cdot \vec{k}_1 = m_1^2, \text{ etc}$$

$$k_1 + k_2 = k_1' + k_2' \text{ (four equations)} \quad (5.5)$$

The invariant [Mandelstam] variables are defined by

$$\begin{aligned} s &= (k_1+k_2)^2 = (k_1'+k_2')^2 \\ &= (\text{Total Energy in C.M})^2 \\ t &= (k_1-k_1')^2 - (k_2 - k_2')^2 \\ u &= (k_1-k_2')^2 = \Sigma m_i^2 - s - t. \end{aligned} \quad (5.6)$$

One can write

$$\langle |T| \rangle^{\text{on shell}} = T(s,t)$$

The relativistic concept of "off-shell is "off-mass-shell". For interacting particles the relation $k_i^2=m_i^2$ no longer holds. In other words, the energy of the ith particle, k_{i_o}, and the momentum, $\vec{k}_i$, of that particle of rest mass m_i do not satisfy the on-shell relationship; i.e.,

$$k_{i_o} \neq \sqrt{\vec{k}_i \cdot \vec{k}_i + m_i^2} \quad \text{(off-mass-shell)}.$$

Therefore, the T-matrix becomes a function of the effective masses of the interacting particles

$$\langle |T| \rangle^{\text{off-mass-shell}} = T(s,t,k_1^2,k_2^2,k_3^2,k_4^2) \tag{5.7}$$

Eq. (5.7) gives the most general off-shell amplitude which we must consider for the two-body problem, neglecting spin.

a.3) Transformation Properties of the T-matrix. It should be mentioned that one of the most pressing reasons for developing a relativistic theory is the ambiguity in defining the T-matrix in the pion-nucleus center-of-mass system if one uses nonrelativistic π-N T-matrices. With the standard definition of the T-matrix, the quantity

$$[k'_{1_o} k'_{2_o} k_{1_o} k_{2_o}]^{1/2} \langle k_1 k_2 |T| k_1 k_2 \rangle \tag{5.8}$$

is an invariant form. I.e., in another coordinate system the $k_i=(k_i^\circ,\vec{k}_i)$ take on new values, but the expression (5.8) with the new values is invariant. This, however, is inconsistent with a nonrelativistic treatment. The consequences of this problem for the pion-nucleus problem have been discussed by Dr. Gibbs. I shall not discuss it further.

b) The Bethe-Salpeter Equation

b.1) The Nonrelativistic Treatment of Scattering. Here we review the definition of the T-matrix and the Born expansion as an aid to understanding the relativistic treatment. The S-matrix for potential scattering of a particle with initial, final momenta $\vec{k},\vec{k}'$ is defined as

$$\langle k'|\,S\,|\vec{k}\rangle = (\Psi^{(-)}_{\vec{k}}, |\Psi^{(+)}_{\vec{k}}), \tag{5.9}$$

where $\Psi^{(+)}_{\vec{k}}$, $\Psi^{(-)}_{\vec{k}}$ are the solution to the Schroedinger (or Lippman-Schwinger) equation with outgoing, incoming wave boundary conditions, respectively. The partial wave S-matrix element is $S_\ell(k) = e^{2i\delta_\ell(k)}$. The T-matrix is defined by

$$\begin{aligned}\langle\vec{k}|S|\vec{k}\rangle &= \delta(\vec{k}'-\vec{k})-2\pi i\;\delta(E_{k'}-E_k)\;\langle\vec{k}'|T|\vec{k}\rangle \\ \langle\vec{k}'|T|\vec{k}\rangle &= \langle\vec{k}|V|\Psi^{(+)}_{\vec{k}}\rangle = \langle\Psi^{(-)}_{\vec{k}}|V|\vec{k}\rangle\end{aligned} \tag{5.10}$$

The Lippman-Schwinger equation (nonrelativstic) for the T-T-matrix is

$$T = V + V\,\frac{1}{E-H_o \mp i\varepsilon}\,T, \tag{5.11}$$

which has the Born expansion

$$T = V + VG_oV+VG_oV\;G_oV+\;\ldots\;, \tag{5.12}$$

where

$$G_o = \frac{1}{E-H_o+i\varepsilon} \tag{5.13}$$

is the nonrelativistic propagator or Green's function. Fig. 4.2 gives a pictoral representation of this series.

For a velocity independent local potential

$$\langle x_1'x_2'|V|x_1x_2\rangle = V(x)\quad\delta(\vec{x}-\vec{x}'), \tag{5.14}$$

where $\vec{x} = \vec{x}_2-\vec{x}_1$, $\vec{x}' = \vec{x}_2'-\vec{x}_1'$, are the relative variables. Here the T-matrix in coordinate space satisfies

$$\begin{aligned}\langle x'|T|x\rangle &= V(x)\,\delta(\vec{x}-\vec{x}')+\;V(x)G_o(\vec{x}-\vec{x}')V(x') \\ &+\int d^3x''V(x')G_o(\vec{x}'-\vec{x}^\circ)V(x'')G_o(\vec{x}''-\vec{x})V(x) \\ &+\;\ldots\end{aligned} \tag{5.15}$$

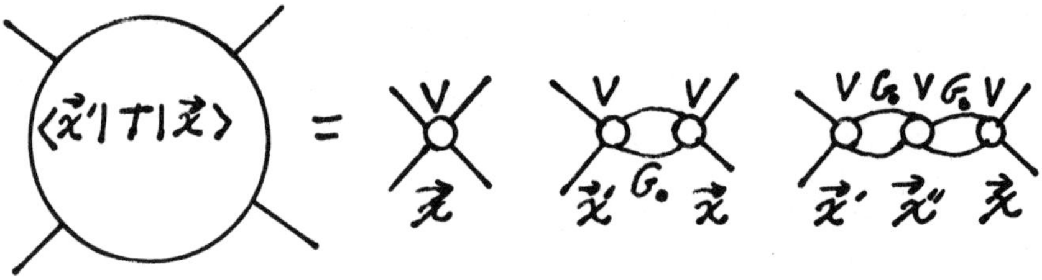

Fig. 5.1 Born Expansion of T-Matrix for Potential Scattering

Thus the propagator takes the system from one interaction point to the next, with suitable integrals.

b.2) <u>A Relativistic Treatment of Scattering</u>: <u>The Bethe-Salpeter Equation</u>. The relativistic definition of the S-matrix for the scattering processes $a(k_1)+k(k_2) \rightarrow c(k_1')+d(k_2')$ is formally the same as the nonrelativistic:

$$\langle k_1'k_2'|S|k_1k_2\rangle = \langle k_1'k_2';\text{out}|k_1,k_2;\text{in}\rangle$$

$$= \langle k_1'k_2';\text{out}|k_1,k_2;\text{out}\rangle$$

$$+ 2\pi i\ \delta^{(4)}(k_1'+k_2'-k_1-k_2)\frac{\langle k_1'k_2'|T|k_1k_2\rangle}{\sqrt{16k_{1_0}k_{2_0}k_{1_0}'k_{2_0}'}}, \qquad (5.16)$$

where the k are all 4-vectors. The formal definition of the in and out states is a bit involved, and we shall not carry out any out the formal manipulation involved with the formal field theory treatment.

The S-matrix $\langle k_1'k_2'|S|k_1k_2\rangle$ is closely related to the two-particle propagator, with which one works in the Bethe-Salpeter method. The two-particle propagator is defined as

$$G^{(2)}(x_1'x_2',x_1x_2) = \langle 0|T\{\psi_c(x_1')\psi_d(x_2')\psi_a^+(x_1)\psi_b^+(x_2)\}|0\rangle \qquad (5.17)$$

with $|0\rangle$ the vacuum, ψ the field operators and where T {} arranges the operators in order of increasing time from right to left. For the scattering problem

$$t_1', t_2' > t_1, t_2$$

$$G^{(2)}(x_1'x_2';x_1x_2) = \langle 0|T\{\psi_c(x_1')\psi_d(x_2')\}T\{\psi(x_1)\psi(x_2)\}|0\rangle \qquad (5.18)$$

Inserting a complete set of states, one has

$$G^{(2)}(x'x_2';x_1x_2) = \sum_n \langle 0|T\{\psi(x_1')\psi(x_2')|n\rangle \langle n|T\{\psi(x_1)\psi(x_2)|0\rangle \qquad (5.19)$$

The quantities $\langle 0|T\{\psi(x_1)\ \psi(x_2)|n\rangle$ are referred to as the Bethe-Salpeter amplitudes. They are analogous to wave functions in the nonrelativistic theory. If one includes only the states equivalent to "ladder diagrams", he obtains a relativistic extension of the Lippman-Schwinger integral equation.

For example, let us consider the scattering of identical scalar mesons with the interaction arising from the exchange of the same type of meson. This is referred to as the ϕ^3 theory. The zeroth order process just consist of the unlinked propagation of two mesons. The first order process consits of meson exchange. The ladder is sum of all the meson exchange processes without crossing lines as shown in Fig. 5.2. The first order diagram (Fig. 5.2b) defines the potential. In coordinate space

$$V(x-x') = g^2\Delta_f(x'-x) = g^2\frac{1}{(2\pi)^2}\int d^4q\frac{e^{q\cdot(x'-x)}}{q^2-\mu^2}, \qquad (5.20)$$

where $\Delta_f(x'-x)$ is the propagator of a meson with mass μ. The sum of the ladder diagrams is obtained from the integral equation for the two-particle propagator,

$$G^{(2)}(x_1'x_2';x_1,x_2) = \Delta_f(x_1-x_1')\ \Delta_f(x_2-x_2') + g^2\iint d^3x_1''d^3x_2''\ \Delta_f(x_1'-x_1'')\ \Delta_f(x_2'-x_2'') \times\Delta_f(x_1''-x_2'')\ G^2(x_1''x_2'';x_1x_2). \qquad (5.21)$$

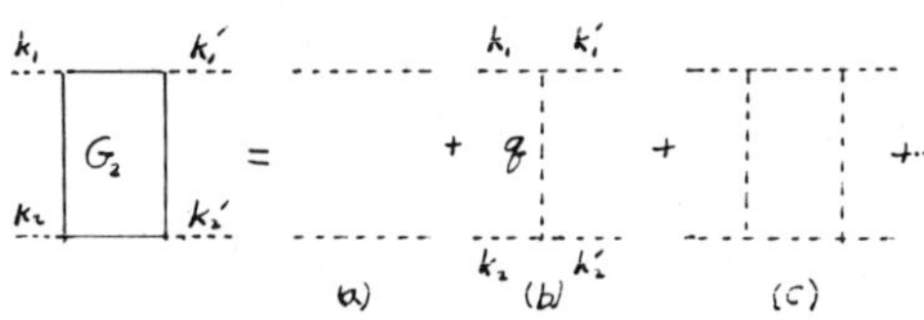

Fig. 5.2 The Ladder Diagrams for Meson Exchange in the ϕ^3 Theory

In terms of the free two-particle propagator

$$\langle x_1' x_2' | G_o | x_1 x_2 \rangle = \Delta_f(x_1 - x_1')\, \Delta_f(x_2 - x_2') \tag{5.22}$$

and the potential defined in Eq. 5.20, Eq. 5.21 is

$$G^{(2)} = G_o + G_o V G^{(2)}. \tag{5.23}$$

One can relate $G^{(2)}$ to the T-matrix with Eq. (5.16). This gives the integral equation for the T-matrix

$$T = V + VG_oT, \tag{5.24}$$

with G_o and V defined in Eqs. 5.22 and 5.20. In momentum space, going to the center-of-mass system (total momentum = 0) we use the coordinates

$$k = (k_1 - k_2)/2$$
$$k' = (k_1' - k_2')/2$$
$$p = k_1 + k_2 = k_1' + k_2'$$
$$s = P^2$$
$$k_1 = k + P/2 \qquad k_1' = k' + P/2$$
$$k_2 = -k + P/2 \qquad k_2' = k' + P/2,$$

Eq. 5.24 becomes

$$\langle k' | T\,(s) | k \rangle = \frac{g^2}{(k'-k)^2 - m^2 + i\varepsilon}$$
$$-i \frac{g^2}{(2\pi)^4} \int d^2k'' \, \frac{1}{(k''-k)^2 - m^2 + i\varepsilon} \qquad \frac{1}{(k''+P/2)^2 - m^2 + i\varepsilon} \tag{5.25}$$
$$\times \frac{1}{(k''-P/2)^2 - m^2 + i\varepsilon} \langle k'' | T(P) | k \rangle$$

The expansion for T is pictured in Fig. 5.3

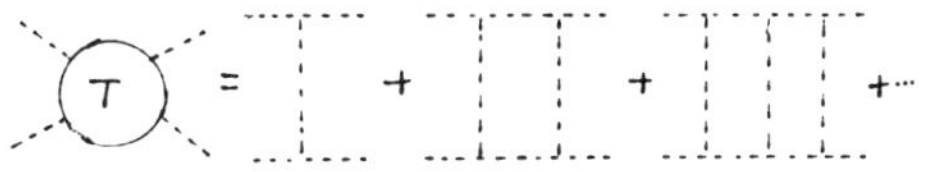

Fig. 5.3 Expansion of the T-matrix in the ϕ^3 Theory

Introducing the partial wave expansion in terms of the scattering angle $[\cos\theta = \vec{k}\cdot\vec{k}'/|\vec{k}||\vec{k}'|]$,

$$\langle k'|T(s)|k\rangle = [2|k||k'|]_{\ell}^{-1}(2\ell+1)\ P_{\ell}(\cos\theta) \times T_{\ell}(|\vec{k}|,\omega,|\vec{k}'|,\omega' s), \tag{5.26}$$

and making use of the relation between Legendre function of the first and second kind

$$Q_{\ell}(1+\frac{t}{2s}) = s\int_{-1}^{1} dz\ P_{\ell}(z)/[t+2s(1-z)],$$

one has the integral equation for the partial wave amplitude

$$T_{\ell}(|\vec{k}|,\omega'|\vec{k}'|,\omega';s) = V_{\ell}(|\vec{k}|,\omega'|\vec{k}'|,\omega') - \frac{i}{(2\pi)^3}\int_{-0}^{\infty} d|\vec{k}''|\int_{-\infty}^{\infty} d\omega''\ V_{\ell}(|\vec{k}|,\omega'|\vec{k}''|,\omega'')\ G_0(|\vec{k}''|,\omega'',s) \times T_{\ell}(|k''|,\omega'';|k'|,\omega';s), \tag{5.27a}$$

where

$$V_{\ell}(|k|,\omega;|k'|,\omega') = \frac{g^2}{(2\pi)^3}\, Q_{\ell}\left(\frac{|\vec{k}|^2+|\vec{k}'|^2-(\omega-\omega')^2+m^2-i\epsilon}{2|\vec{k}||\vec{k}'|}\right) \tag{5.27b}$$

$$G_0(|\vec{k}''|,\omega'',s) = \{[|\vec{k}|^2-(\omega+\frac{\sqrt{s}}{2})^2+m^2-i\varepsilon][|\vec{k}|^2-(\omega-\frac{\sqrt{s}}{2})^2+m^2-i\varepsilon]\}^{-1}$$

Obviously, Eq. (5.27) is more complicated than the corresponding integral equation for the partial wave amplitude in the Schroedinger theory, in that these is an integral one the energy variable $\omega''=k_0''$, the fourth component of the momentum variable of interaction. The variables ω and ω' are the initial and final relative energy variables, so $\omega = \omega' = 0$ for the on=shell amplitude for the equal-mass case which is being treated. The problem is far worse than simply carrying out an extra integration as one can see from Eq. 5.27 defining G_o and V_ℓ. There are four poles in the propagator G_o, and four branch points in the potential V_ℓ (recall that the Legendre functions $Q_\ell(z)$ have branch outs starting at branch points $z=\pm 1$). The singularities are most troublesome in that a pinch of the contour of integration occurs.

A number of people have worked on the ϕ^3 theory for the scattering problem[12]. Although the realistic cases needed for pion-nucleus physics have not yet been treated in a satisfactory way, we can learn a great deal from the present work. We shall return to this below.

I should be mentioned that a method has been developed by Blankenbecler and Sugar which uses a three-dimensional integral equation, but can include effects of inelasticity and relativity in an orderly way. The Green's function is approximated by

$$G_o(|k''|,\omega'',s) \simeq g_o(k'',s)\delta(\omega''-\alpha). \tag{5.28}$$

This procedure replaces the Bethe-Salpeter integral equation of 5.27 by a three-dimensional one, so that one has essentially returned to the Lippman-Schwinger equation. This method has been applied to the π-N problem using the u-channel nucleon pole diagram of Fig. 2.6b as the during term (a nucleon exchange potential).[14] With suitable cutoffs one can obtain the $\Delta(1232)$ and other N* resonances. However, it is difficult to apply relativity "half-way" to the nuclear problems. See also Cpt. 6.

c) The Bethe-Salpeter Equation for the Bound State

The existence of bound states can easily be handled by the methods described in the previous section. The bound states for the ϕ^3 theory were first treated by Cutkosky and more general cases have been considered by others.[15]

Returning to Eq. (5.19) for the two-particle propagator, the existence of a bound state means that there is a state in the intermediate spectrum at fixed total energy.

$$\langle x_1'x_2'|G^{(2)}|x_1x_2\rangle \xrightarrow[P_o \simeq \sqrt{|\vec{P}|^2+M_b^2}]{} \frac{\chi_b^+(x_1'x_2')\chi_b(x_1x_2)}{P_o-\sqrt{|\vec{P}|^2+M_b^2}} \qquad (5.29)$$

By picking out this pole in Eq. 5.21, one can obtain an integral equation (or alternatively a differential or differential-integral equation) for the B-S amplitude $\chi_b(x_1x_2)$. For the case of two scalar mesons of mass m interacting via the exchange of a scalar meson of mass μ, one finds

$$(\Box_{x_1}+m^2)\ (\Box_{x_2}+m\ )\ \chi_d(x_1x_2^2) = g^2\Delta_f(x_1-x_2)\ \chi_d(x_1x_2), \qquad (5.30)$$

where $\Box_{x_1} = \frac{2^2}{2t^2} -\nabla^2$ and Δ_f is defined in Eq.(5.20).

The differential operators which appear in Eq. 5.30 come from the free two-particle propagator, G_o, and Eq. 5.30 can be defined by multiplying both sides of Eq. 5.21 by $(\Box_{x_1}+m^2)\ (\Box_{x_2}+m^2)$ and going to the pole.

In a similar way one can obtain the differential equation for a π-N bound state using the 6-model of sec. 2.a. The Free Green's function now consists of a product of a pion and a nucleon propagator.

$$G_o \sim \frac{1}{q^2-\mu^2+i\varepsilon} \quad \frac{1}{P_o\gamma^o-\vec{P}\cdot\vec{\gamma}-M+i\varepsilon} \qquad (5.31)$$

The same derivation discussed above gives for the differential equation

$$(\Box_{x_1}+m^2)\ (i\gamma^o\frac{2}{2t}+i\gamma\cdot\nabla-M)\ \Psi_b(x_1,x_2)=g^2\Delta_f(x_1-x_2)\Psi_b(x_1,x_2). \qquad (5.32)$$

This equation for the interaction of a fermion and a boson is being treated for the π^-P atom by R. Kwon, F. Tabakin, and myself. For the deuteron one would have an equation similar to (5.32) but with two Dirac Hamiltonian operators, rather than one Klein-Gordon and one Dirac Hamiltonian. As we shall see in Cpt. 6, there is a crutial need for the knowledge of solutions of such equations, and I expect that there will be progress in the next few years.

d) The π-N Low Equation

The S-matrix can be treated in a more nearly general way than the B-S approach discussed above. The B-S approach is accurate if one or a few irreducible diagrams (the driving terms) are adequate, for it just does the ladder sum. As was stressed in Cpt. 2, the potential approach to the π-N or π-nucleus problem does not follow from the microscopic theory in so far as meson exchange does not give an adequate description of medium energy π-N scattering.

Returning to the definition of the S-matrix (Eq. 5.16) for the π-N problem one can make use of

$$\begin{aligned} &\langle \pi(k')N(p')|S|\pi(k)N(p)\rangle \\ &= \langle \pi(k'),N(p'),\text{out}|\pi(k),N(p),\text{in}\rangle \\ &= \langle \pi(k'),N(p'),\text{out}|b_k^{+\text{in}}|N(p)\rangle \quad , \end{aligned} \tag{5.33}$$

where $b_k^{+\text{in}}$ is the creation operator of an incoming pion. It is defined in terms of an "in-field", $\phi^{\text{in}}(x)$, for the pion by a relationship similar to Eq. 2.18. The standard approach makes use of axioms relating the "in" fields to the "out" fields by the S-matrix using an interpolating field ϕ_π for the pion to relate asymptotic times $t \to \pm\infty$.

The asymptotic field can be used to define the equivalent of the interaction Hamiltonian used for the Chew-Low theory in Sec. 2a. One can show, using translational invariance and the axioms that

$$\begin{aligned} &\langle (k'),N(p'),\text{out}\ b_k^{+\text{in}}\ N(p)\rangle = \\ &2\pi i\delta^{(4)}(k'+p'-k-p)\ \frac{1}{\sqrt{2k_o'}}\ \langle \pi(k')N(p')|j(o)|N(p)\rangle \end{aligned} \tag{5.34}$$

or from Eqs. 5.16, 5.33 and 5.35

$$\langle k'p'|T(s)|kp\rangle \alpha \langle k'p'|j(o)|p\rangle ,$$

where j(x) is the nucleon source of pions

$$(\Box_x + m_\pi^2)\ \phi_\pi(x) = j_\pi(x).$$

For the Chew-Low theory of Sec. 2a

$$j_t(o) = ig\Psi_N^+(o)\vec{\sigma}\cdot\vec{k}\tau_t\Psi_N(o) \equiv V_{k,t}, \tag{5.36}$$

the operator which we have we used in Sec. 2.c. Carrying out the "reduction" technique outlined in Eqs. 5.33-5.35 one arrives at the Low equation

$$\langle k't',p's'\tau'|t(s)|kt,ps\tau\rangle = \sum_n \frac{\langle p'\tau'|V_{k't'}|n\rangle\langle n|V_{kt}|p\tau\rangle}{P_o+k_o - E_n+it} + \frac{\langle p'\tau'|V_{kt}|n\rangle\langle n|V_{k't'}|p\tau\rangle}{P_o-k_o - E_n+i\varepsilon} \tag{5.37}$$

where the sum is over all physical states which can be connected. The details of the derivation can be found, e.g., in Ref. 2 and 6.

There have been many approximate solutions to the Low equation published over the past two decades. I shall not review these here.

One recent formal development has been the deviation of a linear, Lippman-Schwinger type equation from the truncated Low equation, Eq. 5.37 limited to one-meson intermediate states.[16] It has the form

$$t(E) = v+v\frac{E}{H_o}\frac{1}{E-H_o+i\varepsilon}\frac{E}{H_o}\ t(E)\ , \tag{5.38}$$

where v is obtained by inserting the one-nucleon states for in> in the intermediate state sum Eq. 5.37. The main difference from the ordinary norelativistic treatment is the appearance of the operators $\frac{E}{H_o}$, which serve as rather long-range cutoffs for the interaction. This in turn is related to the effective $g_\ell(k)$, discussed in Cpt. 2, which are needed for the optical potential. We shall return to this for the π-nucleus optical potential in Sec. 6c.

e) The Off-Mass-Shell π-N Amplitude

The off-mass-shell π-N amplitude, defined in Eq. 5.8 can only be obtained from a dynamic theory. Experiments on the π-N scattering, regardless of their accuracy, range of energy, and so forth just give information about the on-mass-shell amplitude. One needs the basic interaction or its equivalent in order to extrapolate in the masses of the pions and nucleons. There have been only a few attempts to carry out such a program, including recent work on the S-wave amplitudes[3] and on the p-wave amplitudes within a Chew-Low type theory.[17]

I shall discuss some of the work carried out in the late 1960's based on the B-S equation and on the peripheral model for NN and π-N reactions. I believe that this is the most reliable treatment which we now have available. The model used[18] treats the off-mass-shell behavior of the π-N amplitudes at resonance. This is accomplished by deriving the off-shell behavior of the πNN* vertex function,

$$\langle \pi N | V | N^* \rangle \sim \Gamma(M^2, M_\pi^2, q^2), \tag{5.39a}$$

where on shell

$$\Gamma^{\text{on-shell}} = \Gamma(q^2). \tag{5.39b}$$

The on-shell value can be determined for physical q^2 from the width of the N^*, which is determined from scattering experiments.

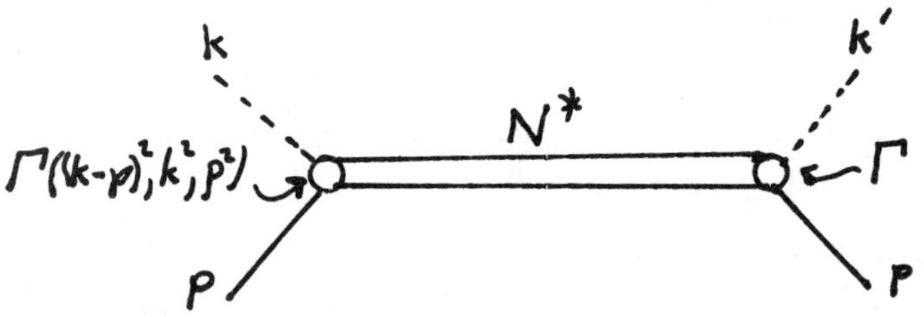

Fig. 5.3 The π-N amplitude for a N^* resonance and the off-mass shell π-N-N* vertex function

The occurrence of the resonance can be determined by the Fredholm method applied to Eq. (5.27). The integral equation for the partial wave amplitude

$$T_\ell = V_\ell + V_\ell G_o T_\ell, \tag{5.40}$$

has a Fredholm solution

$$T_\ell = V_\ell + \frac{V_\ell G_o V_\ell}{1-\mathrm{Tr}\{G_o V_\ell\}} \tag{5.41}$$

The resonance condition corresponds to the vanishing of the denominator. The resonance is made to occur at the correct energy by the choice of the coupling constant. This enables one to obtain an off-mass-shell extrapolation. See Ref. 18 for details.

Although the dynamics of Ref. 18 are not expected to be suitable to calculate the off-mass-shell behavior of the t-matrix reliably, it suggests functional forms which are reasonable. These have been used[19] for studying a variety of reactions which are interpreted as N^* dominated in the peripheral model. Typical reactions are shown in Fig. 5.4.

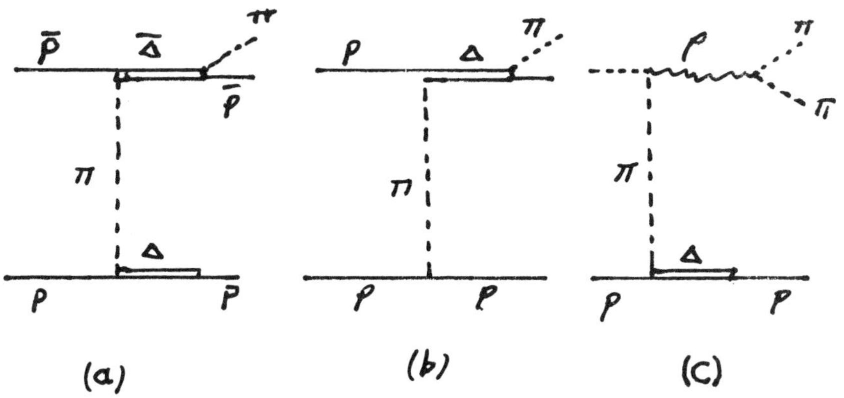

Fig. 5.4 Various reactions used with a peripheral model to determine the π-N off-shell t-matrix; a) pp→ππpp; b) pp→πpp; b) pp→πpp; a) πN→πππp.

The results are a phenomenological representation of the π-N off-mass-shell behavior which can be used for extrapolating in the mass of the pion. Refering to Sec. 5.a.1, the off-shell behavior is defined as

$$t_\ell^{\pi N}\ (s,k^2,\ p^2,\ k_1{}^2,p'^2) = t_\ell(s,k_1{}^2) \qquad (5.42)$$

since the nucleons and one of the pions are on mass shell (except for higher order [distortion] phenomena).

In the model used by Wolf[19] based on the B-S work one has the ansatz

$$t_\ell(s,k_1{}^2) \quad = \quad V_\ell(k_1{}^2)\ t_\ell(s) \qquad (5.43)$$

where t_ℓ (s) is the on-shell partial wave amplitude. The off-shell factor for a positive parity resonance with J-L+1/2 (e.g. the Δ(1232)) is

$$V_\ell(k_1{}^2) = \left[\frac{k}{k_r}\ \frac{(M_r+M)^2 - k_1{}^2}{(M_r+M)^2 - m_\pi{}^2}\right]^{\frac{1}{2}} \sqrt{\frac{u_\ell(k_rR)}{u_\ell(k\ R)}} \qquad (5.44)$$

with

$$u_\ell(kR) = \frac{1}{2(QR)^2}\ Q_\ell\ (1 + \frac{1}{2(kR)^2}), \qquad (5.45)$$

where k, k_r are the center of mass momentum at the energy s and at resonance energy $\sqrt{s_r} = M_r$. The Q_ℓ are the Legendre functions of the second kind defined earlier. R is a range parameter fit for each resonance. Values of this parameter are given in Ref. 19. Therefore, we have available a possible off-mass shell amplitude for use in pion=nuclear physics. The job is by no means complete, but this is a good start.

6. RECENT DEVELOPMENTS IN THE THEORY OF THE PION-NUCLEUS OPTICAL POTENTIAL

In this section I shall briefly review some of the current work in pion-nucleus physics which is related to the theoretical ideas of the previous chapters. This is in no sense a complete review, but is a sample restricted to considerations of the optical potential, which has been introduced by Dr. Gibbs, to illustrate how the concepts enter into nuclear physics. This is an active area of research, and one of the most stimulating in theoretical physics. I shall discuss the second order optical potential in Cpt. 8.

a) Multiple Scattering Theory

The multiple scattering approach for deriving the optical potential has as its basic ingredients the two-body T-matrix, and the nuclear density and nuclear correlation functions. In the impulse approximation, the first order optical potential is

$$\langle \vec{k}'|V^{opt}|\vec{k}\rangle \approx \langle k'|t^{\pi N}|k\rangle \rho(k'-k) \tag{6.1}$$

In the past several years there has been a large body of work giving corrections to the simple expression 6.1. The work which deals most directly with the relativistic considerations which I have discussed is that of Celenza, Liu, and Shakin.[20] They work with a potential of the same general type as that of Eq. 6.1, as in the conventional theory; however, they define it in terms of an off shell π-N amplitude. The basic correction which is involved in their attempt is a correction to the factorization approximation of Eq. 6.1, which has been shown to be in error by perhaps 20% at large angles.[21] Thus they carry out an integral over an internal nuclear momentum variable, Q,

$$\langle \vec{k}'|V^{opt}|\vec{k}\rangle = \int d^3Q \langle k',p-Q|T_{\pi N}(s)|k,p-q-Q\rangle \, F(P,Q), \tag{6.2}$$

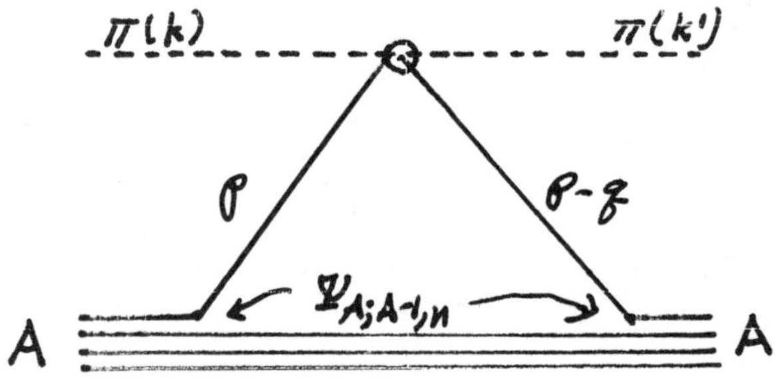

Fig. 6.1 Basic process for the first order optical potential.

where F(p,Q) is a nuclear form factor. In the formal expression of Eq. 6.2 one can allow the nucleon to go off-mass-shell, and thus attempt to introduce some of the general considerations discussed in Cpt. 5. However, to really introduce the nucleon mass as a new variable, one needs a relativistic extension of the nuclear wave function. Since this is not done, one does not have the general form for the off-mass-shell amplitude or the new variables of expected in the optical potential. Thus the expression 6.2 has as its main new element an improved treatment of the question of factorization with an improved kinematics.

The optical potential given in Eq. 6.2 is used in an integral of the Blankenbecler-Sugar type

$$\langle\vec{k}'|T|\vec{k}\rangle = \langle\vec{k}'|V|\vec{k}\rangle + \langle\vec{k}'|VG_0T|\vec{k}\rangle,$$

where G_0 is obtained by one of a variety of ansatz's of the form of Eq. 5.28. Therefore, although the equation formally has the proper relativistic transformation properties, one does not deal with the full complications of the off-mass-shell problem. The general problem is difficult, and will be an open area of research for some time.

There has been a good deal of attention paid to the higher order processes for the optical potential. I shall discuss the second order terms in Cpt. 8, emphasizing questions related to field theory and relativity. There has also been some interesting progress in studying higher order processes in a self-consistent formalism,[22] but since this is done in the approximation of fixed scattering centers it avoids most of the questions which I have been discussing.

b. The Isobar-Doorway Theory

The projection operator method which was used in Cpt. 3 to represent the isobar can also be applied for the pion scattering and reactions in a nucleus.[23] For the optical potential one has the general form

$$V^{opt} = V^{N.R.} + V^{Res} + V^{abs} . \tag{6.3}$$

The $V^{N.R.}$ is the optical potential one obtains from the nonresonant part of $t^{\pi N}$. The absorptive potential, V^{abs}, contains contributions to elastic scattering from processes with no-pion intermediate states

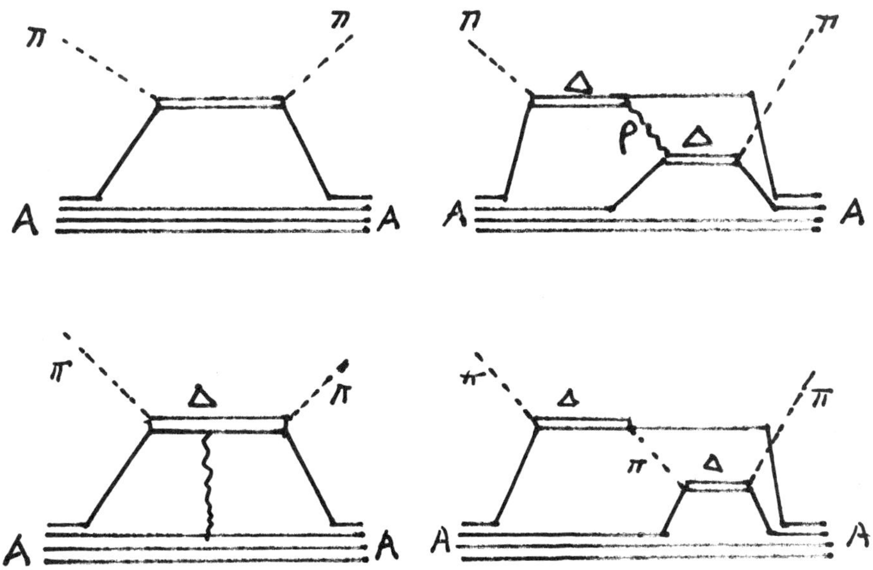

Fig. 6.2 Various processes included in the Isobar-Doorway Theory by V^{Res}.

which do not arise from isobar processes. V^{Res} contains all processes in which one enters and leaves the nucleus via an N^*. Thus the processes illustrated in Fig. 6.2 are included. This theory extends the strong interaction by allowing isobar interactions in nuclei to be included. Microscopic calculations in a π-nucleon hole state representation have been carried out and applied to the calculation of π-nucleus scattering.[24] A. Saharia and I have also derived a form for V^{abs} and are applying a momentum space version of the form of Eq. 6.3 to pi-nucleus elastic scattering. I will return to this briefly in Cpt. 8.

Dr. Moniz will discuss this general approach to π-nucleus physics and the accomplishments to date.

c. The Low Equation for π-Nucleus Physics

Let us return to the Low equation (Eq. 5.37), keeping only the one nucleon and one nucleon-one meson intermediate states. The one nucleon contribution, V_1, is

$$V_1 = \int d^3p'' \frac{\langle p'\tau'|V^+_{k't'}|p''\rangle\langle p''|V_{kt}|p\tau\rangle}{p_0 + k_0 - p_0'' + i\varepsilon} + \text{c.t.}\,, \qquad (6.4)$$

where $|p''\rangle$ is a one-nucleon state. For the one nucleon and one meson term we can use Eq. 5.35. This gives us the form

$$\langle k',p'|T(s)|k,p\rangle = V_1 + \sum_{p'',k''} \frac{\langle k'p'|T^+|p''k''\rangle\,\langle p''k''|T|kp\rangle}{p_0 + k_0 - p_0'' - k_0'' + i\varepsilon} + \text{c.t.} \qquad (6.5)$$

The term called c.t. in Eqs. 6.4 and 6.5 comes from the "crossed term", the second term in Eq. 5.37.

Eq. 6.5 is the Low equation in the approximation which has been considered for nuclear physics. It was first proposed by Dover and Lemmer.[26] The derivation of Eq. 6.5 using the pion interpolating field as sketched in sec. 5.c. was carried out by Cammarata and Banerjee,[27] who stressed the importance of the crossing relation, a property arising from the self-conjugate nature of the pion fields. Recently, Miller has derived this form starting from a field theory of the Chew-Low type,[27] and has derived a linear equation similar to Eq. 5.38. As discussed after this equation the most important difference from standard multiple scattering theory is the appearance of the E/H_o factors, which serve as long-range cutoffs. These results have been derived in a static model. It will be most interesting when more general treatments become available. The long-range cutoff and the treatment of crossing are the most specific results from the Low equation approach which have been included in practical calculations up to the present time.

7. TRUE ABSORPTION: $pp \leftrightarrow \pi d$

As has been emphasized throughout these lectures, meson-nucleon and meson-nucleus processes proceed mainly through absorption. Processes in which a meson does not appear in an intermediate state are called "true absorption" in conventional multiple scattering theory and must be added separately. Since much of the absorption at medium energy can be represented by the creation of an isobar from a nucleon, the "true absorption" potential which must be added in the Isobar-Doorway Model is quite different from that of the conventional multiple scattering theory.

In any case, the prototype absorption process in nuclei is the process π + deuteron $\rightarrow$ two nucleons, and it must be understood if one is to make progress with the problem of absorption in nuclei.

The simplest process for $\pi d \rightarrow pp$ is illustrated in Fig. 7.1. Neglecting initial and final state interactions, this process is easily evaluated using the methods which we have discussed. It is a good exercise. The T-matrix element for the nucleon transfer process for $\pi d \rightarrow pp$ is

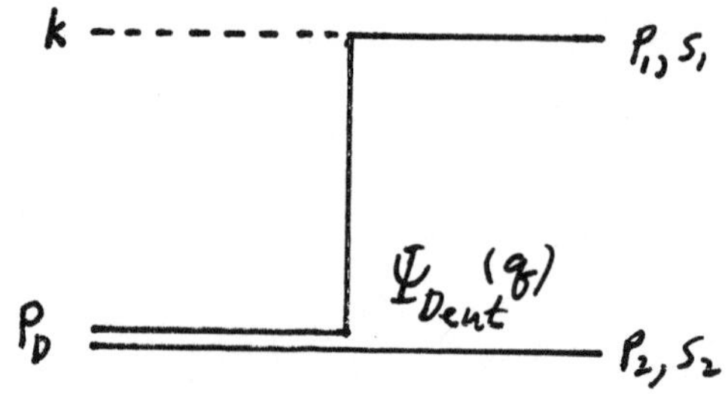

Fig. 7.1 Nucleon Pole (exchange) Process for $\pi d \to pp$

$$T^1_{s_1 s_2 M_D} = \frac{g}{2M} \langle u^+_{s_1} u^+_{s_2} \, e^{-\vec{P}_2 \cdot \vec{r}_2} \, e^{i\vec{P}_1 \cdot \vec{r}_1} \, \sigma \cdot (k - \frac{\mu}{2M} \vec{P})$$

$$\times \frac{1}{\sqrt{2\omega k}} (e^{i\vec{k} \cdot \vec{r}_1} \tau^1_+ + e^{i\vec{k} \cdot \vec{r}_2} \tau^2_+) \frac{e^{iP_D \cdot (\frac{\vec{r}_1 + \vec{r}_2}{2})}}{(2\pi)^{3/2}} \tag{7.1}$$

$$|\Psi(r_2 - r_1), M_D\rangle$$

Introducing relative and center-of-mass coordinates, $\vec{r} = \vec{r}_2 - \vec{r}_1$, $\vec{R} = (\vec{r}_1 + \vec{r}_2)/2$, $\vec{P} = \vec{P}_1' - \vec{P}_1$, $\vec{P} = \vec{P}_1' + \vec{P}_1$ and treating the case of near forward scattering, $\hat{k} \approx \hat{P}_1$, one finds that

$$T' \propto \Psi_D\left(\frac{\vec{k} - \vec{P}}{2}\right) \tag{7.2}$$

where Ψ_D (q) is the Fourier transform of $\Psi_D(r)$ for momentum q. In the low-medium energy region $|\vec{k} - \vec{P}|/2$ is rather large with $\frac{k-p}{2} \approx P_1 \approx 1 \text{ fm}^{-1}$. Therefore the deuteron wave function has fallen to a rather small value, and the first order contribution is not large. It turns out that the mechanism of Fig. 7.1 is perhaps a ten percent contribution to the scattering in the energy region of the $\Delta(1232)$. By this I mean energies such that the π and the transfered neutron at the upper vertex of Fig. 7.1 form a $\Delta(1232)$.

The $\pi d \leftrightarrow pp$ reaction is one of those peculiar cases where the lowest order process is less important than a higher order one. Actually, this is not as unusual in reactions at high momentum transfer. Recall that in the multiple scattering series for the T-matrix for elastic scattering the higher order terms become more

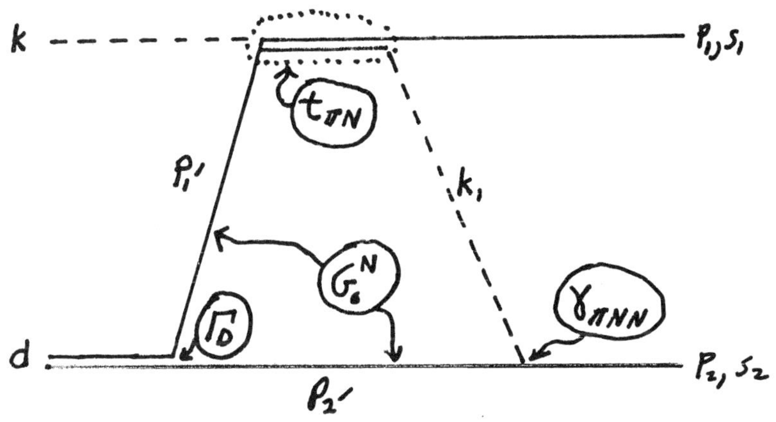

Fig. 7.2 The one-meson-exchange or pion rescattering process for $\pi d \leftrightarrow pp$.

important at larger angle. This is because for high momentum transfer it pays to share the momentum among two or more nucleus, since the single-particle nuclear wave functions decrease rapidly with momentum. The fact that the second order process, which we refer to as "pion rescattering", dominates the first order nucleon transfer process was recognized two decades ago.[28] As illustrated in Fig. 7.2, the resonance scattering of the pion by one of the nucleons enables the two nucleons to efficiently absorb the pion mass and share in the final momentum, without requiring large momentum components from the deuteron.

This problem has been studied by a large number of theorists with a variety of methods. I shall give the relativistic treatment used by B. Keister and myself.[29] We study the process as a Feynman diagram with physical elements. The elements of the diagram are encircled in Fig. 7.2. Schematically,

$$T_{\pi d \to pp} = u(P_2)^+ \gamma_{\pi N} G_o^N \Gamma_D G_o^N t_{\pi N} u(P_1) G_o^\pi$$

which is interpreted as "reading" the diagram starting with the proton spinor $u(P_2)$, then $\gamma_{\pi N}$, the π-N vertex operator, the nucleon (P_2') propagator, G_o^π, then the deuteron-two nucleon vertex function, Γ_D, the nucleon (P_1') propagator, G_o^N, the π-N off shell T-matrix, and the pion (k_1) propagator, G_o, finishes the loop. We restrict ourselves to forward scattering, $\hat{k}=\hat{p}_1$.

The kinematics is an important part of the problem. We shall work in the deuteron rest system. The deuteron center-of-mass and relative momenta are

$$P_D = P_1'+P_2' = (M_D,0) \text{ (deuteron mass shell)}$$

$$q = (P_2'-P_1')/2 = (q_o,\vec{P}_1')$$

Since there is a single loop, there is one 4-dimensional integral to be done. We shall use the variable q as the variable of integration. Then

$$T_{\pi d\to pp} = \int d^4q \;\; u^+_{s_2}(P_2)\; \gamma_{\pi N}(k_1^2,P_2'^2)\frac{1}{\not{P}_2'-M}\,\Gamma_D(q,P_2'^2,P_1'^2)$$

$$\times \frac{1}{\not{P}_1'-M}\;\; t_{\pi N}(s,ts,P_1'^2,k_1^2)\; u_{s_1}(P_1)\;\frac{1}{k_1^2-M_\pi^2} \qquad (7.2)$$

Let us look at the elements of this integral. We study the integral in the fully relativistic form of Eq. 7.2, and look for kinematic conditions under which we can

i) Place the intermediate nucleons on shell: $P^2_{2'} \simeq P^2_{1'} \simeq M^2$.

ii) Drop the q_o integration. I.e., introduce $\Gamma_D(q,P^2_{2'},P^2_{1'}) \simeq \delta(q_o)\;\Gamma_D(\vec{q})$.

This latter approximation permits us to introduce the nonrelativistic deuteron wave function:

iii) $\frac{1}{\not{P}_2'-M}\;\Gamma_D\;\frac{1}{\not{P}_1'-M} \to \Psi_D(\vec{q})$. This follows from, e.g., the Bethe-Salpeter eq. (see sec. 5b) for the deuteron

$$(\not{P}_1-M)\;(\not{P}_2-M)\,|\Psi_D^{B.S.}\rangle = V^{eff}\,|\psi_D^{B.S}\rangle, \text{ and}$$

$$\langle P_1P_2|\; V^{eff}|\psi_D^{B.S.}\rangle = \Gamma_D$$

iv) $u_s(p) \to X_s e^{i\vec{p}\cdot\vec{r}}/(2\pi)^{3/2}$. The Dirac spinors can be replaced by Pauli spinors, and final state distortion is neglected for the moment.

v) t_N: The π-N T-matrix in Eq. 7.2 is defined as a function of the variables appropriate to the π-N system,

$$s=(k+p_1')^2 = (k_1+P_1)^2;\ t_s=(k-k_1)^2$$

Since $P_1'^2 \simeq M^2$,

$t_N = t_N(s,t_s,k_1^2)$, an extrapolation of $t_{\pi N}$ off-mass-shell in the exchanged pion.

vi) In the p.s. theory, with $P_2'^2 \simeq M^2$, referring to Sec. 4b

$$\gamma_{\pi N}(k_1^2 ' P_2'^2) = ig\,\gamma_5\tau_{t_1} \simeq \frac{g}{2M}\vec{\sigma}\cdot\left(\nabla_\pi - \frac{M_\pi}{2M}\nabla_N\right)\tau_{t_1}$$

for a pion of charge t_1 exchanged. However, we introduce a structure function[30] for the π-N vertex to take into account the finite extent of these hadrons.

$$\gamma_{\pi N} \approx F(k_1^2)\,\frac{g}{2M}\,\vec{\sigma}\cdot\left(\nabla_\pi - \frac{m_\pi}{2M}\nabla_N\right)\tau_{t_1},$$

we use the form $F(k_1) = \alpha/(\alpha+k_1^2-M^2)$.

Putting all of this together we have

$$T_{\pi d\,pp} = \frac{g}{2M}\int d^3q\ F(k_1^2)\,\frac{1}{k_1^2-M^2}\,t_{\pi N}(s,\cos\theta_s,\,k_1^2)\,\Psi_D(\vec{q}), \tag{7.3}$$

where $\cos\theta_s = \vec{k}\cdot\vec{k}_1/|\vec{k}||\vec{k}_1|$. The variables s, k_1^2, and $\cos\theta_s$ are functions of $\vec{q}$. The relations are a little complicated, so I shall not give them. The t-matrix is evaluated in the π-N center of mass system and transformed to the deuteron rest system. This is done by making a partial wave expansion and using Eq. 5.31 to get the k_1^2, off-mass-shell behavior, and using the invariant expression 5.8.

One of the most important questions in all of this is the convergence of the integral 7.3. In the first work that was done with the relativistic form 7.2, it was assumed that because of the rapid decrease of $\Psi(\vec{q})$ with q, that the integral could be factorized thus

$$T_{\pi d\to pp} \rightarrow \left[u^+\gamma_{\pi N}\,t_{\pi N}\,u\,\frac{1}{k_1^2-m_\pi^2}\right]_{q=0}\Gamma, \tag{7.4}$$

where Γ is a factor arising from $\int \Psi_d(q) d^3q$. However, note that

$$\int d^3q \quad \Psi_D(q) = \Psi_D(r=0),$$

which is a completely unknown quantity. For a hard core interaction $\Psi_d(r=0)=0$, and for the type of repulsive interactions ex--pected, it is a small, but unknown quantity. In the quark model it is presumably finite. Therefore, although the deuteron wave function is the most important factor in cutting off the integral the rest of the integrand plays a vital role in determining the convergence properties of the integrand.

The kinematic information given in Table 7.1 helps one to evaluate the contribution of the various parts of the integral for the energy shown, $E_{cM} \approx 3.0$GeV. For each value of $|\vec{q}|$, one sees the values of $p_1'^2 = p_2'^2$ and the maximum and minimum values of k_1^2 over the angular range of $\vec{q}$. The numerical evaluation of the integrand shows that by the value $|\vec{q}| = .36$ the integral has converged. Note first that P_1^2 is not very different from the mass-shell value of 0.88 over the range $0<|\vec{q}|<0.36$. This is the main justification for out taking the intermediate nucleons on shell and setting $q_o=0$. Another important feature to note is that

Table 7.1

Kinematics Associated with the $\vec{q}$ Integral. $E_{c\mu}=3.$GeV

		direct		exchange	
$\lvert\vec{q}\rvert$	$p_1'^2$	$k_{1\,min}^2$	$k_{1\,max}^2$	$k_{1\,min}^2$	$k_{1\,max}^2$
.09	.87	-2.03	-4.45	-.38	-.18
.18	.85	-2.34	-1.19	-.50	-.11
.27	.81	-2.68	- .94	-.64	-.05
.36	.75	-3.02	-..70	-.78	-.01
.45	.68	-3.38	- .49	-.97	-.013
.54	.59	-3.76	- .29	-1.16	-.022
.63	.48	-4.16	- .10	-1.36	-.015
.72	.36	-4.57	- .066	-1.58	-.009
.81	.22	-5.00	- .22	-1.81	-.048

k_1^2 at some points approaches 0.02, which is the mass shell value, and that $\frac{1}{k_1^2 - m_\pi^2} \to \infty$ for these values. In other word, "real" rather than virtual mesons can participate in the exchange. This effect tends to slow the convergence of the integral and lead to larger $|\vec{q}|$. It also has the important physical consequence of leading to a very long-range interaction, which could make higher order terms more important.

There is a dynamic feature which also has important consequences for the matematics involved in the integral. For the forward direction which we are considering, for $\vec{q}=\vec{p}_1{}'=0$, $s = 1232$ at $E_{cm} = 2.17$ GeV. Thus at characteristic values of the incoming pion energy, one forms N* resonances for no relative momentum in the deuteron ($\vec{q}=0$). At these values the integrand converges especially rapidly, for as $|\vec{q}| \neq 0$, not only is $\psi_D(q)$ decreasing rapidly, but the $t_{\pi N}$ amplitude also decreases. For these values the main effect of relativity is the off-mass-shell behavior of $t_{\pi N}$, which we treat. For all values of the energy <u>except</u> near-threshold values and $2.3 \lesssim E_{cm} \lesssim 2.6$ GeV this is the case, and we have a reliable treatment of the process. For $2.3 \lesssim E_{cm} \lesssim 2.6$ GeV the integral does not converge for reasonable values of q. This means that the step of putting $q_o=0$ and the nucleons on mass shell is invalid. At $\vec{q} \approx 1$ GeV, $p_1{}'^2 \approx 0$, so that the nucleons are massless, a completely relativistic phenomena. This is a region in which one should be using relativistic methods, such at the Bethe-Salpeter equation.

At the energies of concern here, the eikonal approximation should be valid for the intial and final state absorption. We use the Sopkovich approximation, in which each partial wave is taken as

$$T_\ell^{\pi d \to pp} = \sqrt{S_\ell^{\pi d}}\; T_\ell^{p.\omega.}\; \sqrt{S_\ell^{pp}}, \tag{7.5}$$

where $T^{p.w}$ is the Born T-matrix element which we have just been treating (plane wave), and $S_\ell^{\pi d}$ and S_ℓ^{pp} are the elastic partial wave S-matrix elements.

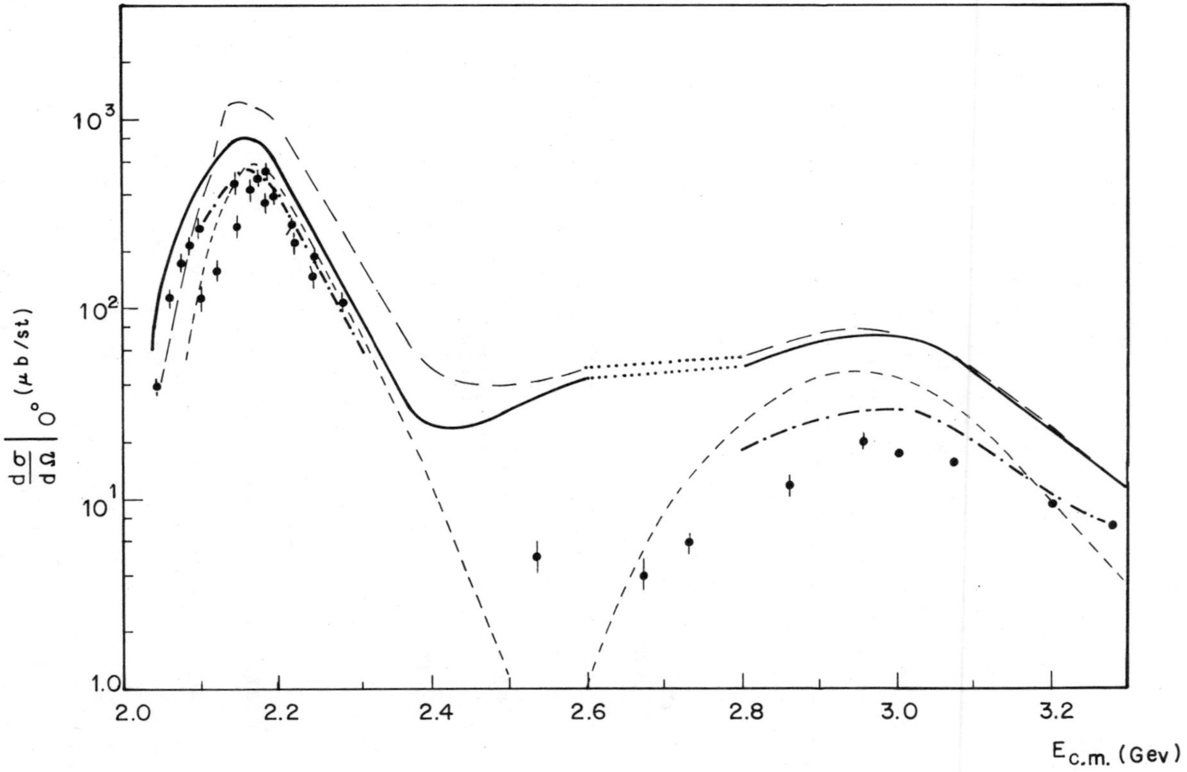

Fig. 7.2 The pp→πd forward cross section. The curves are explained in the text.

The results for the 0° cross section are shown in Fig. 7.2. The long-dashed curve is the plane wave result without the off-mass-shell factor ($V_\ell(k_1)=T_m$ Eq. 5.31). The solid curve gives the theoretical cross-section with the off-mass-shell effect included, again in the plane wave approximation. The importance of the off-shell effect is readily seen in the region of the Δ(1232). At higher energies the relative π-N momentum is large compared with the pion mass, and the off-mass shell corrections become small. One can also observe that the peaks are considerably broader than the free π-N resonance peaks. This is a form of collision broadening due to the internal motion. In the factorized approximation one just has the free width, as shown by the short-dashed curve. Note that the short-dashed curve has an arbitrary normalization, as explained above.

Finally, the dash-dotted curve shows the result of the calculation with both off-mass-shell and distortion effects included. The results are quite satisfactory at the resonances. In the energy region such that the internal π-N energy is between the Δ(1232) and the higher resonances (the I=3/2 resonances are most important for the π^+d reaction), the entire calculation fails. The integral does not converge until q values reach enormous values, so that the $q_o=0$ approximation is not reasonable. Not only does one need relativistic wave functions, but the internal mesons can become real, as explained above. A much more sophisticated calculation is needed, one using methods such as those I have been discussing. It should prove most interesting.

Before I leave this subject I would like to mention a most interesting recent development for the B=2 system. In the experiment[32]

$$\gamma + d \rightarrow p + n$$

there has been observed an interesting energy dependence of the polarization. In Fig. 7.3 the polarization at 90° scattering angle as a function of photon energy is shown. The arrows shown the points

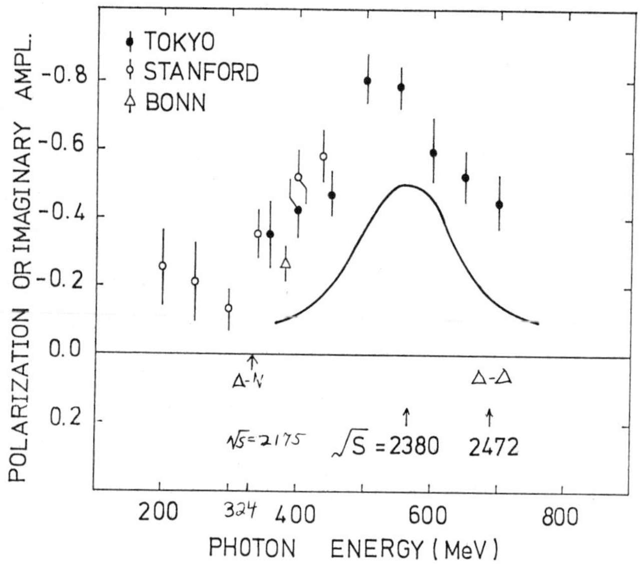

Fig. 7.3 Polarization as a function of γ energy for $\gamma d \to pp$.

points $\sqrt{S}$ = 2.175, 2.38, and 2.472, corresponding to the Δ-N, experimental peak, and the Δ-Δ energies. Note that the peak is about 100 MeV below the Δ-Δ energy. This has been interpreted by Kamae and coworkers[32] as a deeply bound state of the Δ-Δ system, as illustrated in Fig. 7.4a, alternatively, as a new six quark object, which has been named baryonium, as illustrated in Fig. 7.4b. This

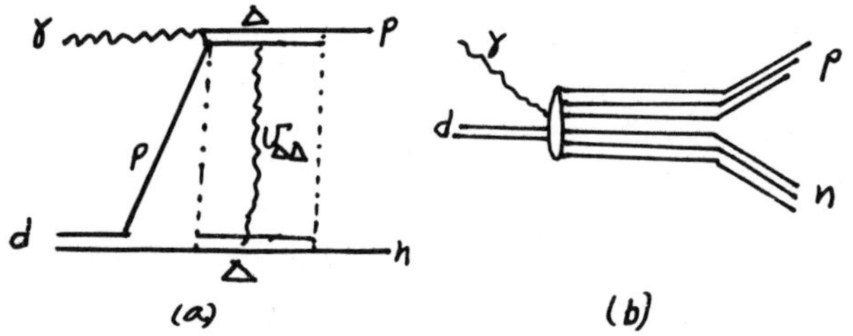

Fig. 7.4a) The interpretation of $rd \to p+n$ with a 2Δ intermediate strongly bound by a potential $v_{\Delta\Delta}$. b) The barionium, 6-quark interpretation.

would be an important object, as it means that six quarks can be treated as being confined as an "elementary" object. That would be an important state for the quark model to deal with.

However, I would like to point out that there is another, and much less interesting interpretation. Note from Figs. 7.2 and 7.3 that the energy of the "bariyonium" peak in the $\gamma d \to pn$ polarization occurs in the minimum of the $\pi d \to pp$ cross section plotted as a function of energy (Fig. 7.2). A natural interpretation is that the polarization arises from the interference of two N*-N systems, a Δ-N at s<2.38 and N*-N at s<2.38. The calculation showing that his can explain the peak shown in Fig. 7.3 has not been completed partly because I am giving these lectures instead of working at my desk, and partly because this is a region which can only be truely satisfactorily treated using a relativistic deuteron wave function.

8. THE OPTICAL POTENTIAL: SECOND ORDER TERMS AND PHENOMENOLOGICAL DEVELOPMENT

a. The Second-Order Optical Potential:

As has been discussed by Dr. Gibbs, the second order term in the expansion for the optical potential is (in momentum space)

$$\langle k'|V^{(2)}|k\rangle = \sum_j \sum_{k\neq j} \langle\Psi_o,\vec{k}'|t_j \frac{1-\mathcal{P}_o}{E\ H_o^+ + it}\ t_j|\Psi_o,\vec{k}\rangle , \qquad (8.1)$$

where t_j is the π-N T-matrix for scattering from the jth bound nucleon, and $\mathcal{P}_o$ is the projection operator onto the ground state. Introducing intermediate states $|\Psi_N,\vec{k}''\rangle$, with inelastic nuclear states Ψ_N and mesons $|\vec{k}''\rangle$, one can write (assuming factorization)

$$\langle \vec{k}'|V^{(2)}|\vec{k}\rangle = \sum_{Nk''j} \sum_{k\neq j} \langle \vec{k}'|t,|k'',N\rangle \langle N_{k''}|t_k|k\rangle \frac{1}{E_o-E_{N,k''}}$$

$$\times \iint d^3r\delta^3r' e^{i\vec{r}\cdot(\vec{k}''-\vec{k}')} e^{i\vec{r}'\cdot(\vec{k}''-\vec{k})}$$

$$\{\langle\Psi_o|\delta(\vec{x}_j-\vec{r})\delta(\vec{x}_k-\vec{r}')|\Psi_o\rangle$$

$$- \langle\Psi_o|\delta(\vec{x}_j-\vec{r})|\Psi_o\rangle\langle\Psi_o|\delta(\vec{x}_k-\vec{r}')|\Psi_o\rangle\} \qquad (8.2)$$

The factor in brackets is related to the nuclear two-particle density and two-particle correlation function. The standard definitions are

$$\rho^{(2)}(\vec{r},\vec{r}') \equiv \frac{1}{A(A-1)} \langle\Psi_o| \sum_{j,k\neq j=1}^{A} \delta(\vec{x},-\vec{r})\delta(\vec{x}_k-\vec{r}')|\Psi_o\rangle$$

$$\rho^{(1)}(\vec{r}) = \frac{1}{A}\sum_{j=1}^{A}\langle\Psi_o|\delta(\vec{x}_j-\vec{r})|\Psi_o\rangle ,$$

and

$$\rho^{(2)}(\vec{r},\vec{r}') - \rho^{(1)}(\vec{r})\ \rho^{(2)}(\vec{r}') = \rho^{(1)}(\vec{r})\ \rho^{(2)}(\vec{r}')g(\vec{r},\vec{r}') , \qquad (8.3)$$

which defines the two particle correlation function. In coordinate space we find, neglecting off-mass-shell behavior,

$$\langle\vec{r}|V^{(2)}|\vec{r}'\rangle = \frac{A(A-1)}{(2\pi)^3} \int d^3k''d^3kd^3k'\langle k'|t|k''\rangle\rho(\vec{x})\rho(\vec{x}')g(\vec{x},\vec{x}')$$

$$\langle\vec{k}''|t|\vec{k}\rangle\frac{1}{E-E_{k''}}\ e^{i\vec{k}'\cdot\vec{r}}\ e^{i\vec{x}\cdot(\vec{k}''-\vec{k}')}$$

$$e^{-i\vec{x}'\cdot(\vec{k}''-\vec{k})}\ e^{-i\vec{k}\cdot\vec{r}'}\ d^3xd^3x' \qquad (8.4)$$

The appearance of correlation functions is characteristic of expansions of optical potentials, and produces much faster convergence than expansions of the many-body T-matrix. There have been a number of calculations for $V^{(2)}$, but all with rather drastic simplifying assumptions. Eisenberg, Hüffner, and Moniz[33] have i) neglected all off-mass-shell behavior which up to now is standard practice in the field, ii) used a factorized t-matrix, $\langle k'|t|k\rangle = b(E)\vec{k}''\cdot\vec{k}v(k)v(k')$,

iii) used the factorization approximation to evaluate the $t_{\pi N}$ at free values, neglected higher-order correlations, and iv) dropped tensor-type forces which appear in pseudo-scalar meson exchange, and v) used a fixed scattering center approximation, which involves a closure approximation. With approximations i), ii) (Let v = 1 for simplicity), iii) and v),

$$V^{(2)} \sim \int d^3x d^3x' \; t(x) \frac{e^{ik|\vec{x}-\vec{x}'|}}{|\vec{x}-\vec{x}'|} t(x')\rho(\vec{x})\rho(\vec{x}')g(\vec{x},\vec{x}') \;, \quad (8.5)$$

where the non-relativistic propagator $G_o(x,x') \alpha \; e^{ik|\vec{x}-\vec{x}'|}/|\vec{x}-\vec{x}'|$ appears. Dropping the exponential (phase) factor for G_o and using the simple form $t \sim \vec{k}'\cdot\vec{k}$, one has

$$t(x)G_o(x,x')t(x')\alpha\nabla_x\nabla_x\cdot\frac{1}{|x-x'|} = C_1\delta(\vec{x}-\vec{x}') + C_2 S(\vec{x}-\vec{x}') \quad (8.6)$$

where $S(x,x')$ is a tensor-like force. In Ref. 33 the tensor-like force is neglected (approximation iv), giving a local potential

$$V^{(2)}_{(x)} = \frac{4\pi A(A-1)b(E)}{2\mu 3} g(0)\nabla\cdot\rho(x)^2\nabla \;. \quad (8.7)$$

With assumptions i)...v) one can iterate the optical potential to all orders.

$$V^{opt}_{(x)} = -\nabla \cdot \frac{\frac{4\pi}{2\mu} b_1(E)A\rho(x)}{1 + \frac{4\pi}{3} b_1(E)(A-1)\xi_\Delta\rho(x)} \nabla \;, \quad (8.8)$$

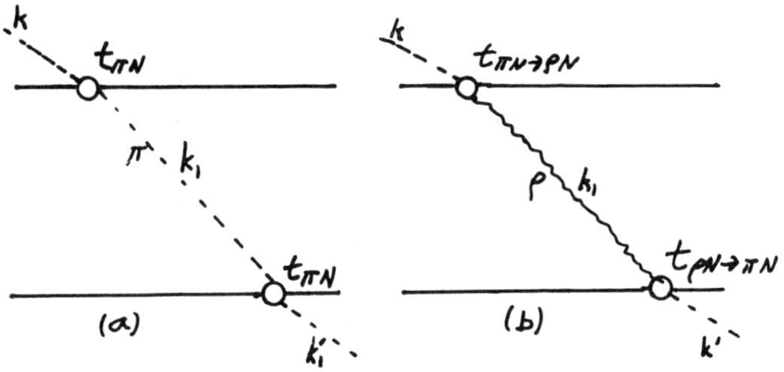

Fig. 8.1 Processes contributing to the second-order pion optical potential: a) pion exchange; and b) rho exchange

a form given by Ericson and Ericson.[34] The parameter ξ_Δ of that paper for the "Lorentz-Lorentz" effect is found to be very small in Ref. 33. This suggests that higher-order corrections to the optical potential arising from multiple pion exchange are very small.

b. Virtual ρ-Mesons, Correlations, and δ-Functions

The second order process just discussed is pictured in Fig. 8.1 (a). One can see that the exchange of the intermediate meson resembles the process associated with the N-N interaction, as discussed in Sections 1 and 2. It has been suggested[35] that the ρ-meson exchange process is very important, giving a much larger second-order pion optical potential (and "Lorentz-Lorentz" effect) than the one pion exchange process considered in Section 8 (a) above. Also, it has been suggested that the ρ-exchange is of the same order as π-exchange for the $\pi d \to pp$ reaction.[36] (Replace the π by a ρ intermediate meson in Fig. 7.1.) However, the estimate for a large ρ-meson contribution in processes like Fig. 8.1(b) is done by using a δ-function trick. As in Eq. 8.7, the exchange of the ρ-meson generates a ρ-potential of the form

$$t_{\pi\rho} G_o^\rho(k_1) t_{\rho\pi} \alpha g_1 \delta(\vec{x}-\vec{x}') + g_2 \times \text{finite range force} \,, \quad (8.9)$$

where g_1 and g_2 are functions of spin, isospin, and momentum, and $\vec{x},\vec{x}'$ are the coordinates of the nucleons. The assumption[35] is made to drop the δ-function term due to the short-range repulsion between nucleons. This is the opposite assumption compared to Ref. 33, where only the δ-function term was kept. Recently it has been shown[37] that

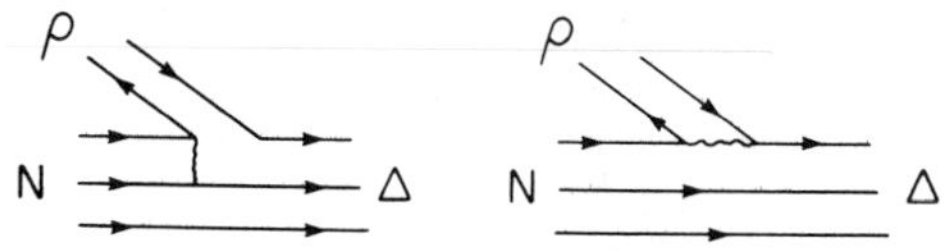

Fig. 8.1a The ρ-N-Δ vertex in the quark model.

the δ-function force cannot be dropped when calculating processes with large momentum transfer. The physical argument is that the nucleons are such extended objects that one cannot make use of the δ-function trick. E.g., the meson exchange takes place over such a short range that it is within both extended nucleons, and one does not really have a well-defined δ-function procedure. It has also been shown by considering the momentum-space treatment,[38] that the effect of the ρ-meson is greatly overestimated by the δ-function trick, and that its effect for the pion optical potential is probably not large. However, a quantitative treatment has not yet been carried out. Miller has also estimated that the effect of the ρ is small in his static Chew-Low Model.[39]

An important question with regard to determining the contribution of the ρ meson to pionic phenomena concerns the magnitude of the ρ-N-Δ coupling constant, f^*_ρ. An important technical point which I would like to make is that the static quark model which has been used overestimates this coupling constant. The current estimates use a π-quark coupling $\sigma^q \cdot \nabla$ or ρ-quark coupling $\sigma^q \times \nabla$. However, the proper procedure for the quark model is to let a $\bar{q}$ in the meson destroy a q in the N or Δ, giving two quark operators $[\sigma^{q_1} \times \sigma^{q_2}]^1$, among other operators. Thus one encounters the matrix element

$$<[q\ q\ q]N\ [\sigma^{q_1} \times \sigma^{q_2}]^1\ [q\ q\ q]\Delta> .$$

For a completely symmetric 3q configuration this matrix element vanishes. The Δ is completely symmetric, while the N has mixed spin-isospin symmetry. On the average the N is 3/4 in the spin symmetric state, giving the result that $f^* \rightarrow 1/4\ f^*$. This will result in ρ-contributions to the pion optical potential, $\pi d \rightarrow pp$, and so forth being very small indeed.

c. Phenomenological Models

I have stressed in these lectures the need to use field theory and relativity in order to understand the interactions of mesons with nuclei. Some of the processes and theoretical modifications give rise to new forms in the optical potential. Considering the uncertainty in the theory, the best we can hope for now is a theoretical prediction of the form of the potential, and some limits on the range of the various parameters. As an example of a phenomenological form, the potential

$$V_{opt}(r) = C_1(E)\rho(r) + C_2(E)\nabla\cdot\rho(r)\nabla + C_3(E)\rho^2(r) + C_4(E)\nabla\cdot\rho^2(r)\nabla \tag{8.11}$$

has been used to fit the low energy (30-50 MeV) data.[40] The first two terms are of the standard form, but the parameters $C_1(E)$ and $C_2(E)$ contain various modifications from effects such as the transformation from the π-N system to the π-nucleus system. The third term is used to simulate true absorption and the last term is the contribution from the second order potential. Typical results are shown in Fig. 8.2. The data is from Dytman et al.[41] It is interesting to compare this to a fit with the gradient potential

$$V_{opt}(r) = b_o\rho(r) + b_1\nabla\cdot\rho\nabla , \tag{8.11}$$

where b_o and b_1 are taken as free parameters.[41] This fit is shown in Fig. 8.3. This fit shows the main problem which has shown up at low energies. The phenomenological values of b_o vs. the theoretical value at 50 MeV is

$$b_o^{exp} = -3.5 - 0.6\,i$$

$$b_o^{theory} = -0.83 + 0.51\,i\,. \tag{8.12}$$

Presumably, the form (8.11) with the phenomenological parameter b_o^{exp} can take into account some of the corrections which come into play at low energies. It is interesting to note that the modification for the low energy elastic scattering seems to enable one to fit the inelastic scattering (in DWBA). There have been other theoretical-phenomenological fits to the data.[42]

One feature which is not contained in these models is the possibility of a long-range nonlocality arising from many-body Δ-

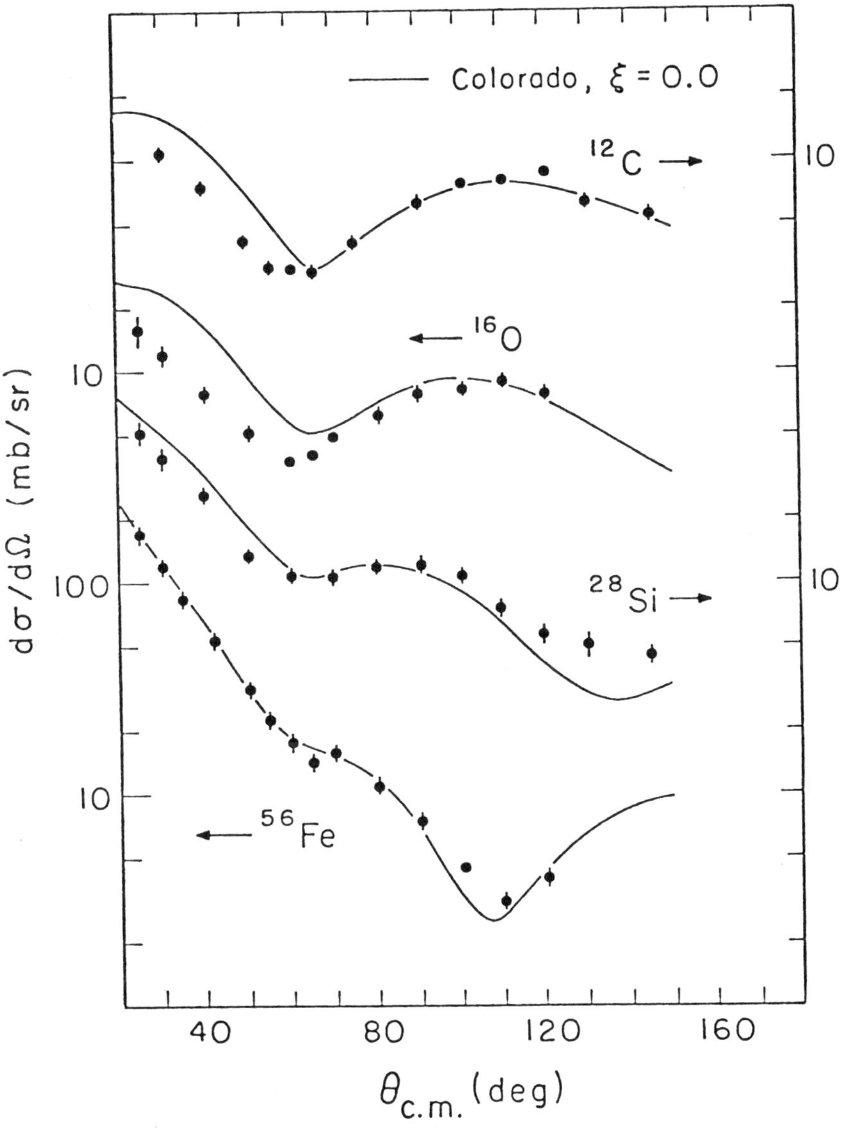

Fig. 8.2 Fit to 50 MeV π-nucleus elastic scattering by the Colorado group.

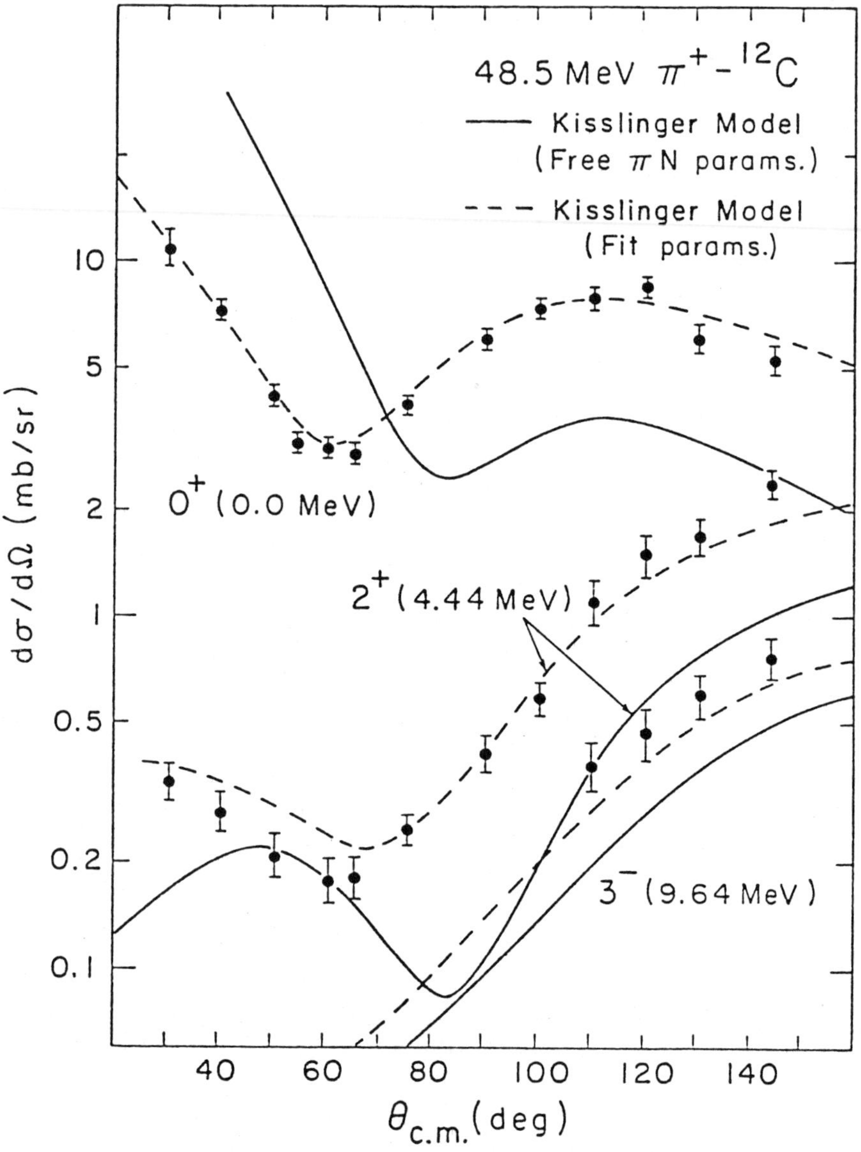

Fig. 8.3 Fit to 50 MeV -nucleus elastic scattering with phenomenological gradient potential.

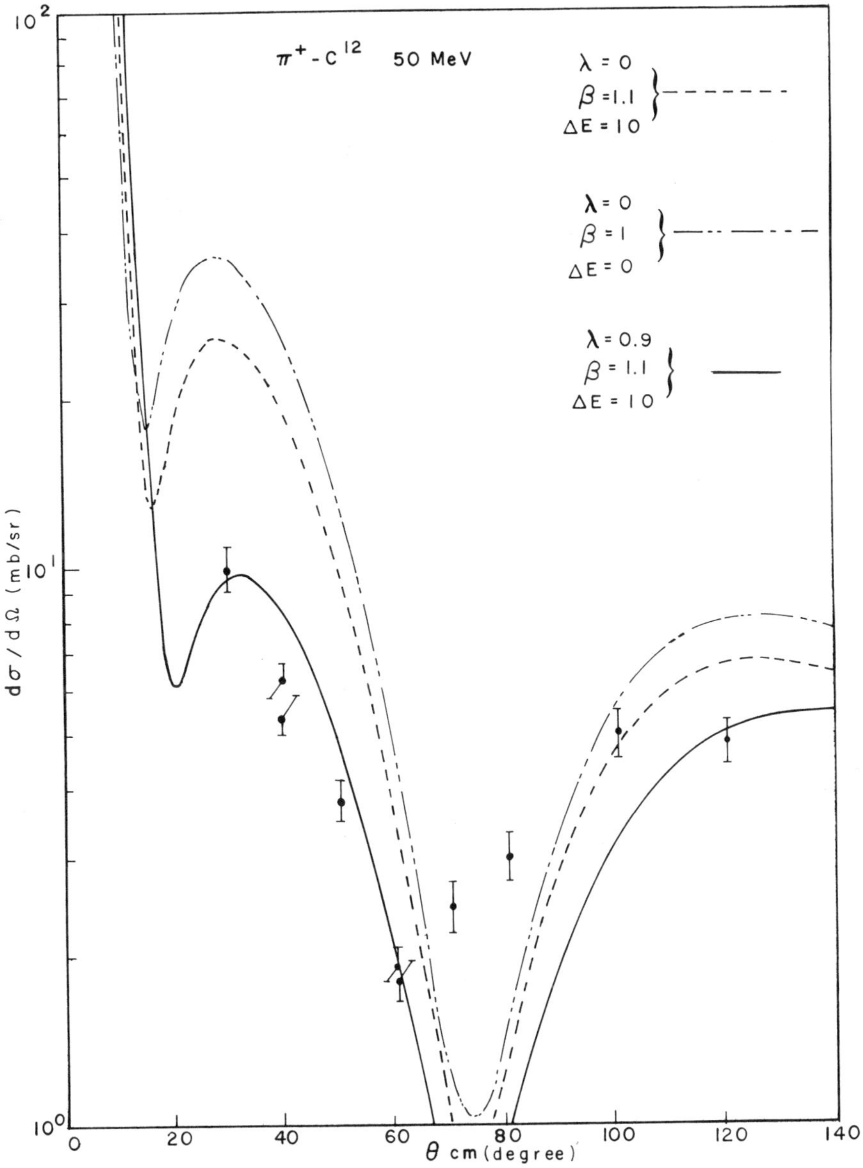

Fig. 8.4 Effects of some Isobar-Doorway parameters for 50 MeV π-nucleus elastic scattering.

propagation effects in nuclei.[23] This has been used in a phenomenological version of the Isobar-Doorway theory in the form

$$\langle r'|V|r\rangle = b_o\rho(r)\delta(\vec{r}'-\vec{r}) + \frac{E-M_\Delta+i\Gamma_\Delta(E)/2}{E-M_\Delta-\Delta E+i\beta\Gamma_\Delta(E)/2}\, t^\Delta_{\pi\rho}\rho_\lambda\ (\vec{r}-\vec{r}') \tag{8.12}$$

where λ is a parameter characterizing this feature, while ΔE and β give the difference in binding energy of a Δ and a nucleon, and β is the ratio of the width of the Δ in the nuclear medium to the free width. Typical results[43] are shown in Fig. 8.3. The most striking result is the forward scattering, as the nonlocality affects the way the pion enters the nucleus and thus the forward scattering is strongly modified.

The phenomenological models are important in that they can be used to form a bridge between the large amount of systematic experimental data and the large, still not so systematic theoretical questions.

REFERENCES

1) J. D. Bjorken and S. D. Drell, "Relativistic Quantum Mechanics" (McGraw-Hill, N.Y. (1964). Chapter 10 and the Appendices are especially useful for this course.

2) See, eig., C. Kållén, "Elementary Particle Physics" (Addison-Wesley, Reading, MA. (1964))

3) M. K. Banerjee and J. B. Cammarata, Nuc. Phys.

4) B. Ditta-Roy, I. R. Lapidus, and M. J. Tansnev, Phys. Rev. 181, 2091 (1969)

5) C. F. Chew and F. Low, Phys. Rev. 101, 1570, 1579 (1956)

6) S. Schweber, "An Introduction to Relativistic Quantum Field Theory" (Row Peterson, Evanston, IL. (1961)

7) M. Bawin, Nucl. Phys. B28, 109 (1971); R. Landau and F. Tabakin, Phys. Rev. D5, 2746 (1972); J. T. Londergan, K. W. McVoy, and E. J. Moniz, Ann. Phys.(N.Y.) 86, 147 (1974)

8) H. Feshbach, A. K. Kerman, and R. Lemmer, Ann. Phys. (N.Y.) 41, 230 (1967)

9) M. V. Barnhill III, Nuc. Phys. A131, 166 (1969)

10) G. A. Miller, Nuc. Phys. A224, 269 (1974); M. Bolsterli, W. R. Gibbs, B. F. Gibson, and L. J. Stephenson, Jr., Phys. Rev. C10, 1225 (1974)

11) J. M. Eisenberg, J. V. Noble, and H. J. Weber, Phys. Rev. C11, 1048 (1975)

12) C. Schwartz and C. Zemach, Phys. Rev. 141, 1454 (1966); M. J. Levine, J. Wright, and J. A. Tjon; these references contain references to the earlier work

13) R. Blankenbecler and R. Sugar, Phys. Rev. 142, 1051 (1966)

14) R. Thompson, Ph.D. dissertation (1969), Case-Western Reserve University, unpublished

15) R. E. Cutkosky, Phys. Rev. 96, 1135 (1954); E. Zur Linden and H. Mitter, Nuovo Cim. 61 B, 389 (1969), contains references to earlier work

16) G. A. Miller, Phys. Rev. C14, 2230 (1976)

17) W. T. Nutt and C. M. Shakin, Phys. Lett. 69B, 290 (1977)

18) J. Benecke and H. P. Durr, Nuovo Cim. 56 A, 269 (1968)

19) G. Wolf, Phys. Rev. 182, 1538 (1969)

20) L. Celenza, L. C. Liu, and C. M. Shakin, Phys. Rev. C11, 1593 (1975)

21) K. Nakano and Chi-Shiang Wu, Phys. Rev. C11, 1505 (1975)

22) M. Johnson and B. Keister, LASL preprint LAUR-77-2496 (1977)

23) L. S. Kisslinger and W. L. Wang, Phys. Rev. Lett. 30, 1071 (1973); Ann. Phys. 99, 374 (1976)

24) M. Hirata, J. H. Koch, F. Lenz, and E. J. Moniz, Phys. Rev. Lett. B70, 28 (1977)

25) L. S. Kisslinger and A. Saharia, "Meson-Nuclear Physics", AIP Conference Proceedings No. 33 (P. Barnes et al, Eds.) p 159

26) C. B. Dover and R. H. Lemmer, Phys. Rev. C7, 2312 (1973)

27) G. A. Miller, Phys. Rev. C16, 2324 (1977)

28) S. Mandelstam, Proc. Roy. Sol. A244, 491 (1958)

29) B. D. Keister and L. S. Kisslinger, Carnegie-Mellon University preprint (1978); contains references to earlier work

30) E. Ferrari and F. Selleri, Phys. Rev. Lett. 7, 387 (1961)

31) T. Yao, Phys. Rev. 134B, 454 (1964)

32) T. Kamae et al, Phys. Rev. Lett. 38, 468; 38, 471 (1977)

33) J. M. Eisenberg, J. Huffner, and E. J. Moniz, Phys. Lett. 47B, 385 (1973)

34) M. Ericson and T. E. O. Ericson, Ann. Phys. (N.Y.) 36, 323 (1966)

35) G. Baym and G. E. Brown, Nucl. Phys. A247, 395 (1975)

36) M. Brack, D. D. Riska, and W. Weise, Nucl. Phys. A287, 425 (1977)

37) L. S. Kisslinger, Carnegie-Mellon University preprint (1978)

38) L. S. Kisslinger, Carnegie-Mellon University preprint (1978); E. Levin and J.M. Eisenberg, Nuc. Phys. A292, 459 (1977).

39) G. A. Miller, University of Washington preprint (1978)

40) N. J. DiGiacomo, A. S. Rosenthal, E. Rost, and D. A. Sparrow, Phys. Lett. 66B, 421 (1977)

41) S. A. Dytman et. al., Phys. Rev. Lett. 38, 1059 (1977); 39, 53 (1977)

42) R. H. Landau and A. W. Thomas, Phys. Lett. 61B, 361 (1976); L. C. Liu and C. M. Shakin, Phys. Rev. C16, 333 (1977); K. Stricker, H. McManns, and J. Carr, Michigan State University preprint (1978)

PION-NUCLEUS SCATTERING IN THE ISOBAR FORMALISM[†]

E. J. Moniz

Massachusetts Institute of Technology

Cambridge, Massachusetts 02139

I. INTRODUCTION

Pion-nucleus interactions play a central role in our attempts to gain a deeper understanding of nuclear systems. In some ways, this role is fulfilled by providing another "conventional" nuclear probe. For example, the pion may be used to measure nuclear matter distributions and, because of considerable variation in the mean free path over the accessible energy range, can be used to emphasize surface or volume effects. Because it comes in three charge states, double charge exchange can be used to study double analog transitions or proton rich nuclei. Because of its quantum numbers $J^{\pi}T = 0^{-}1$, reactions such as radiative capture may be used to study isovector M1 states. However, in addition to these, rather more unique opportunities are also presented. These originate in the boson nature of the pion and in its intimate connection to the nuclear force.

A brief reminder of the role played by the pion in the dynamics of the two nucleon system is in order. Of course, the long range nuclear force has been known for a long time to be given by one pion exchange (Fig. 1a). The nonrelativistic pion-nucleon vertex is dictated by the pseudoscalar nature of the pion. This same coupling led Chew and Low to explain very simply the origin of the attractive "force" in the J = 3/2, T = 3/2 πN scattering channel which leads to the famous resonance at an invariant energy of 1232 MeV. The interaction in this channel is the

[†]This work is supported in part through funds provided by the U.S. DEPARTMENT OF ENERGY (DOE) under contract EY-76-C-02-3069.

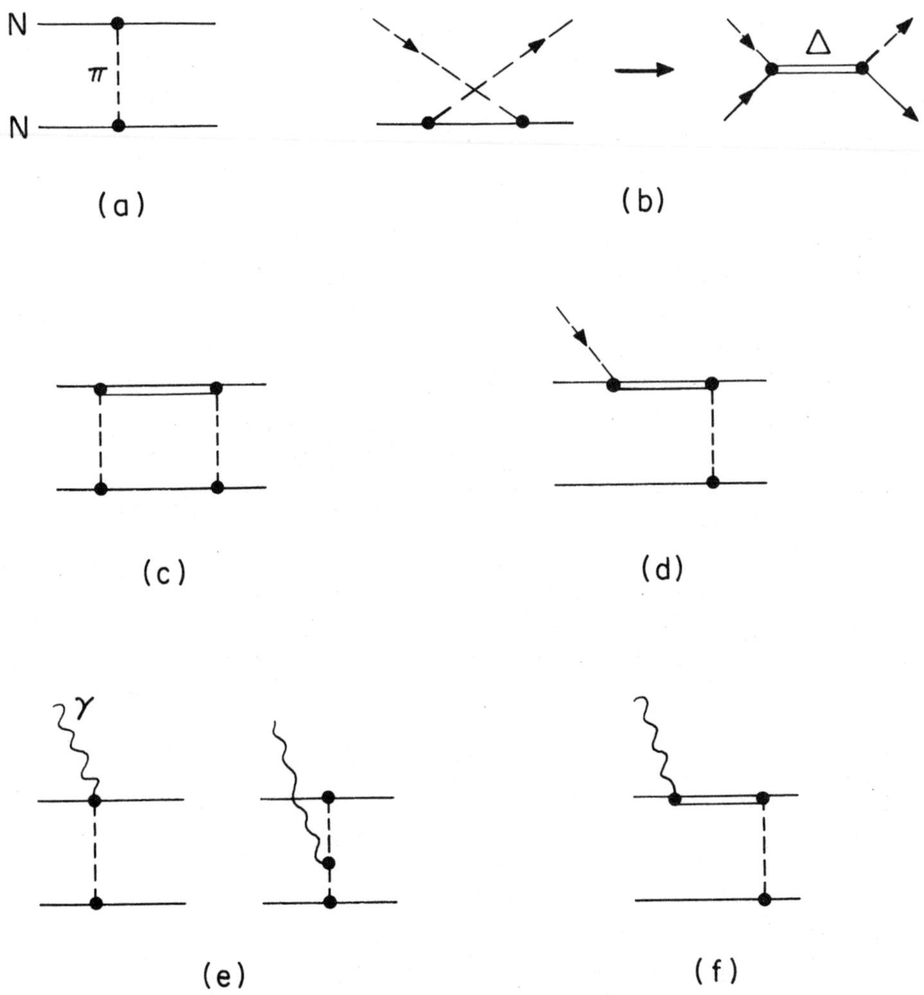

Figure 1: Pions and isobars in the nuclear force, in pion scattering and absorption, and in the nuclear electromagnetic current.

crossed πN Born term in Fig. 1b and we can think of the resonance as intermediate excitation of the (unstable) Δ-isobar. This is the object predicted in the quark model as the "first excited state" of the nucleon, although the fundamental connection between the Chew-Low and the quark model dynamics is not understood. The Δ in turn plays a very important role in the NN force; while not understood nearly as well, a large part of the intermediate range attraction which binds nuclei appears to be due to intermediate excitation of the Δ (Fig. 1c).

The possibility of pion absorption is implicit in the πNN vertex discussed above. Energy/momentum conservation requires that absorption of a physical (timelike) pion occurs on at least two nucleons. The cross section for $\pi^+ + d \to p + p$ shows a "resonance bump" just like that seen in the πN total cross section, so that it is empirically clear that the Δ is again playing a central role in the relevant energy regime. This is indicated in Fig. 1d. Finally, the pion and isobar degrees of freedom in the nucleus are known to play an important role in determining nuclear electromagnetic current distributions. Examples are indicated in Figs. 1e and 1f and are usually called exchange current contributions. The lesson in all this is that we understand at least semi-quantitatively the important role of pions and isobars in the nuclear force, in pion reactions with one and two nucleons, and in the "nuclear structure" of the two nucleon system. In these lectures, we will be concerned with using this understanding as input for a theoretical approach to the pion-nucleus many-body problem.

The conventional theoretical framework for high energy hadron-nucleus reactions is the multiple scattering or optical potential approach. Here, one focuses upon the pion degree of freedom as it works its way through the nucleus. However, this approach certainly does not emphasize the role of isobar dynamics in the nuclear medium and, in practice, this is often "frozen" out of the problem. Yet these features may be expected to be quite important. The isobar propagation distance in free space is not small compared to other length scales of interest in the nucleus. That is, if we write the Δ-propagator crudely as

$$G_\Delta(E,p) = (E - E_R + i\Gamma/2 - p^2/2m_\Delta)^{-1} \quad , \tag{1.1}$$

where E_R and $\Gamma(\approx 115 \text{ MeV})$ are the resonance position and width, respectively, then the coordinate space propagator is

$$G_\Delta(r) \sim \frac{e^{i\kappa r}}{r} \; , \; \kappa^2 \equiv 2m_\Delta(E - E_R + i\Gamma/2) \quad . \tag{1.2}$$

Near resonance, we have

$$\mathrm{Im}\kappa \sim \sqrt{\frac{M_\Delta \Gamma}{2}} \sim (.7\ \mathrm{fm})^{-1} \tag{1.3}$$

while the usual mean free path in nuclear matter is

$$\lambda = (\rho\ \sigma_T)^{-1} \approx \frac{1}{2}\ \mathrm{fm} \quad . \tag{1.4}$$

at resonance. Clearly, isobar propagation should be important for a quantitative description of pion interactions at intermediate energies. Furthermore, pion absorption (which goes through the Δ) is known to constitute a very large part of the pion reaction cross section for $\pi - {}^{12}C$ scattering at 130 MeV and an even larger part at lower energies.[1] Again, this is obviously a very important part of the dynamics and must be incorporated in the formalism.

In these lectures, the isobar-hole formalism for pion reactions will be described. This is an alternate theoretical approach in which the focus is upon the isobar degree of freedom and its dynamics in the nucleus. As will become clear, the physics is not fundamentally different from that in the multiple scattering approach. However one is led to a rather different set of approximations, and we find the isobar formalism a natural way to incorporate the physics discussed above. The concept of isobar-hole doorway states was discussed first by Kisslinger and Wang[2] and has formed the basis of microscopic calculations of pion scattering.[3,4,5,6] Application of the model to $\pi - {}^{16}O$ scattering will be discussed in considerable detail; these calculations have been performed in collaboration with M. Hirata, J. H. Koch and F. Lenz.[4]

The lectures are structured as follows. We shall first discuss more precisely the introduction of the isobar as an explicit degree of freedom and the connection of such a picture to one involving only pions and nucleons. The isobar-hole formalism and its relation to the usual pion optical potential will then be described. The goal here is to evaluate the isobar-hole propagator in a finite nucleus by diagonalizing the Hamiltonian in an appropriate basis. The result is a set of isobar-hole eigenstates and eigenvalues, with strong collectivity a feature of the results. This collectivity will lead to introduction of an extended schematic model for pion scattering.[4] This is basically a generalization of the familiar Brown-Bolsterli nuclear particle-hole schematic model for excited states and allows considerable streamlining of the calculations. The collectivity will also facilitate interpretation of the role of different aspects of the isobar dynamics. In order to see explicitly how the calculations "work", a simple spinless s-wave model will be discussed.[7] Analytic results will be obtained in the high energy limit and some insight into the collectivity gained. Finally, we discuss application of the model to π - ^{16}O scattering and compare with recent high precision elastic scattering data. We thus hope to show the extent to which microscopic treatment of the many-body dynamics explains the data and the extent to which additional physical input is required. Another test of the model is provided by various inelastic processes and we discuss briefly inclusive reactions. Hopefully, the impression left will be one in which the successes of the model can be appreciated while not disguising the need for considerably more development in the microscopic calculations.

II. THE ISOBAR

There has been considerable confusion over the introduction of the Δ isobar as an explicit degree of freedom, so a fairly detailed discussion of this point is worthwhile. On the one hand, some argue that the Chew-Low model for the 3-3 resonance indicates the absence of an elementary Δ in the strong interaction Hamiltonian; others point to the quark model as providing a three-quark Δ on a footing completely equal to that of the nucleon. In fact, this distinction is practically irrelevant for pion-nucleus physics. The point is that we can always introduce an "elementary" isobar or quasiparticle to describe a two-body interaction.[8,9] This is not different fundamentally from the game played with the NN force of introducing separable potentials (of various rank) to describe the interaction very well in some limited energy range, with a residual interaction made up of the difference between the true and separable interactions. With a proper choice of the separable form, the residual interaction may be treated in perturbation theory. The isobar approximation for the πN interaction is more physically motivated because of the clear resonance behavior at invariant mass 1232 MeV; the separable behavior of the transition matrix close to a resonance is inherent in the isobar model.

Let us start by introducing a Δ particle with bare mass m_Δ. This particle must couple to the open πN channel, as in Fig. 2. We now have a coupled channel situation for the πN and Δ channels, and we can express the πN transition operator with either channel projected out:

$$T_{\pi N} = V_{\pi N} + V_{\pi N} G^0_{\pi N} T_{\pi N} \tag{2.1}$$

$$= (g_{\Delta\pi N})^\dagger G_\Delta g_{\Delta\pi N} \tag{2.2}$$

$$V_{\pi N} = (g_{\Delta\pi N})^\dagger G^0_\Delta g_{\Delta\pi N} \tag{2.3}$$

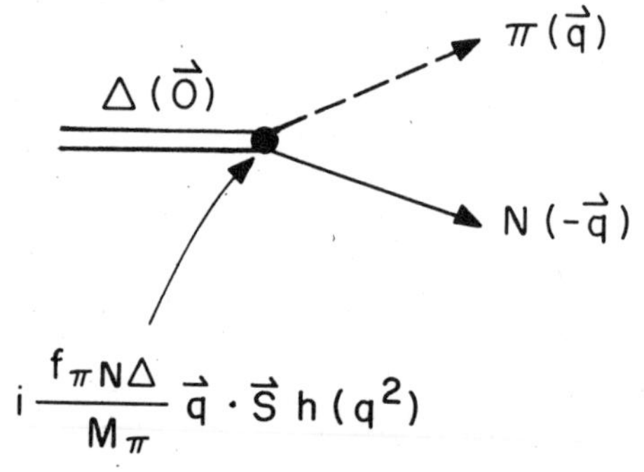

Figure 2: The πNΔ vertex.

$$G_\Delta = G^0_\Delta + G^0_\Delta \Sigma_\Delta G_\Delta \tag{2.4}$$

$$\Sigma_\Delta = g_{\Delta\pi N} G^0_{\pi N} (g_{\Delta\pi N})^\dagger \quad . \tag{2.5}$$

Equation 2.1 expresses $T_{\pi N}$ in the usual Lippmann-Schwinger form, but the interaction $V_{\pi N}$ is provided by intermediate coupling to the isobar (Equation 2.3). These equations are represented diagrammatically in Figure 3. The free isobar propagator has the usual form $G^0_\Delta(E,p) = (E - m_\Delta - p^2/2m_\Delta)^{-1}$, where m_Δ is again the bare mass; note that we have used nonrelativistic notation here for simplicity, although this will be modified shortly. An entirely equivalent expression is given in Equation 2.2, where the picture is now one of exciting the Δ through the vertex operator $g_{\Delta\pi N}$ and de-exciting it with $g_{\Delta\pi N}{}^\dagger$; G_Δ is the full or dressed isobar propagator obtained by including the self-energy (Equation 2.4) generated by intermediate coupling to the πN channel (Equation 2.5). The equivalence of the two expressions for $T_{\pi N}$ can be checked by direct expansion in powers of $g_{\Delta\pi N}$. The different approaches to the pion-nucleus problem, pion multiple scattering versus isobar-hole dynamics, can already be seen in Equations 2.1 and 2.2: the first approach focuses upon iteration of the pion-nucleon interaction, and the second upon evaluation of the isobar propagator. However, if the basic coupling is the same, the physics is identical.

To proceed further, we must be more quantitative in describing the coupling. One concern is a proper treatment of the spin and isospin. Recall that the interaction Hamiltonian describing the πNN vertex is written as $i(f_{\pi NN}/m_\pi)f(q^2)\vec{q} \cdot \vec{\sigma}\, \tau_\alpha$ where $\vec{q}$ is the

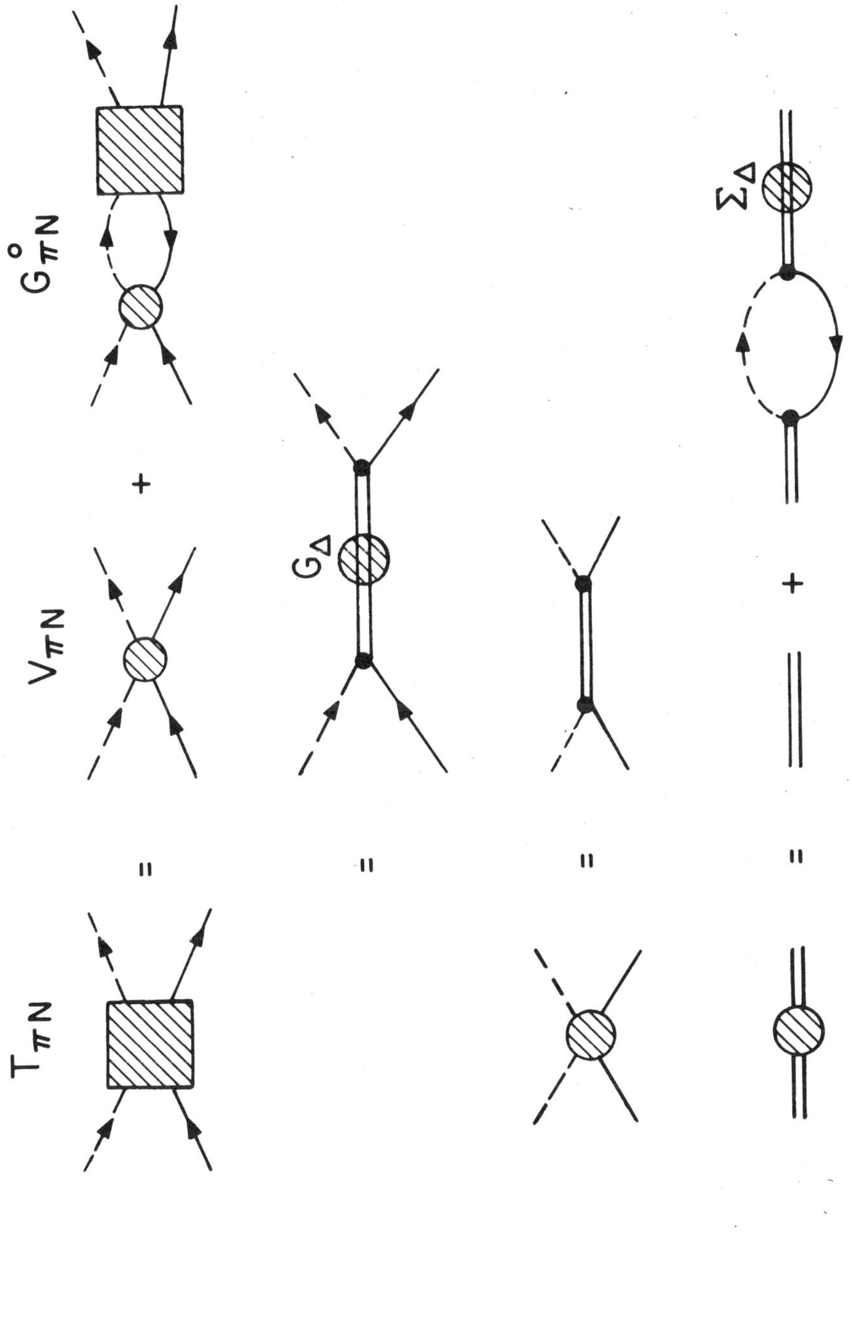

Figure 3: πN scattering in the isobar model.

pion momentum, α is the isospin label, and $f(q^2)$ is a vertex function which cuts off the coupling for large momentum transfer. We introduce an analogous description of the $\pi N\Delta$ vertex $i(f_{\pi N\Delta}/m_\pi)h(q^2)\vec{q}\cdot\vec{S}\,T_\alpha$, where $\vec{S}$ and $\vec{T}$ are the spin and isospin transitions operators.[10] These are defined according to

$$<3/2,\ m_\Delta|\vec{S}|1/2,\ m_N> \equiv \sum_\lambda <1\lambda\ ,\ 1/2m_N|3/2\ m_\Delta>\ \hat{\varepsilon}_\lambda \tag{2.6}$$

where the $\hat{\varepsilon}_\lambda$ are the spherical components of the unit vectors. This definition is obviously similar to the definition of the Pauli spin matrices, except that $\vec{S}$ connects spaces of spin 1/2 and 3/2. Obviously, these operators are not self-adjoint (in matrix representation, they have dimensionality 2 x 4). A similar definition applies to the isospin operator. The important point is that the operator $\vec{S}_i{}^\dagger\vec{S}_j$ operates in the nucleon space and is nothing other than the p-wave J = 3/2 projection operator:

$$\vec{p}\cdot\vec{S}^\dagger\vec{q}\cdot\vec{S} = \frac{2}{3}\vec{p}\cdot\vec{q} - \frac{i}{3}\vec{p}\times\vec{q}\cdot\vec{\sigma} \quad . \tag{2.7}$$

The first and second terms are the spin nonflip and spin flip terms, respectively. With the use of the vertex functions, the effective πN interaction in the center of mass frame is, for invariant energy $\sqrt{S}$,

$$V_{pq}(s) = \frac{f_{\pi N\Delta}}{m_\pi}\ \frac{\vec{p}\cdot\vec{S}^\dagger h(p^2)\vec{q}\cdot\vec{S}\ h(q^2)}{S - m_\Delta{}^2}\ \frac{f_{\pi N\Delta}}{m_\pi} \tag{2.8}$$

where $(S - m_\Delta^2)^{-1}$ is the relativistic bare propagator. With a Blankenbecler-Sugar prescription[11] for the relativistic generalization of the Lippmann-Schwinger equation, the transition operator is

$$T_{pq}(s) = \frac{V_{pq}(s)}{D(s)}$$

$$D(s) = G(s)^{-1}$$

$$= S - m_\Delta{}^2 + \frac{1}{3}\left(\frac{f_{\pi N\Delta}}{m_\pi}\right)^2 \int_0^\infty \frac{dq\ q^2}{(2\pi)^2}\ \frac{\omega_q + E_q}{\omega_q E_q}\ \frac{q^2h^2(q^2)}{(\omega_q+E_q)^2-S-i\varepsilon} \tag{2.9}$$

$$\omega_q = (m_\pi{}^2 + q^2)^{1/2} \qquad E_q = (m_N{}^2 + q^2)^{1/2} \quad .$$

The basic structure of the self energy is obvious, involving an integral over the $\pi N\Delta$ vertex functions and the free πN propagator. The exact form of the propagator and the additional factors involving ω_q and E_q are special to the Blankenbecler-Sugar prescription, but any choice will lead to the form

$$D(S) = S - m_\Delta^2 + \int_0^\infty dq\, q^2 f(q,k_s) \frac{q^2 h^2(q^2)}{q^2 - k_s^2 - i\varepsilon} \qquad (2.10)$$

where k_s is the center of mass on-shell momentum for energy $\sqrt{s}$ and $f(q,k_s)$ is a slowly varying function of the off-shell and on-shell momenta. It is important to note that the prescription of the specific form of the self-energy cannot be accomplished without specification of the dynamics (i.e., it is not a kinematic question). We shall return to this point below.

The next requirement is specification of the parameters in Equation (2.9). Contact with experiment comes from the fact that the transition operator, and therefore the dressed propagator, carries the phase of the πN scattering amplitude:

$$G(s) = |G(s)| \exp(i\delta_{33}(s)) \quad . \qquad (2.11)$$

Below pion production threshold, $\delta_{33}(s)$ is the πN scattering phase shift in the J = 3/2 T = 3/2 channel. Using the analytic properties of $G^{-1}(s)$ in the complex s-plane, we can solve the inverse scattering problem using a modified Muskelishvili-Omnes procedure.[12] The basic idea is that the discontinuity of $G(s)^{-1}$ across the unitarity branch cut is given in terms of the vertex function $h(q^2)$. After subtracting out the asymptotic behavior of the bare propagator, one can write a dispersion relation for $\ell n\, G^{-1}(s)$; the integrand in the dispersion integral then contains the difference

$$\ell n\, G^{-1}(s^+) - \ell n\, G^{-1}(s^-) = -2i\, \delta_{33}(s) \quad . \qquad (2.11)$$

The result[2] is an expression for the vertex function in terms of an integral over the phase shift:

$$\left(\frac{f_{\pi N\Delta}}{m_\pi}\right)^2 h^2(k_s^2) = -\frac{24\pi\sqrt{s}}{k_s \xi(k_s)} \{S - (m_\pi + m_N)^2\} \sin\hat{\delta}_{33}(s) \qquad (2.12)$$

$$\otimes \exp\left\{-\frac{\mathcal{C}}{\pi}\int_{(m_\pi+m_N)^2}^{\infty} ds' \frac{\hat{\delta}_{33}(s')}{S'-S}\right\}$$

where the phase $\hat{\delta}$ is determined by

$$\xi(k)\, f(k) = \xi(k) \frac{\eta e^{2i\delta} - 1}{2ik} \equiv \frac{e^{i\hat{\delta}} \sin\hat{\delta}}{k} \qquad (2.13)$$

and

$$\xi(k) \equiv \mathrm{Im} \frac{-1}{kf(k)} \qquad (2.14)$$

Note that $\xi(k)$ is one for elastic scattering, and the $\hat{\delta}$ is then the scattering phase shift. Further, the bare mass is given by[12]

$$m_\Delta{}^2 = S_R + \frac{1}{3}\left(\frac{f_{\pi N\Delta}}{m_\pi}\right)^2 \mathcal{P}\int_0^\infty \frac{dq\, q^2}{(2\pi)^2}\, \frac{\omega_q+E_q}{\omega_q E_q}\, \frac{q^2 h^2(q^2)\xi(q)}{(\omega_q+E_q)^2-S_R} \qquad (2.15)$$

where $\sqrt{S_R}$ = 1232 MeV (i.e., the renormalized isobar "mass"). The numerical results, calculated from the experimental phase shifts, are given in Reference 12. For our purposes, it is sufficient to note that the simple parametrization[13] (no inelasticity)

$$\frac{f_{\pi N\Delta}}{m_\pi}\, h(q^2) = \frac{g}{1+q^2/\alpha^2} \qquad (2.16)$$

with $\alpha = 1.8\ \mathrm{fm}^{-1}$, $m_\Delta = 6.83\ \mathrm{fm}^{-1}$ = 1348 MeV and g = 25.5 reproduces the experimental phase shifts up to 300 MeV pion kinetic energy. Note that the self-energy interactions shift the bare mass by over a hundred MeV; when inelasticity is included, the shift is even larger. As an aside, this role of the dynamical self-energy shifts raises interesting questions about the meaningfulness of static quark model calculations of hadron spectroscopy.

It is important to note that the inverse scattering procedure tells us that we can describe the scattering phase shifts exactly within the isobar model even if the fundamental interaction $V_{\pi N}$ has nothing whatsoever to do with a bare isobar. Clearly, if the πN scattering amplitude did not display resonant behavior, such a parametrization would not make much sense. We might very well expect to have a rather poor off-shell extension of the amplitude. However, given the existence of the resonance and the associated time-delay, the isobar model is simply a convenient way to describe the center of mass motion of the interacting πN system. The isobar model automatically builds in the separability of the transition operator required near any pole in the energy variable. Further, there is always the opportunity to eliminate the Δ by letting the bare mass go to infinity; i.e., in this limit, Equation (2.8) becomes

$$\begin{aligned} V_{pq}(s) &\to -\left[\tilde{f}\,\vec{p}\cdot\vec{S}^\dagger\, h(p^2)\right]\left[\tilde{f}\,\vec{q}\cdot\vec{S}\, h(q^2)\right] \\ &= -\left(\tfrac{2}{3}\,\vec{p}\cdot\vec{q} - \tfrac{i}{3}\,\vec{p}\times\vec{q}\cdot\vec{\sigma}\right)\tilde{f}^2 h(p^2)h(q^2) \end{aligned} \qquad (2.17)$$

which is simply a static, rank-one, separable potential. The off-shell πN amplitude has also been described within this framework via the inverse scattering procedure.[14]

In closing this chapter, two comments relating to use of the isobar model in the nuclear problem are in order. The first involves use of the transition operator in reference frames other than the center of mass frame. In non-relativistic potential scattering, there is of course no problem; one simply evaluates the transition operator using relative momenta of the πN system and the energy of relative motion (i.e., total energy minus center of mass kinetic energy). Relativistically, the situation is more complicated. This problem must be addressed since the pion is already quite relativistic at the resonance ($p/E \sim .9$), although the nucleon can be treated nonrelativistically to a good approximation. Within the isobar model, in which the pion is effectively projected out, both isobars and nucleons are essentially nonrelativistic, but the $\pi N\Delta$ vertex itself demands a relativistic treatment for consistency; this is because $m_\Delta \neq (m_\pi + m_N)$, so that Galilean invariance cannot be preserved at the vertex. However, the "frame transformation" in a relativistic Hamiltonian theory again cannot be specified independent of a dynamical theory for the many-body physics.[15] The problem is not kinematic and there is no real meaning to "transforming the off-shell πN transition matrix". The origin of the problem is that, in the relativistic theory, the interaction (defined as the difference between the full and noninteracting Hamiltonian), does not commute with the generators of Lorentz boosts.[15] Only within a specific theoretical framework (e.g., a Blankenbecler-Sugar treatment of the πd problem[13,16]) is the prescription for evaluating interactions of the πN subsystem unambiguous. Such a theory prescribes the relative momenta to be used in evaluating vertex functions in various frames and the Jacobian of the transformation between πN variables and relative variables (the square root of which multiplies the transition operator in different frames).

The situation is not so bleak in practice. To lowest order in ω/m, where ω is the relativistic pion energy, the relative momentum at the πN interaction vertex both in the relativistic isobar models and in direct interaction theories[17] is

$$\vec{\kappa} = \vec{q} - \frac{\omega_q}{m+\omega_q} \vec{P} \tag{2.18}$$

where $\vec{q}$ is the pion momentum and $\vec{P}$ is the πN center of mass momentum. A truly relativistic treatment is impractical for anything more than the two nucleon target, so the simple prescription in Eq. 2.18 will be adopted in the work described in later chapters.

The second point involves antisymmetrization of nucleons. In the nuclear medium, the "Pauli blocking" modification of the πN transition matrix will be somewhat different in isobar models or field theories. In all models, the Δ self energy arising from the πN coupling has the form given in Equation 2.10. At this level, the only difference in Pauli effects is connected to the off-shell extension of the transition matrix; i.e., different models will favor a somewhat different range of intermediate nucleon momenta. Such differences are not terribly great if all theories yield the same phase shifts as solutions to a Lippmann-Schwinger-like scattering equation. Field theories may have an additional effect. For example, the Chew-Low driving term has a 2π - N intermediate state, which is sensitive to Pauli effects. However, the effects incorporated in the πN self energy are far more important (the nucleons have lower intermediate momenta), so that effectively the various models are not intrinsically different.

In summarizing this chapter, we have seen how to construct a representation of the πN transition operator focusing upon the isobar degree of freedom. This is not dependent on the existence of a bare Δ. The real motivation for introducing the model is convenience. It provides a mechanism for focusing attention in the many-body problem upon various important aspects of the pion-nucleus dynamics.

III. ISOBAR-HOLE FORMALISM FOR PION SCATTERING

1. The Projection Operator Formalism

The basic idea behind the isobar-hole formalism can be understood quite compactly with projection operators. Starting with the pion plus one nucleon, let P project onto the πN space and D onto the isobar space (this is a re-writing of some results from the last chapter). In this language, the interaction term (Figure 2) is an off-diagonal channel coupling H_{PD}, and the coupled-channel Lippmann-Schwinger equations read

$$
\begin{aligned}
T_{PP} &= H_{PD}\, G_D^0\, T_{DP} \\
T_{DP} &= H_{DP} + H_{DP}\, G_P^0\, T_{PP}
\end{aligned}
\tag{3.1}
$$

Equations (2.2), (2.4) and (2.5) then read

$$
\begin{aligned}
T_{PP} &= H_{PD}\, G_D\, H_{DP} \\
G_D^{-1} &= (G_D^0)^{-1} - H_{DP}\, G_P^0\, H_{PD}
\end{aligned}
\tag{3.2}
$$

Naturally, we can evaluate Equation (3.2) with any complete set of isobar wavefunctions. In the nuclear case, it will prove convenient to use eigenfunctions ψ_i in a shell-model potential:

$$\langle \vec{p}',\vec{\pi}'|T_{PP}(E)|\vec{p},\vec{\pi}\rangle = \sum_{ij} F_i^*(\vec{p}',\vec{\pi}')\, G_{ij}(E) F_j(\vec{p},\vec{\pi})$$

$$G_{ij}(E) = \langle i|G_D(E)|j\rangle = \int \frac{d\vec{p}}{(2\pi)^3} \frac{\psi_i^*(\vec{p})\psi_j(\vec{p})}{G_D^{-1}(E,\vec{p})} \tag{3.3}$$

$$F_i(\vec{p},\vec{\pi}) = \langle i|H_{DP}|\vec{p},\vec{\pi}\rangle$$

$$= i\,\frac{\tilde{f}_{\pi N\Delta}}{m_\pi}\, h(\kappa^2)\,\psi_i^*(\vec{P})\vec{\kappa}\cdot\vec{S}\,\chi_p$$

$$\vec{P} \equiv \vec{p} + \vec{\pi} \qquad\qquad \vec{\kappa} \equiv \vec{\pi} - \frac{\omega}{m+\omega}\vec{P} \quad .$$

The coupling constant $\tilde{f}_{\pi N\Delta}$ contains the "frame transformation" factor described in the last chapter and χ_p is the nucleon spinor. The F_i are now vertex functions for creating the isobar in the shell model state ψ_i and G_{ij} is the isobar propagator connecting various shell model states.

Finally we turn to the nucleus. The P-space will now represent the pion plus the nucleus in the ground state, which we assume to be described by a closed shell harmonic oscillator Slater determinant. The doorway or D space includes states consisting of an isobar plus one nucleon hole. The Q-space includes all the rest, such as nuclear multihole states. In elastic scattering, we wish to calculate the projected transition matrix T_{PP}, and formal solution of the Lippmann-Schwinger equation gives[2,1,18]

$$T_{PP} = H_{PD}\,\{E + i\varepsilon - H_{DD} - H_{DD}^{\uparrow} - H_{DD}^{\downarrow}\}^{-1}\, H_{DP}$$

$$H_{DD}^{\uparrow} = H_{DP}\,[E + i\varepsilon - H_{PP}]^{-1}\, H_{PD} \tag{3.4}$$

$$H_{DD}^{\downarrow} = H_{DQ}\,[E + i\varepsilon - H_{QQ}]^{-1}\, H_{QD}$$

This result uses the doorway state hypothesis[19] $H_{PQ} = 0$, meaning that the complicated Q-space is reached only via the doorway states. The term in curly brackets is the isobar-hole propagator in the nucleus. The interaction $H_{DD}^{\uparrow}$ includes intermediate scattering through the nuclear ground state and therefore corresponds to pion propagation in the elastic channel. Iteration of T_{PP} in powers of $H_{DD}^{\uparrow}$ generates the multiple scattering series. The term H_{DD} is just the doorway space diagonal interaction and includes effects such as isobar kinetic energy and binding. The spreading interaction $H_{DD}^{\downarrow}$ summarizes intermediate coupling to the

Q-space. Comparing Equations (3.4) and (3.2), it is evident that the nuclear transition matrix has been expressed in terms of self-energy operators for the isobar-hole states. The evaluation will proceed by again inserting a complete set of shell model isobar states in Equation (3.4) and then diagonalizing the Δ-hole propagator.

2. Connection to Pion Multiple Scattering

In order to relate directly the isobar-hole and pion multiple scattering theories, we present a brief and rather crude review of the latter. We assume that the pion-nucleus Hamiltonian can be written as

$$H = k_\pi + V + H_N, \quad V \equiv \Sigma_i^A V_{\pi N}(i)$$
$$H_N = \Sigma_i^A k_i + \Sigma_{i<j}^A U_{ij} \tag{3.5}$$

where k is the kinetic energy operator and H_N is the nuclear Hamiltonian (assuming two-body forces). The pion-nucleus transition operator satisfies the Lippmann-Schwinger equation

$$T(E) = V + V\, G(E^+) T(E)$$
$$G(E) = (E - k - H_N)^{-1} \tag{3.6}$$

and can be solved formally to give[20]

$$T(E) = \Sigma_i^A \tau_i(E)\psi_i(E)$$
$$\tau_i(E) = V_{\pi N}(i) + V_{\pi N}(i) G(E)\tau_i(E) \tag{3.7}$$
$$\psi_i(E) = 1 + G(E)\Sigma_{j(\neq i)}^A \tau_j(E)\psi_j(E)$$

Note that τ_i is not the two-body πN transition matrix since the propagator G(E) contains the full nuclear Hamiltonian. The free t-matrix is defined by

$$t_i(\varepsilon) = V_{\pi N}(i) + V_{\pi N}(i)\; g_i\,(\varepsilon)\; t_i\,(\varepsilon)$$
$$g_i(\varepsilon) = (\varepsilon - k_\pi - k_i)^{-1} \tag{3.8}$$

and can be related to τ_i according to

$$\tau_i(E) = t_i(\varepsilon)[1 + (g_i(\varepsilon) - G(E))t_i(\varepsilon)]^{-1}$$
$$= t_i(\varepsilon) - t_i(\varepsilon)[g_i(\varepsilon) - G(E)]t_i(\varepsilon) + \ldots \tag{3.9}$$

Expansion of T(E) in powers of τ or of t yields the multiple scattering series. The optical potential $\mathcal{V}(E)$ is defined as that potential which describes correctly pion-nucleus scattering when used in the two-body Lippmann-Schwinger equation for the pion plus ground state nucleus:

$$T(E) \equiv \mathcal{V}(E) + \mathcal{V}(E) g_o(E) T(E)$$
$$g_o(E) \equiv \frac{P_o}{E - k_\pi - K_N} \tag{3.10}$$

where K_N is the nuclear kinetic energy operator and $P_o \equiv |\bar{\psi}_o\rangle\langle\bar{\psi}_o|$ projects onto nuclear ground states. Again, we can solve formally for $\mathcal{V}$ and obtain

$$\mathcal{V}(E) = T(E)\,(1 + g_o(E)T(E))^{-1}$$
$$= \Sigma_i^A\, \tau_i(E) + [\Sigma_{i\neq j}^A\, \tau_i(E)G(E)\tau_j(E) \tag{3.11}$$
$$- \Sigma_{ij}^A\, \tau_i(E) g_o(E) \tau_j(E)] + (\text{order } \tau^3)$$

$$= \Sigma_i^A\, t_i(\varepsilon) + \Sigma_{i\neq j}^A\, t_i(\varepsilon)[G(E) - g_o(E)]t_j(\varepsilon)$$
$$+ \Sigma_i^A\, t_i(\varepsilon)[G(E) - g_o(E) - g_i(\varepsilon)]t_i(\varepsilon) \tag{3.12}$$
$$+ (\text{order } t^3)$$

Of course, other expansions could be written, but the question is always one of convergence. Equation (3.11) provides a simpler expansion, but the τ-matrix involves nuclear dynamics. Equation (3.12) involves the two-body t-matrix but is more complicated in having an additional self-correlation term and in requiring choice of the parametric energy ε. Note that choice of this parameter requires examination of the second order optical potential and adoption of some convergence criterion. Note also that no questions of coherent approximation versus closure approximation enter in Equation (3.12) until second order.

Most multiple scattering calculations to date have been performed with the first order optical potential. With the isobar model for the elementary πN interaction, we can write[3,1]

$$\mathcal{V}^{(1)}(\vec{p},\vec{q};E) = A\langle\vec{p};0|\tau(E)|\vec{q};0\rangle$$

$$= A\langle\vec{p};0|g_{\Delta\pi N}^{\dagger}\ \frac{1}{D(E-H_{\Delta})}\ g_{\Delta\pi N}|\vec{q};0\rangle \tag{3.13}$$

$$H_{\Delta} \equiv T_{\Delta} + V_{\Delta} + H_{A-1}$$

To obtain this, we have written

$$\begin{aligned} H_N &= H_{A-1} + k_i + \Sigma_{j(\neq i)}U_{ij} \\ &\equiv H_{A-1} + k_i + V_{\Delta} \end{aligned} \tag{3.14}$$

where it is assumed that the ith nucleon is struck by the pion. The interaction energy of the nucleon constituent of the isobar has been transformed into a binding potential V_{Δ} acting on the isobar. The pion and nucleon kinetic energies have been decomposed into center of mass motion, given by the isobar kinetic energy operator, and relative motion, which is implicit in the calculation of D. Clearly, the resonance denominator $D(E - H_{\Delta})$ accounts for isobar propagation and binding effects; in the closure approximation, the operator H_{Δ} is taken as a constant (see Chapter VI). Assuming a shell model wavefunction for the target and inserting a complete set of isobar shell model states, we have (more details can be found in Reference 3):

$$\mathcal{V}^{(1)}(\vec{p},\vec{q};E) = \sum_{\substack{NN' \\ (occ)}} \sum_{\Delta\Delta'} F^{*}_{\Delta'N'}(\vec{p})\langle\Delta'N'^{-1}|D^{-1}(E - H_{\Delta})|\Delta N^{-1}\rangle \otimes F_{\Delta N}(\vec{q}) \tag{3.15}$$

$$\begin{aligned} F_{\Delta N}(\vec{q}) &= \langle\Delta|g_{\Delta\pi N}|\vec{q};N\rangle \\ &= \int \frac{d\vec{k}}{(2\pi)^3}\, \psi_{\Delta}^{*}(\vec{k} + \vec{q})\, \frac{\tilde{f}_{\pi N\Delta}}{m_{\pi}}\, h(\kappa^2)\vec{\kappa} \cdot \vec{S}\, \psi_N(k) \quad . \end{aligned} \tag{3.16}$$

As before, $\vec{\kappa}$ is the πN relative momentum. In performing calculations, we shall linearize the resonance denominator

$$\begin{aligned} D(E - H_{\Delta}) &\approx D(E) - \gamma(E)H_{\Delta} \\ \gamma(E) &\equiv \frac{\partial}{\partial E}\, D(E) = 1 - \frac{\partial}{\partial E}(E_R(E) - \frac{i}{2}\Gamma(E)) \quad . \end{aligned} \tag{3.17}$$

The last form is obtained by writing the resonance denominator explicitly in Breit-Wigner form; this is not an approximation since the proper energy dependence is retained.

The next step is to calculate the nuclear elastic scattering amplitude using Equations (3.15) and (3.10). The formal summation obviously yields

$$T(\vec{k}',\vec{k};E) = \sum_{\substack{NN' \\ (occ)}} \sum_{\Delta\Delta'} F^*_{\Delta'N'}(\vec{k}') \otimes \langle\Delta'N'^{-1}|[D(E) - \gamma(E)H_\Delta - \mathcal{W}]^{-1}|\Delta N^{-1}\rangle F_{\Delta N}(\vec{k}) \tag{3.18}$$

$$\langle\Delta'N'^{-1}|\mathcal{W}|\Delta N^{-1}\rangle = \langle\Delta'N'^{-1}|g_{\Delta\pi N}\, g_o(E)\, g_{\Delta\pi N}^{\dagger}|\Delta N^{-1}\rangle = \int \frac{d\vec{q}}{(2\pi)^3} \frac{F^*_{\Delta'N'}(\vec{q})\, F_{\Delta N}(\vec{q})}{(E-q^2/2Am_N)^2-\omega_q^2+i\varepsilon} \tag{3.19}$$

Clearly, $\mathcal{W}$ plays the role of a pion "exchange" interaction between isobar-hole states and enters as an isobar-hole self energy. Comparing Equations (3.18) and (3.4), we see that we have exactly the same form. The diagonal interaction is generated by H_Δ while the elastic channel interaction $H_{DD}^{\uparrow}$ is given by $\mathcal{W}$. However, the first order optical potential corresponds to a complete absence of the spreading interaction $H_{DD}^{\downarrow}$. This is expected since the latter arises from coupling to more complicated states. We shall return to this point later.

3. Isobar-Hole Calculations

The calculations are carried out by diagonalizing the isobar-hole Hamiltonian in the shell model basis. As is usual in similar Tamm-Dancoff calculations of nuclear excited states, the basis is reduced to subspaces of fixed angular momentum. Thereby, the pion-nucleus partial wave with orbital angular momentum L is described by coupling to Δ-hole states with the same angular momentum. Since the ^{16}O ground state has isospin zero, the relevant Δ-hole states all have T=1. The reduction is accomplished using standard shell model techniques; for example, the pion exchange term $\mathcal{W}$ has the Δ-hole matrix elements

$$\langle\Delta'h'^{-1};JT\|\mathcal{W}|\Delta\, h^{-1};JT\rangle = -\sum_{J'T'}(2J'+1)(2T'+1)\begin{Bmatrix} J_{\Delta'} & J_{h'} & J \\ J_\Delta & J_h & J' \end{Bmatrix}\begin{Bmatrix} 3/2 & 1/2 & 1 \\ 3/2 & 1/2 & T' \end{Bmatrix} \otimes \langle\Delta'h;J'T'|\mathcal{W}|\Delta\, h';J'T'\rangle \tag{3.20}$$

$$\langle\Delta' h|\mathcal{W}|\Delta h'\rangle = \int \frac{d\vec{k}\, d\vec{k}'\, d\vec{q}}{(2\pi)^9} h(\kappa'^2) h(\kappa^2) \left(\frac{\tilde{f}_{\pi N\Delta}}{m_\pi}\right)^2$$

$$\otimes \frac{[\psi^*_{\Delta'}(k')\vec{\kappa}' \cdot \vec{S}^\dagger \phi_{h'}(\vec{k}' - \vec{k} + \vec{p})][\phi^*_h(\vec{p})\vec{\kappa} \cdot \vec{S}\, \psi_\Delta(\vec{k})]}{k_o^2 - (\vec{k} - p)^2 + i\varepsilon} \qquad (3.21)$$

where k_o is the external pion wavenumber. In Equation (3.21), the ψ and ϕ are isobar and nuclear shell model wavefunctions, respectively, and must still be coupled to J', T' with Clebsch-Gordan coefficients. The expansion in Δ wavefunctions is, of course, truncated; typically 8 to 10 $\hbar\omega$ is sufficient for the energies of interest. Diagonalization produces a set of Δ-hole eigenstates $|D_i^L\rangle$ and complex eigenvalues ε_i^L for each angular momentum:

$$T_L(E) = g^\dagger_{\Delta\pi N} \Sigma_i \frac{|D_i^L\rangle\langle\tilde{D}_i^L|}{E-E_R+i\Gamma/2-\varepsilon_i^L(E)} g_{\Delta\pi N} \qquad (3.22)$$

The dimensionality of the space is determined by the truncation of the isobar basis set and the $\langle\tilde{D}_i|$ are the biorthogonal wavefunctions which must be introduced in the treatment of non-Hermitian Hamiltonians. The eigenstates are represented as a sum over the Δ-hole basis states

$$|D_i\rangle = \Sigma_j\, C_{ij}|(\Delta\ h)j\rangle \quad . \qquad (3.23)$$

Equation (3.22) constitutes the isobar-hole representation of the pion-nucleus transition operator.

Before applying the formalism, at least two improvements in the physical input are needed. First, Pauli blocking of the isobar decay should be incorporated. That is, the free-space width of the Δ is generated by decay into a pion and nucleon, given by the imaginary part of the Δ self-energy in Equation 2.9. In the nuclear medium, this width will be quenched since some of the "decays" will lead to already filled nucleon orbitals. Consequently, we add a term $\delta\mathcal{W}$ to the isobar-hole Hamiltonian[3]

$$\langle\Delta' N'^{-1}|\delta\mathcal{W}|\Delta N^{-1}\rangle = -\,\delta_{NN'} \sum_{\substack{m\\(occ)}} \int \frac{d\vec{k}}{(2\pi)^3}$$

$$\otimes \frac{F^*_{\Delta' m}(k) F_{\Delta m}(\vec{k})}{(E-\varepsilon_{Nm}-k^2/2Am_N)-\omega_k{}^2+i\varepsilon} \qquad (3.24)$$

where $\varepsilon_{NM} = \varepsilon_M - \varepsilon_N$. This term is diagonal in the hole and therefore is simply a Δ self-energy term. There is an obvious similarity to the pion exchange term $\mathcal{W}$. This term can be expected to be quite important at the lower energies since the Pauli blocking effect in pion quasifree scattering is substantial for small momentum transfers.[1]

Another important ingredient is the spreading interaction $H_{DD}^{\downarrow}$. As remarked earlier, this is not present in any treatment based upon the first order optical potential. Nevertheless, the very large cross section observed for pion absorption in the resonance region points to the importance of this damping mechanism for the isobar-hole states. The states reached by absorption are clearly in the Q-space, involving for example a nuclear 2p - 2h state with no pion present. Unfortunately the various Q-space couplings are too complicated to evaluate quantitatively at this time, so we resort to a phenomenology. We are guided by the expectation that absorption will be the dominant spreading interaction (at least for energies at and below the resonance) and by the fact that this process goes through the Δ. Consequently, we introduce a phenomenological isobar self-energy given by a Δ spreading potential which is local and proportional to the nuclear ground state density:

$$\mathcal{V}_{sp}(r) = \mathcal{V}_0\, \rho(r)/\rho(0) \tag{3.25}$$

The complex strength of the potential $\mathcal{V}_0$ will be taken as a free parameter fit to the forward elastic amplitude. It is important to make clear the nature of the approximation made here. First, we have assumed that the dominant Q-space effects can be incorporated as isobar self-energies, diagonal in the hole label. In principle, vertex corrections arising from Δ-h potential interactions or from multihole states should be included. We shall test this assumption to some extent later on by calculating total absorption and inclusive (π,π') cross sections from the spreading width. Second, even within this framework we have assumed a local spreading potential. Clearly, the isobar self-energy interactions should be nonlocal. Nevertheless, we do not expect to be very sensitive to this approximation since only inclusive quantities are being evaluated and the pion wavelength is still quite long at intermediate energies. The danger lies less in describing elastic scattering than in possibly extracting unphysical parameters because of a poor phenomenology. We shall test this assumption in Chapter VI when comparing the detailed calculations with data on π - ^{16}O scattering.

It is worth collecting the various parts of the isobar dynamics now included. The Δ-hole propagator which is to be diagonalized, Equation 3.18, now reads

$$G_{\Delta h}^{-1} = D(E) - \mathcal{H}_\Delta$$
$$= D(E) - [\gamma(E)H_\Delta + \mathcal{W} + \delta\mathcal{W} + \mathcal{V}_{sp}] \quad . \tag{3.26}$$

This reflects not only pion multiple scattering, as would be included in a standard first order optical potential calculation, but also isobar propagation and binding, Pauli blocking, and coupling to more complicated states (e.g., absorption) through the phenomenological spreading potential. Additional effects such as a ρ-meson effective isobar-hole force could also be included quite easily. The eigenvalues reflect directly the importance of the various dynamical effects.

Before applying the model, we summarize the approximations which have been introduced:

(i) shell model for the nuclear ground state wavefunction
(ii) retention of only the resonance part of the πN interaction
(iii) neglect of Coulomb effects
(iv) retention of H_Δ to first order only
(v) truncation of the isobar shell model space
(vi) neglect of vertex corrections
(vii) local density approximation for the spreading interaction.

Approximations (ii) and (iii) will not be made in Chapters VI and VII, where quantitative comparison with data will be made. Approximations (iv) and (v) are amenable to quantitative evaluation and will be quite suitable for the applications discussed. We examine in Chapter VI the effects of relaxing approximation (vii) with a reasonable extension to nonlocal form. However, approximations (i) and (vi) are far more difficult to evaluate meaningfully without a significant increase in calculational complexity. The shell model approximation is probably quite good for our calculations of elastic scattering, although some uncertainty in the interpretation of the strength of the spreading potential may arise from the neglect of short range correlations. The neglect of vertex corrections may become significant at the higher energies, where multinucleon knockout by pions is known to increase. This will be discussed in Chapters VI and VII.

We now show some results[4] for $\pi - {}^{16}O$ at 140 MeV (close to the peak of the total cross section). The calculations discussed here do not include Coulomb or nonresonant πN interaction, so comparison with data is deferred to Chapter VI. In Table 1, we give the contribution of each 0^- eigenstate to the partial wave transitiom matrix (see Equation 3.22) and the associated eigenvalues. The basis was 17 dimensional with truncation at 8 $\hbar\omega$ (the dimensionalities are considerably greater for higher L).

i	ε_i [MeV]		t_i	
1	-38.3	-i175.2	-.124	i.675
2	+28.7	- i43.1	.043	-i.066
3	+55.3	- i5.9	.034	-i.027
4	+41.8	+ i2.2	.025	-i.016
5	+68.3	+ i4.8	-.005	-i.008
6	+11.6	- i24.9	.004	-i.007
7	+23.3	+ i8.9	.001	i.003
8	+46.2	- i35.9	-.001	-i.003
9	+46.0	+ i11.0	-.004	i.002
10	+61.3	+ i23.2	.002	-i.001
11	+28.1	+ i8.3	.002	i.001
12	+51.1	- i36.1	.000	i.001
13	+45.6	+ i19.3	.001	i.001
14	+46.5	+ i14.6	.000	-i.001
15	+39.4	+ i10.3	-.000	i.000
16	+53.0	+ i13.3	-.000	-i.000
17	+28.3	+ i11.3	.001	-i.000

Table 1: Eigenvalues ε_i of the Δ-hole eigenstates and contributions to the nuclear amplitude $T = \Sigma_i t_i = \frac{i}{2}(1 - \eta e^{2i\delta})$ for the 0^- partial wave at T_π = 140 MeV. With the truncation described in the text, the 0^- Δ-hole space is 17-dimensional.

The interesting result here is the strong "collectivity" of one eigenstate[4]; i.e., one eigenstate provides by far the largest contribution to the transition matrix and has an eigenvalue significantly split off from the rest of the spectrum. Note that this splitting occurs principally in the imaginary part of the eigenvalue. Indeed, $\mathrm{Im}\,(\Sigma_i\, \varepsilon_i) = -193.9$ MeV, so that $\mathrm{Im}\,\varepsilon_1 \approx 0.9\ \mathrm{Im}\,(\mathrm{Tr}\mathcal{H}_\Delta)$. This collectivity is expected through analogy with the nuclear Brown-Bolsterli particle-hole schematic model; this will be discussed in detail in the next chapter. In any case, it is clear from Equation 3.22 that the various eigenstates contribute to the elastic transition matrix only to the extent that they have reasonable overlap with the (normalized) state

$$|\Delta_o^L\rangle \equiv \frac{g_{\Delta\pi N}|0;(\vec{k})_L\rangle}{[\langle(\vec{k})_L 0|g^{\dagger}_{\Delta\pi N}\, g_{\Delta\pi N}|0;(\vec{k})_L\rangle]^{1/2}} \qquad (3.27)$$

This state is found to have an overlap of 0.9 with the leading eigenstate and represents the linear response of the nuclear ground state to the incoming pion. It is a coherent superposition of Δ-hole states. For example, the coefficients of the Δ-hole basis states in the doorway state $|\Delta_o\rangle$ are given in Table 2.[4] Note that all the allowed isobar angular momentum states have appreciable contributions in the doorway, with the s-shell isobar having the largest weight. Convergence comes about in the expansion in the Δ principal quantum number. We note that the same oscillator parameter was used for the isobar and nucleon wavefunctions, and it is conceivable that convergence could be aided by using a different parameter or basis set for the isobar.

Some insight into the dynamics can be obtained by examining the decomposition of the expectation value of $\mathcal{H}_\Delta$ in the doorway state:

$$\begin{aligned} \varepsilon_A &\equiv \langle\Delta_0|\mathcal{H}_\Delta|\Delta_0\rangle \\ &= \langle\Delta_0|\mathcal{W}|\Delta_0\rangle + \langle\Delta_0|\delta\mathcal{W}|\Delta_0\rangle + \gamma(E)\langle\Delta_0|H_\Delta|\Delta_0\rangle \qquad (3.28) \\ &+ \langle\Delta_0|\mathcal{V}_{sp}|\Delta_0\rangle \end{aligned}$$

In these calculations, we have used a spreading potential of central strength $\mathcal{V}_o \approx -70i$ MeV. We show the different contributions[4] to Im ε_A in Figure 4, together with the value for the leading collective 0^- eigenstate. The dominant width Im ε_π comes from the pion rescattering term $\mathcal{W}$. This is basically the origin of the collectivity: H_{DP} enters in Equation 3.4 both in generating the doorway state and, in second order, the elastic channel interaction $\mathcal{W} = H_{DD}^{\dagger}$. However, the other aspects of the isobar dynamics are far from negligible. The spreading interaction width Im ε_Δ is larger than the free width, providing an important damping mechanism for the isobar-hole states. Further, the Pauli term Im ε_P quenches the free width substantially. These terms will be seen in Chapter VII to have important consequences for the inelastic reactions.

4. Optical Potential Games

Before developing further and applying the isobar-hole formalism, it is instructive to translate crudely the isobar-hole doorway state parameters back into an optical potential language.

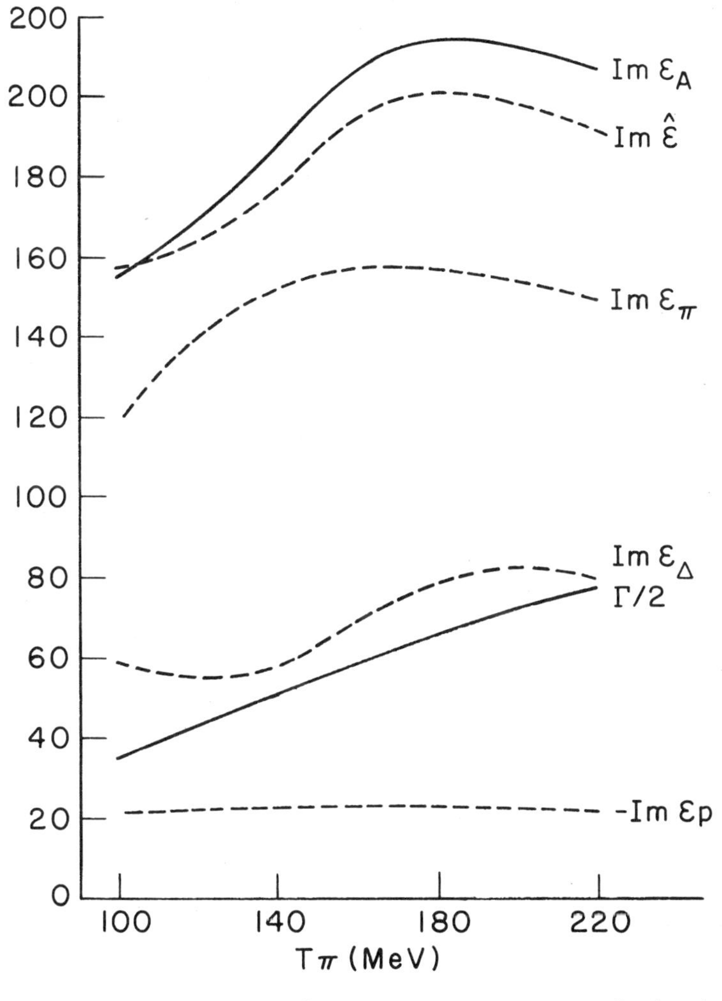

Figure 4: Decomposition of the imaginary part of the doorway expectation value of the isobar-hole Hamiltonian. The eigenvalue of the leading eigenstate is denoted by $\hat{\varepsilon}$, and Γ is the free-space isobar width.

	Δ			hole			
i	n	ℓ	j	n	ℓ	j	c_i
1	0	1	1/2	0	0	1/2	.413
2	1	1	1/2	0	0	1/2	.241
3	2	1	1/2	0	0	1/2	.082
4	3	1	1/2	0	0	1/2	.019
5	0	2	1/2	0	1	1/2	.337
6	1	2	1/2	0	1	1/2	.273
7	2	2	1/2	0	1	1/2	.114
8	3	2	1/2	0	1	1/2	.031
9	0	0	3/2	0	1	3/2	-.578
10	1	0	3/2	0	1	3/2	-.118
11	2	0	3/2	0	1	3/2	.084
12	3	0	3/2	0	1	3/2	.060
13	4	0	3/2	0	1	3/2	.019
14	0	2	3/2	0	1	3/2	.337
15	1	2	3/2	0	1	3/2	.273
16	2	2	3/2	0	1	3/2	.114
17	3	2	3/2	0	1	3/2	.031

Table 2: Decomposition of the doorway state $|\Delta_o\rangle$ into Δ-hole shell model basis states for the 0^- partial wave at 140 MeV.

This also gives a semiquantitative preview of the results to be obtained later. We anticipate the results of Chapter VI to the extent that the collectivity displayed above persists in the more refined calculations and that a good description of the data is obtained. Then, since the Pauli and spreading interactions give isobar (as opposed to isobar-hole) self-energies, and since the doorway state itself already gives a reasonable approximation to the transition matrix, we carry these self-energies over into the resonance denominator in the first order pion optical potential. More specifically, we shall be extremely crude and use the high energy approximation[21] for the nuclear amplitude

$$F(q) = \frac{ik}{2\pi}\int d^2b\, e^{i\vec{q}\cdot\vec{b}}\Gamma(b)$$

$$\Gamma(b) = 1 - e^{i\chi(b)}$$

$$\chi(b) = \frac{-1}{2k}\int_{-\infty}^{+\infty} dz\,\mathcal{V}(\vec{b} + \hat{k}Z) \tag{3.29}$$

$$\mathcal{V}(r) = -4\pi\rho(r)f_{\pi N}(0)$$

where $\chi(b)$ is the semiclassical phase shift for impact parameter b and v(r) is the Glauber first order optical potential. Assuming resonance dominance of the πN amplitude, we have

$$f_{\pi N}(0) = -\frac{k\,\sigma_T^{\,R}\Gamma/8\pi}{E-E_R+i\Gamma/2} \tag{3.30}$$

where $\sigma_T^{\,R}$ = 140 mb is the average of the πp and πn total cross sections at resonance. Motivated by the isobar-hole results, we modify the resonance denominator by the addition of density-dependent self-energies:

$$f_{\pi N}^{(0)} = \frac{-k\sigma_T^{\,R}\Gamma/8\pi}{E-\tilde{E}_R+i\Gamma/2+\frac{i}{2}[\Gamma_{Abs}\tilde{\rho}(r)+\Gamma_{Pauli}\tilde{\rho}(r)^{1/3}]} \tag{3.31}$$

$$\tilde{\rho}(r) \equiv \rho(r)/\rho(0)$$

In the resonance region, the widths are most important and we put $\tilde{E}_R = E_R$. The parameters Γ_{Abs} and Γ_{Pauli} are fixed according to the s-wave doorway state expectation values of the spreading and Pauli interactions, respectively (see Fig. 4):

$$\Gamma_{Abs}/2 \approx 70\text{ MeV}\ ;\ \Gamma_{Pauli}/2 \approx -\ 20\text{ MeV} \tag{3.32}$$

Of course, the elastic width Im ε_π does not enter since we work here with the optical potential rather than the transition matrix (see Equations 3.15 and 3.18). We have assigned the same local density dependence to the absorption width as was used in the

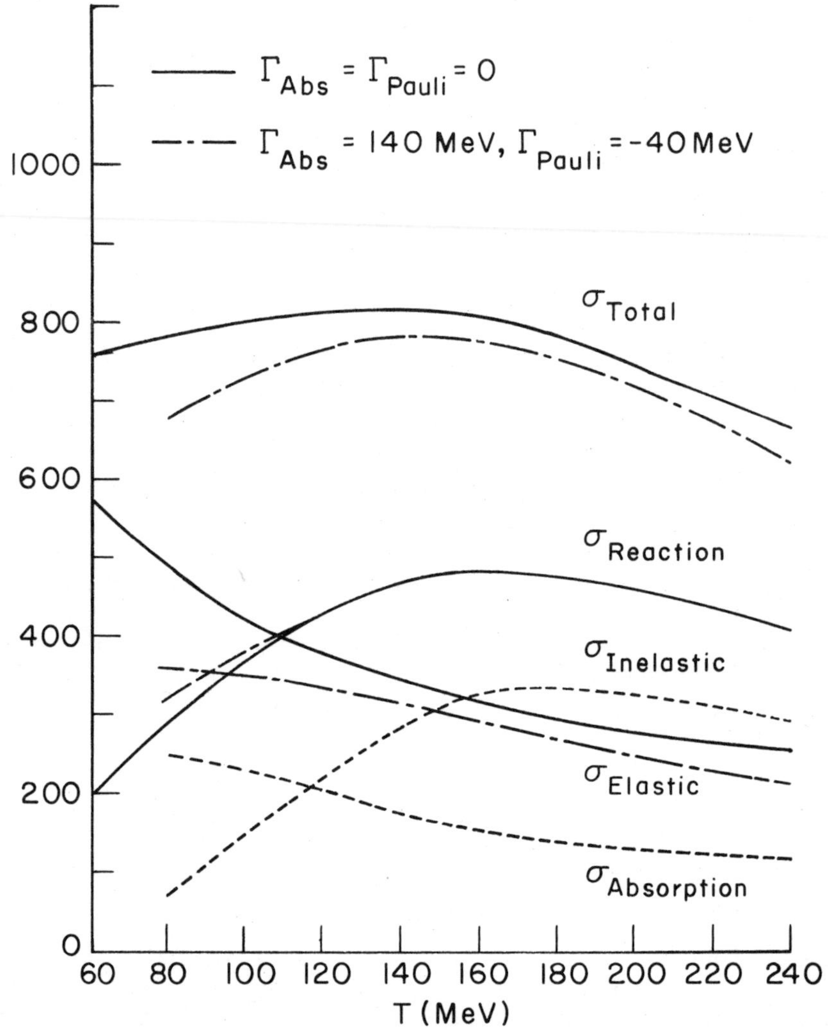

Figure 5: Elastic, reaction and total cross sections for $\pi - {}^{16}O$ scattering, calculated in the eikonal approximation. Solid curves result from using the free πN amplitude, Equation 3.30; dot-dash curves result from including absorption and Pauli widths, Equation 3.31. Dashed curves represent decomposition of reaction cross section into total absorption and total inelastic pion scattering cross section.

spreading potential and take the Pauli width proportional to the local Fermi momentum. The results for the total elastic and reaction cross sections for $\pi - {}^{16}O$ are shown in Figure 5, where it is seen that the main effect is a suppression of the elastic cross section at the lower energies. An excessively large elastic cross section is characteristic of first order optical potential calculations. Furthermore, the reaction cross section can be written as directly proportional to the imaginary part of the optical potential. We interpret $(\Gamma + \Gamma_{Pauli})$ and Γ_{Abs} as partial widths into quasifree and absorption channels, respectively, and thereby are able to decompose the reaction cross section into these two channels. This is also shown in Figure 5. The absorption cross section is seen to dominate below about 120 MeV, in good agreement with the limited data available. These results are encouraging for thinking that the model incorporates the important physics and will emerge also in the more quantitative results given below.

In calculating the absorption and inelastic (i.e., quasifree) cross section, it is interesting to observe the impact parameter dependence of the associated profile functions $\Gamma(b)$. As can be seen from Equations (3.29) and (3.31), the quasifree profile function is suppressed strongly in the nuclear interior. The reaction cross section is changed only little because the absorption profile function is large for small impact parameter. Consequently the rough picture which emerges is of comparatively strong central absorption and peripheral quasifree knockout.

One last exercise is to calculate the elastic differential cross section given by Equation 3.39. In particular, analysis of the differential cross section up to the first minimum has been proposed as a good means for extracting the matter radius of nuclei. We show in Figure 6 the first order Glauber results for two different values of the ${}^{16}O$ harmonic oscillator parameter. As expected, the larger radius leads to a narrower forward peak in the differential cross section. On the other hand, inclusion of the Δ self-energy modifications also changes the shape of the cross section. The lesson appears to be that the "higher order effects" in the optical potential need very careful treatment before a meaningful extraction of neutron radii at the level of say 0.1 fm can be made with pions in the resonance region. This holds also for π^+/π^- comparisons on $N \neq Z$ nuclei, since the absorption term has strong isospin dependence.

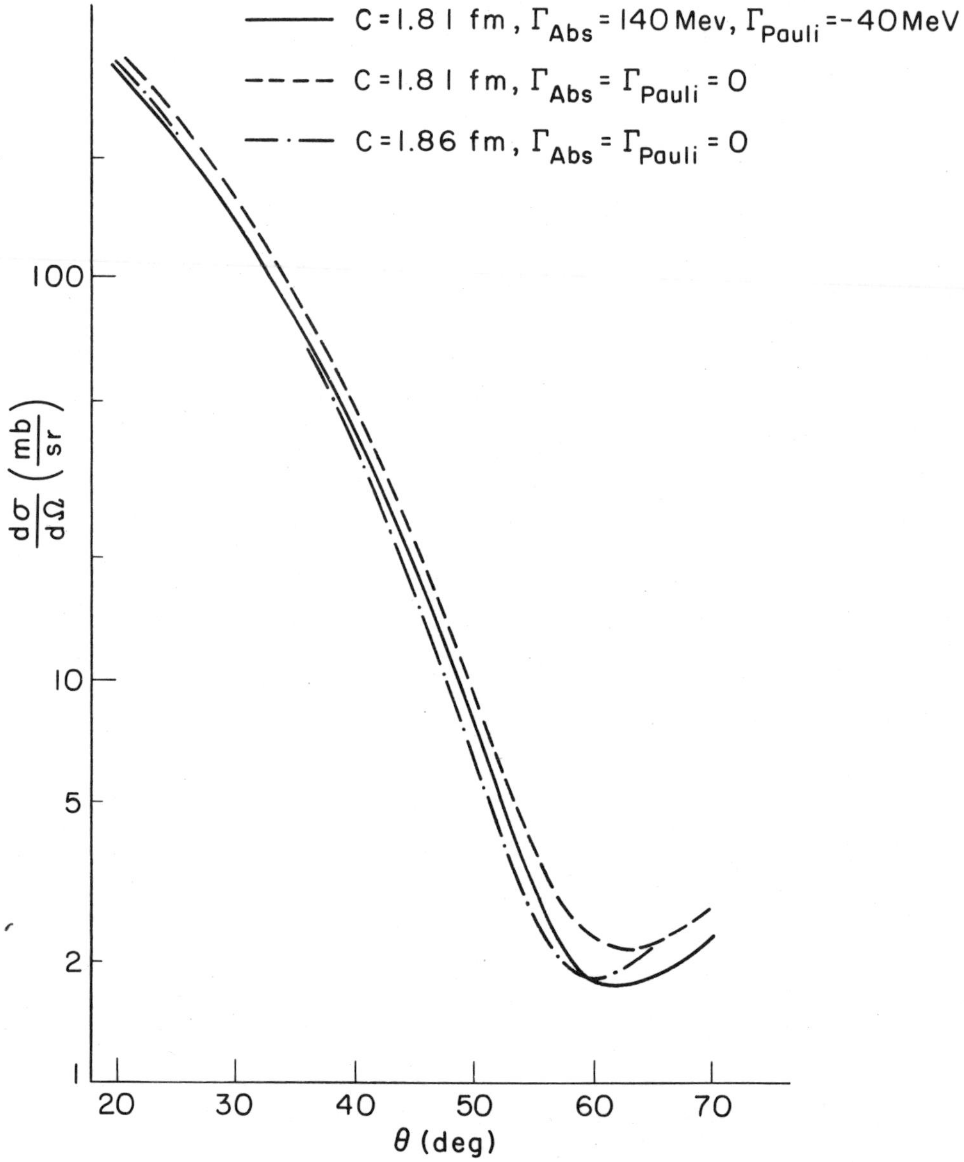

Figure 6: Elastic differential cross section in the eikonal approximation. Solid and dashed curves give results for harmonic oscillator parameter c = 1.81 fm and for πN amplitude with and without absorption and Pauli widths, respectively. Dot-dashed line is first order result with a larger oscillator parameter.

IV. EXTENDED SCHEMATIC MODEL

We remind the reader of the important result obtained in the isobar-hole framework in the last chapter: the isobar-hole "spectrum" is collective in the sense that one or two eigenstates, described as coherent superpositions of Δ-hole shell model basis states, dominate the nuclear partial wave transition matrix. This specifies that the leading eigenstate is very similar to the doorway state. This state is split off from the rest of the Δ-hole spectrum in the imaginary part of the eigenvalue. These features led to reformulation of the isobar-hole formalism in Reference 4. The approach, called the extended schematic model, uses the results summarized above to achieve rapid convergence without diagonalizing in the full isobar-hole space.

1. The Brown-Bolsterli Model

It is useful to begin with a brief review of the Brown-Bolsterli degenerate schematic model[22] for nuclear particle-hole states. We might imagine as an application the calculation of collective states contributing to a low energy photoreaction, as indicated in Figure 7a. Absorption of the photon leads to a doorway state described as a linear superposition of the particle-hole basis states:

$$\begin{aligned} &\langle (ph)i|H_\gamma|0\rangle \equiv \gamma_i \\ &\overline{\Psi}_D = N\,\Sigma_i\,\gamma_i\,\phi_i = N\,\vec{\gamma}\cdot\vec{\phi} \end{aligned} \tag{4.1}$$

where $\phi_i = |(ph)i\rangle$ represents the ith p-h basis state and ψ_D is the doorway state. We also need the nuclear eigenstates and, in the degenerate schematic model, the interaction in the p-h basis is assumed to have the form

$$V_{ij} = g\,D_i\,D_j \tag{4.2}$$

This structure emerges naturally from p-h force of the type depicted in Fig. 7b, where the dashed line represents the interaction (e.g., meson exchange). The D_i play the role of vertex functions and g incorporates coupling constants and the propagator of the (static) exchanged particle. With the further assumption of degeneracy of the unperturbed p-h states ϕ_i, the Hamiltonian is diagonalized easily. There is one collective state

$$\begin{aligned} &\overline{\Psi}_o = N\,\Sigma_i\,D_i\,\phi_i = N\,\vec{D}\cdot\vec{\phi} \\ &\varepsilon_o = g\,\Sigma_i\,D_i^2 = \text{Trace}\,(V) \quad . \end{aligned} \tag{4.3}$$

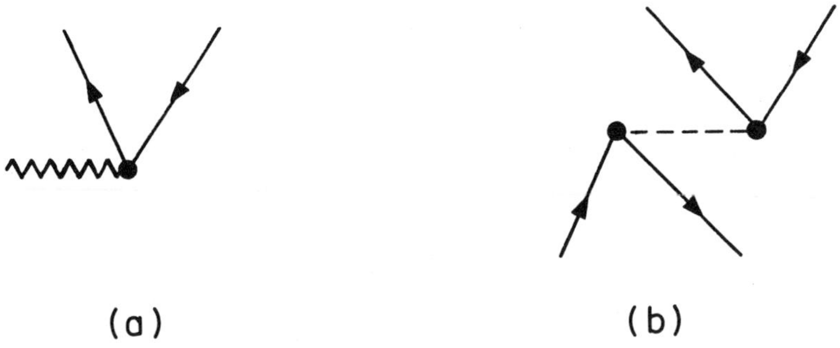

Figure 7: (a) Photoexcitation of a particle-hole state. (b) Exchange diagram for the particle-hole interaction.

The remaining eigenstates are orthogonal to $\overline{\psi}_o$ and remain degenerate ($\varepsilon_i = 0$, $i > 0$). Clearly, if the γ_i are proportional to the D_i, the collective state $\overline{\psi}_o$ becomes identical to the doorway state and concentrates all the photoexcitation strength. This state has been shifted "maximally" from its unperturbed position. This simple model was used to show simply the origin of giant resonance states in a shell model picture.

One might imagine that further residual p-h interactions will weaken the collectivity. The simplest example of this is obtained by adding an additional "separable" interaction

$$V_{ij} = g\, D_i\, D_j + h\, E_i\, E_j \quad . \tag{4.4}$$

Note that the spectrum in an N-dimensional basis can always be reproduced trivially by going to a rank N separable Hamiltonian, so that this procedure is very analogous to the construction of low rank separable potentials to reproduce matrix elements of more complicated forces. In the degenerate limit, this problem is also solved easily[23]; the eigenstates and eigenvalues are

$$\overline{\Psi}_\alpha = \vec{C}^\alpha \cdot \vec{\phi}$$

$$\vec{C}^\alpha = \begin{cases} N_\alpha [g\, \vec{D}\, \vec{D} \cdot \vec{E} - (\varepsilon_\alpha - g\, \vec{D}^2)\vec{E}], & \alpha = 1,2 \\ N_\alpha\, \vec{D} \times \vec{E}\ , & \alpha \geq 3 \end{cases} \tag{4.5}$$

$$\varepsilon_\alpha = \begin{cases} \frac{1}{2}[g\, \vec{D}^2 + h\, \vec{E}^2 \pm \{(g\, \vec{D}^2 - h\, \vec{E}^2)^2 + 4gh(\vec{D} \cdot \vec{E})^2\}^{1/2}], & \alpha = 1,2 \\ 0\ , & \alpha \geq 3 \end{cases}$$

The residual interaction has now fragmented the strength into two states. Still assuming that $\vec{\gamma} \propto \vec{D}$, the doorway state is no longer an eigenstate but has overlap with $\overline{\Psi}_{1,2}$; therefore, each of these will contribute to the photoreaction. Note that the new state mixed in "points" in the direction $\vec{E}$ in the particle-hole Hilbert space. This is to be expected since operation of the Hamiltonian on the doorway state and orthogonalization leads to a new state

$$\overline{\Psi}_D{}' \equiv V\, \overline{\Psi}_D - \overline{\Psi}_D \langle \overline{\Psi}_D, V\, \overline{\Psi}_D \rangle$$
$$\sim \vec{E} \tag{4.6}$$

In other words, diagonalization of the Hamiltonian leads to a rotation in the $\vec{D}$, $\vec{E}$ plane of the Hilbert space.

How does this apply to pion scattering? The analogy between pion scattering and the Brown-Bolsterli model is clear from a comparison of Figures 7 and 8. The isobar-hole doorway state is created by absorption of an on-shell pion. The degeneracy assumption corresponds to the closure limit ($H_\Delta \to$ constant) and the residual interaction corresponds to nonstatic pion "exchange" between Δ-hole states. For on-shell pion exchange (i.e., on-shell pion multiple scattering), the Δ-hole interaction has the same relation to the doorway excitation vertex as does Equation (4.2) to Equation (4.1). Note however that the on-shell condition implies that the interaction is imaginary, so that the eigenvalue splitting is in the imaginary part. In this limit, the doorway state has all the pion excitation strength in pion-nucleus scattering. The residual interaction which fragments the strength then includes the off-shell pion rescattering as well as binding effects, the Pauli blocking, and the spreading interaction. The collectivity will remain if the imaginary part of the pion exchange interaction $\mathcal{W}$ contributes the largest self-energy in the isobar-hole propagator. This is the case for the central par-

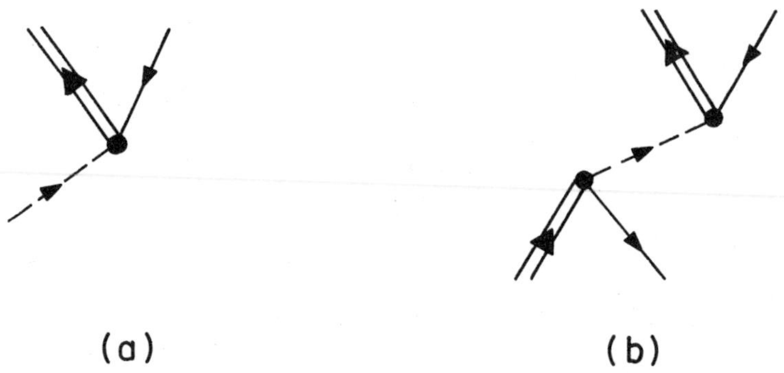

Figure 8: (a) Pion excitation of a Δ-hole state. (b) Pion "exchange" between Δ-hole states.

tial waves over the entire energy range of interest, as seen in Figure 4.

2. The Extended Schematic Model for Pion Scattering

To have an effective theory built explicitly upon the collectivity in the isobar-hole approach, one needs a systematic approach for evaluating corrections at any level of approximation. We accomplish this by introducing a change of basis in the Δ-hole space.[4] Clearly, the doorway state plays a special role in the theory and we use this as a "starting vector". The construction of the rest of the basis is motivated by Equation 4.6 of the degenerate p-h model: the "next" contributing state is obtained by applying the Hamiltonian to the doorway state and orthonormalizing. We continue this procedure (in principle) until the entire Δ-hole space is spanned:

$$|\Delta^L_\alpha\rangle = N_\alpha\{\mathcal{H}_\Delta|\Delta^L_{\alpha-1}\rangle - \Sigma^{\alpha-1}_{\beta=0}|\Delta^L_\beta\rangle\mathcal{H}_{\beta,\alpha-1}\},\ \alpha > 0$$
$$\mathcal{H}_{\alpha,\beta} \equiv \langle\tilde{\Delta}^L_\alpha\mathcal{H}_\Delta|\Delta^L_\beta\rangle. \tag{4.7}$$

Note that this procedure will span only that part of the Hilbert space which has some overlap with the doorway state, but this is anyway the only part which can contribute to the nuclear transition matrix. This construction has the tridiagonal property

$$\mathcal{H}_{\alpha\beta} = 0 \quad \text{for } |\alpha - \beta| > 1 \tag{4.8}$$

and generates a continued fraction representation of the doorway matrix element of the Δ-hole propagator:

$$\langle\tilde{\Delta}_0^L|G_{\Delta h}|\Delta_0\rangle = \langle\tilde{\Delta}_0|(E - E_R + i\Gamma/2 - \mathcal{H}_\Delta)^{-1}|\Delta_0\rangle$$

$$= \cfrac{1}{E - E_R + \frac{i}{2}\Gamma - \mathcal{H}_{00} - \cfrac{\mathcal{H}_{01}^2}{E - E_R + \frac{i}{2}\Gamma - \mathcal{H}_{11} - \cfrac{\mathcal{H}_{12}^2}{E - E_R + \frac{i}{2}\Gamma - \mathcal{H}_{22} - \cdots}}} \tag{4.9}$$

The procedure described by Equations 4.7 through 4.9 is familiar in many branches of physics, variously passing under the names method of minimized iterations, Lanczos method, coupled quasi-particle method, the method of Padé approximants. It has been applied in both nuclear structure and nuclear reaction theory. Extensive shell model calculations[24] have used the construction of Equation 4.7, starting from an ansatz for $|\Delta_0\rangle$, to calculate low-lying collective states. The continued fraction representation Equation (4.9) is familiar from reaction theory[19], reflecting sequential coupling to states of higher and higher complexity. We shall refer to the states $|\Delta_\alpha\rangle$ as the doorway basis.

The point of introducing this basis is the expectation that the continued fraction will converge very rapidly; i.e., we may expect that construction of only the first few doorway states will be sufficient for high accuracy. This can be checked by systematically extending the basis. Further, within a doorway basis of dimensionality N, the eigenvalue problem can be seen from Equation (4.9) to consist of solving for the roots of an Nth order polynomial in $(E - E_R + i\Gamma/2)$. Also, it is worth noting that the first off-diagonal matrix element corresponds directly to the fluctuation of the Hamiltonian in the doorway state

$$\mathcal{H}_{01} = [(\mathcal{H}^2)_{00} - (\mathcal{H}_{00})^2]^{1/2} \tag{4.10}$$

Obviously this reflects the extent to which the doorway state is an eigenstate.

A test of the convergence is shown in Table 3. The exact results refer to the transition matrix obtained by diagonalizing in

	$\varepsilon(J^\pi)$ [MeV]		$T(J^\pi)$	
	0^-	4^-	0^-	4^-
Exact	-38.4 - i175.2	18.2 - i28.0 46.1 - i22.9	-.021 + i.553	.108 + i.189
Schematic N=1	-25.8 - i185.2	20.85 - i15.90	-.022 + i.640	.114 + i.198
N=2	-38.2 - i175.4	11.6 - i29.2 41.9 - i10.9	-.019 + i.549	.108 + i.188
N=3	-38.4 - i175.2	17.3 - i27.1 49.3 - i27.2	-.021 + i.553	.108 + i.188

Table 3: Eigenvalues of the dominant eigenstates and partial wave transition matrix for $L^\pi = 0^-, 4^-$ at 140 MeV. "Exact" refers to the results of the full isobar-hole calculations; "schematic" refers to the extended schematic model calculation using N = 1, 2 and 3 doorway basis states. Taken from Reference 4.

the full Δ-hole shell model basis and to the eigenvalues of the leading eigenstates. For the $L^\pi = 4^-$ partial wave, two eigenstates have substantial contributions. The schematic results refer to calculations done in the doorway basis truncated at dimensionality N. Clearly, the transition matrix for both the central and the peripheral partial wave is essentially exact in the two-dimensional doorway basis. The eigenvalues converge slightly more slowly, but even here it is clear that the "eigenstates" obtained by diagonalizing the first few doorway states are very similar to the "true" eigenstates calculated in the full Δ-hole space. The extended schematic model simplifies the calculations significantly and facilitates insight into the reaction dynamics. This is the calculational approach used for obtaining the results in Chapters VI and VII.

V. A MODEL PROBLEM

We conclude this formal part of the lectures by examining a simple model problem in the isobar-hole framework. This model has been worked out in collaboration with F. Lenz and K. Yazaki[17] and affords an opportunity to see explicitly the connection between multiple scattering and isobar dynamics. The model consists of the simplest, non-trivial physical situation: pion scattering from an uncorrelated nucleus (A >> 1) in the fixed nucleon approximation, with zero-range s-wave πN interactions. The nuclear multiple scattering series for the elastic amplitude is then

$$F(\vec{k}',\vec{k}) = A \int d\vec{x}_1 \rho(\vec{x}_1) e^{i(\vec{k}' - \vec{k}) \cdot \vec{x}_1} f(\vec{k}',\vec{k})$$
$$+ A^2 \int d\vec{x}_1\, d\vec{x}_2\, \rho(\vec{x}_2)\rho(\vec{x}_2) \int \frac{d\vec{p}}{(2\pi)^3} e^{i(\vec{k}'-\vec{p}) \cdot \vec{x}_1} f(\vec{k}',\vec{p}) \frac{4\pi}{p^2-k^2-i\varepsilon}$$
$$\otimes\, e^{i(\vec{p}-\vec{k}) \cdot \vec{x}_2} f(\vec{p},\vec{k}) + \ldots \tag{5.1}$$

Here, $f(\vec{p},\vec{q})$ is the off-shell πN scattering amplitude and we assume

$$f(\vec{p},\vec{q};E) = - \frac{g^2/4\pi}{E-E_R+i\Gamma(E)/2} \tag{5.2}$$

With such a simple assumption, the equivalent optical potential would be local and proportional to the ground state density.

In performing isobar-hole calculations, we are effectively evaluating the Δ-hole spectrum for various values of the total angular momentum. This leads us to consider the partial wave decomposition of the nuclear amplitude:

$$F(\vec{k}',\vec{k}) = \Sigma_{\ell m}\, 4\pi\, Y_{\ell m}^{*}(\hat{k}')\, Y_{\ell m}(\hat{k})\, F_\ell(k)$$

$$F_\ell(k) = \int_0^\infty dx_1\, x_1^2 \frac{[g\,\phi(x_1)\, j_\ell(k\,x_1)]^2}{E-E_R+i\Gamma/2}$$

$$+ \int_0^\infty dx_1\, x_1^2 \int_0^\infty dx_2\, x_2^2 \frac{[g\phi(x_1) j_\ell(kx_1)]\mathcal{W}_\ell(x_1,x_2)[g\phi(x_2) j_\ell(kx_2)]}{(E-E_R+i\Gamma/2)^2}$$

$$+ \ldots \tag{5.3}$$

$$\mathcal{W}_\ell(x_1,x_2) \equiv 4\pi \int \frac{d\vec{p}}{(2\pi)^3} \frac{j_\ell(px_1) j_\ell(px_2)}{p^2-k^2-i\varepsilon}\, g^2\phi(x_1)\phi(x_2)$$

$$= ik\,[g\,\phi(x_1)]\, h_\ell^{(1)}(k\,x_>) j(k\,x_<)\,[g\,\phi(x_2)] \tag{5.4}$$

We have defined $A\rho(x) \equiv \phi^2(x)$, so that $\phi(x)$ plays the role of a nucleon orbital wavefunction in this model. We can now sum the entire multiple scattering series to obtain

$$F_\ell(k) = -\int_0^\infty dx_1 x_1^2 \int_0^\infty dx_2 x_2^2\, \psi_0^\ell(x_1)\Delta_\ell(x_1,x_2)\psi_0^\ell(x_2)$$

$$\psi_0^\ell(x) = g\,\phi(x) j_\ell(kx) \tag{5.5}$$

$$\Delta_\ell = [E - E_R + i\Gamma/2 + \mathcal{W}_\ell]^{-1}$$

Therefore, we have an expression entirely equivalent to the multiple scattering series but with the multiple scattering incorporated into an isobar self-energy $\mathcal{W}_\ell$. It is clear from Equation (5.4) that $\mathcal{W}_\ell$ is the non-static pion exchange interaction and that the on-shell piece is

$$\mathrm{Im}\,\mathcal{W}_\ell(x_1,x_2) = k\,\psi_0^\ell(x_1)\psi_0^\ell(x_2) \quad . \tag{5.6}$$

This is again the condition for the degenerate schematic model, since the problem is to evaluate the expectation value of Δ_ℓ in the doorway ψ_0^ℓ. In other words, if only on-shell rescattering is kept, we have

$$[F_\ell(k)]_{\text{on-shell}} = \frac{-\int_0^\infty dx\, x^2 [\psi_0^\ell(x)]^2}{E-E_R+i\Gamma/2+ik\int_0^\infty dx\, x^2[\psi_0^\ell(x)]^2} \tag{5.7}$$

Clearly, Im$\mathcal{W}$ generates the partial wave elastic width and the isobar free width the inelastic width.

We now apply the extended schematic model to this problem. We drop the angular momentum superscripts and start by constructing the doorway basis. The normalized doorway state is

$$\Phi_0(x) \equiv N\,\psi_0(x)$$
$$\int_0^\infty dx\, x^2 [\Phi_0(x)]^2 = 1 \quad . \tag{5.8}$$

The remaining states are given by

$$\Phi_n = N_n\,[\mathcal{W}\Phi_{n-1} - \Sigma_{i=0}^{n-1}\,\Phi_i\,\mathcal{W}_{i,n-1}] \tag{5.9}$$

and the partial wave amplitude is

$$F_\ell(k) = \frac{-1}{k}\;\frac{\text{Im}\,\mathcal{W}_{00}}{E-E_R+i\Gamma/2+\mathcal{W}_{00}+\cfrac{\mathcal{W}_{01}^{\,2}}{E-E_R+i\Gamma/2+\mathcal{W}_{11}+\ldots}} \tag{5.10}$$

These expressions are carried over directly from Chapter IV.

All the integrals can be evaluated in closed form in the high energy limit.[7] The result for the doorway state is

$$\Phi_n^{(\ell)}(x) = \left[\frac{2(2n+1)}{f(\infty)}\right]^{1/2} \phi(x)P_n(\rho) \begin{cases} k\; j_\ell(kx) & , \quad n \text{ even} \\ -k\; n_\ell(kx) & , \quad n \text{ odd} \end{cases}$$

$$\rho \equiv \frac{f(x)}{f(\infty)} \quad , \qquad f(x) \equiv \int_0^x dy\; A\rho(y) \quad . \tag{5.11}$$

The nonvanishing expectation values of $\mathcal{W}_\ell$ (Equation 5.4) are

$$\mathcal{W}_{n,n} = i\,\gamma\,\delta_{no}$$
$$\mathcal{W}_{n,n+1} = \frac{\gamma}{\sqrt{4(n+1)^2-1}} \tag{5.12}$$

where we have defined

$$\gamma \equiv -\frac{g^2 f(\infty)}{2k(E-E_R+i\Gamma/2)} \tag{5.13}$$

Comparison with Equation (3.29) reveals that γ is just the semi-classical phase shift. Substituting the results for $\mathcal{W}_{i,j}$ into Equation 5.10, one finds that the continued fraction in the latter is essentially the expansion of the exponential:

$$F_\ell(k) = \frac{i}{2k}(1 - e^{2i\gamma})$$

(5.14)

This is, of course, just the high energy scattering amplitude which could have been calculated directly from the optical potential. It should be clear that the multiple scattering and isobar approaches are completely equivalent.

The convergence of the continued fraction representation of the exponential is quite rapid even for large values of the exponent. Convergence is obtained by keeping $n \sim \gamma$ terms in the fraction. For $\pi - {}^{16}O$ scattering, the semiclassical phase shift $\gamma \sim 2$, so that very few doorway states are needed for convergence. Further, we can expect the number of states needed to grow roughly like $A^{1/3}$ for larger nuclei.

One last exercise is to write the pion wavefunction in the isobar representation. The pion outgoing scattering wavefunction is conveniently written in terms of the half-off-shell nuclear scattering amplitude as

$$\psi_\ell^{(+)}(k;r) = j_\ell(kr) + \frac{2}{\pi}\int_0^\infty dp\, p^2\, \frac{j_\ell(pr)}{p^2-k^2-i\varepsilon}\, F_\ell(p,k;E_k)$$

(5.15)

$$F_\ell(k,k;E_k) \equiv F_\ell(k) \equiv \frac{\eta_\ell\exp(2i\delta_\ell)-1}{2i}$$

We now write the amplitude in the isobar representation

$$F_\ell(p,k;E_k) = -\sum_n \langle p|n;k\rangle\langle n;k|\frac{1}{E-E_R+\frac{i}{2}\Gamma+\mathcal{W}_\ell}|0;k\rangle$$

(5.16)

$$\otimes\ \langle 0;k|k\rangle$$

$$\langle p|n;k\rangle \equiv f_n(p)$$

$$= \int_0^\infty dx\, x^2 g\, \phi(x) j_\ell(px) \Phi_n^{(\ell)}(x)$$

(5.17)

$$f_0(k) = \frac{g}{k}\sqrt{\frac{f(\infty)}{2}}$$

Note that the vertex function $f_n(p)$ is non-zero for all the doorway states (in general). Consequently, reactions other than elastic scattering are sensitive to off-diagonal doorway expectation values of the isobar-hole propagator. Thus, the fact that the doorway expansion of the elastic on-shell nuclear scattering amplitude converges rapidly does not necessarily mean that other amplitudes (such as the half-off-shell amplitude) will do so. We now have for the pion wavefunction

$$\psi_\ell^{(+)}(kr) = j_\ell(kr)[1 + ik\, F_\ell(k)]$$
$$-g^2\sqrt{\frac{f(\infty)}{2}} \sum_{n=0}^{\infty} \int_0^\infty dr\, dr'\phi(r')\overline{\Phi}_n^{(\ell)}(r')n_\ell(kr_>)\,j_\ell(kr_<)\langle n|G_{\Delta n}|0\rangle \tag{5.18}$$

Of course, only the n=0 term in Equation (5.18) contributes to the wavefunction outside the nuclear density (i.e., the asymptotic wavefunction can be written in terms of the scattering phase shift). For our simple model, the wavefunction does not converge rapidly in the high energy limit.[7] We shall return to this point in Chapter VII in calculating the absorption cross section.[4]

It is amusing to comment on the phenomenological implications of having both the on- and off-shell amplitude converge rapidly in the doorway expansion. This would imply that a rank N separable potential, with N equal to the number of doorways, would suffice for the nuclear (partial wave) optical potential. Of course, the extreme nonlocality implicit in a very low rank separable potential already leads to the conclusion that the number of states needed should grow with the nuclear dimension. For reactions with very large momentum mismatches, there will generally be strong coupling to higher doorway states. For this case, convergence may be supplied more readily by the off-diagonal matrix elements of the Δ-hole propagator rather than the overlap integral in Equation 5.18. Detailed consideration of these questions is in progress.

VI. π-^{16}O ELASTIC SCATTERING

We now apply the isobar-hole formalism to elastic π - ^{16}O scattering over a wide energy range. All these calculations have been carried out in collaboration with M. Hirata, J. Koch and F. Lenz.[4] Comparison will be made with recent high precision data from S.I.N.[25] over the energy range 79 to 343 MeV and with data from Los Alamos[26] at 50 MeV. In making such a comparison, it is important to include nonresonant πN interactions and the Coulomb interaction. We include these within the two-potential framework:

$$T \equiv T_{background} + T_{resonant} \tag{6.1}$$

with $T_{background}$ calculated from the background interaction and

$T_{resonant}$ calculated with the resonant interaction and pion waves distorted by the background potential. For the nonresonant πN interaction, we take a conventional first order static optical potential.

The first step is to determine the phenomenological spreading potential strength by fitting the forward elastic amplitude (i.e., total cross sections and/or forward differential cross sections); where available data in the Coulomb/nuclear interference region have been used. The resulting spreading potential is shown as a function of energy in Figure 9. The error bars indicate very crudely the uncertainty in the total cross section data. We note that the fit is made effectively to the quantity $(\gamma V_\Delta + V_{sp})$, since these have the same density dependence. We have taken the depth of the binding potential as 55 MeV, and this choice obviously affects somewhat the spreading potential parameter (especially the real part). In any case, the result is that the spreading potential is roughly -50i MeV, with a slightly attractive real part, over a wide energy range centered close to the resonance energy. On the other hand, there are clearly peculiarities in the fit at the highest energy; not only is the spreading potential changing rapidly, but the imaginary part is even slightly positive. While this does not violate unitarity, we nevertheless are caused difficulty in Chapter VII because of this. This will be discussed below. From now on, all calculations will use the spreading potential given in Figure 9. It should be noted that the same spreading potential is found for π^+ and π^- scattering whenever the data for both charges exists at more or less the same energy.

The central spreading potential strength $\mathcal{V}_{sp}(0) \simeq -50i$ MeV may seem at first a bit large, but in fact it is consistent with the absorption widths in pionic atoms. If we write the usual p-wave pionic atom optical potential as

$$\begin{aligned} \mathcal{V}(r) &\equiv 4\pi \vec{\nabla} \cdot \alpha(r)\vec{\nabla} \\ \alpha(r) &= c_o\rho(r) + C_o\rho^2(r) \quad , \end{aligned} \tag{6.2}$$

then phenomenological fits give[27]

$$\begin{aligned} c_o &= 0.21\ m_\pi^{-3} \\ \mathrm{Im}\ C_o &= (.04 \text{ to } .07) m_\pi^{-6} \end{aligned} \tag{6.3}$$

The real part of C_o is generally prescribed to be $\mathrm{Re}C_o = -\mathrm{Im}C_o$, although this is in fact quite arbitrary. In the threshold situation appropriate to pionic atoms, the imaginary part of the

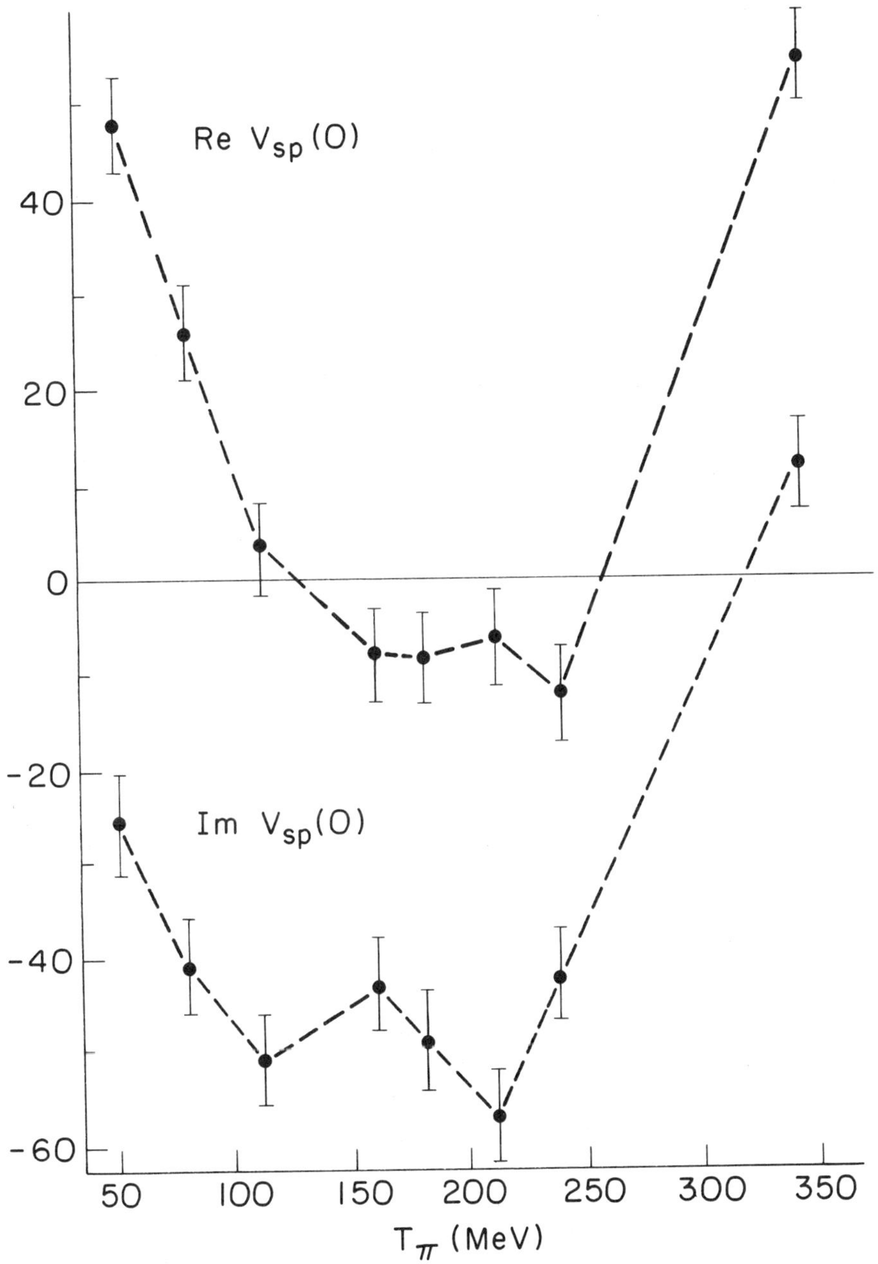

Figure 9: Isobar spreading potential (in MeV) fit to forward elastic scattering amplitude. Dashed lines are intended only as a guide to the eye.

optical potential is generated only by pion absorption. For nuclear matter densities, the ratio of the second to first order terms in the phenomenological optical potential is given by

$$\frac{C_o}{c_o} \rho \approx (.10 \text{ to } .17)(i - 1) \tag{6.4}$$

We now want to calculate this same quantity assuming that the spreading potential represents p-wave absorption mediated by the isobar (which we assume to be the dominant p-wave absorption mechanism). We write for the first order term

$$\alpha(r) = \rho(r)[c_{NR} + c_\Delta] \tag{6.5}$$

where c_Δ is the contribution to the zero-energy p-wave scattering volume coming from the "Δ-pole", while c_{NR} represents the remainder (e.g., a crossed pion graph). Various authors[28] give $c_\Delta \approx .55\, c_o$, and we adopt this value. It should be clear that we do not pretend to supply here more than an order-of-magnitude estimate, and the uncertainty in even defining c_Δ must be weighed at the end. Given this, we now modify the resonant part in a way similar to that used in the Glauber calculation at the end Chapter III:

$$\begin{aligned} \alpha(r) &\equiv \rho(r)\left[c_{NR} + c_\Delta \frac{E-E_R}{E-E_R+\frac{i}{2}\Gamma_{Abs}\tilde{\rho}(r)}\right] \\ &= \rho(r)c_o\left[1 - \frac{c_\Delta}{c_o}\,\frac{i(\Gamma_{Abs}/2)\tilde{\rho}(r)}{E-E_R+i(\Gamma_{Abs}/2)\tilde{\rho}(r)}\right] \end{aligned} \tag{6.6}$$

Taking Γ_{Abs} = 100 MeV (recall that E = 0 for the pionic atom situation), we have from the second term in square brackets

$$\frac{C_o}{c_o} \rho \approx .14(i - .3) \tag{6.7}$$

which obviously is in rather close agreement with Equation 6.4. We repeat that the exercise is intended only to demonstrate that the spreading potential strength is in the right ballpark.

1. The Resonance Region

We start our detailed comparison with the data close to the free resonance energy. Figure 10 shows the differential cross section for $\pi^+ - {}^{16}O$ scattering at 163 MeV[25] together with the Δ-hole theoretical prediction. Also shown is the calculation

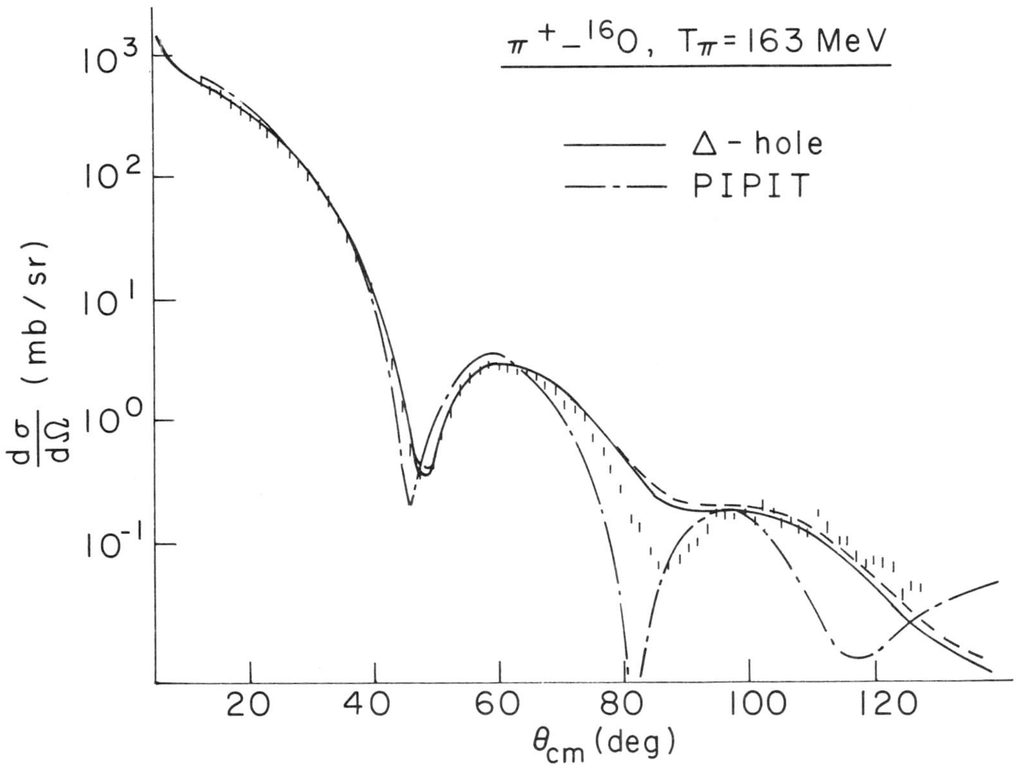

Figure 10: Elastic differential cross section for $\pi^+ - {}^{16}O$ scattering at 163 MeV. Solid and dot-dashed curves result from the isobar-hole and PIPIT calculations, respectively. The dashed curve is the isobar-hole result modified by inclusion of "higher order" Coulomb effects (see discussion on pg. 46). Data from Reference 25.

given by PIPIT[29]; this is a widely available momentum-space, first order optical potential code which uses a nonlocal separable πN transition matrix extended off-shell via the inverse scattering solution.[14] It includes the "frame transformation" of the elementary amplitude and some correction for Fermi averaging.[30] Obviously, the isobar-hole results provide a very good description of the data except in the vicinity of the second minimum. Nevertheless, this is not really very instructive, and we turn to an analysis of the partial wave amplitudes.

Figure 11 is the Argand plot for the partial wave amplitude $T_L \equiv (\eta_L \exp(2i\delta_L) - 1)/2i$ at 163 MeV. Again, the Δ-hole and PIPIT results are given. The deep diffraction structure in the latter arises because of the small real parts of the PIPIT partial wave amplitudes. Also shown is the result of a partial wave fit to the data (χ^2/degree of freedom = 1.1). An important point is the comparatively good agreement between the Δ-hole calculations and the partial wave fit in the peripheral partial waves. For these partial waves, the spreading potential and Pauli blocking terms are comparatively unimportant, but a good treatment of the isobar propagation and binding effects is very important.

Table 4 gives the decomposition of the doorway expectation value of the isobar-hole Hamiltonian

$$<\tilde{\Delta}_o|\mathcal{H}_\Delta|\Delta_o> = <\mathcal{W}> + <\delta\mathcal{W}> + \gamma<H_\Delta> + <\mathcal{V}_{sp}> \tag{6.8}$$

for each partial wave. The first column gives the root-mean-square radius for the isobar in the doorway state and therefore a measure of "peripherality"; recall that the root-mean-square radius of the ^{16}O ground state density is 2.7 fm. In the closure limit (which is appropriate to PIPIT), $<H_\Delta>$ is a constant for all partial waves (approximately 55 MeV in this case). On the other hand, this quantity obviously varies considerably with L in the Δ-hole calculations. The attraction for L = 0 is understood by recalling that the biggest component in the 0^- doorway state (see Table 2) has the isobar in the $0S_{3/2}$ shell model state. Clearly, the differences in $<H_\Delta>$ between various partial waves are not negligible compared to other isobar self-energies.

The central partial waves are more complicated and the Δ-hole results do not agree with the partial wave fit in detail. The dominant interaction is generated by pion exchange. The real part of $<\mathcal{W}>$, which roughly corresponds to off-shell propagation, is substantially smaller than the imaginary part for the central partial waves, implying that the scattering amplitude is fairly insensitive to reasonable changes in the off-shell extension of the πN amplitude. For the peripheral partial waves, $<\mathrm{Re}\mathcal{W}>/<\mathrm{Im}\mathcal{W}>$

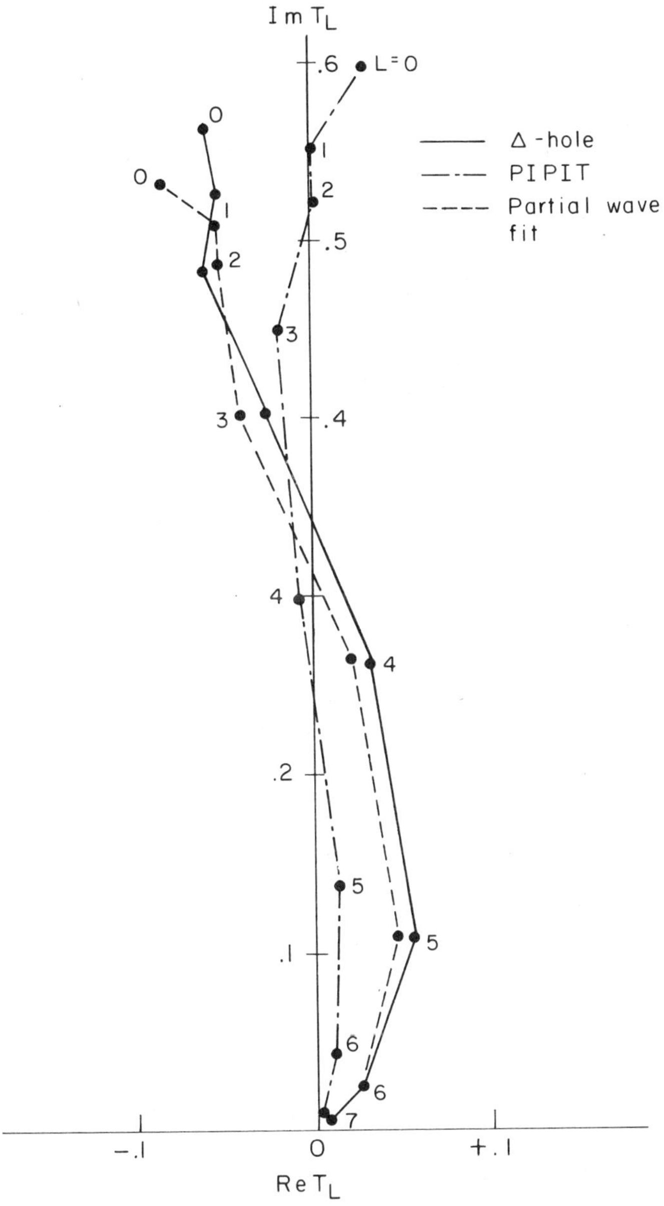

Figure 11: Argand plot corresponding to Figure 10. The points are the partial wave amplitudes, with the lines serving only to guide the eye. The dashed lines correspond to a partial wave fit to the data.

L^{π}	$\bar{R}_{\Delta}$(fm)	$\langle\mathcal{W}\rangle$(MeV)	$\langle\delta\mathcal{W}\rangle$(MeV)	$\langle H_{\Delta}\rangle$(MeV)	$\langle\mathcal{V}_{sp}\rangle$(MeV)
0^-	2.24	-48 - 140i	2 + 25i	-.8 + .6i	-7 - 40i
1^+	2.31	-36 - 99i	10 + 20i	8.6 + .8i	-7 - 38i
2^-	2.49	-28 - 76i	7 + 20i	10.7 + .4i	-7 - 36i
3^+	2.90	-28 - 47i	7 + 14i	23.5 + .1i	-5 - 26i
4^-	3.36	-23 - 22i	6 + 8i	37.3 + .1i	-3 - 17i
5^+	3.80	-14 - 8i	4 + 4.5i	49.3	-2 - 10i
6^-	4.19	- 7 - 2i	3 + 2i	58.6	-1 - 6i
7^+	4.63	- 3 - .5i	1.5 + .7i	65.1	-.5 - 3i

Table 4: Decomposition of the doorway state expectation value of the isobar-hole Hamiltonian for π^+ _ ^{16}O scattering.

is no longer small, but here single scattering provides a good approximation to the amplitude so that again off-shell effects are not too important. Returning to the central partial waves, the "higher order effects" of Pauli blocking and the spreading potential modify the dynamics in an important way. The Pauli term $\langle\delta\mathcal{W}\rangle$ provides a significant quenching of the free isobar decay width ($\Gamma/2 \approx 59$ MeV). Since the latter corresponds to the inelastic width for quasifree scattering, this effect increases the ratio of elastic/inelastic pion scattering. On the other hand, the spreading interaction provides a large damping mechanism, suppressing the ratio of elastic to inelastic scattering. The net result is a suppression of elastic scattering and considerable loss of flux into the Q-space (see Figure 5 and Chapter VII).

Another interesting quantity is the off-diagonal doorway coupling. This determines convergence of the extended schematic model. We give the Hamiltonian matrix among the first two doorway states in Table 5 for the 0^- and 4^- partial waves. Recall that $\langle\mathcal{H}_{\Delta}\rangle_{01}$ is the fluctuation of the Hamiltonian in the doorway state (Equation 4.10). For the central 0^- partial wave, $|\langle\mathcal{H}_{\Delta}\rangle_{01}/\langle\mathcal{H}_{\Delta}\rangle_{00}| \approx 1/3$, implying that the collective doorway state is "close" to an eigenstate. This leads to the rapid convergence of the doorway expansion. On the other hand, the fluctuation in the 4^- doorway state is not small compared to the diagonal value $\langle\mathcal{H}_{\Delta}\rangle_{00}$. However, $\langle\mathcal{H}_{\Delta}\rangle_{01}$ is small compared to

	$L^\pi = 0^-$	$L^\pi = 4^-$
$\langle \mathcal{H}_\Delta \rangle_{00}$	-53 - 154i	13.9 - 14.4i
$\langle \mathcal{H}_\Delta \rangle_{01}$	59 + 8.2i	22.1 + 10i
$\langle \mathcal{H}_\Delta \rangle_{11}$	8.2 + 3.4i	25 + 4i

Table 5: Hamiltonian matrix $\langle \tilde{\Delta}_i | \mathcal{H}_\Delta | \Delta_j \rangle$ in MeV among the first two doorway states for $L^\pi = 0^-, 4^-$ at 163 MeV.

the free half width and this leads again to convergence of the doorway expansion. Clearly, the nature of the convergence is quite different for the central and peripheral partial waves. In the former, the isobar-nuclear dynamics is extremely important and leads to a "true" collective state. For the peripheral partial waves the isobar-hole states simply are damped weakly except for the usual quasifree decay; i.e., impulse approximation eventually becomes adequate for large angular momenta.

Within the Λ-hole framework, it is relatively simple to examine the importance of various dynamical modifications. Because of the collectivity, evaluation of the associated self-energies in the doorway state provides an immediate gauge of the role played by each effect. For example, one can calculate the effects of "higher order" Coulomb effects on π^+/π^- differences. These can arise from mass differences for the different Δ charge states, differences in the binding energy of protons and neutrons in the ^{16}O ground state, and Coulomb energy differences for the different Δ charge states. The latter is the most important, and the Coulomb energy for Δ^{++} at the center of the residual nucleus (i.e., ^{16}O minus one proton) is easily calculated to be 8 MeV. Comparing this to the scale fixed by the L = 0 results in Table 4, it is clear that such effects will not be large for the differential cross section. This can be seen in Figure 10. On the other hand, the effects are not completely negligible for π^+/π^- differences and, in fact, make the theory/experiment comparison virtually the same for π^+/π^- - ^{16}O scattering at 163 MeV (with the same spreading potential). Similarly, a nuclear structure correction would be inclusion of hole widths for the s-shell. Taking a half width $\Gamma_s/2 \approx 8$ MeV, the magnitude of the effect on the differential cross section is roughly the same as that of the Coulomb effects just described, although the cross section is now suppressed slightly at the first minimum.

To summarize this subsection, it is clear that a fairly successful description of pion scattering in the resonance region is offered by the Δ-hole calculations. There are clearly discrepancies in the central partial waves. Here, the physics is complicated, and our simple assumption that the Q-space coupling proceeds through a local density-dependent spreading potential may be inadequate.

2. Scattering at Lower Energies

In Figure 12, we show the differential cross section for π^+- ^{16}O scattering at 79 MeV[25] together with Δ-hole and PIPIT results. The PIPIT results give a much too large elastic cross section (recall the similar results in Figure 5). The isobar-hole calculations, while fitting the total elastic cross section quite well because of the spreading potential, do not do terribly well in the differential cross section. The source of the difficulty is seen by examining the Argand diagram, Figure 13. Comparing the Δ-hole partial wave amplitudes with those from the partial wave fit to the data, it is clear that the central partial wave amplitudes (i.e., 0^- and 1^+) must be reduced considerably. This is not surprising. At the lower energies, the background interaction is comparatively more important and we have used a simple first order optical potential. Phenomenological optical potential fits[31] indicate the need for a substantial, repulsive second order, s-wave optical potential. We have included such a term of central strength 10 MeV in the background potential and show the resulting partial wave amplitudes in Figure 13. This clearly goes in the right direction and in fact results in a good fit to the data except in the region of the first minimum. It is interesting that an important effect of introducing the background potential is to reduce the resonant part of the amplitude. This is because the repulsive potential "pushes out" the pion wavefunction and thus reduces the isobar excitation strength.

Further improvement is obtained by including a binding energy shift and Pauli suppression in the first order background potential.[32] This is seen most clearly at 50 MeV, where the "background" interaction is very much in the foreground. In Figure 14 we show the data[26] and a theoretical calculation. The latter includes the same type of ρ^2-background potential described above, with central strength (20 - 3i) MeV. In addition, the energy at which the first order optical potential is evaluated has been shifted by about 20 MeV and the imaginary part has been quenched (to simulate the role of the Pauli principle). The last modifications were important for obtaining the right ratio of

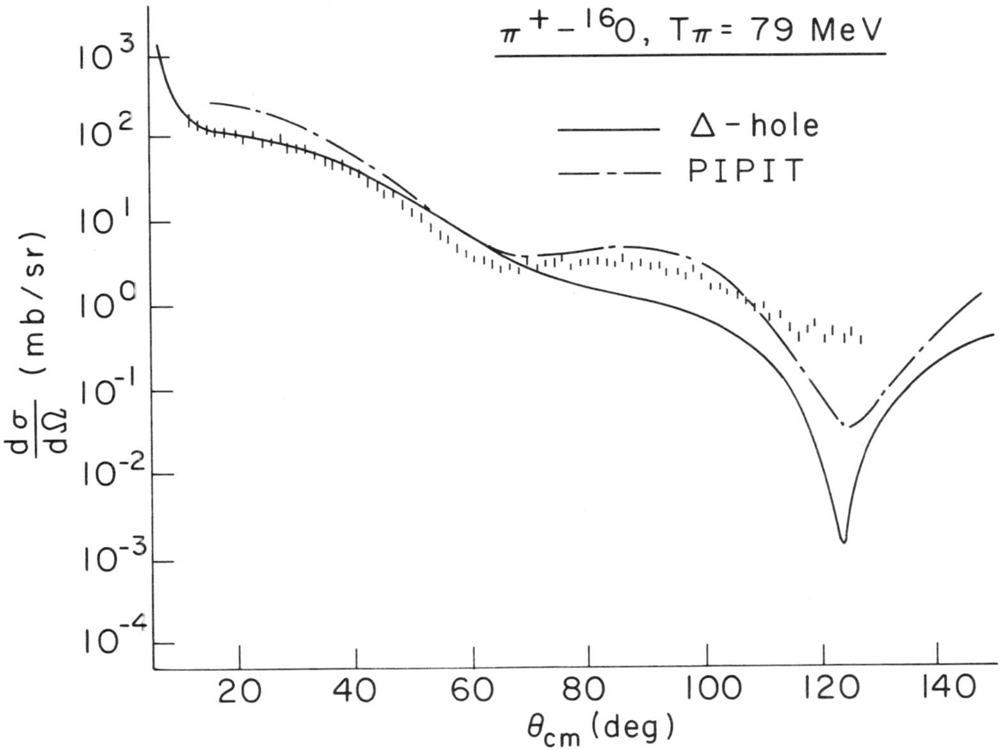

Figure 12: Elastic differential cross section for $\pi^+ - {}^{16}O$ scattering at 79 MeV; data from Reference 25.

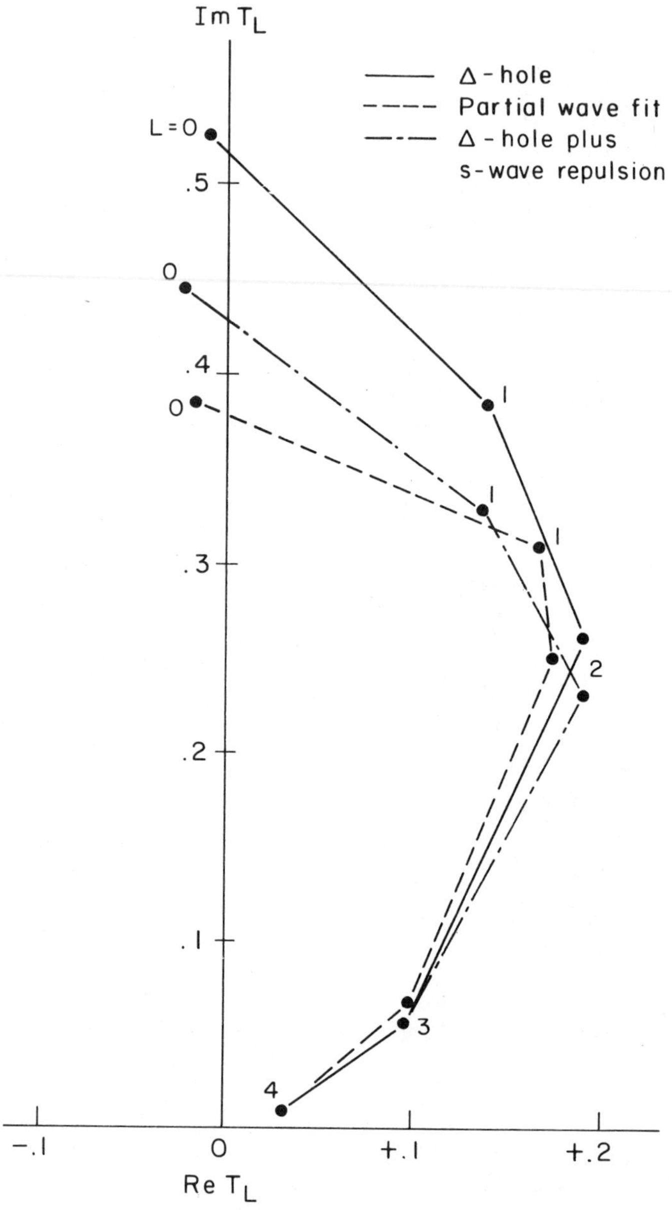

Figure 13: Argand plot corresponding to Figure 12. Dot-dashed line represents the isobar-hole results modified by including an s-wave repulsive second order optical potential of central strength 10 MeV in the background interaction.

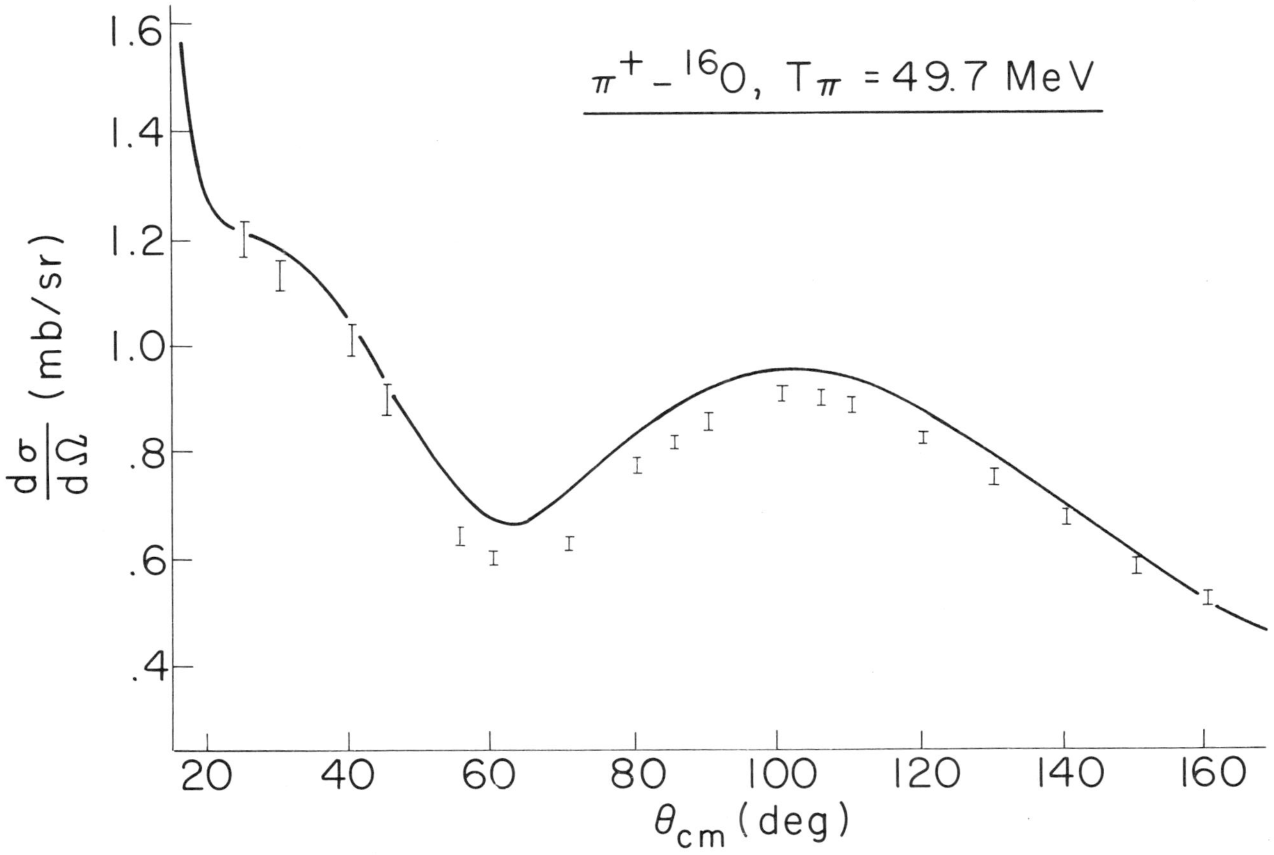

Figure 14: Elastic differential cross section for π^+ - ^{16}O scattering at 49.7 MeV; data from Reference 26.

forward to back angle scattering. The ingredients necessary for reproducing the data are obviously present.

3. Scattering Above the Resonance

Figure 15 shows the differential cross section for $\pi^+ - {}^{16}O$ elastic scattering at 240 MeV.[25] The theoretical curves are again the isobar-hole and PIPIT results. The latter obviously appear to be better, but a rather different light is cast on the comparison by examining the Argand plot, Figure 16. Here, we show also the result of a partial wave fit (χ^2/degree of freedom $\approx$ 1.1). Notice that the peripheral partial waves ($L \geq 5$) are quite different. We stress that the isobar propagation and binding effects needed for a good calculation of the peripheral partial waves are not "speculative"; they are calculated just as reliably in the nuclear shell model framework as are the pion rescattering and Pauli corrections. Their absence in the PIPIT calculation is simply a shortcoming of static optical potential calculations. However, the absence of deep minima in the PIPIT differential cross section, which obviously improves the appearance in comparison with the data, is aided significantly by the too large real parts of the peripheral amplitudes.

The isobar-hole calculations clearly have a significant problem with the central and semi-peripheral partial waves. In particular, the relative position of the L = 0 and 1 amplitudes from the partial wave fit cannot be reproduced in the Δ-hole approach without some appreciable change. It is certainly true that the partial wave fit is far from unique, but all good fits which we found had some "non-smooth" L-behavior for the first few partial waves. We note that some phenomenological optical potential fits which do include binding effects[31,33] do seem to get better descriptions of the data. However, these calculations also give total cross sections about 15% too big. If we allow the total cross section to be this big in fitting the spreading potential, the Δ-hole results also appear quite good for the angular distribution. However, such values for the total cross section are not reasonable and point to the importance of including this constraint when evaluating the success of theoretical calculations. It also points to the need for accurate total cross section measurements.

Since the πN resonance interaction clearly dominates in this energy regime, we must assume that our treatment of the spreading potential needs refinement. As stressed earlier, this can take two forms: introduction of vertex corrections and/or inclusion of nonlocality in the isobar spreading potential. Apart from

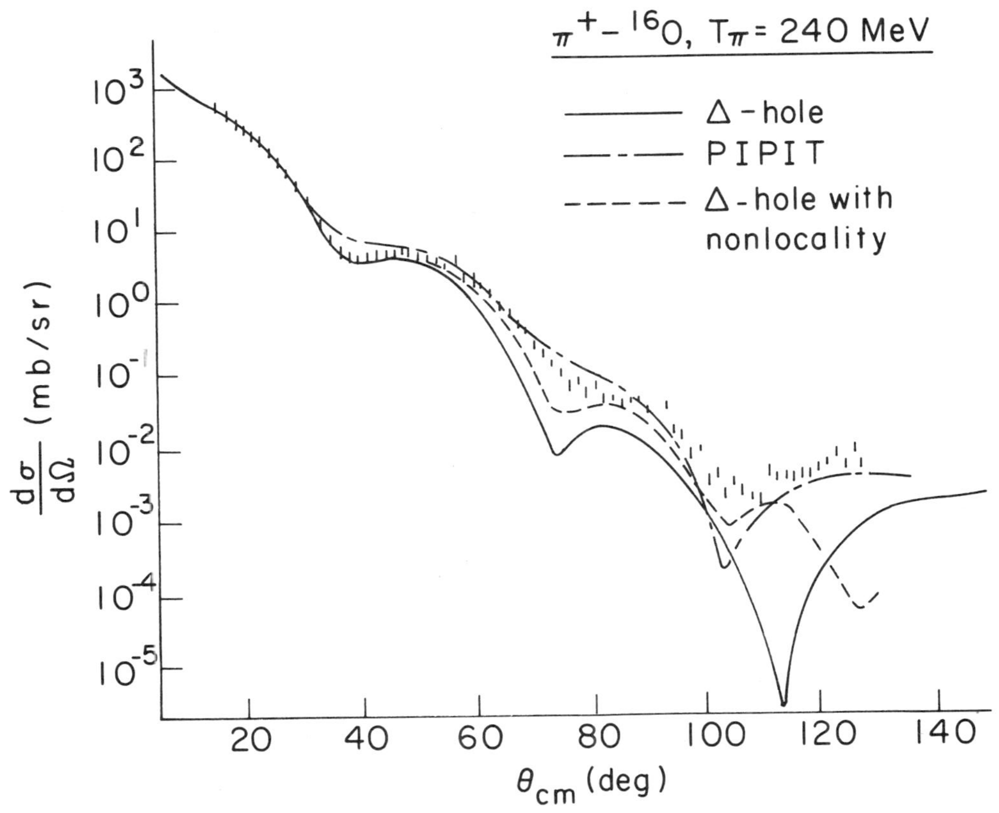

Figure 15: Elastic differential cross section for $\pi^+ - {}^{16}O$ scattering at 240 MeV; data from Reference 25. The dashed curve is the isobar-hole result modified by inclusion of nonlocality in the spreading potential (see discussion on pg. 53).

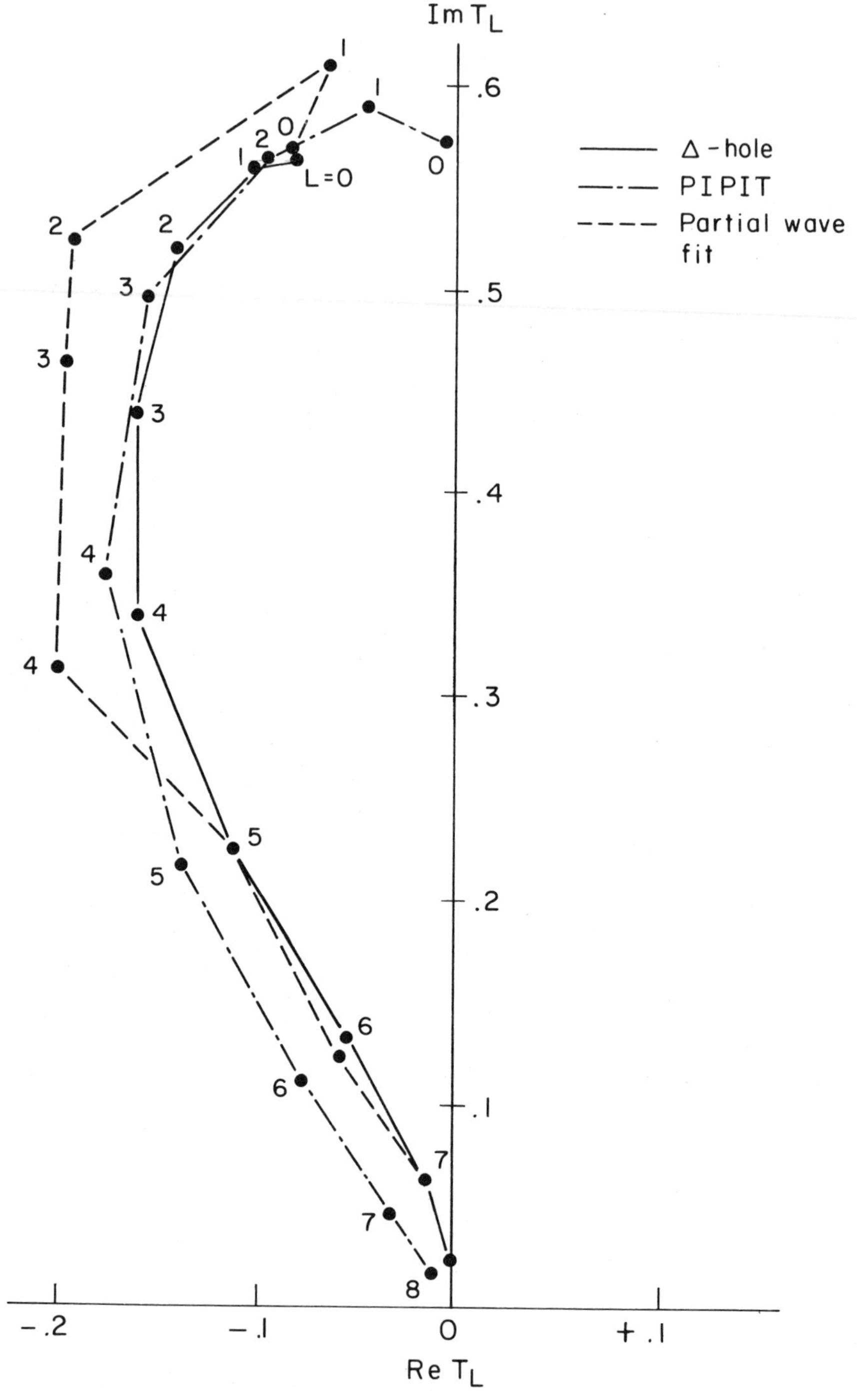

Figure 16: Argand plot corresponding to Figure 15.

simple Δ-hole "potential" interactions (which is a vertex correction), microscopic calculation of such terms requires consideration of two-hole diagrams. Simple examples of isobar self-energy terms are given in Figure 17, corresponding to pion absorption and pion multiple reflection (local field effects). Some examples of vertex corrections are given in Figure 18; these are obtained by hole exchange of the diagrams indicated in Figure 17. Quantitative evaluation of these terms may be expected to be very difficult; for example, NN and ΔN short-range forces are probably very important for quantitative accuracy.[34] Nevertheless, we can expect this "two-hole physics" to become very important above the resonance. For example, multinucleon knockout certainly increases strongly in this energy regime.[1] This is not surprising for several reasons: with more energy available, multinucleon knockout is easier; above the resonance, loss of energy in one quasifree interaction takes the pion into the resonance region, where another scattering is highly probable, the backward peaking of the πN scattering amplitude in the resonance region makes substantial energy transfer to the nucleon more likely.

The self energy corrections are the easiest to incorporate phenomenologically. We have estimated the role of nonlocality by expanding

$$\mathcal{V}_{sp}(r) = (1 + \tfrac{1}{6}\ell^2\nabla^2 + \ldots)\,\mathcal{V}_{sp}^{(o)}(r) \tag{6.9}$$

where ℓ plays the role of a nonlocality range. This range can be estimated from the absorption diagram (Figure 17a) to be $\ell \approx 1$ fm.[34] The resulting modification of the differential cross section is shown in Figure 15. There is clearly substantial improvement in raising the cross section at back angles, and this appears encouraging for examining Δ-dynamics in detail. Unfortunately, the total cross section is increased by about 8% and, as mentioned above, such a modification of the angular distribution could be achieved with the original spreading potential if a larger total cross section is allowed. Therefore, the lessons from this exercise are that smooth (or long-range) modifications of the dynamics do not change substantially the angular distribution while maintaining the constraint offered by the forward amplitude but that the interpretation of the phenomenological spreading potential strength may be altered somewhat. It is conceivable that very different forms for the spreading interaction could have substantial effects, but it is very difficult to think of any physical process which might generate these. This takes us to the point of evaluating the vertex corrections, but no progress has been made here yet.

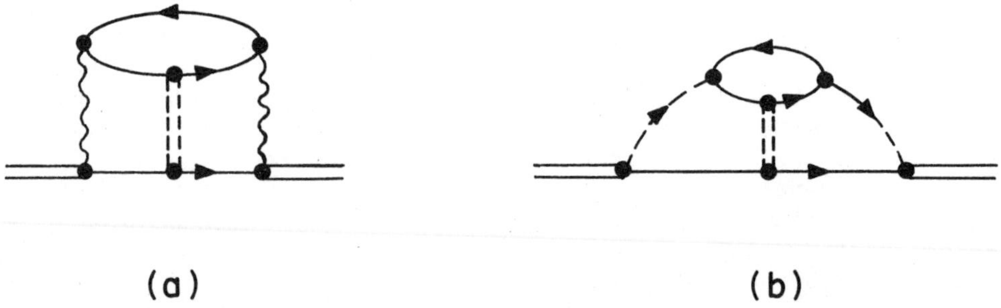

Figure 17: Simple examples of isobar self-energies: (a) absorption; (b) multiple reflection. The double-dashed lines represent particle-particle scattering.

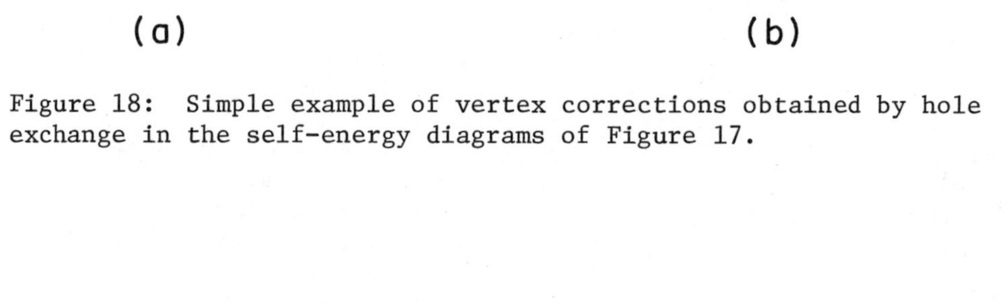

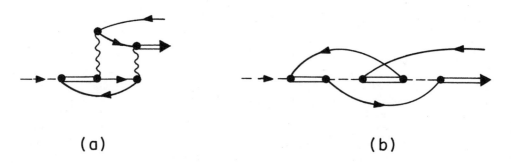

Figure 18: Simple example of vertex corrections obtained by hole exchange in the self-energy diagrams of Figure 17.

At the highest energy studied, 343 MeV, the isobar-hole calculations yield excellent agreement with the differential cross section data.[25] However, the spreading potential has changed drastically (See Figure 9) and even has a slightly positive imaginary part. We repeat that this does not violate unitarity (the imaginary part is still small in magnitude compared to the free Δ half-width), but physical interpretation is difficult and the phenomenology is therefore poor.

VII. THE REACTION CROSS SECTION

We have seen that the phenomenological spreading potential plays a crucial role in the description of pion scattering throughout the resonance region. One test of the model is provided, of course, by the description provided for the elastic angular distribution. This was reproduced reasonably well at all energies except 240 MeV. A more direct test came from the comparison with pionic atom level widths. However, this test is only semi-quantitative. We provide a further test by calculating the partial cross section into the Q-space. Further, we shall assume that this cross section is predominantly that for "true" pion absorption. As discussed already, inelastic data implies that this should be a good assumption from threshold up to at least the resonance energy.

The basic idea is understood easily in the limit that the scattering amplitude is given by the leading doorway state. We then have

$$T_L(E) = \frac{-\Gamma_{Elas}/2}{E-E_R+i\Gamma/2-\langle\tilde{\Delta}_o|\mathcal{H}_\Delta|\Delta_o\rangle}$$
$$= \frac{-\Gamma_{Elas}/2}{E-\tilde{E}_R+\frac{i}{2}[(\Gamma-\Gamma_{Pauli})+\Gamma_{Elas}+\Gamma_{Abs}]} \tag{6.1}$$

where the Pauli, elastic and absorption half-widths are given by the imaginary part of the expectation value of $\delta\mathcal{W}$, $\mathcal{W}$, and $\mathcal{V}_{sp}$, respectively. The real part of the expectation value is incorporated into $\tilde{E}_R$. The partial wave contribution to the total cross section is given by

$$\sigma^L_{Total} = \frac{4\pi}{k^2}\,\mathrm{Im}\,T_L \tag{6.2}$$
$$= \frac{4\pi}{k^2}\,\frac{\frac{1}{4}\Gamma_{Elas}((\Gamma-\Gamma_{Pauli})+\Gamma_{Elas}+\Gamma_{Abs})}{(E-\tilde{E}_R)^2+\frac{1}{4}((\Gamma-\Gamma_{Pauli})+\Gamma_{Elas}+\Gamma_{Abs})^2}$$

The interpretation in terms of partial cross sections is now clear. The contributions proportional to Γ_{Elas}, $(\Gamma - \Gamma_{Pauli})$ and Γ_{Abs} give total elastic, total inelastic (quasifree), and total absorption cross sections, respectively. This decomposition is very similar to that performed in the simple Glauber calculation in Chapter III.

For quantitative results, the one doorway approximation is not sufficient. We evaluate the reaction cross section as the expectation value of the imaginary part of the pion optical potential between pion distorted waves:

$$\sigma^L_{Reaction} = \frac{4\pi}{k} \langle 0, \chi_k^{(-)} | \mathrm{Im}\,\mathcal{V}_{PP} | 0, \chi_k^{(+)} \rangle$$
$$= \frac{4\pi}{k} \langle 0, (\vec{k})_L | (1+T_{PP}^{\dagger} G_{PP}^{\dagger}) \mathrm{Im}\,\mathcal{V}_{PP} (1+G_{PP}T_{PP}) | 0, (\vec{k})_L \rangle \tag{6.3}$$

The elastic pion transition matrix operator has the isobar-hole representation

$$T_{PP} = g_{\Delta\pi N}^{\dagger} [D(E) - \mathcal{H}_{\Delta}]^{-1} g_{\Delta\pi N} \tag{6.4}$$

Finally, we relate the isobar spreading potential to the term in the optical potential generated by pion absorption:

$$\mathrm{Im}\,\mathcal{V}_{sp} \equiv g_{\Delta\pi N} G_{PP}^{\dagger} \mathrm{Im}\,\mathcal{V}_{PP}^{Abs} G_{PP} g_{\Delta\pi N}^{\dagger} \tag{6.5}$$

Combining Equations (6.3) through (6.5), we have[4]

$$\sigma^L_{Abs} = \frac{4\pi}{k} \sum_{ij} \alpha_i^{*} \langle \tilde{\Delta}_i | \mathrm{I_m}\,\mathcal{V}_{sp} | \Delta_j \rangle \alpha_j \tag{6.6}$$

$$\alpha_j = \langle \tilde{\Delta}_j | \frac{1}{D(E)-\mathcal{H}_{\Delta}} g_{\Delta\pi N} | 0, (k)_L \rangle$$
$$= \frac{\langle \tilde{\Delta}_j | [D(E)-\mathcal{H}_{\Delta}]^{-1} | \Delta_o \rangle}{\langle 0, (\vec{k})_L | g_{\Delta\pi N}^{\dagger} g_{\Delta\pi N} | 0, (\vec{k})_L \rangle} \tag{6.7}$$

Note that the vertex function α_j is just the off-diagonal expectation value of the isobar-hole propagator in the doorway basis. This is the same quantity which entered in the pion wavefunction or half-off-shell elastic transition matrix (Equations 5.16 and 5.18). The vertex functions α_j provide a rapid convergence for the calculation of the absorption cross section; for example, $|\alpha_3/\alpha_0| \lesssim 10^{-2}$ for both central and peripheral partial waves. Consequently the absorption cross section, which involves the half-off-shell transition matrix, converges almost as rapidly as the on-shell elastic scattering amplitude.

The decomposition of the reaction cross section into absorption and inelastic scattering cross sections is shown in Figure 19. Note that the inelastic (quasifree) cross section peaks at about 200 MeV, reflecting the binding energy shift discussed previously. The absorption cross section peaks at about 100 MeV and falls slowly with increasing energy. Unfortunately, there is essentially no data for the total absorption cross section on ^{16}O. However, an estimate based upon results available with other targets indicates that the cross section is about 200 mb between 80 and 220 MeV. Our calculation is more or less consistent with this.

At the highest energies, there are problems. Since the imaginary part of the spreading potential is positive at 343 MeV, the interpretation of the absorption cross section given above is no longer possible. In fact, even at 240 MeV, the calculated absorption cross section is likely too small. We have discussed already the problems at high energy in discussing the elastic cross section. Presumably, all of these problems could be related to a breakdown of the assumption that Q-space coupling can be summarized via an isobar self-energy. This is difficult to evaluate quantitatively and would require consideration of dynamics such as those indicated in Figure 18. Unfortunately, the needed reaction data which might resolve the problem are not available. For example interpretation of the spreading potential above the resonance would be aided by experimental resolution of the question as to whether absorption occurs after quasifree nucleon knockout.

VIII. CONCLUDING REMARKS

We conclude by summarizing the main features of the isobar-hole formalism and of the comparison with data. It is important to stress again that the descriptions of pion-nucleus reactions in the multiple scattering/optical potential language and in the isobar-hole language contain, in principle, the same physics. The fundamental advantage of the isobar-hole formalism is that the nuclear degrees of freedom are not "frozen out" as they are in most multiple scattering calculations. Thus, isobar propagation and binding effects and Pauli-blocking effects are calculated within the same framework and as reliably as the pion multiple scattering itself. In the optical potential language, this corresponds to evaluation, within the framework of the

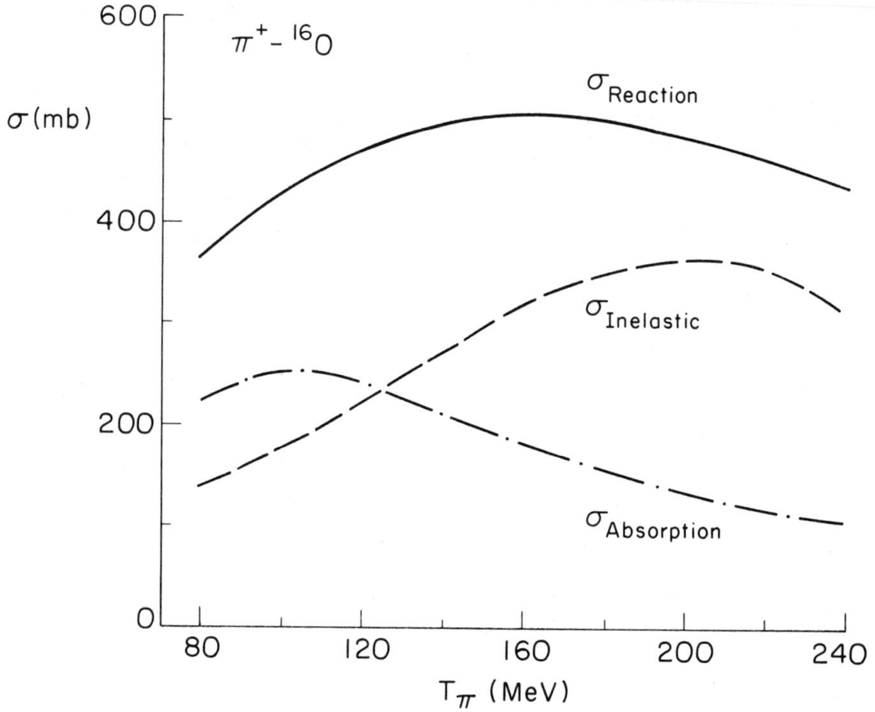

Figure 19: Decomposition of the reaction cross section into absorption and inelastic scattering cross sections.

nuclear shell model, of the πN transition operator in the medium (i.e., $\tau(E)$ in Equation 3.7). The higher order effects are handled phenomenologically, as they are in optical potential calculations. We believe that the most important "spreading interaction", at least for energies at and below the 3-3 resonance, is pion absorption. The isobar dominance of this process leads to a "natural" phenomenology for incorporating these effects as an isobar self-energy. This is reinforced by the good description of pion elastic scattering and by the evaluation of the absorption cross section and pionic atom level widths in terms of the phenomenological spreading interaction. We note that this phenomenology does not correspond to adding a simple local ρ^2 term in the optical potential. Above the resonance, there are unresolved problems involving the elastic angular distribution at 240 MeV and the abrupt change in the spreading potential strength at 340 MeV.

With regard to the calculational technology, we find the similarity to the many body techniques used in nuclear structure calculations and to the concepts employed in low energy nuclear reaction theory very satisfying. In particular, the extended schematic model or doorway expansion makes this similarity quite clear and somewhat simplifies the calculations. This doorway approach is based explicitly on the "collectivity" expected by analogy with the Brown-Bolsterli model. The dominance of the elastic transition matrix by one or two Δ-hole doorway states also allows clear insight into the role of the various aspects of pion-nuclear dynamics in determining the scattering. Further, to the extent that calculations of various "off-shell" reactions converge similarly in the doorway expansion, a simple unification of these reactions is afforded in terms of only a few parameters (doorway expectation values and coupling strengths). The absorption cross section offered an example of this rapid convergence.

Returning to the dynamics, the isobar propagation and binding effects were crucial for calculation of the peripheral partial wave elastic scattering amplitudes. The results seem to be in good agreement with the data. The Pauli blocking effects were important for suppressing the inelastic or quasifree pion scattering cross section. This increases the ratio of elastic to reaction cross sections. However, this is more than compensated by the spreading interaction which summarizes loss of flux into the Q-space. This finally allowed for a good description both of the elastic to reaction cross section ratio and of the inelastic to absorption cross section ratio. At the lower energies, our original treatment of the nonresonant interactions through a static first order optical potential was not adequate, but addition of a repulsive s-wave term and inclusion of a binding energy shift and Pauli quenching led to a very good description of the data. All of these features are known to be important from optical potential fits to the data. Above the resonance, the situation was not so good. Our simple treatment of the Q-space coupling needs refinement, and this presumably requires some serious consideration of various "two-hole diagrams". However, these are very difficult to evaluate quantitatively, since this would require consideration of the strong short-distance interactions among nucleons and isobars. There is no reason to believe that a simple treatment based upon static NN correlation functions is sufficient. Considerable theoretical effort is needed here, whether in the isobar-hole or multiple scattering approach.

Finally, we note that much of the "testing" of the model came from a consideration of inelastic reactions. For example, the binding and Pauli blocking effects are reflected directly in the (π,π') inclusive spectrum; the spreading potential strength is reflected in the total absorption cross section (although this connection is far from model-independent). The most urgent theoretical need for further development of the microscopic description of pion-nucleus interactions is more reaction data. The nuclear response function for pions is still very poorly known. Systematic studies of the pion quasielastic spectrum throughout the resonance region are essential; this is the channel which leads to attenuation of the elastic channel wavefunction in a first order optical potential. Good measurements of the total absorption cross section are needed, as well as more exclusive reactions such as π(in flight) $\rightarrow$ NN. The role of multinucleon knockout is crucial for interpretation of the second order optical potential. This was relevant directly to our discussion of the spreading interaction above the resonance. The goal is to understand the reaction mechanism by which the pion transfers energy and momentum to the nucleus, particularly in the absorption process. The role of photoinduced reactions as a complementary probe of the isobar-nuclear dynamics should not be overlooked. Hopefully, the rather meager amount of such reaction data will be expanded greatly in the next few years and will provide guidance for the theoretical developments. An understanding of the behavior of pions and Δ's in the nucleus will both allow detailed study of nuclear bound states with pions and provide some measure of unification between descriptions of the strong force and of nuclear structure. This is a fundamental challenge of intermediate energy physics.

REFERENCES

1. For recent reviews, see the lecture notes of E.J. Moniz in the Proceedings of the Les Houches Summer School on Nuclear Physics with Heavy Ions and Mesons, R. Balian, editor, to be published, and the lecture notes of F. Lenz in the Proceedings of the Topical Meeting on Intermediate Energy Physics (Zuoz, 1976).

2. L. Kisslinger and W. Wang, Ann. Phys. 99 (1976) 374.

3. M. Hirata, F. Lenz and K. Yazaki, Ann. Phys. 108 (1977) 116.

4. M. Hirata, J.H. Koch, F. Lenz and E.J. Moniz, Phys. Lett. 70B (1977) 281 and to be published.

5. K. Klingenbeck, M. Dillig, H.M. Hofmann, and M.G. Huber, in Abstract Volume of the Proceedings of the 7th International Conference on High Energy Physics and Nuclear Structure (Zurich, 1977).

6. W. Weise, Nucl. Phys. A278 (1977) 402.

7. F. Lenz, E.J. Moniz and K. Yazaki, to be published.

8. M.T. Vaughn, R. Aaron and R. Amado, Phys. Rev. 124 (1961) 1258.

9. S. Weinberg, Phys. Rev. 130 (1963) 776.

10. H. Sugawara and F. Von Hippel, Phys. Rev. 172 (1968) 1764.

11. R. Blankenbecler and R. Sugar, Phys. Rev. 142 (1966) 1031.

12. J.T. Londergan and E.J. Moniz, Phys. Rev. C16 (1977) 721.

13. R.M. Woloshyn, E.J. Moniz and R. Aaron, Phys. Rev. C13 (1976) 286.

14. J.T. Londergan, K.W. McVoy and E.J. Moniz, Ann. Phys. 86 (1974) 147.

15. For a review, see L. Heller in Proceedings of the International Conference on Meson-Nuclear Physics, P.D. Barnes et al, editors (Carnegie-Mellon, 1976) 93.

16. R. Aaron, R.D. Amado and J.E. Young, Phys. Rev. 174 (1968) 2020.

17. L. Heller, G.E. Bohannon and F. Tabakin, Phys. Rev. C13 (1976) 742.

18. W.A. Friedman, Phys. Rev. C12 (1975) 1294.

19. H. Feshbach, Reaction Dynamics (Gordon and Breach, New York, 1973).

20. K.M. Watson, Phys. Rev. 89 (1953) 575; A.L. Fetter and K.M. Watson in Advances in Theoretical Physics, K. Brueckner editor (Academic Press, New York, 1965).

21. R.J. Glauber, in Lectures in Theoretical Physics, vol. 1 (Interscience, New York, 1959) 315.

22. G.E. Brown and M. Bolsterli, Phys. Rev. Lett. 3 (1959) 472.

23. A.M. Lane, Nuclear Theory (W.A. Benjamin, Inc., New York, 1964).

24. R.R. Whitehead, A. Watt, B.J. Cole and I. Morrison in Advances in Nuclear Physics, vol. 9, M. Baranger and E. Vogt, editors (Plenum Press, 1977) 123.

25. J.P. Albanese et al, Phys. Lett. 73B (1978) 119.

26. D.J. Malbrough et al, Phys. Rev. C17 (1978) 1395.

27. J. Hüfner, Phys. Reports 21 (1975) 1.

28. For example, see S. Barshay, G.E. Brown and M. Rho, Phys. Rev. Lett. 32 (1974) 787.

29. R.A. Eisenstein and F. Tabakin, Comp. Phys. Comm. 12 (1976) 237.

30. R. Landau and F. Tabakin, Phys. Rev. D5 (1972) 2746.

31. L.C. Liu and C.M. Shakin, to be published.

32. A.W. Thomas "Application of Three-Body Techniques to Pion-Nucleus Scattering", lectures appearing elsewhere in this volume.

33. J.P. Maillet, J.P. Dedonder and C. Schmidt, to be published.

34. F. Lenz and E.J. Moniz, to be published.

APPLICATION OF THREE-BODY TECHNIQUES TO PION-NUCLEUS SCATTERING*

A. W. Thomas

TRIUMF, U.B.C.
Vancouver, B.C., Canada V6T 1W5

Lecture I

There is a considerable number of nuclear physicists who would argue that the title of these lectures is very misleading. In particular, the few-body problem has long been considered a small, very specialized, area of nuclear physics that could safely be ignored by the majority. I will have succeeded in my task if in the course of this presentation I can convince you that in pion physics, the pion few-nucleon system has provided several striking theoretical insights, and may continue to do so.

Our guiding theme taken from few-body physics, is an emphasis on exact unitarity--Amado's remark that unitarity is "the hobgoblin of small minds" not withstanding! This concept is carefully introduced in the introductory section on the non-relativistic three-body problem. (Those readers strictly interested in pion-nucleus phenomena could skip §A completely on first reading.) The significance of the two-body t matrix, which is the exact solution of the two-body scattering problem, is also stressed there. Later it will be seen that some of the sharpest divisions in the theory of π-nucleus scattering centre on the choice of off-shell behavior for this object.

After the elimination of a two-body potential for the relativistically preferred scattering amplitude (or t-matrix), we show how (guided by unitarity) one can obtain practical relativistic three-body equations. Such equations have been applied to the

*Research supported by the National Research Council of Canada (Grant #A3233)

πNN system, and particularly πD scattering with some success and we shall briefly review those results. In comparison with the π-many-nucleon problem, where one must make many approximations in order to carry out a calculation, the πD system allows essentially exact calculation once the πN t-matrices are given. Thus agreement with data in the πD system should be regarded as a prerequisite for trust in any theory of π-nucleus scattering! Of course, agreement with experiment aside, from a purely theoretical point of view, the more rigorous treatment of relativistic effects in the three-body systems can tell one a good deal about directions for improving the usual, essentially non-relativistic, π-nucleus scattering theories.

We have already hinted at the controversial question of the πN range--that is, the most reliable off-shell behavior of the πN t-matrix. This question has recently been asked much more clearly by Myhrer and Thomas, who consider the πN interaction in a system with an extra nucleon--and particularly the implications of two and three-body unitarity in such a system for the off-shell behaviour of $t_{\pi N}$. We shall present these arguments rather briefly in §3, relying on Professor Kisslinger for the πN field theory.

Having justified our off-shell model for the πN t-matrix, we turn to the formulation of the π-many-nucleon problem in terms of this object. The general theory outlined in §4 is that of Kerman et al., and Watson. From the general theory it is relatively easy to write down all the conventional π-nucleus optical potentials (Kisslinger, Ericson-Ericson, Landau-Phatak-Tabakin, etc.). Although there are alternative ways to obtain the same result, we show how few-body ideas, and particularly the formulation of a three-body model for the first-order optical potential, have led to a deeper understanding of the π-nucleus system. In fact, many of the conventional theories drastically overestimate the cross-section for quasi-elastic scattering, and omit real pion absorption. While there are possibly important consequences for other reactions whose description needs distorted pion waves (as we mention), the most dramatic consequences may still await discovery in the (3,3) resonance region. Towards this end we mention the self-consistency question first raised by Ericson and Hüfner.

Finally, if time permits, we shall describe the efforts made in the πNN system to formally separate the "effects of absorption" on elastic scattering, from the usual multiple scattering series. The implication of these formal results for the π-nucleus case will be discussed briefly.

A. Non-Relativistic Three-Body Scattering Theory

For background material we refer the interested reader to Refs. 1-6, and particularly the classic paper of Lovelace and the most recent text (Ref. 4).

(i) Integral equation for scattering--the two-body t-matrix

We assume that the reader is familiar with the Lippmann-Schwinger formulation of the two-body scattering problem as a single integral equation equivalent to both the Schrodinger equation and appropriate boundary conditions. For a local potential V(r), this integral equation is

$$\psi_{\vec{k}}^{(+)}(\vec{r}) = (2\pi)^{-3/2}\, e^{i\vec{k}\cdot\vec{r}} - \frac{\mu_{12}}{2\pi}\int d\vec{r}\,'\, \frac{e^{ik|\vec{r}-\vec{r}\,'|}}{|\vec{r}-\vec{r}\,'|}\, V(r')\psi_{\vec{k}}^{(+)}(\vec{r}\,'), \tag{A.1}$$

with $\mu_{12}[\equiv m_1m_2/(m_1+m_2)]$ the reduced mass and $\psi_{\vec{k}}^{(+)}$ the full solution of the two-body scattering problem for an incident projectile of momentum $\vec{k}$. Eq. (A.1) may be shown to have the following asymptotic behaviour

$$\psi_{\vec{k}}^{(+)}(r) \underset{r\to\infty}{\longrightarrow} (2\pi)^{-3/2}\, e^{i\vec{k}\cdot\vec{r}} + f(k,\theta)\,\frac{e^{ikr}}{r}, \tag{A.2}$$

with

$$f(k,\theta) = -4\pi^2\,\mu_{12}\int d\vec{r}\,'\,\frac{e^{-i\vec{k}'\cdot\vec{r}}}{(2\pi)^{3/2}}\, V(r')\psi_{\vec{k}}^{(+)}(\vec{r}\,'), \tag{A.3}$$

the scattering amplitude $|\vec{k}'| = |\vec{k}|$, $\cos\theta = \hat{k}'\cdot\hat{k})$. That is, the differential cross-section for two-body scattering is

$$\frac{d\sigma(\theta)}{d\Omega} = |f(k,\theta)|^2 . \tag{A.4}$$

If we write Eq. (A.1) in operator form as

$$|\psi_{\vec{k}}^{(+)}\rangle = |\vec{k}\rangle + G_o(k^2/2\mu_{12} + i\varepsilon)\, V|\psi_{\vec{k}}^{(+)}\rangle , \tag{A.5}$$

it is in fact valid for any short-range potential (local or non-local). The free Green's function $G_o(E)$ is defined as usual

$$G_o(E) = (E-H_o)^{-1} , \tag{A.6}$$

with H_o the kinetic energy operator, and the fully interacting

Green's function G(E) is

$$G(E) = (E-H_o-V)^{-1} \tag{A.7}$$

The so-called resolvent equation for G(E) is readily derived using the operator identity

$$B - A = A(A^{-1}-B^{-1})B, \tag{A.8}$$

and hence

$$G(E^+) = G_o(E^+) + G_o(E^+)\ V\ G(E^+). \tag{A.9}$$

The two-body t-matrix, "t(E)", corresponding to the potential V is defined in terms of this Green's function as

$$G(E) = G_o(E) + G_o(E)\ t(E)\ G_o(E)\ . \tag{A.10}$$

At this stage it is worth pointing out that the easiest derivation of Eq. (A.5) follows from (A.9). Since one can define the scattering wave function as

$$\psi_{\vec{k}}^{(+)} = \lim_{\varepsilon\to 0+}\ i\varepsilon\ G(E+i\varepsilon)\ |\vec{k}> \quad , \tag{A.11}$$

(with $E \equiv k^2/2\mu_{12}$), and

$$\lim_{\varepsilon\to 0+}\ i\varepsilon\ G_o(E+i\varepsilon)\ |\vec{k}> = |\vec{k}> \tag{A.12}$$

multiplying (A.9) to the right by k , and throughout by iε, leads to Eq. (A.5). Using Eq. (A.10), however, leads to the alternative expression

$$|\psi_{\vec{k}}^{(+)}> = |\vec{k}> + G_o(E^+)\ t(E^+)\ |\vec{k}> \ . \tag{A.13}$$

In the momentum representation, using momentum states normalized so that

$$<\vec{p}|\vec{p}\,'> = \delta(\vec{p}-\vec{p}\,')\ , \tag{A.14}$$

Eq. (A.13) becomes

$$\psi_{\vec{k}}^{(+)}(\vec{p}) = \delta(\vec{p}-\vec{k}) + (k^2+i\varepsilon-p^2)^{-1}\ <\vec{p}|t(E)|\vec{k}> \ . \tag{A.15}$$

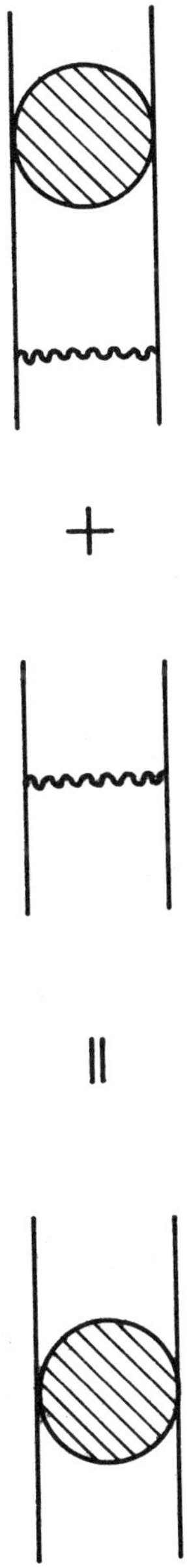

Fig. 1 Graphical representation of the two-body Lippmann-Schwinger equation (A.18). The circle represents the t-matrix, and the wiggly line a potential.

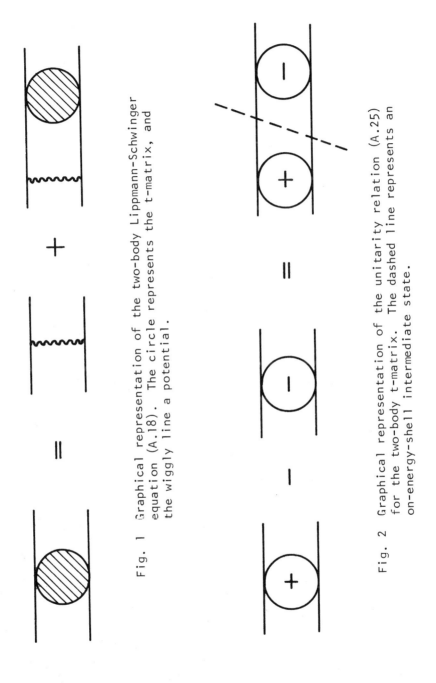

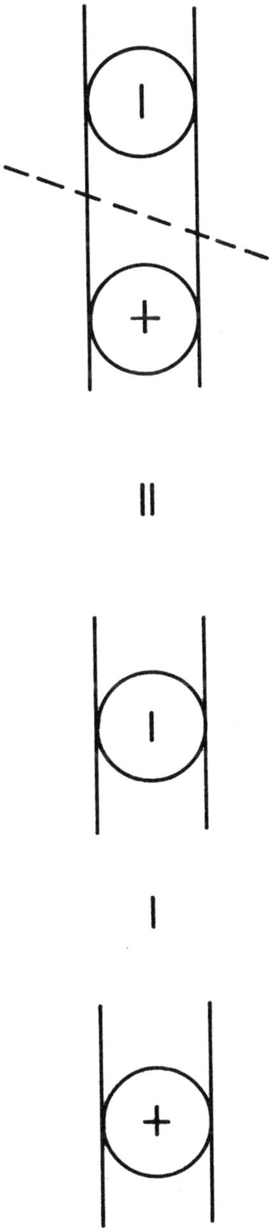

Fig. 2 Graphical representation of the unitarity relation (A.25) for the two-body t-matrix. The dashed line represents an on-energy-shell intermediate state.

(Alternatively, one may replace

$$(k^2+i\varepsilon-p^2)^{-1} = P/(k^2-p^2) - i\pi\, \delta(k^2-p^2), \qquad (A.16)$$

where "P" means principal value integral.)[7] The matrix element of the operator t(E) appearing in Eq. (A.15) is often written as $t(\vec{p},\vec{k};E)$ and is said to be "half off-shell", because $|\vec{p}| \neq |\vec{k}| = (2\mu_{12}E)^{1/2}$. The close relationship between the short range behaviour of the two-body wave function (i.e., Fourier components $p \neq k$), and the off-shell behaviour of the two-body t-matrix is made explicit through (A.15).

In practice, it is more common to write an integral equation for the t-matrix itself, rather than working through the wave function. In fact, multiplying (A.13) to the left by V, and [from Eqs. (A.5) and (A.13)] identifying

$$V|\psi_{\vec{k}}^{(+)}> = t(E^+)|\vec{k}> , \qquad (A.17)$$

one finds

$$t(E^+) = V + V\, G_o(E^+)\, t(E^+) . \qquad (A.18)$$

This is often represented diagrammatically as in Fig. 1.

Because of the simple form taken by the kinetic energy operator in momentum space, this is by far the most common representation of Eq. (A.18) in use--that is

$$t(\vec{k}',\vec{k};E) = V(\vec{k}',\vec{k}) + \int d\vec{k}''\, V(\vec{k}',\vec{k}'')(E^+-k''^2/2\mu_{12})^{-1} \times t(\vec{k}'',\vec{k};E) \qquad (A.19)$$

Equation (A.19) defines elements of the t-matrix for arbitrary values of (k',k;E), or in the usual jargon "fully off-shell". Unfortunately experiment only provides a constraint on one line in that three-dimensional space, namely $|\vec{k}| = |\vec{k}'| = (2\mu_{12}E)^{1/2}$. In fact, comparing Eqs. (A.3) and (A.17) we see that the experimental scattering amplitude is simply

$$\begin{aligned} f(k,\theta) &= -4\pi^2\mu_{12}\, t(\vec{k}',\vec{k};k^2/2\mu_{12}) \;:\; k=k', \\ &= -4\pi^2\mu_{12}\, t^{(on)}(k,\theta) \quad . \end{aligned} \qquad (A.20)$$

Using the partial wave decomposition

$$t(\vec{k}',\vec{k};E) = \sum_{\ell m} Y_{\ell m}(\hat{k}')t_\ell(k',k;E)Y_{\ell m}(\hat{k}), \qquad (A.21)$$

and the usual relation of a scattering amplitude to phase shifts implies

$$t_\ell(k,k;k^2/2\mu_{12}) \equiv t_\ell^{on}(k) = -e^{i\delta_\ell} \sin\delta_\ell/(\pi\mu_{12}k). \tag{A.22}$$

(ii) Unitarity of the two-body t-matrix

In the introduction we mentioned that the concept of unitarity was fundamental to these lectures. Perhaps the simplest example of a unitarity relation is provided by the two-body t-matrix. In fact, multiplying Eq. (A.18) to the left and right by V^{-1} and t^{-1}, respectively, implies

$$t^{-1}(Z) = V^{-1} - G_o(Z) , \tag{A.23}$$

for Z an arbitrary complex parameter. Setting $Z = E \pm io$ and subtracting leads to the formula for the discontinuity of t across the real axis

$$t^{-1}(E^-) - t^{-1}(E^+) = 2\pi i \; \delta(E-H_o) , \tag{A.24}$$

and hence

$$\text{disc } t(E) \equiv t(E^+) - t(E^-) = -2\pi i \; t(E^+) \; \delta(E-H_o) \; t(E^-). \tag{A.25}$$

This represented diagrammatically in Fig. 2, where the dashed line represents "$-2\pi i \; \delta(E-H_o)$" -- that is, an on-energy-shell intermediate state.

The most commonly seen form of Eq. (A.25) is the optical theorem, which relates the imaginary part of the forward scattering amplitude to the total cross section. Indeed, taking the on-shell matrix element of (A.25) leads to (k'=k)

$$\text{Im } t^{(on)}(k,0^o) = [\text{disc } t^{on}(k,0^o)/2i] = -\pi\mu_{12} \; k \int |t(\vec{k}',\vec{k};k^2/2\mu_{12})|^2 d\hat{k}', \tag{A.26}$$

and using (A.20)

$$\text{Im } f(k,0^o) = \left(\frac{k}{4\pi}\right) \sigma_{total} \tag{A.27}$$

We shall see very soon that a similar relation holds even in the far more complicated three-body problem, where many inelastic processes are included in σ_{total}.

While the unitarity of the scattering amplitude is an extremely significant observation, equivalent to conservation of probability, the operator relation (A.25) is actually far more powerful. It places constraints on the analytic structure, in the energy variable, of the completely off-shell t-matrix. In a given partial wave, this is simply an algebraic constraint, namely

$$\text{disc } t_\ell(k',k;E) = -2\pi i\mu_{12}\, k_E\, t_\ell(k',k_E;E)\, t^*_\ell(k_E,k;E),$$
$$k_E = (2\mu_{12}\, E)^{1/2} . \tag{A.28}$$

Clearly then, knowledge of the half-shell t-matrix is sufficient to construct the fully off-shell t-matrix by dispersion relation. Lack of space prevents us from pursuing the consequences of (A.28) further, but the interested reader is referred to the work of Lovelace,[3] which is still one of the best discussions available.

For later discussion it is useful to record one more result here concerning the half-shell t-matrix. In fact, it can be shown that in any partial wave, the half off-shell t-matrix is simply related to the fully on-shell t-matrix by

$$t_\ell(k',k_E;E) = t_\ell^{on}(k_E)\; f_\ell(k',k_E), \tag{A.29}$$

where $f_\ell(k',k_E)$ is unity for $k' = k_E$, and for a real potential is purely real. The function $f_\ell(k',k_E)$ is called the Kowalski-Noyes half-shell function.[5,8]

(iii) Two-body t-matrix in the three-body system

A minimal knowledge of the usual notational conventions in three-body theory is necessary at this stage:

(a) It saves on subscripts if any two-body property is labeled by the spectator particle's number. Thus (V_3, t_3) are the potential and t-matrix for the pair (12).

(b) Unless otherwise stated we work in the center of mass (c.m.) of the three particles, so that

$$\vec{p}_1 + \vec{p}_2 + \vec{p}_3 = 0. \tag{A.30}$$

Therefore only two momenta are needed to specify any state. These are usually the relative momentum of any pair [with (i,j,k) cyclic]

$$\vec{q}_i = (m_k\vec{p}_j - m_j\vec{p}_k)/(m_j + m_k) , \tag{A.31}$$

and the momentum of the third particle relative to that pair (which is simply p_i in the three-body c.m.).

(c) The kinetic energy operator is then

$$H_o = p_i^2/2\mu_i + q_i^2/2\mu_{jk} , \tag{A.32}$$

with

$$\mu_i = m_i\, m_{jk}/M;\ \mu_{jk} = m_j\, m_k/m_{jk}\ , \qquad \text{(A.33)}$$
$$m_{jk} = m_j + m_k;\ M = m_i + m_j + m_k\ .$$

(d) It is convenient to define the channel Hamiltonian and Green's function

$$H_i = H_o + V_i;\ G_i(E) = (E-H_i)^{-1}\ i\varepsilon(0,1,2,3). \qquad \text{(A.34)}$$

corresponding to just two particles interacting in the presence of a spectator. The fully interacting Green's function is

$$G(E) = (E-H)^{-1} \equiv (E-H_o - \sum_i V_i)^{-1} \qquad \text{(A.35)}$$

Consider now the scattering process shown in Fig. 1 with a third particle line disconnected from the other two representing the interaction of particles 1 and 2 in the presence of a non-interacting spectator (3). The appropriate Lippmann-Schwinger equation is

$$t_3(E^+) = V_3 + V_3\, G_o(E^+)\, t_3(E^+)\ , \qquad \text{(A.36)}$$

in analogy with (A.17), but the meaning has changed a little. In fact, t_3 and V_3 are operators in a space of three particles, so that the matrix elements are to be taken between states like $|\vec{p}_3\vec{q}_3\rangle$ and $|\vec{p}_3^{\,\prime}\ \vec{q}_3^{\,\prime}\rangle$. Clearly the two-body potential in three-body space is simply

$$\langle \vec{p}_3^{\,\prime}\ \vec{q}_3^{\,\prime}|V_3|\vec{p}_3\vec{q}_3\rangle = \delta(\vec{p}_3^{\,\prime} - \vec{p}_3)V_3(\vec{q}_3^{\,\prime},\vec{q}_3)\ . \qquad \text{(A.37)}$$

Note that we have not bothered to make the formally necessary discrimination between the operators V_3 on the left and right of (A.37) which act in different spaces. Eq. (A.37) says simply what we expect, namely that a pair-wise potential only changes the relative momentum of the pair, not their c.m. motion.

Let us take the momentum representation of Eq. (A.36), and use (A.37). This leads to the equation

$$t_3(\vec{p}_3^{\,\prime}\ \vec{q}_3^{\,\prime};\vec{p}_3\vec{q}_3,E) = \delta(\vec{p}_3^{\,\prime} - \vec{p}_3)V_3(\vec{q}_3^{\,\prime},\vec{q}_3)$$
$$+ \int d\vec{p}_3^{\,\prime\prime}\ d\vec{q}_3^{\,\prime\prime}\ \frac{\delta(\vec{p}_3^{\,\prime}-\vec{p}_3^{\,\prime\prime})V_3(\vec{q}_3^{\,\prime},\vec{q}_3^{\,\prime\prime})t_3(\vec{p}_3^{\,\prime\prime}\ \vec{q}_3^{\,\prime\prime};\vec{p}_3\vec{q}_3;E)}{\{E^+-p_3^{\prime 2}/2\mu_3\} - q_3^{\prime\prime 2}/2\mu_{12}} \qquad \text{(A.38)}$$

which clearly [c.f. Eq.(A.19)] has the solution

$$t_3(\vec{p}_3'\,\vec{q}_3';\,\vec{p}_3\vec{q}_3;E) = \delta(\vec{p}_3' - \vec{p}_3)\,t_3(\vec{q}_3',\,\vec{q}_3;\,E-\vec{p}_3^{\,2}/2\mu_3) \qquad (A.39)$$

The δ-function in (A.39) simply says that the non-interacting particle goes through with its momentum unchanged, and that the two-body t-matrix should be evaluated at the two-body sub-energy. (That is, the total three-body energy is E, and the energy of relative motion of the pair and spectator is $p_3^2/2\mu_3$, leaving $(E-p_3^2/2\mu_3)$ in the two-body sub-system. This is a direct consequence of the form of G_o).

While Eq. (A.39) as derived above is almost obvious, its consequences are quite profound! In any three-body scattering problem one needs the t-matrix for arbitrarily high spectator momentum p_3, and thus the two-body energy variable covers the range $(-\infty,E)$. The first consequence of this is that any two-body pole (bound state or resonance) becomes a cut in the three body energy variable. Secondly one needs the two-body t-matrix at negative values of the energy parameter. Since Eq. (A.19) defines the t-matrix for arbitrary (complex) values of the parameter "E", negative values obviously present no difficulty of principle. Indeed, for a real potential, the negative energy t-matrix defined by (A.19) is obviously real.

We shall see very soon that the form of (A.39) is essential to the proof of three-body unitarity. However, the concept of a negative energy scattering amplitude seems to disturb many people. Perhaps a word on how it relates to more familiar objects in scattering theory will be worthwhile. As we mentioned in part (ii) above, the unitarity relation (A.28) which gives the discontinuity across the only cut the two-body t-matrix has [from $(0,\infty)$], can be used to write a dispersion relation for the t-matrix at arbitrary energy. In fact, in the absence of a bound state, one finds (using the contour of Fig. 3)

$$t_\ell(k',k;Z) = V_\ell(k',k) + (2\pi i)^{-1}\int_0^\infty dE\ \mathrm{disc}\{t_\ell(k',k;E)\}/(Z-E). \qquad (A.40)$$

with "$V_\ell(k',k)$" the limit of $t_\ell(k',k;Z)$ as $Z \to \infty$ (it can be removed by a subtraction). Now using Eq. (A.28) one finds the fully off-shell t-matrix for arbitrary "Z" (e.g. on the negative real axis) is directly related to the half off-shell t-matrices at positive energy (i.e., in the scattering region)

$$t_\ell(k',k;Z) = V_\ell(k',k) - \int_0^\infty dk_E\ k_E^2\ \frac{t_\ell(k',k_E;E)\,t_\ell(k_E,k;E)}{(Z-E)} \qquad (A.41)$$

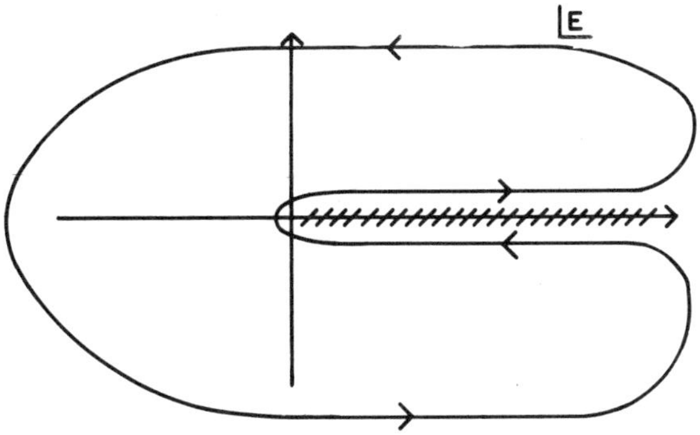

Fig. 3 Contour used in the proof of Eqs. (A.40) and (A.41).

But we have seen in (A.15) that knowledge of the half-shell t-matrix at positive energy, is equivalent to knowing the scattering wave function at that energy. Clearly, then, one could rewrite any system of equations which require negative energy t-matrices, entirely in terms of more familiar scattering wave functions. Such a system of three-body equations has indeed been developed,[9] but will not be considered here.

(iv) The Necessity for Faddeev's Equations

The standard technique for the numerical solution of an equation like (A.19), after partial wave expansion, is to approximate the integral by some quadrature formula e.g.

$$t_\ell(k_i,k_j;E) = V_\ell(k_i,k_j) + \sum_{n=1}^{N} W_n \frac{V_\ell(k_i,k_n)\; t_\ell(k_n,k_j;E)}{E^+ - k_n^2/2\mu_{12}} \; . \qquad (A.42)$$

Equation (A.42) can then be solved as a set of coupled linear equations, with the accuracy improving as N is increased. A sufficient condition that this numerical solution is possible (leading to a unique solution), is that the kernel (G_0 V) of the integral equation should be square integrable (or L^2 in the usual terminology). The L^2 condition means in fact that

$$T_r\ [(G_0\ V)^\dagger\ G_0\ V] < \infty \; . \qquad (A.43)$$

(For a much more complete discussion of these ideas we refer to the work of Lovelace.)[3,10] For the usual short range potentials, and E^+ not exactly on the real energy axis, it is relatively easy to show that the kernel of the two-body scattering equation (A.19) satisfies Eq. (A.43). The proof that a numerical solution is still possible in the limit where E^+ becomes real is more difficult,[10] but the result is not surprising.

However, the kernel of Eq. (A.38) which describes two-body scattering in the three-body system contains a δ-function, representing the non-interacting spectator. Now when one constructs the trace (A.43) one finds the combination

$$T_r \left\{(G_o\ V_3)^{\dagger}\ (G_o\ V_3)\right\} \sim \int d\vec{p}_3 \int d\vec{p}_3{}'\ \delta(\vec{p}_3-\vec{p}_3{}')\ \delta(\vec{p}_3{}'-\vec{p}_3)\ , \tag{A.44}$$

which is infinite! This may seem very much like making a mountain out of a mole-hill, since we could obviously factor out $\delta(p_3-p_3')$ and solve the purely two-body problem as before. Unfortunately this is not so in the true three-body system with more than one interaction. Indeed "G_o V" in that case is $G_o(V_1+V_2+V_3)$ which contains three distinct δ-functions. The trace of this "G_o V" inevitably contains combinations like $\{(G_o\ V_i)^+\ G_o\ V_i\}$, and is therefore not L^2. While this fact is not sufficient to show that the three-body Lippmann-Schwinger equation

$$T = V + V\ G_o\ T \tag{A.45}$$

does not have unique solutions, it does indicate that there may be severe difficulties. The problem which we have discussed is known as the "disconnectedness problem" (because the δ-functions which lead to problems are represented diagrammatically by a third line not connected to either of the lines representing the interacting nucleons).

The failure of the Lippmann-Schwinger equation for three (or more) particle scattering was a real formal difficulty in the late 50's. However like many formal problems in physics its existence was ignored for practical purposes. Indeed the formal manipulations which we shall carry out in deriving the optical model in Lecture 2 are in this class—i.e. mathematically meaningless, but the end result happens to be correct! The correct solution to this problem by the Russian mathematician Faddeev not only solved the formal problem but opened the door to considerable new physical understanding.

(v) Faddeev's Equations

The solution given by Faddeev had essentially two steps. First all the series of purely two-body, disconnected potential diagrams which could be summed, were. This is most clearly seen in a perturbation expansion of Eq. (A.45). (While this is not always guaranteed to converge, the reader can rest assured that a more formal operator proof along the lines of the next section is possible.) That is,

$$T = V_1 + V_2 + V_3 + (V_1 + V_2 + V_3)\ G_o\ (V_1 + V_2 + V_3) + \ldots \tag{A.46}$$

and the subseries like

$$V_i + V_i G_o V_i + V_i G_o V_i G_o V_i + \dots \tag{A.47}$$

can be identified (through the perturbative expansion of Eq. (A.36) as t_i. This leads to the series

$$T = (t_1 + t_2 + t_3) + t_1 G_o t_2 + t_1 G_o t_3 + t_2 G_o t_1 + \dots \tag{A.48}$$

where all the disconnected diagrams which can be trivially summed have been. Thus we see that the present day emphasis on the two-body t-matrix takes considerable impetus from Faddeev.

Equation (A.48) describes the rather unlikely process where three free particles scatter to three free particles (the so-called $3 \to 3$ amplitude). The more realistic case of scattering from a bound state is considered in the next section. The second step in Faddeev's solution was to break T into three parts $\sum_i T^i$, where T^i is defined so that the last pair to interact were the pair (jk). Thus, we find

$$T^i = t_i + t_i G_o (T^j + T^k), \tag{A.49}$$

which is a set of three coupled equations with kernel

$$\tilde{\tilde{K}} = \begin{pmatrix} 0 & t_1 & t_1 \\ t_2 & 0 & t_2 \\ t_3 & t_3 & 0 \end{pmatrix} G_o . \tag{A.50}$$

Let $\tilde{T}$ and $\tilde{t}$ be the obvious column vectors, then Eq. (A.49) can be re-written as

$$\tilde{T} = \tilde{t} + \tilde{\tilde{K}} \tilde{T}, \tag{A.51}$$

and when iterated once

$$\tilde{T} = \tilde{t} + \tilde{\tilde{K}} \tilde{t} + \tilde{\tilde{K}}^2 \tilde{T}, \tag{A.52}$$

the new kernel $\tilde{\tilde{K}}^2$ is *completely connected*. That is all its elements have the form $t_i G_o t_j$ $(i \neq j)$. This condition (namely $i \neq j$) is sufficient to avoid the infinity caused by two identical δ-functions coming together [as in Eq. (A.44)].

The formal proof that for reasonable short range potentials Eq. (A.52) has a compact kernel in the limit $\varepsilon \to 0+$, is very difficult, and need not concern us here [c.f. Refs. (2, 11)]. It is enough that such a proof exists, and was made possible by the *separation of* T *into three parts according to a proper choice of boundary conditions*. This realization that the formal problems of three-body theory arose from the ambiguity of boundary conditions [e.g. particle 1 hitting 2 and 3 can lead to any of the three possible combinations 1 (23), 2 (31) or 3 (12)] was Faddeev's second crucial contribution.

(vi) The Amplitudes U_{ji}

The equations derived immediately above were not very practical as they applied only to the $3 \to 3$ scattering process. By far the most commonly examined experimental situation is, however, the scattering of a projectile i from a bound state "u" of the pair (jk) leading to either elastic scattering or re-arrangement

$$\begin{aligned} i + (jk)_c &\longrightarrow i + (jk)_c \\ &\longrightarrow j + (ki)_d \,. \end{aligned} \tag{A.53}$$

The t-matrix for such a process can be proven by formal scattering theory[3,12] to be

$$\begin{aligned} T_{j(d);i(c)}(E^+) &= \langle \chi_{j;d} \,|\, V^j \,|\, \psi^{(+)}_{i;c} \rangle \equiv \langle \chi_{j;d} \,|\, U^{(+)}_{ji}(E^+) \,|\, \chi_{i;c} \rangle \\ &= \langle \psi^{(-)}_{j;d} \,|\, V^i \,|\, \chi_{i;c} \rangle \equiv \langle \chi_{j;d} \,|\, U^{(-)}_{ji}(E^+) \,|\, \chi_{i;c} \rangle \,, \end{aligned} \tag{A.54}$$

where $|\chi_{i;c}\rangle$ represents the asymptotic state of (jk) in the bound state c with a non-interacting spectator i, satisfying the equation

$$H_i \, \chi_{i;c} = E \, \chi_{i;c} \,, \tag{A.55}$$

and

$$V^i \equiv V - V_i = V_j + V_k \,. \tag{A.56}$$

The wave functions $\psi^{(\pm)}$ are the full three-body scattering wave functions,[4] with ingoing or outgoing boundary conditions, which we shall not discuss further. Instead we concentrate on $U_{ji}^{(\pm)}$ which are formally

$$U^{(+-)}_{ji}(E) = V^{(ji)} + V^j \, G(E) \, V^i \,. \tag{A.57}$$

These operators were first introduced by Lovelace who derived Faddeev-like equations for them.

To prove this we need the operator identity

$$G = G_k + G \, V^k \, G_k \,, \tag{A.58a}$$

$$\;\;= G_k + G_k \, V^k \, G \,, \tag{A.58b}$$

which is easily proven using Eqs. (A.8), (A.34) and (A.56). Now, inserting this in the definition (A.57) for $U_{ji}^{(+)}$ and substituting for V^i implies

$$U_{ji}^{(+)} = V^j + \sum_{k\neq i}\left\{V^j G_k V_k + V^j G V^k G_k V_k\right\}$$

$$= V^j + \sum_{k\neq i} U_{jk}^{(+)} G_o t_k , \qquad (A.59)$$

which is of the Faddeev form. [The last step follows from the definition of $U_{ji}^{(+)}$, and the fact that [from Eqs. (A.10) and (A.18) —c.f. also (A.17)].

$$G_k V_k = G_o t_k . \qquad (A.60)$$

Equation (A.59) is clearly of the Faddeev form. Similarly one can show

$$U_{ji}^{(-)} = V^i + \sum_{k\neq j} t_k G_o U_{ki}^{(-)} . \qquad (A.61)$$

While Eqs. (A.59) and (A.61) are of the Faddeev form it does seem somewhat superfluous to have two different transition operators which give identical on-shell results. Therefore we follow Alt *et al.* in introducing the operator U_{ji},[13] defined by

$$G(E) = G_j(E)\, \delta_{ij} + G_j(E)\, U_{ji}(E)\, G_i(E) , \qquad (A.62)$$

[with G_i the channel Green's function (A.34)]. The relationship of U_{ji} to $U_{ji}^{(+)}$ is best shown by substituting (A.58a) into (A.58b)

$$G = G_j + G_j V^j G_i + G_j V^j G V^i G_i$$

$$= G_j + G_j U_{ji}^{(+)} G_i , \qquad (A.63)$$

and comparing (A.63) with (A.62) we find

$$U_{ji}(E) = \bar{\delta}_{ji}(E-H_i) + U_{ji}^{(+)}(E) , \qquad (A.64)$$

[where $\bar{\delta}_{ij} \equiv (1-\delta_{ij})$]. Similarly one can show

$$U_{ji}(E) = \bar{\delta}_{ji}(E-H_j) + U_{ji}^{(-)}(E) . \qquad (A.65)$$

An integral equation for U_{ji} is now obtained by substituting (A.64) into (A.59), with the result (after some algebra)

$$U_{ji}(E) = \bar{\delta}_{ji}(E-H_o) + \sum_{\ell \neq i} U_{j\ell}(E)\ G_o(E)\ t_\ell(E)\ ,$$

$$= \bar{\delta}_{ji}(E-H_o) + \sum_{\ell \neq j} t_\ell(E)\ G_o(E)\ U_{\ell i}(E)\ . \qquad (A.66)$$

Equation (66) satisfies both of the criteria emphasised by Faddeev. The potentials have been completely eliminated in favour of t-matrices, and the kernel of the three-body equations is completely connected after one iteration [c.f. Eq. (A.52) above]. Finally we stress that the on-shell (physical) transition amplitudes $T_{j(d);i(c)}$ calculated with U_{ji} are equal to those calculated from $U_{ji}^{(\pm)}$ [the proof requires only Eq. (A.55)].

(vii) The Break-Up Amplitude and Cross Section

If the kinetic energy of the projectile is greater than the binding energy of bound state 'c', the additional possibility of a three-body final state must be considered in addition to those shown in Eq. (A.53). [This is the 2→3 amplitude '$T_{0;i(c)}(E)$'.] In fact, the general equations (A.54) and (A.57) are quite valid even in this case, and

$$T_{0;i(c)}(E) = \langle 0 \mid U_{0i}^{(+)}(E) \mid \chi_{i;c} \rangle$$

$$= \langle 0 \mid \sum_\ell t_\ell\ G_o\ U_{\ell i} \mid \chi_{i;c} \rangle \text{ on-shell} \qquad (A.67)$$

(We have substituted for $U_{0i}^{(+)}$ using Eq. (A.59) with $V^o \equiv 0$.)

For completeness we record here how to calculate the scattering cross section using the operators described above. In fact our transition operators have precisely the normalizations of Goldberger and Watson[14] so that, if V_{in} ($\equiv p_i/\mu_i$) is the relative velocity in the initial state, and the final state contains N_f particles with momenta $\{\vec{p}_1 \ldots \vec{p}_{N_f}\}$, the c.m. differential cross section is

$$d\sigma_{fi} = (2\pi)^4\ V_{in}^{-1}\ |T_{f;i}|^2\ \delta(E_f-E_i)\ \delta\left[\sum_{j=i}^{N_f} \vec{p}_j\right] \times$$

$$\times \prod_{j=1}^{N_f} d\vec{p}_j\ . \qquad (A.68)$$

For a detailed discussion of the case of identical particles we refer to §1.7 of Ref. 4 and to Ref. 14.

(viii) Unitarity of the Three-Body Equations

In the section above dealing with the two-body t-matrix we discussed its singularity structure in considerable detail—particularly that structure associated with two-body unitarity. As part of our discussion of pion-nucleus optical potentials will be concerned with their unitarity content (in a three-body model), it is important to clearly establish here the unitarity structure of the three-body equations.

The neatest derivation is that of Schmid and Ziegelmann[15] which we shall follow. Let us write Eq. (A.66) as

$$\tilde{\tilde{U}} = \tilde{\tilde{U}}^o + \tilde{\tilde{U}}^o \tilde{\tilde{\Gamma}} \tilde{\tilde{U}} , \tag{A.69}$$

in an obvious matrix notation. Then multiplying to the left by $\tilde{\tilde{U}}^{o-1}$ and to the right by $\tilde{\tilde{U}}^{-1}$ we find

$$\tilde{\tilde{U}}^{-1}(E^+) = \tilde{\tilde{U}}^{o-1}(E^+) - \tilde{\tilde{\Gamma}}(E^+) . \tag{A.70}$$

Subtracting the similar expression for E^- yields a relation very similar to (A.24), except that the driving term is no longer hermitian, *viz:*

$$\tilde{\tilde{U}}^{-1}(E^-) - \tilde{\tilde{U}}^{-1}(E^+) = [\tilde{\tilde{U}}^{o-1}(E^-) - \tilde{\tilde{U}}^{o}(E^+)^{-1}] + [\tilde{\tilde{\Gamma}}(E^+) - \tilde{\tilde{\Gamma}}(E^-)] . \tag{A.71}$$

The inverse of the operator $\tilde{\tilde{U}}^o$ can be readily found by writing out the appropriate 3×3 matrix.

$$[(\tilde{\tilde{U}}^o)^{-1}]_{ij} = (\frac{1}{2} - \delta_{ij})\, G_o . \tag{A.72}$$

Thus we find the discontinuity of $\tilde{\tilde{U}}^{-1}$ is

$$\text{disc}\ [U^{-1}(E)]_{ij} = \frac{1}{2}\, \text{disc}\ G_o(E) - \delta_{ij}\ \text{disc}\ G_i(E) \tag{A.73}$$

(where we used Eq. (A.10) to replace $G_o\, t_i\, G_o$ in $\tilde{\tilde{\Gamma}}$ by the channel Green's function G_i).

From Eq. (A.34) one sees that

$$\text{disc}\ G_i(E) = -2\pi i\ \delta(E-H_i), \tag{A.74}$$

which may mean a little more if we multiply by a complete set of eigenstates of H_i—including bound states ($\chi_{i;c}$ with binding energy ε_c), and scattering states (ψ_i), for two particles in the presence of a non-interacting spectator.

$$\text{disc } G_i(E) = \sum_c \int d\vec{p}_i \ (-2\pi i)\,|\chi_{i;c}(\vec{p}_i)>\delta(E-\vec{p}_i^{\,2}/2\mu_i-\varepsilon_c) \times$$

$$\times < \chi_{i;c}(\vec{p}_i)|$$

$$+ \int d\vec{p}_i \ d\vec{q}_i \ (-2\pi i)\,|\psi_i(\vec{p}_i\vec{q}_i)>\delta(E-\vec{p}_i^{\,2}/2\mu_i-\vec{q}_i^{\,2}/2\mu_{jk})<\psi_i(\vec{p}_i\vec{q}_i)| \quad (A.75)$$

Now, with the definitions

$$\Delta_{i;c} = \int d\vec{p}_i\,|\chi_{i;c}(\vec{p}_i)>\delta(E-\vec{p}_i^{\,2}/2\mu_i-\varepsilon_c)<\chi_{i;c}(\vec{p}_i)|\,,$$

$$\Delta_o = \int d\vec{p}_i \ d\vec{q}_i \ |\vec{p}_i\vec{q}_i>\delta(E-\vec{p}_i^{\,2}/2\mu_i-q_i^2/2\mu_{jk})<\vec{p}_i\vec{q}_i|\,, \quad (A.76)$$

and identifying [from Eq. (A.13)]

$$|\psi_i(\vec{p}_i\vec{q}_i)> = (1+G_o\ t_i)\,|p_iq_i>\,, \quad (A.77)$$

equation (A.71) can be written as

$$\text{disc}\left[\tilde{\tilde{U}}^{-1}(E)\right]_{ij} = -\pi i\ \Delta_o(E) + 2\pi i\ \delta_{ij}\left\{\sum_c \Delta_{i;c}(E)\right.$$

$$\left. + \left[1+G_o(E^+)t_i(E^+)\right]\ \Delta_o(E)\ \left[1+t_i(E^-)G_o(E^-)\right]\right\}. \quad (A.78)$$

To obtain the conventional expression for the right-hand discontinuity we next multiply to the left and right by $\tilde{\tilde{U}}(E^+)$ and $\tilde{\tilde{U}}(E^-)$ respectively.

$$\text{disc } U_{ji}(E) = \pi i \sum_{k,\ell} U_{jk}(E^+)\ \Delta_o\ U_{\ell i}(E^-)$$

$$- 2\pi i\left\{\sum_{k;c} U_{jk}(E^+)\ \Delta_{k;c}\ U_{ki}(E^-)\right.$$

$$\left. + \sum_k U_{jk}(E^+)\left(1+G_o(E^+)t_k(E^+)\right)\Delta_o(E)\left(1+t_k(E^-)G_o(E^-)\right)U_{ki}(E^-)\right\}. \quad (A.79)$$

Also summing Eq. (A.66b) over j implies that

$$\sum_k U_{ki} = 2\ U_{oi}\,, \quad (A.80)$$

and adding $t_j\ G_o\ U_{ji}$ to both sides of (A.66b) implies

$$(1 + t_k\ G_o)\ U_{ki} = (\overline{\delta}_{ki}-1)(E-H_o) + U_{oi}\,. \quad (A.81)$$

Equations (A.80) and (A.81) can be used to eliminate U_{ki} from all but the bound-state part (second term) of (A.79). Finally, using the fact that $(E-H_0)\Delta_0$ is zero one finds the complete expression for the unitarity of the Alt, Grassberger Sandhas operators U_{ji}

$$\text{disc}\left\{U_{ji}(E)\right\} = -\,2\pi i\left\{\sum_{k,c} U_{jk}(E^+)\ \Delta_{k;c}(E)\ U_{ki}(E^-) + U_{jo}(E^+)\ \Delta_o\ U_{oi}(E^-)\right\}. \tag{A.82}$$

This result allows us to write down the optical theorem for particle i scattering from a bound state c. Indeed using (A.54) and (A.67) we obtain

$$\text{disc}\left\{T_{i(c);i(c)}\left(\vec{p}_i^{\,(E)},\ \vec{p}_i^{\,(E)};E\right)\right\} = -\,2\pi i\left\{\sum_{j,d}\int d\vec{p}_j{}'\ \left|T_{j(d);i(c)}\left(\vec{p}_j{}',\ \vec{p}_i^{\,(E)};E^+\right)\right|^2\ \delta\left(E-p_j{}'^2/2\mu_j-\varepsilon_{j;d}\right)\right.$$
$$\left. + \int d\vec{p}_i{}'\ d\vec{q}_i{}'\ \left|T_{0;i(c)}\left(\vec{p}_i{}',\vec{q}_i{}';\vec{p}_i^{\,(E)};E+\right)\right|^2\ \delta\left(E-p_i{}'^2/2\mu_i-q_i{}^2/2\mu_{jk}\right)\right\}, \tag{A.83}$$

which together with Eq. (A.68) gives the promised generalisation of (A.26,27), *viz:*

$$\text{Im}\left\{T_{i(c),i(c)}(E,\theta = 0^{\circ})\right\} = -\ \pi(2\pi)^4\ V_{in}\left\{\sum_{jd}\sigma_{j(d)\leftarrow i(c)} + \sigma_{\text{break-up}}\right\}. \tag{A.84}$$

We have examined the unitarity relations for the three-body equations in considerable detail. It has been shown that the right-hand unitarity cut is actually a series of cuts, one starting at each $\varepsilon_{j;c}$—corresponding to each allowed re-arrangement channel—in addition to the break-up cut beginning at E = 0. We stress that even though one usually only calculates explicitly the elastic (or re-arrangement) amplitude for comparison with data, the information about all other open channels is implicitly contained in it, and gives important constraints on its analytic behaviour. This point will be stressed again when in Lecture 3 we examine the reaction content of theories of pion-nucleus scattering.

(ix) Three-Body Interactions for a Separable Two-Body t-Matrix

Historically the acceptance of separable interactions was very closely connected with the development of three-body scattering theory. Over many years of dealing with local Yukawa potentials, based on one pion exchange, the nuclear physics community developed a prejudice which ruled the highly non-local separable N-N interactions unrealistic. Even the demonstration by Lovelace that the *t-matrix* for *any* potential is separable in the vicinity of a bound state or resonance[3] would probably have had little impact were it not for the incredible agreement with low energy (14.1 MeV) n-d elastic data obtained at about the same time by Aaron, Amado and Yam.[16]

In general, a potential is *separable* in the ℓth partial wave if it can be written in the form

$$V_\ell^{(Sep)}(p',p) = \lambda_\ell \, \tilde{g}_\ell^T(p') \, \tilde{g}_\ell(p), \tag{A.85}$$

where $\tilde{g}_\ell$ is a linear array of form-factors. A rank-one potential, of the type used by Aaron *et al.* has just one element in $\tilde{g}_\ell$—for example in the $^3S_1$ N-N state

$$g_0(p) = C/(p^2+\beta^2). \tag{A.86}$$

While (A.85,86) may look quite different from a Yukawa potential

$$V_0^{(Yukawa)}(p',p) \sim \ell n\left(\frac{p^2+p'^2+2pp'+\mu^2}{p^2+p'^2-2pp'+\mu^2}\right)\Big/ 4\mu pp' , \tag{A.87}$$

the key point is that at the deuteron pole the *t-matrices* given by *both* have the form[3]

$$t(p',p;E) \xrightarrow[E\to-\varepsilon_d]{} \frac{(p'^2+m\varepsilon_d)\; \psi_d(p')\; \psi_d(p)\; (p^2+m\varepsilon_d)}{E + \varepsilon_d} , \tag{A.88}$$

and for $p \lesssim 1$ fm^{-1}, Eq. (A.86) with $\beta \sim 1.4$ fm^{-1} gives a reasonable representation of the deuteron wave function $[\psi_d(p)]$.

Thus, even when a separable potential may seem quite unrealistic, it is possible that the resulting t-matrix can be a good approximation to the real thing. However the crucial factor in favour of separability (and this is what motivated Lovelace) is the simplification it gives in the Faddeev equations. To see this, we need to re-write the equations of Alt *et al.* for separable t-matrices. In general, in the three-body system a separable two-body potential has the form

$$V_i = \sum_a \lambda_a \int d\vec{p}_i \; | \, i\vec{p}_i;a \rangle \langle i\vec{p}_i;a \, | , \tag{A.89}$$

where "a" labels possible quantum numbers, such as

$$\langle \vec{p}_i{}' \vec{q}_i \mid i\vec{p}_i; a\rangle = \delta(\vec{p}_i - \vec{p}_i{}')\ g_\ell(q_i)\ Y_{\ell m}(\hat{q}_i{}'), \tag{A.90}$$

in the spinless-isospinless case where $a \equiv (\ell m)$.

Substituting (A.89) into the two-body Lippmann-Schwinger equation, the reader should be able to prove in a couple of lines that the solution is

$$t_i(E^+) = \sum_a \int d\vec{p}_i \mid i\vec{p}_i; a\rangle\ D_{ia}^{-1}(E^+ - p_i{}^2/2\mu_i) \langle i\vec{p}_i; a \mid, \tag{A.91}$$

with

$$D_{ia}(E^+) = \lambda_a^{-1} - \int d\vec{q}\,'\ \frac{|\langle a|\vec{q}\,'\rangle|^2}{E^+ - q'^2/2\mu_{jk}}. \tag{A.92}$$

Equation (A.91) demonstrates the essence of the separable t-matrix in relation to resonance formation ($\langle a \mid \vec{q}_i \rangle$), propagation ($D_{ia}{}^{-1}$) and decay ($\langle \vec{q}_i{}' \mid a \rangle$) all independent of each other.

If one substitutes (A.91) into Eq. (A.66b) for U_{ji} and multiplies to the left and right by $\langle j\,\vec{p}_j{}'; b| G_0$ and $G_0 | i\,\vec{p}_i; a\rangle$ respectively one finds

$$X_{jb;ia}(\vec{p}_j{}', \vec{p}_i; E) = Z_{jb;ia}(\vec{p}_j{}', \vec{p}_i; E)$$

$$+ \sum_{\ell, c} \int d\vec{p}_\ell{}''\ Z_{jb;\ell c}(\vec{p}_j{}', \vec{p}_\ell{}''; E)\, D_{\ell c}{}^{-1}(E^+ - p_\ell{}''^2/2\mu_\ell)\, X_{\ell c;ia}(\vec{p}_\ell{}'', \vec{p}_i; E), \tag{A.93}$$

where we have defined

$$X_{jb;ia}(\vec{p}_j{}', \vec{p}_i; E) \equiv \langle j\,\vec{p}_j{}'; b|G_0(E)\ U_{ji}(E)\ G_0(E)|i\,\vec{p}_i; a\rangle,$$

$$Z_{jb;ia}(\vec{p}_j{}', \vec{p}_i; E) = \bar{\delta}_{ij} \langle j\,\vec{p}_j{}'; b|G_0(E)|i\,\vec{p}_i; a\rangle. \tag{A.94}$$

Although we did not write out Eqs. (A.66) in detail in momentum representation, a little reflection will show that in general they are coupled integral equations in *two* momentum variables. These are extremely difficult to solve in practice, although several heroic efforts using Padé techniques seem to have succeeded for the three-nucleon problem.[17] On the other hand, Eq. (A.93) is a set of coupled equations in only one variable, and therefore presents little more difficulty than a coupled channels two-body problem. [The numerical difficulties associated with the more complicated (three-body) singularity structure of the driving term can be overcome by contour rotation[11]—see particularly Ref. 15]. This is an extremely important practical simplification of the formal three-body theory!

For on-shell momenta

$$|\vec{p}_i\,'| = |\vec{p}_i| = 2\mu_i(E-\varepsilon_{i;c}), \tag{A.95}$$

the X-amplitude is the elastic (or re-arrangement) t-matrix, up to a normalisation constant. In fact, the form-factor in any two-body channel with bound state is intimately related to the bound-state wave function [in analogy with (A.86,88)], i.e.

$$<\vec{p}_i\,'\vec{q}_i\,'|G_o(E)|i\,\vec{p}_i\,;a> = -\delta(\vec{p}_i-\vec{p}_i\,')(\varepsilon_{i;a}+q_i\,'^2/2\mu_{jk})^{-1}g_a(\vec{q}_i\,')$$

$$\equiv <\vec{p}_i\,'\vec{q}_i\,'\,|\,\chi_{i;a}(\vec{p}_i)>, \tag{A.96}$$

up to a normalisation constant. [To verify this, the reader should consider the two-body Schrödinger equation $(H_o + V_i)\psi_B = \varepsilon_{i;B}\,\psi_B$ with the separable interaction (A.89).] Now, assuming the bound-state normalisation has been included in the definition of the two-body form-factor we obtain

$$X_{jb;ia}(\vec{p}_j\,',p_i;E) = <\chi_{j;b}(\vec{p}_j\,')\,|\,U_{ji}(E)\,|\,\chi_{i;a}(\vec{p}_i)>_{\text{on-shell}}, \tag{A.97}$$

which is exactly the (elastic or) re-arrangement scattering amplitude $T_{j(b);i(a)}(E)$. The break-up amplitude is readily shown [from Eqs. (A.67) and (A.91)] to be

$$T_{o\leftarrow i(a)}(\vec{p}\,',\vec{q}\,';\vec{p}_i) = \sum_{\ell,c} <\vec{q}_\ell\,'\,|\,\ell;c>D_{\ell c}^{-1}(E^+-p_\ell\,'^2/2\mu_\ell)$$

$$X_{\ell c;ia}(\vec{p}_\ell\,',\vec{p}_i;E). \tag{A.98}$$

The diagrammatic representation of both Eqs. (A.93) and (A.98) will be considered in the next section.

§B SCATTERING THEORY FOR THE πNN SYSTEM

At first sight the πNN system in the energy region where the (3,3) resonance can play a significant role seems to be ideally suited to a description using the Faddeev equations. Particularly πd scattering involves a bound state in the NN sub-system, and a resonance in each πN sub-system, so that the separable potential equations [§A.(ix) above] should be suitable. This considerably reduces the computational difficulties over the general case, as we have already indicated.

For the moment we choose to ignore the additional complications caused by the pion being a boson, and therefore strictly requiring a field theoretic description. [Some of the implications of standard field theories for the type of t-matrix necessary in a multiple scattering formalism will be considered in part C (below)

—see also Dr. Kisslinger's lectures.] Even without this problem, the very low mass of the pion unavoidably presents us with a relativistic scattering problem. Thus pion physics immediately takes us one step beyond standard nuclear physics.

In the early days of three-body theory much of the optimism surrounding the discovery of the Faddeev equations was related to the fact that because the essentially non-relativistic concept of potential had been replaced by the t-matrix, or scattering amplitude, the generalisation to relativistic systems did not seem so obscure. Those heady days of extremely relativistic calculations (such as generating the pion as a three pion bound state) have long been history. Nevertheless, the equations developed in the late 60's still provide an excellent framework for the far more reasonable πNN problem. (We consider it more reasonable, because in the intermediate energy region only one of the three particles is significantly relativistic—as opposed to all three being far off-mass-shell in the case mentioned above.)

The advantages of working first with a three-body system should be apparent here. There are no nuclear structure complications, and the possibility exists to develop a complete, consistent theory—without the necessity to make many-body approximations. For comparison it is interesting to read Heller's criticisms of the inconsistencies of many pion-nucleus calculations performed so far—consistencies related to the ad hoc introduction of relativity into a non-relativistic formalism.[18] Unfortunately the solution he proposed does not seem practical, because the theory (a relativistic potential theory based on the work of Bakamjian and Thomas)[19] inevitably generates many body forces, which render numerical work essentially impossible. We are not aware, for example, of a complete formulation of the three-body problem in this approach (including a proof of unitarity).

One of the most significant early formulations of the relativistic three-body problem[20] dealt with the properties of resonant solutions to the two-body Bethe-Salpeter equation as a first step towards generalising the work of A(ix). This was ambitious work aiming at a complete formulation of a practical technique for solving three-body problems in field theory. More recent work has tended to be much more phenomenological,[21,22] with the limited objective of finding a practical, unitary, Lorentz invariant generalisation of the two-body Blankenbecler-Sugar equation to three-body systems. Essentially for lack of space, we tend to follow the latter approach.

The content of this section is therefore a presentation of the relativistic two-body problem as it appears for a πN pair. It will be seen that almost all commonly used πN wave equations follow as approximations to the general form. Having established a suitable two-body equation, we derive a practical system of relativistic three-body equations for the πNN system. To conclude, we briefly compare some very recent theoretical results with available πd data.

(i) Relativistic Integral Equations for the πN System

All practical formulations of the relativistic two-body problem begin with the Bethe-Salpeter equation, which can be represented diagrammatically as in Fig. 1 (although the meaning has changed considerably from part A). Now the driving term represents the sum of all irreducible diagrams (in some field theory) which result in pion-nucleon scattering. This set of irreducible diagrams plays the role of a potential in the theory (and hence the notation V for it). Algebraically the scattering equation is*

$$M(K',K;P) = \overline{V}(K',K) + (2\pi)^{-4}\int d^4K''\ \overline{V}(K',K'';P)\ \overline{G}(P,K'')\ M(K'',K;P), \quad (B.1)$$

where, if $(p,k),(p',k'),(p'',k'')$ are the initial, intermediate and final nucleon and pion four-momenta, the conserved total momentum is $P[P = (p+k) = (p''+k'') = (p'+k')]$, and the relative four momentum is $K,K'',K'[K \equiv (k-p)/2$ etc.]. The invariant mass (squared) of the system is $s = P^2$.

Neglecting the complications of spin, the propagator is simply

$$\overline{G}^{(+)}(P,K'') = +i[(P/2-K'')^2-m_N{}^2+i\varepsilon)^{-1}[(P/2+K'')^2-m_\pi{}^2+i\varepsilon)^{-1}, \quad (B.2)$$

where the factor i is included to match other conventions. (In fact, all our considerations will involve a non-relativistic treatment of the nucleon spin.) Using Eq. (A.16) then implies that the discontinuity of this Green's function across the (right-hand) unitarity cut is

$$\mathrm{disc}\{\overline{G}(P,K'')\} = -4\pi^2 i\ \delta^+[(P/2-K'')^2-m_N{}^2]\delta^+[(P/2+K'')^2-m_\pi{}^2], \quad (B.3)$$

(where as usual $\delta^+(k^2-m^2)$ is $\delta[k_0-(\vec{k}^2+m^2)^{1/2}]/2k_0)$. Note that the unitarity cut involves only physical states so that both the pion and nucleon must be on-mass-shell.

The Blankenbecler-Sugar procedure aims to preserve this crucial property (B.3) in a new Green's function $[G(P,K'')]$, which has

---

*Note that in this section only we conform to the practice of using momentum states normalised so that $\langle\vec{p}'|\vec{p}\rangle = (2\pi)^3\ \delta(\vec{p}'-\vec{p})$, which is conventional in the literature on this subject.

a much simpler structure—essentially freezing out one degree of freedom of the real system. The easiest way to preserve (B.3) is to write the following dispersion relation

$$G(P,K'') = \frac{1}{2\pi i}\int_{(m_N+m_\pi)^2}^{\infty} \frac{ds'}{s'-s}\ \mathrm{disc}\ \bar{G}(\sqrt{s'/s}\ P,K'')\Big\} \qquad (B.4)$$

Now, if we define the on-mass-shell values of the nucleon and pion energies

$$E_{\vec{K}} = (\vec{K}^2+m_N^2)^{1/2};\ W_{\vec{K}} = (\vec{K}^2+m_\pi^2)^{1/2}\ , \qquad (B.5)$$

and remember the standard relation for the Dirac δ-function

$$\delta(a+b)\ \delta(a-b) = \delta(a)\ \delta(b)/2\ , \qquad (B.6)$$

Eq. (B.4) is easily evaluated in the three-body c.m. system, where $P = (\sqrt{s},\ \vec{0})$. Then we find

$$\begin{aligned}\mathrm{disc}\ \bar{G}\left(\sqrt{s'/s}\ P,K''\right) &= -4\pi^2 i\ \delta^+\left((\sqrt{s'}/2 - K_0'')^2 - E_{\vec{K}''}^2\right)\\ &\qquad \delta^+\left((\sqrt{s'}/2 + K_0'')^2 - W_{\vec{K}''}^2\right)\\ &= -\frac{i\pi^2}{E_{\vec{K}''}W_{\vec{K}''}}\ \delta\left(\sqrt{s'}/2-K_0''-E_{\vec{K}''}\right)\delta\left(\sqrt{s'}/2+K_0''-W_{\vec{K}''}\right)\\ &= -\frac{2i\pi^2}{E_{\vec{K}''}W_{\vec{K}''}}\ \delta\left(\sqrt{s'} - E_{\vec{K}''} - W_{\vec{K}''}\right)\delta\left(2K_0'' - W_{\vec{K}''}+E_{\underline{K}''}\right).\end{aligned} \qquad (B.7)$$

The first δ-function in (B.7) allows us to perform the dispersion integral (B.4)

$$G(P,K'') = \frac{2\pi(E_{\vec{K}''} + W_{\vec{K}''})\ \delta(2K_0'' - W_{\vec{K}''} + E_{\vec{K}''})}{E_{\vec{K}''}W_{\vec{K}''}\ [s^+ - (E_{\vec{K}''} + W_{\vec{K}''})^2]}\ . \qquad (B.8)$$

Finally, the δ-function in (B.8) makes it possible to remove the K_0''-integral in the original Bethe-Salpeter equation (B.1). The new equation which results is called the Blankenbecler-Sugar equation

$$T(\vec{K}',\vec{K};s) = \bar{V}(\vec{K}',\vec{K}) + (2\pi)^{-3}\int \frac{d\vec{K}''}{2E_{\vec{K}''}\ W_{\vec{K}''}}\ \bar{V}(\vec{K}',\vec{K}'')$$

$$\frac{(E_{\vec{K}''} + W_{\vec{K}''})}{[s+i\varepsilon - (E_{K''}+W_{K''})^2]}\ T(\vec{K}'',\vec{K};s)\ . \qquad (B.9)$$

Whereas Eq. (B.1) presents major problems for numerical solution even after partial wave decomposition (because it involves two variables $|\vec{K}''|$ and K_o''), Eq. (B.9) reduces to a one variable integral equation just like the Lippmann-Schwinger equation. Thus we have certainly obtained a more practical relativistic equation, which can be solved using the technique described in A (iv). Unfortunately the equation is not unique, because in Eq. (B.4) we could have multiplied $(s'-s)^{-1}$ by any function $f(s,s')$ which is unity when $s = s'$. Alternatively we can write

$$\overline{G} = G + R, \tag{B.10}$$

where R is a residual function which has no two-particle singularities.

The difference in the analytic behaviour of the solutions to (B.1) and (B.9) for a scalar boson exchange interaction $\overline{V}$ was discussed by Blankenbecler and Sugar. In particular the left-hand structure is only approximately reproduced. More recently the tendency is not to use a potential $\overline{V}$ calculated from field theory, but rather to adjust $\overline{V}$ so that the resulting (on-shell) t-matrix fits the experimental phase shifts over some energy range.

$$t_\ell^{(on)}\ [K,K;s(K)] = -\frac{32\pi^2\ s^{1/2}}{K}\ e^{i\delta_\ell}\ \sin\delta_\ell\ . \tag{B.11}$$

While this is a more phenomenological approach, it seems necessary at the present time, if one wants reliable calculations of pion reactions in more complicated systems of two or more nucleons.

Many πN interactions have been constructed to fit the phase shifts in the intermediate energy region, but very few use Eq.(B.9). The most common form is probably the relativistic Lippmann-Schwinger equation, which is "derived" by simply replacing the kinetic energy operators in Eq. (A.18) by a relativistic (on-mass-shell) form. That is,

$$t_\ell^{LS}\ (p',p;\sqrt{s}) = V_\ell^{LS}(p',p) + \int_0^\infty dp''p''^2\ \frac{V_\ell^{LS}(p',p'')\,t_\ell^{LS}(p'',p;s)}{\sqrt{s} - E_{p''} - W_{p''}} \tag{B.12}$$

with the on-shell relationship

$$t_\ell^{LS}\ [p,p;\sqrt{s(p)}] = -e^{i\delta_\ell}\ \sin\delta_\ell\,/\pi\,\mu(p)\,p,\ \mu(p) = E_p W_p/(E_p+W_p)\ . \tag{B.13}$$

It is rather instructive to see what approximations must be made to reduce Eq. (B.9) to (B.12).

The factor $(2\pi)^3$ in (B.9) is removed by using δ-function normalised momentum states in (B.12). The linearity of the propagator in (B.12) can be obtained by relying on the fact that the most important part of the integral in Eq. (B.12) corresponds to $(E_{K''} + W_{K''}) \sim \sqrt{s}$, so that

$$s - (E_{K''}+W_{K''})^2 \simeq 2(E_{K''}+W_{K''})[\sqrt{s} - E_{K''} - W_{K''}]. \qquad (B.14)$$

(Notice that this approximation does not affect unitarity!) Finally, let us make the definitions

$$t^{LS}(\vec{k}',\vec{k};\sqrt{s}) = (2W_{\vec{k}'}2E_{\vec{k}'})^{-1/2}\, T(\vec{k}',\vec{k};s)\,(2W_{\vec{k}}\, 2E_{\vec{k}})^{-1/2}, \qquad (B.15)$$

which explicitly show the relativistic phase space factors,* which relate the Lorentz invariant quantity 'T' to the frame dependent "t". If we now multiply Eq. (B.9) to the left and right by $(2W_{\vec{K}'}2E_{\vec{K}'})^{-1/2}$ and $(2W_{\vec{K}}2E_{\vec{K}})^{1/2}$ respectively, Eqs. (B.13) and (B.15) lead directly to (B.12).

To some extent the discussion leading to Eq. (B.12) was slightly misleading since we could have written the original dispersion relation (B.4) using $\sqrt{s}$ instead of s as the approximate variable. (Indeed such an approach has been carried through for pion-nucleus scattering by Shakin *et al.*)[24] However, in a relativistic theory certain constraints, such as analytically continued time reversal invariance,[25] suggest that s is the most natural variable and we prefer (B.9) on these grounds.

Probably the most widely used πN interactions based on Eq. (B.12) are the rank-one separable interactions of Landau and Tabakin[26] and later Londergan *et al.*[27] Including spin and isospin the partial wave decomposition of the potential is [generalising Eq. (A.21) to include spin and isospin]

$$V(\vec{p}',\vec{p}) = \sum_{\ell jmI} Y^m_{\ell 1/2 j}(\hat{p}')\; V_{\ell jI}(p',p)\; Y^{m*}_{\ell 1/2 j}(\hat{p})\; P_I, \qquad (B.16)$$

with "P_I" an isospin projection operator. Let us denote the combination $(\ell j I)$ as α then

$$V_\alpha(p',p) = \lambda_\alpha\, g_\alpha(p')\, g_\alpha(p), \qquad (B.17)$$

[with $\lambda_\alpha = +1\ (-1)$ for a repulsive (attractive) interaction.] By the same reasoning used to prove Eq. (A.92), the solution to Eq. (B.12) using (B.17) is

*This is the source of the factors which are usually introduced into pion-nucleus scattering theory in a very ad hoc way—c.f. §D below.

$$t_\alpha(p',p;\sqrt{s}) = g_\alpha(p')\ D_\alpha^{-1}(\sqrt{s})\ g_\alpha(p), \tag{B.18}$$

$$D_\alpha^{-1}(\sqrt{s}) = \lambda_\alpha^{-1} - \int_0^\infty dp''\ \frac{p''^2\ g_\alpha^2(p'')}{\sqrt{s} + i\varepsilon - E_{p''} - W_{p''}} \tag{B.19}$$

For details of the method used to uniquely determine the phenomenological form-factors $g_\alpha(p)$ we refer to Refs. 26, 27. The main difference between the two approaches is that in the Landau-Tabakin work no account was taken of opening inelastic channels. The complex πN phases in the inverse scattering formalism then lead to form-factors which are complex for all momenta. This does not destroy the unitarity of the on-shell t-matrix (by construction), however the half-shell function [c.f. Section A (ii)] for a separable t-matrix is just $\{g_\alpha(p')/g_\alpha(p_{on})\}$, and will in general be complex even below the first inelastic threshold. This is a violation of off-shell unitarity, and would certainly destroy the unitarity relations in a three-body system. On the other hand, Londergan *et al.* managed to avoid this problem by introducing extra energy dependence in the D-function corresponding to the coupled inelastic channels. The form-factors in their formalism are real.

The shapes of the form-factors produced in both formulations were very similar. If mapped into co-ordinate space, the apparent range of the (highly non-local) interaction is of the order of 0.6 fm, which is quite a long range! This finding has important implications for pion-nucleus scattering theory, and we shall see later that there is by no means unanimous agreement on this result. However, as explained briefly in part C (Lecture 2), recent theoretical studies in the πNN system have tended to confirm that the πN interaction should be quite long range.

The relativistic form of nucleon kinetic energy was retained in Eq. (B.12). However, this may lead to inconsistencies when calculating pion-nucleus interactions where the nucleons are usually treated non-relativistically. For this reason, an alternative to Eq. (B.12) in which E_p goes to $(m_N + p^2/2m_N)$ has also found practical use *viz:*

$$t_\alpha(p',p;E) = V_\alpha(p',p) + \int_0^\infty dp''\ \frac{p''^2\ V_\alpha(p',p'')\ t_\alpha(p'',p;E)}{E + i\varepsilon - p''^2/2m_N - Wp''}\ . \tag{B.20}$$

In fact, Thomas has found analytic form-factors

$$g_\ell(p) = p^\ell\left[S_1/\left(p^2+\beta_1^2\right)^{\ell+1} + S_2\ p^{2\ell}/\left(p^2+\beta_2^2\right)^{\ell+1}\right], \tag{B.21}$$

which reproduce the experimental phase shifts quite well up to

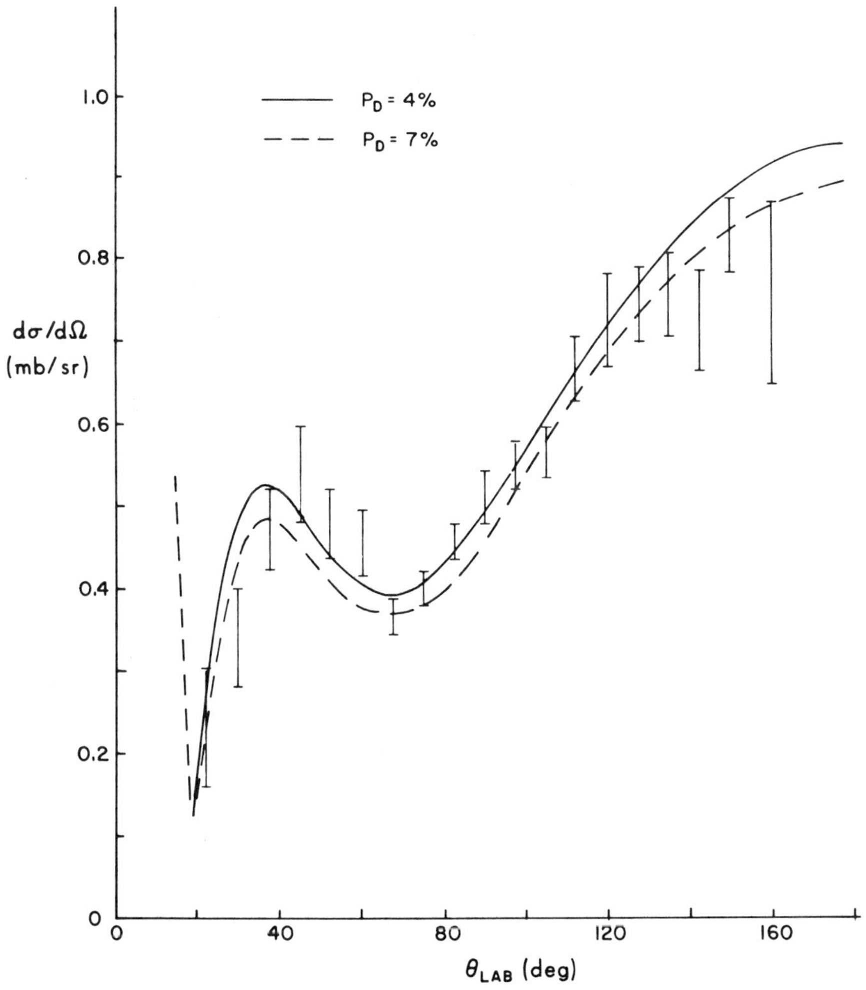

Fig. 4 Comparison of the 50 MeV π^+d data of Axen *et al.*, with the results of a semi-relativistic three-body calculation—corresponding to the two-body equation (B.20)—see Ref. [28].

300 MeV[28] when Eq. (B.20) is used. Although we have not discussed the solution of the three-body scattering equations, it should be mentioned that this involves continuation of the form-factors into the complex momentum plane, so that analytic expressions like (B.21) are essential.

(ii) Relativistic Three-Body Equations for the πNN System

In view of the discussion immediately above, the simplest semi-relativistic three-body theory could be obtained from Eqs. (A.93) by replacing $p^2/2m_\pi$ everywhere by W_p. These equations (called RPK for "relativistic pion kinematics") have been used quite successfully in a calculation of low energy (50 MeV) π^+d scattering as shown in Fig. 4. However, in order to test such an approximate set of equations it is necessary to construct something better. The main purpose of this section is to explain how that is done, and and what numerical effects the improvements have. Perhaps the most significant fact about the RPK calculation now, is that two quite independent groups[28,30] have obtained agreement on the partial wave πD amplitudes in this model (using the separable πN interactions of Thomas) to better than 1/2%. This system of equations can therefore serve as the test case requested in Ref. 31.

We have already mentioned that the solution of the full two-body Bethe-Salpeter equation (BSE) is very difficult. Clearly then, the full three-body BSE is not a good starting point for practical three-body equations. Instead, the experience gained in part A with separable interactions is used to guess a suitable set of quasi-two-body relativistic equations. Finally the Blankenbecler-Sugar techniques are used to preserve unitarity in a new (calculable) set of one-dimensional integral equations.

Figure 5 illustrates the three-body equations derived in part A for separable two-body t-matrices. For simplicity of notation, we consider only the case of *three identical spinless particles* in the formal discussion, and only show the πNN case explicitly at the end of this section. The approach of Aaron *et al.*[32,33] is to make the *ansatz* that a similar equation can represent the relativistic case of separable pair-wise interactions. That is, one writes the integral equation for scattering of a projectile from a two-body bound (or resonant) state as

$$T(p',p;P) = B(p',p;P) + (2\pi)^{-4}\int d^4p''\, B(p',p'';P)\, \tau[(P-p'')^2]\, T(p'',p;P). \tag{B.22}$$

With a slightly different mathematical interpretation, this can also be represented by Fig. 5. As in the non-relativistic case, the driving term "B" is an effective potential generated by the exchange of one particle between the initial and final pairs.

After partial wave decomposition, Eq. (B.22) leads to coupled two-dimensional integral equations, of complexity similar to the two-body Bethe-Salpeter equation. To obtain more practical one-dimensional equations we apply the Blankenbecler-Sugar procedure to Eq. (B.22). In order to apply unitarity constraints we need the break-up $(2 \to 3)$ amplitude, and consistent with the isobar idea [c.f. Eq. (A.98)], that is given by

$$T_{3\leftarrow 2}(p_1'p_2'p_3';p;P) \sim g(p_i',p_j')\; S[(P-p_k')^2]\; T(p_k',p;P), \qquad (B.23)$$

where S describes the isobar propagation, and g the dissociation of the isobar (see Fig. 6).

It is now possible to write down two expressions for the discontinuity of the scattering amplitude T across its right hand cut. The first is simply the general expression of two- and three-body unitarity which is *assumed* to be obeyed. That is[32,33]

$$T(p',p;s^+) - T(p',p;s^-) = i \sum_{n=2,3} d\rho_n\; T_{n\leftarrow 2}(n,p';s^-)\; T_{n\leftarrow 2}(n,p;s^+), \qquad (B.24)$$

where $d\rho_n$ is the two- or three-body relativistic phase space

$$d\rho_n = (2\pi)^4\; \delta^4(P - \sum_{i=1}^{n} p_i) \prod_{i=1}^{n} [d^4p_i (2\pi)^{-4}\; 2\pi\; \delta^+(p_i^2-m_i^2)]. \qquad (B.25)$$

The second expression for the discontinuity is given directly by the integral equation (B.22)—as first proven by Freedman *et al.*[20] (see also Appendix A of Ref. 33). When the momenta (p',p) are on-energy-shell (that is correspond to physical momenta for scattering from a bound state at energy s), this discontinuity is

$$\begin{aligned} T(p',p;s^+) - T(p',p;s^-) = {} & (2\pi)^{-4}\int d^4p''\; T(p',p'',s^+)\,[\tau(\sigma^+_{p''}) \\ & - \tau(\sigma^-_{p''})]\; T(p'',p;s^-) \\ & + (2\pi)^{-8}\int d^4p''\; d^4p'''\; T(p',p'';s^+)\;\tau(\sigma^+_{p''}) \\ & [B(p'',p''';s^+) - B(p'',p''';s^-)]\;\tau(\sigma^-_{p'''})\; T(p''',p;s^-), \end{aligned} \qquad (B.26)$$

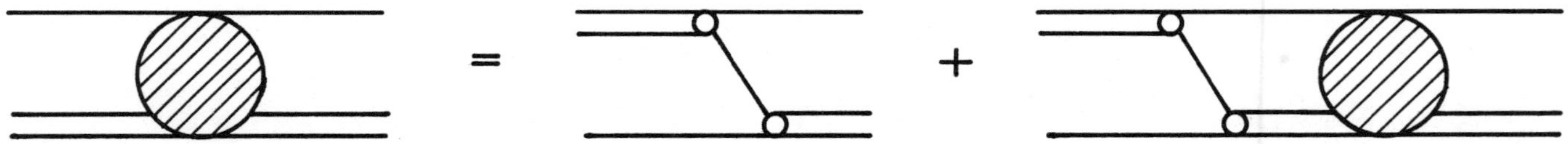

Fig. 5 Graphical representation of either the non-relativistic three-body equations (A.93), or the relativistic equations (B.22) or (B.35), in the case of separable interactions. In all cases the driving term represents the free exchange of one particle between the initial and final quasi-two-body states.

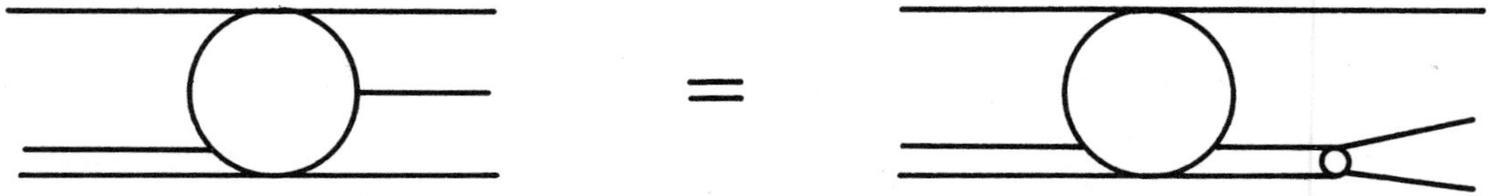

Fig. 6 Illustration of the isobar assumption (B.23) for the $2 \to 3$ (break-up) amplitude.

where [from Eq. (B.22)] we have defined

$$\sigma_p = (P - p)^2 \ , \tag{B.27}$$

the invariant mass of an interacting pair. Physically the first term on the right has two parts. First the bound state pole in "τ" (if any) gives the contribution of elastic and re-arrangement scattering to the unitarity relation. The continuum discontinuity of τ, and the second term on the right of (B.26), both contribute to the break-up part of the unitarity relation.

Equating the r.h.s. of (B.26) and (B.24) [using (B.23)]—that is demanding exact two and three-body unitarity—leads to equations for the right hand discontinuities of B(s) and τ(σ). With one more assumption, namely that the isobar propagators in the two and three-body equations are essentially the same

$$\tau\,(\sigma_p) = 2\pi\ \delta^+(p^2-m^2)\ S(\sigma_p) \ , \tag{B.28}$$

this gives (for details we refer to Ref. 33)

$$B(p',p;s^+) - B(p',p;s^-) = i\ g[(P-p'-2p)^2]\ 2\pi\ \delta^+[(P-p'-p)^2-m^2]$$
$$g[(P-p-2p')^2]:\ p'^2 = p^2 = m^2 , \tag{B.29}$$

$$S(\sigma^+) - S(\sigma^-) = 2\pi i\ \delta^+(\sigma - \mu^2)$$
$$+ \frac{S(\sigma^+)S(\sigma^-)}{2(2\pi)^4}\ i \int d^4q_i\ g^2(4q_i^2)\,(2\pi)^2\ \delta^+(p_j^2 - m^2)\ \delta^+(p_k^2 - m^2). \tag{B.30}$$

We stress that the assumption (B.28) means that Eq. (B.29) applies only for on-mass-shell particles ($p^2 = p'^2 = m^2$). Thus we can construct only the on-mass-shell born term by writing a dispersion relation with (B.29) as the right-hand discontinuity.

As in the two-body case, this dispersion relation is most easily evaluated in the three-body c.m., where [$\varepsilon_{\vec{k}} = (\vec{k}^2+m^2)^{1/2} = w_{\vec{k}}$]

$$\delta^+\left[(\sqrt{s'/s}\ P-k-k')^2 - m^2\right] = \left(2\varepsilon_{\vec{k}+\vec{k}'}\right)^{-1} \delta\left(\sqrt{s'} - \varepsilon_{\vec{k}} - \varepsilon_{\vec{k}'} - \varepsilon_{\vec{k}+\vec{k}'}\right) \tag{B.31}$$

(remember that $k^0 = \varepsilon_{\vec{k}}, k^{0\prime} = \varepsilon_{\vec{k}'}$, because of the on-mass-shell condition). Performing the dispersion integral analogous to Eq. (B.4) leads to the result

$$B(p',p;s) = \frac{g(Q'^2)(w_{\vec{p}}+w_{\vec{p}'}+w_{\vec{p}+\vec{p}'})\ g(Q^2)}{w_{\vec{p}+\vec{p}'}[s+i\varepsilon-(w_{\vec{p}}+w_{\vec{p}'}+w_{\vec{p}+\vec{p}'})^2]} . \qquad (B.32)$$

Similarly the isobar propagator can be constructed from Eq. (B.30) (where μ is the bound state mass, when there is a bound state). In the absence of a bound state, it can be proven that the solution of the two-body Blankenbecler-Sugar equation (B.9) with a separable potential like (B.17)

$$D_\alpha(\sigma) = \lambda_\alpha^{-1} - (2\pi)^{-3}\int_0^\infty \frac{dp\ p^2}{w_{\vec{p}}}\ \frac{g^2(p)}{\sigma+i\varepsilon-4w_{\vec{p}}^2} , \qquad (B.33)$$

satisfies Eq. (B.30), provided

$$S_\alpha(\sigma) = D_\alpha^{-1}(\sigma). \qquad (B.34)$$

[We have specialised Eq. (B.33) to the equal mass case.] This is a very satisfying consistency test on the procedure we have used. The key assumption which led to this result was implicit in the very first three-body equation (B.22), where the isobar propagator was only allowed to depend on the invariant mass of the two-body sub-system.

If we now put Eqs. (B.34), (B.28) and (B.32) into (B.22), we obtain the following relativistic three-body equation (in the three-body c.m.)

$$T(\vec{p}',\vec{p};s) = B(\vec{p}',\vec{p};s) + (2\pi)^{-3}\int \frac{d^3\vec{p}''}{2w_{\vec{p}''}}\ \frac{B(\vec{p}',\vec{p}'';s)T(\vec{p}'',\vec{p};s)}{D(\sigma_{p''})} , \qquad (B.35)$$

$$B(\vec{p}',\vec{p};s) = \frac{g(Q'^2)(w_{\vec{p}'} + w_{\vec{p}} + w_{\vec{p}+\vec{p}'})g(Q^2)}{w_{\vec{p}+\vec{p}'}[s+i\varepsilon-(w_{\vec{p}'}+w_{\vec{p}}+w_{\vec{p}+\vec{p}'})^2]} . \qquad (B.36)$$

Notice that the on-mass-shell *assumption* for the spectator particle in intermediate states [Eq. (B.28)] results in equations for only the on-mass-shell scattering amplitudes. Fortunately this is what we need for calculating cross sections. (The deeper question of whether this assumption is actually necessary has only recently been asked.[34] It may well be that the apparent solution of the clustering problem through (B.28) is not strictly correct. This is certainly a topic for further research.) Thus, the theory which we have described is an on-mass-shell, off-energy-shell theory, analogous to the coupled channels Lippmann-Schwinger equation. The resulting scattering amplitudes are nevertheless Lorentz invariant.

The generalisation of these equations to the πd system (including spin and isospin) has been given by Rinat and Thomas[35] (see also Ref. 36). (N.B. Henceforth $w_{\vec{p}} = (\vec{p}^2 + m_\pi^2)^{1/2}$, $E_{\vec{p}} = (\vec{p}^2 + m_N^2)^{1/2}$.)

$$X_{d,d}(\vec{p},\vec{p}_o;s) = 2\int \frac{d^3p'}{16\pi^3 E_{\vec{p}'}} Z_{d,\Delta}(\vec{p},\vec{p}';s)\; D_\Delta^{-1}(\sigma_{\vec{p}'}) X_{\Delta,d}(\vec{p}',\vec{p}_o;s)$$

$$X_{\Delta,d}(\vec{p},\vec{p}_o;s) = Z_{\Delta,d}(\vec{p},\vec{p}_o;s) + \int \frac{d^3p'}{16\pi^3 E_{\vec{p}'}} Z_{\Delta,\Delta}(\vec{p},\vec{p}';s)\; D_\Delta^{-1}(\sigma_{\vec{p}'})\, X_{\Delta,d}(\vec{p}',\vec{p}_o;s)$$

$$+\int \frac{d^3p'}{16\pi^3 w_{\vec{p}'}} Z_{\Delta,d}(\vec{p},\vec{p}';s)\; D_d^{-1}(\sigma_{\vec{p}'})\; X_{d,d}(\vec{p}',\vec{p}_o;s)\,. \tag{B.37}$$

In addition to indicating whether the bound or resonating pair is a deuteron or P_{33} state, the labels "d" and "Δ" (respectively) include spin and isospin projections. After a partial wave decomposition which is rather more complicated than usual [see Ref. 35 Eqs. (2.21) and (3.7)], this leads to a set of one dimensional coupled integral equations which can be solved by the usual technique of contour rotation and numerical quadrature.

On-shell, the Lorentz invariant πd scattering amplitude X_{dd} $[\vec{p}_o,\vec{p}_o;s(p_o)]$ is related (in a diagonal partial wave "L") to the phase shifts by

$$X^{J\pi}_{Ld,Ld}\,[p_o,p_o;s(p_o)] = -\frac{32\pi^2\; s^{1/2}}{p_o}\, e^{i\delta_{LJ}}\, \sin\delta_{LJ}\,. \tag{B.38}$$

The driving terms in Eq. (B.37) are specified entirely by kinematic terms and the $d\to NN$ and $\Delta\to\pi N$ form factors [c.f. (B.36)]

$$Z_{\Delta,d}(\vec{p},\vec{p}';s) = \frac{g_\Delta(\vec{Q})\,(E_{\vec{p}} + w_{\vec{p}} + E_{\vec{p}+\vec{p}'})\,g_\Delta(\vec{Q}')}{E_{\vec{p}+\vec{p}'}\,[s+i\varepsilon - (E_{\vec{p}} + w_{\vec{p}'} + E_{\vec{p}+\vec{p}'})^2]}\,, \tag{B.39}$$

$$Z_{\Delta,\Delta}(\vec{p},\vec{p}';s) = \frac{g_\Delta(\vec{Q})\,(E_{\vec{p}} + E_{\vec{p}'} + w_{\vec{p}+\vec{p}'})\,g_\Delta(\vec{Q}')}{w_{\vec{p}+\vec{p}'}\,[s+i\varepsilon - (E_{\vec{p}} + E_{\vec{p}'} + w_{\vec{p}+\vec{p}'})^2]} \tag{B.40}$$

So far the only quantities not defined are the relative three-momenta $(\vec{Q},\vec{Q}')$. The correct expressions for these were first derived by Aaron *et al.*[32] showing that with a careful choice the three-vector dot-product $\vec{Q}\cdot\vec{Q}$ could be made Lorentz invariant. The answer was later shown[37] to be equivalent to applying a Lorentz boost to the c.m. of the pair of particles. That is, if two particles have (on-mass-shell) momenta (k_1,k_2) in some frame, then defining the total momentum $K = k_1 + k_2$, and $L(K)$ the Lorentz transformation which boosts to the frame where K is time-like, one finds

$$\pm\,\vec{Q}_i = L_{i\mu}\,(K) \begin{cases}(k_1)^\mu \\ (k_2)^\mu\end{cases}, \tag{B.41}$$

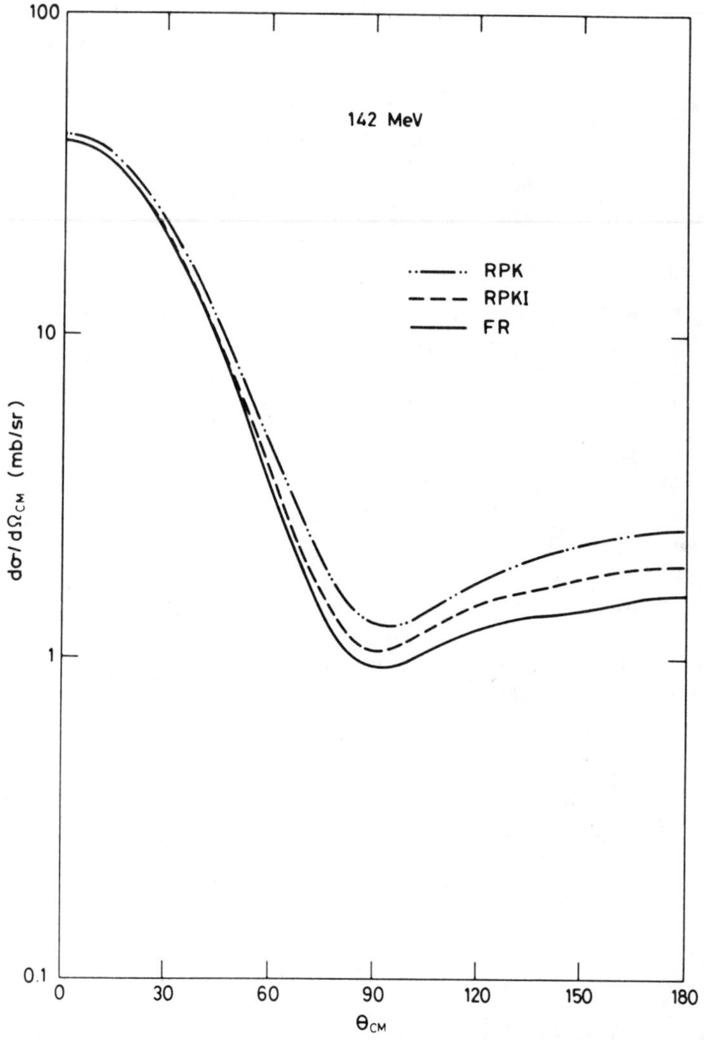

Fig. 7 Pion-deuteron scattering at 142 MeV with only the P_{33} interaction. This shows the effect of various theoretical approximations (a) RPK means that relativistic kinematics are used for the pion only (as in Fig. 4), (b) RPKI shows the effect of using the correct relativistic relative momenta (B.42), (c) FR is the result of the fully relativistic calculation using Eq.(B.37).[35]

which are the momenta of 1 and 2 in their c.m. system. More explicitly this relativistic relative momentum is

$$\vec{Q}(k_1,k_2) = \vec{q} - (\vec{q}\cdot\vec{K})\ \vec{K}/\left(K_0[K_0+(K^2)^{1/2}]\right) . \tag{B.42}$$

(It is the rather complicated dependence of $\vec{Q}$ on the energies of the particles which complicates the partial wave decomposition, as we mentioned earlier.) Note that consistent with the on-mass-shell off-energy-shell nature of these equations, the four-momentum of (say) the exchanged nucleon in $Z_{\Delta,d}$ [Eq. (B.39)], used to calculate $(\vec{Q},\vec{Q}')$, is

$$p_{ex} = (E_{\vec{p}+\vec{p}'},\ -\vec{p}-\vec{p}') . \tag{B.43}$$

(iii) Applications to πd Scattering

The relativistic three-body equations whose derivation was sketched in the preceding section, represent the minimal equations which guarantee unitarity within an isobar-like theory. We refer to the original paper [35] for details of the deuteron and Δ form-factors used. Fig. 7 illustrates the effect at 142 MeV laboratory energy (the lowest energy studied), of gradually including the relativistic extension of the non-relativistic expression for the pion kinetic energy. (These results have now been confirmed completely independently by the Lyon group.)[30] In this model the πN relative momentum was simply

$$\vec{Q}^{RPK}_{\pi N} = (m_N\ \vec{k}_\pi - m_\pi\ \vec{k}_N)/(m_N + m_\pi) . \tag{B.44}$$

The effect of using the relativistic relative momentum (B.42) is shown as RPKI (improved). This is obviously quite a significant effect beyond ~70°. Unfortunately no check has been made of the expression

$$\vec{Q}(\text{semi-relativistic}) = (m_N\ \vec{k}_\pi - w_{\vec{k}\pi}\ \vec{k}_N)/(m_N + w_{\vec{k}\pi}) , \tag{B.45}$$

which can be shown analytically to be closer to Eq. (B.42), and which has found common use in pion-nucleus calculations. Finally the curve labelled FR (fully relativistic) shows the effect of using the non-linear form of propagator (B.39) [c.f. $(\sqrt{s} - w_{\vec{p}'} - E_{\vec{p}} - E_{\vec{p}+\vec{p}'})^{-1}$ for RPK], as well as retaining the relativistic phase space factors ($w_{\vec{p}+\vec{p}'}$ etc). This improvement clearly produces significant differences at all angles.

The comparison of theory with experiment is still in a very early stage. There is no available polarisation data, and while the general trends of the differential cross section have been mapped out by a number of groups, there is as yet no high precision data (i.e. errors of order 5%). Hopefully this will soon be reme-

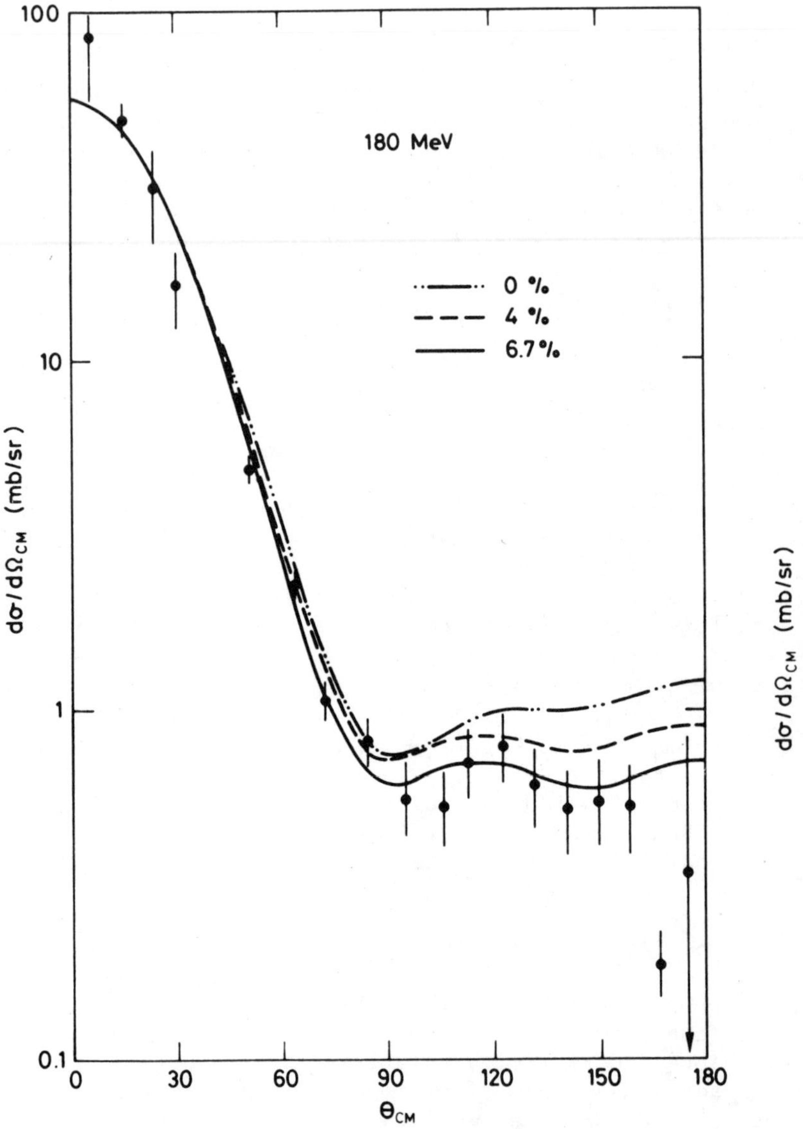

Fig. 8 A comparison of the fully relativistic πd calculation of Rinat and Thomas[35]) with the data of Norem[39]) at 180 MeV for various deuteron d-state probabilities. (N.B: The deviation at the furthest forward angles occurs because no account is taken of the coulomb interaction.)

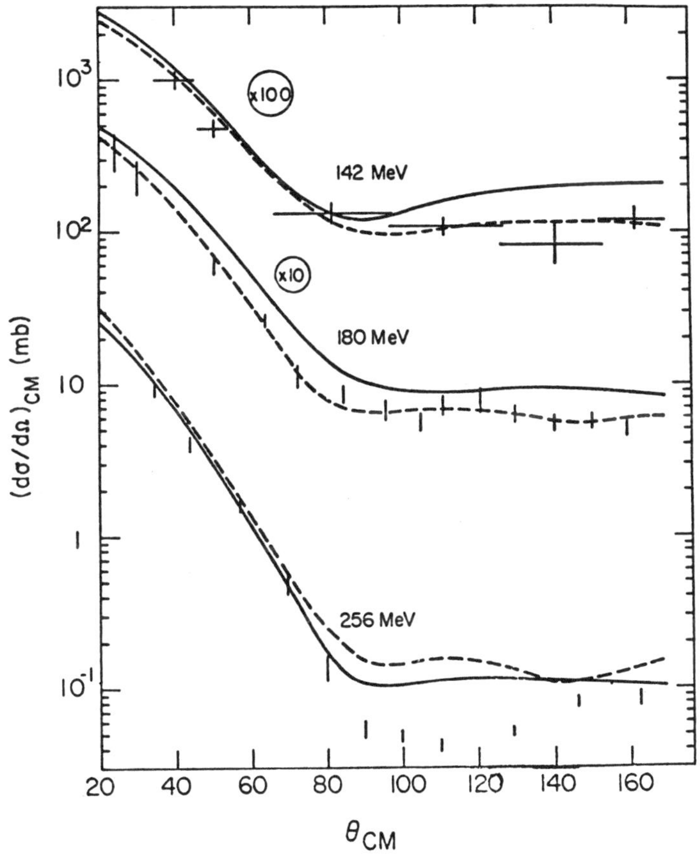

Fig. 9 Comparison of the Glauber calculations of Hoenig and Rinat[41] (solid curve), with data and the relativistic three-body calculations of Rinat *et al.*[40] (dashed curve), for πd scattering at several energies through the resonance region.

died by the group using the pion spectrometer SUSIE at SIN—at least through the resonance region.

If one includes only the d and Δ two-body interactions, the agreement with existing data[39] is quite good at 180 MeV (see Fig. 8). This is because the πN interaction has almost no real part at this energy, so that the interference effect of other non-resonant πN channels is negligible. Away from resonance, however, the interference effects of other partial waves are very important. Fig. 9 compares the recent relativistic three-body-calculations of Rinat *et al.*[40] with Glauber calculations[41] and the available data at 142, 180, 234 and 256 MeV. The agreement at 142 MeV is very good, and the non-resonant background is essential to this. At 180 and 234 MeV the situation is also quite good. However, the CERN data at 256 MeV[42] is very much below the theory in the (80°-160°) region. This experimental trend seems to have been confirmed by recent LAMPF data.[43]

For more detail of the results of recent calculations, and the latest experiments we refer to the latest review articles.[44] To summarize briefly, however, the theory seems good (as far as it has been tested) below about 230 MeV. Above this energy some new process seems to set in which is not contained within the present model.

LECTURE 2

The first lecture was devoted to the development of a very solid basis in the non-relativistic theory of three-body scattering, and its relativistic generalisation. If we have succeeded in communicating some understanding of the two-body t-matrix, the advantages of the momentum representation, and the fundamental role of unitarity in few-body problems our aims have been met.

In this section we first describe (in Section C) a very recent application of three-body scattering theory to the πNN system which has thrown considerable new light on the question of the range of the πN interaction. As we shall see later, this is a very important question in developing a theory of pion-nucleus scattering. The major part of this lecture will then be used to explain some aspects of the treatment of pion-nucleus scattering by an optical model. As well as deriving the conventional optical potentials from the general formalism, we will explain the very recent three-body model of the first-order optical potential in some detail.

C. The Range of the πN Interaction

The field theory of the πN system has been discussed in considerable detail by Dr. Kisslinger, so we will only recall the aspects

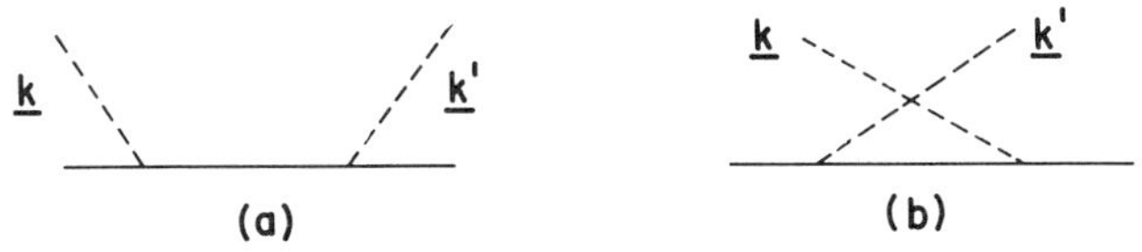

Fig. 10 The Born approximation diagrams for πN scattering in the Chew-Low model.

directly relevant to the present discussion. In the static Chew-Wick field theory the $N \to N\pi$ vertex is schematically

$$H_I \sim (f/m_\pi)\ \nu(k)\ \vec{\sigma}\cdot\vec{k}\ , \tag{C.1}$$

where $\vec{k}$ is the momentum of the pion produced, and $\nu(k)$ is a form-factor $[\nu(k) \sim \theta\ (m_N - k)]$ for the vertex. The lowest order contribution to πN scattering is therefore $(|\vec{k}| = |\vec{k}'|)$

$$t_{\pi N}^{(Born)}(\vec{k}',\vec{k}) \sim \left(\frac{f}{m_\pi}\right)^2 \frac{\nu(k')\nu(k)}{\sqrt{4w_{k'}\ w_k}} \left\{\frac{\vec{\sigma}\cdot\vec{k}'\ \vec{\sigma}\cdot\vec{k}}{w_k} - \frac{\vec{\sigma}\cdot\vec{k}\ \vec{\sigma}\cdot\vec{k}'}{w_k}\right\}\ . \tag{C.2}$$

This is shown diagramatically in Fig. 10, where the direct (s-channel) nucleon pole term (Fig. 10a) is the first term on the r.h.s. of Eq. (C.2), and the crossed term (Fig. 10b) is the second. In a theory with no anti-nucleons, Fig. 10a can only contribute in a channel with the quantum numbers of the nucleon, namely $(J^\pi = \frac{1}{2}^+,\ I = \frac{1}{2})$. In the standard notation of πN scattering this is the P_{11} channel.

The crossed term, represented by Fig. 10b, contributes in all πN P-waves. In fact, the amplitude (C.2) (including isospin) is crossing symmetric, and one may look at the crossed term as being a necessary addition to the nucleon pole [clearly Eq. (C.2) has a pole at w=0, which is the nucleon mass] to guarantee crossing symmetry— this is its 'raison d'être'. Using the conventional normalisation, and including isospin, the correct expression for the (on-shell) πN Born amplitude is

$$t^{(Born)}(\vec{k}',\vec{k};w) = \frac{4\pi\ \nu(k')\ \nu(k)}{(4\ w_{k'}\ w_k)^{\frac{1}{2}}} \sum_\alpha P_\alpha(\lambda_\alpha/w)\ , \tag{C.3}$$

where P_α is a projection operator for channel α [c.f. Eq.(B.15)], and $\lambda_\alpha = 2/3\ (f^2/m_\pi^2)\ (-4,\ -1,\ -2)$ for $\alpha = (P_{11},\ P_{13} = P_{31},\ P_{33})$.

It is a fundamental, but often neglected, problem of multiple scattering theory, how to incorporate the pole at w=0 evident in Eqs. (C.2) and (C.3). The direct pole in the P_{11} channel (with strength $\lambda_{11}^{(Dir)} = -3\ f^2/m_\pi^2$ - see Fig. 10a) is associated with pion absorption in a system with more than one nucleon, as we shall mention again in Lecture 3. However, all the other πN P-waves also have a pole at w=0, or at the nucleon mass. This is a little more difficult to interpret because it seems to be merely a reflection of the direct P_{11} pole into the other channels as a result of crossing symmetry—through the relationship[46]

$$h_\alpha(-w) = A_{\alpha\beta}\, h_\beta(+w) \quad . \tag{C.4}$$

In Eq.(C.4) $h_\alpha(w)$ is related to the full Chew-Low scattering amplitude by

$$t(\vec{k}',\vec{k};w) = \frac{4\pi\ \nu(k')\ \nu(k)}{(4\ w_{\vec{k}'}\ w_{\vec{k}})^{\frac{1}{2}}} \sum_\alpha P_\alpha\, h_\alpha(w) \quad , \tag{C.5}$$

and in Born approximation h_α equals (λ_α/w).

In fact, the correct fully off-shell energy denominator corresponding to this pole in the on-shell P_{33} interaction is

$$G_0 \sim (w+i\varepsilon-w_{\vec{k}} - w_{k'})^{-1} \tag{C.6}$$

which for real momenta $\vec{k}$ and $\vec{k}'$ cannot vanish for $w < 2m_\pi$. That is, the first unitarity cut associated with this diagram begins at the threshold for pion production. Therefore, in potential language the pole in the on-shell P_{33} amplitude is left-hand structure, and properly incorporated with the potential singularities (i.e. in the N-function in an N/D decomposition).

This is where the Chew-Low representation of the πN scattering amplitude is therefore a little misleading. Equation (C.5) is identical to the form of a separable t-matrix [c.f. Eq.(B.17)], so that one is tempted to identify $D_\alpha^{-1} \equiv h_\alpha$. However, as we have shown, h_α contains a 'nucleon pole' even for $\alpha \neq P_{11}$. An additional problem, which we have not mentioned, is that h_α also has a left-hand cut $w\ \varepsilon\ (-\infty, -m_\pi)$, resulting from the application of the crossing relation (C.4) to the standard right-hand (unitarity) cut. This too is properly incorporated amongst the potential singularities in a potential model.

In case this argument sounds like an irrelevant exercise in changing the name associated with a singularity, we remind the reader that in any many nucleon system, and particularly in a three-body system, the energy argument of each two-body t-matrix is needed over the range $(-\infty, E_{c.m.})$, with $E_{c.m.}$ the three-body c.m. energy (c.f. part (iii) of Section A). In a separable model, this means performing an integral over the D-function for this energy range. The naive identification of D_α^{-1} with h_α would then mean that the pole at $w_{\pi N}=0$, and the cut for $w_{\pi N}\ \varepsilon\ (-\infty, -m_\pi)$, would both contribute to the discontinuity of a three-body (e.g. πNN) scattering amplitude for any positive energy (actually any $E > -m_\pi$ and $E > -2m_\pi$ respectively). This would destroy the three-body unitarity relation (or optical

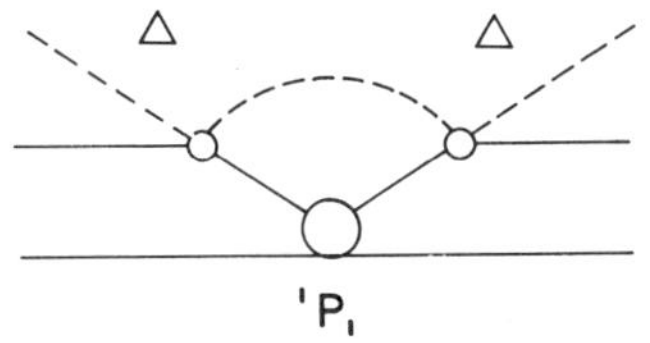

Fig. 11 One triple scattering contribution to πd scattering involving the πN resonant interaction (Δ) and the N-N $^1P_1$ interaction. The model dependence of results obtained in Ref.[51] for this graph, led to the consideration of the πN range problem based on three-body unitarity.

theorem), because as we discussed near Eq.(C.6), Fig. 10b cannot describe a real intermediate state unless the two body kinetic energy is greater than m_π. This observation was first made by Myhrer and Thomas,[51] in connection with the triple scattering contribution to πd elastic scattering shown in Fig. 11.

To solve this problem we could divide h_α ($\alpha \neq P_{11}$) into two parts $h_\alpha{}^L$ and $h_\alpha{}^R$ (for left and right), so that

$$h_\alpha{}^L(w) = \frac{\lambda\alpha}{w} + (2\pi i)^{-1}\int_{m_\pi}^{\infty} \frac{\text{disc}\,\{h_\alpha(-w')\}\,dw'}{w' + w^+} \quad ,$$

$$h_\alpha{}^R(w) = (2\pi i)^{-1}\int_{m_\pi}^{\infty} \frac{\text{disc}\,\{h_\alpha(+w')\}\,dw'}{w' - w^+} \quad ,$$

$$h_\alpha = h_\alpha{}^L + h_\alpha{}^R \;;\; \alpha \neq P_{11} \;. \tag{C.7}$$

Here, $h_\alpha{}^R$ contains only the right-hand (s-channel unitarity) cut and is properly identified with the D-function of a separable t-matrix. On the other hand, the full 'potential structure' in the Chew-Low model is $\{h_\alpha{}^L(w_k)\,\nu^2(k)\}$, which is approximated by the form-factors of a separable model $g_\alpha^2(k)$. Now we are in a position to understand why Ernst and Johnson find $\nu(k)$ to have a very short range of order $m_N{}^{-1}$ [actually $\nu(k) \sim \exp(-30\,k^2)$ with k in units of m_π], whereas all separable models of the πN t-matrix have form factors with quite a long range [$\sim \frac{1}{2}$ fm - see the discussion after Eq.(B.18)]

What the separable models simulate is the nearby left-hand structure in $h_\alpha{}^L(w_k)$, namely a pole at $k = \pm i m_\pi$ and a cut starting at $k = \pm 2i m_\pi$. With an analytic form like Eq.(B.21), for example, one gets poles in the on-shell t-matrix at k equals $i\beta_1$ and $i\beta_2$. Clearly if $h_\alpha{}^L(w_k)$ has its most significant strength at small values of w_k (e.g. at w_k=0 from the nucleon pole), at least one of the fitted ranges β_1 and β_2 could be quite small—of order one or two times m_π. Thomas, for example, using the semi-relativistic scattering Eq. (B.20), found that (β_1,β_2) equals (1.475, 3.4) fm^{-1} gave a good representation of the P_{33} phase shifts below 300 MeV.[28]

Our conclusion therefore, is that the range of a πN t-matrix suitable for use in a multiple scattering treatment of pion scattering from more than one nucleon, is quite long (in co-ordinate space). The implications for pion-nucleus scattering will be discussed in the following sections, but essentially it reduces the importance of N-N short-range correlations, and implies that Bèg's theorem[47,48)] is not applicable. (That is, one cannot calculate pion-nucleus scattering from purely on-shell πN information.) Our conclusion concerning the relationship of the πN range to the singularity structure of Chew-Low theory is closely related to other recent work.[49,50] Nevertheless we feel that the clear statement that unitarity is violated in pion scattering from more than one nucleon if the separation (C.7) is not made,[51)] is the most forceful demonstration yet presented. It indicates the power of considerations based on few body systems where exact unitarity is a fundamental consideration.

D. Pion - Nucleus Scattering Theory

In recent times a considerable use has been made of the fixed scatterer approximation (f.s.a.) as an approach for calculating pion-nucleus elastic scattering. Indeed the f.s.a. is an ideal pedagogical tool,[52)] and may be especially useful in obtaining a qualitative understanding of previously unexplored density regions.[53)] We have also seen in Dr. Gibbs lectures that, at least for light nuclei ($A \lesssim 12$), once the f.s.a. has been made the complete pion-nucleus multiple scattering series can be summed exactly.

The alternative approach, which we prefer, is to retain the nucleon motion, while expanding the scattering amplitude in series depending on the degree of nuclear excitation. Formally such an approach reduces the full (A+1) body problem to an effective two-body problem, where the *optical potential* contains all the many-body complications. Although the formulation of the optical model which we follow was first given in 1959 by Kerman *et al.* (KMT),[54)] we shall see that three-body ideas have recently deepended our understanding of it (and helped to explain recent low energy data).

(i) Three-body model of the first-order optical potential. The intimate connection of the equations derived here with three-body scattering theory was first emphasised by Tandy *et al.*[55-57] (for proton-nucleus scattering), but similar equations had been found earlier by Revai.[58)] The starting point is non-relativistic scattering theory for a system of a pion (particle o) and A nucleons, with only pairwise potentials [V_{oi} and V_{ij}, $(i,j)\varepsilon(1 \ldots A)$]. Then the channel Hamiltonian for a free pion incident on A bound nucleons is

$$H_o = K_o + H_A \doteq K_o + \sum_{i=1}^{A} K_i + \sum_{i<j} V_{ij} \; . \qquad (D.1)$$

The interaction which scatters the pion is

$$V^o = \sum_{i=1}^{A} V_{oi} \ , \tag{D.2}$$

and the appropriate scattering equation is the (A+1)-body Lippmann - Schwinger equation

$$T = V^o + V^o G_A T \ , \tag{D.3}$$

$$G_A(E^+) = E^+ - H_o \ . \tag{D.4}$$

[Note that we ignore for the moment the formal difficulties of the non-compactness of Eq.(D.3) as discussed below Eq.(A.44).]

The KMT formalism has the advantage that only physically allowed anti-symmetric nuclear intermediate states occur. This is signified by carrying the anti-symmetrization operator '$\mathcal{A}$' explicitly for each intermediate state. With the understanding that all the following operator equations are to be evaluated between anti-symmetric nuclear states one can replace $\sum_i V_{io}$ by AV. Equation (D.3) then becomes

$$T = AV \ (1 + \mathcal{A}G_A T) \ . \tag{D.5}$$

Let us define the operator 'τ', which we shall see below is closely related to the free πN t-matrix

$$\tau = V \ (1 + \mathcal{A}G_A \tau) \ . \tag{D.6}$$

Equation (D.6) may be inverted to obtain

$$V = \tau(1 + \mathcal{A}G_A\tau)^{-1} \ . \tag{D.7}$$

Similarly Eq.(D.5) becomes

$$T = AV(1 - \mathcal{A}G_A \ AV)^{-1} \ , \tag{D.8}$$

and substituting for V from (D.7) implies after a little algebra

$$T = A\tau[1 - (A-1) \ \mathcal{A}G_A\tau)^{-1} \ . \tag{D.9}$$

This can be converted to the integral equation

$$T' = U^{(o)} + U^{(o)}\mathcal{A}G_A T' \ , \tag{D.10}$$

with the definitions

$$T' = [(A-1)/A] \ T \ , \tag{D.11}$$

$$U^{(o)} = (A - 1)\tau \ . \tag{D.12}$$

Unfortunately, Eq.(D.10) still involves nuclear excited states explicitly. As KMT observed, one expects that in general matrix elements of U leading to an excited state should be smaller ($\sim 1/A$) than diagonal ones. One therefore introduces the ground and excited state projection operators

$$P = |o\rangle\langle o| \ ; \ Q = \sum_{n \neq o} |n\rangle\langle n|$$

$$= A - P \ . \tag{D.13}$$

Inserting (P + Q) into Eq.(D.10) we find

$$T' = U^{(o)} + U^{(o)}(P + Q)\ G_A T'$$

$$= U + UPG_A T' \ , \tag{D.14}$$

provided that the new operator U satisfies the equation

$$U = U^{(o)} + U^{(o)} QG_A U \ . \tag{D.15}$$

Equation (D.14) is ideally suited for the description of pion-nucleus elastic scattering. Indeed taking the ground state expectation value of Eq.(D.14) we obtain a two-body scattering equation

$$T'_{oo} = U_{oo} + U_{oo}\ G_A^{(o)} T'_{oo} \ , \tag{D.16}$$

where of course $G_A^{(o)}$ is the ground state expectation value of the channel Green's function (D.4). The effective potential U_{oo}, which is non-local and energy dependent in general, is called the optical potential. Clearly all the difficulties associated with virtual nuclear excitation have been hidden in it [c.f. the Q operator in Eq.(D.15)].

It is unfortunate that there is no unique definition of the *first order optical potential* '$U^{(1)}$'. Such freedom of choice is one of the wonderful things in nuclear physics! The definition which we propose is designed to incorporate as far as possible only the one-body properties of the nucleus. First we discard all but the first term in Eq.(D.15), so that there is no explicit inclusion of nuclear excitation. Second, we approximate H_A by the shell model form $\sum_i (K_i + u_i)$, neglecting residual N-N interactions. Third, all the nuclear wave functions are assumed to be given by the independent particle shell model (IPSM).

With these approximations Eq.(D.6) is very close to the scattering equation for a pion from one nucleon in a binding potential 'u'. This is only spoilt by the anti-symmetrisation operator which can involve more than one nucleon in fourth and higher order through exchange effects. Such terms involve higher nuclear excitation, that is larger energy denominators and will be neglected. With this last approximation the operator 'τ', now labelled '$\tau^{(1)}$' to indicate that it is the first order approximation, obeys the equation

$$\tau^{(1)} = V + V \frac{Q^{(P)}}{E^{+} - K_{o} - K_{A-1} - K - u - \varepsilon_{B}} \tau^{(1)} . \qquad (D.17)$$

(We have assumed that only the ground state expectation value of $\tau^{(1)}$ is needed.) Here E is the c.m. kinetic energy in the pion-nucleus system, K_{A-1} the kinetic energy operator for the c.m. motion of the core, K and u the nucleon kinetic energy operator and binding potential respectively, and $\varepsilon_B(>0)$ the single particle binding energy. (We set the energy scale so that $H_A \psi_A^{(g.s.)} = 0$, which means $H_{A-1}^{(int)} \psi_{A-1} = \varepsilon_B \psi_{A-1}$.) Finally, $Q^{(P)}$ is a Pauli operator which does not allow the nucleon into an occupied level of the (A-1) system—this is equivalent to the operator A in the approximation outlined above.

If we omit the Pauli operator for the moment, Eq.(D.17) represents the simple three-body problem shown in Fig. 12. Notice particularly that *there is no projectile-core interaction in the first-order potential*. Let us define the πN and N-core t-matrices as usual

$$t_{\pi N}(z) = V + V G_o(z) t_{\pi N}(z) , \qquad (D.18)$$

$$t_{NC}(z) = u + u G_o(z) t_{NC}(z) , \qquad (D.19)$$

$$G_o(z^+) = (z + i\varepsilon - K_o - K - K_{A-1})^{-1} . \qquad (D.20)$$

Then solving (D.18) for V we find

$$V = (1 + t_{\pi N} G_o)^{-1} t_{\pi N} . \qquad (D.21)$$

Substituting (D.21) into Eq.(D.17) (without $Q^{(P)}$) implies that

$$\tau^{(1)}(E^+) = t_{\pi N}(E^+ - \varepsilon_B) + t_{\pi N}(E^+ - \varepsilon_B)[G_u(E^+ - \varepsilon_B) - G_o(E^+ - \varepsilon_B)] \times \tau^{(1)}(E^+) , \qquad (D.22)$$

$$G_u(z) = (z^+ - K_o - K - K_{A-1} - u)^{-1} . \qquad (D.23)$$

Now Eq.(A.10) allows us to replace $(G_u - G_o)$ by $G_o t_{NC} G_o$. The resulting form of Eq.(D.22) is equivalent to the following coupled integral equations

$$\tau^{(1)}(E^+) = t_{\pi N}(E^+ - \varepsilon_B) G_o(E^+ - \varepsilon_B) \tau_C(E^+) ,$$

$$\tau_C(E^+) = G_o^{-1}(E^+ - \varepsilon_B) + t_{NC}(E^+ - \varepsilon_B) G_o(E^+ - \varepsilon_B) \tau^{(1)}(E^+) . \qquad (D.24)$$

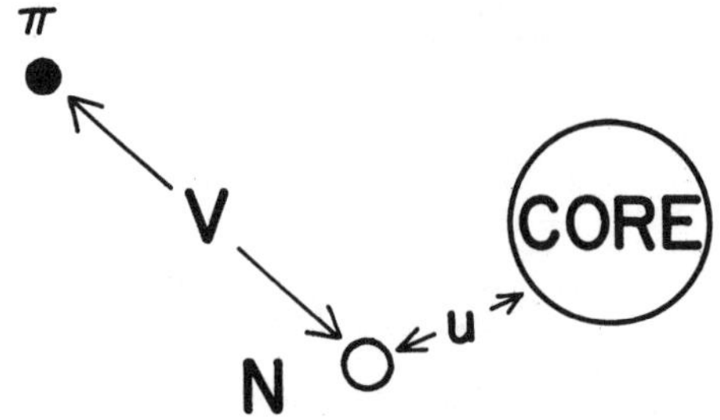

Fig. 12 Illustration of the simple three-body problem (in which only two of the three pairs interact) whose solution defines $\tau^{(1)}$.

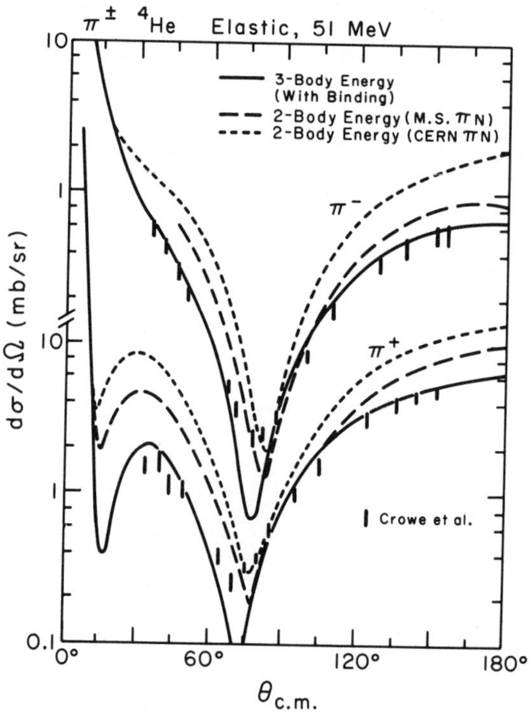

Fig. 13 A comparison of various first order optical potential calculations[61,64]) with the 51 MeV ^{4}He data of Crowe *et al.* The difference between the short and long dashed curves illustrates the sensitivity to poorly known low energy πN phase shifts, while the long dashed to solid comparison shows the importance of the three-body choice of πN interaction energy.

Comparison with Eq.(A.65) shows that Eq.(D.24) has exactly the form of the three-body equations of Alt *et al.* for the problem shown in Fig. 12, where there are only two non-zero interactions [e.g. $U_{11}(E-\varepsilon_B) \equiv \tau^{(1)}(E)$, $U_{21}(E-\varepsilon_B) \equiv \tau_C(E)$, $t_1 = t_{\pi N}$, $t_2 = t_{NC}$, $t_3 = 0$]. While we have omitted $Q^{(P)}$ from equations (D.18) through (D.24) they can all be corrected simply by re-inserting it wherever there is a Green's function. The effect is merely to restrict the range of momenta the nucleon can have in each intermediate state.

The most obvious approximation to Eq.(D.24) is to iterate it once, so that $\tau^{(1)}$ is approximately the πN t-matrix in the <u>three-body system</u> evaluated at $(E^+-\varepsilon_B)$. From the discussion in section <u>A (iii)</u> we know that the πN interaction energy therefore covers the range $(-\infty, E^+-\varepsilon_B)$ because of the recoil energy of the core and the πN c.m. motion. In terms of $\tilde{t}_{\pi N}$, the πN t-matrix in the two-body c.m. the first order optical potential is therefore

$$U_{oo}^{(1)}(E^+) \simeq (A-1) \langle o|\tilde{t}_{\pi N}(E^+-\varepsilon_B - K_{A-1} - K_{\pi N}^{c.m.})|o\rangle \; . \tag{D.25}$$

(The corrections to (D.25) arising from the pion-nucleus to πN c.m. transformation are discussed in the next section.) Since typical nucleon binding energies are of the order of (10-20) MeV and so are the kinetic energy terms, there is a considerable downward shift of the πN interaction implied by Eq.(D.25). This effect was first observed in a non-relativistic model of pion-nucleus scattering,[59] but essentially the same result has also been derived in a fully relativistic theory.[60] A very important result of the shift is that this optical potential is purely real for $E < \varepsilon_B$. This is sufficient to guarantee elastic unitarity for T_{oo} (to order 1/A), if there is not sufficient energy to knock-out a nucleon.

The difficulty with this naive use of Eq.(D.25) is of course that t_{NC} is not weak—indeed it supports a bound state. From the fact that u is attractive, the original equation (D.17) suggests that the effect of the N—core interaction might be to compensate to some extent the full energy shift in (D.25). For this reason Landau and Thomas[61] treated ε_B as a parameter 'E_B' to vary between the real binding energy and zero, in an attempt to explain the dramatic failure of conventional optical models[62] for 50 MeV $\pi^{\pm}$ ^{4}He and π^+ ^{12}C scattering. (The conventional models are briefly discussed below, but all ignore the three-body shift completely.) Figure 13 shows the improvement of a factor of (2-3) obtained in the fit to the 50 MeV ^{4}He data when this shift was introduced (with E_B = 5 MeV) into a momentum space calculation using separable πN interactions. This was the first strong empirical evidence for the theory described here (see Refs. 63, 64 for a more detailed presentation, including Pauli effects and absorption).

Clearly the present situation regarding the three-body nature of $\tau^{(1)}$ is unsatisfactory. The assumption in most calculations that a shift of the energy in the two-body t-matrix can approximate the full solution of the three-body problem is untested. Given such an assumption, the work of Landau and Thomas[61] ($E_B \sim 5$ MeV) represents an intermediate result—intermediate between Liu and Shakin[60] who ignore binding effects completely (and thus get an effective downward shift of order 30 MeV), and Schmidt *et al.*[65] who find no shift at all using an approximate Padé technique to solve Eq.(D.17).

The one firm theoretical result on this matter was obtained very recently by Amado *et al.*,[66] who used essentially three-body techniques to extend a theorem of Fuda for n-d scattering to the general projectile - nucleus optical model problem. They established unambiguously that the longest range (peripheral) part of any optical potential can be written entirely in terms of the on-shell projectile-nucleon t-matrix at an energy equivalent to that used in Eq. (D.25). In conclusion, while some hard theoretical work remains to be done, the importance of the three-body picture of the first-order potential, and particularly the choice of πN sub-energy has been established beyond doubt.

(ii) Relativistic aspects of the first-order potential. Almost all theoretical descriptions of pion-nucleus scattering start with non-relativistic scattering theory as we have just done. The relativistic aspects of the problem are then added in at the end (as we do below). A much more satisfying alternative is to follow the same sort of procedure we used for the relativistic three-body problem in part B. That is, one writes down effectively a Bethe-Salpeter equation for the pion-nucleus system and uses the Blankenbecler - Sugar techniques to make it calculable. The latter approach has been carried through by Shakin *et al.*[24] Fortunately both these approaches lead to essentially the same first-order potential.

The first correction arises because the pion-nucleon t-matrix is usually calculated in the πN c.m. system using a relativistic Lippmann-Schwinger equation [e.g. Eq.(B.12)]. This t-matrix is however needed in the pion-nucleus c.m. in Eq.(D.25). To make this transformation we use Eq.(B.14) to relate the t-matrix in the πN c.m. [for scattering from $(\vec{Q},-\vec{Q})$ to $(\vec{Q}',-\vec{Q}')$], to a Lorentz invariant amplitude (T). A similar expression yields the t-matrix in the pion-nucleus system [for scattering $(\vec{k},\vec{p})$ to $(\vec{k}',\vec{p}')$] in terms of T. To summarise then, the first order optical potential including relativistic effects is approximately[64]

$$U^{(1)}(\vec{k}',\vec{k};E^+) = \frac{A-1}{A} F_{c.m.}(\vec{q}) \sum_a \int d\vec{p}\,\psi_a^*(\vec{p}-\vec{q}) \times \left(\frac{w_{\vec{Q}'} E_{\vec{Q}'} w_{\vec{Q}} E_{\vec{Q}}}{w_{\vec{k}'} E_{\vec{p}-\vec{q}} w_{\vec{k}} E_{\vec{p}}}\right)^{1/2} t^{LS}_{\pi N}\left(\vec{Q}',\vec{Q};E^+ - E_B^{(a)} - \frac{(\vec{p}+\vec{k})^2}{2M}\right) \times \psi_a(\vec{p}) \ . \tag{D.26}$$

(Here M is the reduced mass of the (πN) + core system, and $F_{c.m.}(\vec{q})$ is the usual correction factor, based on a harmonic oscillator model, which corrects for the omission of a c.m. constraint in the IPSM.[67]) Here the sum is over all the nucleons in their IPSM states $\{\psi_a\}$, the momentum transfer is $\vec{q} = \vec{k}' - \vec{k}$, $t^{LS}_{\pi N}$ is calculated in the πN c.m. system, and we have allowed E_B to depend on the state of the struck nucleon. As in the relativistic πNN case we choose to keep all particles on-mass-shell, so that the relative momenta $(\vec{Q},\vec{Q}')$ are given by Eq.(B.41) again.

(iii) Comparison with conventional potential models

a) Landau - Phatak - Tabakin (LPT):

Our entire development of two and three-body scattering theory, as well as the optical model (D.26) has been illustrated using the momentum representation. Because of the simple representation of particle energies this is the natural representation for scattering problems. The reader should try to write the transformation factor in Eq.(D.26) in co-ordinate representation to appreciate the problem! It therefore comes as something of a shock to realize that the first momentum space calculation of pion-nucleus scattering was not published until 1973.[68] One reason is that the price for the extra sophistication is a considerable increase in computational difficulty which does require a large modern computer.

The LPT potential was derived using the theory of Kerman *et al.*,[54] and is equivalent to Eq.(D.26) with two approximations. First the choice of interaction energy was not derived from any theoretical considerations, such as those leading to the three-body choice above. Instead the on-mass-shell four-momenta for the pion and nucleon were used to construct the conventional relativistic s and t for the pair as

$$s = (k + p)^2 = (w_{\vec{k}} + E_{\vec{p}})^2 - (\vec{p} + \vec{k})^2$$

$$t = (k' - k)^2 = (w_{\vec{k}'} - w_{\vec{k}})^2 - (\vec{k} - \vec{k}')^2 \quad . \qquad \text{(D.27)}$$

The πN interaction energy (in the two-body c.m. system) is then $(\sqrt{s} - m_N)$ - including the pion mass—and runs above and below the incident pion energy as we integrate over $\vec{p}$. This choice of energy will be called $E_{2\text{-body}}$ from now on. It is not unique, of course, because s′ defined as $(k' + p')^2$ will not be equal to s in general (i.e. for an arbitrary πN collision)!

The second difficulty in the LPT approach is also related to the non-uniqueness of the interaction energy. Their πN t-matrix in the c.m. was parameterised in terms of a scattering angle which was calculated in the standard way from s and t. Unfortunately in an on-mass-shell, off-energy-shell theory (i.e. $s \neq s'$) of this type (c.f. Section B), this does not always lead to a physical scattering angle [$\cos\theta \in (-1,+1)$].

While these two approximations sound rather drastic, in practice the nuclear form-factor [which is contained in (D.26) i.e. $U(\vec{k}',\vec{k}) \sim F(\vec{k}'-\vec{k})$] restricts $|\vec{k}'| \sim |\vec{k}|$ so that the degree by which $s \neq s'$ is not so great. On the positive side, the transformation factor in (D.26) was included for the first time in any pion-nucleus calculation. In addition, the definition of πN scattering angle while not strictly correct, incorporated an extremely important physical effect—later dubbed the angle transformation—into pion-nucleus scattering theory for the first time. The results of the calculations were in excellent agreement with the available data on ^{12}C and ^{16}O in the resonance region.[68,69]

b) Δ - propagation:

The full sophistication of the work by Lenz and others has been described at this meeting by Dr. Moniz. In its simplest form, their approach represents an alternative calculational technique to the first order optical model. Their basic πN interaction in the P_{33} channel is separable, but greater use is made of the separability to identify the 'Δ' as an extra degree of freedom. It should be born in mind that their Δ has exactly the same meaning as that in the relativistic three-body equations (B.36).

The 'propagation of the Δ' refers to the independent realization by Lenz[70] (in the context of pion scattering in the resonance region), that the three-body energy $\sim(E^+-E_B-T_\Delta)$ is the natural argument of the πN D-function in a many nucleon system. Without the kinetic energy term T_Δ the c.m. of the πN sub-system does not move through the nucleus. This clear inclusion of $E_{3\text{-body}}$ is a little hidden when higher order connections are introduced through a complex Δ - core interaction, but the essential idea is there.

The scattering angle in the πN c.m. system is calculated through $\underset{\sim}{Q}\cdot\underset{\sim}{Q}'/|\underset{\sim}{Q}||\underset{\sim}{Q}'|$, with the πN relative momentum

$$\underset{\sim}{Q} = (m_N\vec{k} - w_{\vec{k}}\vec{p})/(m_N + w_{\vec{k}}) \; . \qquad (D.28)$$

The reader should confirm that this is the semi-relativistic limit of Eq. (B.41), and is probably a rather good approximation to it. For details of the results of this approach I defer to the lectures of Dr. Moniz.

c) Co-ordinate space calculations:

The co-ordinate space potentials of Kisslinger,[71] and later (and more sophisticated) of Ericson and Ericson,[72] are still being very widely used to understand pion-nucleus data. Undoubtedly a major reason is the familiarity of the co-ordinate space picture to nuclear physicists, and the existence of many scattering programs which could easily be modified to accept the velocity dependence in this model.

To be fair, the model also provided first an explanation of the role of the nuclear surface in pion scattering, leading to strong back-scattering,[71)] and then (with some parameter adjustment) to an impressive explanation of the strong interaction shifts in pionic atoms.[73)]

The first step in deriving the Kisslinger-type potential is to *factorise* the πN t-matrix out of Eq.(D.26). As we soon show, this is no problem in the usual co-ordinate space picture as the dependence of $t_{\pi N}$ on $\vec{p}$ is usually ignored. However, factorisation has also been used in almost all momentum space calculations.* For a detailed description of the steps involved we refer to Refs. [64] (§3B) and [70]. The crudest factorisation of (D.26) is simply,

$$U^{(1)}(\vec{k}',\vec{k};E) \simeq (A-1)\ F(\vec{q})\gamma t_{\pi N}\ [\vec{Q}'(\vec{k}',\vec{p}_0-\vec{q}),\ \vec{Q}(\vec{k},\vec{p}_0);\ E^+-E_B - (\vec{p}_0+\vec{k})^2/2M] \tag{D.29}$$

where γ is the relativistic transformation factor and $\vec{p}_0$ is the nucleon momentum which optimises the approximation, and $F(\vec{q})$ is the nuclear form-factor, as measured (say) by electron scattering.

At this stage, if we drop the factor γ (which we noted above cannot be calculated in the co-ordinate space representation), and approximate the p-wave πN t-matrix as

$$t_{\pi N}\ (\vec{Q}',\vec{Q};E^+-E_B - (\vec{p}_0+\vec{k})^2/2M) \simeq c(E)\ \vec{k}'\cdot\vec{k}\ , \tag{D.30}$$

we obtain the Kisslinger potential

$$U^{Kiss}(\vec{k}',\vec{k};E) = (A-1)c\ F(\vec{q})\ \vec{k}'\cdot\vec{k}\ . \tag{D.31}$$

When fourier transformed this leads to the velocity dependent potential

$$U^{Kiss}(\vec{r}) = (A-1)c\ \vec{\nabla}\ [\rho(r)\ \vec{\nabla}]\ , \tag{D.32}$$

and it is the emphasis on the gradient of the single particle density [$\rho(r)$] which gives the nuclear surface its importance in pion scattering. Note that in order to derive Eq.(D.31,32), on top of all the other approximations the three-body energy shift, and the very important dependence of $(\vec{Q},\vec{Q}')$ on the nucleon momentum were ignored.

---

*The exceptions are calculations by Lee,[74)] and Liu and Shakin,[75)] both of whom find that the approximation gets worse at lower energy. Nevertheless the agreement between Liu and Shakin,[75)] and the approximation of Landau and Thomas[64)] is quite good even at 50 MeV.

The latter approximation has been shown to be so poor,[68] that recent calculations have attempted to make some improvement on it.[76,77] Using Eq.(D.28) one obtains[76] $[m_N/(m_N+w_{\vec{k}}) \sim 1$, $w_{\vec{k}}/(m_N+w_{\vec{k}}) = \xi]$

$$\vec{Q}'\cdot\vec{Q} \simeq [\vec{k}'- \xi\ (\vec{p}_o-\vec{q})]\ \cdot\ [\vec{k} - \xi\ \vec{p}_o]\ , \tag{D.33}$$

and if p_o is set to zero, to $O(\xi)$ we find

$$\vec{Q}'\cdot\vec{Q} \simeq (1 + \xi)\ \vec{k}'\cdot\vec{k} - \xi\vec{k}^2\ . \tag{D.34}$$

The extra term $\xi\vec{k}^2$ contributes to an effective s-wave πN interaction in the pion-nucleus system. Depending on the way its approximated it gives either a purely local term $\{\vec{\nabla}^2\ \rho(r)\}$,[77] or a mixture of a velocity dependent and a local part.[76] In either case this recoil correction has a dramatic effect on low energy pion elastic scattering.

The p-wave part of the Ericson-Ericson potential is also given by Eq.(D.32). In addition they added a term $\{b_o\ \rho(r)\}$ corresponding to s-wave πN scattering, plus terms dependent on $\rho^2(r)$ representing a phenomenon which really lies outside the scope of potential theory—namely pion absorption. (We shall return to this question in Lecture 3.) Finally Ericson and Ericson noticed that the divergent off-shell behaviour of the postulated πN interaction (D.30) placed a very strong emphasis on the short-range behaviour of the two nucleon wave function. In particular, the double scattering term at zero energy goes like[78]

$$\begin{aligned} T^{(2)}(\vec{k}',\vec{k}) &\sim c^2 \int d\vec{r}_1\ d\vec{r}_2\ \rho(\vec{r}_1,\vec{r}_2) \int d\vec{q}\ \frac{e^{i\vec{q}\cdot(\vec{r}_1-\vec{r}_2)}}{0^+-\vec{q}^2}\ (\vec{k}'\cdot\vec{q})(\vec{q}\cdot\vec{k}) \\ &\sim c^2\ \vec{k}\cdot\vec{k}' \int d\vec{r}_1\ d\vec{r}_2\ \rho(\vec{r}_1,\vec{r}_2) \int d\vec{q}\ e^{i\vec{q}\cdot(\vec{r}_1-\vec{r}_2)} \\ &\sim c^2\ \vec{k}\cdot\vec{k}' \int d\vec{r}_1\ d\vec{r}_2\ \rho(\vec{r}_1,\vec{r}_2)\ \delta(\vec{r}_1-\vec{r}_2)\ , \end{aligned} \tag{D.35}$$

which should vanish in the presence of nucleon-nucleon correlations! This is the simplest derivation of the Lorentz-Lorenz effect, first derived for pions by the Ericsons.[72] It says that for a realistic nuclear density the single-scattering approximation is essentially exact for the p-wave potential (D.32). To subtract out the spurious higher order scattering the p-wave term has a non-linear form,[72,79] and the Ericson-Ericson potential (without absorption) is

$$\overset{E-E}{U}(\vec{r}) = -4\pi b_o\rho(r)+4\pi\ c_o\vec{\nabla}\ [\rho(r)/\{1 + \lambda\frac{4\pi}{3}\ c_o\rho(r)\}]\ \vec{\nabla}\ , \tag{D.36}$$

where $\lambda=1$ is the 'full Lorentz-Lorenz effect'. For an excellent discussion of the origin of (D.36) and the definition of b_o and c_o in terms of πN phase shifts we refer to Ericson's lectures in Ref. [79]. Recent work by Thies[76] and Stricker *et al.*[77] has introduced some improvements, such as the angle transformation (D.34) and some energy dependence for λ, but the essential physics has not changed.

For comparison, we recall the discussion of Section C concerning the range of the πN interaction. Whereas the postulated πN interaction (D.30) has zero range [$\sim \delta(\vec{r}_N - \vec{r}_\pi)$], there we suggested it should be of order 0.6 fm. This results in a tremendous change in emphasis in the theory. Indeed, as Eisenberg *et al.*[80] have shown [and Hirata *et al.* have confirmed[70]], the longer range πN interaction 'smooths over' the effect of correlations, so that in a typical momentum space calculation correlation effects can be ignored in a first approximation. Thus, as we mentioned in part C, the question of the πN range is absolutely fundamental in pion-nucleus scattering theory.

LECTURE 3

'He that riseth late must trot all day.'
Benjamin Franklin.

The difficulty with the topic being discussed here is that by the time enough basic material has been explained, there is little time left for the most interesting questions. Thus our strategy has been to put enough detail into Lecture 1, not all of which could be covered orally, that the serious student can draw on it in places like part E. There we are motivated by the three-body model of $U^{(1)}$, to examine what inelastic reactions are contained in it. Amongst the processes not included is real pion absorption. The final section therefore reviews what recent theoretical understanding the πNN system has provided on this question.

E. The Reaction Content of the First-Order Potential

We have shown throughout our discussion of three-body theory the emphasis placed on exact unitarity. The real advance made by Tandy *et al.*[55] for proton-nucleus scattering therefore, was not that they were able to write three-body-equations for $U^{(1)}$, but that they used these equations to show what approximations concerning the inelastic channels were implicit in $U^{(1)}$. The first application of these ideas to pion-nucleus scattering was by Thomas and Landau.[81]

In this section we illustrate the unitarity relation with a simple example. We then establish the exact unitarity relation for the t-matrix $T^{(1)}$ corresponding to $U^{(1)}$. The practical implications of this for calculations of pion-nucleus cross-sections, as well as calculations involving pion distorted waves, are then discussed. Finally we indulge in a little speculation on 'the way things really are' in the (3,3) resonance region.

(i) The unitarity relation in Born approximation. Consider the Born approximation, $T^{(1)}_{Born} \equiv U^{(1)}$, to the optical model Eq.(D.16) using the first order potential (D.26). (We drop the relativistic transformation and c.m. correction factors for simplicity.) Using the unitarity relation (A.24) for $t_{\pi N}$ implies

$$\text{disc } t_{\pi N}(\vec{Q}',\vec{Q};\varepsilon) = -2\pi i \int d\vec{Q}'' \, t(\vec{Q}',\vec{Q}'';\varepsilon^+) \, \delta[\varepsilon - H_o(\vec{Q}'')] \times t^*(\vec{Q}'',\vec{Q};\varepsilon^+) \ . \tag{E.1}$$

Therefore, the discontinuity of $T^{(1)}_{Born}$ in the forward direction ($\hat{k} = \hat{k}'$) is simply

$$\begin{aligned} \text{disc } \{T^{(1)}_{Born}(E,0°)\} = &-2\pi i(A-1)/A \sum_a \int d\vec{p} \, d\vec{Q}'' \, \psi^*_a(\vec{p}) \\ &\times t^*[\vec{Q},\vec{Q}'';E^+-E_B^{(a)}-(\vec{p}+\vec{k})^2/2M] \; t[\vec{Q}'',\vec{Q};E^+-E_B^{(a)}-(\vec{p}+\vec{k})^2/2M]\psi_a(\vec{p}) \\ &\times \delta[E^+-E_B^{(a)}-(\vec{p}+\vec{k})^2/2M-H_o(\vec{Q}'')]. \end{aligned} \tag{E.2}$$

In terms of diagrams, Eq.(E.2) corresponds simply to cutting the on-energy-shell free πN intermediate states of a πN t-matrix (c.f. Fig.2) in the presence of a non-interacting core. This is the clue to understanding Eq.(E.2). The combination '$\psi_a \; t_{\pi N}$' is the matrix element for nucleon knock-out by the incident pion in *plane wave impulse approximation* [denoted $M^{(a)}_{PWIA}(\vec{k};\vec{p},\vec{Q}'')$]. Thus, Eq(E.2) says simply that

$$\begin{aligned} \text{Im } \{T^{(1)}_{Born}(E(\vec{k}),0°)\} &\propto \int d\vec{p} \, d\vec{Q}'' \sum_a |M^{(a)}_{PWIA}(\vec{k};\vec{p},\vec{Q}'')|^2 \\ &\times \delta(E_{initial} - E_{final}), \\ &\propto \sigma_{PWIA}(\pi,\pi N). \end{aligned} \tag{E.3}$$

That is, the optical theorem for the Born approximation solution to the first order optical potential (D.26) implies that the imaginary part of the forward amplitude is related to the cross-section for pion quasi-free scattering in PWIA. With a little forethought we could have guessed this result, because in the three-body model the core is inert, and the pion does not interact with it (c.f. Fig. 12). Clearly then, the outgoing pion wave is never distorted in the first-order three-body potential. In the next section we show that using the full solution of the pion-nucleus scattering equation $T^{(1)}$, leads to distortion of the incident pion wave only.

The point which this simple demonstration is meant to stress, is that the only inelastic channel whose presence is incorporated (through unitarity) in a first-order optical model is quasi-elastic scattering (one nucleon knock-out).

(ii) The unitarity relation for the first order potential. Let us forget the relativistic aspects of the problem for a moment.

We recall that the three-body equations (D.24) for the operator $\tau^{(1)}$ are of the AGS form. In fact, identifying $\tau^{(1)} = U_{1,1}$, $\tau_C = U_{2,1}$, $t_{\pi N} = t_1$, $t_{NC} = t_2$, $t_3 = 0$, we see that $\tau^{(1)}$ is an elastic scattering operator, and τ_C describes a pick-up process, in which the projectile and nucleon stick together. However, in the present case the πN interaction has no bound state, so the discontinuity relation for $U_{1,1}$ [c.f. Eq.(A.82)] has only one bound state contribution $\Delta_{1;a}$ [where 'a' denotes the IPSM level of the one nucleon involved—this label will be dropped for notational convenience below Eq.(E.4)]. Using Eq.(A.82) we now find the discontinuity of $\tau^{(1)}$ across its right-hand cut

$$\begin{aligned} \text{disc}\ \{\tau_a^{(1)}\} = -2\pi i\ \{&\tau_a^{(1)\dagger}\ \Delta_{1;a}\ \tau_a^{(1)} \\ &+ (\tau_C^{\dagger}\ G_o^{\dagger}\ t_{NC}^{\dagger} + \tau_a^{(1)\dagger}\ G_o^{\dagger}\ t_{\pi N}^{\dagger})\ \text{disc}\ \{G_o\} \\ &\times (t_{NC}\ G_o\ \tau_C + t_{\pi N}\ G_o\ \tau_a^{(1)})\}\ . \end{aligned} \tag{E.4}$$

However $t_{\pi N}\ G_o\ \tau^{(1)}$ can be replaced by just τ_C using Eq.(D.24), and so the second term on the r.h.s. of Eq.(E.4) contains the combination

$$(1 + t_{\pi N}\ G_o)\ \tau_C \equiv \Omega_{\pi N}\ \tau_C\ , \tag{E.5}$$

where $\Omega_{\pi N}$ is the Möller operator converting a plane wave πN state into a full πN scattering wave function [c.f. Eq.(A.13)], in the presence of a non-interacting core. Next we recall the optical model Eq.(D.16). By manipulations similar to those used in Eqs.(A.70,71) the reader should convince him- (her-) self that

$$\begin{aligned} \text{disc}\ \{T'_{oo}\} &= T'^{\dagger}_{oo}\ \text{disc}\ G_A^{(o)}\ T'_{oo} \\ &\quad - T'^{\dagger}_{oo}\{U^{-1}_{oo}(E^+) - U^{-1}_{oo}(E^-)\}\ T'_{oo} \\ &= T'^{\dagger}_{oo}\ \text{disc}\ \{G_A^{(o)}\}\ T'_{oo} + \Omega_o^{\dagger}\ \text{disc}\ \{U_{oo}\}\ \Omega_o\ . \end{aligned} \tag{E.6}$$

In Eq.(E.6) we have defined the elastic distortion operator as usual ($\Omega_o \equiv U^{-1}_{oo}\ T'_{oo}$).

Now the major trick in proving elastic unitarity within the KMT formalism is that to get the physical scattering amplitude (T_{oo}) we must form $[A/(A-1)]\ T'_{oo}$. Thus, ignoring the second term on the r.h.s. of (E.6) we find

$$\text{disc}\ T_{oo} \sim \frac{A-1}{A}\ |T_{oo}|^2\ . \tag{E.7}$$

A comparison with the usual two-body unitarity relation shows that elastic unitarity is not guaranteed to order (1/A) without the first term on the r.h.s. of Eq.(E.4) for disc $\tau^{(1)}$. This in turn requires a full solution of the three-body Eq.(D.24). Therefore the impulse

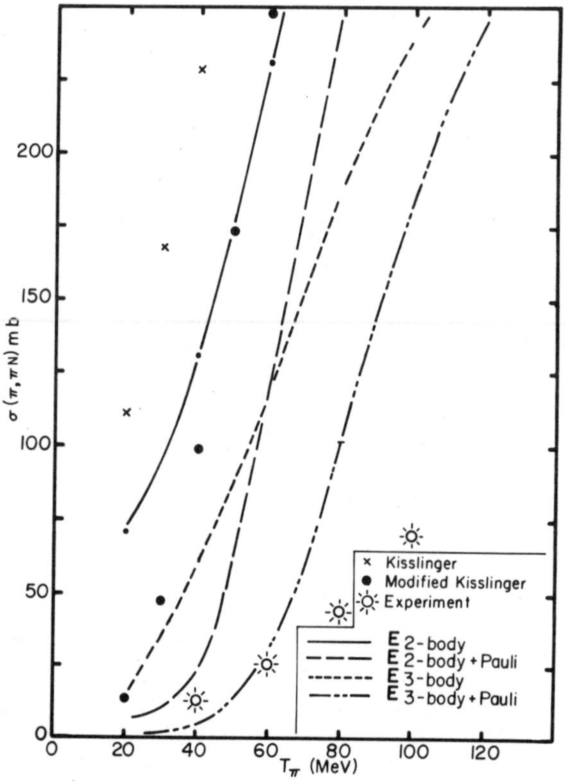

Fig. 14 A comparison of the total cross-sections for quasi-elastic scattering implicit in several standard first-order optical potentials. Note particularly the discrepancy of one to two orders of magnitude in the region below 60 MeV in models that do not use $E_{3\text{-body}}$ and Pauli effects.

high as experiment is obvious from the discussion in part (ii) above. In particular, in the resonance region the pion-nucleus interaction is strongly absorptive, so that the plane wave in the approximate DWIA (E.10) results in a considerable overestimate of '$\sigma(\pi,\pi N)$' in Eq.(E.11).* This is apparently sufficient to (fortuitously) compensate for the many other open channels whose presence is ignored in $U_{00}^{(1)}$.

*The omission of absorption in the nucleon wave function also tends to overestimate '$\sigma(\pi,\pi N)$'.

approximation $\tau^{(1)} \to t_{\pi N}$ formally violates elastic unitarity [to $O(1/A)$]. Having pointed this out, we ignore it [along with other minor details of $O(1/A)$] for clarity of presentation.

Equation (E.6) then becomes

$$\text{disc } T_{oo} = T_{oo}^{\dagger} \text{ disc } \{G_A^o\} \, T_{oo} + \sum_a \Omega_o^{\dagger} \, \tau_c^{(a)\dagger} \, \Omega_{\pi N}^{\dagger} \text{ disc } \{G_o\} \, \Omega_{\pi N} \, \tau_c^{(a)} \, \Omega_o \, . \tag{E.8}$$

That is, the optical theorem relates $T_{oo}(0°)$ to the total elastic cross-section ($\sigma_{elastic}$) plus the cross-section for break-up of the bound nuclear system leading to a π- N- core three-body final state. The combination $\Omega_{\pi N} \, \tau_c^{(a)} \, \Omega_o$ can be rewritten as $\Omega_{NC} \, \tau^{(1)} \, \Omega_o$ (an exercise for the reader), which leads to the partially distorted matrix element

$$T_{\pi N \leftarrow \pi}^{(a)} = < \chi_N^{(-)} \, \phi_\pi \, |\tau^{(1)}| \, \chi_\pi^{(+)} \, \psi_a > . \tag{E.9}$$

Here ϕ_π is a plane wave state and $\chi_N^{(-)}$ is the nucleon-core scattering wave function in the real shell-model potential 'u'. Finally, using the impulse approximation $\tau^{(1)} \sim t_{\pi N}$ this is simply

$$T_{\pi N \leftarrow \pi}^{(a)\,IA} = < \phi_N \, \phi_\pi \, |t_{\pi N}| \, \chi_\pi^{(+)} \, \psi_a > . \tag{E.10}$$

Of course, $\chi_\pi^{(+)}$ is the optical model wave function for the incident pion. In comparison with Eq.(E.3) we see that the only effect of using the full solution of the optical model equation, as opposed to Born approximation, is this extra distortion. In summary, the optical theorem for the solution of the first-order optical model is[81)]

$$\text{disc } T_{oo}(0°) \sim \sigma_{elastic} + \text{'}\sigma(\pi,\pi N)\text{'}, \tag{E.11}$$

where the quotation marks on the quasi-elastic total cross-section imply it is evaluated approximately.

(iii) Comparison with data. Until very recently, the prevailing attitude to the reaction content of the first-order optical model was rather naive. Experimentally it is known that quasi-elastic scattering accounts for only about one-third of the total inelastic cross-section on ^{12}C at 130 MeV.[82)] At lower energies this fraction seems to become even smaller—although to be honest the data are very meagre. This did not prevent (for example) the authors of Ref. [68]—c.f. part (b) of section D(iii) above—making a comparison of the total cross-sections calculated using their first-order potential with the experimental total cross-sections. Surprisingly the agreement was quite good in the resonance region—far better than it had a right to be!

The reason why the theoretical total cross-section can be as

In view of the importance of the insight which the unitarity relation provides into the content of $U^{(1)}$, it is worth while to state clearly what has and has not been proven, using low energy pion scattering from ^{12}C as an example.[81]

a). The purpose of the derivation of Eqs.(E.9-11) was not to obtain a better DWIA for quasi-elastic pion scattering—as suggested recently by Eisenberg.[83] Indeed, as we have stressed here, Eq. (E.10) should overestimate the 'true' DWIA matrix element more and more as the pion energy rises.

b). Therefore, if one were interested in quasi-elastic scattering per se, the best agreement with experiment would (hopefully) be obtained by calculating a full DWIA using distorted initial and final pion wave functions calculated by an optical model code. On the other hand, the calculation of '$\sigma(\pi,\pi N)$' *(implicit in all the approximations necessary to obtain a calculable form of* $U^{(1)}$*)* via the optical theorem [Eq.(E.11)] is an important test of the internal consistency of the theory.

c). One problem in comparing this result with data is of course that Eq.(E.11) is only exact for $U^{(1)}$. If one included an extra term (e.g. proportional to $\rho^2(r)$—as we mention in part F) in the potential to describe the effect of real pion absorption, this unitarity relation would become much more complicated. For example, both the initial and final pion waves in Eq.(E.10) would include the effects of absorption, and $\sigma(\pi,\pi N)$ would be lowered. Fortunately, at (say) 50 MeV the pion mean free path for absorption is greater than $\lambda_{abs} \approx 6$ fm in ^{12}C (a number calculated from the experimental total cross-section which includes real absorption). Since a pion needs to travel only of the order (1-2) fm to escape from carbon, the reduction in $\sigma(\pi,\pi N)$ from this effect is expected to be no more than $O(e^{-4/6})$, or less than a factor of two. Thus the omission of true pion absorption should not alter our qualitative conclusions.

d). Figure 14 shows the comparison of the implicit quasi-elastic cross-sections in several of the first-order potentials discussed in Section D, with appropriate 'data' constructed from the $\pi^{\pm}$ measurements of Dropesky *et al.*[84] - see Ref. [81]. The key point is that the simple first-order potentials with essentially free πN strengths give values of '$\sigma(\pi,\pi N)$' which lie between one and two orders of magnitude above the data near 40 MeV. The crude inclusion of Pauli effects (which restrict the range of momenta allowed to the outgoing nucleon) by Stricker *et al.* is a slight improvement. However, only the calculations which include the three-body energy and a more sophisticated treatment of Pauli effects are within a factor of two of the data!

e). We stress that the 'agreement' between the '$E_{3\text{-body}}$+Pauli' calculation and the data in Fig. 14 is excellent (at least below

60 MeV) from our present perspective! Remember we cannot expect that the approximations necessary to obtain a calculable form of $U^{(1)}$ will ever allow an accurate prediction of $\sigma(\pi,\pi N)$ through the optical theorem. It is enough that the $(\pi,\pi N)$ cross-section is 'in the right ball park'—namely a small part of the total inelastic cross section.

f). The reason why the use of $E_{3\text{-body}}$ lowers $\sigma(\pi,\pi N)$ is very instructive. In fact following through the derivation of Eq.(E.2) again, it will hopefully be obvious that the argument of the energy conserving δ-function is directly related to that of the two-body t-matrix. In particular, *the correct phase space factor in* $\sigma(\pi,\pi N)$ *is only obtained when* $E_{3\text{-body}}$ *is used.* The reader is invited to prove that when $E_{2\text{-body}}$ (c.f. Eq.(D.27) is used only the $\vec{Q}''$ integral is restricted by the energy-conserving δ-function in Eqs.(E.2,3). That is, the $\vec{p}$-integral is unrestricted, and the unitarity relation contains a possibly quite large unphysical contribution! In any case, the use of $E_{2\text{-body}}$ leads to a very poor representation of the real π-N-core phase space.

The 'Pauli effects' which are referred to in Fig. 14 are calculated as described in Ref. [64]. Their origin is simply that we retain the $Q^{(P)}$ operator (i.e. anti-symmetry) in Eq.(D.17) and the following equations. This effect reduced the quasi-free cross-section because it essentially forbids (occupied) low momentum states to the knocked-out nucleon. The rather severe drop near 30 MeV in Fig. 14 is just an artifact of the nuclear matter approximation for $Q^{(P)}$.

It is an interesting and as yet unanswered question whether the differences between the potential models in Fig. 14 have observable consequences in other reactions. From studies in the resonance region on reactions like (π^+,π^0), (γ,π^0), etc. it seems that the conventional optical models may be too absorptive. This energy region will be mentioned again in Section E(iv). However, the low energy region ($E_\pi < 100$ MeV) where we have found the most dramatic (order of magnitude) overestimates of the implicit quasi-elastic cross-section has hardly been investigated at all. We have calculated the pion co-ordinate space wave function using (case I) the pure $E_{2\text{-body}}$ model (essentially Ref. [68]), and the more sophisticated model (case II) with $U^{(1)}$ including $E_{3\text{-body}}$ and Pauli, plus a term $[U^{(abs)}]$ representing the effect of true absorption. The latter term will be described in part F below, but for now we note that it restores σ_{total} to near experiment ($\sim$120 mb on ^{12}C at 50 MeV), but has a $\rho^2(r)$ dependence, and can therefore produce a different wave function.

Indeed, at 50 MeV the p-wave function in case II is far less strongly distorted by the nuclear surface than case I. Preliminary calculations of the (γ,π^0) reaction[85] using these wave functions indicate a significant qualitative difference (of order 300%) in the back-angle cross-sections. Further tests in experimentally more accessible reactions like charge exchange and quasi-free scattering are essential.

(iv). Higher order corrections - self-consistency. A major reason for the overestimate of the implicit $(\pi,\pi N)$ cross section, as we have mentioned, is the omission in $U^{(1)}$ of any pion-core interaction while the 'active' nucleon is in a continuum state. This is the major characteristic of the first-order potential and resulted in the appearance of ϕ_π in Eq.(E.10)—rather than $\chi_\pi^{(-)}$. It is therefore tempting to try to remedy this defect by incorporating some higher order corrections into the theory through a pion-core interaction. Because the interaction of the pion with the core will not be drastically different from its interaction with the original nucleus, this approach is closely related to the question of the 'self-consistency' of the pion-nucleus interaction.

The first investigation of the effect of imposing self-consistency on the pion-nucleus interaction was an investigation by Ericson and Hüfner[86] of total cross-sections in the resonance region. In particular, if a pion is traversing nuclear matter of density ρ, the dispersion relation relating its momentum k in the medium and its energy is $(w^2 = k^2 + \mu^2)$

$$-\nabla^2 + \mu^2 + 2wV - w^2 = 0 \ , \tag{E.12}$$

or explicitly for a factorised form of $U^{(1)}$

$$K^2 + \mu^2 - 4\pi\rho \ f_{\pi N}(k,w) = w^2 \ . \tag{E.13}$$

In Eq.(E.13) $f_{\pi N}(k,w)$ is the forward πN scattering amplitude. Ericson and Hüfner suggested that what should really appear as the momentum in the πN scattering amplitude is the local pion momentum K. This leads in essence to a self-consistency relation for $K(w)$, or the refractive index $n(w) \equiv (K(w)/k)$

$$K^2 + \mu^2 - 4\pi\rho \ f_{\pi N}(K,w) = w^2 \ . \tag{E.14}$$

Using a reasonable extrapolation of the resonant (3,3) scattering amplitude into the near off-shell region

$$f_{\pi N}(k,w) = \frac{C \ k^2}{w-w_R+i\Gamma/2} \ , \tag{E.15}$$

they showed that the effect of self-consistency in the form (E.14) is to lower the resonance position [i.e. where $\mathrm{Re}(n(w)-1)$ goes through zero] by (30-40) MeV.

We have mentioned this work in some detail, because it represents a first attempt to introduce this new concept. However, it should be clear from our discussion of the three-body model of $U^{(1)}$ that we disagree with the approach. The natural place to introduce a pion-core interaction $V_{\pi C}$ is in fact the denominator of Eq.(D.17). In the crude approximation where $V_{\pi C}$ is a constant, the effect is to modify the interaction energy in $\tau^{(1)}$, not the momentum dependence! For the nuclear matter case this would still involve a half-off-shell πN amplitude, but the off-shell extrapolation is different. We leave it as an interesting exercise for the student to carry this simple calculation through.

On the phenomenological level, something very similar to what we have just described has been carried out by Hirata *et al.*[70] In part (b) of section D(iii) we have already mentioned that their simplest calculation includes the three-body energy shift through the argument of the D-function—namely $D_\Delta(E^+-E_B-T_\Delta)$, with T_Δ the kinetic energy of the πN c.m. However, as a means of including some higher order effects, in practice they replace T_Δ by $\{T_\Delta + V_\Delta\}$, where the complex Δ-core potential is adjusted to fit the pion-nucleus scattering data. Clearly then V_Δ is a phenomenological representation of $\{u + V_{\pi C}\}$ in our language (and possibly includes some effects of true pion absorption as well).

To conclude this section I would like to use a little poetic licence in describing the way things really are in the resonance region. A considerable amount of evidence has been amassed showing that optical models are far too absorptive there. Even the qualitative features of single charge exchange, for example, are wrong—with distorted wave calculations producing a dip near the resonance where experiment has a slight bump![87]

One indication of this problem from the theoretical point of view is the well known observation (see e.g. Ref. [88]) that adding terms representing real absorption has essentially no effect on the pion scattering results in conventional optical models—in spite of the fact that real absorption represents of the order of 50% of the experimental total reaction cross-section.[82] Clearly one needs a mechanism to reduce the total cross-section produced by '$U^{(1)}$' (i.e. the potential without absorption) to a value of the order of the observed elastic plus quasi-elastic cross-sections.

One possible mechanism suggested by the three-body model is the inclusion of the pion-core interaction. From the point of view of a first principles calculation (rather than phenomenology), it is important to observe that the energy shift means that this potential is needed at an energy considerably lower than the starting potential. Since the optical potential grows rapidly weaker as the energy falls this sort of self-consistency procedure may converge very rapidly.... tuum est.

F. The Effect of Pion Absorption

L'urgent est déjà fait,
Le difficile est en train de se faire,
Pour l'impossible on demande plusiers jours.
Inconnu.

A discussion of the effect of absorption does not fit strictly into our development of multiple scattering theory, starting from the Faddeev equations. Nevertheless, our considerations of unitarity alone for the π-nuclear system have shown that true absorption is a very important inelastic channel. Until recently the inclusion of absorption was very phenomenological. In this final section we shall briefly outline the development of the phenomenology and show that the effects on the differential cross-section can be very large in the low energy region.

The real theoretical advances in understanding this phenomenon have however been made in a few body system (namely NNπ) which has served as an excellent theoretical laboratory. The results obtained in this system and the implications for the pion-nucleus problem will also be mentioned briefly.

(i) Phenomenological treatment of absorption. The new distinct feature of pion-nucleus scattering, which was evident even in the early 50's in the formation of stars in photographic emulsions, is that the pion can disappear giving its rest mass into a violent breaking apart of the nucleus. A single nucleon with 140 MeV kinetic energy has a momentum of greater than 500 MeV/c, which is very difficult to find in a single particle wave function. If on the other hand the pion is absorbed on two nucleons which are emitted in almost opposite directions, little momentum transfer is required from the nuclear wave function. This was the original reason why Ericson advocated the use of the (π, 2N) reaction to measure the N-N correlation function.[89] While distortion effects have made this essentially impossible, there is now considerable evidence that absorption on two nucleons is the basic mechanism underlying pion absorption.[90,91]

This evidence also suggests that absorption is more likely on a deuteron-like pair (i.e. n-p in I=0, S=1). Thus the original model proposed by the Ericsons to explain the widths of pionic atom levels (remember $U^{(1)}$ is real in their model at threshold),[72,73] is still

*The phrase 'true absorption' is not meant as a religious catchphrase. This term is used to distinguish processes in which the pion actually disappears, from inelastic processes [e.g. (π, πN)] which result in a loss of flux (absorption) from the elastic channel.

in keeping with current ideas. Using a zero-range two nucleon operator (as an approximation to the real operator, which should have a short range because of the high relative momentum involved), they derived the cross-section for [$\pi d \to NN$] and [$\pi A \to NN\ (A-2)$] in terms of two constants—namely the absorption strengths of s- and p-wave pions (with respect to the NN c.m.).

In the πA system this gave rise to a new term in the optical potential proportional to $\rho^2(r)$ (because of the zero-range approximation). This term is supposed to represent a phenomenon not contained at all in the pion-nucleus multiple scattering series—the interaction with a cluster. Incorporating this into the potential means that by solving the pion-nucleus scattering equation we allow for processes where (e.g.) after interaction with a cluster the pion can scatter from other nucleons.

The Ericson-Ericson absorptive potential was

$$U^{(abs)}_{E-E}(\vec{r}) = -(4\pi/2m_\pi)\ [-B_o\ A^2\ \rho^2(r) + \vec{\nabla}(C_o\ A^2\ \rho^2(r))\vec{\nabla}]\ , \quad (F.1)$$

with $\rho(r)$ the single particle density. In practice the calculated values of Im B_o and Im C_o were a factor of two too small to fit pionic atom level widths. In retrospect this seems to have been more because of uncertainties in the zero-range approximation [e.g. the need for $\psi_d(\vec{r}=0)$?], than a failure of the quasi-deuteron model. Indeed, a very recent calculation by Bertsch and Riska gave 80% of the pionic atom value of Im B_o.[92]

The intensive study of low energy pion elastic scattering in the last two years has shown the necessity for extending Eq.(F.1) above $E_\pi=0$. This is most reasonably done within the quasi-deuteron model by scaling the s- (B_o) and p-wave (C_o) strengths[64] according to the elementary $\pi d \to NN$ cross-section.[93] In addition, since the conventional absorption mechanism involves scattering of the pion from one nucleon before its absorption on a second, it is reasonable to introduce form-factors (g_o and g_1) into the momentum space generalisation of (F.1), with a range typical of the πN system (c.f. Section D). This leads to the generalisation of Eq.(F.1),[64]

$$U^{abs}(\vec{k}',\vec{k}) = -4\pi\ [2m_\pi\ (2\pi)^3]^{-1}\ A(A-1) \left\{ B_o(k_o)\ \frac{g_o(K')\ g_o(K)}{g_o^2(K_o)} + C_o(k_o)\ \frac{g_1(K')\ g_1(K)}{g_1^2(K_o)}\ \cos\theta_{\vec{k}\,\vec{k}'} \right\} \tilde{\rho^2}(\vec{k}'-\vec{k}), \quad (F.2)$$

where K is the πN c.m. momentum and $\tilde{\rho^2}$ the fourier transform of $\rho^2(r)$. The values of Im $B_o(k_o=0)$ and Im $C_o(k_o=0)$ were fixed by the appropriate pionic atom values [Im $B_o(0) = 0.02\ m_\pi^{-4}$, Im $C_o(0) = 0.04\ m_\pi^{-6}$ corresponding to a full Lorentz-Lorenz correction—or finite size πN interaction].

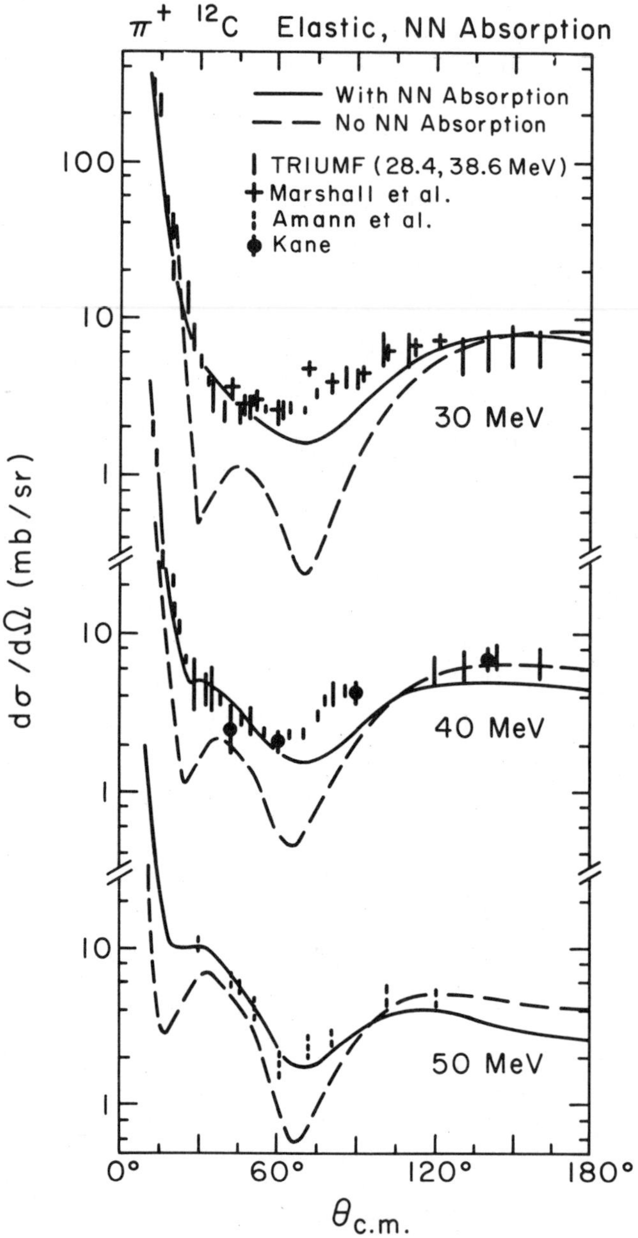

Fig. 15 A comparison with available low energy pion-nucleus elastic scattering data, of the results of Landau and Thomas using $U^{(1)}$ alone (including $E_{3\text{-body}}$)—dashed curve—or including the effect of absorption through the phenomenological term $U^{(abs)}$ [c.f. Eq.(F.2)].[64] Clearly absorption is essential for agreement at 30 and 40 MeV.

$$\text{Im } B_o(k_o) \simeq \text{Im } B_o(0) \left\{ \frac{k_o \; \sigma_{\ell=o} \; (\pi^+ d \to pp;\; k_o)}{\lim\limits_{k\to o} k \; \sigma_{\ell=o} \; (\pi^+ d \to pp;\; k)} \right\} , \qquad (F.3)$$

$$\text{Im } C_o(k_o) \simeq \text{Im } C_o(0) \left\{ \frac{k_o \; \sigma_{\ell=1} \; (\pi^+ d \to pp;\; k_o)}{\lim\limits_{k\to o} k \; \sigma_{\ell=1} \; (\pi^+ d \to pp;\; k)} \right\} . \qquad (F.4)$$

For further discussion of the details of Eqs.(F.3) and (F.4) we refer to Ref. [64]. (Note particularly that the separation into ℓ=0 and ℓ=1 parts of the $\pi^+ d \to pp$ cross-section is quite ambiguous.)

Figure 15 shows the calculation of 30, 40 and 50 MeV π^+ ^{12}C scattering in comparison with data from TRIUMF (at 30 and 40 MeV),[94] and old data of Carnegie-Mellon (at 50 MeV). The dominant feature of low energy π^+ scattering in the forward direction is the destructive interference between the repulsive coulomb and attractive strong interactions. Because of the use of $E_{3\text{-body}}$ in the first order potential of Landau and Thomas,[64] this leads to a very low cross-section—particularly at 30 MeV. (Both the coulomb, and strong amplitude calculated with $U^{(1)}$, were almost real.) As the figure shows, the inclusion of some imaginary part in the strong amplitude through $U^{(abs)}$ is essential to fit the data! This was the first demonstration of the importance of absorption in elastic scattering.

(ii) Theoretical approach to the effect of absorption. While the use of $U^{(abs)}$ given by Eqs.(F.2-4) is physically reasonable, the need for a deeper theoretical understanding has been felt for many years. Progress has been made recently by considering a simple field theory of the πNN system (where the nuclear physics uncertainties are minimal). The same result has now been derived by four methods.[95-98] We shall follow the diagrammatic method, which is the fastest way to establish the correct result. However, the serious student is urged to also read carefully the work of Koltun and Mizutani (using Feshbach projection operators),[97] and Rinat (using the reduction techniques of Ballot and Becker).[98]

The power of the diagrammatic technique is that we do not have to specify exactly the Hamiltonian involved. It is sufficient that we have somewhere in the (renormalized) theory an operator which produces and absorbs pions. For simplicity we neglect all couplings to negative energy nucleon states. Finally we assume that the full $\pi NN \to \pi NN$ scattering amplitude (denoted $\langle 1|T|1\rangle$, where $|1\rangle$ shows the number of pions present) is given by the set of all renormalized diagrams connecting one one-meson two-nucleon state with another.

The set of all such diagrams falls naturally into two (topological) subsets:

(a) those in which every intermediate state has one or more pions (denoted $<1|T|1>_1$, with the subscript giving the minimum number of pions)
(b) those in which there is at least one intermediate state with no mesons.
If we cut the diagrams in part (b) at the last two-nucleon-only intermediate state we find

$$<1|T|1> = <1|T|1>_1 + <1|T|o>_1 \, G_o \, <o|T|1> \, . \qquad (F.5)$$

(Again the subscript '1' on $<1|T|o>_1$ means at least one pion in every intermediate, and G_o is the propagator for two non-interacting nucleons.)

Next one may cut $<o|T|1>$ at the first two-nucleon-only intermediate state, so that

$$<1|T|1> = <1|T|1>_1 + <1|T|o>_1 \, \{G_o + G_o <o|T|o> G_o\} <o|T|1>_1 \, . \qquad (F.6)$$

Now $<o|T|o>$ is simply the sum of all diagrams which take us from the initial to the final N-N state—i.e. the N-N t-matrix t_{NN} in this theory. From Eq.(A.10) we recognise the term in braces as G_{NN}, the interacting N-N Green's function. Using the spectral representation of G_{NN} we find that (F.6) becomes

$$<1|T|1> = <1|T|1>_1 + \sum_n <1|T|o>_1 \frac{|\psi_n^{(+)}><\psi_n^{(+)}|}{E^+ - \varepsilon_n} <o|T|1>_1 . \qquad (F.7)$$

The combination '$<f| \; <1|T|o>_1 \; |\psi_n^{(+)}>$', where $<f|$ is a two-nucleon final state (e.g. the deuteron) is loosely speaking the complete matrix element for pion production.*

Thus, the full elastic amplitude in this model is the sum of a term which could have been generated by some multiple scattering theory—namely $<1|T|1>_1$—plus a term which involves virtual pion absorption. Of course $<1|T|1>_1$ contains many complicated intermediate states with more than one meson, but just as Chew picked out the one meson diagrams for πN as those relevant to unitarity, here we expect also that the approximation

$$<1|T|1>_1 \simeq T^{(Fadd)} \, , \qquad (F.8)$$

---

*The caution here is because strictly to get the $NN\to\pi D$ amplitude one should pick out the residue of the deuteron pole term in $<1|T|o>_1$. Having done this, the condition on $\tilde{T}$ in $M_{\pi d, NN} \equiv <\psi_d| \; <1|\tilde{T}|o>_1 \; |\psi_n^{(+)}>$, is simply that the last interaction be a πN scattering—not N-N scattering!

should be a good starting point.

The second term in Eq.(F.7) also involves $T^{(Fadd)}$, in the same level of approximation, as we see by cutting $<o|T|1>_1$ at the last place where there is one-pion-only intermediate state. That is,

$$<o|T|1>_1 = <o|T|1>_2 \, [1 + G_o <1|T|1>_1] \,, \tag{F.9}$$

and we recognise the term in brackets as the Möller operator for the πNN system with one pion always present [$\Omega^{(Fadd)}$ - c.f. Eq.(A.13)]. In such a theory $<o|T|1>_2$ is essentially the elementary pion production vertex, given to lowest order by Eq.(C.1), for example. To summarise therefore, the contributions of multiple scattering and absorption can be separated quite clearly in such a simple model as

$$T(E^+) = T^{(Fadd)}(E^+) + \Omega^{(Fadd)\dagger}(E^-) <1|T|o>_2 \, G_{NN}(E^+) \, <o|T|1>_2 \, \Omega^{(Fadd)}(E^+) \; . \tag{F.10}$$

Equation (F.10) is agreed upon by each of Refs. [95-98]. It is currently being applied to the πd problem between zero and 250 MeV.[99,100] The key to avoiding double counting of absorption somewhere in the Faddeev term is the subscript '1' in $<1|T|1>_1$. That is, every intermediate state in the first term in Eq.(F.10) must contain at least one pion. Therefore the nucleon pole term (c.f. Fig. 10a) in the P_{11} interaction must be excluded from the Faddeev calculation. This is the only caution necessary to avoid double counting problems.

The formal generalisation of Eq.(F.10) to the pion-nucleus case is straightforward. However, when one tries to connect that with the phenomenology discussed in part (i), the picture is very unclear. Hopefully the near future will see some clarification of this very important problem in pion physics.

ACKNOWLEDGEMENTS

It is a pleasure to thank R.H. Landau, F. Myhrer, E.F. Redish, A.S. Rinat and P.C. Tandy for their invaluable assistance in many stages of the work described here. The long and difficult task of preparing this manuscript would have been impossible without the help of Ada Strathdee, Teresa Diaz and Ingrid Duelli, to whom I am very grateful. I would like to thank my wife Joan for her understanding and patience while these lectures were prepared. Finally I would like to thank K. McVoy and W. Friedman for providing all the participants with an excellent opportunity for the exchange of ideas and stimulation essential to progress in fundamental physics.

References

1. L.D. Faddeev, Soviet Physics JETP 12, 1014 (1961).
2. L.D. Faddeev, Mathematical Aspects of the Three-body Problem (Daniel Davey and Co. Inc., New York, 1965).
3. C. Lovelace, Phys. Rev. B135, 1225 (1964).
4. I.R. Afnan and A.W. Thomas, Fundamentals of Three-body Scattering Theory in Modern Three-Hadron Physics, ed. A.W. Thomas (Springer-Verlag, Berlin, 1977) Chapter1.
5. K.M. Watson and J. Nuttal, Topics in Several Particle Dynamics (Holden - Day Inc., San Francisco, 1967).
6. E.W. Schmid and H. Ziegelmann, The Quantum Mechanical Three-body Problem (Permagon Press, Oxford, 1974).
7. P.A.M. Dirac, The Principles of Quantum Mechanics, 4th. ed. (Oxford University Press, 1958).
8. K.L. Kowalski, Phys. Rev. Lett. 15, 798 (1965).
   H.P. Noyes, Phys. Rev. Lett. 15, 538 (1965).
9. B.R. Karlsson and E.M. Zeiger, Phys. Rev. D11, 939 (1975)
   T.A. Osborn and K.L. Kowalski, Ann. Phys. (N.Y.) 68, 361 (1971).
10. C. Lovelace, in Strong Interactions and High Energy Physics, ed. R.G. Moorhouse (Oliver and Boyd, London, 1964).
11. A.T. Stelborics, Nucl. Phys. A288, 461 (1977).
    L.R. Dodd, Chapter 2 in Ref. [4].
12. H. Eckstein, Phys. Rev. 101, 880 (1956).
    E. Gerjuoy, Ann. Phys. (N.Y.) 5, 58 (1958).
13. E.O. Alt, P. Grassberger and W. Sandhas, Nucl. Phys. B2, 167 (1967).
14. M.L. Goldberger and K.M. Watson, Collision Theory 1st. ed. (John Wiley and Sons Inc., New York, 1964).
15. E.W. Schmid and H. Ziegelmann, Ref. [6].
16. R. Aaron, R.D. Amado and Y.Y. Yam, Phys. Rev. 136, B650 (1964).
17. W.M. Kloet and J.A. Tjon, Ann. Phys. 79, 407 (1973).
18. L. Heller, in Proc. Int. Conf. on Mesons in Nuclei (AIP Conference Series, #33).
    L. Heller, G.E. Bohannon and F. Tabakin, Phys. Rev. C13, 742 (1976).
19. B. Bakamjian and L.H. Thomas, Phys. Rev. 92, 1300 (1953).
20. D. Freedman, C. Lovelace and J.M. Namyslowski, Nuovo Cim. 43A, 258 (1966).
21. R. Aaron, R.D. Amado and J.E. Young, Phys. Rev. 174, 2022 (1968).
22. R. Aaron, Chapter 6 in Ref. [4].
23. R. Blankenbecler and R. Sugar, Phys. Rev. 142, 1051 (1966)
24. L.S. Celenza, L.C. Liu, W. Nutt and C.M. Shakin, Phys. Rev. C14, 1090 (1976).
25. H.J. Weber, Nucl. Phys. A264, 365 (1976).
26. R.H. Landau and F. Tabakin, Phys. Rev. D5, 2746 (1972).
27. J. Londergan, K. McVoy and E. Moniz, Ann. Phys. (N.Y.) 86, 147 (1974).
28. A.W. Thomas, Nucl. Phys. A258, 417 (1976).

29. D. Axen *et al.*, Nucl. Phys. A256, 387 (1976).
30. N. Giraud, C. Fayard and G.H. Lamot, Phys. Rev. Lett. 40, 438 (1978).
31. A.W. Thomas, in Ref. [18], p. 375.
32. R. Aaron *et al.*, Ref. [21].
33. R. Aaron, Ref. [22], p. 142.
34. U. Weiss, Nucl. Phys. B44, 573 (1972).
J.M. Namyslowski, Nuovo Cim. 57A, 355 (1968).
35. A.S. Rinat and A.W. Thomas, Nucl. Phys. A282, 365 (1977).
36. I.R. Afnan and A.W. Thomas, Phys. Rev. C10, 109 (1974).
37. J.M. Namyslowski (communication via I.R. Afnan).
38. A.J. MacFarlane, Rev. Mod. Phys. 34, 41 (1962).
39. J.H. Norem, Nucl. Phys. B33, 512 (1971).
40. A.S. Rinat, E. Hammel, Y. Starkand and A.W. Thomas, Weizmann Institute Preprint, WIS-78/17-Ph; and to be published.
41. M. Hoenig and A.S. Rinat, Weizmann Institute Preprint (1977).
42. K. Gabathuler *et al.*, Nucl. Phys. B55, 397 (1973).
43. R.C. Minehart *et al.*, Phys. Rev. C17, 681 (1978).
44. A.W. Thomas, Proc. 7th Int. Conf. on High Energy Physics and Nuclear Structure, ed. M. Locher (Birkhäuser Verlag, Basel, 1977), p. 109; Proc. Graz Conference, Aug. 1978 (to be published).
45. D.J. Ernst and M.B. Johnson, Phys. Rev. C17, 247 (1978).
46. G.F. Chew and F.E. Low, Phys. Rev. 101, 1570 (1956).
47. M.A.B. Bèg, Ann. Phys. 13, 110 (1961).
48. J. Hüfner, Phys. Reports 21C, 1 (1975).
49. G.A. Miller, Phys. Rev. C16, 2325 (1977).
50. S. Gasiorowicz, Elementary Particle Physics (Wiley, N.Y., 1966) p. 401.
51. A.W. Thomas and F. Myhrer, TRIUMF preprint: TRI-PP-78-7 (1978).
52. L.L. Foldy and J.D. Walecka, Ann. Phys. 54, 447 (1969).
53. H.A. Bethe and M.B. Johnson, Los Alamos Report, LA-UR-76-1844 (1976).
54. A.K. Kerman, H. McManus and R.M. Thaler, Ann. Phys. (N.Y.) 8, 551 (1959).
55. P.C. Tandy, E.F. Redish and D. Bollé, Phys. Rev. Lett. 35, 921 (1975).
56. P.C. Tandy, E.F. Redish and D. Bollé, Phys. Rev. C16, 1924 (1977).
57. E.F. Redish, in Ref. [4], Chapter six.
58. J. Revai, Nucl. Phys. A208, 20 (1973).
59. E. Kujawski and M. Aitken, Nucl. Phys. A221, 60 (1974).
60. L.C. Liu and C.M. Shakin, Phys. Rev. C16, 333 (1977)
61. R.H. Landau and A.W. Thomas, Phys. Lett. 61B, 364 (1976).
62. J.F. Amann *et al.*, Phys. Rev. Lett. 35, 426 (1975).
63. R.R. Johnson *et al.*, Nucl. Phys. A296, 444 (1978).
64. R.H. Landau and A.W. Thomas, TRIUMF preprint TRI-PP-77-4 (accepted for publication in Nucl. Phys. A).
65. J. Maillet, J.P. Dedonder and C. Schmit, Nucl. Phys. A271, 253 (1976).

66. R.D. Amado, F. Lenz and K. Yazaki, SIN preprint PR-77-014.
67. D. Jackson, Nuclear Reactions (Chapman and Hall, London, 1970).
68. R.H. Landau, S.C. Phatak and F. Tabakin, Ann. Phys. (N.Y.) 78, 299 (1973).
69. S.C. Phatak, F. Tabakin and R.H. Landau, Phys. Rev. C7, 1803 (1973).
70. F. Lenz, Ann. Phys. 95, 348 (1975).
M. Hirata, F. Lenz and K. Yazaki, Ann. Phys. (N.Y.) 108, 116 (1977).
71. L.S. Kisslinger, Phys. Rev. 98, 761 (1955).
72. M. Ericson and T.E.O. Ericson, Ann. Phys. (N.Y.) 36, 323 (1966).
73. M. Krell and T.E.O. Ericson, Nucl. Phys. B11, 521 (1969).
74. T.-S.H. Lee, Phys. Lett. 67B, 282 (1977).
75. L.C. Liu and C.M. Shakin, see Ref. [60].
76. M. Thies, Phys. Lett. 63B, 43 (1976).
77. K. Stricker, H. McManus and J. Carr, MSU report π 100T (1977).
78. F. Scheck and C. Wilkin, Nucl. Phys. B49, 541 (1972).
79. T.E.O. Ericson, Proc. Banff Summer School on Intermediate Energy Nuclear Physics (U. Alberta, 1970).
80. J. Eisenberg, J. Hüfner and E.J. Moniz, Phys. Lett. 47B, 381 (1973).
81. A.W. Thomas and R.H. Landau, LBL report LBL-7166 (1977) (revised Feb. 1978). (Accepted for publication (Phys.Lett.B,1978)
82. E. Belotti, D. Cavalli and C. Matteuzzi, Nuovo Cim. 18A, 75 (1973).
83. J. Eisenberg, Tel Aviv preprint (May 1978).
84. B.J. Dropesky *et al.*, Phys. Rev. Lett. 34, 821 (1975).
85. R. Woloshyn, R.H. Landau and A.W. Thomas, unpublished.
86. T.E.O. Ericson and J. Hüfner, Phys. Lett. 33B, 601 (1970).
87. Y. Shamai *et al.*, Phys. Rev. Lett. 36, 82 (1976).
88. M. Thies, Phys. Lett. 63B, 39 (1976).
89. T.E.O. Ericson, Int. Conf. on High Energy Physics and Nuclear Structure (1963), p. 39.
90. H.K. Walter, 7. Int. Conf. on High Energy Physics and Nuclear Structure (Birkhäuser - Verlag, Basel, 1977), p. 225.
91. R.H. Landau and A.W. Thomas, Phys. Reports C (to be published).
92. G.F. Bertsch and D.O. Riska, 'Threshold Pion Absorption in Nuclei', MSU preprint (1978).
93. J. Spuller and D.F. Measday, Phys. Rev. D12, 3550 (1975).
94. R.R. Johnson *et al.*, see Ref. [63] Fig. 12.
95. A.W. Thomas, Thesis, Flinders University of South Australia (1973); A.W. Thomas, Proc. Int. Conference on Few Body Problems (Les Presses de l' Université Laval, 1974), p. 287.
96. T. Mizutani, Thesis, University of Rochester (1975).
97. D.S. Koltun and T. Mizutani, Ann. Phys. (N.Y.) 109, 1 (1978).
98. A.S. Rinat, Nucl. Phys. A287, 399 (1977).
99. N. Giraud, Y. Avishai, C. Fayard and G.H. Lamot, Lyon preprint.
100. A.S. Rinat *et al.*, see Ref. [40] Figs. 1, 2 and 3.

INDEX